中国盐穴储气第一库

金坛储气库二十周年科技成果汇编

(2003—2023)

李　龙　李建君　陈加松◎等编

内 容 提 要

本书汇编了金坛储气库建设团队在 2003—2023 年国际国内刊物上正式发表的论文、获得的省部级及以上科技奖励、授权的专利、发布的标准、出版的专著以及获得的软件著作权等，反映了金坛储气库建设团队 20 年来在前沿趋势探究、地质设计、老腔改建、钻采工程、造腔工程、注采运行、安全监测、地面工程和经济评价等领域所取得的代表性成果。

本书可为从事盐穴储气库建设和运营的工程技术人员、研究人员和管理人员提供参考。

图书在版编目(CIP)数据

中国盐穴储气第一库金坛储气库二十周年科技成果汇编：2003—2023 / 李龙等编 . —北京：石油工业出版社，2024.3

ISBN 978-7-5183-6566-1

Ⅰ.①中… Ⅱ.①李… Ⅲ.①地下储气库-科技成果-汇编-常州-2003-2023 Ⅳ.①TE972

中国国家版本馆 CIP 数据核字(2024)第 043224 号

出版发行：石油工业出版社
　　　　　(北京市朝阳区安华里二区 1 号楼　100011)
　　网　　址：www.petropub.com
　　编辑部：(010)64523687　图书营销中心：(010)64523633
经　　销：全国新华书店
印　　刷：北京中石油彩色印刷有限责任公司

2024 年 3 月第 1 版　2024 年 3 月第 1 次印刷
787×1092 毫米　开本：1/16　印张：48.75
字数：1216 千字

定价：360.00 元
(如出现印装质量问题，我社图书营销中心负责调换)
版权所有，翻印必究

《中国盐穴储气第一库金坛储气库二十周年科技成果汇编(2003—2023)》编委会

主　任：李　龙

副主任：李建君　余国平　赵廉斌　程　林　张青庆

成　员：(按姓氏笔画排序)

　　　　王一单　王元刚　王文权　王成林　王春亮　王桂九
　　　　王桂林　王晓刚　井　岗　云少闯　方　亮　巴金红
　　　　邓　琳　付亚平　兰　天　成　凡　任众鑫　刘　岩
　　　　刘　春　刘玉刚　刘浩生　刘继芹　齐得山　闫凤林
　　　　安国印　孙军治　杨海军　杨清玉　杨普国　李　祥
　　　　李小明　李文斌　李海伟　李淑平　肖罗坤　肖恩山
　　　　何　俊　汪会盟　张　幸　张志胜　张新悦　陆守权
　　　　陈加松　陈维帅　范丽林　罗天宝　周冬林　房维龙
　　　　屈丹安　孟　君　赵　岩　贡建华　柳　雄　侯　磊
　　　　姚　威　敖海兵　贾建超　徐宝财　翁小红　郭　凯
　　　　黄　玮　黄盛隆　龚　璨　焦雨佳　谢　楠　管　笛
　　　　薛　雨　霍永胜　魏东吼

前　言

金坛储气库被誉为"中国盐穴储气第一库"，位于江苏省常州市金坛区直溪镇，是我国首座盐穴储气库，设计库容约26亿立方米，工作气量约17亿立方米。项目于2003年10月启动，2007年2月注气投产，承担着长三角及华中、华南地区季节调峰和应急供气重任。作为国家石油天然气管网集团有限公司（以下简称"国家管网集团"）管理唯一已投产的盐穴储气库，金坛储气库具有"随注随采"的独特功能，投产后多次高质量完成应急保供硬任务。截至2023年底，金坛储气库工作气量突破10亿立方米，最高日采气量达2200万立方米，有效发挥了管网"心脏起搏器"的重要作用，有力保障了管道沿线企业和人民平稳用气。

金坛储气库作为西气东输工程重要组成部分，从建设之初就承载了保障天然气管网安全平稳运行的重任，建设者们勇于担当、敢于创新、精心组织、克服重重困难，确保金坛储气库工程建设及生产运行平稳，为西气东输筑牢安全平稳运行防线。2003年2月，中国石油西气东输管道公司成立储气库管理处（项目部），开启了金坛储气库建设与运营工作。2018年7月，金坛储气库由中国石油西气东输管道公司移交至中国石油华北石油管理局管理。2020年10月，按照国家油气管网运营机制改革部署，金坛储气库相关业务由中国石油华北石油管理局移交至国家管网集团西气东输公司，2022年12月13日，国家管网集团成立江苏储气库分公司负责金坛储气库建设运营。2023年12月，按照国家管网集团"四化"改革部署，国家管网集团储能技术有限公司挂牌成立，专业化负责储气库建设与运营，提速地下储能发挥作用。

金坛储气库建设之初专业技术人员严重缺乏，关键技术装备全部依靠国外引进，经过20年的探索创新，金坛储气库建设团队聚焦储气库瓶颈技术难题，攻坚克难、创新突破、立足现场大量工程及生产数据，技术发展经历了从无到有、从技术探索到成熟推广，克服国内盐层薄、夹层多、杂质含量高、建库条件差、造腔设计难度大等问题，建立完善了集勘探、钻井、造腔、运行、监测于一体的盐穴储气库建库技术体系，研发了具有自主知识产权的一系列新技术和新装备，取得多项"国内首次"的技术成果，在国内多夹层、高不溶物盐矿复杂地质条件下建成国内首座盐穴储气库。金坛储气库已成为我国在盐穴储气库建设和运营领域的科技创新基地、人才培养基地和学术交流基地。

在地质工程方面，形成了符合国内盐岩地质条件下的盐穴储气库选址评价技术，建立了盐穴储气库选址评价标准，并针对国内金坛、淮安、楚州等 10 多个盐矿进行了重点筛选和评价，金坛、淮安、楚州、平顶山等 4 个盐穴储气库已陆续进入建设阶段；开展建库区域三维地质精细化建模，攻关盐层成分测井评价技术，深度刻画盐层、夹层、断层空间展布，为井位部署和造腔设计奠定基础；完成不同岩性岩石力学性能测试评价技术，建立了盐穴储气库稳定性评价体系，明确了腔体运行压力上限和下限的评价方法；优化盐穴储气库安全矿柱比由 2.5 至 2.0，进一步提升盐矿资源利用效率。

在老腔改建储气库方面，建立了国内老腔改建储气库的筛选评价标准和程序，形成了老腔筛选、检测、评估和改建等一整套老腔改建储气库技术，金坛储气库在 2005 年首次利用盐化 5 口老腔改建储气库，形成工作气量 5000 万立方米，达到国际领先水平；引入德国 SOCON 和法国 Flodim 声呐检测仪器和设备对老腔的腔体形状进行精确检测；发明了老腔卤水试压判断腔体密封性的工艺流程和评价方法；提出了"弃井用腔"老腔改建储气库思路；首次针对盐化公司对流井复杂连通老腔开展检测和评价，探索"两注一排""两注两排""一注一排"等改建工艺。

在钻采工程方面，针对盐穴储气库钻井井眼尺寸大、顶替效率低、固井质量要求高等特点，开展优快钻井技术、固井水泥浆体系、优化井身结构等关键核心技术攻关，形成一整套盐穴储气库钻完井工艺技术，成功应用于金坛、淮安、楚州、平顶山、云应等盐穴储气库近 80 口井；在金坛储气库大规模使用五段制"S"型井建库技术，降低土地征用范围、实现储气库井集约化管理；优化了盐穴储气库注采完井及注气排卤工艺方法，金坛储气库已完成注气排卤 40 口井，注气排卤顶替效率由 93%提升至 96%以上；开展地应力测试及评价技术研究，形成了一整套压力上限配套评价技术、现场提压工艺技术、储气库完整性监测技术，增加了盐穴储气库压力上限，储气库工作气量可提升 5%。

在造腔工程方面，通过引进、消化、吸收国外造腔工程技术，结合金坛储气库建设实践经验，创新形成了一整套造腔工程配套技术和造腔跟踪预测技术，通过对成果应用前后的效果对比，根据声呐测量结果比对，腔体单腔体积提升了 37.2%，单位厚度盐层溶腔体积较之前提高了 27.9%，在复杂多夹层盐岩溶腔建库技术领域达到国际领先水平；研制并大规模应用了光纤实时监测垫层界面系统，确保了反循环造腔顺利进行；金坛储气库全部实现一次循环，同等卤水处理条件下，造腔速度提高 20%，造腔成本节约 10%以上；开发了造腔动态跟踪分析软件（GSDMAS）、夹层垮塌控制处理技术和造腔故障处理应对措施，造腔事故事件明显降低、造腔腔体形状更加规则；推广应用天然气阻溶造腔和氮气阻溶造腔工艺技术和配套设施，并开展了工程试验，试验效果良好，盐穴储气库造腔柴油垫层开始向气体垫层转变，更加经济和环保。

在注采运行与安全监测方面，建立健全了盐穴储气库注采运行安全监测技术体系，探索并推广应用多项监测手段，对盐穴储气库注采运行进行全方位的监测，主要包括温度、压力和流量监测、地面沉降监测、腔体变形声呐形状监测、井筒泄漏监测、微地震腔体垮塌监测、示踪剂泄漏监测等，有力保障了盐穴储气库注采运行过程中井筒和腔体的完整性，有效防止腔体垮塌、井筒变形，导致盐穴密封性破坏事件发生，提升了金坛储气库安全运行水平。

在数字化方面，研发出国内首款具有自主知识产权的盐穴储气库建库设计系统，该系统集成了国内首套集地质模型—造腔模拟—稳定性评价—注气排卤—投资决策—数据管理为一体的优化设计平台（CavBen）；独立自主开发国内首套具有独立知识产权的造腔模拟软件（Cav-Simu），集井腔管理、盐穴建模、造腔设计、注采模拟功能为一体，造腔匹配度 95%以上，造腔设计符合率大大提升，造腔故障发生率大大降低，实现造腔模拟软件国产化；为了优化造腔管理流程和提升造腔管理效能，自主开发盐穴储气库造腔管理分析系统（GSDMAS），可以实现生产数据的组织和存储、造腔跟踪分析及作业计划安排、造腔异常监测和故障诊断以及库区造腔配产配注方案优化；建立了盐岩破坏预测模型，提出了安全评价指标并分配权重建立了安全性定量评价标准；建立了注气排卤参数设计计算模型，并开发了设计软件，形成了高效注气排卤实施工艺；开发了优化经济投资决策模块，针对各个造腔方案进行综合评价，确定最优方案。

本书是上述成果的总结，收录了金坛储气库建设团队从 2003 年至 2023 年 20 年来在国内外重要权威刊物和学术会议上发表的 84 篇论文，获得的 18 项省部级及以上科技奖励，出版的 3 部专著，授权的 19 个专利，发布的 14 个标准等，反映了金坛储气库建设团队在地下储气库前沿趋势探究、地质设计、老腔改建、钻采工程、造腔工程、注采运行、安全监测、地面工程和经济评价等领域所取得的研究成果。本书可以为从事盐穴储气库建设和运营的工程技术人员、研究人员和管理人员提供参考。

金坛储气库所取得的成就离不开所有参与设计、施工、监理、检测、运维等单位及科研院所、高校的奉献与支持，在此谨代表金坛储气库建设方对所有参与金坛储气库建设与运行的管理人员、工程师、运维人员等表示衷心感谢。由于本次科技成果汇编主要为金坛储气库建设方阶段性成果总结，受限于篇幅要求未能将行业内各单位的高水平论文收纳其中，在此表示遗憾。

特别感谢杨春和院士在此书编制过程中给予的指导与支持，作为中国地下储能领域的科学家和"能源卫士"，为本书提供了地下储能前瞻性报告《我国深地储能机遇、挑战与发展建议》，指明了地下储能技术的发展方向。

鉴于编著者水平与认识有限，书中难免存在不妥之处，敬请广大读者批评指正。

编者
2024 年 2 月

目　录

论文篇

前沿综述

我国深地储能机遇、挑战与发展建议 ………………………… 杨春和　王同涛（4）
中国盐穴地下储气库技术现状及发展趋势 ……………………………… 李　龙（14）
中国地下储气库发展现状及展望 ………………………………………… 李建君（23）
中国盐穴储气库建设关键技术及挑战 …………………………………… 杨海军（32）
国内盐穴储气库建库关键技术研究进展 ………… 孙军治　陈加松　井　岗　等（41）
盐穴地下储气库在我国的发展动力和趋势 …………………… 房维龙　屈丹安（52）

地质设计

盐穴地下储气库盐岩力学参数的校准方法 ……… 李建君　陈加松　吴　斌　等（58）
金坛盐穴储气库地质力学评价体系研究进展 …… 陈加松　李建君　井　岗　等（68）
含夹层盐岩孔隙特征及非线性渗透模型 ………… 刘继芹　寇双燕　李建君　等（80）
楚州盐穴储气库建库地质条件分析 ……………… 薛　雨　王元刚　周冬林（88）
盐穴储气库建槽工程实践与顶板极限跨度分析 …… 屈丹安　施锡林　李银平　等（94）
层状盐层中水平腔建库及运行的可行性 ………… 杨海军　王元刚　李建君　等（104）
基于分形理论的盐岩储气库腔底堆积物粒度分布特征
　　………………………………………………… 任众鑫　李建君　汪会盟　等（114）
盐穴储气库造腔巨厚隔层处理的新思路 ………… 何　俊　赵　岩　井　岗　等（122）
盐穴储气库表征渗透率研究 ……………………… 王元刚　薛　雨　李心凯（130）
盐岩地层地应力测试方法 ………………………… 周冬林　杨海军　李建君　等（136）
Thermal analysis for gas storage in salt cavern based on an improved heat transfer model
　　………………………………… Youqiang Liao　Tongtao Wang　Long Li　et al（143）

I

老腔改建

采盐井腔改建储气库和声呐测量技术的应用 ………… 屈丹安　杨海军　徐宝财(166)
利用现有采卤溶腔改建地下储气库技术 ……………………………………… 丁建林(173)
复杂对流井连通老腔改建储气库技术 …………… 刘继芹　乔　欣　李建君　等(180)
复杂老腔改建储气(油)库可行性分析 …………………………… 杨海军　闫凤林(189)
盐化对流井老腔改建水平腔储气难点分析 ……… 薛　雨　王元刚　周冬林(194)
盐穴地下储气库对流井老腔改造工艺技术 ……… 薛　雨　王元刚　张新悦(199)
云应地区采盐老腔再利用的可行性 …………… 周冬林　李建君　王晓刚　等(207)
CSAMT法在盐化老腔形状检测中的勘探试验 ……… 井　岗　何　俊　翁小红　等(215)
Feasibility analysis of using abandoned salt caverns for large-scale underground energy storage in
　　China …………………… Chunhe Yang　Tongtao Wang　Yinping Li　et al(222)

钻采工程

中俄东线楚州盐穴储气库配套钻井液技术 ………… 薛　雨　张新悦　王立东(250)
楚州盐穴储气库钻井工程难点及对策 …………… 薛　雨　王立东　王元刚　等(258)
金坛地下储气库盐腔偏溶与井斜的关系 ………………………… 杨海军　于胜男(268)
平顶山盐穴储气库固井技术 ……………………… 张　幸　覃　毅　李海伟　等(276)
盐腔声呐测量前的井下作业施工 ………………… 杨清玉　李海伟　张　幸　等(283)
中俄东线楚州盐穴储气库固井难点及对策 ……… 薛　雨　齐奉忠　王元刚　等(286)
盐岩储气库堆积物注气排卤试验研究 …………… 任众鑫　巴金红　任宗孝(293)
盐岩储气库注气排卤期剩余可排卤水分析 ……… 陈　锋　杨海军　杨春和(300)
盐岩储气库注气排卤工艺参数的数值模拟 ……… 任众鑫　李建君　朱俊卫　等(308)
盐穴储气库气卤界面检测技术 …………………… 付亚平　陈加松　李建君(313)
金坛盐穴储气库新溶腔井注气排卤情况分析 …… 李　龙　杨海军　刘玉刚　等(321)
Modeling debrining of an energy storage salt cavern considering the effects of temperature
　　…………………………… Dongzhou Xie　Tongtao Wang　Long Li　et al(325)
Temperature distribution of brine and gas in the tubing during debrining of a salt cavern gas
　　storage ………………… Dongzhou Xie　Tongtao Wang　Long Li　et al(349)

造腔设计与分析

井间距对盐穴储气库小间距对井造腔的影响 ………… 李 龙 侯 磊 李建君 等(372)
盐穴储气库水溶造腔工艺优化研究与现场应用 …… 王元刚 周冬林 邓 琳 等(386)
光纤技术在盐穴储气库油水界面监测中的应用 ……… 付亚平 吴 斌 敖海兵 等(393)
淮安盐穴储气库厚夹层盐层造腔工艺设计 …………… 齐得山 李建君 巴金红 等(401)
盐穴储气库水溶造腔的影响因素 …………………………… 王元刚 高 寒 薛 雨 等(409)
金坛盐穴储气库精细造腔腔体体积优化 ……………… 齐得山 李建君 赵 岩 等(416)
金坛盐穴储气库腔体偏溶特征分析 ……………………… 齐得山 李淑平 王元刚(424)
盐穴储气库造腔设计与跟踪 ……………………………… 刘 春 高云杰 何邦玉 等(433)
盐穴储气库反循环造腔的试验研究 ……………………… 李 龙 屈丹安 李建君 等(443)
盐穴储气库回溶造腔技术研究 …………………………… 刘继芹 焦雨佳 李建君 等(450)
盐穴储气库快速造腔方案 ………………………………… 何 俊 井 岗 赵 岩 等(458)
盐穴储气库天然气阻溶回溶造腔工艺 …………………… 李建君 陈加松 刘继芹 等(466)
盐穴储气库溶腔参数优化方案的研究 …………………… 李建君 李 龙 屈丹安(478)
盐穴储气库巨厚夹层垮塌控制工艺 ……………………… 王元刚 陈加松 刘 春 等(484)
盐穴储气库溶腔排量对排卤浓度及腔体形态的影响 … 王文权 杨海军 刘继芹 等(492)
盐穴储气库造腔过程动态监控数据分析方法 ………… 齐得山 巴金红 刘 春 等(499)
盐穴储气库造腔过程夹层处理工艺——以西气东输金坛储气库为例
  …………………………………………………………… 郭 凯 李建君 郑贤斌(507)
盐穴储气库天然气阻溶恒压运行技术 …………………… 何 俊 井 岗 陈加松 等(513)
盐岩储库腔底堆积物空隙体积试验与计算 ……………… 任众鑫 杨海军 李建君 等(522)
金坛盐穴储气库腔体畸变影响因素 ……………………… 李建君 王立东 刘 春 等(534)
盐岩储气库腔底堆积物扩容及工艺应用 ………………… 任众鑫 李建君 巴金红 等(542)
金坛盐穴储气库现场问题及应对措施 …………………… 李建君 巴金红 刘 春 等(547)
Applications of numerical simulation in fault recognition during dissolving mining for gas
  storage caverns ………………………………………… Huiyong Song Jianjun Li (553)

注采运行与安全监测

盐穴储气库运行损伤评价体系 …………………………… 敖海兵 陈加松 胡志鹏 等(564)
金坛盐穴储气库上限压力提高试验 ……………………… 井 岗 何 俊 陈加松 等(577)

金坛盐穴储气库 JZ 井注采运行优化 ················· 陈加松　程　林　刘继芹　等（587）
盐穴储气库单腔长期注采运行分析及注采压力区间优化——以金坛盐穴储气库西 2 井
　　腔体为例 ····································· 杨海军　郭　凯　李建君（598）
盐穴储气库注采运行对邻近铁路安全影响分析 ········ 张　幸　李文斌　谢　楠　等（608）
盐穴储气库稳定性的实时微地震监测 ················ 井　岗　何　俊　陈加松　等（615）
热应力对盐穴储气库稳定性的影响 ·················· 李建君　敖海兵　巴金红　等（623）
金坛盐穴地下储气库地表沉降预测研究 ·············· 屈丹安　杨春和　任　松（631）
气密封检测技术在储气库井应用研究 ················ 李海伟　李梦雪　孟凡琦　等（641）
气体示踪技术在盐穴地下储气库微泄漏监测中的应用
　　 ·· 王建夫　张志胜　安国印　等（647）
Geomechanical investigation of roof failure of China's first gas storage salt cavern ············
　　 ································· Tongtao Wang　Chunhe Yang　Jiasong Chen　et al（658）
Failure analysis of overhanging blocks in the walls of a gas storage salt cavern：A case study
　　 ································· Tongtao Wang　Chunhe Yang　Jianjun Li　et al（677）

地面工程

盐穴储气库造腔地面工艺技术 ······································ 杨清玉　庄清泉（696）
盐穴储气库注采集输系统优化 ······················ 吕亦瑭　黄　坤　方　亮　等（699）
电磁流量计在盐穴储气库造腔过程中的应用 ·········· 成　凡　焦雨佳　张青庆　等（707）
节能技术在储气库地面工程中的应用 ················ 柳　雄　云少闯　黄　玮（712）
盐穴储气库地面工程技术要点研究 ·································· 刘　岩　程　林（715）
盐穴储气库注水站整体造腔参数优化 ················ 耿凌俊　李淑平　吴　斌　等（722）
盐穴储气库造腔管理分析系统设计与应用 ············ 李淑平　刘继芹　齐得山　等（731）

经济评价

考虑垫底气回收价值及资金时间价值的盐穴型地下储气库储气费计算方法
　　 ·· 王元刚　李淑平　齐得山　等（738）
我国地下储气库垫底气经济评价方法探讨 ············ 罗天宝　李　强　许相戈（747）
盐穴地下储气库建设项目经济评价应注意的几个问题 ······ 罗天宝　程风华　孟少辉（752）

成果篇

获得省部级及以上科技奖励 …………………………………………………………（759）

授权专利 ……………………………………………………………………………（761）

发布标准 ……………………………………………………………………………（762）

出版专著 ……………………………………………………………………………（763）

获得软件著作权 ……………………………………………………………………（764）

论文篇

前沿综述

我国深地储能机遇、挑战与发展建议

杨春和　王同涛

（中国科学院武汉岩土力学研究所，岩土力学与工程国家重点实验室）

Opportunities, challenges, and development suggestions for deep underground energy storage in China

Yang Chunhe　Wang Tongtao

(State Key Laboratory of Geomechanics and Geotechnical Engineering, Institute of Rock and Soil Mechanics, Chinese Academy of Sciences)

Deep underground energy storage(DUES)is defined as using deep underground spaces(such as depleted reservoirs, aquifers, salt caverns, and mining cavities)for the storage of oil, natural gas, hydrogen, compressed air, CO_2, and helium. It is a significant strategic option for improving the efficiency of clean energy utilization, ensuring national energy security, and ensuring the security of the strategic material supply. The large-scale application of hydrogen energy, CO_2 emission reduction, strategic oil reserves, and natural gas peakshaving have stimulated the rapid development of the DUES industry. DUES has made significant progress in China, especially in the use of depleted oil, gas reservoirs and salt caverns to build underground gas storage, which have reached international leading levels. In 1999, China built its first commercial gas storage using depleted condensate gas reservoirs in Dagang Oilfield. Till now, underground natural gas storage has been significantly developed, and a total of 24 gas storages have been built, forming a working gas capacity of 19 billion m^3. Underground oil storage in salt caverns offer the following advantages：(1)Significant economic benefits, (2)high security, (3)small footprint and(4)oil quality assurance. The 3rd phase of the underground oil reserve project is in the planning stage, with a focus on deep underground oil storage, including salt cavern oil storage depots in Huai'an, Jiangsu, and Yunying, Hubei. Compressed air energy storage power stations using deep underground spaces are being developed rapidly. Compressed air energy storage for power generation has been achieved through a grid connection in Jintan, Jiangsu, and Feicheng, Shandong. Nineteen projects are underconstruction or are being planned in Yingcheng, Hubei, Huai'an, Jiangsu, etc., with a total installed capacity of 5.38 GW. Compressed air energy storage is expected to form a new industry. The geological disposal and utilization of CO_2 have been tested on an industrial scale. Critical technologies and industrialization experiments for geological storage of hydrogen, helium, and oil are being developed. It can be expected that these will all become reality. The challenges of the DUES in China mainly include(1)lack of basic research in theoretical and technical aspects such as multi-field, multi-phase, and multi-scale coupling, *in-situ* biological chemical and physical reactions, small molecule gas seepage, and evolution of physical and mechanical properties of energy storage geological bodies;(2)lack oftop-level design planning for collaborative utilization of mining and energy storage; and(3) lack of professional and technical personnel for a field involving geotechnical engineering, engineering geology, geochemistry, oil and gas storage and transportation, energy engineering. To our best knowledge and based on a review of the literature, it is suggested to focus on(1)basic theory and key technology for DUES, targeting application scenarios such as oil storage, natural gas, hydrogen, helium, compressed air, and CO_2 disposal;(2)laboratory and field experimental research to address the challenges faced by engineering implementation;(3)survey and location of energy storage resources by combining the differences in physical properties and functions of different reserve substances; and(4)the top-level design of the strategic layout of energy storage. Priority should be given to engineering practice in salt cavern strategic oil storage, compressed-air energy storage power stations using deep underground spaces, geological hydrogen storage, and CO_2 geological disposal and utilization. Planning and management organization for the DUES should be set up. Professional and technical personnel training for the DUES should also be given priority.

能源清洁化和低碳化是全球能源发展的不可逆趋势[1]。图 1 给出了全球可再生能源装机容量及其增长速率(2017—2022 年)[2]，年平均增长速度为 8.88%。其中，风电和光伏发电装机容量增加幅度最为显著，其次为生物质发电，水力发电增加不显著。中国作为能源消费大国，能源消费主要以煤炭为代表的化石能源为主，但能源清洁化已经成为国家能源发展重大的战略部署。以天然气和风能、光能等为代表的清洁能源规模化应用是实现"双碳"战略和能源结构升级转型的关键。2016—2022 年，清洁能源在我国能源消费占比从 19.1%增加到 25.9%，年均增长速率达到 5.2%，其中风能、光能累计装机容量从 $2.46×10^8$ kW 增加到 $7.60×10^8$ kW。清洁能

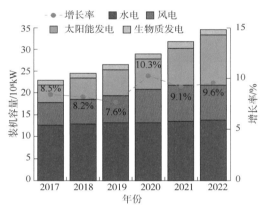

图 1 全球可再生能源累计装机容量及增长速率(2017—2022 年)[2]

源已经成为我国能源重要构成部分，并且是我国能源今后开发利用的重点，预计到 2060 年清洁能源在我国能源消费占比将会达到 80%。

以天然气、风能和太阳能为代表的清洁能源分布上具有很强的地域性和时域性，制约了清洁能源的平稳供给和高效利用。我国天然气主产区在西北而消费区在东南，生产区和消费区通过上千千米的管道连接，且消费存在较强的季节性和时段性，给天然气平稳安全供给带来了巨大的挑战。例如，2017 年席卷全国的"气荒"严重影响了国民经济发展和民生。将风能、光能等清洁能源转换为可存储的能源(压缩空气、氢气等)是克服其地域性和时域性缺点的必由之路，而大规模储能设施建设是基础和前提，建设深地储能库是最佳选择。另外，我国石油对外依存度高，但储备不足且主要以地面储备为主，容易受到战争、恐怖主义和自然灾害等影响，严重威胁国家能源和国防安全，加大储备、提高储备安全势

在必行[3]。21世纪是人类开发利用深部地下空间的世纪。因此,资源化利用我国深部地下空间进行石油、天然气、氢气、压缩空气及CO_2等能源或能源物质和氦气等战略稀缺物质的储备,对优化我国能源结构、快速提高能源储备能力和确保国家能源及战略物资安全具有重要意义,是我国能源储备重要的发展方向。

1 深地储能概念及其意义

深地储能是指利用深部地下空间(地质体)实施石油、天然气、氢气、压缩空气及CO_2等能源或能源物质和氦气等战略稀缺物质的储备,保障能源平稳供给和战略物资安全、提高清洁能源利用效率和减少CO_2排放等[1]。深地储能地质体类型包括:枯竭油气藏、含水层、盐穴和矿洞等(图2)。

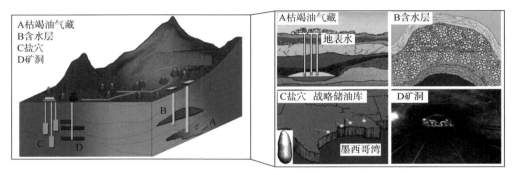

图 2 深地储能地质体类型示意图

枯竭油气藏是指已经开采衰竭或者开采到一定程度的退役油气藏,常被用来建设地下天然气储备库。截至2021年,全世界已经建成油气藏型储气库523座、工作气量约$3500×10^8 m^3$,建库数量和工作气量分别占到储气库总数和总工作气量的73%和79%(https://www.cedigaz.org/shop-with-selector/?type=publications)。用于储气的含水层一般由含水砂层和不透气覆盖层组成,全球已经建成含水层型储气库87座、总工作气量达到$467×10^8 m^3$(https://www.cedigaz.org/shop-with-selec-tor/?type=publications)。盐穴是指利用水溶法在盐岩地层中为储能专门建造的或者采矿形成的地下腔体。利用盐穴进行地下能源储备已经较为成熟,全世界目前已经建成盐穴储气库约100座、形成工作气量超$330×10^8 m^3$(https://www.cedigaz.org/shop-with-selector/?type=publications)。同时,盐穴石油储备库、盐穴压缩空气储能发电站和盐穴储氢库等都已经有成功实施的案例。我国已经建成江苏金坛和湖北江汉盐穴储气库,累计投产腔体超过40个,形成工作气量超过$10×10^8 m^3$。矿洞作为矿产资源开发完成后形成的地下空间,具有资源量大、分布广等优势,1963年在美国科罗拉多DENVER附近首次建成废弃矿洞储气库,目前世界上有3座废弃矿洞储气库[4]。同时,德国计划利用废弃的煤矿矿洞建设抽水蓄能发电站[5,6],以提高清洁能源利用效率。利用深部地下空间进行石油和天然气储备已经具有较长的发展历史并形成相关产业。随着能源低碳和清洁化发展以及对提高能源供给安全可靠度的需求不断增加,深地储能已经被赋予新的内涵并进入了快速发展阶段。

图3是利用枯竭油气藏、含水层、盐穴和矿洞等深部地下空间储能在能源互联网中作

用的示意。第一，利用深地地下空间实施压缩空气和氢储能，可以有效提高清洁能源利用效率和丰富清洁能源应用场景。以风能、太阳能和潮汐能为代表的可再生能源，存在不能连续供给和分布地域性强等缺点，给电网安全运行带来了巨大挑战，造成弃风弃光严重[3]。通过将风能、光能等可再生能源转换为压缩空气势能和氢能(电解水制氢)是解决其上述缺点的有效手段。由于压缩空气能量密度低，大规模存储压缩空气的空间是制约压缩空气储能经济性、快速发展中的重要一环。绿电制氢并进入管网是实现绿氢大规模、高效利用的关键。利用深部地下空间建设储氢库是氢能"制—输—储—销—用"产业链中重要的一环。第二，截至2022年，全世界有超过$10×10^8$ bbl原油和成品油、$4000×10^8 m^3$天然气被存储在深部地下空间中[1]，对保障全球能源安全和能源贸易发挥了巨大的作用。第三，氦气作为重要的战略物资，广泛地应用在航空航天、高端制造和精密医疗等领域。我国对外依存度长期保持在95%以上，利用地下空间实施氦气储备对保障我国氦气供给安全和大宗氦气国际采购博弈具有重要意义。第四，利用深部地下空间进行CO_2地质封存(包括利用)是实现化石能源减碳和零碳的有效途径之一。我国已经在CO_2驱油和煤层气等领域得到长足发展。因此，深地储能是能源互联网中的关键节点和重要基础设施。

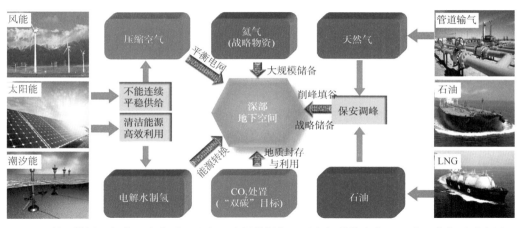

图3 利用枯竭油气藏、含水层、盐穴和矿洞等深部地下空间储能在能源互联网中作用示意图

2 新形势下深地储能机遇

2.1 清洁能源大规模应用和CO_2减排

以风能和太阳能为代表的清洁能源在我国能源整体消费占比从2016年的19.1%增加到2022年25.9%(图4)。根据预测，2030年我国清洁能源消费量占比将要达到30%以上。由于风能、光能等清洁能源具有不能持续供给和地域性分布强等天然缺点，2022年新疆、西藏、山东和内蒙古等地出台相应政策要求风光发电项目需要配备1~2h 5%~40%装机容量的储能。储能作为风能、光能等清洁能源项目上马的先决条件，客观上促进了储能产业的发展。依托深部地下空间建设大型压缩空气储能发电站具有储能规模大、经济性好和安全性高等优势，已经在行业内达成共识，例如德国Huntorf和美国Mcintosh压气蓄能电站[7]等都是以深部地下盐穴作为储能载体，并已经实现了商业化运行。压缩空气储能已经被列入

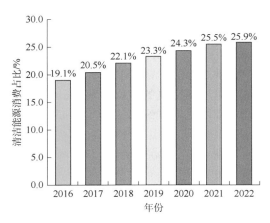

图4 2016—2022年中国清洁能源消费占比

国家"十四五"储能重点发展方向[8]。2021年，中国压缩空气储能实现了跨越式增长，新增投运规模170MW，累计装机规模为183.35MW，规划建设的包括：江苏淮安465MW/2600MW·h压缩空气储能项目、湖北应城300MW压缩空气储能项目和辽宁朝阳300MW压缩空气储能项目等。

将风能、光能等清洁能源产生的富裕电能转换为氢能进行储能也是大规模储能的重要方式，是国际上储能领域研究的热点和难点，也是我国大规模储能优先发展的方向之一。根据《中国氢能及燃料电池产业白皮书》预测[9]，2030年中国氢能年需求量将高达3500×10^4t，2050年氢能将在我国终端能源体系中占比10%，年需求量6000×10^4t（折$6600 \times 10^8 m^3$）。面对如此规模化氢能利用，必须建立相应的保安调峰储备。利用深部地下空间储氢是实现大规模储氢的经济和安全手段，可以实现万吨级规模的氢能储备，满足太瓦和季节性调峰需求。美国和英国等开展了利用深部地下盐穴实施大规模氢能储备方面的现场试验[10]，单个盐穴最大储氢量可以达到近万吨。德国、法国和美国等已经开展了H_2和CO、CH_4等混合气体在地下含水层和衰竭油气藏中大规模存储的工程试验[11]，并取得了相关工程经验和基础理论认识。

CO_2地质封存是实现化石能源低碳甚至零碳排放的有效途径，已经被国际社会列为固碳的主要技术手段[12]。化石能源作为我国能源主要构成部分（图5），今后很长一段时间内仍将是我国能源消费的主体，减碳和固碳压力巨大。CO_2封存地质构造包括衰竭油气藏、咸水层、盐穴和深部煤层等。根据测算，我国2030年和2060年CO_2地质封存需求量分别为$(0.2 \sim 4.08) \times 10^8$t和$(10 \sim 18.2) \times 10^8$t[13]。超临界状态下，封存1t CO_2需要地下空间体积为$1 \sim 1.4 m^3$，则2030和2060年CO_2地质封存需要地下空间将会达到$4 \times 10^8 m^3$和$18 \times 10^8 m^3$，给深部地下空间利用带来新的机遇。

2.2 战略石油和成品油储备

1993年我国从石油出口国转换为石油净进口国，近年来我国石油对外依存度超过70%，而石油储备量与石油消费量严重不匹配且石油储备主要以地面储罐为主，严重威胁到我国石油供给安全。石油具有易燃易爆的特征，火灾等隐患巨大。例如，1989年黄岛油库火灾、2011年日本地震诱发千叶县地面储罐火灾[3]和2022年古巴雷击导致大量战略储备石油被烧毁等（http：//news.sohu.com/a/678463926_121123871）。同时，地面储油罐也是军事打击和恐怖袭击的主要对象（https：//k.sina.cn/article_6435187353_17f912a9900100sd3w.html?mod=wpage&r=0&tr=381）。地面储罐占地面积大。以建设500×10^4t地面石油储备库为例，大约需要2000亩土地，且在距离油库一定范围内不能从事人员密集型生产和活动。利用深部地下空间，尤其是深部地下盐穴进行战略石油和成品油储备，具有经济性好、安全性高和占地少等优势，已经在欧美多个国家得到推广和应用。目前全世界有超过10×10^8bbl原油和成品油存储

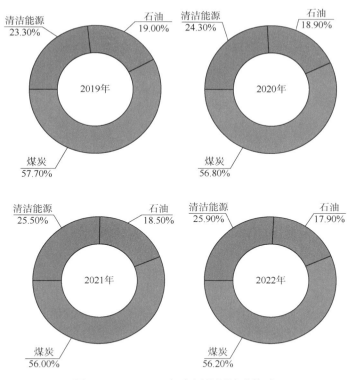

图 5 2019—2022 年中国能源消费构成

在深部地下盐穴中[1],为保障相关国家能源安全作出了巨大贡献。根据国际能源组织建议,一个国家战略石油储备量应不少于其 90 天的消费量才能保障其石油供给安全。2022 年,我国成品油消费量达到了 $3.26×10^8$ t,将成品油存储于深部地下空间也是我国成品油存储的重要发展方向。这为深部地下空间储能带来了新的需求。

2.3 天然气调峰与战略储备

天然气作为一种清洁能源,是替代煤炭和石油等化石能源的一种过渡能源,在我国能源消费占比中逐年增加。2017 年,"气化中国"战略实施以来天然气消费在中国呈现出爆发式增长。2022 年,中国天然气表观消费量达到了 $3663×10^8 m^3$,其中进口天然气为 $1532×10^8 m^3$,占比 41.8%,加大储备是保障天然气供给安全的迫切需求。同时,我国天然气主产区在西南和西北地区,而天然气主要消费区在中东部地区,且消费量随季节性变化大,大规模天然气存储是保证长输管道天然气平稳供给的关键。目前我国大规模天然气储备主要包括:地下油气藏型储气库、盐穴储气库和 LNG 储罐。根据国际惯例,储备工作气量达到年消费量的 15% 左右才能保证管网运行和天然气供给安全[1,3]。我国天然气储备严重滞后天然气产业发展需求,供需不平衡问题突出,导致大规模"气荒"屡次发生。2020 年 4 月 10 日,国家发展和改革委员会等 5 部门联合下发《关于加快推进天然气储备能力建设的实施意见》,要求加快天然气储备能力建设,提升天然气储备能力。2023 年 4 月 12 日,国家能源局印发《2023 年能源工作指导意见》,要求着力增强能源供应保障能力,以地下储气库为主,沿海 LNG 储罐为辅,推进储气设施集约布局。根据预测,2030 年我国天然气消费量将

会达到 $6000×10^8m^3$，保安调峰需要的储备工作气量将会达到 $900×10^8m^3$。加快天然气战略储备制度建设已经在全世界形成共识。根据我国天然气消费市场规模及发展前景预测，战略储备规模将会达到 $600×10^8m^3$。目前，我国累计建成储气库24座，形成工作气量约 $190×10^8m^3$，距离实际需求还具有很大的差距，这为利用深部地下空间储能带来了新的机遇。

3 我国深地储能现状

（1）储油库。

2008年我国石油对外依存度突破50%，2019年超过70%。我国的战略石油储备相对滞后，战略石油储备量占比消费量的比例远低于美国、德国、法国等欧美发达国家。我国石油储备基地大部分建在沿海地区，且以地面储备为主。截至2014年，国家战略石油储备一期工程已经建设完成、形成原油储备能力 $1243×10^4t$，储备基地均位于沿海港口。石油储备第二期建设正在进行，包括：辽宁锦州、山东青岛和甘肃兰州等。较第一期更具有战略纵深和安全性，储备方式由传统的地面储罐转为水封花岗岩硐室、盐穴等多种方式。石油储备第三期正在规划中，建设重点为深部地下石油储库。

（2）储气库。

20世纪60年代末，在大庆油田开始尝试利用废弃油气藏改建为地下储气库。1999年，我国首次在大港油田利用枯竭凝析气藏建成了第一座商业化的调峰储气库——大港大张坨地下储气库，并于2000年投产运行。2007年，金坛储气库开始投产运营，是我国第一个盐岩地下储气库，也是世界上第一个利用已有溶腔改建而成的储气库。截止到2022年底，我国累计建成储气库24座，形成工作气量约为 $190×10^8m^3$，其中衰竭油气藏型储气库20座、盐穴储气库4座。储气库工作气量约占天然气消费量（2022年）的5%，距离保安调峰所需的15%消费量仍具有较大差距。中东部地区是中国经济最发达区域之一，对天然气等能源消费需求旺盛，具有丰富的盐矿资源，例如江苏金坛、江苏淮安和河南平顶山等大型盐矿，为建设盐穴储气库提供了便利。目前金坛盐穴储气库已经建成投产盐穴40余个、形成工作气量超过 $10×10^8m^3$。依据国家总体战略部署，我国将形成四大区域性联网协调的储气库群，2025年全国储气能力达 $(550~600)×10^8m^3$，2030年达到 $(600~700)×10^8m^3$，2035年达到 $(700~800)×10^8m^3$。

（3）压缩空气储能。

2009年，利用盐穴实施压缩空气储能发电被提出并结合金坛盐矿地质条件开展了地质可行性评价，筛选出10余个采卤盐穴作为建设压缩空气储能电站的潜在库址[14]。2021年，在前期盐穴选址基础上建成了我国首座压缩空气储能发电站——金坛盐穴压缩空气储能发电站。该储能发电站设计功率为60MW，远期设计功率为300MW，于2022年5月并网发电[15]。山东肥城盐穴压缩空气储能发电站2021年9月实现并网发电，设计储能发电功率为10MW。由于利用深部地下空间进行压缩空气储能具有安全、经济和高效等优势，在我国呈现出井喷式的发展。例如，湖北应城300MW盐穴压缩空气储能项目、江苏淮安400MW压缩空气储能项目、山东肥城300MW压缩空气储能示范项目二期等正在建设和规划建设的项目共有19个，规划总装机量达到5.38GW（https：//www.163.com/dy/article/HUBUU2NH0552DME7.html）。根据预测，我国2025年压缩空气储能发电装机量达到6.76GW，2030年达到43.15GW（https：//

chuneng. ofweek. com/news/2022-10/ART-180221-8120-30575757. html），有望形成新的产业。

（4）储氢库。

2023 年，湖北大冶绿电绿氢制储加用一体化氢能矿场综合建设制氢工厂项目正式开始建设，其中包括利用采矿矿洞进行氢气存储的工程试验，有望成为中国地质储氢第一库（http：//www. hbdaye. gov. cn/xwzx/tpxw/202303/t20230326_1000793. html）。中国科学院武汉岩土力学研究所主要负责采矿矿洞储氢可行性分析及现场工程试验关键技术等方面工作。同时，中国科学院武汉岩土力学研究所正在与中国平煤神马集团联合开展利用盐穴进行大规模储氢相关研究工作。

4 存在的问题

深地储能在我国已经取得了长足发展，尤其是在利用枯竭油气藏和盐穴建设地下储气库方面已经达到了国际领先水平，深地压缩空气储能、盐穴储油、盐穴储氦、地质储氢和 CO_2 地质处置等都在蓬勃发展，深地储能产业逐渐形成，但是仍然存在以下问题。

（1）深地储能领域前沿基础研究不足，尚未形成系统的理论体系，无法对深地储能关键新技术策源提供支撑。利用深地储氢、压缩空气和氦气等涉及多场多相和多尺度耦合、原位生物—化学和物理反应、小分子气体渗流、储能地质体物理力学性能演化等理论和技术方面的新问题和新挑战，是国内外深地储能最新研究前沿。由于前期未涉及类似工程情景，导致已有的基础理论和关键技术无法满足深地储能发展新形势的需求。

（2）深地储能缺乏顶层设计规划，采矿和储能等相关领域协调不足，降低了深部采矿地下空间利用效率，滞后了深地储能实施进程。利用采矿形成的深部地下空间进行储能，可以变废为宝，实现采矿空区的资源化利用，显著降低储能成本、加快储能库建设进度。由于缺乏顶层规划设计，未能实现采矿和深地储能一体化协同发展，导致采空区安全性、密封性和耐久性等无法满足深地储能要求，造成深部地下空间资源的大量浪费。例如，我国已有盐矿水溶开采形成的盐穴超过 2000 个，而可以满足储能等需求的不足 50 个。

（3）深地储能专业化人才缺乏，人才培养有待于进一步加强。深地储能涉及岩土工程、工程地质、地球化学、油气储运、能源工程等领域，是典型的交叉学科，是经济发展和能源消费遇到的新情况，导致相关科技人才储备不足，无法满足大规模深地储能快速发展需求。

5 加快深地储能发展建议

5.1 加强基础理论和关键技术研究

（1）针对利用衰竭油气藏建设储气库，突破油藏建库技术瓶颈，拓展建库新领域，优先开展库容高效动用关键技术研究提高在役库容利用效率。开展推动储气库智能化建设关键技术和拓展储气库新功能等方面的研究。

（2）针对我国层状盐岩地质特征，重点开展高杂质含盐地层建库新技术研究，提高建库效率，实现层状盐岩多建库、建大库和低成本建库。

（3）聚焦地质储氢相关理论和关键技术研究，揭示地质储氢原位生物、化学和物理反应机制和机理以及对储能地质体性能演化影响规律，形成地质储氢系列理论体系。

（4）加快利用深部地下空间实施压气蓄能方面的研究工作，攻克高频率、高强度注采和剧烈温度变化等对储能库安全演化影响的调控关键技术。

5.2 加强室内和现场试验研究

（1）研发模拟深地储能过程中高温、高压、渗流、生物—化学—物理反应等室内系统，实现对实际深地储能工况的精准模拟，为深地储能实施提供实验手段。

（2）利用深部地下空间建立深地储能原位实验室，开展多尺度的深地储能模拟实验，获得跨尺度条件下深地储能库安全演化规律，研发深地储能安全调控关键技术。

（3）结合实际深地储能工程建设观测站和实验室，开展现场专项实验和反演分析，优化储能库运行参数和监测方案，提高储能库运行效率和安全水平。

5.3 开展深部地下空间资源普查与储能库选址工作

（1）开展衰竭油气藏精细化普查工作，划分出优先建库区、可建库和待建库区，并结合国家储气库建设规划推荐出建库库址。

（2）开展盐矿和盐穴资源普查工作，摸清盐矿和盐穴资源家底，并结合储油、储气、储氢和压气蓄能等技术特征，推荐出优先建库目标库址。

（3）开展含水层有利圈闭筛查评价技术研究并形成相应的评价技术体系，对全国含水层资源进行普查，为含水层储能库建库推荐出潜在库址。

（4）开展 CO_2 驱油、咸水层地质封存和深煤层封存 CO_2 等地质封存 CO_2 地质体普查工作，结合区域 CO_2 封存需求及运输成本，推荐出 CO_2 处置库址。

（5）开展煤矿等非金属矿和铁矿等金属矿采空区普查工作，结合储油和压缩空气储能等技术特征，给出潜在建库库址。

5.4 开展储能库建库战略布局顶层设计

（1）加强油气藏开发与建设天然气储库、压缩空气储能发电站、CO_2 处置和储氢库等顶层设计，兼顾油气开发与储能需求，最大限度实现地面设备设施、井筒和地下空间等安全高效利用。

（2）加强盐矿开采规划设计，因地制宜实施新型采矿方法，实现采矿—储能一体化发展，带动盐矿开采向能源储备方向发展。

（3）加强以煤矿为代表的非金属矿和以铁矿为代表的金属矿开采的规划设计，积极对接国家和地方储能规划，实现采矿储能协调发展。

（4）将深地储能作为确保国家能源和国防安全的重要措施之一，设立国家深地储能规划管理部门，定期制定并发布"国家深地储能"规划，加强我国深地储能的顶层设计。设置深地储能学科，加强深地储能专业技术人才培养。

6 总结与展望

随着我国能源清洁化和低碳化发展不断加速、石油和天然气对外依存度保持高位，深地储能在提高清洁能源利用效率、保障国家能源安全和确保战略物资供给等方面优势显著，为应对新形势下能源安全挑战提供了新的解决思路和方法，但是深地储能仍然存在一系列理论和技术挑战亟须解决。在基础理论方面，建议针对深部地下空间储油、天然气、氢、氢、压缩空气和CO_2处置等应用场景开展具有针对性的研究工作,形成深地储能基础理论体系,为深地储能新技术研发和重大工程落地提供理论和技术支撑。在工程实践和人才培养方面,建议优先开展战略石油盐穴储备、深部地下空间压缩空气储能发电、地质储氢和CO_2地质处置等方面的技术开发和工程实践工作,设置国家深地储能规划管理部门,加大相关人才培养力度,促进深地储能形成新的产业和经济增长点。

致谢

感谢中国科学院战略性先导项目(XDC10020300)、国家自然科学基金(42072307)和湖北省杰出青年基金(2021CFA095)资助。本文部分观点和信息根据第748次香山科学会议"地下储能"与会李术才院士、赵文智院士、李阳院士、李根生院士、孙焕泉院士和马新华教授级高工等报告和发言总结获得,在此一并表示最诚挚的感谢!

参 考 文 献

[1] 杨春和,王同涛.深地储能研究进展.岩石力学与工程学报[J].2022,41:1729-1759.

[2] 水电水利规划设计总院.中国可再生能源发展报告2022[R].2023.

[3] Yang C, Wang T, Chen H. Theoretical and technological challenges of deep underground energy storage in China[J]. Engineering, 2023, 25: 168-181.

[4] Muhammed N S, Haq B, Al Shehri D, et al. A review on underground hydrogen storage: Insight into geological sites, influencing factors and future outlook[J]. Energy Rep, 2022, 8: 461-499.

[5] 何涛,王传礼,高博,等.废弃矿井抽水蓄能电站基础建设装备关键问题及对策[J].科技导报,2021, 39: 59-65.

[6] Morabito A. Underground cavities in pumped hydro energy storage and other alternate solutions. In: Cabeza L F, ed. Encyclopedia of Energy Storage[J]. Netherlands: Elsevier, 2022, 3: 193-204.

[7] Soltani M, Kashkooli F M, Jafarizadeh H, et al. Diabatic compressed air energy storage (CAES) systems: State of the art. In: Cabeza L F, ed. Encyclopedia of Energy Storage[J]. Netherlands: Elsevier, 2022, 2: 173-187.

[8] 国家发展改革委国家能源局关于印发《"十四五"新型储能发展实施方案》的通知.发改能源[2022]209号[EB/OL].(2022-01-29)[2023-07-15]. https://www.gov.cn/zhengce/zhengceku/2022-03/22/content_5680417.htm.

[9] 中国氢能联盟.中国氢能源及燃料电池产业白皮书[M].北京:人民日报出版社,2020.

[10] Crotogino F, Schneider G S, Evans D. Renewable energy storage in geological formations[J]. J Power Energy, 2018, 232: 100e14.

[11] 周庆凡,张俊法.地下储氢技术研究综述[J].油气与新能源,2022,34: 1-6.

[12] Li Q Y, Chen Q, Zhang X. CO_2 long-term diffusive leakage into biosphere in geological carbon storage[J]. Chin Sci Bull, 2014, 59: 3686-3690.

[13] 蔡博峰,李琦,张贤,等.中国二氧化碳捕集利用与封存(CCUS)年度报告(2021)——中国CCUS路径研究,2021.

[14] 杨花.压气蓄能过程中地下盐岩储气库稳定性研究[C].武汉:中国科学院武汉岩土力学研究所,2009.

[15] 陈海生,李泓,马文涛,等.2021年中国储能技术研究进展[J].储能科学与技术,2022,11: 1052-1076.

中国盐穴地下储气库技术现状及发展趋势

李 龙

（国家管网集团西气东输公司）

【摘 要】 本文从国内能源结构与地区条件方面论证了国家管网公司发展盐穴储气库的必要性。以我国21年的盐穴储气库工程建设实践为基础，分析了国内盐穴储气库建设在建库技术、运行管理和建设平台等方面的发展优势。此外，国内盐穴储气库的发展存在建库地质条件复杂、造腔周期长、腔体形状控制难度大、复杂老腔检测与改造困难、高不溶物盐层成腔率低等瓶颈问题，面临安全、环保等外部因素和市场化程度、政策机制方面的挑战。针对这些问题和挑战，提出了加快储气库关键技术研究、优化整体建设方案、完善政策机制和加快数字化转型等对策与建议。

【关键词】 盐穴储气库；优势；瓶颈问题和挑战；科技创新

Technical status and development trend of salt-cavern UGS in China

Li Long

(PipeChina West East Gas Pipeline Company)

Abstract From the aspects of global energy structure and domestic and regional conditions, the necessity of PipeChina to develop salt cavern gas storage is demonstrated in this paper. Based on the 21 year construction practice of salt cavern gas storage engineering in China, this paper analyzes the development advantages of domestic salt cavern gas storage construction in terms of construction technology, operation management, and construction platform. In addition, the development of salt cavern gas storage in China has some bottleneck problems, such as complex geological conditions, long cavity construction period, difficult cavity shape control, arduous detection and transformation of complex old cavity, low cavity formed rate of salt layer with high insoluble, and challenges from external factors such as safety and environmental protection, as well as from marketization degree and policy mechanism. In view of these problems and challenges, countermeasures and suggestions are put forward, such as accelerating the research on critical technologies of salt-cavern UGS, optimizing the overall construction plan, improving the policy mechanism and accelerating the digital transformation.

作者简介：李龙，男，高级工程师，1966年10月13日生，毕业于西南石油大学采油工程专业，主要从事地下盐穴储气库工程建设及生产运营技术管理的研究工作。地址：江苏省镇江市南徐大道60号商务A区A座，212000。电话：021-50954899，E-mail：lilong@pipechina.com.cn。

Keywords Salt-cavern UGS; Advantages; bottleneck problems and challenges; Scientific and technological innovation

2002年，我国首次开展盐穴储气库的建设和运营工作。现已建成了中国盐穴储气库第一库——金坛储气库，形成了盐穴储气库的库址选取、初步设计、钻井、水溶造腔、注采运行和安全监测等全套建库技术。在已建成的诸多种类的储气库中，盐穴储气库具有吞吐量大、注采频次高、损耗小、灵活性高等显著优势，在保障长输管道平稳运营和天然气持续供应方面起到了"压舱石"作用。盐穴储气库是天然气生产调峰和资源储备的最佳选择，不仅可以满足年、季度、月、日，甚至小时调峰的需求，而且可以应急采气，有助于输气管网和生产系统的运行与优化[1]。

1 国内盐穴储气库现状与技术积累

1.1 发展现状

国内盐穴地下储气库建设起步较晚，21年来经历了从无到有，从技术探索到技术创新的过程。目前，国内有4座盐穴储气库，如表1所示，分别是国家管网金坛储气库、中国石化金坛储气库、港华金坛储气库(一期)和中国石化江汉储气库，其中国家管网金坛储气库是国内第一座投产的盐穴储气库[2]。

1.2 技术积累

我国的盐穴储气库经过21年的探索和发展，在引进国外技术的基础上进行自主创新，目前已形成了一整套工程建设技术和生产运行管理体系，包括选址[3-4]、勘探、钻完井、造腔、注气排卤、带压作业、注采运行等，同时研发出了一整套拥有自主知识产权的装备和技术，经过多年技术积累，国内盐穴储气库建库技术体系已基本成形，部分单项特色技术达到国际先进水平。

表1 中国盐穴储气库储气能力情况

名称	设计库容/$10^8 m^3$	设计工作气量/$10^8 m^3$	已形成库容/$10^8 m^3$	已形成工作气量/$10^8 m^3$	设计最大日注气量/$10^4 m^3$	设计最大日采气量/$10^4 m^3$
国家管网金坛储气库	26.39	17.14	15.6	10	820	1500
中国石化金坛储气库	12	7	4.6	2.8	450	600
港华金坛储气库	10	6	3.6	2.2	150	600
中国石化江汉储气库（一期阶段）	2.37	1.4	0.25	0.14	250	200

1.2.1 老腔改建储气库技术

老腔改建储气库是一种投资低、周期短的建库方式[5-6]。如图1所示，西1井是金坛储气库中盐化老腔改建储气库的第一口井，主要采用"弃井用腔"的方式进行改建，开创了国内盐穴储气库老腔建库的先河。第一批共改建5口老腔，新增工作气量约$5000×10^4 m^3$，至今已累计安全运行16年。在淮安储气库建设中，初步研究了对流井连通老腔改建储气库技术，可采用"两注一排"或"一注一排"的注气排卤方式并在现场开展了验证试验[7]。

（a）金坛储气库老腔改建储气库群

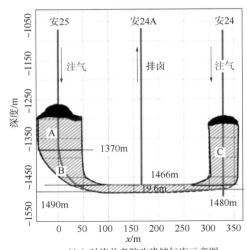
（b）对流井老腔改建储气库示意图

图 1 老腔改建储气库

1.2.2 造腔工程配套与监控技术

通过引进、消化、吸收国外造腔工程技术，结合金坛储气库建设实践经验，创新形成了一整套造腔工程配套技术和造腔跟踪预测技术（GSD）[8]。独立自主开发国内首套具有独立知识产权的造腔模拟软件（CavSimu），完全取代国外造腔软件，集井腔管理、盐穴建模、造腔设计、注采模拟功能为一体，造腔匹配度95%以上，造腔设计符合率大大提升，造腔故障发生率大大降低。可实现排卤浓度与垫层压力的实时模拟，对在建盐穴的腔体扩容进度、泥质夹层对腔体形态的影响、库区内腔体的生长情况及相邻腔体间的相互关系等实现动态监控，随着自主研发的溶腔模拟软件 CavSimu 在金坛储气库全面推广应用，造腔跟踪技术的持续进步，金坛储气库造腔效率实现了较大提升。通过对成果应用前后的效果对比，根据声呐测量结果比对，腔体单腔体积提升了37.2%，单位厚度盐层溶腔体积较之前提高了27.9%。在复杂多夹层盐岩溶腔建库技术领域达到国际领先水平[9]。

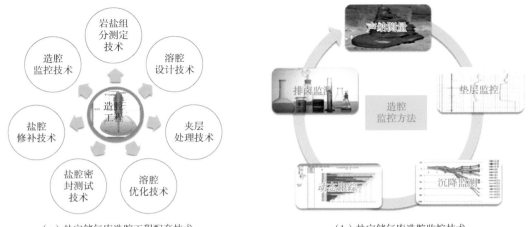

（a）盐穴储气库造腔工程配套技术　　　　（b）盐穴储气库造腔监控技术

图 2 造腔工程配套与监控技术

1.2.3 盐穴储气库气体阻溶造腔技术

在造腔过程中由于地质资料有限、工艺参数不合适或人为误操作等因素导致局部腔体形态未达到设计要求，可以使用易回收、成本低且无污染的天然气作为阻溶剂在注气排卤阶段进行回溶，不仅可以对原不规则腔体进行修复，还可以实现原腔体体积的扩容。如图3所示，金坛储气库开展国内首次天然气阻溶造腔工程试验，并形成一整套配套技术[10]。

氮气阻溶技术采用氮气取代原来的柴油作为阻溶剂，具有高效、健康、环保的特点。金坛储气库JK7-1井开展了国内首次氮气阻溶造腔工程试验并形成配套工艺[11]，利用液氮泵车或地面注氮橇装设备实现液氮气化和氮气注入，实现氮气阻溶造腔，如图4所示。

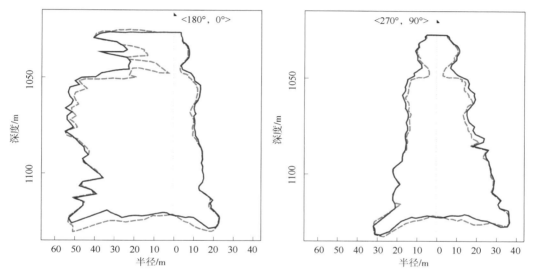

图3 腔体修复前后声呐检测形状对比

红色虚线—修复前形状；蓝色实线—修复后形状

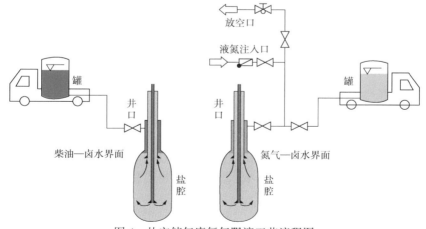

图4 盐穴储气库氮气阻溶工艺流程图

1.2.4 光纤油(气)水界面检测技术

造腔过程中，阻溶剂界面位置控制对腔体形状至关重要。利用集成了感温光缆与电加热单元的复合光缆,开发了国内首套光纤油(气)水界面检测仪[12],如图5所示。较中子测井,大大降低了成本,同时可实现连续检测,保障了造腔成功率和安全性。

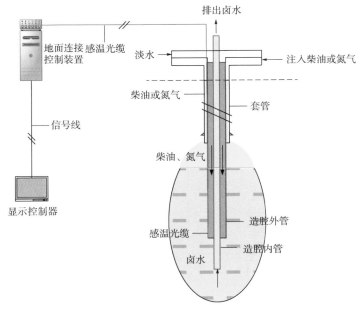

图5 光纤界面检测示意图

1.2.5 盐穴储气库安全监测技术

为确保盐穴储气库注采运行过程中井筒和腔体的完整性,防止腔体垮塌、井筒变形,导致盐穴密封性破坏事件发生,提高储气库运行的安全性,金坛储气库通过探索并推广应用了多项监测手段,对盐穴储气库注采运行进行全方位的监测,主要包括温度、压力和流量监测、地面沉降监测[13]、腔体变形声呐形状监测、井筒泄漏监测[14]、微地震腔体垮塌监测[15-16]、示踪剂泄漏监测等,从而形成了系统的盐穴储气库注采运行监测技术体系。

2 中国盐穴储气库发展的困难

2.1 地质条件复杂,建设难度大

2.1.1 建库地质条件复杂

我国的盐矿主要为湖相沉积盐层,与海相盐矿[17]相比,湖相沉积形成的盐体矿床规模和分布面积相对较小,沉积过程中盐层单层厚度小、层数多,多含有不溶夹层且盐岩品位相对较低,因此利用陆相盐矿建设储气库时,需着重对盐层的密闭性和稳定性进行分析。另外还受到与主输气管道距离和所在区域天然气消费量的制约,优质库址将更为稀少[18]。

2.1.2 水溶造腔周期长

受地质条件、造腔方式、管柱尺寸、盐化企业卤水接收能力等多方面因素影响,盐穴储气库造腔周期长。金坛储气库平均单腔有效体积约$20 \times 10^4 m^3$,造腔周期一般为6~8年。造腔期间需多次调整造腔管柱深度位置,另需在每个造腔阶段结束后对腔体进行形状检测,单腔达容时间长。

2.1.3 腔体形状控制难度大

为满足盐腔稳定性和容积尽量大的需求,一般将腔体设计成"梨形"。但是,由于造腔层段内隔夹层发育,非均质强,各向异性显著,岩盐溶漓时侧溶速率小于上溶速率等原因,腔体存在不同程度的偏溶[19],部分腔体顶部形成了"大平顶",或因阻溶剂界面失稳,导致腔体横向发展未达到设计要求,造成盐腔形状较差、岩盐利用率变低。

2.1.4 高不溶物盐层成腔率低

盐岩层中的不溶物主要是指泥质、钙芒硝和石膏等不溶于水的物质。金坛区块盐层的不溶物含量约为15%,楚州、淮安、平顶山等储气库不溶物含量分别为42.76%、27.4%、23%,不溶物在卤水中膨胀后占据了部分盐腔有效体积,使得成腔率降低,如图6所示。

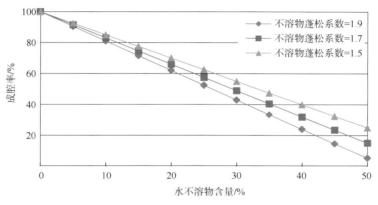

图6 成腔率与水不溶物含量关系图

2.2 外部因素影响

储气库作为系统工程,受安全、环保等外部因素影响较大。储气库建设涉及规划、用地、安评、环评等各类专项评价,办理周期长,协调难度大,且随着环保政策日益收紧,新建项目与生态保护红线、自然保护区、永久基本农田等相冲突的情况逐渐增多。同时,库区附近盐化工厂卤水处理能力有限,极大制约了循环排量,限制了造腔的进度,直接影响储气库项目的建设工期。

3 对策与建议

结合我国盐穴储气库发展的瓶颈问题和挑战,急需开展复杂建库关键技术研究、管理创新和完善相关的政策机制。

3.1 加快盐穴储气库关键技术研究

3.1.1 造腔速度提高和过程优化管理

卤水浓度对岩盐的溶漓速率影响很大,优选大尺寸井眼匹配大排量可提高造腔速度[20]。另外,优化造腔阶段和循环方式、加强阻溶剂界面监测与控制,做好造腔与井下作业的转换衔接,可提高造腔效率、缩短造腔周期。加强造腔各阶段的监测,针对未达到设计要求的腔体,制定针对性修复措施,统一安全、速度与质量的对立关系,辩证看待"存量"与"增量"问题,思想上由"建成腔"向"建好腔"转变。

3.1.2 水平多步法造腔关键技术

对于厚度较小的盐层采用单井单腔的方式建库形成的腔体体积小，单位工作气量投资大，岩盐资源动用量小。因此，水平多步法造腔是一种较合理的薄盐层建库方式。如图7所示，主要方法是钻1口直井（B井）和1口水平井（A井），先溶直井形成腔体后连通水平井，再进行水平井段多步造腔，形成水平腔，再通过B井完成注气排卤作业。腔体的体积可以通过调整水平段的长度进行确定[21-22]。

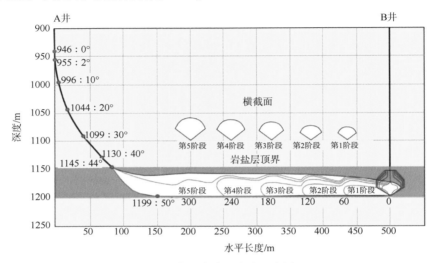

图7 水平多步法造腔示意图

3.1.3 盐腔空间扩展利用技术

井下采盐后形成的腔体主要被蓬松后的水不溶物和卤水充满，注气排卤后腔体底部被水不溶物和部分卤水占据了一定的体积，不溶物含量越高，其成腔率越低。因此，若能对沉渣空隙进行利用，将大大提高库容和工作气量[23-24]。腔底的水不溶物残渣就如同一块吸饱水的海绵，必须要将其内部的卤水驱替出来，才能实现储气利用。急需开展不溶物水化膨胀机理研究，通过物理化学方法抑制其膨胀，另外，需验证气驱水的可行性。

3.2 优化储气库整体建设方案，打造集约型储气库

在我国，储气库主要是作为长输管道的附属设施进行规划建设，主要是为了解决峰谷供气问题，确保天然气管道安全平稳运行，提高管道运行效率。储气库地面建设主要受到规划用地、安全、环保等多方面因素制约[25]。为解决这一问题，必须要优化储气库整体建设方案，比如采用丛式井建库[26]，减少地面井场数量和占地面积，打造集约型储气库。同时要提高储气库的安全性和环保能力，采用成熟工艺技术，确保安全。

3.3 加快数字化转型，提速智慧储气库建设

盐穴储气库注采速度远大于普通气藏，以及由渗透性储层改造形成的储气库，在注采周期交变载荷下，盐穴储气库的完整性会受到一定影响[27]。同时，由于我国地下盐岩资源多为陆相沉积盐丘，其中多含有薄厚不等的泥质夹层，对盐穴的成腔形状及密封性会产生重要影响，因此，在设计选址、施工建造、运行管理等各环节需要加强技术攻关，构建地上地下一体化的盐穴储气库数字化模拟系统，形成盐穴储气库数字化平台。

利用盐穴储气库数字化平台，可以实现统筹建库各环节与生产运行工作，构建地质、井筒、地面设施三位一体的信息化支持平台，提升盐穴储气库的自动化和数字化水平，进一步提速智慧储气库建设，提升工作效率，降低人工成本。

3.4 构建市场化储气库运营模式

按照2017年发布的《中长期油气管网规划》，为缓解供气紧张，2025年中国地下储气库工作气量将达到$300×10^8 m^3$。天然气储气库的大力发展，有利于促进天然气贸易，进一步保障供气稳定，但储气设施的建设需要大量投资，储气资源的利用及运营模式需要合理优化。根据2020年发布的《关于加快推进天然气储备能力建设的实施意见》，要避免储气设施小型化、分散化，鼓励合建共用储气设施，形成区域性储气调峰中心。建立健全储气设施运营模式，由于储气库建设周期较长，应支持各类主体及资本参与储气设施的建设与运营，科学整合自建、租赁储气设施资源，提高储气设施运行管理水平。

因此，需要根据建库资源分布情况，形成统一的协调、统筹管理体系，将距离较近的小型储气库科学配置组成储气库群，进行统一规划、统一调配，通过合理整合库容资源增大有效调峰规模，同时可以大幅度降低投资成本。

4 结论

（1）我国盐穴储气库历经21年的发展，实现了从无到有，从引进合作到自主创新，从金坛单点突破到楚州、淮安、平顶山、云应多点开花的快速发展阶段，形成选址、造腔、运行、监测、老腔改造一体化技术，研发出具有自主知识产权的一系列新技术和新装备，建立了专业化的技术和管理团队和功能强大的生产运行管理系统。随着国内天然气市场需求量的逐步扩大，加上一系列在建和待建盐穴储气库的建设需求，我国的盐穴储气库发展十分必要且优势明显。

（2）从地质条件、造腔工艺与管理、老腔改建等方面分析了制约国内盐穴储气库发展的瓶颈问题，为下一步技术攻关指明了方向。

（3）面对国内建库条件复杂、优质库址较少的现状，必须依靠科技创新的驱动发展。生产建设单位应与国内高校、研究院和设计院进一步开展合作，共同技术攻关，破解高不溶物盐矿复杂地质条件下建库难题。同时，要加快储气库数字化转型、完善储气库政策机制，为储气库建设与运营管理注入更多活力。

参 考 文 献

[1] 完颜祺琪，安国印，李康，等.盐穴储气库技术现状及发展方向[J].石油钻采工艺，2020，42(4)：444-448.

[2] 谢卫炜.中盐金坛盐化有限责任公司盐穴储气库的发展与思考[J].中国盐业，2018，22：11-13.

[3] 郑雅丽，完颜祺琪，邱小松，等.盐穴地下储气库选址与评价技术[J].天然气工业，2019，39(6)：123-130.

[4] 完颜祺琪，丁国生，赵岩，等.盐穴型地下储气库建库评价关键技术及其应用[J].天然气工业，2018，38(5)：111-117.

[5] 石悦，郭文朋，徐宁，等.采卤老腔改建盐穴储气库关键技术及应用[J].特种油气藏，2021(5)：134-139..

[6] 巴金红，康延鹏，姜海涛，等.国内盐穴储气库老腔利用现状及展望[J].石油化工应用，2020，39

（7）：1-5.

［7］薛雨，王元刚，张新悦．盐穴地下储气库对流井老腔改造工艺设计［J］．天然气工业，2019，39（6）：131-136.

［8］刘春，高云杰，何邦玉，等．盐穴储气库造腔设计与跟踪［J］．油气储运，2019，38（2）：220-227.

［9］Yuan G J, Shen R C, Tian Z L, et al. Review of underground gas storage in the bedded salt deposit in China［C］. Calgary: SPE Gas Technology Symposium, 2006: SPE100385.

［10］何俊，井岗，陈加松，等．盐穴储气库天然气阻溶恒压运行技术［J］．油气储运，2020，39（11）：1298-1303.

［11］王建夫，安国印，王文权，等．盐穴储气库氮气阻溶造腔工艺技术及现场试验［J］．油气储运，2021，40（7）：802-808.

［12］付亚平，吴斌，敖海兵，等．光纤技术在盐穴储气库油水界面监测中的应用［J］．油气储运，2019，38（11）：73-78.

［13］Morgan J L, Snider-lord A C, Ambs D C, et al. InSAR for monitoring cavern integrity: 2D surface movementover Bryan Mound［C］. Solution Mining Research Institute Spring 2018 Technical Conference, Salt Lake City, Utah, USA, 2018: 1-12.

［14］Amer A H, Mohd A. Theoretical and experimental basics for a new tightness test method to enable testing of gas storage caverns during gas storage operation［C］. San Antonio: Solution Mining Research institute Spring Conference, 2014: 130-149.

［15］Patrick R, Eric F, Christophe M, et al. Microseismicity induced within hydrocarbon storage in salt caverns, manosque, france, hazard review and event relocationin a 3D velocity model［C］. Solution Mining Research Institute Fall 2013 Technical Conference, Avignon, France, 2013: 1-10.

［16］Hosseini Z, Collins D, Shumila V, et al. Inducedmicroseismic monitoring in salt caverns［C］. 49th U.S. Rock Mechanics/Geomechanics Symposium, San Francisco, California, 2015: 1-9.

［17］Quirijn H, Henk D, Fritz W, et al. Gas storage in salt caverns Zuidwending-The Netherlands［C］. Leipzig: Solution. Mining Research Institute Fall Conference, 2010: 241-250.

［18］完颜祺琪，冉莉娜，韩冰洁，等．盐穴地下储气库库址地质评价与建库区优选［J］．西南石油大学学报（自然科学版），2015，37（1）：57-64.

［19］李建君，王立东，刘春，等．金坛盐穴储气库腔体畸变影响因素［J］．油气储运，2014，33（3）：269-273.

［20］何俊，井岗，赵岩，等．盐穴储气库快速造腔方案［J］．油气储运，2020，39（12）：1435-1440.

［21］唐海军，李晓康，刘伟，等．盐穴储气库水平井造腔扩展规律试验探究［J］．地下空间与工程学报，2019，15（3）：762-768.

［22］陈涛，施锡林，李金龙，等．盐穴储气库水平井造腔模拟试验［J］．油气储运，2019，38（11）：1257-1264.

［23］郑雅丽，邱小松，丁国生，等．盐穴储气库残渣空间利用实验研究［J］．盐科学与化工，2019，48（11）：14-19.

［24］王自敏，袁光杰，班凡生．对接井盐穴储气库沉渣空间利用实验研究［J］．中国井矿盐，2020，51（5）：27-30.

［25］刘岩，程林．盐穴储气库地面工程技术要点研究［J］．油气田地面工程，2019，2：71-75，82.

［26］朱健颖，钱彬，赵云松，等．应用丛式井技术建设盐穴储气库的优势［J］．煤气与热力，2021，41（5）：1-3，17，44.

中国地下储气库发展现状及展望

李建君

（国家管网集团西气东输公司江苏储气库分公司）

【摘　要】 中国天然气消费每年以近10%的增速增长，能源平稳供应的重要性日益凸显，地下储气库对于天然气调峰保供和战略储备至关重要。通过调研中国天然气需求增量及目前储气库区域分布特点，提出当前储气库发展面临的建设、运营及技术短板，并预测储气库选址、建库范围将更加广泛。在中国致力于打造天然气"全国一张网"的背景下，将形成联网协调的储气库群，未来储气库将向着集约、高效、智能、科技和经济等方向发展。研究结果可为进一步推动储气库建设提供借鉴。

【关键词】 地下储气库；调峰；关键技术；建库资源；卤水消化；发展趋势

Development status and prospect of underground gas storage in China

Li Jianjun

(Jiangsu Gas Storage Sub-Company, PipeChina West-East Gas Pipeline Company)

Abstract As the natural gas consumption increases at 10% annual rate of growth in China, the importance of stable energy supply is becoming increasingly prominent. Thus, the underground gas storage becomes very critical to peak shaving, stable supply and strategic reserve of natural gas. Herein, the shortcomings of construction, operation and technology in the current development of gas storage were put forward by investigating the increment of gas demand in China and the regional distribution characteristics of gas storage at present. The scope of site selection and construction of gas storage was expected to be more extensive, and thereby to form the networked and coordinated gas storage group under the background of striving to build a unified national network of natural gas in China. In the future, underground gas storage will develop towards the direction of intensive, efficient, intelligent, technological and economic development. The research results can provide reference to promote the construction of gas storage in the next step.

Keywords underground gas storage; peak shaving; key technology; construction resources of gas storage; brine treatment; development tendency

---

作者简介：李建君，男，1982年生，高级工程师，2008年硕士毕业于清华大学动力机械及工程专业，现主要从事地下储气库的研究工作。地址：江苏省镇江市润州区南徐大道商务A区D座，212000。电话：0511-212004。E-mail：lijj@pipechina.com.cn。

目前，全球经济整体处于快速复苏阶段，随着能源供给低碳化加速转型和碳减排，天然气作为过渡型能源的需求持续增长。2021年，中国天然气表观消费量为 $3726×10^8m^3$，相比2020年和2019年，分别增长了12.7%和19.5%。预计2040年前天然气消费达峰，峰值约 $6500×10^8m^3$，消费增长主体为天然气发电与集中供热[1]。

在天然气供需调峰方面，基本形成以地下储气库和沿海LNG接收站储罐为主，其他调峰方式为补充的综合调峰体系。截至2021年底，中国已形成储气能力 $264×10^8m^3$，其中地下储气库工作气量为 $164×10^8m^3$，占全国天然气消费总量的7.1%，远低于12%～15%的国际平均水平。中国LNG接收站数量、规模相比LNG进口大国日韩两国也相差甚远[2]。储气基础建设滞后、储备能力不足等问题与天然气行业迅速发展、气量消费需求快速增长不匹配，成了制约中国天然气安全稳定供应和行业健康发展的短板。

1 发展历程

中国地下储气库相较国外建设起步较晚但发展迅速，大致经历了形成阶段、发展阶段、成熟阶段。

（1）建设形成阶段。20世纪60年代末，大庆油田首次开展储气实验，1975年建成投产了中国首座气藏型地下储气库——喇嘛甸地下储气库。1999年建成第一座商业地下储气库——大张坨地下储气库，承担京津冀地区天然气"错峰填谷"任务和北京地区冬季调峰保供，自此中国储气库已形成较成熟的建设运行体系。

（2）发展阶段。2005年开始建设的江苏金坛地下储气库，是中国同时也是亚洲第一座盐穴型储气库，填补了中国利用盐穴改建储气库的历史空白，为长三角地区调峰保供起到重要作用，自此储气库进入了发展并逐步提速阶段。2010年以来，中国相继建成新疆呼图壁、西南相国寺等一批气藏型储气库，取得了复杂储气库重大关键技术突破，培养了专业化科研队伍和技术人才。

（3）成熟阶段。目前，中国在油气藏型、盐穴型储气库建库理论和技术方面基本成熟，创造了世界储气库建设的多项第一，创新成果整体达到国际先进水平。2021年10月，冀东油田南堡储气库增注项目顺利投产，标志着中国储气库建设完成了从陆路到海上、从油气藏型到油藏型的成功转型，拓宽了中国储气库建库思路，为今后油藏改建储气库奠定了理论基础，具有重要的借鉴和指导意义。储气库建设整体处于成熟并向全国范围内辐射阶段，将在中国的油气消费、油气安全领域发挥更加重要的作用，建库目标将从调峰型向战略储备型延伸与发展，建库技术水平也将在实践中不断得到提高。

2 主要成就

2.1 储气库建设高速发展

截至2021年底，中国建成地下储气库（群）17座，设计总工作气量 $247×10^8m^3$（表1），其中2013年6月建成的新疆呼图壁储气库，是目前中国最大的气藏型储气库，仅用两年时间全面建成投产，创造了世界储气库建设史上的奇迹。

表1 中国储气库(群)主要设计参数表

储气库(群)	地理位置	设计库容/$10^8 m^3$	工作气量/$10^8 m^3$ 设计	工作气量/$10^8 m^3$ 已建成	企业主体
大港库群	天津大港	68.98	30.30	20.00	国家管网集团
华北库群	河北永清	17.40	7.80	4.10	国家管网集团
国家管网金坛	江苏金坛	26.39	17.14	7.80	国家管网集团
江苏刘庄	江苏刘庄	4.55	2.45	2.45	国家管网集团
中原文23一期	河南濮阳	84.31	32.67	19.60	国家管网集团
新疆呼图壁	新疆呼图壁	107.00	45.10	34.90	中国石油、国家管网集团
西南相国寺	重庆渝北区	42.60	26.00	22.80	中国石油、国家管网集团
辽河双6	辽宁盘锦	57.50	32.20	25.00	中国石油、国家管网集团
辽河雷61	辽宁盘锦	5.25	3.40	1.00	中国石油、国家管网集团
大庆喇嘛甸	黑龙江大庆	35.10	1.00	0.10	中国石油
大庆四站库群	黑龙江肇东	5.17	3.05	0.10	中国石油
华北苏桥	河北永清	67.38	23.32	11.00	中国石油
大港板南	天津滨海新区	10.47	5.57	4.30	中国石油
长庆陕224	靖边气田	10.40	3.70	3.80	中国石油
中原文96	河南濮阳	5.88	2.95	2.35	中国石化
中石化金坛	江苏金坛	11.79	7.23	3.00	中国石化
港华金坛	江苏金坛	5.23	3.07	1.65	港华储气有限公司

储气库建库地域分布不均,气库主要集中在中西部地区,而主消费城市位于东南部地区,对于此供需分离现状,国字号战略工程"西气东输"与储气库同步规划,实现了气能源供给与需求的无缝衔接[3]。储气库在保障骨干管网安全平稳运行方面发挥了不可替代的作用。

2.2 关键技术自主创新

经过20多年的储气库建设实践,中国克服了建库地质条件差、卤水消化慢等外部困难,通过自主创新和引进国外先进技术,形成了勘探、钻井、设计、建设、运行和监测一体化建库技术,研发出一系列具有自主知识产权的新技术和新装备,部分单项技术达到国际同行领先水平,形成多项特色技术,为中国储气库的高效、快速建设打下了坚实基础。

2.2.1 突破复杂地质条件下油气藏型储气库建设壁垒

复杂地质条件下气藏型储气库建设取得多项创新成果,整体达到国际先进水平,创造了断裂系统最复杂、储层埋藏最深、地层温度最高、注气压力最高和地层压力系数最低等多项世界纪录。目前,已形成一系列适合中国气藏型储气库建库核心技术,包括储气地质体动态密封理论、气藏型储气库库容动用理论及优化设计方法、储气库建设关键技术、储气库运行风险预警与管控技术等,扩宽了选址范围,提高了库容动用程度,解决了强注强采条件下储气库井注采工艺技术难题,形成了地质体—井筒—地面"三位一体"的风险管控体系等,为大规模建设油气藏型储气库奠定了基础[4-8]。

2.2.2 形成盐穴储气库特色技术

针对盐穴储气库建设过程中出现的腔体形态控制难、造腔速度慢和成腔率低等难点进行技术攻关,形成了一系列建库关键技术(图1)。

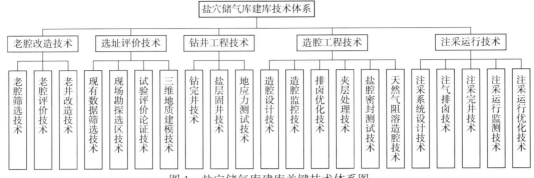

图 1　盐穴储气库建库关键技术体系图

（1）统一腔体气密封检测方法和标准。在研究和借鉴国外气密封检测技术的基础上，充分考虑中国盐层的实际情况及特点，提出适用于中国的盐穴腔体密封测试方法，具有现场操作性强、试压费用低、评价结果准确合理等特点。

（2）研发低成本高精度造腔油水界面监测技术。改进的光纤DTS测温技术可实时连续测量垫层界面，界面深度差在0.5m之内，取代了传统的地面观察法、电阻率及中子法相结合的方式进行垫层界面的监测，解决了传统方法成本高、稳定性差的问题，完全满足造腔垫层监控需要，单次界面检测费用由10万元降至1000元。

（3）拓展腔体形态控制技术。自主研发造腔软件，能够准确模拟盐层建腔形状和体积，并能对盐穴底坑形状扩展进行预测，腔体体积吻合度达到95%左右，腔体平均体积由原先的 $20\times10^4 m^3$ 提升至 $26\times10^4 m^3$，同时可计算耗能指标，保证盐穴腔体的安全性、经济性。以氮气、天然气为阻溶剂代替传统柴油作为垫层，形成新的造腔控制技术，开展了中国首次氮气阻溶和天然气阻溶造腔试验并取得突破，盐穴储气库建库逐步向高效、环保建库转型，同时可对原畸形腔体进行修复和扩容（图2）。提出复杂盐层条件下水平多步法造腔、单井双腔、双井单腔等造腔新技术，并开展了工程试验验证，可加快新井造腔速度、实现薄盐层建腔、实现多盐层一次利用建库[9-11]。

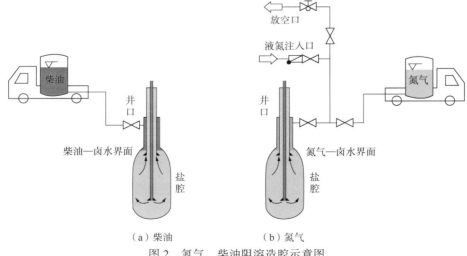

（a）柴油　　　　（b）氮气

图 2　氮气、柴油阻溶造腔示意图

3 存在问题

3.1 优质地下建库资源匮乏

储气库库址选择不仅要考虑地质条件,还需综合考虑资源分布、交通运输与管网建设、环境保护、区域发展、国家需求及战略部署等多方面因素。目前所建17座储气库(群)中,14座为油气藏型,3座为盐穴型,含水型尚未涉及。

3.1.1 油气藏型储气库

中国地下储气库建库资源分布不均衡。油气资源主要集中在长江以北,分布在东北、西北、西南、中西部及环渤海湾地区,可供选择建库的地层类型多。国外地下储气库建库埋深较浅,基本在2000m以内,储层物性好且非均质性弱,但是中国地下储气库建库条件较差,90%地下储气库构造为复杂断块、埋藏深、储层物性较差、非均质性强、建库气藏采出程度高,增加了建库难度。随着储气库建库需求的增长,未来建库方向将由砂岩气藏、碳酸盐岩气藏、带油环气藏、含硫气藏向低渗透率、岩性、超高压、油藏转变,给选址、设计、建设、运行带来巨大挑战,需建立配套的地质工程技术,如特殊储层孔隙空间动用评价技术、特殊储层改造工艺技术等。

3.1.2 盐穴型储气库

长三角、中南及东南部地区作为天然气消费市场主体,油气资源匮乏,含水层勘探程度低,但盐矿资源发育,盐穴储气库是建库的首选目标。据调查,中国南方井矿盐主要分布在江苏、江西、湖北、湖南、广东、云南、山东和安徽等省,盐岩资源丰富,分布范围广,埋深几十米至几千米不等。一般以层状为主,具有矿层多、单层薄、夹层多、夹层厚、埋藏过浅或过深等特点。除金坛建库条件良好以外,其余地方或厚度小或夹层厚或建成层段水不溶物含量高,优质建库资源缺乏(表2)。

表2 长三角、中南及东南部地区部分盐矿基本信息表

盐矿	埋深/m	含矿地层总厚/m	含矿地层占比/%	平均品位或可溶物占比/%	夹层特征及其他
金坛	860~1300	80~260	>80	73~85	两个主要夹层,不超过5m
淮安	800~2200	>200	—	>60	夹层多且厚
平顶山	1100~1900	300~460	63	>90	夹层多,但多数较薄
丰县	800~1700	200~500	45~51	75.10	上盐层存在超10m厚夹层
清江	800~1100	250	40~50	60~70	夹层多且薄
三水	1210~1450	200~450	—	69.23	一般小于5m
潜江	1900~3600	>800	—	>85	存在厚夹层
肥城	700~1200	250	50	60	夹层多,厚度多为5~9m
衡阳	800~1100	250	—	>60	夹层较少较薄,品位一般
小板	800~1500	800~1500	—	>90	夹层较少
长宁	3500	300~500	>90	>90	基本无夹层,埋深过深
宁晋	2500~2700	150~227	85~90	—	位于邢台地震带

续表

盐矿	埋深/m	含矿地层总厚/m	含矿地层占比/%	平均品位或可溶物占比/%	夹层特征及其他
云应	400~800	—	—	50~77	夹层杂质多
安宁	369~730	—	—	58.86	不溶物含量高
威西	800~1800	15~44	—	>95	单层盐层
桐柏	840~930	2.3	—	—	不具备成腔条件
周田	320~600	200~250	—	—	局部塌陷及地面沉降
长寿	2700~3000	30~62	—	—	埋藏较深，需进一步调研
高峰	2900~3000	60	—	—	夹层较少

通过前期调研分析，淮安和平顶山建库条件相对较好，丰县、清江和三水可作为下一步研究的方向。未来需加大对低品位盐层快速造腔及扩大空间动用的技术攻关，如复杂老腔连通改造技术、薄盐层水平腔体建造技术、高残渣空隙水排出技术、厚夹层垮塌技术等。

3.2 盐穴储气库建设周期长

盐穴储气库建设周期的长短很大程度上受限于当地卤水接收规模、接收浓度限制等影响。根据金坛储气库造腔经验，直井单腔造腔 $20×10^4 m^3$，常规用时3~5年，难以快速适应中国天然气调峰需求，因此提升造腔速度刻不容缓。目前大井眼造腔设计的管柱组合将原有直径177.8mm的造腔外管+直径114.3mm的造腔内管对应提高为直径273mm的造腔外管+直径177.8mm的造腔内管，增大了管流通道面积，提升了造腔速率并降低了泵能损耗，易于快速形成工作气并投入使用。该项技术正在河南平顶山叶县储气库开展，目前已完成1口井的大井眼钻井先导试验。

目前，中国储气库井固井质量未达到现行行业标准事件时有发生，针对水平井、大井眼和易漏失地层等固井提出更高的要求，提高和完善钻井液配套技术、固井工艺及水泥浆配套技术是未来储气库固井研究的主要方向[12-13]。

3.3 储气库市场化运营模式不成熟

储气库作为长输管道附属设施，由于其建设周期长、投资大、风险高和矿权归属不明晰等原因，长期以来一直由石油公司投资建设，外来资本对储气库投资热情不足，因此储气库建设略滞后。

由于中国储气库建设存在市场化程度低、天然气价格机制不完善、产权界定不清晰和商业运行模式不成熟等问题，影响了参与建设储气库的积极性。未来需要不断探索运行管理模式及供储销价格机制，激发投资建设主体的积极性，加快中国储气库建设和运营[14-17]。

4 发展趋势

依据国家总体战略部署，中国将形成四大区域性联网协调的储气库群，储气库发展已步入黄金期，其将在推进中国能源生产和消费革命、保障能源安全方面发挥重大作用，储气库未来将向着集约、高效、智能、科技和经济等方面发展。

4.1 储气库联动保障

中国天然气产供储销体系建设已取得阶段性成效，四大进口战略通道全面建成，中国油气管网骨架基本形成，干线管道互联互通基本实现。"十三五"时期累计建成长输管道 $4.6×10^4$ km，全国天然气管道总里程约达到 $11×10^4$ km，未来的储气库建设将依托天然气"全国一张网"而深度融合、联动，统一参与全国范围内调峰应急，并将在天然气产业链中发挥越来越大的作用，通过储气库独立运营管理和单独的市场化定价机制，实现储气库运营管理和利用效率双提升。

4.2 合资合作建库

2020年4月23日，国家发展和改革委员会、财政部、自然资源部、住房和城乡建设部、国家能源局5部门联合印发的《关于加快推进天然气储备能力建设的实施意见》指出：要多方合资建设储气设施，加大各大储气库建设方、政府投资平台及其他注入资金企业的合作力度，落实各方主体责任，建立健全定期考核制度，共同提高储气库建设能力。

在盐穴储气库建设方面，可由国家统筹规划盐岩资源综合利用，将储气库项目建设用地列入国家整体规划，各区域政府为主导，以合资公司方式整合当地盐企和储气库多元投资方。一方面，有利于打破单一来源卤水消化量受限问题，避免各自主导建设造成的地上地下空间规划、上游天然气资源分配和下游天然供给、站场和管道系统建设等方面出现交错、重复和综合利用率低下等一系列问题。另一方面，合作桥梁的架设，将使投资方在盐企采盐期考虑腔体形态控制，使传统意义的"老腔"变为"新腔"；可充分共享建设资源，在源头设计中，投入匹配的注水系统、电力系统等设备设施，减少独立管理模式下投资浪费；同时可充分利用盐企已建区域，集约土地使用、矿权使用和卤水消化等瓶颈性问题，实现采、建、供、销一体化，加快天然气战略储备能力。目前，江苏苏盐井神股份有限公司与中国石油集团储气库有限公司出资设立的江苏国能石油天然气有限公司正进入合作建设的快车道。

4.3 加快储气库建设布局

在长输管道较完善但天然气供需矛盾突出区域、东部及沿海地区积极开展油气藏储气库及盐穴储气库库址筛选和建设，同时加快油藏及含水层建库技术攻关，开展油藏及含水层库址普查及评价，寻找合适目标建设储气库[18]。在天然气消费快速增长、天然气管道"全国一张网"的背景下，补足储气能力短板，构建中国储气区域中心，促进产供储销体系健康发展势在必行。

4.4 储气库建设数字化转型

在现有工程建设及生产运行数字化成果的基础上，建立地质体—井筒—地面"三维一体"化的平台，通过对气藏、注采井、管网和设备进行全方位的管理和监测，对实时数据进行分析、故障预判和预测报警，实现储气库安全高效生产[19]。

4.5 储气库节能及空间利用多元化

利用储气库调峰采气降压产生的压差实现机械能与电能的转换，增加经济效益，实现节能降耗、绿色发展以及提质增效。

"盐链"延伸，带动转型。鉴于盐穴储气库具有高稳定性与封闭性，未来将不局限于天然气储存，地下空间将用于实现压缩空气储能、储油、储氢和储氦等，参与各能源系统的调峰，应用前景广阔[20-21]。

4.6 应急保供与市场化运营同步发展

中国天然气消费指数持续上涨，为储气库的能源应急保供角色需求提供了更大的施展舞台，以中国石油、中国石化、国家管网为代表的央企将在储气库能力建设中大力投入，能源需求高度刺激也必将使得储气库产品市场化，从而逐步形成以应急保供为基础、以市场竞价为辅助的局面。储气库可以参考北美、欧洲更成熟的产品模式，根据工作气量、注采能力、注采周期和稳定性等方面匹配组合固定储气产品、可中断储气产品、寄存或暂借服务产品等用户服务模式[22]。

5 结束语

（1）立足"双碳"目标，在中国天然气消费持续增长、"全国一张网"基本成型的背景下，积极发展储气库建设，形成联网协调的储气库群，增强储气能力，满足社会经济发展对清洁能源的需求。

（2）加快储气库技术革新，在拓宽建库方向、提高建设速度的同时，加大储气库空间多元化利用，积极发展压缩空气储能、储油、储氢和储氦等新技术，促进能源结构转型，实现储气库的绿色发展；同时，加快储气库向数字化转型、智能化发展，为储气库的建设及运行管理提供决策支持。

（3）加快储气库商业化运行模式转变，保证储气库在调峰应急正常运行的同时，针对多元化市场需求，优化资源配置，达到收益最大化。激发更多投资主体参与储气库建设，加快中国储气库建设和运营步伐。

参 考 文 献

[1] 中国石油经济技术研究院．2060年世界与中国能源展望（2021版）[R/OL]．（2021-12-26）[2022-04-10]．https：//max.book118.com/html/2022/0115/8141063102004055.shtm.

[2] 国家能源局石油天然气司，国务院发展研究中心资源与环境政策研究所，自然资源部油气资源战略研究中心．中国天然气发展报告（2021）[M]．北京：石油工业出版社，2021.

[3] 刘烨，何刚，杨莉娜，等．"十四五"期间我国储气库建设面临的挑战及对策建议[J]．石油规划设计，2020，31(6)：9-13，62.

[4] 丁国生，魏欢．中国地下储气库建设20年回顾与展望[J]．油气储运，2020，39(1)：25-31.

[5] 马新华，郑得文，申瑞臣，等．中国复杂地质条件气藏型储气库建库关键技术与实践[J]．石油勘探与开发，2018，45(3)：489-499.

[6] 刘国良，廖伟，张涛，等．呼图壁大型储气库扩容提采关键技术研究[J]．中外能源，2019，24(4)：46-53.

[7] 孙军昌，胥洪成，王皆明，等．气藏型地下储气库建库注采机理与评价关键技术[J]．天然气工业，2018，38(4)：138-144.

[8] 金根泰，李国韬．油气藏型地下储气库钻采工艺技术[M]．北京：石油工业出版社，2015：25-135.

[9] 张博，吕柏霖，吴宇航，等．国内外盐穴储气库发展概况及趋势[J]．中国井矿盐，2021，52(1)：

21-24.
[10] 郑雅丽, 完颜祺琪, 垢艳侠, 等. 盐矿地下空间利用技术[J]. 地下空间与工程学报, 2019, 15(增刊2): 534-540.
[11] 完颜祺琪, 安国印, 李康, 等. 盐穴储气库技术现状及发展方向[J]. 石油钻采工艺, 2020, 42(4): 444-448.
[12] 谢丽华, 张宏, 李鹤林. 枯竭油气藏型地下储气库事故分析及风险识别[J]. 天然气工业, 2009, 29(11): 116-119.
[13] 袁光杰, 张弘, 金根泰, 等. 我国地下储气库钻井完井技术现状与发展建议[J]. 石油钻探技术, 2020, 48(3): 1-7.
[14] 张福强, 曾平, 周立坚, 等. 国内外地下储气库研究现状与应用展望[J]. 中国煤炭地质, 2021, 33(10): 39-42, 52.
[15] 马华兴. 国内盐穴储气库发展现状初探[J]. 盐科学与化工, 2021, 50(11): 1-4.
[16] 龚承柱, 张康, 潘凯. 不同运营模式下天然气地下储气库的投资价值评估[J]. 油气与新能源, 2021, 33(3): 84-90, 96.
[17] 杨义, 陈进殿, 王露, 等. 中国储气库业务发展前景及运营模式发展路径探析[J]. 油气与新能源, 2021, 33(4): 11-15, 38
[18] 李国永, 徐波, 王瑞华, 等. 我国天然气地下储气库布局建议[J]. 中国矿业, 2021, 30(11): 7-12.
[19] 叶康林. 地下天然气储气库信息化建设现状与探讨[J]. 信息系统工程, 2019(7): 124-126.
[20] 朱力洋, 熊波, 王志军, 等. 天然气压差发电技术在地下储气库的应用[J]. 天然气工业, 2021, 41(3): 142-146.
[21] 王文权. 盐穴地下储气库产业链增值增效技术思路[J]. 天然气工业, 2021, 41(3): 127-132.
[22] 李锴, 郭洁琼, 周韬. 中国地下储气库市场化产品研究与经济分析[J]. 油气储运, 2021, 40(5): 492-499.

本论文原发表于《油气储运》2022年第41卷第7期。

中国盐穴储气库建设关键技术及挑战

杨海军

(中国石油西气东输管道公司储气库项目部)

【摘 要】 中国盐穴储气库的建设历程划分为技术研究与探索、技术消化与形成、技术成熟与发展3个阶段,在建设过程中形成了老腔改造、腔体气密封测试、光纤测试油水界面以及高效注气排卤等多项特色技术。为保证盐穴储气库的安全稳定运行,提出了一套在注采运行过程中监测井口温度压力、地面沉降、腔体形态和井筒完整性在内的体系及方法。指出中国盐穴储气库建设存在造腔速度慢、老腔改造难、适建库址少等问题,现有工艺技术尚不能完全满足中国层状盐层地质条件下盐穴储气库建设的需要。针对盐穴储气库建设过程中面临的技术挑战,提出了相应的攻关研究内容,为中国盐穴储气库的技术发展指明了方向。

【关键词】 盐穴储气库;溶盐造腔;老腔改造;管道调峰

Construction key technologies and challenges of salt-cavern gas storage in China

Yang Haijun

(Gas Storage Project Department, PetroChina West-East Gas Pipeline Company)

Abstract The construction course of salt-cavern gas storage in China is divided into three stages, i.e., technology research and exploration, technology digestion and formation, and technology maturity and development. In the process of salt-cavern gas storage construction, a number of distinguished technologies are developed, such as old cavern rebuilding, gas leak detection of cavern, oil-brine interface monitoring based on optical fiber and high-efficiency gas injection debrining. In this paper, a complete set of system and technologies was developed to monitor wellhead temperature and pressure, ground subsidence, cavity shape and wellbore integrity in the process of injection and production to ensure the safe and stable operation of salt-cavern gas storage. The problems involved in the construction of salt-cavern gas storage were described, such as slow solution mining, difficult old cavern rebuilding and fewer sites available for gas storage building, and it was demonstrated that existing technologies can't meet the con-

---

基金项目:中国石油储气库重大专项"地下储气库关键技术研究与应用"子课题"盐穴储气库加快建产工程试验研究",2015E-4008。

作者简介:杨海军,男,1961年生,高级工程师,1983年毕业于大庆石油学院钻井工程专业,现主要从事地下储气库的运营管理工作。地址:江苏省镇江市润州区南徐大道商务A区D座中国石油西气东输管道公司储气库管理处,212000。电话:13951201923,E-mail:yhjun@petrochina.com.cn。

struction requirements of salt-cavern gas storage in layered salt beds in China. To cope with the technological challenges in the process of salt-cavern gas storage construction, a series of research contents were proposed correspondingly, and the technology development direction of salt-cavern gas storages in China was pointed out.

Keywords salt-cavern gas storage; solution mining; old cavern rebuilding; peak shaving

盐穴储气库采用溶盐采矿的方式在地下盐层或盐丘中开采盐岩形成的地下洞穴中储存天然气，具有注采灵活、单井吞吐量大、储气无泄漏、工作气量比例高等优点，非常适合用于储气调峰[1-5]。自 2001 年将西气东输管道工程金坛储气库作为中国首个盐穴储气库建库目标以来，中国盐穴储气库的建设至今已有 15 年历史，经历了从无到有、从技术探索到成熟的建设历程[6-8]。在引进、吸收国外相关技术的基础上，通过自主攻关，目前中国已初步形成一套具有自身特色的盐穴储气库建设与运行技术。

1 建设历程

（1）2001—2005 年，技术研究与探索阶段。随着西气东输管道工程的建设与运营，中国盐穴储气库工程的建设拉开序幕。该阶段以国内研究单位为主体，与国外公司联合开展技术研究与攻关，针对重大技术难点问题开展专题研究，同时引进、吸收国外技术，完成了金坛储气库可行性研究和初步设计，为工程建设奠定了理论基础。

（2）2006—2010 年，技术消化与形成阶段。基于西气东输金坛储气库的初步设计方案，金资 1 井于 2006 年开展造腔技术先导试验，中国盐穴储气库的工程建设正式付诸实施。该阶段引进与消化吸收国外造腔技术、声呐检测技术、不压井取排卤管柱等盐穴储气库建设技术，并改进、应用了多项油气田勘探开发技术。2010 年 12 月，西气东输金坛储气库第一口井金资 1 井注气投产，标志中国盐穴储气库建设技术基本形成。

（3）2010—2015 年，技术成熟与发展阶段。在金坛储气库建设的基础上，开始在云应、淮安、平顶山、楚州等地开展盐穴储气库的建设工作，建设规模不断扩大，表明中国盐穴储气库建设开启了新的里程。

通过 10 多年的建设实践，克服了中国建库地质条件差和卤水消化慢等困难，通过不断改进和创新，形成了勘探、钻井、设计、造腔、运行及监测一整套盐穴储气库建库技术[9-17]，并研发了具有自主知识产权的一系列新技术和新装备，中国盐穴储气库建库技术已基本成熟，其中老腔改造、盐腔修补及光纤监测等技术达到国际先进水平。

2 主要技术成果

2.1 老腔改建储气库成套技术

中国盐化企业采卤过程中形成大量老腔，如果能对这些老腔进行筛选评价，并按储气库标准改造成库，将大大加快中国盐穴储气库投产进度，并节约投资。金坛储气库建设初期，为了加快储气库建设，尽快在西气东输管道运行中发挥调峰和应急作用，尝试利用中盐采卤过程中形成的老腔改建储气库。2003 年开始进行筛选、检测、评价及改造等工作，2007 年改造完成 5 个老腔，形成腔体体积 $100\times10^4 m^3$，库容达到 $8900\times10^4 m^3$，调峰气量

$5000×10^4 m^3$。由此，形成了老腔筛选、声呐检测、腔体稳定性及密封性评价、井筒改造、注采完井、MIT测试及注气排卤等配套技术[18]，为之后老腔的利用做好了技术储备。

2.2 造腔工艺技术

中国盐穴储气库造腔技术虽然起步较晚，但是水溶法采盐工业发展历史悠久。采盐与造腔虽然目的不同，但在工程技术方面具有相同之处，综合中国盐矿采盐工艺，目前中国掌握的造腔工艺包括：无油垫单井单管造腔、有油垫单井单管造腔、双井水平对流造腔、双井单腔加油垫造腔及单井双管加油垫造腔。

前3种造腔方法具有腔体形状不易控制、声呐测腔困难等缺陷；双井单腔加油垫造腔方法虽然排量大、腔体形状可控，但排卤质量浓度低、需钻两口井、硬件投资高。因此，目前国际上普遍采用的造腔工艺技术是单井双管加油垫造腔，即钻一口单井，在生产套管固井后，下入内外两层造腔管柱，循环造腔。该工艺具有操作相对简单、腔体形状易控制、盐岩使用充分、声呐测腔方便的优点。

中国采用的是直井双管加油垫造腔法，是在引进消化国外技术的基础上进行了改进和再创新。自2006年以来，西气东输管道工程已经累计造腔超过$850×10^4 m^3$，22个盐腔投产，26个盐腔正在建设中，造腔成功率达到100%。同时，在淮安储气库开展了10m厚夹层溶蚀的先导性试验，在云应储气库开展了双井单腔造腔先导性试验，均获得成功。

2.3 多项特色技术

2.3.1 腔体气密封检测

盐穴储气库腔体密封检测在中国尚无先例，在国外也没有统一的方法和检测标准。目前，国外在腔体试压方面主要是向生产套管中注入柴油，通过检测油水界面是否移动来判断盐腔的密封性。基于上述方法，并充分考虑中国盐层及井腔的实际情况，西气东输金坛储气库对盐腔气密封检测方法进行了改进和创新：向生产套管中注入氮气至指定深度，定时测量气液界面深度及记录井口压力，并依据上述数据计算井筒内的气体漏失率，从而判断盐腔的密封性是否达标。采用氮气测试腔体密封性的技术具有现场操作性强、试压费用低、评价结果准确等特点。

2.3.2 造腔油水界面监测

国外造腔过程中油水界面的监测主要采用中子法，测量精度高，但单次测量的费用也较高。由于中国主要是在层状盐岩中造腔，需要精细控制油水界面，监测的频率比国外要高，因而致使采用中子法监测的成本很高。西气东输金坛储气库造腔过程中采用地面观察法、电阻率与中子法相结合的方式进行垫层的监测，但是成本仍然较高，且稳定性差，不能完全满足油水界面监控的需要。

基于此，西气东输金坛储气库研发了采用光纤实时连续测量油水界面的技术：将包含电缆和光纤的外铠固定在造腔外管上，并随之一起下入井筒中；在井口通过电缆可以对井筒中外铠周围的柴油或卤水加热，而光纤则可以将不同深度上液体的温度上传至井口温度测试仪；由于柴油和卤水的导热性不同，在相同的加热条件下，二者的温度变化也不同，因而致使油水界面上下的温度差异较明显，温度产生明显差异的深度就是油水界面深度。该技术成本低、可靠性好，已成功应用于数口造腔井，并长时间保持良好的工作状态，界面深度误差小于 0.5m。

2.3.3 高效注气排卤

国外盐穴储气库注采井通常采用 339.7mm 的井身结构,既加大了造腔的卤水排量,加快造腔速度,又增大了注气排卤管柱直径,提高注气排卤的速度。而中国造腔过程受到排出卤水质量浓度必须符合盐化企业要求的限制,大直径井身结构不仅不能提高造腔速度,而且会导致钻井工艺复杂、钻井成本增加,因此中国均采用生产套管 244.5mm 的井身结构。

注气排卤完井的常规方法是注气管柱、排卤管柱分别选用 177.8mm、114.3mm 的油管,注气排卤完毕后带压起出 114.3mm 排卤管柱,导致注气排卤速度慢,如体积 $20×10^4 m^3$ 的腔体排卤时间超过 4 个月,且排卤期间排卤管柱易堵塞。经过多年研究与实践,金坛储气库拟采取简化的注气排卤完井管柱,即取消 114.3mm 的排卤管柱,直接通过 244.5mm 油管注气、177.8mm 油管采卤。177.8mm 油管下部安装有丢手,注气排卤完毕,将丢手以下的油管直接丢落于腔体内,丢手以上的油管作为注采管柱(图 1)。

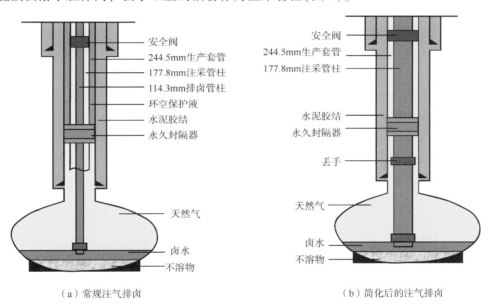

（a）常规注气排卤　　　　　　　　（b）简化后的注气排卤

图 1　常规注气排卤与简化后的注气排卤技术对比示意图

与常规注气排卤尺寸相比,简化后的注气排卤工艺的注气和排卤管柱尺寸均增大,既可以提高排卤速度,又可以解决排卤管柱的堵塞问题。同时,由于取消了带压起 114.3mm 的排卤管柱作业,降低了管柱投资和作业成本。

2.3.4 连续油管排卤

盐穴储气库在注气排卤过程中,由于排卤管柱不能下入腔体的最深处,因此在完成注气排卤后,在腔体的底部留有超过 $1×10^4 m^3$ 的卤水,由于腔体和管柱直径大小不等,使得腔体有效体积造成浪费。为了在注气排卤后最大限度地排出剩余卤水,在完成注气排卤后,再利用连续油管插入腔体底部,抽出剩余卤水,提高腔体利用率。使用 76.2mm 连续油管,在 15MPa 井口压力的情况下,每小时可排出卤水 $50~60 m^3$,既可以排出腔内残余卤水,又可以避免常规注气排卤因气液界面低于排卤管柱而产生井喷事故，安全高效。

2.3.5 天然气回溶腔体修复

在造腔过程中，由于地质、工艺或人为因素导致腔体局部未达到设计要求，特别是接近腔底位置，可以采用回溶技术在注气排卤阶段进行修补。回溶过程中改用天然气作为阻溶剂，不仅易回收、成本低，且对环境基本无污染。西气东输金坛储气库以库区内的一口井进行天然气阻溶回溶试验，专门开发了回溶造腔模拟软件。在回溶试验前先进行模拟分析，防止出现超溶问题。此外，在回溶试验过程中研发了实时监测气液界面仪器，预防腔顶破坏。试验结束后，腔体体积增加了 $1.4 \times 10^4 m^3$，取得了显著效果。

2.3.6 夹层处理

中国盐穴储气库造腔层段夹层发育，其厚度几十厘米至十几米不等，因此夹层处理工艺对造腔过程具有重大影响。夹层溶蚀效果的好坏直接决定腔体的形状，如果夹层未被溶蚀，会在夹层上部形成二次造腔，造成盐层段体积损失。夹层类型不同，造腔过程中所采取的应对措施也不相同。针对薄夹层，主要考虑腔顶油垫厚度不能太大，油水界面、造腔管柱距薄夹层必须有一定距离[14]。针对厚夹层，当夹层中含有大量裂缝且被盐岩填充时，主要采用使夹层自然垮塌的方法。首先在夹层下部建造足够大的腔体，然后上提造腔管柱和油水界面在夹层上部造腔，当夹层裂缝内填充的盐岩被溶解时，夹层就会在重力的作用下发生垮塌。

2.4 盐穴储气库运行监测技术

2.4.1 温度、压力和流量的监测

单井温度、压力和流量的监测对于保证盐穴储气库正常运行、及时发现运行过程中存在的问题具有重要作用[19]。每天记录井口温度、压力及流量计流量可用于单井盘库计算，通过监测套压可以判断井筒的完整性，而停井期间的井口压力监测则可用于判断盐穴是否发生漏气。

2.4.2 地面沉降监测

在注采运行过程中，为了防止腔体可能产生地面沉降而对地面建筑物造成损害，西气东输金坛储气库每年对库区内 10 多个监测点的累计沉降值和沉降速率进行监测，监测时间已超过 10 年。结果表明：金坛储气库中西部和北部地区沉降量较大，但仍在正常水平以内，且沉降速率无增大趋势，不会对地面建筑物及设施造成影响。

2.4.3 腔体带压监测

在盐穴储气库注采运行间歇期，对储存有天然气的带压腔体进行声呐测腔，可以直观识别腔体的垮塌和蠕变现象[20]。基于测腔结果，通过对腔体做稳定性评价分析，重新调整注采方案，以确保腔体安全平稳运行。每个腔体每 5 年带压监测一次声呐，西气东输金坛储气库已实施 2 次声呐带压测腔作业，共测腔 4 个。结果表明：腔体蠕变正常，但局部有垮塌现象。

2.4.4 井筒泄漏监测

利用新型光纤测井技术，将光纤固定在管柱外侧，并下入腔体中，通过测量腔体和井筒内的温度分布情况，判别井筒和固井水泥环是否发生泄漏。目前，西气东输金坛储气库研发了相关仪器，并进行了总装测试，准备开展现场工程试验。采用该技术，可以根据储气库实际运行情况,按照现场需求监测井筒的泄漏情况。

2.4.5 腔体泄漏监测

向盐穴内注入可微量检测、地层中不含成本较低且安全环保的气体示踪剂,然后在井筒和库区范围内采样进行化验分析。利用气相色谱仪检测样品中的烃类组分及示踪剂含量,判断盐穴存储的天然气是否泄漏,进而分析储气库盐穴地区地层的完整性。

2.4.6 腔体垮塌及裂缝监测

就西气东输管道工程,与东方地球物理勘探公司合作,利用微地震技术实时监测腔体垮塌和壁面裂缝情况,已在金坛储气库某井区附近开展了现场测试,效果良好。

3 技术挑战与应对措施

3.1 建设进度有待加快

2000年以来,以西气东输管道工程建成投产为标志,中国天然气的消费规模迅速增长,由2000年的$245×10^8 m^3$增至2014年的$1830×10^8 m^3$,年均增长达15.44%。经测算,2020年中国天然气消费量可达$4000×10^8 m^3$,显示了强劲的天然气消费需求。

"十三五"期间,中国规划在环渤海、东北、长三角、西南及中南5大区域建设储气库群,2020年形成有效工作气量超过$200×10^8 m^3$。但目前全国已建成地下储气库只有25座,2014年调峰量仅为$30×10^8 m^3$,与2020年目标工作气量尚有较大差距。因此,加快储气库建设是当前储气库工程建设的最大挑战[21]。

为了加快盐穴储气库的建设进度,从工程建设方面应当采取以下应对措施:(1)充分利用盐矿现有老腔。中国盐矿大规模开采已有50多年历史,在开采过程中形成了大量老腔,据不完全统计,仅在已选中的盐穴储气库建库区域以内就有老腔300多口,体积超过$3000×10^4 m^3$,可形成库容$40×10^8 \sim 50×10^8 m^3$。如果这部分老腔得以充分利用,对于加快盐穴储气库建设具有非常重要的意义。(2)加快造腔新技术的应用。采用高效的排卤新工艺,提高建设速度;在厚度超过600m的盐层,采用双井叠加造腔的方式,可以提高造腔速度;采用双井单腔造腔工艺技术,既可以加快造腔速度,又可以加快注气排卤的速度;开展大直径井身结构的利用,可以提高造腔和注气排卤的速度。(3)建立畅通的卤水消化渠道。由于中国造腔速度受到卤水消化渠道的限制,因此,建库选址应当靠近沿海地区,建立造腔使用的淡水和排出卤水的渠道,以期加快储气库建设速度。

3.2 建设过程中不断出现新问题

3.2.1 复杂老腔利用难度大

盐化企业采盐形成的老腔数量多,但由于腔体形态复杂、检测困难,因此利用难度较大[17],具体原因如下:(1)对流井形成的老腔,其水平连通腔体的形态无法进行声呐检测;(2)有的老腔对流段通过人工压裂形成,地层可能已经受到破坏,是否可以利用,尚无成功经验;(3)有的老腔形状不规则或多个腔体已连通,是否可以利用也无评价标准。

针对上述问题,今后应当加强以下研究:(1)开展新型的声呐检测设备研究,如人工智能型声呐检测设备或潜艇型的声呐检测设备;(2)开展复杂老腔稳定性评价研究,制定老腔可利用的评价标准。

3.2.2 腔体容易偏溶

在造腔的过程中经常出现腔体偏溶的情况,即容易造腔,但腔体形状不规则,稳定性较差,两个腔体间距变小,导致造腔体积达不到设计指标[10-11]。因此,需要积极开展偏溶控制技术研究。

3.2.3 注气排卤管柱易结晶

在盐穴储气库注气排卤过程中,管柱经常出现盐结晶堵塞排卤管柱的情况,严重时导致注气排卤中断,需要重新注入卤水排出已经注入的气体,给工程建设带来较大麻烦。因此,需要开展注气排卤防结晶技术研究,以及改善排卤管柱工艺技术研究。

3.2.4 节流采气管段易冻堵

在盐穴储气库采气初期,由于腔体压力高,需要节流采气,在节流管段易出现冻堵,一旦在站场或管道低压端出现冻堵,将有发生管道破裂爆炸的风险。

3.2.5 注水造腔泵站效率低

盐穴储气库的造腔注水站通常是用相同扬程的多台注水泵对应多口造腔井,当各个井的注水排量要求不同时,需要通过阀门调节排量,因而形成压力损失,降低注水泵工作效率[22]。当金坛储气库25口井同时造腔时,泵效率仅为29%,因此,有必要开展相应的节能造腔技术研究。

3.2.6 阻溶剂成本高且易污染环境

中国造腔过程中阻溶剂全部使用柴油,成本较高,单井造腔周期内平均至少消耗柴油250m³,且废弃柴油处置困难,易污染环境。如果使用国外氮气阻溶造腔技术,则对环境无影响、价格低廉、易控制,且作业时可以直接排空(图2)。目前,西气东输金坛储气库正在对该技术进行地面设备设计、工程试验工作。

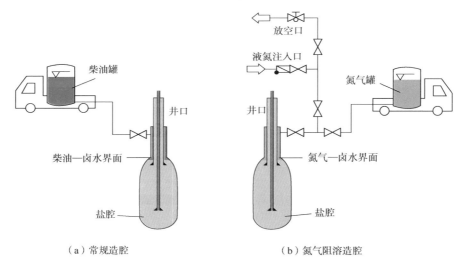

(a)常规造腔 (b)氮气阻溶造腔

图2 常规造腔与氮气阻溶造腔技术对比示意图

3.3 现有工艺技术不能满足建库需要

中国境内虽然含盐盆地众多,岩盐资源丰富,分布范围广,埋深从几十米到几千米不等,但盐矿普遍表现为以层状为主,具有矿层多、单层薄、夹层多、夹层厚、埋层过浅、埋藏过深等特点，缺乏适合建设盐穴储气库的优质盐矿资源。现有工艺技术不能满足中国层状盐层地质条件下盐穴储气库造腔的需要。为此，需在以下几个方面开展技术攻关：(1)开展水平多步法造腔研究，解决薄盐层造腔问题；(2)开展多夹层条件下造腔研究，解决多夹层地质条件造腔；(3)开展厚隔层条件下造腔技术研究，解决隔层上下两段造腔或隔层垮塌问题；(4)开展丛式井造腔或小井距布井技术研究，有效利用盐矿资源；(5)开展2000~2500m深井造腔技术研究，解决埋藏深盐层的造腔问题。

4 结论

中国盐穴储气库建设技术经历了研究与探索、消化与形成、成熟与发展3个阶段，现已基本成熟，具备了自主建设盐穴储气库的能力。在盐穴储气库建设实践过程中，中国取得了一系列的技术成果，尤其是在造腔、注气排卤及注采运行方面，形成了氮气气密封测试、光纤测试油水界面、高效注气排卤等一系列特色技术。但仍存在造腔速度慢、老腔改造难和适建库址少等问题，为了探索形成一套更适合中国盐矿地质条件的盐穴储气库建设和运行技术，针对工程建设中面临的技术挑战，尚需深入开展技术攻关。

参 考 文 献

[1] Seto M, Nag D K, Vutukur V S. In-situ rock stress measurement from rock cores using the emission method and deformation rate analysis[J]. Geotechnical and Geological Engineering, 1999, 17: 241-266.

[2] Bérest P, Brouard B, Durup J G. Tightness tests in saltcavern wells[J]. Oil & Gas Science and Technology, 2002, 56(5): 451-469.

[3] Liang W G, Yang C H, Zhao Y S, et al. Experimental investigation of mechanical properties of bedded salt rock[J]. International Journal of Rock Mechanics and Mining Sciences, 2007, 44(3): 400-411.

[4] 杨春和, 梁卫国, 魏东吼, 等. 中国盐岩能源地下储存可行性研究[J]. 岩石力学与工程学报, 2005, 24(24): 4409-4417.

[5] 周学深. 有效的天然气调峰储气技术——地下储气库[J]. 天然气工业, 2013, 33(10): 95-99.

[6] 魏东吼. 金坛盐穴地下储气库造腔工艺技术研究[D]. 北京: 中国石油大学(北京), 2008: 49-65.

[7] 丁国生, 张昱文. 盐穴地下储气库[M]. 北京: 石油工业出版社, 2010: 179-185.

[8] 肖学兰. 地下储气库建设技术研究现状及建议[J]. 天然气工业, 2012, 32(2): 79-82.

[9] 杨海军, 于胜男. 金坛地下储气库盐腔偏溶与井斜的关系[J]. 油气储运, 2015, 34(2): 145-149.

[10] 李建君, 王立东, 刘春, 等. 金坛盐穴储气库腔体畸变影响因素[J]. 油气储运, 2014, 33(3): 269-273.

[11] 王文权, 杨海军, 刘继芹, 等. 盐穴储气库溶腔排量对排卤浓度及腔体形态的影响[J]. 油气储运, 2015, 34(2): 175-179.

[12] 刘继芹, 焦雨佳, 李建君, 等. 盐穴储气库回溶造腔技术研究[J]. 西南石油大学学报(自然科学版), 2016, 38(5): 122-128.

[13] 郑雅丽, 赵艳杰, 丁国生, 等. 厚夹层盐穴储气库扩大储气空间造腔技术[J]. 石油勘探与开发, 2017, 44(1): 1-7.

[14] 耿凌俊, 李淑平, 吴斌, 等. 盐穴储气库注水站整体造腔参数优化[J]. 油气储运, 2016, 35(7): 779-783.

[15] 郭凯, 李建君, 郑贤斌. 盐穴储气库造腔过程夹层处理工艺——以西气东输金坛储气库为例[J]. 油

气储运,2015,34(2):162-166.

[16] 李建君,陈加松,吴斌,等.盐穴地下储气库盐岩力学参数的校准方法[J].天然气工业,2015,35(7):96-101.

[17] 井岗,杨海军,李建君,等.二维地震勘探在楚州盐矿勘探中的应用[J].中国井矿盐,2016,47(2):22-24.

[18] 杨海军,闫凤林.复杂老腔改建储气(油)库可行性分析[J].石油化工应用,2015,34(11):59-61.

[19] 杨海军,霍永胜,李光,等.TJ盐穴地下储气库注采气运行动态分析[J].石油化工应用,2010,29(9):61-65.

[20] 杨海军,李建君,王晓刚,等.盐穴储气库注采运行过程中的腔体形状检测[J].石油化工应用,2014,33(2):22-25.

[21] 丁国生,李春,王皆明,等.中国地下储气库现状及技术发展方向[J].天然气工业,2015,35(11):107-112.

[22] 耿凌俊,李淑平,吴斌,等.盐穴储气库注水站整体造腔参数优化[J].油气储运,2016,35(7):779-783.

本论文原发表于《油气储运》2017年第36卷第7期。

国内盐穴储气库建库关键技术研究进展

孙军治　陈加松　井　岗　杨普国　王一单　孟　君

(国家管网集团西气东输公司)

【摘　要】　盐穴储气库是天然气管网的重要配套设施,在天然气应急调峰和战略储备中发挥着重要作用。目前国内盐穴储气库建库面临缺乏优质库址、建成投产少、对流老腔改造技术不成熟、造腔速度慢等问题。针对这些问题,总结归纳了国内盐穴储气库建库关键技术的研究进展,认为选址评价技术、对接连通采卤老腔改造技术、水溶造腔模拟技术是目前的研究热点。给出了未来国内盐穴储气库的研究方向与建议,为我国盐穴储气库建库提供借鉴作用。

【关键词】　盐穴储气库;造腔技术;选址评价;研究进展

Research progress on key technologies of salt cavern gas storage construction in China

Sun Junzhi　Chen Jiasong　Jing Gang　Yang Puguo　Wang Yidan　Meng Jun

(PipeChina West-East Gas Pipeline China)

Abstract　Salt cavern gas storage is an important supporting facility of natural gas pipeline network, and plays an important role in natural gas emergency peak regulation and strategic reserve. At present, the construction of salt cavern UGSs in China faces problems such as lack of high-quality storage sites, few completions and commissioning, immature convective cavity reconstruction technology, and slow cavity construction. In view of these problems, this article summarized the research progress of the key technologies for the construction of salt-cavern gas storage in China. It is considered that the site selection evaluation technology, the reconstruction technology of the docking and connecting brine mining old cavity, the water-soluble cavity modeling technology are the current research hotspots. The research directions and suggestions of future domestic salt-cavern gas storage are given, which can provide reference for the construction of salt-cavern gas storage in China.

Keywords　salt cavern gas storage; cavity making technology; site selection evaluation; research progress

　　地下储气库是天然气管输系统的重要基础设施,承担着天然气应急调峰和战略储备的重要任务。地下储气库的发展和进步,不仅能够减轻天然气管输的压力,还能够解决天然

作者简介:孙军治(1995—),男,硕士研究生,工程师,现从事地下盐穴储气库建库技术研究工作。联系方式:17660451912,E-mail:1142960645@qq.com。

气供应夏季产能过剩、冬季供不应求的主要矛盾[1]。近年来，随着中俄东线、川气东送和西气东输西三线等多个大型工程的开工建设，储气库的建设也加快了步伐，尤其是在西北地区和华北地区，先后建设了多个大型油气藏型地下储气库。我国长三角、中南以及东南沿海地区作为主要的天然气消费市场虽然油气资源匮乏，但是岩盐资源丰富，非常适合建设盐穴储气库。盐穴储气库是利用地下盐丘或盐层，通过钻井等技术手段，水溶盐岩形成腔体。其具有注采速度快，吞吐量大，垫底气少且能回收利用等优点。这主要得益于盐岩的低渗透性、良好的蠕变性和自修复能力[2]。然而，我国盐矿大都是陆相湖泊层状沉积而来，具有夹层多，盐层薄的特点，非常不利于建库。近年来，国内学者针对我国盐穴储气库建库特点，从选址评价技术、老腔改造技术、造腔模拟技术几个方面展开研究。文章对最近几年国内文献进行调研，总结归纳了盐穴储气库建库关键技术的研究进展与发展方向，旨在为我国盐穴储气库建设发展提供借鉴作用。

1 选址评价技术

1.1 选址评价原则

库址筛选是地下盐穴储气库建库流程的第一步，选择满足建库条件的库址是成功建设地下盐穴储气库的必要条件。目前，我国唯一建成投产的盐穴储气库是位于江苏常州的金坛储气库，金坛储气库在天然气管网季节性调峰中发挥着重要作用。然而，随着对天然气需求量的逐渐增大，仅依靠金坛储气库已经远远不够。利用综合选址评价技术以寻找新的适合建设地下盐穴储气库的目标区域是极其重要的。盐穴储气库选址评价的基本原则可分为地质条件、地表条件两个方面(表1)。

表1 地下盐穴储气库选址基本原则

	构造	构造较简单，断层不发育
地质条件	埋藏深度	介于500~1500m，不宜超过2000m
	含盐地层厚度	不低于100m，含盐率大于70%
	氯化钠含量	综合含量大于60%
	顶板	大于30m且分布稳定
	储量	含盐地层面积大，有扩展余地
地表条件	周围环境	无大型工厂、建筑物、居民生活区及敏感区
	水源	有充足的水源保障
	卤水消化能力	有足够的卤水消化能力
	与管网距离	满足市场需求和调峰需要

1.2 建库条件评价

根据盐穴储气库建库选址的基本原则，综合利用建库条件评价技术，才能筛选出有利于建库的目标区域。建库条件评价技术主要采用数据筛选、地震勘探、地质建模、数值和物理实验等方法，从盐矿地质构造、含盐层特征、盐层顶底地层的密闭性、稳定性、储气规模以及地表条件几个方面进行综合评价。如刘凯等[3]2013年从地表条件和地质特征等方面，分析了盐穴储气库建设的影响因素；郑雅丽等[4]2019年提出了盐穴储气库的选址筛选的原则，创建了建库条件评价体系。

1.2.1 地质构造评价

地质构造评价通过对目标区域的地震资料进行精细处理和解释，确认目标区域的地质构造形态，识别地层空间展布特征与断层分布特征。重点分析断层发育带位置、走向特征、断距大小以及其对密封性的影响[5]。值得注意的是，如果地震资料不满足盐穴储气库建库评价的精度，需要重新进行地震资料采集，以识别出5~10m的小断层和微构造。最后优选出地质构造比较简单，起伏不大，盐岩厚度大，断层分布少的区域进行井位部署。

1.2.2 含盐层特征评价

含盐层特征是盐穴储气库建库选址的重要指标。我国盐矿不同于国外海相沉积，主要来源陆相湖泊层状沉积，常呈现夹层多且厚，不溶物含量高，盐岩品位低的特点。这些特点直接导致了造腔速率低、溶腔不规则发展、不溶物膨胀占用储气空间，甚至造成造腔管柱堵塞等情况，严重制约了造腔工程的实施。因此，在建库选址阶段分析目标区域的含盐层特征，优选出利于建库的盐岩区至关重要。根据区域地质特征资料，采用层序划分方法，进行小层划分与对比，编制连井剖面图，结合地震解释成果，确定含盐层以及顶底面埋藏深度、厚度、夹层分布特征、含盐率以及综合氯化钠含量等信息。如垢艳侠等[6]2021年通过引入人工神经网络模型，对不同组分含量盐岩的溶蚀速率进行了分析，张博等[7]2021年提出将重构曲线定量识别法与BP神经网络反演相结合的方法，识别了平顶山地区的盐层厚度与盐岩品味。

1.2.3 密闭性及稳固性评价

在地下盐穴储气库建设和运行过程中，良好的盖夹层是保证密闭性和安全运行的基础。盖夹层的评价需要结合构造解释成果，通过对样品进行岩性分析、渗透率测试以及地应力测试等方法进行综合评价。储气库盖层要求具有良好的岩性，低渗透率，扩散能力弱，突破压差大等特性。盖层稳固性的评价一般通过临界深度法进行判定[8]。另外，盖层中有地下水与地面水系连通会增加天然气泄漏的风险，因此还应充分了解目标区域的水文地质特征[9]。对于盐岩夹层，为了保证良好的密封性，要求塑性要强，毛细管压力要高，渗透率要低，扩散能力要弱，夹层数量越少越好[10]。如张耀平等[11]2009年提出双重介质固气耦合模型的数值模拟方法，对盐穴储气库夹层的渗透性进行了分析。

1.2.4 储气规模

单个腔体的有效体积的估算可以利用腔体梨形的特点，将低槽、主体和顶部简化为上下圆锥和中圆台的计算。单腔储气量和工作气量的估算可以根据气体状态方程来计算[12]，计算的关键在于上下限压力的取值，上限压力的取值可参考国内外储气库经验、最小主应力测试以及稳定性评估的结果，下限压力参考管网的压力以及稳定性评估的结果。王建夫等[13]2021年提出一种排卤质量浓度测定，结合盐腔体积守恒原理的盐腔有效体积计算公式。

1.2.5 地表条件

根据国外储气库运行的经验，盐穴储气库运行过程有安全性事故出现的可能[14]，我国盐矿通常埋藏浅、夹层多、品位低，出现天然气泄漏和地表沉降的可能性更大。因此，在选址建库评价阶段,基于安全性的考虑是必不可少的。盐穴储气库应选择没有大型工厂、建筑物、居民生活区以及敏感带的区域建库,避免发生不可挽回的事故,危害生命财产安全。盐穴储气库是采用水溶解盐岩形成储气空间进行储存天然气的,充足的淡水源对于建造储气库是必不可少的,一般造腔需要盐穴设计体积 7~10 倍的淡水资源。同时,建库目标区域需要有一定能力对水溶造腔形成的卤水进行处理。盐穴储气库可以有效保障长输管道稳定运行,并且具备季节调峰功能。为了满足下游用户的稳定供气,盐穴储气库要建设在天然气用户集中、距离长输管道较近的地区,一般调峰半径为 150km,距离长输管道不应超过 100km。

2 采卤老腔改造技术

近年来,天然气消费量快速增长、调峰需求逐渐增大与盐穴储气库工作气量低、建库达产速度慢的矛盾逐渐凸显出来[15]。主要原因是盐穴储气库造腔时间成本较高,往往需要 3~5 年才能形成腔体。事实上,建设盐穴储气库除了常规新钻井水溶造腔之外,还可以利用已有的采卤老腔改建。利用已有老腔改建储气库可以大大加快盐穴储气库投产速度、提升经济效益以及缓解卤水处理的压力。采卤老腔类型有单井单腔和连通对接井老腔(图1)。

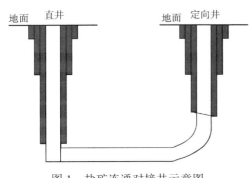

图 1 盐矿连通对接井示意图

2.1 预选评价

采卤老腔的预选遵循以下原则:处于构造稳定区域,断层与裂缝发育较少;含盐层埋深适中,厚度较大,分布稳定,品位较高;盐岩盖层厚度大,岩性好,渗透率低;腔体具备一定体积,矿柱厚度满足要求;腔体密封性、稳定性均要满足要求;井口与周围建筑物具备一定安全距离,尽量靠近天然气管网和天然气消费市场[16]。

2.2 腔体评价

腔体形态和体积是决定采卤老腔有没有改造价值的重要评价指标。对于单井老腔,为了符合力学稳定性要求,腔体应该是近似球体、圆柱体、梨状体等几何形态。腔体体积应该大于 $8×10^4 m^3$,腔体顶部盐层应该大于 15m,底部盐层大于 5m,不同腔体之间的矿柱厚度应该大于腔体直径的 2.5 倍[17]。对于连通对接老腔,近些年进行了大量研究。如杨海军等[18] 2015 年提出对声呐测腔仪器增加转向棒,测量出对接井中间的水平腔体;垢艳侠等[19] 2019 年采用声呐测腔和物质平衡法,结合老腔井眼轨迹,预测了连通对接老腔的腔体形态和体积,同时提出一种排出老腔残渣中卤水增加储气空间的方法;施锡林等[20] 2020 年总结了高杂质盐矿建库技术的研究进展,针对连通对接腔体形态探测难题,采用声呐与电法联合测腔、声呐与地震联合测腔两套物探探测技术,并给出了该技术造腔的具体流程以及风险防控措施。

2.3 密封性和稳定性评价

密封性是采卤老腔改建储气库的重要因素,通常采用饱和卤水试压的方法进行测试。

试压前根据腔体情况和卤水资料确定试压压力，要求不超过套管鞋所处地层最小主应力的80%，还需将其折算为井口处压力。对于不确定最小主应力值的，可以按照经验压力梯度计算。测试过程中，为了保持压力恒定不变，需要不定期补充卤水，如果井口压力下降幅度随着注入卤水次数的增多而减小，最后保持稳定压力，说明试压合格。如周冬林等[21] 2020年采用卤水试压法对云应、淮安地区老腔密封性进行测试，通过监测注入卤水量和井口压力变化计算漏失量。

2.4 井身结构改造

大部分采卤老井存在生产套管直径小，固井质量差和变形腐蚀严重等问题，不能满足储气库注采运行的要求，有必要对这些井身结构进行改造。老井改造遵循安全性、密封性和满足注采运行要求三个原则。根据金坛储气库对老腔井身结构改造的经验，单井老腔有三种不同改造工艺。第一种是直接将原有老井封堵，重新按照储气库标准钻一口新井；第二种是锻铣老井部分套管和水泥，进行封堵扩大井眼形成新井；第三种是直接进行全井套铣，扩大井眼下入大直径套管。对流井采卤老腔的改造有两种：一种是封堵原有两口老井，然后按照储气库标准新钻一口注采井，最后在通道中新钻一口排卤井[22]；另一种是在两口老井之间新钻一口新井，将排卤管柱下到水平连通通道内，老井经改造合格后可用于注采气，通过两口老井注气，盐腔以及溶渣中的卤水从新钻井排卤管中排出。

2.5 气密性评价

气密性测试是检验采卤老腔改造成功与否的评判标准。基本原理是在生产套管和测试管柱的环空中注入氮气，直至气水界面稳定在套管鞋下5~10 m，使套管鞋处压力恒定，不断监测气水界面，记录压力、温度以及流量等参数，绘制气体泄漏曲线，根据气体泄漏曲线评价腔体气密性。测试过程中气水界面变化不超过1 m，或者气体泄漏曲线随着时间的推移泄漏量越来越小视为气密性合格。目前我国盐矿企业大多采用连通对接井开采的方式采卤制盐，并且连通对接老腔具有产能高、事故极少、服务年限长以及采收率高等优点。因此未来采卤老腔改造的研究方向必定是对于连通对接老井的改造。如石悦等[23] 2021年总结了采卤老腔改建储气库的经验，并分析了连通对接老井对改造技术的适应性，确定了适合对接老腔改造的工艺技术。

采卤老腔改造是目前盐穴储气库建库的主要手段，经过多年的研究探索，单井老腔改造技术已经逐渐成熟，对流连通老井的改造还需要进一步加强研究。对流连通老井改造的核心问题在于如何确定腔体形态，目前提出的声呐测腔结合物质平衡法等方法都是基于预测盐腔形态的思路，难免会与真实的腔体形态产生偏差。因此，准确探明腔体形态，以便更好地进行后续的评价改造是目前需要解决的问题。

3 造腔技术

3.1 基本原理

盐穴型地下储气库的基本原理是采用水溶造腔的方式在地下溶解形成一个储气空间。以最常见的单井双管造腔为例，其工艺原理是通过钻井将生产套管下到盐层中，然后在生产套管内下入造腔内管和造腔外管。根据造腔设计参数，往井下注入淡水，水溶解盐岩形成卤水排出。同时在生产套管和造腔外管形成的环空中注入阻溶剂，控制腔体上溶速率，自下而上逐步溶解形成腔体。为了控制腔体形状，造腔过程中不断调整造腔内外管柱的相对位置，通过采出盐量和声呐测腔确定腔体形状，达到设计形状后开始注气排卤，卤水基本采完继续注气达到设计压力后关井，造腔完成。其中往造腔内管注入淡水，造腔外管排除卤水，称为正循环；往造腔外管注入淡水，造腔内管排除卤水，称为反循环(图2)。国内针对造腔技术的研究颇多，主要集中在参数优化、数值模拟、阻溶剂以及厚夹层垮塌几个方面。

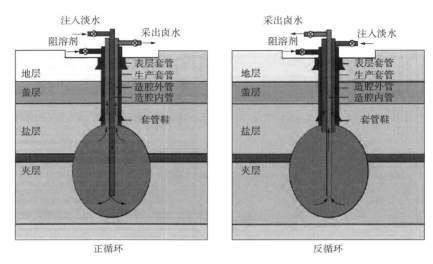

图 2 水溶造腔示意图

3.2 造腔参数优化

传统方法的水溶造腔所耗费时间长，经济成本高。因此，有必要对造腔工艺过程进行优化，改变造腔工艺参数，降低经济成本，提高造腔速率。近年来，通过室内实验、数值模拟以及生产试验等手段，众多学者围绕着提升造腔速率、控制腔体形态等方向进行了研究。如肖恩山等[24]2017年引入非线性规划理论，建立了多井水溶造腔工艺参数优化模型，简化地面工艺流程，减少地面设备设施维护保养成本，降低了单位造腔体积能耗；王建夫等[25]2020年结合现场数据，从注水循环方式、油垫提升高度、注水排量、造腔管距四个参数对造腔进行优化，有效地提升了造腔速度的扩大腔体体积；张敏等[26]2020年通过对平顶山、淮安、安宁、楚州四个地区的岩石进行水溶实验，分析其水溶特征及其对造腔过程中的影响，并且根据各个地区的岩性给出了水溶造腔过程中的对策；王元刚等[27]2020年结合金坛储气库造腔实例，研究了盐穴储气库水溶造腔过程中导致盐腔有效体积减小的因素，总结了井下异常情况和地面临井相互影响两大类因素，并给出了避免水溶造腔过程中出现有效体积减小的建议措施；徐贵春等[28]2020年分析了卤水中颗粒的沉降规律，研究了颗粒沉浮速度和卤水流速之间的关系，提出了改善卤水排量方法，有效地避免造腔管柱堵塞。

3.3 数值模拟

最大程度利用地下含盐层，尽可能造出空间大、形态稳定的腔体是水溶造腔的目标。因此，充分利用造腔模拟技术是至关重要的。造腔模拟技术最早由 Jessen 等[29-31]1964 年提出，随后众多学者在此基础上进行完善补充，逐渐形成了较为成熟的水溶造腔模拟技术。造腔模拟技术分为物理模拟和数值模拟，其中物理模拟因为其成本较高，难以实现等特点目前研究较少，而数值模拟技术因为其是利用计算机资源进行计算，实现较为简单，受到了国内外学者的青睐。水溶造腔数值模拟技术大致可以分为四步[32]（图 3）：建立目标区域的地质模型；给定初始造腔参数和腔体设计目标；优化造腔参数；达到设计腔体目标，模拟完成。

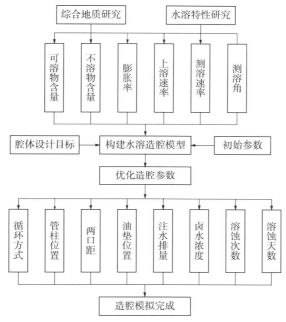

图 3 造腔模拟流程图

目前造腔数值模拟总共有四种算法：纳维—斯托克斯方程、达西流、浮羽流和平衡法。国外造腔模拟主要是针对计算流体力学，实现更高精度的造腔技术。国内层状盐岩不溶物含量多，水溶机理复杂，因此不能照搬国外已有技术。国内主要关注含盐地层水溶机理等问题进行了深入探讨[33]。针对国内盐岩复杂岩性，造腔模拟技术应采用物理模拟和数值模拟相结合的方式，通过室内实验，不断优化数值模拟技术，提升造腔模拟精度。同时，发展定向井和水平井造腔模拟技术，以更好地适应国内盐穴储气库造腔地质特征。如陈涛等[34]2019 年，王文权等[35]2020 年分别通过物理实验的方法发展了水平井造腔模拟技术。

3.4 阻溶剂

在水溶造腔前中期，如果上溶速度过快，会影响腔体形状，使腔体达不到设计体积。

为了更好地控制腔体形状，需要在造腔过程中注入阻溶剂，防止腔体上溶。在盐穴储气库建设初期，一般采用柴油作为阻溶剂，安全性高，可以很好地控制油水界面。然而柴油成本较高且不环保，为了改善柴油阻溶的不足，国内外学者开始考虑用氮气作为阻溶剂[36]，如董建辉等[37]2009年采用氮气作为阻溶剂进行水溶造腔，对比分析了氮气和柴油作为阻溶剂的经济成本。肖恩山等[38]2020年研究了建槽期不同工矿对气水界面的影响，并计算了不同工矿下应补充氮气的体积。王建夫等[39]2021年通过改造地面工艺流程，改善注氮设备，提出气水界面监测技术，有效改善了氮气作为阻溶剂易泄漏，气水界面难控制等问题。经过多年的理论研究和现场试验，氮气阻溶造腔技术已经趋于成熟，在我国金坛储气库造腔应用中取得了不错的效果。近年来，有学者提出使用天然气作为阻溶剂，天然气直接来源于管道，又可重新回采入管道，经济成本较低。如何俊等[40]2020年等提出一种快速造腔技术，在造腔的同时注入天然气，利用天然气作为阻溶剂，有效缩短了造腔周期。

3.5 厚夹层垮塌

我国盐岩矿床中常存在较厚夹层，如硬石膏层、钙芒硝层、泥岩层等，这些厚夹层给水溶造腔带来巨大挑战。厚夹层的存在会影响腔体中卤水正常流速，从而降低盐岩溶蚀的速度，并且不利于腔体形状的控制。另外，造腔过程中，厚夹层突然垮塌会造成造腔管柱受损、套管被卡等工程事故。国内外学者从厚夹层的水溶特性、力学性质以及对腔体稳定性的影响等方面进行了深入探讨[41-43]。如郑雅丽等[44]2017年研究了夹层的水溶机理，水浸力学参数变化规律，通过研究夹层的不同参数对垮塌的影响，建立厚夹层垮塌临界跨度数学模型，对夹层垮塌进行预测，进行造腔扩大储气空间可行性分析，并且通过实验验证；齐得山等[45]2020年针对淮安地区通过设计造腔方案，有效处理了造腔过程中厚夹层问题；垢艳侠等[46]2020年提出单井双腔造腔方案，并且通过造腔模拟验证了该方案的可行性，该方案能够扩大盐岩利用率，提高经济效益。

造腔技术的研究一直是盐穴储气库建库技术的热点，然而目前还存在一些问题：氮气造腔理论研究已经相对成熟，下一步应重点关注实际应用效果；盐岩夹层水溶特性的不同给造腔模拟技术带来了一定困难，相应的物理实验研究较少；厚夹层的力学性质以及对腔体稳定性影响的研究还需进一步深入。

4 结论

通过对国内盐穴储气库建库现状进行分析，针对存在的问题，从选址评价技术、采卤老腔改造技术、造腔技术等方面总结归纳了近些年国内盐穴储气库建库技术的研究进展。得出以下结论：

（1）天然气消费量快速增长、调峰需求逐渐增大与盐穴储气库工作气量低、建库达产速度慢的矛盾是国内盐穴储气库现阶段的主要矛盾。

（2）随着盐穴储气库的发展，国内盐穴储气库建库技术也日趋成熟，逐渐从单一简单化向复杂多元化转型，为我国天然气管网调峰供气提供重要保障。下一步应加强含盐目标层特征分析，造腔物理数值模拟，对流连通井腔体评价以及厚夹层垮塌等关键技术的研究。

（3）目前国内盐穴储气库存在缺乏优质库址、建成投产少、对流老腔改造技术不成熟、造腔速度慢等问题，未来几年盐穴储气库的研究应针对这些问题，从选址评价、扩大储气空间、优化造腔参数以及对流连通井改造等几个方面进行攻关。

参 考 文 献

[1] 丁国生，魏欢．中国地下储气库建设20年回顾与展望[J]．油气储运，2020，39（1）：25-31．

[2] 杨春和，梁卫国，魏东吼，等．中国盐岩能源地下储存可行性研究[J]．岩石力学与工程学报，2005（24）：4409-4417．

[3] 刘凯，宋茜茜，蒋海斌．盐穴储气库建设的影响因素分析[J]．中国井矿盐，2013，44（6）：24-27．

[4] 郑雅丽，完颜祺琪，邱小松，等．盐穴地下储气库选址与评价新技术[J]．天然气工业，2019，39（6）：123-130．

[5] 完颜祺琪，冉莉娜，韩冰洁，等．盐穴地下储气库库址地质评价与建库区优选[J]．西南石油大学学报（自然科学版），2015，37（1）：57-64．

[6] 垢艳侠，完颜祺琪，李康，等．不同盐岩组成含量对盐岩溶解速率影响规律研究[J]．地下空间与工程学报，2021，17（z1）：108-113．

[7] 张博，安国印，刘团辉，等．平顶山地下盐穴储气库建库盐层分布预测[J]．盐科学与化工，2021，50（2）：5-9，13．

[8] 吴建春，张定柱．江苏省某盐矿水溶开采工程效应研究[J]．西部探矿工程，2013，25（4）：132-134，137．

[9] 井文君，杨春和，李银平，等．基于层次分析法的盐穴储气库选址评价方法研究[J]．岩土力学，2012，33（9）：2683-2690．

[10] 袁光杰，田中兰，袁进平，等．盐穴储气库密封性能影响因素[J]．天然气工业，2008（4）：105-107，148-149．

[11] 张耀平，曹平，赵延林，等．双重介质固气耦合模型及含夹层盐穴储气库渗漏研究[J]．中南大学学报（自然科学版），2009，40（1）：217-224．

[12] 丁国生，谢萍．西气东输盐穴储气库库容及运行模拟预测研究[J]．天然气工业，2006（10）：120-123，185-186．

[13] 王建夫，王娜，许开志，等．盐穴储气库盐腔有效体积计算方法[J]．油气储运，2021，40（8）：909-913．

[14] Evans D J. An appraisal of underground gas storage technologies and incidents for the development of risk assessment methodology[R]．RR605 Research Repot．

[15] 丁国生，李春，王皆明，等．中国地下储气库现状及技术发展方向[J]．天然气工业，2015，35（11）：107-112．

[16] 巴金红，康延鹏，姜海涛，等．国内盐穴储气库老腔利用现状及展望[J]．石油化工应用，2020，39（7）：1-5，19．

[17] 田中兰，夏柏如，苟凤．采卤老腔改建储气库评价方法[J]．天然气工业，2007（3）：114-116，160．

[18] 杨海军，闫凤林．复杂老腔改建储气（油）库可行性分析[J]．石油化工应用，2015，34（11）：59-61，65．

[19] 垢艳侠，完颜琪琪，丁国生，等．高效利用复杂连通老腔新方法与效果分析[J]．科技创新导报，2019，16（29）：89-93，97．

[20] 施锡林,马洪岭,章雨豪.高杂质盐矿已有溶腔大规模储气技术研究进展[J].山东科技大学学报(自然科学版),2020,39(4):55-65.

[21] 周冬林,杜玉洁,肖恩山.利用卤水试压法评价老腔的密封性[J].油气储运,2020,39(2):183-187.

[22] 薛雨,王元刚,张新悦.盐穴地下储气库对流井老腔改造工艺技术[J].天然气工业,2019,39(6):131-136.

[23] 石悦,郭文朋,徐宁,等.采卤老腔改建盐穴储气库关键技术及应用[J].特种油气藏,2021,28(5):134-139.

[24] 肖恩山,刘继芹,王晓刚,等.盐穴储气库造腔节能优化技术[J].石油化工应用,2017,36(4):77-81,88.

[25] 王建夫,朱阔远,李友才,等.金坛盐穴储气库造腔关键参数优化[J].石油钻采工艺,2020,42(4):471-475.

[26] 张敏,朱华银,武志德,等.盐穴储气库不同类型盐岩溶蚀特性实验研究[J].盐科学与化工,2019,48(2):31-35.

[27] 王元刚,高寒,薛雨.盐穴储气库水溶造腔的影响因素[J].油气储运,2020,39(6):662-667.

[28] 徐贵春,江永强.盐穴储气库造腔过程中堵管原因分析及措施[J].盐科学与化工,2020,49(5):12-15.

[29] Durie R W, Jessen F W. Mechanism of the Dissolution of Salt in the Formation of Underground Salt Cavities[J]. Society of Petroleum Engineers Journal, 1964, 4(2):183-190.

[30] Durie R W, Jessen F W. The Influence of Surface Features in the Salt Dissolution Process[J]. Society of Petroleum Engineers Journal, 1964, 4(3):275-281.

[31] Kazemi H, Jessen F W. Mechanism of Flow and Controlled Dissolution of Salt in Solution Mining[J]. Society of Petroleum Engineers Journal, 1964, 4(4):317-328.

[32] 李康,完颜祺琪,王立献,等.盐穴储气库水溶造腔数值模拟技术研究进展[J].盐科学与化工,2017,46(12):1-5.

[33] 刘中华.钙芒硝矿床原位水溶开采的溶解—渗透耦合数学模型及数值模拟[D].太原:太原理工大学,2007.

[34] 陈涛,施锡林,李金龙,等.盐穴储气库水平井造腔模拟实验[J].油气储运,2019,38(9):1257-1264.

[35] 王文权,苗胜东,邢红艳,等.注水排量对水平井造腔腔体形态影响实验研究[J].石油钻采工艺,2020,42(4):476-480.

[36] Holschumacher F, Saalbach B, Mohmeyer K U. Nitrogen blanket initial operational experience[C]. Hannover: SMRI Fall 1994 Technical Conference 3-6 October 1994, 1994.

[37] 董建辉,袁光杰,申瑞臣,等.盐穴储气库腔体形态控制新方法[J].油气储运,2009,28(12):35-37,79.

[38] 肖恩山,杜玉洁.盐穴储气库建槽阶段工况变化对气水界面的影响[J].石油化工应用,2020,39(11):56-59.

[39] 王建夫,安国印,王文权,等.盐穴储气库氮气阻溶造腔工艺技术及现场试验[J].油气储运,2021,40(7):802-808.

[40] 何俊,井岗,赵岩,等.盐穴储气库快速造腔方案[J].油气储运,2020,39(12):1435-1440.

[41] 李银平,刘江,杨春和.泥岩夹层对盐岩变形和破损特征的影响分析[J].岩石力学与工程学报,2006,25(12):2461-2466.

[42] Bauer S J, Ehgartner B L, Levin B L, et al. Waste disposal in horizonal solution mined caverns. 1998.

[43] Bekendam R, Paar W. Induction of subsidence by brine removal[C]. Proceedings of the SMRI Fall Meeting, 2002: 1-12.

[44] 郑雅丽, 赵艳杰, 丁国生, 等. 厚夹层盐穴储气库扩大储气空间造腔技术[J]. 石油勘探与开发, 2017, 44(1): 137-143.

[45] 齐得山, 李建君, 巴金红, 等. 淮安盐穴储气库厚夹层盐层造腔工艺设计[J]. 油气储运, 2020, 39(8): 947-952.

[46] 垢艳侠, 白松, 贾建超, 等. 厚夹层盐穴储气库单井双腔可行性分析[J]. 石油钻采工艺, 2020, 42(4): 449-453.

本论文原发表于《盐科学与化工》2022年第51卷第10期。

盐穴地下储气库在我国的发展动力和趋势

房维龙　屈丹安

（中国石油西气东输管道公司）

【摘　要】 本文从天然气产业发展的角度，分析了天然气生产、运输和市场对大型地下储气库存储设施的需求动力，根据不同类型地下储气库的运行特点，分析了未来10~20年盐穴地下储气库在我国的发展趋势。通过分析，我们能很清楚地看到大容量和高注采速率的盐穴地下储气库在我国南方地区的需求将会持续增加，但由于其建设周期较长，与当地盐化工产业的发展紧密相关，一般应提前3~5年先于管道工程建设，才能与管道系统的投产和运行紧密衔接，保证管道系统的调峰和安全运行，提高管道系统的运行效率和经济效益。

【关键词】 天然气；输气管道；存储设施；盐穴地下储气库

随着全球石油价格的不断上涨，各种替代能源的开发和应用受到了人们的关注，我国在面临经济快速发展、石油供应紧张的形势下，正在加快天然气能源利用的发展。天然气作为一种清洁、优质、高效的能源和化工原料，对改善能源结构、保护大气环境、缓解石油供应紧张、提高能源利用效率、促进工商业的现代化和实现国民经济的可持续发展具有重要的保障作用。从国家能源政策上讲，也已明确提出在今后一个时期要大力开发和利用天然气资源，以弥补石油缺口，不断改善我国能源利用结构。

目前，我国天然气资源主要分布在四川盆地、鄂尔多斯盆地、塔里木盆地和海上区域，主要和潜在的天然气消费市场却分布在人口稠密的东北、华北、长江三角洲和我国南方经济发达地区，因此，输气管道和地下储气库等储运基础设施的建设是保证我国天然气产业稳定健康发展的关键，随着时间的推移和我国天然气产业的逐渐成熟，天然气存储在生产、运输和市场销售中所起的作用将愈发明显。本文主要从天然气产业发展的角度，分析了天然气生产、运输和市场对大型地下储气库存储设施的需求动力，根据不同类型地下储气库的运行特点，分析了未来10~20年盐穴地下储气库在我国的发展趋势。

1　天然气市场

从国外天然气市场发展历史和目前国内天然气市场发展需求来看，天然气市场主要分为四类：居民用气、商业用气、工业用气和发电用气，每个市场都有其独特的消费模式。

居民用户早晨用气较多，主要是为了准备早餐，并为白天的日常活动做准备，傍晚时候用气也较多，主要是为了准备晚餐，并为晚上的日常活动做准备。另外，居民用户的消费也是季节性的，冬季用气较多，夏季较少。因此，居民用气需求比较像两条叠加的正弦曲线。第一条曲线每天有两个峰值，它叠加在季节用气的另一条正弦曲线上。

商业用气是一个相对稳定的消费模式。营业时用气量增加，停业时用气量减少，因此，商业用气需求更像是一个阶梯函数，营业时，需求从最小值升至峰值，停业后，又从峰值降至最小值。

工业用气一般非常稳定，每周七天、每天 24h 持续使用，但在北方地区，由于一些工厂要向职工供热，工业用气在冬季时需求也会增加。

发电用气可以细分为三个不同的组成部分：基本负荷发电设备，中间负荷发电设备和高峰期发电设备。基本负荷发电设备每周 7 天、每天 24h 连续发电，这部分发电站中，大部分是核能发电或燃煤发电，只有少部分是燃气发电，基本负荷燃气发电设备全天乃至全年的耗气量非常稳定。中间负荷发电设备全年向市场提供电能，这些电站每天运行 4~18h，一般在早晨和下午达到完全负荷，这些电站基本上是燃煤发电、燃油发电和燃气发电，由于受天气的影响，中间负荷燃气发电设备的全天耗气量非常不稳定，在冬季和夏季，这些发电站的负荷系数会达到 40%~75%，以满足供热和空调用电，在相邻季节的一段时间内，这些设备的负荷系数为 20%~50%，因此，天然气需求量在一天中和各个季节里的变化都很大。

居民用气、商业用气、工业用气和发电用气的消耗量形成了一个复杂且高度变化的需求曲线，这就要求向市场供应天然气的管道公司或市场销售商具有一定的天然气供需调配能力，少量的天然气调控可以使用管道充填量或建设小型地面存储设施来满足高峰需求，但大量天然气的调控，尤其是季节性、燃气发电需求变化和其他巨大的潜在气量变动，需要使用大型的地下储气库存储设施。

2 天然气生产

天然气产自地表下各种不同的地质构造中，不同的地质构造特性确定了其不同的开采模式。某些储层的天然气采出速率快、产量高，如我国塔里木盆地的超高压、特高产克拉 2 号大气田，美国墨西哥湾的许多高产气田；与之相反，煤层甲烷气的开采速率较低，如我国山西许多煤层气井的开发。每个气田在制定开发和生产计划时，首先需要根据储层情况来优化天然气开采速率，以保证气田有较长的开采寿命和较高的采收率。但是，由于天然气市场消费模式的不同，天然气的需求量总是在不断变化和波动中，对气田的稳定开发带来不利影响。因此，要经济、稳定和高效地开发一个气田，既要考虑气田的最大生产能力、不断变化的市场需求，还要考虑天然气生产能力过剩情况下的存储能力，这种存储能力既能满足长期的季节生产需求变化，又能满足短期的生产需求波动，使气田的生产能力保持在一个稳定的最佳水平上。

3 天然气管道输送

目前，大多数国外天然气管道公司的运作模式是为天然气供应商和市场销售商提供天然气输送服务。一般情况下，市场销售商根据消费需求确定天然气需求量，与天然气供应商签订供应合同，规定了天然气的接收点和购买的数量，然后，天然气供应商再与管道公司签订输送合同，并通知管道公司从哪里接收天然气进入管道系统，接收多少天然气并把它们输送到什么地方。一旦合同签订并确定了天然气的接收量和分输量后，管道公司为了保证输气管道的平稳运行，将对进入管道和分输出去的天然气量进行精确的计算和调度，但实际输送过程中，每个合同预计的接收量和分输量很难精确执行，因此，管道公司会在输送合同条款中对接收量和分输量发生的差异制定出非常明确的处罚措施。许多市场销售商采用复杂的天然气供应(需求)模型来确定需要消费的天然气量，但是，许多变化如温度波动、未在计划内的发电站用气消耗等是销售商无法控制的，这会导致合同约定的分输量和实际的消费量之间产生巨大的差异，天然气储存是解决这种困难的有效方法，销售商可以从储气库中采气来解决供应短缺或过高的问题，同样，供应商也可以向储气库中注气以解决输送过多的问题，并且对未按预计发展的市场情况做出调整。因此，针对一段时间内天然气销售商从管道系统过度或过低采集天然气，供应商向管道系统过高或过低输送天然气的情况，管道公司开发了一系列服务措施，使用户可以寄存正向失衡、并向逆向失衡的用户提供调峰服务。为了免受平衡处罚，用户必须获得储存条件，而管道公司为了提供这些服务也必须拥有足够的天然气储存能力，地下储气库能够提供这种能力，并对每小时和每天的天然气供应波动、市场需求和管道系统失衡做出高度反应。

目前，我国正处于天然气发展的起始阶段，上游正在不断加大天然气资源的勘探和开发，中游以西气东输和陕京线为代表的主干线管道系统，基本还处于单线、长距离和大输量的运行状态，下游天然气消费市场尚在不断培育和发展中，远未达到成熟阶段，作为连接上下游的天然气管道公司，其运作模式在借鉴国外成功经验的基础上，具有我国自己的特色。以西气东输管道公司为例，目前，西气东输管道公司管理着我国天然气输送距离最长、输送量最大的天然气管道系统，其运作模式是从上游拥有天然气生产能力的油气田公司购买天然气，根据照付不议合同，向下游用户提供天然气。由此可见，与国外发达国家不同的是，在目前我国天然气市场总体需求大于供应的环境下，下游用户不可能去上游订购天然气后再委托给管道公司运输，即便能到上游去订购天然气，也只能委托一家管道公司运输，因此，在国家统一指导价格下，下游用户在门站(分输站)以到站价接收天然气显得更加方便。按照这种运作模式，西气东输管道公司在满足下游多种用户不同需求的基础上，既要考虑上游气田的稳定供气需求，又要考虑管道系统的最佳运输能力，因此，拥有足够的天然气存储能力，将可为管道系统的高效、安全运行提供较大的操作空间和弹性：第一，可在每年两个供气淡季里，将管道系统过剩的天然气注入地下储气库，在每年两个用气高峰期，将储气库存储的天然气投入市场，不仅保证了上游气田、整个管道系统的稳定生产和运行，减少了生产运行调整和维护保养费用，还保证了下游市场的季节调峰，在此基础上，还提高了管道系统的运输效率、增加了管道公司的销售量和经济效益；第二，在下游市场销售时，按照照付不议、照供不误的对等原则，足够的天然气存储能力，可将某些尚未培育起来的市场用户用不完的天然气按照一定的存储价格寄存在储气库，对某些发育过头的市场用户的超量需求以一定的超送价格进行供应，即可消除市场用户的担忧，保证天然气消费市场健康有序发展，又可利用这种存储能力提高天然气销售量；第三，足够的储气库存储能力，还能够解决整个管道系统的高管存运行风险、解决各种用户的临时用气高峰问题、降低市场用户对储气设施的投资和管理；最后，也是最重要的一点，足够的储气库存储能力，对管道系统各个地方的应急供应、特别是对整个管道系统的停气应急供应是一个非常重要的安全保障。

4 天然气存储

大型天然气地下储气库主要有以下三种类型：含水层、枯竭油气藏和盐穴。

单个含水层构造储气库一般可容纳 $3\times10^8 \sim 45\times10^8 m^3$ 的天然气，工作气量一般占其总库容的 35%，每天容许的采气能力约为工作气量的 0.5%，因此，含水层构造地下储气库通常每年只有 1 个注采周期。

单个枯竭油气藏构造储气库一般可容纳 $3\times10^8 \sim 45\times10^8 m^3$ 的天然气，工作气量一般占其总库容的 50%，每天容许的采气能力约为工作气量的 1%，因此，枯竭油气藏构造地下储气库通常每年可有 1~2 个注采周期。

单个盐穴一般可容纳 $0.05\times10^8 \sim 2.5\times10^8 m^3$ 的天然气，众多盐穴组成的储气库群可容纳数十亿立方米的天然气，工作气量一般占其总库容的 70%~80%，每天容许的采气能力约为工作气量的 10%，因此，盐穴地下储气库通常每年可有 12 个注采周期以上。

5 天然气存储的发展趋势

在当今世界天然气工业中，地下储气库在满足市场调峰和保证稳定安全供应方面起着非常重要的作用。同时，地下储气库也被用于管理天然气供应成本，天然气销售商可以在用气淡季以低价购买天然气，在需求高峰期以高价卖出，取得可观的经济效益。从世界各国的能源消费结构看，在发达国家的经济活动中，天然气的消费比例在 80 年代初为 19%，2000 年达到 22%，2020 年将达到近 28%，随着全球天然气需求的继续增长，地下储气库项目开发商将会选择高注入、高采出速度的项目，来满足不断变化且愈加复杂的市场需要，盐穴储气库因其独有的高注采能力，成为市场优选的一个方向，开发商会根据市场需求来确定最经济可行的造腔方案，包括盐穴腔体的大小、深度和数量、造腔工程总费用以及运行成本等，因此，在世界范围内，盐穴储气库造腔工程项目在未来十年内的需求将会非常强劲。

目前，我国能源需求正在快速增长，以煤炭为主导的能源受各种条件的制约，供应压力越来越大，石油消费迅速增长，对进口的依存度越来越高，石油供应安全已成为影响国家经济安全的重要因素，因此，加快我国天然气工业发展，是改善中国能源结构的战略选择，对确保我国经济增长向着高质量迈进具有重要的意义。目前我国的能源结构中，天然气的消费比例比世界发达国家的平均水平低近 20 个百分点，天然气在我国一次能源中的发展比例还有很大潜力，根据国家发改委牵头制定的《天然气管网布局及"十一五"发展规划》，2006—2010 年，我国还将规划建设天然气管道大约 $1.6\times10^4 km$，到 2010 年我国天然气管道总长将达到 $4.4\times10^4 km$，基本形成覆盖全国的天然气基干管网。随着西气东输、忠武线、涩宁兰线、陕京一线和二线等管道系统的建成投产，以及西气东输二线、川气东送等项目的开工建设，一个覆盖全国的天然气管网正在逐步形成，为提高管道系统的运行效率，保证管道系统的调峰和安全可靠运行，急需加快与之配套的地下储气库建设。然而，建设地下储气库需要相应的地质条件，而且距离天然气主要消费地区又不能太远，这对地下储气库的建设提出了很高的要求。从目前我国天然气主要消费地区、管道建设和适宜建设地下储气库的地质资源来看，在我国北方地区的天然气主要消费区域，陕京一线和二线依托大港油田、华北油田的油气藏建设了地下储气库群，基本满足了京津地区及华北地区管道沿线的调峰保安需求。

目前，在我国南方地区的地下储气库建设规划中，西气东输建设的金坛盐穴地下储气库，其工作用气量规模占总设计的85%以上；西气东输二线规划建设的3个地下储气库，其中2个是盐穴地下储气库；川气东送也规划在两湖地区、长江三角洲地区建设盐穴地下储气库；筹划中的中缅天然气能源大通道也计划在西南地区建设盐穴地下储气库。由此可见，未来10~20年内，我国南方地区对盐穴地下储气库建设的需求将会持续增加。

与油气藏地下储气库建设相比，盐穴地下储气库的建设周期较长。由于采用的是水溶法采矿工艺，单个盐穴腔体的建造一般需要数年时间，主要是受单井注水量和排卤量的制约；另外，造腔过程中需要消耗大量淡水，同时又生产大量卤水，通常建设$100×10^4m^3$的盐穴体积，要生产$1000×10^4m^3$左右的卤水，卤水的消化处理能力是影响造腔速度的另一个主要因素，与当地盐化工产业的发展紧密相关。因此，根据西气东输金坛盐穴地下储气库的建设经验，盐穴地下储气库的建设一般应提前3~5年先于管道工程建设，才能与管道系统的投产、运行紧密衔接，保证管道系统的调峰和安全运行，提高管道系统的运行效率和经济效益。

参 考 文 献

[1] 王希勇，罗雨香，龙刚，等．中国天然气供应安全战略研究[J]．中国能源，2006，28(2)：23-25.
[2] 吴刚，刘兰翠，魏一鸣．能源安全政策的国际比较[J]．中国能源，2004，26(12)：36-41.
[3] 丁国生，谢萍．中国地下储气库现状与发展趋势[J]．天然气，2005(1)：74-77.
[4] 姚建军，华爱刚．国内外天然气利用现状及发展趋势对我们的启示[J]．天然气，2006(4)．

本论文原发表于《科技资讯》2008年第29卷。

地质设计

盐穴地下储气库盐岩力学参数的校准方法

李建君[1]　陈加松[1]　吴　斌[1]　汪会盟[1]　王晓刚[1]　敖海兵[1]　陈　锋[2]

（1. 中国石油西气东输管道公司储气库项目部；2. 中国科学院武汉岩土力学研究所）

【摘　要】　盐穴地下储气库具有调配灵活、垫层气量需求量少、吞吐能力强等优点，但同时也面临着诸如地表沉陷、盐岩破坏、气体渗漏以及腔体收缩过快等安全稳定性问题。盐穴稳定性评价中一些关键力学参数的选择和校准极为重要，评价所涉及的岩石力学参数主要包括盐岩弹性参数和黏塑性参数。在盐穴腔体设计阶段，关键的力学参数的选择只能通过室内岩心试验来确定，其数值的大小可能与实际的原岩参数有较大出入，这在其他的地下岩体开挖工程中也经常碰到，需要进一步对盐岩力学参数进行校准。为此，在国内盐穴储气库造腔和注采运行的多年经验和数据积累的基础上，应用造腔和注采运行回归法，利用现场数据校准蠕变参数，建立了盐穴地下储气库岩石力学参数校准的试验方法。应用该方法对江苏金坛盐穴地下储气库的盐岩的弹性和黏塑性参数进行了校准，所获得的力学参数现场应用效果良好。校准后的力学参数对优化已运行腔体注采方案、提高储气库运行经济性和安全性都有重要作用。

【关键词】　金坛盐穴地下储气库；盐岩力学参数；弹性参数；黏塑性参数；数值模拟；校准方法

A calibration method for salt rock mechanics parameters of salt-cavern gas storage

Li Jianjun[1]　Chen Jiasong[1]　Wu Bin[1]　Wang Huimeng[1]
Wang Xiaogang[1]　Ao Haibing[1]　Chen Feng[2]

（1. Gas Storage Project Department, PetroChina West-East Gas Pipeline Company；
2. Institute of Rock and Soil Mechanics, Chinese Academy of Sciences）

Abstract　The salt-cavern gas storage has the advantages of flexible deployment, less cushion gas demand, and strong throughput, etc. However, it also has safety and stability problems, such as surface subsidence, salt rock damage, gas leakage, and cavity shrinkage. The selection and calibration of some key parameters in the stability evaluation of salt-caverns are very important. The rock mechanical parameters in the evaluation mainly include the elastic parameters and the visco-plastic parameters of the salt rocks. During the salt-cavern design stage, the selection of key mechanical parameters can only be

---

作者简介：李建君，1982年生，工程师；2008年毕业于清华大学并获硕士学位；主要从事盐穴地下储气库建设工作。地址：（212000）江苏省镇江市中国石油西气东输管道公司储气库项目部。电话：13814795191。E-mail: cqklijianjun@petrochina.com.cn。

determined by laboratory core test. Their values may have a large discrepancy with the actual parameters of the original rocks, which is common in other underground rock excavation engineering, thus the mechanical parameters of salt rocks should be further calibrated. Therefore, based on many years of experiences and data accumulation in the domestic salt-cavern gas storage cavity construction and injection-production operation, with the cavity construction and injection-production operation regression method and on-site data to calibrate creep parameters, a test method for calibrating rock mechanics parameters of salt-cavern gas storage was established. This method was applied to calibrate the elastic and visco-plastic parameters of salt rocks in Jintan salt-cavern gas storage, showing good effects of such mechanics parameters. The calibrated mechanics parameters are very important in optimizing the operating chamber injection scheme, and in improving the economy and safety of the gas storage operation.

Keywords jintan salt-cavern gas storage; mechanics parameters of salt rock; elastic parameter; visco-plastic parameter; numerical simulation; calibration method

随着西气东输管道及其他管道线路的陆续建成，作为油气管道的配套设施，地下储气库在安全平稳供气、季节调峰以及国家能源战略储备等方面起到重要作用[1-2]。盐岩因其良好的蠕变性能、超低渗透性、损伤自我恢复性能以及具有一定的地层压力从而成为盐穴储气库的良好介质[3]，盐穴地下储气库具有调配灵活、垫层气量需求少和吞吐能力强等优点，但同时也面临着诸如地表沉陷、盐岩破坏、气体渗漏以及腔体收缩过快等安全稳定性问题，特别是国外一般在巨型盐丘中建设储气库，而我国在地质条件更加复杂的层状盐岩层中建设储气库[4]。因此开展盐岩的力学研究尤其重要。

国外关于盐岩力学性质的研究开展了半个多世纪，第一次盐岩力学大会在美国举行，此后定期开展盐岩讨论会，会议从理论、模拟方法和实验方法等方面分别对盐岩储气库以及盐岩力学特性进行了探讨，解决了诸多技术问题。我国关于盐岩的力学研究起步较晚但是进展很快。国内学者首先对各盐矿区的盐岩进行了常规的力学试验研究，如陈锋等对江苏金坛[5]、李银平等对湖北云应[6-7]、王芝银等对江苏淮安[8]、梁卫国等对四川同庆[9-10]等诸多矿区盐岩样品进行了单轴压缩试验、三轴压缩试验和剪切试验，获得了包括弹性参数、强度参数等在内的基本力学参数。盐岩具有良好的流变特性，杨春和、梁卫国等通过大量的盐岩样品蠕变试验数据[11-12]，认为盐岩蠕变分为瞬时蠕变阶段、稳态蠕变阶段和加速蠕变阶段，并建立了盐岩的蠕变本构方程。关于盐岩储气库的模拟方法目前比较常用的是利用 FLAC3D、Abaqus 等有限元软件进行数值模拟[13]，陈卫忠等利用 Abaqus 软件对金坛废弃盐矿老腔进行了稳定性研究并对储气库的注采压力和套管鞋高度进行了探讨[14]。杨春和等利用 FLAC3D 软件对金坛不同洞形、不同注采压力下进行了数值模拟和稳定性评价[3,15]，但以上学者的研究多偏重于理论和室内试验研究，未得到现场实际数据验证。

笔者建立了一套盐穴地下储气库力学参数的校准方法，并在此基础上，利用 Abaqus 有限元软件校准了金坛盐穴储气库腔体围岩的力学参数，所得结果可以用于今后盐穴腔体的力学稳定性评价，特别是对优化已运行腔体注采方案、提高运行的经济性与安全性具有重要作用。

1 校准方法

为了确保盐穴地下储气库注采运行的安全，注采方案必须使腔体围岩处于弹性变化范围内，再加上岩盐的蠕变特性，国际上对盐穴进行稳定性评价时大多采用 Lemetre 蠕变模型[16]，该模型将盐岩的总应变分为弹性应变和黏塑性应变，其应力—应变关系如下：

$$\varepsilon_\mathrm{T} = \varepsilon_\mathrm{E} + \varepsilon_\mathrm{VP} = \frac{\sigma}{E} + \varepsilon_\mathrm{VP} \tag{1}$$

$$\varepsilon_\mathrm{VP} = \left(\frac{\sigma}{K}\right)^\beta t^\alpha \exp\left[-A_0\left(\frac{1}{T} - \frac{1}{T_\mathrm{r}}\right)\right] \tag{2}$$

式中：ε_T 为总应变；ε_E 为弹性应变；ε_VP 为黏塑性应变；σ 为应力差；E 为杨氏模量；α、β、K 为 3 个 Lemaetre 参数；t 为时间，d；T 为温度；T_r 为温度基准；A_0 为 Arrhénius 系数。

为了在有限元软件 Abaqus 使用该模型，需要对模型中的黏塑性应变部分进行进一步简化如下：

$$\dot{\varepsilon}_\mathrm{VP}^\mathrm{eq}(t) = A\sigma_\mathrm{eq}^n t^m \tag{3}$$

$$A = 10^{-6} \exp\left[-A_0\left(\frac{1}{T} - \frac{1}{T_\mathrm{r}}\right)\right] \Big/ K^\beta$$

其中，

$$m = a - 1$$

$$n = \beta$$

式中：A、m、n 分别为简易蠕变模型中的蠕变参数。

通过现场溶腔试验以及长期注采运行的生产数据来校准弹性模量（E），泊松比（μ），以及黏塑性模型中的 A 参数，黏塑性参数（m、n）采用试验值，使得模拟计算与现场实际试验或运行数据相匹配。

1.1 弹性参数确定方法

盐穴储气库腔体在短时间内进行升压和降压操作，并使得压力在较小区间变化，基本上能够保证腔体围岩处于弹性变形状态，可以借此推导围岩的弹性参数。由于受计算机计算能力的限制只能建立轴对称模型来进行模拟计算。因此在试验选井时要选择形状轴对称性较好的腔体。

现场试验最好使用浓卤水来进行，但是由于浓卤水获取困难，且注水升压和排卤降压施工麻烦。因此实际选用造腔排出的淡卤水作为试验工质，这样在计算过程中需要考虑淡卤水继续溶腔带来的影响。

试验的基本原理是向盐穴腔内注入或排出一定体积卤水，腔体的体积变化包括黏塑性变形、弹性变形和腔体溶蚀 3 部分，与腔内卤水的体积变形相等。试验时间较短所以可以忽略与时间相关的黏塑性变形，剩余的弹性变形可以用腔体的弹性压缩系数表示。具体推导如下：

$$\Delta V_{腔} = \Delta V_{蠕} + \Delta V_{弹} + \Delta V_{溶} = \Delta V_{注/排} + \Delta V_{水} \tag{4}$$

式中：$\Delta V_{腔}$ 为腔体体积变化量，m^3；$\Delta V_{蠕}$ 为腔体蠕变体积收缩量，m^3；$\Delta V_{弹}$ 为腔体弹性体积变化量，m^3；$\Delta V_{溶}$ 为注入淡卤水继续造腔体积，m^3；$\Delta V_{注/排}$ 为注入或者排出卤水体积，m^3；$\Delta V_{水}$ 为腔内卤水体积变化量，m^3。

$$\Delta V_{水} = V \Delta p / K_v \tag{5}$$

式中：$\Delta V_{水}$ 为卤水的压缩量，m^3；Δp 为应力增量，MPa；V 为腔内初始卤水体积，m^3；K_v 为卤水的体积模量，取 3.56×10^3 MPa（中科院试验数据，50℃下）。

$$Z_{腔} = \Delta V_{弹} / (\Delta p V) \tag{6}$$

式中：Δp 为应力增量，MPa；V 为溶腔体积，m^3；$\Delta V_{弹}$ 为腔体体积弹性变化量（不包括继续造腔体积增加部分），m^3；$Z_{腔}$ 为腔体的弹性压缩系数。

$$\gamma_{腔} = \Delta V / V \tag{7}$$

式中：$\gamma_{腔}$ 为腔体收缩率；ΔV 为腔体体积变化量，m^3；V 为腔体初始体积，m^3。

现场试验获得腔体的弹性压缩系数后，使用 Abaqus 有限元计算软件对选中的腔体进行校准模拟计算，从而确定与试验相匹配的弹性参数。

1.2 黏塑性参数确定方法

由于黏塑性参数与时间相关，因此必须参考较长时间的实际数据。黏塑性参数有 A、m、n，通过现场试验确定这3个数据非常复杂，需要大量的运行数据和声呐测量数据支持，只考虑了保持 m、n 值使用室内试验数据不变，通过现场数据校准 A 值，使得模拟的腔体收缩率与现场试验或实际运行结果相匹配。具体有以下两种方法。

1.2.1 造腔回归法

顾名思义，造腔回归方法分析的对象是处于造腔过程中的盐穴腔体，分析的对象最好是声呐测量完毕后就长期处于停井状态，腔体内卤水处于饱和状态，不会进一步溶蚀，可以排除溶蚀影响。停井时间超过7d，时间越长越好。通过调整 A 值来使得模拟的腔体收缩率与实际的体积收缩率相匹配。

1.2.2 注采运行回归法

该方法需要实际的注采运行数据，利用数值模拟软件对盐穴腔体长期变形进行模拟，通过调整 A 值来使得模拟的腔体收缩率与实际的腔体体积收缩率相匹配。该方法需要对注采运行中腔体进行声呐测量，对比完腔时的声呐测量结果，计算得到实际的腔体收缩率作为参考。

2 现场试验

水溶造腔指通过管柱往盐岩层中注入淡水溶解盐岩形成近饱和卤水后排出，在地下盐岩层中形成的洞穴。造腔的循环方式主要有两种：正循环和反循环（图1）。正循环造腔指淡水从溶腔内管进入，卤水从溶腔内管与外管的环形空间返回地面；反循环造腔指淡水从溶腔外管的环形空间进入，卤水从溶腔内管返回地面。为了有效控制盐腔形态和保护盐岩顶板的密封性，金坛盐穴储气库从生产套管与溶腔外管的环空中注入柴油阻溶剂，从而在卤水和腔体顶部之间形成一层保护层，确保顶部盐层免遭溶蚀破坏。通过对井口注水压力、井口排卤压力、井口油垫压力进行监控可以掌握腔内造腔活动和管线运行情况。

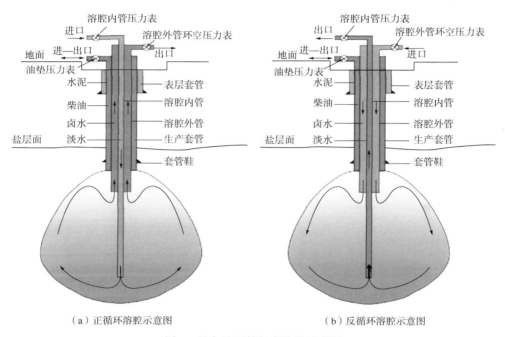

（a）正循环溶腔示意图　　　　　　　（b）反循环溶腔示意图

图1　盐穴地下储气库溶腔示意图

2.1　A腔体的油垫压力升压及降压试验

A腔体在某轮溶腔阶段结束后进行了声呐测腔，测得腔体体积为140339m^3。升压试验中将油垫压力通过注入淡卤水的方式从4MPa升高至7MPa，注入淡卤水量为284.98m^3，其中注入的淡卤水还可以继续溶腔。降压试验将油垫压力通过排出卤水的方式从7MPa降至6.45MPa，排出卤水量为56m^3，2次试验的时间都较短，因此可以忽略腔体的蠕变体积收缩变化，近似认为腔体的体积变化值为腔体的弹性收缩变化。关井7d后油垫压力从6.45MPa升至6.61MPa，其主要原因是由于盐岩的蠕变导致腔体蠕变收缩，卤水压缩使得油垫压力升高。

2.2　B腔体停井期间油垫压力上升试验

B腔体2012年9月2日进行了声呐测腔，腔体体积为92430.7m^3，从2012年9月15日停井，此时的油垫压力为3.79MPa，到2012年12月30日油垫压力上涨至7.49MPa，其主要原因为腔体周围的盐岩的蠕变性导致腔体体积收缩压缩卤水导致油垫压力上涨。

2.3　C、D腔体两期声呐测腔试验

C、D腔体为同采同注腔组，腔体距离较近，不到30m。2005年1月11日对C、D腔体进行了第一次声呐测腔，测得体积分别为105664m^3和129830m^3，腔体从2007年6月30日排卤完毕后，腔体停止作业，腔内压力平衡近4个月，2007年10月22日开始注采气运

行，工作气压介于 8.0~13.5MPa，运行方式为注气 1 个月，采气 2 个月，最高压停留 3 个月，期间伴有一些应急采气等突发状况，2013 年 6 月又进行了一次声呐测腔，C、D 腔体的体积分别为 103890m^3 和 128305m^3，说明腔体在运行期间存在不同程度的收缩。

3 数据分析

3.1 A 腔体计算过程及结果

A 腔体计算过程及结果见表 1。

表 1 A 腔体现场试验数据计算结果表

试验内容	腔体体积/m^3	试验前油垫压力/MPa	试验后油垫压力/MPa	腔内卤水变化量/m^3	腔体体积变化量/m^3	腔体弹性压缩系数/%	腔体体积收缩率/%
升压试验	140339.00	4.00	7.00	165.38	165.38	0.0434	0.1180
降压试验	140504.38	7.00	6.45	34.05	34.05	0.0440	0.0242
停井试验	140470.33	6.45	6.61	6.31	6.31	—	0.0045

3.1.1 升压试验腔体弹性压缩系数（Z_1）的计算

（1）根据地层压力计算注入淡卤水后腔内的压力约为 15.9142MPa，根据式（5），注入腔体内的卤水体积为 283.68m^3；（2）4MPa 时腔体卤水体积为声呐测腔体积，根据式（5），升压至 7MPa 时腔体内卤水体积变化量为 118.3m^3；（3）根据现场测得卤水浓度差计算，注入的 284.98m^3 淡卤水可以继续造腔 17.35m^3；（4）根据式（4），腔体体积变化量为 165.38m^3；（5）根据式（6），Z_1 为 0.0434%。

3.1.2 降压试验腔体弹性压缩系数（Z_2）的计算

（1）根据地层压力计算排出卤水后腔内的压力约为 15.3642MPa，根据式（5），腔体压力下排出的卤水体积为 55.76m^3；（2）7MPa 时腔体卤水体积为 140504.38m^3，降压至 6.45MPa 时根据式（5），腔体内卤水体积变化量为 21.71m^3；（3）根据式（4），腔体体积变化量为 34.05m^3；（4）根据式（6），Z_2 为 0.044%。

3.1.3 腔体停井期间体积收缩率（γ）的计算

（1）油垫压力为 6.45MPa 的腔体体积为 140470.33m^3；（2）根据式（5），关井 7d 后腔内卤水变化量为 6.31m^3；（3）根据式（7），关井 7d 后腔体体积变化量即为卤水体积变化量，为 6.31m^3，其包括蠕变体积变化和弹性体积变化，该段时间内腔体体积收缩率为 0.0045%。

3.2 B 腔体计算过程及结果

B 腔体计算过程及结果见表 2。

（1）根据式（5），卤水的变化体积为 96.06m^3；（2）根据式（4），腔体的变化体积为卤水的变化体积，结果为 96.06m^3，根据式（7），则腔体的体积收缩率就等于 0.1039%。

表 2 B 腔体试验数据计算结果表

声呐体积/m^3	停井初油垫压力/MPa	停井末油垫压力/MPa	停井时间/d	卤水体积变化量/m^3	腔体体积变化量/m^3	腔体体积收缩率/%
92430.7	3.79	7.49	106	96.06	96.06	0.1039

3.3 C、D腔体计算结果

根据C、D腔体两期声呐测腔所得体积和式(7)算出腔体体积收缩率，结果见表3。

表3 C、D腔体体积收缩计算结果表

腔体	2005年声呐测腔体积/m³	2013年声呐测腔体积/m³	运行期间体积损失量/m³	腔体体积收缩率/%
C	105664	103890	1774	1.68
D	129830	128305	1525	1.17

4 试验数据与数值模拟数据对比

4.1 数值模拟计算模型

根据单个腔体声呐数据及相关地质数据，利用数值模拟软件建立腔体的二维轴对称几何模型，选取腔体的一个纵剖面的一半作为研究对象，计算模型为一个2000m×2000m的正方形，平面为xy坐标系平面，竖直方向为y轴，坐标原点为模型顶部左侧顶点，整个模型考虑了0~2000m的地层信息，考虑到泥岩夹层较少且厚度较薄，因此暂不考虑夹层的影响，其几何模型如图2所示，根据腔体的受力情况，分别施加地应力和内部拉张力，地应力场采用三向等压自重应力场，模型底边、左边和右边分别加以法向约束。采用有限元数学方法对建立的物理模型进行模拟计算，需要对几何模型划分网格，网格划分质量的好坏直接决

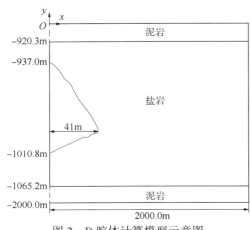

图2 D腔体计算模型示意图

定了模拟的准确度和精度，通常腔体周围的网格需要加密以分析腔体周围的物理过程。

4.2 盐岩弹性参数校准

A腔体升压和降压试验过程中由于试验时间极短，因此可以认为是由于盐岩的弹性收缩导致腔体体积发生变化，弹性参数主要为弹性模量(E)和泊松比(μ)，根据实验室对盐岩的弹性参数测量结果，弹性模量值介于2~18GPa，泊松比取值范围变化不大(0.30~0.35)，弹性参数模拟结果见表4，根据模拟结果泊松比在0.30~0.35范围内对模拟结果的影响有限，弹性模量对模拟结果影响较大。因此根据模拟结果与实际数据的契合度，将弹性模量定为5.8GPa，泊松比定为0.35。

表4 A腔体盐岩弹性参数校准表

项目	弹性模量/GPa	泊松比	升压腔体弹性压缩系数/%	降压腔体弹性压缩系数/%	备注
试验结果	2.0~18.0	0.30~0.35	0.0434	0.0440	根据岩心测试，弹性模量2~18GPa，泊松比0.30~0.35

续表

项目	弹性模量/GPa	泊松比	升压腔体弹性压缩系数/%	降压腔体弹性压缩系数/%	备注
模拟结果	13.0	0.30	0.0168	0.0133	中国科学院武汉岩土力学研究所数据
模拟结果	6.0	0.30	0.0403	0.0422	—
模拟结果	6.0	0.35	0.0417	0.0432	—
模拟结果	5.8	0.30	0.0418	0.0436	—
模拟结果	5.8	0.35	0.0431	0.0441	模拟结果与试验数据符合

4.3 盐岩黏塑性参数校准

4.3.1 A腔体盐岩黏塑性参数校准

盐岩的黏塑性参数主要为 A、n 和 m，A 值与岩性矿物组分和温度有关，n 和 m 值与盐岩的矿物组成有关，根据实验室试验结果 A 值变化范围较大数量级介于 $10^{-6} \sim 10^{-9}$，n 值变化范围较小介于 $3 \sim 4$，此次数值模拟保持 n、m 值不变，根据中国科学院武汉岩土力学研究所对金坛盐岩岩心的力学试验结果 $n = 3.75$，$m = -0.525$，模拟结果见表5，根据模拟结果，$A = 3.0 \times 10^{-7}$，$n = 3.75$，m 为 -0.525 时的模拟结果与试验结果契合较好。

表5 A腔体盐岩黏塑性参数校准结果表

项目	A	n	m	腔体收缩率/%	备 注
试验结果	$10^{-6} \sim 10^{-9}$	$3 \sim 4$	−0.525	0.0045	根据简化公式(1)至公式(6)，岩心力学试验 A 值范围为 $10^{-6} \sim 10^{-9}$，n 值为 $3 \sim 4$
模拟结果	2.0×10^{-7}	3.75	−0.525	0.00042	中国科学院武汉岩土力学研究所利用 Lemaitre 公式(1)至公式(5)拟合金坛储气库岩心力学数据的结果
模拟结果	3.0×10^{-7}	3.75	−0.525	0.00419	该模拟结果与现场试验结果契合较好

4.3.2 B腔体试验数据验证盐岩力学参数

为了验证A腔体校准出来的盐岩力学参数的准确性和实用性，选取B腔体现场试验数据(表2)进行模拟验证，从表6中可以看出模拟处理的腔体体积收缩率与现场试验实际腔体收缩率十分接近，其主要原因是B腔体和A腔体距离较近，约250m，腔体的深度以及腔体周围盐岩的性质变化不大，同时也说明该方法模拟出来的金坛储气库盐岩力学参数具有一定的实用性和指导意义。

表6 B腔体体积收缩模拟结果表

项目	弹性模量/GPa	泊松比	A	n	m	腔体收缩率/%	备 注
试验结果	—	—	—	—	—	0.1039	
模拟结果	5.8	0.35	3.0×10^{-7}	3.75	−0.525	0.0971	模拟结果与B腔体现场试验数据近似

4.3.3 C、D腔体试验数据验证盐岩力学参数

为了进一步验证校准的盐岩力学参数，选取了距离A腔体较远的C、D腔体，为同采

同注腔体，其现场试验数据见表3，根据其实际的注采过程进行数值模拟，其模拟结果见表7，当模拟过程使用B腔体和A腔体的盐岩力学参数时，腔体收缩率的模拟结果与实际情况存在一些差距，但其结果误差在可接受范围内。究其主要原因是C、D腔体距离B腔体和A腔体较远，盐岩层深度相差几十米，盐岩的地层温度以及矿物组成等方面的差别导致其力学性质也存在一些差异。

表7 C、D腔体体积收缩模拟结果

项目		弹性参数		黏塑性参数			腔体收缩率/%	备 注
		E/GPa	μ	A	n	m		
C腔体	试验数据	—	—	—	—	—	1.68	
	模拟结果	5.8	0.35	3.0×10^{-7}	3.75	−0.525	2.27	B腔体、A腔体校准的力学参数
	模拟结果	5.8	0.35	2.0×10^{-7}	3.75	−0.525	1.74	改变A值模拟结果更贴近真实值
D腔体	试验数据	—	—	—	—	—	1.17	
	模拟结果	5.8	0.35	3.0×10^{-7}	3.75	−0.525	2.17	B腔体、A腔体校准的力学参数
	模拟结果	5.8	0.35	2.0×10^{-7}	3.75	−0.525	1.66	改变A值模拟结果更贴近真实值

5 结论

笔者在长期的现场工作经验基础上，收集了大量的现场试验数据，理论结合实际，总结出一套行之有效的盐穴储气库围岩力学参数校准方法，并基于该方法对金坛盐穴储气库的盐岩力学参数进行了校准，获得了良好的效果。校准后的力学参数对于该储气库以后造腔和长期注采运行有重要的指导意义。

另外，气腔声呐数据获取昂贵，目前国内仅有两个注采运行的腔体进行过声呐测腔，由于声呐数据的不足，蠕变模型中的黏塑性参数未得到完全校准，目前校准的参数结合室内试验参数已经可以使模拟结果与实际工况良好匹配，今后将在现场数据不断丰富之后作进一步深入研究。

参 考 文 献

[1] 严铭卿，廉乐明，焦文玲，等．21世纪初我国城市燃气的转型［J］．煤气与热力，2002，22（1）：12-14．
[2] 李铁，张永强，刘广文．地下储气库的建设与发展［J］．油气储运，2000，19(3)：1-8．
[3] 杨春和，梁卫国，魏东吼，等．中国盐岩能源地下储存可行性研究［J］．岩石力学与工程学报，2005，24(24)：4410-4417．
[4] 杨春和，李银平，陈锋．层状盐岩力学理论与工程［M］．北京：科学出版社，2009．
[5] 陈锋．盐岩力学特性及其在储气库建设中的应用研究［D］．武汉：中国科学院研究生院（武汉岩土力学研究所），2006．
[6] 李银平，杨春和，罗超文，等．湖北省云应地区盐岩溶腔型地下能源储库密闭性研究［J］．岩石力学与工程学报，2007，26(12)：2430-2436．
[7] 刘江，杨春和，吴文，等．盐岩短期强度和变形特性试验研究［J］．岩石力学与工程学报，2006，25（1）：3105-3109．

［8］唐明明，王芝银，丁国生，等．含夹层盐岩蠕变特性试验及其本构关系［J］．煤炭学报，2010，35（1），42-45．

［9］梁卫国，徐素国，莫江，等．盐岩力学特性应变率效应的试验研究［J］．岩石力学与工程学报，2010，29（1）：43-50．

［10］徐素国，梁卫国，赵阳升．钙芒硝盐岩物理力学特性研究［J］．地下空间与工程学报，2007，3（6）：1054-1059．

［11］高小平，杨春和，吴文，等．盐岩蠕变特性温度效应的试验研究［J］．岩石力学与工程学报，2005，24（2）：2054-2059．

［12］梁卫国，徐素国，赵阳生，等．盐岩蠕变特性的试验研究［J］．岩石力学与工程学报，2006，25（7）：1386-1390．

［13］Colomé Jaime D，Monárdez Christian. Potasio Río Colorado pilot cavern creep modeling using FLAC3D-comparative analysis with values obtained from sonar mapping and pressure measurements inside the cavern［C］//Solution Mining Research Institute Spring 2011 Technical Conference，18-19 April 2011，Galveston，Texas，USA.

［14］陈卫忠，伍国军，戴永浩，等．废弃盐穴地下储气库稳定性研究［J］．岩石力学与工程学报，2006，25（4）：848-854．

［15］赵克烈，杨海军，陈锋，等．深部储气库群盐层蠕变参数优化研究［J］．岩石力学与工程学报，2009，28（2）：3550-3555．

［16］Tijani M，Hadj-Hassen F，Gatelier N. Improvement of Lemaitre's creep law to assess the salt mechanical behavior for a large range of the deviatoric stress［C］//9th International Symposium on Salt，5-7 September 2009，Beijing. DOI：https：//hal-mines—paristech. archives-ouvertes. fr/hal-00566278.

本论文原发表于《天然气工业》2015年第35卷第7期。

金坛盐穴储气库地质力学评价体系研究进展

陈加松 李建君 井 岗 敖海兵 刘春 王元刚 付亚平

（中国石油西气东输管道公司储气库项目部）

【摘 要】 基于盐穴储气库所在目标地层的地质特征、物性参数及运行参数等，建立盐穴储气库的地质力学模型对盐穴围岩响应规律进行研究，是目前盐穴储气库运行参数优化及安全评价的主要方法之一。通过对盐穴储气库地质力学评价中的岩石力学参数获取、地质力学模型建立及安全评价指标体系的国内外研究进展进行综述，并结合金坛盐穴储气库实际情况，对盐穴储气库地质力学评价体系的应用效果进行了验证。同时，根据金坛盐穴储气库后期发展面临的技术难题，对盐穴储气库地质力学评价体系进行了展望。

【关键词】 金坛盐穴储气库；地质力学评价；稳定性判据；数值模拟

Research progress of geomechanical evaluation system used for Jintan salt cavern gas storage

Chen Jiasong Li Jianjun Jing Gang Ao Haibing
Liu Chun Wang Yuangang Fu Yaping

(Gas Storage Project Department, PetroChina West-East Gas Pipeline Company)

Abstract At present, one of the main methods to optimize operation parameters and evaluate the safety of salt cavern gas storage is to establish a geomechanical model of salt cavern gas storage based on the geological characteristics, physical property parameters and operation parameters of the target formation so as to investigate response laws of the rocks around the salt cavern. In this paper, the research progress of geomechanical evaluation systems used for salt cavern gas storage at home and abroad was reviewed from aspects of the obtainment of rock mechanical parameters, the establishment of geomechanical model and the stability evaluation index system. Then, their application effects were verified based on the actual situations of Jintan salt cavern gas storage. Besides, the geomechanical evaluation systems of salt cavern gas storage were prospected based on the technical difficulties encountered in the late development stage of Jintan gas storage.

Keywords Jintan salt cavern gas storage; geomechanical evaluation; stability criteria; numerical simulation

---

基金项目：中国石油地下储气库重大科技专项"地下储气库关键技术研究与应用"，2015E-40。

作者简介：陈加松，男，1986年生，工程师，2013年硕士毕业于中国石油大学（北京）矿产普查与勘探专业，现主要从事盐穴储气库建设及稳定性评价研究工作。地址：江苏省镇江市中国石油西气东输管道公司储气库项目部，212000。电话：18605253656，E-mail：chenjiasong@petrochina.com.cn。

盐穴储气库由于其注采气灵活、运行效率高等优点，在保障安全平稳供气方面起到了关键作用[1-2]。盐穴稳定性评价对于储气库的安全平稳运行具有重要作用，是国内外研究的重点。欧美国家针对盐岩力学性质，进行了大量力学实验，为开展盐穴储气库地质力学研究奠定了基础；有限元数值模拟软件的出现，为盐穴储气库稳定性的整体评价提供了有效的研究方法[3-4]。中国盐穴储气库建设起步较晚，自2000年在江苏金坛建设中国第一座盐穴储气库以来，中国盐穴储气库建设进度明显加快。中国在盐岩基础理论和力学实验方面作了大量研究，利用有限元数值模拟软件对盐穴进行模拟计算，并利用相关评价指标开展稳定性评价研究[4-16]。但目前开展的大量工作还局限于理论研究和设计阶段，缺少足够的现场数据作为支撑，难以应用于现场实践。金坛盐穴储气库已有26口盐穴投入生产运行，随着投入盐穴数量的增加，工作重心从造腔工程转移至盐穴注采运行稳定性监测及分析。对金坛盐穴储气库地质力学评价体系研究方面的成果进行综述，利用地质力学模型等手段，对盐穴储气库的稳定性评价发展趋势进行预测。

1 金坛盐穴储气库概况

金坛盐穴储气库位于江苏省常州市金坛区直溪镇，地处长三角天然气销售市场终端，作为天然气管道的配套工程，在保障天然气管道平稳运行和市场平稳用气方面发挥了重要作用。金坛盐穴具备了建设盐穴储气库良好地质条件，盐岩埋深适中（900~1200 m），盐层厚度大（150~300 m），可溶物NaCl含量高（80%~90%），夹层少且薄，单个厚度小于2 m，断层和裂缝发育相对较少。金坛盐穴储气库设计库容$26×10^8 m^3$，工作气量$17×10^8 m^3$，单腔设计有效体积$25×10^4 m^3$，单腔运行压力为7~17 MPa，单腔工作气量$2700×10^4 m^3$。

2 评价流程

目前，中国对于盐穴储气库完腔后以及注采运行中的力学评价方面的研究较少。基于国际上常用的有限元数值模拟方法，形成了金坛盐穴储气库稳定性评价方法(图1)：(1)利用前期地质勘探获取建库目标地层信息、岩石力学参数、地应力数据及盐穴加载历史等，建立盐穴储气库地质力学模型并求解；(2)通过预设不同盐穴形状、尺寸和运行条件等，对盐穴围岩响应规律进行分析；(3)利用设置好的评价指标对盐穴储气库的安全性进行评价，对不同参数的敏感性进行分析；(4)根据计算结果，对盐穴储气库形状、尺寸和运行参数进行优化，从而达到安全评价的目的。

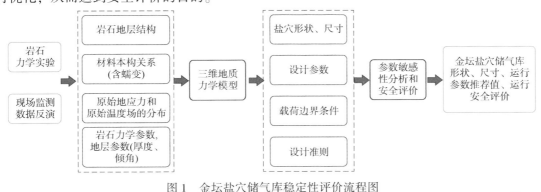

图1 金坛盐穴储气库稳定性评价流程图

3 岩石基本力学参数获取

3.1 岩石力学本构模型

盐岩以外的岩石、泥岩盖层及夹层作为线性弹性材料,主要为弹性变形。盐岩作为弹性—黏塑性材料,变形主要包括弹性变形和蠕变变形,关于盐岩的蠕变变形目前国际上普遍采用 Lemaître 模型[3],该模型很好地描述了温度与蠕变速率的关系:

$$\varepsilon_T = \varepsilon_E + \varepsilon_{VP} = \frac{\sigma}{E} + \varepsilon_{VP} \tag{1}$$

$$\varepsilon_{VP} = \left(\frac{\sigma}{K}\right)^\beta t^\alpha \exp\left[-A_0\left(\frac{1}{T} - \frac{1}{T_r}\right)\right] \tag{2}$$

式中:ε_T 为盐岩总应变;ε_E 为盐岩弹性应变;ε_{VP} 为盐岩黏塑性应变;σ 为盐岩试样受到的应力差,MPa;E 为杨氏模量,MPa;α、β、K 分别为 3 个 Lemaître 参数;t 为时间,d;T 为盐岩实验温度,℃;T_r 为温度基准,℃;A_0 为 Arrhénius 系数。

在不考虑温度对盐岩蠕变速率的影响下,式(2)简化为

$$\dot{\varepsilon}_{VP}^{eq} = A\sigma^n t^m \tag{3}$$

式中:$\dot{\varepsilon}_{VP}^{eq}$ 为盐岩稳态蠕变应变率;A、n、m 均为盐岩的蠕变参数,其主要与盐岩样品的结构组分有关,且参数 A 还与盐岩试样实验温度相关。

唐明明等[8-11]经过实验分析,认为盐岩蠕变主要包括初始蠕变阶段、稳态蠕变阶段以及加速蠕变阶段(图 2,其中 ε_B 为加速蠕变阶段的应变;t_B 为加速蠕变阶段的时间),处于初始蠕变阶段的时间较短,多数时间处于稳态蠕变阶段,而加速蠕变在实际工程基本上不可能出现。因此,忽略掉初始蠕变和加速蠕变的影响,稳态蠕变率可以由实验室获取。当拟合盐岩蠕变参数和数值模拟计算时,多采用 Norton 幂指数函数形式[11-14],但该本构模型无法显示盐岩的初始蠕变阶段:

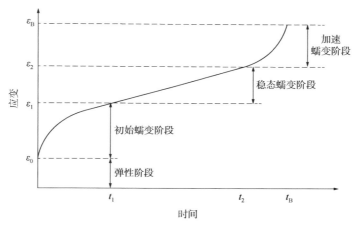

图 2 盐岩蠕变应变不同阶段曲线图

$$\dot{\varepsilon}_{\text{VP}}^{\text{eq}} = A(\sigma_1 - \sigma_3)^n \tag{4}$$

式中：σ_1、σ_3 分别为盐岩试样所受最大主应力、最小主应力，MPa。

式（3）和式（4）为目前常用的盐岩蠕变本构方程，但其无法反映盐岩地层温度变化对于盐岩蠕变速率的影响。在中国，盐岩的蠕变参数多在室内常温条件下获得，与地层条件的真实蠕变参数仍有一定差距，其主要原因是目前中国能够用于盐岩高温蠕变的实验仪器较少，且缺少切实可行的高温盐岩蠕变实验方案。如何模拟金坛盐穴储气库实际地层温度和压力条件，并开展盐岩蠕变实验，从而准确描述盐岩蠕变变形特征，是盐岩力学实验的主要发展方向。

3.2 力学参数获取方式

3.2.1 实验室实验法

常规岩石力学参数主要包括弹性参数、泊松比、黏聚力、内摩擦角、抗压强度、抗拉强度及蠕变参数等。岩石力学参数的获取对于盐穴储气库稳定性评价至关重要，直接决定了评价的精度。岩石力学参数获取需要开展室内力学研究，对不同深度、不同岩性的岩心样品进行分类制样加工，再对不同样品进行单轴压缩实验、三轴压缩实验、劈裂及剪切实验和蠕变长时实验等[6-10]。岩石力学的数值精度受限于实验仪器、实验方法、实验员的经验及责任心。目前，中国能够很好地获取常规盐岩力学参数，但是关于盐岩高温蠕变规律的研究还较少。特别是中国盐穴储气库建造深度有加深趋势，虽然金坛盐穴储气库的深度约为1000m，其深度较适中，但其他外围储气库，如平顶山、淮安、楚州等盐穴深度为1500~2200m，其深度跨度大、地温变化大、蠕变速率快。因此，需要进行力学实验，明确力学参数与温度的变化规律。

3.2.2 工程数据校核法

盐岩弹性参数、泊松比及蠕变参数对于盐穴储气库稳定性评价工作具有重要影响。利用现场监测和反演是校核和验证室内实验数据是否可靠的有效手段。同时，室内实验结果也可以为现场监测反演结果提供佐证。利用室内实验和现场监测反演手段相结合的方法获得盐岩力学性能参数，可以更加准确描述盐穴实际变形和受力情况。目前，现场校核方法主要有卤水井加压、泄压工程实验结果回归法，注采运行盐穴形状对比回归法。李建君等[17]通过现场卤水井快速加载和卸载的方法获得了盐岩的弹性力学参数；杨海军等[18]通过金坛盐穴储气库老腔两期声呐形状对比和校核，获得了盐岩蠕变参数；赵克烈等[19]通过造腔卤水井或观察井长时间停留时腔内压力的变化折算盐穴体积的变化，从而核算盐岩蠕变参数。可见，综合利用室内实验和现场监测数据获得盐岩力学参数，相互验证对比，确定最终力学参数，具有较高的精度，是今后的研究方向之一。

4 地质力学模型建立

4.1 计算模型

目前，国内外采用数值模拟的方法进行地质力学评价，采用的有限元软件有Abaqus、ANSYS及FLAC3D。陈卫忠等[20]利用Abaqus软件对金坛废弃盐矿老腔进行了稳定性研究，并对储气库的注采压力和套管鞋高度进行了探讨[13]。梁光川等[21]利用FLAC3D软件对金坛盐穴储气库不同洞形、不同注采压力开展了数值模拟和稳定性评价。

建立盐穴的计算模型需要综合考虑地层信息和盐穴形状。中国盐岩主要是层状结构，盐岩和非盐夹层交替出现。地层信息由钻井综合录井提供，地质建模时仅考虑厚度大于1.5m的非盐夹层。盐穴形状主要包括初始设计盐穴形状或声呐检测盐穴形状，对形状变化较大盐穴应进行适当的圆滑处理(图3)；二维模型主要选取单腔某个剖面作为研究对象；二维轴对称模型适用于形状较为对称的盐穴；三维模型可以模拟单腔，也可以模拟多腔，并可全面模拟盐穴各个部位的力学性质，但是对计算机硬件的配置要求较高。几何模型建好后，需要对模型进行网格划分，网格划分精度决定了计算结果的精度，通常盐穴周围的网格需要加密，能够更加清楚地显示出盐穴周围的变形情况。

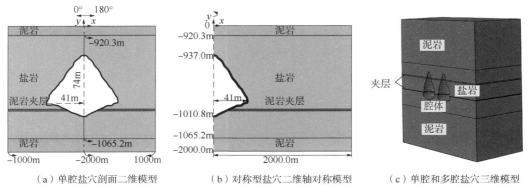

(a) 单腔盐穴剖面二维模型　　(b) 对称性盐穴二维轴对称模型　　(c) 单腔和多腔盐穴三维模型

图3　不同类型盐穴储气库地质力学计算模型图

4.2　岩石力学参数选择

中国科学院武汉岩土力学研究所对金坛盐岩做了相关力学实验[17-19]，根据泥岩、盐岩与泥岩夹层的单轴实验、三轴实验、蠕变实验结果，并通过工程实验反演及校核方法对相关力学参数进行了校核，确定了金坛储气库地质力学评价所需的岩石力学参数(表1)，其中盐岩采用黏塑性本构模型，泥岩和泥岩夹层则采用弹塑性本构模型。

表1　金坛盐穴储气库基本力学参数

岩性	弹性模量/MPa	泊松比	黏聚力/MPa	内摩擦角/(°)	蠕变参数			密度/(g/cm³)
					A	n	m	
盐岩	6000	0.30	2.5	30	$3×10^{-7}$	3.75	-0.525	2.16
泥岩	10000	0.27	2.5	35	—	—	—	2.46
泥岩夹层	4000	0.30	2.0	30				2.35

4.3　地应力加载

地应力是影响储气库稳定性的外部因素。如果无地应力测试数据，可采用三向等压自重应力场，垂向主应力大小为上覆载荷重力，其计算公式为

$$\sigma_v = g\int_0^z \rho(z)\mathrm{d}z \tag{5}$$

式中：σ_v 为垂向主应力，MPa；g 为重力加速度，m/s²；z 为岩石深度，m；$\rho(z)$ 为深度 z 处岩石密度(根据密度测井曲线获得)，kg/m³。

金坛盐穴储气库利用小型水力压裂法获得了最小主应力，以该储气库某井为例，根据其最小主应力和裂纹开裂压力，通过空隙介质弹性力学，可以计算得出最大主应力(表2)。金坛盐岩层具有各向同性的应力状态(3个主应力分量大小相似)，与流体相似，验证了相关学者的结论[22-24]：盐岩具有典型蠕变特征，构造应力在成岩过程中充分释放，现今地应力场分布满足静水压力分布特征。但小型水利压裂法获得的最小主应力，由于无法获知裂缝走向，因此无法判断最小主应力的走向。

表2 金坛储气库某井地应力测试及计算结果

岩性	深度/m	主应力/MPa		
		垂向	最小	最大
盐岩	1042	24.5	23.5	26.0
泥岩	1074	24.5	24.2	27.2
泥岩	1085	25.1	24.5	27.4
盐岩	1104	25.5	24.6	27.0
盐岩	1121	25.3	26.3	26.3

4.4 盐穴内部载荷变化历史

在盐穴建造和运行过程中，其内部载荷的变化历史(图4)是影响储气库稳定性的内在因素。在开挖前，盐穴为静岩压力；在开挖后，造腔过程中静岩压力逐步变为卤水压力；在注气排卤过程中，逐渐演变成气体压力；在运行过程中，腔内运行压力伴随着注采气压力呈现出周期变化，整个盐穴的力学性质随着腔内压力的周期变化也在发生改变。

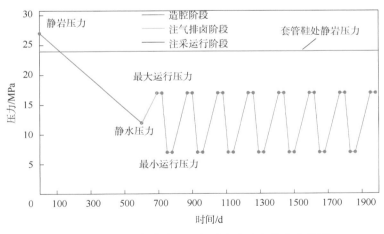

图4 盐穴储气库建造和运行过程中腔内载荷变化图

5 安全性评价判据指标

对于盐穴的稳定性判据指标研究较多[15-16,25-37]，均是预先设置好评价标准，通过计算机模拟分析，确定稳定性状况，但尚无统一的判断标准。目前，中国盐穴储气库稳定性行评价仍停留于定性和半定量状态，综合考虑一个或者多个评价指标对于储气库安全的影响。金坛盐穴储气库选用了无拉张应力、热应力、剪切破坏等8个评价指标，即盐穴的力学指标需要满足所有既定的稳定性判据指标，使得现阶段的生产运行设计方案相对较保守，但可以确保盐穴储气库安全运行。鉴于目前盐穴储气库地质力学评价尚无统一标准，应当选取适用性较好的指标，并根据选取指标的重要性分配权重进行综合评分，划分安全等级，将安全评价指标进行量化。

5.1 无拉张应力

盐岩在拉张应力作用下较脆，其抗拉张强度极低，因此盐穴周围腔壁中不允许有拉张应力的存在，否则将导致盐岩发生破坏，该指标是目前判断盐穴稳定性公认的原则[34-35]。以下5种情况可能产生张应力：腔顶大平台形成的顶层下陷、盐腔运行过程中超压、盐岩层与泥岩夹层之间的变形不匹配、临腔之间的运行压力差效应、快速注采气形成的热应力。当盐岩主应力小于0时，呈挤压状态，则无拉张应力形成。

5.2 热应力

当盐穴储气库长期运行时，盐岩蠕变过程中会形成拉张应力，特别是在注采过程中腔内温度发生变化形成焦耳汤普逊效应产生的热冲击力，会加速腔内拉张应力的形成。但目前对于金坛盐穴储气库注采过程中腔内的温度变化规律尚不清楚，仅开展了一些理论研究[16]，建立以下计算公式：

$$\sigma_t = \frac{E}{1-\nu}\alpha_{\mathrm{lin}}\Delta T_{\mathrm{w}} \tag{6}$$

式中：σ_t 为由于温度改变在盐岩中产生的热应力，MPa；ν 为盐岩的泊松比；α_{lin} 为盐岩膨胀系数，m/℃；ΔT_{w} 为盐穴储气库腔壁温度变化，℃。

Mousset 等[30]认为盐穴腔壁每产生1℃的温度变化就会产生1MPa的热应力，即 $\sigma_t \approx$ 1MPa/℃，腔内温度的变化与注采气速率、压降有关，但尚未得出统一结论，还需进一步研究。目前，模拟热应力主要有两种方法：（1）顺序耦合，即先根据盐穴注采气情况模拟出温度场变化情况，再将模拟结果叠加到力学模拟中[31]；（2）完全耦合，即根据盐穴注采气情况，同时进行温度场和力学模拟耦合计算。顺序耦合方法忽略了力学变化对于温度场的影响，而完全耦合方法考虑了热力耦合，准确度更高[32]。研究热应力对于注采运行腔体的稳定性影响是目前储气库研究重点之一，对于确定最优注采气速率具有重要作用，但金坛盐穴储气库的相关研究较少。

5.3 剪切破坏

岩石的抗剪强度虽高于抗拉强度，但在较高的偏应力作用下仍有可能产生剪切破坏，通常可以使用摩尔库伦准则来判断剪切破坏程度，其计算公式为

$$\frac{\sigma_1-\sigma_3}{2}-c\cos\varphi-\frac{\sigma_1+\sigma_3}{2}\sin\varphi \geq 0 \tag{7}$$

式中：c 为盐岩黏聚力，MPa；φ 为盐岩内摩擦角，(°)。

盐穴储气库小范围、局部的剪切破坏区可以接受，但不可以连接成片，否则可能会导致盐穴发生掉块、片帮等破坏。

5.4 膨胀损伤

盐岩在较小偏应力的作用下会发生体积收缩，当偏应力超过某个值时则会由于微裂缝的形成和扩展导致体积膨胀而损伤，盐岩发生膨胀损伤会使得腔壁发生崩塌和失稳，通常用膨胀损伤指标 η 来描述盐岩产生膨胀损伤时的应力状态，该指标也是目前国内外公认的常用稳定性判断指标。当 $\eta \geqslant 1$ 时，盐岩产生膨胀损伤，关于 η 的计算公式主要有以下3种[33,36-37]。

（1）当膨胀损伤指标为线性损伤指标，其计算式为

$$\eta = \frac{\sqrt{J_2}}{aI_1 + b} \tag{8}$$

式中：J_2 为第二应力偏量不变量，MPa；I_1 为第一应力不变量，MPa；a、b 均为盐岩损伤系数。

（2）当膨胀损伤指标为考虑洛德角影响的线性损伤指标，其计算式为

$$\eta = \sqrt{J_2} \frac{\cos\theta - \frac{1}{\sqrt{3}}\sin\theta\cos\varphi}{c\cos\varphi - \frac{1}{3}I_1\sin\varphi} \tag{9}$$

式中：θ 为盐岩洛德角，（°）。

（3）当膨胀损伤指标为非线性损伤指标，其计算式为

$$\eta = \sqrt{J_2} \frac{\sqrt{3}\cos\theta - D_2\sin\theta}{D_1\left(\frac{I_1}{\sigma_0}\right)^f + T_0} \tag{10}$$

式中：D_1、D_2、f、T_0 为盐岩材料内部参数；T_0 为盐岩抗拉强度，MPa；σ_0 为常数，取 1MPa。

目前，金坛储气库主要使用式(8)、式(9)中线性方程计算盐岩膨胀损伤指标，可以较好地确定储气库的最小运行压力，但尚未针对式(10)进行力学损伤实验获取相关材料参数。

5.5 盐穴体积收缩

盐穴储气库在运行过程中由于盐岩的蠕变会发生体积收缩[38-39]，当盐穴体积收缩过大将会导致盐穴力学性质变差甚至发生失效，且影响储气库的经济性。因此，盐穴在前5年的体积收缩率不能超过5%，整个生命周期内收缩率不能超过20%。

5.6 穿刺渗透

当平行于盐穴壁的切向应力小于作用于盐穴壁的拉张力（即气压）时，需要考虑气体渗透的风险。当腔壁周围平行应力大于垂直应力时，盐穴无渗透风险；反之，则有渗透风险；当盐穴处于最大气压时，最容易发生渗透风险，可以形成宏观裂缝。因此，当腔壁周围出

现小范围的穿刺渗透区域可以接受，但若出现整体性穿刺渗透区时，需要重新评价其稳定性，并可利用穿刺渗透指标确定腔内最大运行压力。

5.7 蠕变应变

盐岩的蠕变性能导致盐穴储气库在运行过程中会发生收缩变形，盐岩蠕变应变不能超过所给定的限值，盐穴的最大蠕变应变率一般要求每年小于1%，整个运行周期内小于10%。对于套管鞋处的盐岩蠕变率一般要求小于0.3%，否则套管鞋会发生塑性变形，存在发生损伤的风险，据此可以确定生产套管鞋与腔顶的安全距离。

5.8 地面沉降

盐穴储气库运行过程中由于盐穴收缩会导致地面沉降。地面沉降过快、过大则会导致地面建筑、构筑物的损坏，因此地面沉降速率必须满足建筑、构筑物稳定指标为0.044mm/d的控制标准要求[40-41]，目前盐穴储气库地面沉降指标只是符合地面建筑物的沉降标准，是否满足盐穴长期稳定运行尚无统一定论。

6 优化设计

6.1 几何形状

金坛盐穴储气库几何形状(图5)的主要参数包括腔顶盐岩顶板厚度、脖颈高度、生产套管鞋下深、盐穴高度、盐穴最大直径、两井距以及矿柱宽度等，各参数根据现场地质条件确定，以确保盐穴的结构稳定性和力学稳定性[18]。按照初始设计方案，金坛盐穴储气库盐岩顶板至少预留30～35m，脖颈长度为15m，生产套管鞋下入盐层深度为15m，盐穴最大直径为80m，矿柱宽度为200m，矿柱宽度比(矿柱宽度与盐穴最大直径比)约为2.5，相邻井距为280m，盐穴顶部锥顶角不超过120°。中国科学院武汉岩土力学所通过对金坛储气库几

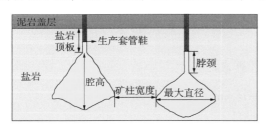

图5 金坛盐穴储气库几何形状示意图

何参数进行优化，研究结果表明金坛储气库最大直径达90m、矿柱比降至2.0，大大增加了单盐穴容积和库区总的储气能力。

在盐穴储气库设计阶段所考虑的地质模型和盐穴形状较理想，但是在实际造腔和运行过程中，由于复杂的地质条件和工艺技术问题导致盐穴呈现偏溶等畸形形状。目前，在金坛储气库遇到的盐穴畸形形状导致力学性质变差的现象有：盐穴偏溶，导致两口盐穴之间的矿柱宽度小于设计宽度；盐穴脖颈处被溶，脖颈高度变小；腔顶收口失败，形成大平顶。盐穴几何形状难以进行修复，因此需要根据盐穴本身形状进行力学模拟分析，通过优化注采运行方式以弥补盐穴形状缺陷；对于周边的盐穴，若已经溶腔结束，则需要优化并调整注采方式；若还在继续溶腔或未进行溶腔的盐穴则需要对初始盐穴形状设计进行优化，以满足稳定性要求。

6.2 注采运行方式

盐穴注采运行参数包括注采运行压力区间、注采频次、注采气速率和最小运行压力停

留时间等。当对某个注采运行参数进行设计或优化时，将其他参数固定，再进行模拟分析对比[18,42]。金坛盐穴储气库按照注采运行压力7~17MPa、每年2次调峰、注采气压降控制在0.5MPa/d进行设计，但在实际过程中往往需要进行6~10次调峰。最小运行压力及最小压力停留时间对于盐穴的力学性质影响极大，但目前最小压力的停留时间尚未形成统一定论，只有尽量缩短最小压力停留时间[43]，其与调峰次数有关。与国外盐穴储气库运行方式相比，金坛盐穴储气库目前的注采运行方式偏于保守，尚未发挥最大调峰能力，关于盐穴储气库注采运行参数的确定方法后续仍需要进行研究和细化，在稳定性评价工作的基础上，最终实现盐穴运行的精细化管理。

盐穴储气库造腔完毕后，需要针对完腔后的盐穴形状进行稳定性评价，初步确定注采运行方式，以后每5年需要进行一次带压声呐检测，跟踪分析盐穴的形状变化情况，并进行稳定性评价分析和注采方式优化。对于形状规则且运行状况良好的盐穴，可以扩大运行压力区间，特别是适当增加最大运行压力上限、库容、工作气量和注采频次，并可以根据实际需要适当提高注采气速率；对于形状不规则盐穴、在运行过程中盐穴部分区域发生垮塌或变形较大的盐穴，则需要缩小运行压力区间，特别是要提高最低压力，减少低压运行时间，降低注采频次和注采气速率，在运行过程中采取慢注慢采模式。

7 结论及建议

盐穴储气库稳定性评价工作是储气库建造及运行过程中重要一环，直接决定了储气库的运行安全及经济效益。目前，中国关于盐穴储气库的稳定性评价工作流程及方法尚无统一定论，根据金坛盐穴储气库稳定性评价研究结果可见：对盐岩地层实际压力、温度等进行模拟，开展力学实验获得盐岩力学性能参数将是今后的主要研究方向；同时，利用现场监测数据对实验数据进行校核和修正将会变得越来越重要。随着中国盐穴储气库的不断建设、发展，如何利用建库、运行等过程中的大数据实现对盐穴储气库的安全状态、运行寿命以及风险等级的预测，将是中国盐穴储气库研究中的重要发展趋势。

参 考 文 献

[1] 周学深. 有效的天然气调峰储气技术——地下储气库[J]. 天然气工业，2013，33(10)：95-99.

[2] 丁国生，梁婧，任永胜，等. 建设中国天然气调峰储备与应急系统的建议[J]. 天然气工业，2009，29(5)：98-100.

[3] Colomé J D, Monárdez C. Potasiorío colorado pilot cavern creep modeling using FLAC3D. Comparative analysis with values obtained from sonar mapping and pressure measurements inside the cavern[C]. Texas: SMRI Spring 2011 Technical Conference, 2011: 1-11.

[4] Tijani M, Hassen F H, Gatelier N. Improvement of Lemaitre's creep law to assess the salt mechanical behavior for a large range of the deviatoric stress[C]. Beijing: 9th International Symposium on Salt, 2009: 135-147.

[5] 杨春和，梁卫国，魏东吼，等. 中国盐岩能源地下储存可行性研究[J]. 岩石力学与工程学报，2005，24(24)：4410-4417.

[6] 陈锋. 盐岩力学特性及其在储气库建设中的应用研究[D]. 武汉：中国科学院研究生院(武汉岩土力学研究所)，2006：54-68.

[7] 任松，白月明，姜德义，等. 温度对盐岩疲劳特性影响的试验研究[J]. 岩石力学与工程学报，2012，

31(9): 1839-1845.

[8] 唐明明, 王芝银, 丁国生, 等. 含夹层盐岩蠕变特性试验及其本构关系[J]. 煤炭学报, 2010, 35(1): 42-45.

[9] 陈剑文, 杨春和, 高小平, 等. 盐岩温度与应力耦合损伤研究[J]. 岩石力学与工程学报, 2005, 24(11): 1986-1991.

[10] 王军保. 不同加载路径下盐岩蠕变力学特性与盐岩储气库长期稳定性研究[D]. 重庆: 重庆大学, 2012: 17-130.

[11] Yang C H, Wang T T, Qu D A, et al. Feasibility analysis of using horizontal caverns for underground gas storage: a case study of Yunying salt district[J]. Journal of Natural Gas Science and Engineering, 2016, 36: 252-266.

[12] Yang C H, Daemen J J K, Yin J H. Experimental investigation of creep behavior of salt rock[J]. International Journal of Rock Mechanics & Mining Sciences, 1999, 36(2): 233-242.

[13] Wang T T, Ma H L, Shi X L, et al. Salt cavern gas storage in an ultra-deep formation in Hubei, China[J]. International Journal of Rock Mechanics and Mining Sciences, 2018, 102(2): 57-70.

[14] 梁卫国, 徐素国, 赵阳升, 等. 盐岩蠕变特性的实验研究[J]. 岩石力学与工程学报, 2006, 25(7): 1386-1390.

[15] 吴文, 侯正猛, 杨春和. 盐岩中能源(石油和天然气)地下储存库稳定性评价标准研究[J]. 岩石力学与工程学报, 2005, 24(14): 2497-2505.

[16] 王同涛. 多夹层盐岩体中储气库围压变形规律研究及安全评价[D]. 东营: 中国石油大学(华东), 2011: 14-18.

[17] 李建君, 陈加松, 吴斌, 等. 盐穴地下储气库盐岩力学参数的校准方法[J]. 天然气工业, 2015, 35(7): 96-102.

[18] 杨海军, 郭凯, 李建君. 盐穴储气库单腔长期注采运行分析及注采压力区间优化——以金坛盐穴储气库西2井腔体为例[J]. 油气储运, 2015, 34(9): 945-950.

[19] 赵克烈, 杨海军, 陈锋, 等. 深部储气库群盐层蠕变参数优化研究[J]. 岩石力学与工程学报, 2009, 28(增刊2): 3550-3555.

[20] 陈卫忠, 伍国军, 戴永浩, 等. 废弃盐穴地下储气库稳定性研究[J]. 岩石力学与工程学报, 2006, 25(4): 848-854.

[21] 梁光川, 王梦秋, 彭星煜, 等. 盐穴地下储气库溶腔形态变化数值模拟[J]. 天然气工业, 2014, 34(7): 88-92.

[22] 张重远, 吴满路, 陈群策, 等. 地应力测量方法综述[J]. 河南理工大学学报(自然科学版), 2012, 31(3): 305-310.

[23] 闫凤林, 翁小红, 李祥, 等. 盐穴储气库地应力测试新方法[J]. 重庆科技学院学报(自然科学版), 2017, 19(3): 63-64.

[24] 周冬林, 杨海军, 李建君, 等. 盐岩地层地应力测试方法[J]. 油气储运, 2017, 36(12): 1385-1388.

[25] Yang C H, Wang T T, Li Y P, et al. Feasibility analysis of using abandoned salt caverns for large-scale underground energy storage in China[J]. Applied Energy, 2015, 137: 467-481.

[26] Avdeev Y, Vlorobyev V, Krainev B, et al. Criteria for geomechanical stability of salt caverns[C]. Hannover: SMRI Fall 1997 Meeting, 1997: 1-11.

[27] Asgari A, Ramezanzadeh A, Jalali S M E. Stability analysis of natural-gas storage caverns in salt formations[C]. Bremen: SMRI Fall 2012 Technical Conference, 2012: 1-18.

[28] Wang T T, Yang C H, Li J J, et al. Failure analysis of overhanging blocks in the wall of a gas storage salt cavern: a case study[J]. Rock Mechanics and Rock Engineering, 2017, 50(1): 125-137.

[29] 任松, 李小勇, 姜德义, 等. 盐岩储气库运营期稳定性评价研究[J]. 岩土力学, 2011, 32(5): 1465-1472.

[30] Mousset C, Charnavel Y, Hévin G. Evaluating and improving the accuracy of salt cavern thermodynamic models using in situ downhole data[C]. Groningen: SMRI Fall 2014 Technical Conference, 2014: 5-10.

[31] 李建君, 敖海兵, 巴金红, 等. 热应力对盐穴储气库稳定性的影响[J]. 油气储运, 2017, 36(9): 1007-1011.

[32] 李景翠, 王思中, 张亚明, 等. 提高盐穴储气库腔体稳定性的设计方法[J]. 油气储运, 2017, 36(8): 964-967.

[33] Vries D, Kerry L. Geo-mechanical analyses to determine the onset of dilation around natural gas storage caverns in bedded salt[C]. Belgium: SMRI Spring 2006 Conference, 2006: 10-15.

[34] Hardy H R. Design and stability monitoring of salt caverns[C]. Manchester: SMRI Fall 1982 Meeting, 1982: 5-10.

[35] Adams J B. Determination of salt cavern operating pressures using rock mechanics and finite element analysis[C]. Ohio: SMRI Fall 1996 Meeting, 1996: 2-23.

[36] Rokahr R B, Staudtmeister K, Schiebenhofer D K. Development of a new criterion for the determination of the maximum permissible internal pressure for gas storage in caverns in rock salt[C]. Hannover: SMRI Fall 1997 Meeting, 1997: 145-155.

[37] 敖海兵, 陈加松, 胡志鹏, 等. 盐穴储气库运行损伤评价体系[J]. 油气储运, 2017, 36(8): 910-915.

[38] 杨海军, 李建君, 王晓刚, 等. 盐穴储气库注采运行过程中的腔体形状检测[J]. 石油化工应用, 2014, 33(2): 23-25.

[39] 丁国生, 张保平, 杨春和, 等. 盐穴储气库溶腔收缩规律分析[J]. 天然气工业, 2007, 27(11): 94-96.

[40] 屈丹安, 杨春和, 任松. 金坛盐穴地下储气库地表沉降预测研究[J]. 岩石力学与工程学报, 2010, 29(增刊1): 2706-2710.

[41] Rokahr R, Staudtmeister K, Schiebenhofer D Z. Rock mechanical determination of the maximum internal pressure for gas storage caverns in rock salt[C]. Rome: SMRI Fall 1998 Meeting, 1998: 2-31.

[42] Wang T T, Yang C H, Yan X Z, et al. Allowable pillar width for bedded rock salt caverns gas storage[J]. Journal of Petroleum Science and Engineering, 2015, 127: 433-444.

本论文原发表于《油气储运》2018年第37卷第10期。

含夹层盐岩孔隙特征及非线性渗透模型

刘继芹[1,3]　寇双燕[2]　李建君[1]　李淑平[1]　陈加松[1]　肖恩山[1]

（1. 中国石油西气东输管道公司储气库管理处工艺技术研究所；
2. 中海油田服务股份有限公司油田生产研究院；3. 中国石油大学（华东）石油工程学院）

【摘　要】 为了有效识别腔体之间以及腔体与断层之间的封闭性、安全性，有必要对盐穴储气库中夹层及盐岩的密闭性及渗透特征进行研究，识别腔体间或腔体与断层间存在的渗漏风险。结合金坛地区的地质情况，分析夹层及盐层孔隙结构，建立气体渗透非线性渗流模型，考虑了气固分子作用力及扩散作用对渗流的影响，基于此模型对金坛储气库两口相邻腔体矿柱存在夹层情况下的密闭性进行分析，并对腔体周边存在断层的情况进行研究。盐岩有很好的密闭作用，是非常好的储库建址选择，但泥岩夹层中气体的渗透性明显。因此，对含夹层盐层进行建腔，确定腔体之间、腔体与断层之间距离具有重要意义。

【关键词】 盐穴储气库；夹层；非线性；渗漏；断层

Pore characteristics and nonlinear permeability model of salt rocks with interlayers

Liu Jiqin[1,3]　Kou Shuangyan[2]　Li Jianjun[1]
Li Shuping[1]　Chen Jiasong[1]　Xiao Enshan[1]

（1. Technology Research Institute, Gas Storage Project Department, PetroChina
West-East Gas Pipeline Company;
2. Oilfield Production Institute, China Oilfield Services Ltd.;
3. School of Petroleum Engineering, China University of Petroleum (East China)）

Abstract　In order to identify the sealing capacity and safety between cavities or between cavity and fault, it is necessary to study the sealing capacity and permeability of the interlayers and salt rocks of salt cavern gas storages, and to identify the risk of seepage between cavities or between cavity and fault. In this paper, the pore structures of interlayers and salt rocks were analyzed based on the geological conditions in Jintan area. Then the nonlinear seepage model of gas permeability was established, and it considers the effects of gas-solid molecular force and diffusion on gas seepage. Finally, this model was applied in Jintan Gas Storage to analyze the sealing capacity between adjacent cavities with interlayers in pillars, and investigate the situations of the faults around the cavities. Salt rocks are very good choice for the site of underground gas storage due to its good sealing capacity, but the gas permeability of mudstone interlayers is relatively obvious. So it is of great significance to the solution mining of salt rocks with interlayers and the optimization and determination of the distance between the cavities or between cavity and fault.

Keywords salt cavern gas storage; interlayers; nonlinear; seepage; fault

中国地下盐层丰富,在河南平顶山、江苏金坛、江苏淮安、湖北云应等地均开展了天然气储气库选址研究及建设,但盐层普遍较薄、夹层较多[1-9],影响造腔过程中腔体形状的控制。盐穴储气库的气密性是储气库的关键要求,在储气库后期注采过程中,夹层的渗透现象不容忽视。腔与腔之间、腔体与周边断层之间夹层的渗透性,决定了腔体长期注采运行的安全性[10-17]。

1 夹层及盐岩的孔隙特征

1.1 夹层

盐层夹层一般由泥岩组成,是气体渗漏的主要通道,在金坛地区,由钻井取心岩样分析,夹层泥岩裂隙相对发育,可见一些被石膏充填的不同方向裂隙。

对选取岩样进行分析,夹层碎屑颗粒以石英为主(体积分数为12%~26%),含少量钾长石和斜长石。通过电子扫描镜观察发现,夹层样品微孔隙相对发育,层理发育的薄片多见微裂隙和微孔隙(图1)。

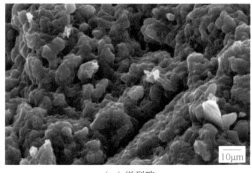

(a)微裂隙　　　　　　　　　　　(b)微孔隙

图1　夹层泥质岩的微裂隙和微孔隙显微图

根据分析,岩样的比表面积大,为5.2~31.8m²/g,均值为13.0m²/g。表明夹层岩石颗粒细,具有较多的微小孔隙。孔隙总量较大,孔径较小,孔径为7.8~14.7nm,选取岩样的平均孔径为11.45nm。样品含盐或样品水平层理发育时,钻取样品易造成微裂缝,物性测试结果普遍偏高。试验分析可得:孔隙度主要分布在3.1%~8.3%范围内,均值为6.4%;常规渗透率较低,介于0.0039~0.09mD之间,均值为0.014mD,渗透率主要分布在0.01~0.001mD范围内;地层渗透率介于0.0033~0.0723mD之间,均值为0.012mD。

1.2 盐岩

盐岩的孔隙相对较少，一般肉眼观察不到，在电子显微镜下可以观察到一些晶间孔和晶体间杂质形成的微孔隙。Cuevas[18]将岩盐中的孔隙分为宏观孔隙(孔径大于7.5μm)、微观孔隙(孔径7.5μm~100nm)及亚孔隙(孔径小于100nm)，试验结果可知亚孔隙占孔隙总体积的60%以上。

在WIPP盐岩场[19]对原始盐岩开展试验，得出渗透率为1×10^{-4} mD。Berest等[20]在一口井进行为期一年的试验，得出渗透率为6×10^{-5} mD。通常情况下，盐岩很难渗透气体，但是盐岩结晶边界形成的网状结构容易受到应力作用而导致孔隙结构破坏及裂缝生成与成长，这在一定程度上导致盐岩渗透率数量级提高。

2 孔隙中气体非线性渗流

2.1 流动形态

孔隙中的渗流常采用达西定律描述，未考虑气体分子与微孔隙的分子作用力，当孔隙为纳微米级别时，气体分子与孔隙壁面的碰撞以及气体分子在孔隙壁面的滑移不能忽略。气体在微孔隙中的运动形态可以通过克努森数K_n来判断。

$$K_n = \frac{\bar{\lambda}}{r} \tag{1}$$

$$\bar{\lambda} = \frac{K_B T}{\sqrt{2}\pi\delta^2 p} \tag{2}$$

式中：$\bar{\lambda}$为气体分子平均自由程，m；r为孔隙喉道半径，m；K_B为玻尔兹曼常数，取1.3805×10^{-23} J/K；T为温度，K；δ为气体分子碰撞半径，m；p为孔隙压力，Pa。

不同K_n对应不同的渗流方式：$K_n \leq 0.001$为连续流动区域，可通过达西定律计算；$0.001 \leq K_n \leq 0.1$为滑移流动区域，存在滑移边界；$0.1 \leq K_n \leq 10$为过渡区域，存在滑移边界，为过渡流；$K_n > 10$为分子自由运动区域，此时孔隙半径为纳米级，此时分子自由流动。Deng等[21]分析了不同孔吼半径下压力与K_n的关系(图2)：盐层及夹层中气体流动主要以滑脱流动为主，孔吼半径较大的夹层中部分流动为连续流动。

2.2 流动模型

对于常规多孔介质，气体流动分析一般采用达西定律。但是对于盐层及其及夹层，其渗透率非常低，孔隙半径为微米或纳米级，微孔吼道迂曲复杂，气体分子与孔隙壁面碰撞强烈，扩散作用与滑脱效应的作用不能忽略，此时达西定律不再适用。

在边续介质中，考虑气体分子的滑脱效

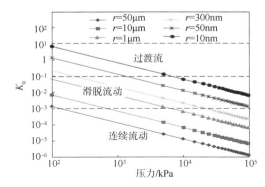

图2 不同孔吼半径下压力与K_n的关系图

应，扩散及黏性流工况下，可以通过 Beskok-Karniadakis 模型[22-23]计算气体的流动速度 v：

$$v = \frac{K_0}{\mu}(1+\alpha K_n)\left(1+\frac{4K_n}{1-bK_n}\right) \quad (3)$$

$$\alpha = \frac{128}{15\pi^2}\tan^{-1}(4K_n^{0.4}) \quad (4)$$

式中：K_0 为盐层或夹层渗透率，D；μ 为气体黏度，Pa·s；α 为稀疏因子；b 为滑移系数。

由连续性方程知，气体在盐岩及夹层中的渗流方程为

$$\frac{\partial}{\partial t}(\rho_g \phi) - \nabla \cdot (\rho_g \vec{v}_g) = q \quad (5)$$

式中：ρ_g 为气体密度，kg/m³；ϕ 为岩石孔隙度，%；$\vec{v}_g$ 为气体孔隙中流速，m/s；t 为时间，s。

气体的密度方程为

$$\rho_g = \frac{M_g p}{RT} \quad (6)$$

式中：M_g 为气体摩尔质量，kg/mol；R 为理想气体常数，取 8.314J/(mol·K)。

将式(6)、式(3)代入式(5)可得

$$\frac{\partial}{\partial t}\left(\frac{\phi M_g}{RT}p\right) - \nabla \cdot \left[p\frac{M_g K_0}{RT\mu}(1+\alpha K_n)\left(1+\frac{4K_n}{1-bK_n}\right)\nabla p\right] = q \quad (7)$$

3 工程应用

基于西气东输金坛储气库盐性参数（表1），考虑存在夹层的两腔体之间及周边气体渗透情况及压力分布，腔体等效为椭圆形，腔体直径80m，高150m，两腔体间矿柱宽度为200m，腔体之间存在2个夹层，厚度分别为3m、4m（图3）。

表1 西气东输金坛储气库夹层及盐岩的物性参数

盐岩孔隙度/%	盐岩渗透率/m²	地层压力/MPa	夹层孔隙度/%	夹层渗透率/D
1	3×10⁻²⁰	10	7	1.4×10⁻¹⁷

正常情况下，每年6—9月为注气阶段，12月至次年2月为用气高峰，为采气阶段（图4），为了便于计算，采用正弦函数等效描述腔内压力变化趋势。

虽然夹层渗透率较低，但其已成为盐穴中气体渗透的主要通道。运行3年后，两腔体间任一点压力均发生变化；腔体运行10年后，压力分布无较大变化，表明在当前矿柱宽度下盐层及夹层密封性较好，能够保证腔体密封性（图5）。

根据两腔矿柱中点盐层上点 $A(0,0)$ 处压力变化（图6），A 点压力逐渐升高，但基本不受腔内压力周期性变化的影响。运行570天时，A 点开始受注采气运行压力变化的影响，压力升高了0.1MPa，运行10年后，A 点最高压力升至11.1MPa。

根据腔矿柱中点夹层上点 $B(0,-40)$ 处压力变化(图7),运行60天时,A 点开始受注采气运行腔内压力变化影响,运行1年后,B 点开始受腔内压力周期性变化影响,B 点压力整体趋势逐渐升高。运行10年后,B 点最高压力升至12.61MPa。

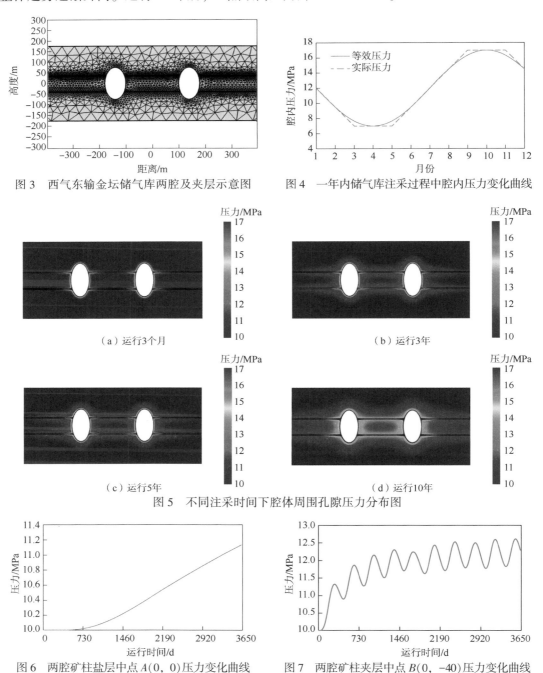

图3 西气东输金坛储气库两腔及夹层示意图 图4 一年内储气库注采过程中腔内压力变化曲线

图5 不同注采时间下腔体周围孔隙压力分布图

图6 两腔矿柱盐层中点 $A(0,0)$ 压力变化曲线 图7 两腔矿柱夹层中点 $B(0,-40)$ 压力变化曲线

模型研究了造腔存在与不存在夹层情况下单腔周围压力分布(图8),假设断层为开放性断层,模型设置为高渗透率条带,腔壁距断层距离为90m。对于不存在夹层的情况,距

腔体90m处断层对腔体的安全运行基本无影响，表明盐岩是极好的储气库封闭介质。对于存在夹层的情况，夹层成为断层与腔体连通的载体，运行3年后，断层内压力开始受腔体注采运行的影响；运行10年后，断层内压力发生较明显的变化，断层极可能成为气体渗漏通道，影响腔体运行及地面安全[24]（图9）。

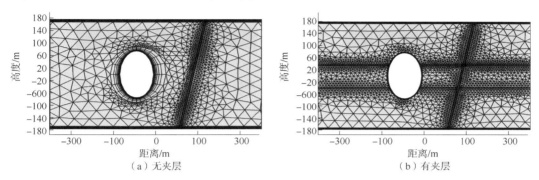

图 8　腔体邻近断层示意图

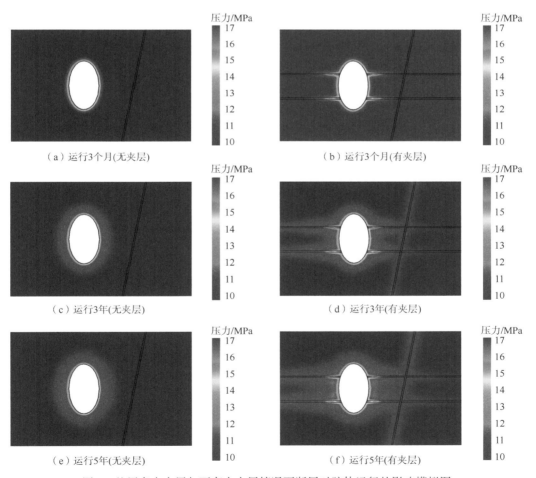

图 9　盐层存在夹层与不存在夹层情况下断层对腔体运行的影响模拟图

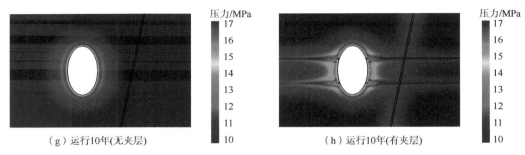

(g) 运行10年(无夹层)　　　　　　　(h) 运行10年(有夹层)

图 9　盐层存在夹层与不存在夹层情况下断层对腔体运行的影响模拟图(续)

4　结论

（1）盐岩及夹层渗透率非常低，属于特低渗透岩层，孔隙中气体渗流已不再遵守达西定律，需要考虑气体分子与孔隙壁面的作用力及扩散作用的影响。

（2）气体在盐岩中渗透极低，但存在夹层的盐岩中泥岩微孔隙相对发育，其渗透率高于盐岩渗透率多个数量级，是腔体渗透的主要通道，尤其是腔体周边存在断层的情况下，夹层与断层的贯通是腔体安全运行的隐患。

（3）对于存在夹层及断层的盐层，腔内气体的周期性压力使腔体周围压力分布复杂，会对腔体的稳定安全运行产生一定影响，同时也可能是气体渗漏的通道。

（4）盐层是极好的储气库建设选择，但泥岩夹层的存在是储气库泄漏分析不可忽略的因素。确定腔体之间及腔体与断层之间的距离时，需要定量分析并保障气体渗透量在安全范围内，同时进行稳定性评价时需要考虑夹层与断层存在造成压力分布的复杂性。

参 考 文 献

[1] 罗天宝，李强，许相戈．我国地下储气库垫底气经济评价方法探讨[J]．国际石油经济，2016，24(7)：103-106.

[2] 成金华，刘伦，王小林，等．天然气区域市场需求弹性差异性分析及价格规制影响研究[J]．中国人口资源与环境，2014，24(8)：131-140.

[3] 王元刚，陈加松，刘春，等．盐穴储气库巨厚夹层垮塌控制工艺[J]．油气储运，2017，36(9)：1035-1041.

[4] 周学深．有效的天然气调峰储气技术——地下储气库[J]．天然气工业，2013，33(10)：95-99.

[5] 杨海军，王元刚，李建君，等．层状盐层中水平腔建库及运行的可行性[J]．油气储运，2017，36(8)：867-874.

[6] 王峰，王东军．输气管道配套地下储气库调峰技术[J]．石油规划设计，2011，22(3)：28-30.

[7] 冉莉娜，完颜祺琪，王立献，等．国外盐穴地下储气库发展趋势及启示[J]．盐科学与化工，2017，46(12)：9-12.

[8] 垢艳侠，完颜祺琪，罗天宝，等．中俄东线楚州盐穴储气库建设的可行性[J]．盐科学与化工，2017，46(11)：16-20.

[9] 郭峒．地下储气库建设技术研究现状及建议[J]．石化技术，2017，24(8)：46.

[10] 袁光杰，田中兰，袁进平，等．盐穴储气库密封性能影响因素[J]．天然气工业，2008，28(4)：105-107.

[11] 赵延林. 层状岩盐储库气体渗漏固气耦合模型及储库稳定性研究[D]. 太原：太原理工大学，2006：1-21.
[12] 刘伟，李银平，杨春和，等. 层状盐岩能源储库典型夹层渗透特性及其密闭性能研究[J]. 岩石力学与工程学报，2014，33(3)：500-506.
[13] 周玉鑫. 含夹层岩盐储库渗透性与密闭性研究[D]. 天津：河北工业大学，2011：5-22.
[14] 陈卫忠，谭贤君，伍国军，等. 含夹层盐岩储气库气体渗透规律研究[J]. 岩石力学与工程学报，2009，28(7)：1297-1304.
[15] 周宏伟，何金明，武志德. 含夹层盐岩渗透特性及其细观结构特征[J]. 岩石力学与工程学报，2009，28(10)：2068-2073.
[16] 武志德. 考虑渗流及时间效应的层状盐岩溶腔稳定分析[D]. 北京：中国矿业大学(北京)，2011：1-25.
[17] 张耀平，曹平，赵延林，等. 双重介质固气耦合模型及含夹层盐穴储气库渗漏研究[J]. 中南大学学报(自然科学版)，2009，40(1)：217-224.
[18] Cuevas C D L. Pore structure characterization in rock salt[J]. Engineering Geology，1997，47(1-2)：17-30.
[19] Shosei S，Kittitep F. Permeability studies in relation to stress state and cavern design[C]. Richmond：Solution Mining Research Institute Spring 2008 Technical Conference，1991：1-56.
[20] Berest P，Blum P A，Durup G，et al. Long term creep test in a salt-caverns[C]. Houston：Solution Mining Research Institute Spring 2008 Technical Conference，1994：1-18.
[21] Deng J，Zhu W，Ma Q. A new seepage model for shale gas reservoir and productivity analysis of fractured well[J]. Fuel，2014，124(15)：232-240.
[22] Michel G M，Sigal R，Civan F，et al. Parametric investigation of shale gas production considering nano-scale pore size distribution，formation factor，and non-Darcy flow mechanisms[C]. Denver：SPE Annual Technical Conference and Exhibition，2011：1-20.
[23] Freeman C M，Moridis G J，Blasingame T A. A numerical study of microscale flow behavior in tight gas and shale gas reservoir systems[J]. Transp Porous Media，2011，90(1)：253-259.
[24] 耿凌俊，李淑平，吴斌，等. 盐穴储气库注水站整体造腔参数优化[J]. 油气储运，2016，35(7)：779-783.

本论文原发表于《油气储运》2018年第37卷第9期。

楚州盐穴储气库建库地质条件分析

薛 雨　王元刚　周冬林

（中国石油西气东输管道公司储气库项目部）

【摘　要】 楚州盐矿毗邻中俄东线天然气管道。在分析地震解释、资料井及老井钻探资料的基础上，为筛选出储气库建库层段，对该地区构造形态、地层特征、断裂发育、盐层横纵向展布、盖层密封性等地质条件进行了综合分析。该地区构造形态相对简单，浦二段上盐亚段第三岩性组合含盐厚度大，品位高，夹层少，盐层分布稳定，适合建设盐穴地下储气库。张兴区块盐岩含量分布稳定，夹层少，不溶物含量小，是建库有利区。通过分析各个盐群的地质情况，选取Ⅲ-7—Ⅲ-9盐群为建库有利层段进行储气库先期建设；杨槐区块目前只有一口资料井，待其地质资料进一步明确后再确定储气库建设层段。

【关键词】 中俄东线；楚州盐矿；盐穴储气库；构造形态；地质分析

Analyses on the geological conditions of building the salt cavern gas storage in chuzhou

Xue Yu　Wang Yuangang　Zhou Donglin

(Gas Storage Project Department, PetroChina West-East Gas Pipeline Company)

Abstract　Chuzhou salt mine is near to the Sino-Russian Eastern Route Gas Pipeline. In order to select the strata and member for the salt cavern construction, on the bases of the seismic interpretation, the information well and old well drilling data, the comprehensive analysis was conducted on the following geological conditions: the structural shapes, stratigraphic characteristics, fault development, horizontal and vertical distributions of the salt layers, seal-ing ability of the caprock and so on. The overall structure shape in the region is relatively simpler, the salt-bearing thickness in Group $K_2 p_{2-3(3)}$ is pretty higher, the quality is higher and with few interlayers, the distribution of the salt group is stable, so it is feasible to construct the gas storage in the underground salt cavern. The salt-concentra-tion distribution is stable in Zhangxing Block characterized by less interlayers and low insolubles content, therefore the region is considered as the favorable area to build the gas storage. By means of the geological analyses on each salt layer, Salt Group Ⅲ-7—Ⅲ-9 were selected to construct first, because there is only one

---

基金项目：中国石油天然气集团公司储气库重大专项"地下储气库关键技术研究与应用"（2015E-40）之课题八"盐穴储气库加快建产工程试验研究"（2015E-4008）。

作者简介：薛雨，男，1987年生，工程师，硕士，从事地下储气库项目管理工作。E-mail：xqdsxueyu@petrochina.com.cn。

exploratory well in Yanghuai Block, The constructed layer and member should be determined after further making clear of the geolog-ical data.

Keywords Sino - Russian Eastern Route Gas Pipeline; Chuzhou salt mine; salt cavern gas storage; structural fea-ture; geological analysis

与枯竭油气田型和含水型地下储气库相比，盐穴储气库具有注采灵活、单井吞吐量大、注采频次高、垫气量小、垫气可回收和储气损失小等优点[1-3]。目前世界各国都在大力开展盐穴储气库建设，美国有22座盐穴储气库、德国24座、法国3座、英国4座、白俄罗斯1座、亚美尼亚1座。2014年5月21日，中国石油天然气集团公司与俄罗斯天然气工业股份公司签署了《中俄东线供气购销合同》，合同约定中国石油天然气集团公司负责中国境内输气管道和储气库等配套设施建设。中俄东线天然气管道拟建的配套储气库有大庆升深2-1和四站、吉林昌10和伏龙泉、辽河雷61、冀东南堡1-29、大港驴驹河和江苏楚州等9个储气库。作为拟建的9座储气库之一，楚州储气库是唯一一座盐穴储气库，其距在建的中俄东线天然气管道约61km，距冀宁联络线楚州分输站约7km，对长三角地区的天然气供应以及中俄东线、冀宁联络线的调峰和应急供气具有非常重要的意义。

盐穴储气库的安全性，运行的稳定性和可靠性对库址要求很高[4-10]，因此需对建库区的地质条件进行系统分析，包括盐矿构造地质特征、含盐地层分布特征、盖层密封性等核心内容[11-15]。

1 区域地质概况

楚州盐矿位于苏北盆地西北端的淮安盆地内，淮安盆地可划分为南部断阶带、中部深凹带和北部斜坡带。其中深凹带沉积深度大于3000m，且盐矿层厚度最大(图1)。

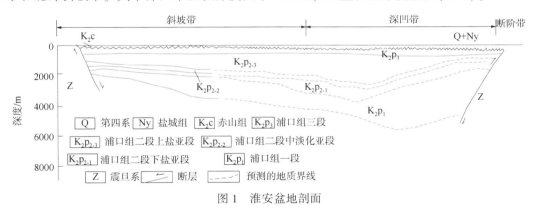

图1 淮安盆地剖面

资料井钻遇4个盐层组中深凹带浦口组上盐亚段第三盐性组合，盐层分布稳定、厚度大、埋藏深度适中、盐岩品位高，是适合储气库建设的主力层段(表1)。

表1 淮安盆地地层划分

地层系统			厚度/m	岩性组合特征
组	亚段	岩性组合		
东台组 Qd			90~294	灰色粉砂质黏土夹粉砂，下部含砾砂层、砂砾层

续表

地层系统			厚度/m	岩性组合特征
组	亚段	岩性组合		
盐城组 Ny 盐二段 Ny_2			281.5~1072.5	灰棕色砂、砂砾与杂色黏土,含粉砂黏土互层
盐城组 Ny 盐一段 Ny_1 三垛组 E_3s 戴南组 E_2d 阜宁组 E_1f 泰州组 E_1t				缺失
赤山组 K_2c	赤二段 K_2c_2			剥蚀
	赤一段 K_2c_1		0~550	砖红色、红棕色粉细砂岩夹泥质粉砂岩与含砾砂岩
浦口组 K_2p	浦三段 K_2p_3		138~916	暗棕色粉砂质泥岩、粉砂岩、泥质粉砂岩,泥岩含石膏芒硝
	浦二段 K_2p_2 上盐亚段 K_2p_{2-3}	第四岩性组合 $K_2p_{2-3(4)}$	139~438	以灰色泥质岩和钙芒硝岩为主,夹少量的盐岩和粉砂岩
		第三岩性组合 $K_2p_{2-3(3)}$	169~647	以灰白色盐岩为主,夹钙芒硝岩、泥岩和粉砂岩
		第二岩性组合 $K_2p_{2-3(2)}$	124~517	以褐灰色钙芒硝岩为主,夹泥岩、盐岩、少量硬石膏岩
		第一岩性组合 $K_2p_{2-3(1)}$	85~517	灰色盐岩和钙芒硝岩为主,夹暗棕、灰绿色粉砂质泥岩和泥质粉砂岩
	中淡化亚段 K_2p_{2-2}		138.0~388.5	以泥质粉砂岩、细砂岩为主,夹粉砂质泥岩
	下盐亚段 K_2p_{2-1}		113.0~758.5	暗棕色灰黑色泥岩、膏质泥岩、粉细砂岩,硬石膏,钙芒硝与盐岩组成韵律的不等厚互层,夹薄层凝灰岩
	浦一段 K_2p_1		619	咖啡色泥岩夹砂岩、云质泥岩、泥质粉砂岩,底部砂岩、砂砾岩

2 建库地质条件分析

2.1 盐层分布特征

楚州盐矿存在大量老腔,在分析盐化老腔以及资料井的基础上,建立楚州储气库的地质模型,根据建立的地质模型分析楚州盐矿的盐层分布特征以及夹层分布情况,确定建库的最佳层段。模型中上盐亚段第三岩性组合顶面埋深909.5~1783.1m,盐层分布稳定,埋藏深度由中部向东西部逐渐变深,其中张兴与杨槐两个区块在该盐性组合的盐层厚度分别为 300m，490m，是盐岩的发育有利区，杨槐区块盐层埋深较大，只有 1 口资料井，而张兴区块埋深较浅且资料井较多(图 2)，能最大程度反映地层情况，因此后续对张兴区块进行重点分析。

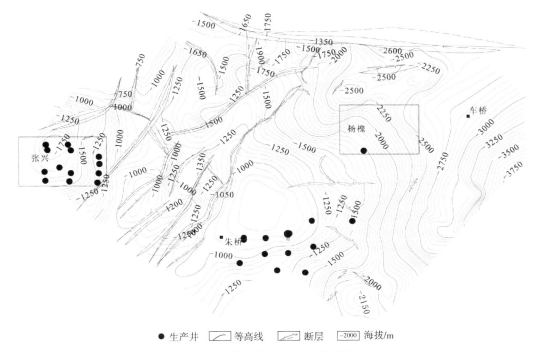

图 2　第三岩性组合顶面埋深

在对比资料井、现有生产井的基础上，结合建立的地质模型，将第三岩性组合由深至浅依次划分为 12 个工业盐群。因此统计结果显示，除Ⅲ-11、Ⅲ-1 盐群平均厚度小于 20m 外，其他盐群平均厚度均大于 20m，其中Ⅲ-9 盐群平均厚度最大，为 119.5m。

Ⅲ-1—Ⅲ-6 盐群平均含矿率为 56.8% 左右，平均总厚度较小，为 194.6m；Ⅲ-7—Ⅲ-9 盐群平均含矿率为 63.9% 左右，且平均总厚度最大，为 235.8m；Ⅲ-10—Ⅲ-12 盐群平均含矿率为 50.5%，平均总厚度最小，为 81.9m(图 3，图 4)。

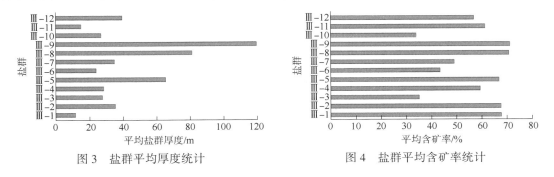

图 3　盐群平均厚度统计　　　　　　图 4　盐群平均含矿率统计

2.2　夹层分布特征

盐层内分布有泥岩、含盐泥岩和含钙芒硝泥岩等夹层，夹层的存在不仅造成盐岩溶解

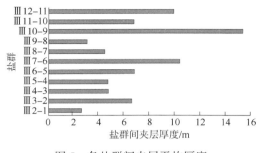

图 5 各盐群间夹层平均厚度

的不均匀性,而且不溶物残渣占据腔体空间,影响盐腔的体积。夹层厚度大,不溶物多,影响程度也就越大,因此建库时应选择夹层厚度小的地区。

对各盐群的夹层进行统计发现,Ⅲ-10盐群底部和Ⅲ-7盐群底部发育的夹层在全区内分布稳定,平均厚度较大,分别为15.5m、10.5m(图5),建议建库层段选择时避开这两个层位。

2.3 顶底板密封性

楚州盐穴储气库建库层段岩性为盐岩与夹层互层,盐岩岩性致密,夹层以泥岩、含盐泥岩为主,顶板厚度为341.2~611.5m,全区分布稳定,厚度大;Ⅲ-1盐矿体的底板围岩较厚,其厚度大于15m。其岩性为灰色、深灰色、棕灰色及咖啡色泥岩、含盐泥岩,泥岩普遍含钙芒硝、硬石膏团块或条带。因此,顶底板岩性致密坚硬,是很好的封盖层。

2.4 建库区块的选择

根据楚州盐矿盐岩的分布特点,结合开采现状及地表建筑物等,确定建库区块选择原则如下:(1)盐岩的蠕变受温度、压力影响,随埋藏深度增加,盐岩的蠕变速率增大,影响盐腔的稳定性,因此建库层段选择在较浅的地层;(2)保证盐腔具有一定的高度、有效体积,能形成一定规模的工作气量。

杨槐区块盐层埋深较大,只有一口资料井,周边地质条件尚不能明确判断,相比之下,张兴区块生产井较多,反映的地质情况较为详细,因此建议先期开发张兴区块,通过对张兴区块的开发,进一步明确第三岩性组合后对杨槐区块进行开发。

2.5 建库层段选择

造腔过程中,夹层的存在对盐腔的溶蚀速度、有效体积都会产生影响,因此选择合理的盐层段造腔是建设地下储气库的关键。楚州盐层段中盐岩的品位相对较好,主要考虑夹层对造腔的影响。本地区夹层岩性以泥岩、含盐泥岩、含钙芒硝泥岩为主,不易溶解,因此造腔层段中夹层的厚度不能过大。

张兴区块的Ⅲ-1—Ⅲ-5盐群地层厚度为114.1~243.5m,平均地层厚度为170.3m,夹层厚度占地层厚度比例大,盐层厚度比例小,平均含矿率为59.5%,夹层数多。

根据夹层分布特点,Ⅲ-12盐群、Ⅲ-10盐群底部和Ⅲ-6盐群顶部发育一套稳定分布、厚度较大的泥岩夹层,造腔过程中是否可溶漓或垮塌还不能确定,因此考虑建库层段避开这几套泥岩厚夹层。

Ⅲ-10—Ⅲ-12盐群地层厚度较小,平均厚度为81.9m,根据盐群厚度与含矿率计算得到的盐岩平均厚度仅41.4m,平均含矿率50.5%,单个夹层厚度>5m的层数很多,不适宜进行储气库建设。

Ⅲ-7—Ⅲ-9盐群平均厚度达到230m以上,经计算该盐群的盐岩平均厚度150.7m,平均含矿率63.9%,单个夹层厚度>5m的层数很少。

综合考虑地层及盐层厚度、夹层分布和夹层数，推荐Ⅲ-7—Ⅲ-9作为储气库建库目的层位。

3 结论

（1）楚州储气库建库区域含盐地层厚度大，且分布稳定。浦二段上盐亚段第三岩性组合平均盐层厚度超过300m，盐矿顶底板岩性致密，盖层厚度大，密封性好，夹层以泥岩为主，夹层数量少，单层厚度较小，适宜建设储气库。

（2）建库区内张兴与杨槐两个区块是储气库建设的有利区，张兴区块埋深较浅且资料井较多，地质情况较详细，可进行先期开发，杨槐区块目前资料井较少，可等地质情况进一步明确后再进行开发。

（3）对张兴区块建库有利区的盐群及夹层进行分析，认为其Ⅲ-7—Ⅲ-9盐群盐层厚度较大，含矿率高，夹层数少，是建库的有利层段，建议在该层位进行储气库建设。

参 考 文 献

[1] 孟浩. 美国储气库管理现状及启示[J]. 中外能源，2015，20(1)：18-23.
[2] 郑雅丽，赵艳杰. 盐穴储气库国内外发展概况[J]. 油气储运，2010，29(9)：652-655.
[3] 杨伟，王雪亮，马成荣. 国内外地下储气库现状及发展趋势[J]. 油气储运，2007，26(6)：15-19.
[4] 丁国生，谢萍. 中国地下储气库现状与发展展望[J]. 天然气工业，2006，26(6)：111-113.
[5] Barron T F. Regulatory technical pressures prompt more US salt-cavern gas storage[J]. Petroleum Society of CIM，1995(2)：169-171.
[6] 李建中. 利用岩盐层建设盐穴地下储气库[J]. 天然气工业，2004，24(9)：119-121.
[7] 刘凯，宋茜茜，蒋海斌. 盐穴储气库建设的影响因素分析[J]. 中国井矿盐，2013，44(6)：24-26.
[8] Vladimir O. Czech Republic 2006-2009 triennium work report, working committee 2: Storage [C]//24th World Gas Conference Argentina, 2009.
[9] Sergey K. Underground Storage of gas, report of working committee 2 [C]//23rd World Gas Conference Amsterdam, Amsterdam, 2006.
[10] 宋桂华，李国韬，温庆河，等. 世界盐穴应用历史回顾与展望[J]. 天然气工业，2004，24(9)：116-118.
[11] 完颜祺琪，冉莉娜，韩冰洁，等. 盐穴地下储气库库址地质评价与建库区优选[J]. 西南石油大学学报(自然科学版)，2015，37(1)：57-64.
[12] 陈波，沈雪明，完颜祺琪，等. 平顶山盐穴储气库建库地质条件评价[J]. 科技导报，2016，34(2)：135-141.
[13] 井文君，杨春和，李银平，等. 基于层次分析法的盐穴储气库选址评价方法研究[J]. 岩土力学，2012，33(9)：2683-2690.
[14] 袁光杰，田中蓝，袁进平. 盐穴储气库密封性能影响因素[J]. 天然气工业，2008，28(4)：1-3.
[15] 丁国生. 金坛盐穴地下储气库建库关键技术综述[J]. 天然气工业，2007，27(3)：111-113.

本论文原发表于《大庆石油地质与开发》2018年第37卷第6期。

盐穴储气库建槽工程实践与顶板极限跨度分析

屈丹安[1,2,3]　施锡林[2]　李银平[2]　杨春和[1,2]　马洪岭[2]　张桂民[2]

（1. 重庆大学，西南资源开发及环境灾害控制工程教育部重点实验室；
2. 中国科学院武汉岩土力学研究所　岩土力学与工程国家重点实验室；
3. 中国石油西气东输管道公司储气库项目部）

【摘　要】 通过分析金坛盐穴储气库的建槽工程案例，探讨建槽施工中遇到的若干问题，总结经验教训并提出工艺改进方案，进而讨论建槽期工艺设计应考虑的关键问题，阐明确定盐腔顶板极限跨度的必要性，并提出建槽期盐腔板极限跨度的确定方法，应用该方法开展基于数值试验的极限跨度分析，获得不同侧溶底角及盐腔高度条件下的盐腔顶板极限跨度，并对其规律进行分析。

【关键词】 岩石力学；盐岩；储气库；水溶造腔；建槽；工程实践；顶板跨度

Engineering practice and analysis of limit roof diameter in building sump of salt cavern for gas storage

Qu Dan'an[1,2,3]　Shi Xilin[2]　Li Yinping[2]　Yang Chunhe[1,2]
Ma Hongling[2]　Zhang Guimin[2]

（1. Key Laboratory for Exploitation of Southwestern Resources and Environmental Disaster Control Engineering, Chongqing University；
2. State Key Laboratory of Geomechanics and Geotechnical Engineering, Institute of Rock and Soil Mechanics, Chinese Academy Sciences；
3. Gas Storage Project Department, PetroChina West-East Gas Pipeline Company）

Abstract　According to the analysis of an engineering case of building sump of Jintan salt cavern, some problems existing in the engineering case are discussed. The experience and lesson are summarized, and some ideas and suggestions are proposed. Then, some key factors that should be considered in the

---

基金项目：国家重点基础研究发展计划（973）项目（2009CB724602，2010CB226701）；国家自然科学基金资助项目（51274187）。

作者简介：屈丹安（1965—），男，1985年毕业于西北大学化工机械与设备专业，现任高级工程师、博士研究生，主要从事盐穴储气库水溶造腔方面的研究工作。E-mail：qudanan@petrochina.com.cn。通讯作者：施锡林（1983—），男，现任助理研究员。E-mail：xlshi@whrsm.ac.cn。

design progress of solution mining are discussed. The necessity of determining limit diameter of cavern roof is expounded. And a method of calculating limit exposed diameter of cavern roof is presented. With this method, the limit roof diameters of some salt caverns are studied basing on simulation test. The limit roof diameters of salt caverns with different lateral solution angles and cavern heights are obtained, and their laws are studied.

Keywords rock mechanics; salt rock; gas storage; solution mining; building sump of salt cavern; engineering practice; diameter of cavern roof

我国盐穴储库采用的建造方法为"单井油垫对流法水溶造腔",本文所述的"建槽"是水溶开采中的广义建槽,具体指盐穴储气库水溶造腔过程中建造不溶物沉渣腔的施工过程。建槽是盐穴储气库水溶造腔的第一步,对于储气库建造的成功具有关键影响。

国外盐穴储气库造腔相关研究,始于20世纪60年代,在工程实践中积累了不少经验与教训,但限于技术保密等方面的原因,一些实质性的技术成果并未公开发表。我国盐穴储气库建设最近几年才刚刚起步,相关的核心技术正在摸索中慢慢建立起来,例如我国第一批新建盐穴储气库16口溶腔已进入造腔后期,将在未来数月内投入储气运行,施工过程中积累了宝贵的技术经验。专门针对盐穴储气库建槽的研究并不多见,与本文研究内容相关的报道简述如下。

S. Bauer等[1]基于梁的弯曲理论,简要研究了盐穴泥质顶板的破坏形式,考虑了顶板的剪切及拉伸破坏形式,研究成果对于单井水溶造腔适用性不强。R. Bekendam和W. Paar[2]研究了盐腔顶板夹层的垮塌问题,并采用FLAC软件对顶板的离层、拉破坏等问题进行了探讨,研究针对的是卤水从盐穴移除后引起的地面沉降问题,未涉及水溶开采过程。Y. Charnavel和N. Lubin[3]研究了难溶夹层对盐穴储气库腔体最终形状的影响,认为腔体段存在大量难溶夹层,会使腔体底部中心出现凸起,导致注气排卤时周边卤水无法排出。K. L. Devries等[4]采用数值模拟方法研究了盐穴储气库运行过程中顶板跨度、顶板深度、护顶盐厚度、泥页岩厚度及刚度等对储气库运行安全的影响,但未涉及盐穴储气库的水溶造腔问题。

余海龙等[5]通过相似模拟试验,探讨了水溶开采中溶腔围岩移动、破坏和应力重新分布特征,得出了不同开采深度岩盐溶腔顶板极限跨距。姜德义等[6]建立了单井溶腔顶板大变形失稳突变分析力学模型,基于该模型推导出了夹层失稳的必要条件,提出了顶板岩石应力释放规律、溶腔失稳判据和溶腔稳定性控制方法。班凡生等[7]根据小挠度薄板弯曲理论,建立了盐岩地层中夹层的数学模型,给出了夹层的强度校核和断裂极限长度的求解方法。施锡林等[8]研究了悬空型难溶夹层垮塌的力学机制,揭示了夹层垮塌的破坏模式及各破坏模式在水溶建腔过程中可能出现的阶段。

本文通过分析我国金坛储气库建槽期盐腔畸形与管柱损伤案例,发现了我国盐穴储库建槽工艺设计与施工中的不足,并提出了改进建议,进而参考前述研究成果,在总结工程经验的基础上,提出了建槽期盐腔极限跨度的确定方法,并通过算例阐明了该方法的具体实施过程,研究将为我国盐穴储气库建槽工程设计与施工提供科学依据。

1 建槽期盐腔畸形与管柱损伤案例

1.1 金坛盐穴储气库建槽期造腔案例

我国某盐穴储气库造腔施工中,为配合声呐测腔及下一阶段溶腔,金坛盐腔于2009年2月底进行管柱调整作业。在起出造腔内管后,发现有2根位于溶腔底部的套管发生严重弯曲变形,计算弯曲套管的所在深度分别为1172~1182m和1144~1153m,并伴随套管接箍严重变形(图1)。

(a) 造腔内管弯曲　　　　　　　　　(b) 套管接箍变形

图1　盐腔造腔内管损伤

该盐腔在此次修井作业前,已进行的4次声呐测腔的腔体边界及地层柱状图如图2所示。最后一次测腔是本次管柱调整期间进行的。4次测腔的主要时间节点为:2006年6月28日,开展第1次测腔,体积为6766.1m³;2007年9月13日,第2次测腔,体积为25387.2m³;2008年6月9日,第3次测腔,体积为43064.7m³;2009年2月2日,第4次测腔,体积为84524.0m³。

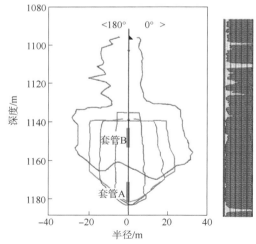

图2　盐腔的地层柱状图及4次声呐测腔的边界轮廓

为方便分析讨论,下文中将深度为1172~1182m的弯曲套管命名为"套管A",深度为1144~1153m的弯曲套管命名为"套管B",这2根套管的位置已标注在图2中。

由前3次的声呐测腔结果,在第3次测腔前盐腔顶板基本未抬升,且盐腔形态扩展趋势非常理想。其主要原因为:在第3次测腔前,造腔外管管鞋始终位于1136~1140m之间,而从地层岩性柱状图可以看出,在此深度以深直到溶腔底板的位置,盐层中的不溶泥质含量仅为3%,属高品位盐层,因而溶腔形态易于控制。

1.2 盐腔畸形的表现形式及原因分析

第 3 次声呐测腔后，提升造腔外管管鞋至 1098m，其下部 20m 范围内均为含泥盐岩，由第 4 次声呐测量边界线可以看出，这次管柱调整之后，溶腔边界的形态就明显差了很多，具体表现形式及原因分析如下：

（1）腔体上部约 35m 范围内的腔体扩展趋势表现出明显不对称性。具体到图 2 中来说，上部腔体向左侧溶蚀的趋势明显大于向右侧的趋势。这主要是由两方面的原因造成的：①套管接箍的损坏后，致使出水口深度及方向发生改变，进而导致左侧卤水流动速度大于右侧；②盐岩层品位水平向分布不均匀。

（2）溶腔上部边界呈锯齿状。这主要是由于泥质夹层的滞后溶蚀造成的。

（3）溶腔底部中心附近呈向上凸起状态。这与法国某多夹层盐矿盐腔底部的边界类似（图 3[3]）。如果在造腔完成后仍然保持这种形态，最终会导致腔底部分卤水无法排出，造成溶腔有效体积的损失。形成这种形态的主要原因为：难溶夹层的滞后溶蚀使腔体形成"瓶颈"（图2，图 3 中的盐腔上部的锯齿状边界），"瓶颈"之上的不溶物在水溶造腔过程会不断沉积到腔体底部的中心附近，最终形成腔底中部凸起的形态。

1.3 造腔管柱损伤原因分析

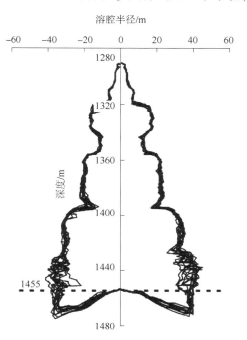

图 3 盐腔底部中心凸起造成的有效储气体积损失[3]

在第 3 次、第 4 次测腔之间的造腔施工过程中，造腔外管管鞋恰好位于一个厚度约 1.5m 的夹层顶部，另外在 1110m 附近还存在一个厚度约 3.0m 的夹层。在此期间造腔内管下入深度为溶腔设计底板深度 1183m。腔体溶蚀期间，上部的夹层，尤其是厚度为 3.0m 和 1.5m 的低含盐率夹层，在卤水浸泡以及腔体溶蚀引起的应力重分布的综合作用下，逐步发生垮塌，期间伴随夹层大范围突然垮塌，岩块下落过程中砸弯套管 B 并造成接箍损坏，最终堆积到腔体底部中心附近，并掩埋下入最深的造腔内管套管 A，造成其产生弯曲变形。

1.4 经验教训与工艺改进

从该盐腔建槽期的工程实践中，得出了如下经验与教训：

（1）必须对大厚度难溶夹层的垮塌给予足够重视，否则可能对溶腔形态控制及施工风险规避造成无法弥补的后果。幸运的是，第 4 次声呐测腔还处于造腔早期，之后的造腔施工中，已经有针对性地调整了造腔工艺，目前该盐腔形态发展状况良好。

（2）对于上述地质特征的盐岩地层，借鉴夹层垮塌动态控制工艺[9]，第 2 次调整造腔外管的管鞋深度，应位于厚度为 3.0m 的夹层的下方，下部溶腔顶板跨度接近极限跨度时，再上提造腔外管至集中分布的夹层带的上部，并适当上提造腔内管，可有效避免大厚度夹层过早垮塌，更加有利于造腔中期盐腔形态的控制。

2 建槽期工艺设计应考虑的关键参数

2.1 盐穴储气库建槽的基本原则

我国金坛储气库某盐腔声呐测量边界与真实溶蚀边界的关系如图 4 所示，采用德国 SO-CON 公司声呐测量仪器获得，图 4(a)为造腔完成后的最后一次声呐测量结果，图 4(b)为依据该盐腔造腔过程中历次声呐测腔结果绘制的。相关试验及工程实践表明，在综合可溶率高达 80%的盐层中开展水溶造腔，造腔完成后，沉渣将在盐腔底部堆积，由于不溶物膨胀及沉渣块体尺寸的影响，沉渣通常会占据 25%左右的腔体体积。沉渣面中心点以下的卤水无法排出到地面，造成了盐腔有效储气库体积的减小。这对于在我国总体厚度较薄的盐层建设储气库，无疑是雪上加霜。

因而，盐穴储气库建槽的基本原则可表述为：在保证盐腔稳定、施工安全的前提下，最大限度地做大沉渣腔，以利于储存更多不溶沉渣，从而提高盐腔有效储气库体积。

2.2 工艺设计应考虑的关键参数

（1）盐腔顶板的极限跨度。

可连续水溶段盐层的总厚度是一定的，要想做大沉渣腔，必须尽量扩大建槽期沉渣腔的横向尺寸。除盐腔顶板自身重力外，建槽期盐腔顶板所承受的外部应力如图 5 所示，包括垂直地应力、水平地应力和卤水浮托力，其中卤水浮托力是经过油垫层间接施加到顶板上的。

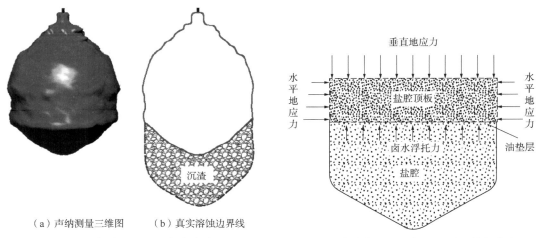

（a）声纳测量三维图　（b）真实溶蚀边界线

图 4　金坛盐腔声呐测量边界与真实溶蚀边界　　　图 5　建槽期盐腔顶板所承受的外部应力

在三维应力场作用下，盐腔顶板的直径过大，就可能发生局部剪切或拉伸破坏，继而诱发顶板大面积突然垮塌。这将对水溶造腔带来两大不利影响：一是造成造腔管柱损坏，影响盐腔形态的扩展趋势，严重时造成盐腔畸形；二是导致盐腔储气能力下降，主要体现为顶板大体积盐岩堆积到溶腔底部后，很难继续溶解，倘若堆积到底部中心位置，还会形成中心凸起，造成注气、排卤困难。

（2）油垫层深度与厚度的调控时机。

建槽期造腔内管的下入深度为一定值，油垫层深度主要由造腔外管控制，油垫层与盐腔底板间的相对距离，决定了盐腔在竖直方向的尺寸，不同尺寸的盐腔的顶板稳定性存在较大差异。因而，设计过程中应注意考虑造腔内管与造腔外管的两口距。

建槽期盐腔总体积较小，在顶板油垫层的作用下，横向尺寸发展很快，油垫层厚度会逐渐变薄，为避免盐腔顶板过早暴露于卤水中发生溶解反应，根据具体生产情况，适时足量向盐腔内补充油垫是十分必要的。

（3）建槽段盐岩层的综合可溶率。

受重力影响，造腔过程中高浓度卤水总是试图向盐腔底部运动，形成上部浓度低、下部浓度高的总体分布趋势，导致盐腔上部盐岩溶解速度大于下部溶解速度。随着溶解时间的延长，盐类矿石中的不溶残渣不断沉积于盐腔底部，覆盖底部未溶的盐岩，盐腔底部将形成一个以钻井为中心的倒圆锥体。圆锥体母线与水平面之间的夹角，叫做"侧溶底角"。

侧溶底角的大小与盐岩层的综合可溶率有关，并影响到盐腔底部盐岩的利用率：综合可溶率高，侧溶底角小，盐腔底部不溶解的矿石损失少、利用率高；综合可溶率低，侧溶底角较大，溶洞底部未溶矿石损失较多、利用率低。

因而，建槽段盐层的综合可溶率，直接影响到溶腔底部边界的扩展趋势，进而影响到整个盐腔形态的发展趋势，在设计建槽期盐层段时应予以重点考虑。

3 建槽期顶板极限跨度的确定方法

盐穴储气库水溶造腔是一项复杂的系统性工程，受地质条件和施工过程的影响，所形成的盐腔顶板的极限跨度，难以通过解析方法获得，需要通过反复试算的方法来确定，工作量大。本文借鉴非线性方程数值解法中的"二分法"求解思想[10]，提出一套极限跨度的快速搜索流程，可有效降低试算工作量。建槽期盐腔顶板极限跨度的搜索流程如图6所示。

单井水溶造腔过程中，受盐腔内水动力及物质扩散和传递规律的制约，盐腔不会沿水平方向无限扩展。生产实践表明，单井水溶开采形成的溶腔直径一般不超过100m[11]，因此，可以推测盐穴储气库建槽期盐腔的顶板跨度也不会超过该值。

令 $D_0 = 100$m，以 D_0 为顶板跨度建立计算模型（所建立的模型将在节5.2中进行论述），开展盐腔稳定性计算。计算完成后，判断建槽期盐腔顶板是否有破损。如果顶板没有破损，说明建槽期溶腔顶板是稳定的；如果有破损，则表明在建槽期设计过程中需考虑盐腔顶板的极限跨度，应继续进行更加深入的分析，以确定极限跨度。

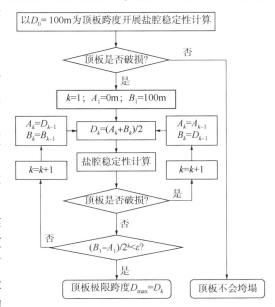

图6 建槽期盐腔顶板极限跨度搜索流程

令 $A_1 = 0$m，$B_1 = 100$m，则极限跨度必然在区间$[A_1,B_1]$上。令 $k=1$，进入第一轮搜索，则 $D_1=(A_1+B_1)/2$ 为区间中点，以 D_1 为顶板跨度建立计算模型，开展盐腔稳定性计算。计算完成后，判断盐腔顶板是否有破损。

若顶板有破损，说明极限跨度在区间$[A_1,D_1]$上，即当进入下一轮搜索后，应令 $A_2=A_1$，且 $B_2=D_1$；若顶板没有破损，则表明极限跨度在区间$[D_1,B_1]$上，即当进入下一轮搜索后，应令 $A_2=D_1$，且 $B_2=B_1$。无论顶板有无破损区，新的极限跨度值所在区间$[A_2,B_2]$仅为$[A_1,B_1]$的1/2。

对减半后的新的极限跨度值所在区间$[A_2,B_2]$，用中点 $D_2=(A_2+B_2)/2$ 再分为两半，以 D_2 为顶板跨度建立计算模型，开展盐腔稳定性计算，从而判断顶板是否有破损，进而确定新的极限跨度值所在区间$[A_3,B_3]$，其长度是$[A_2,B_2]$的1/2。

如此循环下去，即可得到一系列的极限跨度值所在区间，它们的包含关系如下：

$$[A_1,B_1] \supset [A_2,B_2] \supset [A_3,B_3] \supset \cdots \supset [A_k,B_k] \supset \cdots \tag{1}$$

其中每个区间都是前一个区间的1/2。

搜索过程中，将产生一个顶板极限跨度近似解的序列：$\{D_1,D_2,D_3,\cdots,D_k,\cdots\}$，这个序列必然以盐腔的真实极限跨度 $D_{\max}^{\text{true}}$ 为极限。

然而，在实际工程中，不需要完成这个无限的计算过程。因为上述搜索过程存在以下关系式：

$$|D_{\max}^{\text{true}}-D_k| \leq (B_k-A_k)/2 = (B_1-A_1)/2^k \tag{2}$$

假定以 ε 作为预定试算精度，经过若干次试算后，便有

$$|D_{\max}^{\text{true}}-D_k| \leq \varepsilon \tag{3}$$

可见，经过5次试算搜索，便有

$$|D_{\max}^{\text{true}}-D_5| \leq (100-0)/2^5 = 3.125 \tag{4}$$

根据理论与工程实践，实际设计过程中，试算精度 $\varepsilon=3.125$m 已足够，即：若令顶板的极限跨度 $D_{\max}=D_k$，当 $k=5$ 时，$D_{\max}$ 的取值可达到精度要求，试算5次便可解决问题。

4 基于数值试验的极限跨度分析

4.1 试验方法

基于节4所述的极限跨度搜索流程，利用FLAC3D建立不同跨度的盐腔，然后开展盐腔稳定性数值试验，观测盐腔顶板的塑性区发展情况，进而确定顶板极限跨度。

4.2 几何模型

单井油垫对流水溶造腔形成的地下盐腔，可视为以钻井中轴线为旋转轴的旋转体[图7(a)]。考虑到几何形态的对称性，计算过程中选取1/4腔体及其周边扩展地层作为研究对象，以减少数值计算工作量、降低计算时间。

参考节2所述的金坛盐腔的地质资料，建立了如图8所示的几何模型及数值计算有限元网格。几何模型为长方体，尺寸为300m×300m×460m。计算过程中，将含盐系地层内的

夹层和盐层概化为"盐岩层",厚度230m;将含盐系地层上方在计算范围内的岩层概化为"上覆岩层",厚度120m;将含盐系地层下方在计算范围内的岩层概化为"下伏岩层",厚度110m。

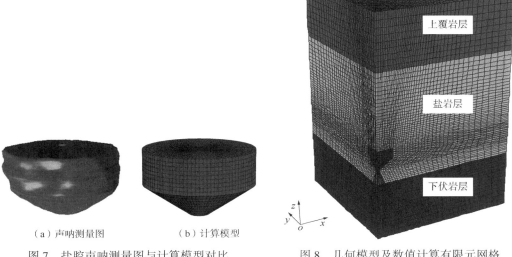

（a）声呐测量图　　（b）计算模型

图7　盐腔声呐测量图与计算模型对比　　图8　几何模型及数值计算有限元网格

图7为盐腔声呐测量图与计算模型的对比图,其中,图7(b)所示的盐腔模型,是利用图8中所示的4个1/4盐腔模型拼合而成的,目的是方便直观对比。

4.3 边界条件与地应力场

模型的上表面施加均布应力,模拟模型上部边界至地表范围内岩体自重作用。分别约束模型的4个侧面及1个底面的法向位移。

盐岩具有很强的蠕变变形能力,计算涉及的盐岩矿床短时期内未发生较大的构造运动,且盐层厚度及埋深较大(含盐系地层埋深为984~1217m),因而,地应力场可视为静水压状态。初始地应力场的获取过程为:将所有岩体的本构模型设置为弹性模型,求解过程中体积模量及剪切模量设置为很大的值,然后求模型的弹性解,平衡后将位移场及速度场清零,得到初始地应力场。

获取初始地应力场后,将盐腔所属单元的模型设置为null,模拟腔体开挖。实际上,盐腔的形成过程是一个与时间相关的渐进过程,建槽期通常会持续数月,计算过程中用一次性开挖模拟这一过程,所得出的顶板极限跨度偏于安全。

卤水压力是通过面力的方式施加到溶腔边界上的,同时考虑了造腔过程中盐腔内卤水的压力梯度。卤水压力取对顶板稳定性的最不利工况,即:井口压力为0;盐腔内的卤水全部为淡水,密度为1.0g/cm³。这一做法也将使所得出的顶板极限跨度偏于安全。盐腔内某一深度的卤水压力通过下式计算得出:

$$p_{\text{brine}} = \rho_{\text{brine}} g h \tag{5}$$

式中：p_{brine} 为盐腔内某点对应的卤水压力，MPa；ρ_{brine} 为盐腔内卤水的密度，g/cm³；g 为当地重力加速度，取 9.8m/s²；h 为从井口到盐腔内某点的深度，m。

4.4 岩石力学本构模型及参数

盐腔顶板稳定性数值试验过程中，采用 Mohr-Coulomb 弹塑性模型，所采用的基本力学参数见表 1[12]。

表 1　数值试验中采用的基本力学参数[12]

地层名称	弹性模量/GPa	泊松比	黏聚力/MPa	内摩擦角/(°)	抗拉强度/MPa
上覆岩层	10	0.27	1	35	1
盐岩层	18	0.30	1	45	1
下伏岩层	10	0.27	1	35	1

4.5 数值试验结果分析

通过数值试验，获得了不同侧溶底角及盐腔高度条件下盐腔顶板的极限跨度，试验结果见表 2，从表 2 中可以得出如下结论。

表 2　建槽期盐腔顶板极限跨度

| 盐腔高度/m | 极限跨度/m | | | | |
	35°	40°	45°	50°	55°
10	—	—	—	—	—
20	—	—	—	—	—
30	71.875	—	—	—	—
40	65.625	71.875	78.125	—	—
50	59.375	65.625	71.875	78.125	—
60	53.125	59.375	65.625	71.875	78.125

注：(1)"—"代表数值试验中盐腔顶板未发生破损，相应工况下的顶板在建槽期处于稳定状态，不存在极限跨度；(2)35°等为侧溶底角。

（1）盐腔设计高度较小的工况中，受侧溶底角限制，盐腔横向尺寸扩展能力有限，不可能形成较大直径的顶板，因而，在建槽期顶板始终处于稳定状态，不存在极限跨度。

（2）盐腔高度相同的情况下，侧溶底角越大，顶板的极限跨度越大。分析出现这种趋势的原因为：侧溶底角越大，盐腔底部未溶解的盐岩越多，所形成的盐腔整体抵抗外力的能力越强，顶板也就越趋于稳定。

（3）侧溶底角相同的情况下，盐腔设计高度越大，顶板极限跨度越小。分析其原因为：盐腔设计高度越大，其整体抵抗外力的能力越弱，顶板也就越趋于不稳定。

5　结论

我国盐矿具有"夹层多、盐层薄"的突出特点，这一地质特征给盐穴储气库造腔带来了极大挑战，建槽质量的好坏更是关系到整个盐穴储库造腔的成败。本文针对建槽过程中的科技问题开展研究，提出了建槽工艺的改进方案，探讨了建槽期工艺设计应考虑的关键问题，进而提出了建槽期盐腔极限跨度的确定方法，并应用该方法研究了不同侧溶底角及盐腔高度条件下盐腔顶板的极限跨度的分布规律。研究成果可为我国盐穴储气库建槽期设计与施工提供参考。

参 考 文 献

[1] Bauer S，Ehgartner B，Levin B，et al. Waste disposal in horizontal solution mined caverns[C]//Proceedings of the SMRI Fall Meeting.［S. l.］：［s. n.］，1998：1–10.

[2] Bekendam R，Paar W. Induction of subsidence by brine removal[C]//Proceedings of the SMRI Fall Meeting. ［S. l.］：［s. n.］，2002：1–12.

[3] Charnavel Y，Lubin N. Insoluble deposit in salt cavern–test case[C]//Proceedings of the SMRI Fall Meeting. ［S. l.］：［s. n.］，2002：1–8.

[4] Devries K L，Mellegard K D，Callahan G D，et al. Cavern roof stability for natural gas storage in bedded salt [R]. Rapid City，South Dakota，USA：RESPEC，2005.

[5] 余海龙，谭学术，鲜学福，等. 岩盐溶腔稳定性模拟试验研究[J]. 矿山压力与顶板管理，1995，4(3)：156–159.

[6] 姜德义，任松，刘新荣，等. 岩盐溶腔顶板稳定性突变理论分析[J]. 岩土力学，2005，26(7)：1099–1103.

[7] 班凡生，高树生，单文文. 夹层对岩盐储气库水溶建腔的影响分析[J]. 辽宁工程技术大学学报，2006，25(增)：114–116.

[8] 施锡林，李银平，杨春和，等. 盐穴储气库水溶造腔夹层垮塌力学机制研究[J]. 岩土力学，2009，30(12)：3615–3620.

[9] 施锡林，李银平，杨春和，等. 多夹层盐矿油气储库水溶建腔夹层垮塌控制技术[J]. 岩土工程学报，2011，33(12)：1957–1963.

[10] 李庆扬，王能超，易大义. 数值分析[M]. 北京：清华大学出版社，2008：212–215.

[11] 王清明. 盐类矿床水溶开采[M]. 北京：化学工业出版社，2003：214–269.

[12] 杨春和，李银平，陈锋. 层状盐岩力学理论与工程[M]. 北京：科学出版社，2009：101.

本论文原发表于《岩石力学与工程学报》2012年第31卷增2。

层状盐层中水平腔建库及运行的可行性

杨海军　王元刚　李建君　薛　雨　井　岗　齐得山　陈加松

（中国石油西气东输管道公司储气库项目部）

【摘　要】　淮安储气库、楚州储气库等西气东输外围储气库纯盐段厚度较小，若建库采用常规直井单腔造腔工艺，则必须钻深井连通不同深度的盐层。随钻井深度的增加，储气压力升高，地温升高，投资增加，蠕变效应也将增加。而采用水平造腔技术，可以在浅层选择相对较厚的纯盐段建造水平腔，从而最大化地利用盐岩体积。通过对水平造腔工艺、水平盐腔稳定性以及经济性进行分析，论证了水平造腔技术在中国的可行性。结果表明：水平造腔成本较小，且腔体稳定性能够满足注采运行条件，但是腔体形状的控制与检测尚存在一定困难，可以通过开展水平造腔试验和调整声呐参数确定腔体形状控制方法，并精确检测腔体形状。

【关键词】　水平盐腔；水平造腔工艺；稳定性分析；经济性分析

Feasibility of construction and operation of strata aligned cavern in bedded salt

Yang Haijun　Wang Yuangang　Li Jianjun　Xue Yu
Jing Gang　Qi Deshan　Chen Jiasong

（Gas Storage Project Department, PetroChina West-East Gas Pipeline Company）

Abstract　The pure salt sections in the salt-cavern gas storages in the periphery of West-to-East Gas Pipeline (such as Huai'an and Chuzhou gas storages) are thinner, so deep wells shall be drilled to connect the salt layers at different depth if the conventional vertical single cavern leaching technology is used. With the increasing of drilling depth, the reservoir pressure, reservoir temperature and investment as well as the creep effects increase. If the strata aligned cavern leaching technology is adopted, strata aligned caverns can be built in the thicker pure salt sections in the shallow layers, and the utilization of salt rock volume can be maximized. The feasibility of strata aligned cavern leaching technology in China was demonstrated by analyzing the strata aligned cavern leaching technology and the stability and economy of strata aligned cavern. It is indicated that this strata aligned cavern leaching technology costs

---

基金项目：中国石油天然气集团公司储气库重大专项"地下储气库关键技术研究与应用"，2015E-40；中国石油天然气集团公司储气库重大专项"盐穴储气库加快建产工程试验研究"，2015E-4008。

作者简介：杨海军，男，1961年生，高级工程师，1983年毕业于大庆石油学院钻井工程专业，现主要从事地下储气库的运营管理工作。地址：江苏省镇江市润州区南徐大道60号商务A区D座中石油西气东输管道公司储气库管理处，212000。电话：13951201923，E-mail：yhjun@petrochina.com.cn。

less and its cavern stability can satisfy the injection and withdraw conditions, but the control and detection of cavity shape are in a way difficult. Cavity shape controlling method can be determined by using the test of strata aligned cavern leaching and adjusting sonar parameters to detect cavity shape precisely.

Keywords strata aligned cavern; strata aligned cavern leaching technology; stability analysis; economic analysis

目前，盐穴储气库建设以常规直井单腔造腔为主，中国第一座盐穴储气库——金坛储气库的建成及投产运行，标志着垂直型盐穴储气库建设技术已经趋于成熟。但是大量盐岩层地质资料表明，中国盐岩矿床普遍为近水平层状分布，江苏淮安、河南平顶山等盐穴储气库存在大量的厚夹层，且厚度小的石膏层、钙芒硝层、泥岩层等难溶夹层相对较多，夹层厚度多在2m以上[1]，若在该类盐层中建垂直储气库，建库过程中不可避免要穿越单个甚至多个夹层[2]。盐岩矿床中的石膏和泥岩等难溶夹层在建腔过程中长期浸泡于卤水中，在一定的力学条件下会失稳坍塌[3]，影响建腔进程；在储气库周期性注采气过程中，腔体围岩将受循环载荷作用，对注采运行的稳定性及安全性均有一定影响[4-5]。

基于中国盐岩层的地质特点，在纯盐段开展水平盐穴储气库建设，避免穿越夹层[6]，建造稳定的腔体，可以大幅降低运行期间天然气从盐岩和夹层交界面处渗漏的风险[7]。为此，从水平造腔工艺与腔体稳定性等方面，分析在中国层状盐层中建设水平腔储气库的可行性。

1 研究现状

国外对于水平腔[8]的研究最早可以追溯至20世纪60年代，德国、法国和俄罗斯等国家均针对盐岩地层水平造腔技术开展了试验研究。1993年，法国在东部Mulhouse附近25m厚的盐层中分5个阶段进行水平多步法造腔[9]。1995年，俄罗斯Podzemgasprom公司开展了水平型盐岩造腔试验，确定了造腔模式与腔体形状，水平腔的截面形状顶大、底小，呈倒水滴状，底部形状易受侧溶角的影响[10]。1979年，俄罗斯在室内模拟了水平多步法造腔后，认为可以通过控制每步的采盐量而使腔体截面积基本相同。俄罗斯伏尔加格勒地下储气库开展了多步法水平腔建设试验，采用水平井注、直井采的方式造腔。造腔初期溶蚀直井附近盐岩，后续盐岩溶蚀段逐步向水平井靠近，腔体形状较规则，盐层利用率较高（图1、图2）。

中国盐矿大多为层状盐层，对水平腔的需求较大，一般采用对井开采，目前尚未开展多步法造腔试验。以江苏淮安某盐矿的一对井为例，该对井采用水平井注、直井采的产盐方式，由于两井间距约为276.3m，在水平井的产出卤水过程中，与裸眼段盐层充分接触，很快达到饱和，因此直井附近大部分盐层未被溶蚀，成腔率较差。声呐测腔结果显示，两井连线上盐层有效溶蚀长度为39.5m，远小于两井间距，无法形成水平腔。

江苏金坛某盐矿采用一对直井开采，造腔过程中两口井单独开采造腔。声呐测腔结果显示，两口井几乎无连通区域，未能形成有效水平腔，后续无法储气注采运行。

根据国内外造腔试验结果可知：(1)单步法造腔，腔体只存在于注水井附近，排卤井附近盐层未得到有效溶蚀，不能形成水平腔，且两口直井单独开采也不能形成水平腔；(2)水

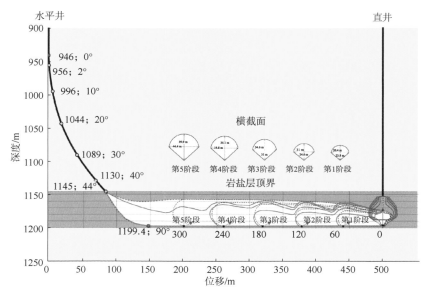

图 1 伏尔加格勒地下储气库水平腔造腔模拟图

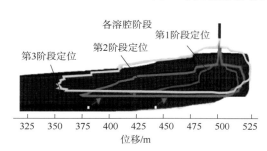

图 2 伏尔加格勒地下储气库水平腔实际腔体示意图

平腔主要是上溶和侧溶,向下不溶;(3)水平腔整个水平段截面形状基本对称,注水排量变化不会对截面形状产生明显影响,腔顶呈穹隆状,高宽比约为一定值;(4)双井多步水平造腔,盐层利用率显著提高,水平腔横截面形状、大小基本相同。

西气东输外围储气库(淮安储气库、楚州储气库)均有厚夹层。根据测井数据,淮安储气库造腔盐岩段内有 10m 以上的不溶夹层,楚州储气库整个盐岩段内厚度超过 5m 的不溶夹层有 10 个。针对盐层中含有厚夹层的情况,如果在该类盐层中开展常规直井储气库建设,则不利于储气库的稳定,可以考虑在较纯的盐层段开展水平多步法造腔,并调整造腔工艺保证排出卤水浓度、控制腔体形状,以期最大化地利用盐层。

2 水平造腔工艺

与常规直井单腔工艺不同,水平造腔工艺主要分为双井钻井、双井连通、多步造腔 3 项工艺(图 3,L_1、L_2、L_3 分别为第 1 步、第 2 步、第 3 步造腔管柱的上提长度)。

2.1 双井钻井

双井钻井的主要步骤:(1)钻垂直井 B,按普通造腔井完井,设计深度略深于水平井;

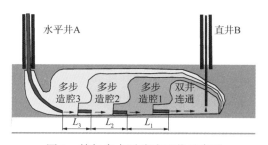

图 3 储气库水平造腔工艺示意图

（2）根据设计体积确定水平段长度，钻水平井 A 确保水平井与垂直井连通或接近连通，根据稳定性和密封性评价结果确定固井套管的下入深度。

2.2 双井连通

双井钻井完成后，在垂直井中下入造腔内、外管柱开始造腔，在水平井中下入造腔外管不进行任何作业。垂直井先造腔，一是为了形成一定腔体作为集卤井，二是为了确保双井连通。垂直井造腔模式与普通井相同，为了保证腔体有较好的形状，需要多步造腔，尽可能不出现 A 形底坑，防止注气排卤时残余卤水过多而损失较多腔体体积。

2.3 多步造腔

完成集卤井后，取出造腔内管，采用水平井注、直井采的方式进行大井眼造腔。精确计算各造腔阶段长度，保证腔体水平段不出现锥形，同时确保排卤浓度达到要求。根据产盐量计算腔体体积，达到设计的阶段体积后从 B 井进行声呐测量，确定腔体形状[11]。若腔体出现锥形，则需要调整下一阶段的造腔方案，如减小造腔阶段长度等；若腔顶即将突破设计高度，则下一阶段需要将 A 井中的管柱上提，使当前腔顶不再继续上溶；若水平井在造腔过程中被堵塞，则可以通过管柱切割或射孔进行管柱调整。

声呐在卤水中能够探测 200m 以内的范围，为了保证腔体形状的可确定性，建议水平段长度最大不超过 400m[12]。当腔体接近 A 井时，声呐测量需要从 A 井、B 井两口井进行，再将测量结果合并，从而确认腔体的最终形状。

3 稳定性数值模拟

淮安储气库典型井的测井结果显示，该井存在约 70m 厚的纯盐段，是开展水平腔建设的优良选择。后续水平盐腔的稳定性分析[13-17]与经济性分析均以淮安储气库典型井为例，稳定性分析常采用 Abaqus、FLAC3D、ANSYS 等有限元数值模拟软件，在此采用 Abaqus 软件。

3.1 模型建立

水平井造腔主要依靠淡水的自然上浮溶蚀盐层，因此，形成的腔体横切面形状与无阻溶剂单井造腔相似。采用水溶造腔软件 WinUbro 对水平腔截面形状进行模拟，结果显示：截面最大直径为 60m，不溶物顶面以上腔体截面面积为 1586m^2（图4）。参考俄罗斯水平腔的建造经验，水平腔长度设置为 300m，则腔体净体积约为 $47×10^4$m^3，造腔时间约为 970 天。

根据水平腔截面的近似形状建立有限元分析模型。由于平面应变模型模拟的水平腔是对称的，为了提高计算效率，计算时只需要计算一半腔体形状（图5）即可，为了确切了解腔体附近应力场的变化细节，对腔壁附近的网格进行加密。

3.2 力学参数

国际上对盐穴储气库稳定性评价的盐岩蠕变变形本构模型大多采用 Lemetre 模型，模型中盐岩总应变包括弹性应变和黏塑性应变：

$$\varepsilon_\mathrm{T}=\varepsilon_\mathrm{E}+\varepsilon_\mathrm{V}=\frac{\Delta\sigma}{E}+\varepsilon_\mathrm{V} \tag{1}$$

式中：ε_T 为总应变；ε_E 为弹性应变；ε_V 为黏塑性应变；$\Delta\sigma$ 为应力差，MPa；E 为杨氏模量，MPa。

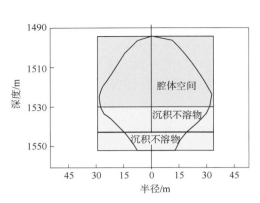

图 4　淮安储气库典型井水平腔截面形状图

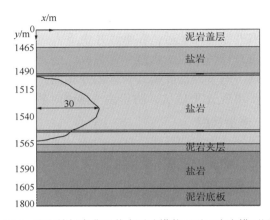

图 5　淮安储气库典型井水平腔横截面平面应变模型图

在 Abaqus 软件中，黏塑性应变可以简化为

$$\varepsilon_V = A\sigma_{eq}^n t^m \tag{2}$$

式中：A、m、n 均为模型蠕变参数；σ_{eq} 为等效偏应力，MPa；t 为时间，d。

根据现场岩心试验确定出不同岩石样块的力学特性参数以及蠕变参数[18]（表 1），将相关参数输入 Abaqus 软件即可计算出不同的应力场变化，从而对盐穴储气库的稳定性进行评价。

表 1　淮安储气库典型井岩石力学参数

岩石种类	弹性模量/GPa	泊松比	黏聚力/MPa	内摩擦角/(°)	重度/(kN/m³)	抗拉强度/MPa	蠕变参数		
							A	n	m
盐岩	6	0.33	3.6	32	21	1	4.0×10⁻⁷	3.4	-0.525
泥岩夹层	10	0.30	4.0	31	22	3	1.3×10⁻⁶	2.4	-0.525
泥岩	16	0.27	5.0	35	25	3	2.5×10⁻⁶	1.7	-0.525

3.3　载荷施加方法

盐穴腔体开挖后主要受外部挤压力与内部拉张力的影响，外部挤压应力主要包括各岩层段的自身重力及水平应力场，水平应力场无特殊情况下采用 3 向等压自重应力场（表 2）。盐穴内部载荷根据建腔或注采气运行不同阶段的实际情况加载（表 3）。

表 2　淮安储气库典型井力学分析载荷计算表

盐岩的不同层段	折算压力梯度/(MPa/m)
泥岩盖层	0.0220
泥岩夹层以上纯盐段	0.0216
泥岩夹层	0.0240

续表

盐岩的不同层段	折算压力梯度/(MPa/m)
泥岩夹层以下纯盐段	0.0211
泥岩底板	0.0248

表3 淮安储气库典型井盐穴不同阶段载荷加载所需的基本参数

分析步骤	时间/d	腔壁载荷	备注
造腔	970	卤水静压	根据腔壁所在深度的静卤水压力折算
注气排卤	100	气体压力	由静卤水压力线性变化为气体内压
注采气运行	3650	循环内压 方案1：12~26MPa 方案2：20~26MPa	按一年注采2次计算，30天注满，30天采完

3.4 稳定性判据

根据稳定性评价结果，当典型井的运行压力介于12~26MPa之间时，淮安储气库水平腔体可以满足安全稳定性条件，适当增加最小内压有利于提高腔体的稳定性[19]。

3.4.1 拉应力准则

拉应力破坏准则要求腔壁周围不出现拉应力，否则将导致腔体掉块和垮塌。由腔体在方案1(运行压力区间为12~26MPa)和方案2(运行压力区间为20~26MPa)两种运行方案下注采运行30年的模拟计算结果可知：除地面最大主应力接近0外，其他区域均小于0；方案2腔壁附近最大主应力距离安全值更远，腔壁周围未出现拉应力，符合注采运行安全要求。

3.4.2 管鞋应变准则

由于盐岩有很强的蠕变特性，为了防止套管在注采运行过程中由于拉应力过载而损坏，套管鞋位置的应变一般控制在0.3%以内。根据两种方案运行30年管鞋处的蠕变应变曲线(图6)可知：随最小运行压力的升高，套管鞋处(约1480m)蠕变应变降低，方案1中套管鞋处蠕变应变为1.2%，方案2中套管鞋处蠕变应变为0.46%；套管鞋距离腔顶越近，套管鞋处应变越大，因此，在造腔过程中需要适当增加腔顶与套管鞋的距离，减小套管鞋处的应变。

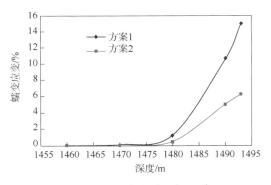

图6 两种方案注采运行30年套管鞋处的蠕变应变曲线

3.4.3 剪切破坏准则

盐岩与含夹层盐岩具有相似的强度特性，其破坏形式主要是拉应力破坏或剪切破坏。剪切破坏采用莫尔剪切破坏准则，计算公式如下：

$$\frac{\sigma_1-\sigma_3}{2}-c\cos\varphi-\frac{\sigma_1+\sigma_3}{2}\sin\varphi \geq 0 \tag{3}$$

式中：σ_1、σ_3 分别为第 1 主应力、第 3 主应力，MPa；c 为黏聚力，MPa；φ 为内摩擦角，(°)。

根据淮安储气库典型井力学数值模拟结果可知：盐穴处于低压状态时容易出现剪切破坏，方案 1 和方案 2 处于低压状态下，两种方案盐穴周围均无塑性区，但方案 2 的剪切破坏风险更小。

3.4.4 膨胀破坏准则

由盐岩的应力应变曲线（图 7）可知：弹性阶段应力随应变的增大而直线增大，之后随应力的增大呈指数形式增大，直到曲线出现拐点后，应力随应变增大而减小，盐岩发生膨胀破坏，为了保证腔体稳定性，必须保证岩体不发生膨胀破坏。

当盐岩发生膨胀破坏时，可以参考公式（4）：

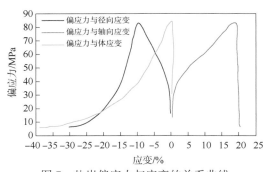

图 7　盐岩偏应力与应变的关系曲线

$$\frac{\sqrt{J_2}}{aI_1+b} \geqslant 1 \quad (4)$$

式中：I_1 为应力张量第一不变量，MPa；J_2 为应力偏量第二不变量，MPa^2；a、b 均为膨胀系数，由试验拟合可得 $a=0.328$，$b=1.3155$。

膨胀破坏计算结果显示，方案 1 运行 30 年后在腔底局部范围内产生膨胀破坏，方案 2 未产生膨胀破坏，增大最小内压可以降低膨胀破坏的可能性。

3.4.5 腔体蠕变准则

蠕变是指盐岩在很小的恒定压差条件下，应变随时间的增加而增大[20]。盐岩的蠕变特性与盐岩的温度、偏差应力相关。国际上一般要求腔壁附近 1 年的蠕变应变为 1%，30 年为 10%，注采运行 30 年后由于蠕变应变导致的体积收缩率均为 20%[21-22]。

根据蠕变应变计算结果可知：方案 1 运行 30 年后，在腔顶和腔底非常小的区域内蠕变应变均为 5%，其他区域均小于 5%；而方案 2 腔体壁面附近和远离壁面区域的蠕变应变都很小。方案 1 运行 30 年体积收缩率为 8.6%，方案 2 运行 30 年体积收缩率为 3.9%。

3.4.6 地面沉降准则

在造腔及长期注采运行过程中，腔体收缩会导致地面沉降[23-26]，根据地面建筑的沉降速率标准，地下工程导致地面沉降的速率不能超过 0.04mm/d。地面沉降的中心理论上在井口附近，方案 2 注采运行 30 年后井口累计沉降 7mm，沉降速率约为 0.0006mm/d，完全在可控范围内。

4　造腔费用计算

（1）钻井费：钻井折算单价为 6100 元/m，水平井钻井费用按直井的 2 倍计算，则深度 1900m 水平井钻井费用约 2318 万元。

（2）注水耗能费：根据造腔数据统计结果进行估算，建设 $47×10^4 m^3$ 的腔体需要注入约

$470×10^4 m^3$ 淡水，注入 $1 m^3$ 水耗电约 $1.4 kW·h$，造腔效率按 0.8 计算，则造腔所需要电费约 822 万元。

(3) 井下作业费：水平造腔需采用多步造腔工艺，根据造腔经验，造腔管柱每次可上提 30m，则水平造腔需要 10 次管柱调整作业，直井需要一次管柱调整作业。造腔过程中共需 11 次井下作业，每次作业按 20 万~30 万元/次计算，井下作业费约 330 万元。

(4) 声呐测腔费：每个造腔阶段均需进行声呐检测，当水平段较短时，可以只进行单井测量，当水平段较长时，需要双井同时测量才能反映腔体真实情况。由于声呐测量距离在 200m 以内，因此前 5 次采用单井测量，后 6 次采用双井测量，共 17 次声呐检测作业，按 6 万元/次计算，声呐测腔共需 102 万元。

(5) 造腔维护费：双井造腔维护费用按直井的 2 倍计算，约 60 万元/年，造腔周期约为 2.7 年，维护费共需 162 万元。

(6) 柴油费：水平造腔完全靠预留盐层和控制上溶速度来防止盐岩顶板太薄，无需进行腔顶保护，只有两口井的裸眼中需要注柴油，每口井约需 $40 m^3$ 柴油，考虑到造腔过程中柴油的损失率约为 50%，因此需 $160 m^3$ 柴油，共需约 107 万元。

根据上述费用分析，一口水平腔的总造腔费用约为 3841 万元，单位体积造腔费用约为 81 元。

5 存在的问题

相对于直井造腔，目前水平造腔尚存在诸多问题：(1) 钻井难度大，盐层厚度分布不均，水平井钻井设计需要基于精细的地质研究。(2) 水平腔长度较大，根据声呐检测原理，水平腔的检测比垂直腔难度大、费用高。目前声呐在卤水中的最大探测半径为 200m，受测量范围限制，水平段长度为 300m 左右时，需要从两口井进行声呐测量，从而确定腔体形状。此外，由于测量腔顶时声波不能垂直于腔体壁面，测量精度会受一定影响，需要专业公司对声呐参数进行调整。(3) 腔体形状控制困难，由于水平造腔工艺无法使用柴油控制腔顶上溶，水平段腔体形状完全由造腔管柱位置和造腔时间控制。由于造腔过程中不同位置的上溶速率不同，因此沿水平段不同位置腔体的截面形状不同。如果盐层上溶速率大，侧溶速率小，水平截面易形成瘦高拱形，盐层利用率受影响；反之，盐层利用率提高，但形成的腔顶较平，腔体稳定性受影响，因此需要通过现场试验确定如何通过调整管柱位置、造腔时间、排量来控制腔体形状。(4) 管柱移动困难，由于水平井在盐层下部，造腔过程中水平段管柱极易被不溶物填埋，导致井下作业过程中水平段管柱无法起出。(5) 注气排卤，水平腔同样存在 A 形底坑问题，注气排卤会有大量卤水残留，在钻井时需要考虑该问题，可使水平井略微倾斜，集卤井处于较深部位，这样造腔完成后集卤井可以排出较多卤水。

6 结论

(1) 通过总结俄罗斯、法国等水平腔储气库的成功建设经验，指出中国在薄岩层进行水平腔建设是可行的。

(2) 水平造腔技术适用于较纯的层状盐岩，与常规直井造腔工艺相比，水平造腔技术可分为双井钻井、双井连通及水平多步造腔等工艺。

（3）水平腔数值模拟结果显示，淮安储气库典型井在12~26MPa压力区间内注采运行30年后，腔壁周围未出现拉应力，无塑性区，不存在剪切破坏，盐穴周围井口累计沉降在可接受范围内，符合安全要求。为了避免套管鞋处出现蠕变损伤以及腔壁蠕变应变过大，建议适当增加套管鞋与腔顶的距离，并提高腔体的最小运行压力。

（4）水平造腔单位体积费用约81元，费用较低，后续层状盐层储气库建设可以考虑水平造腔。

（5）目前，水平造腔在形状控制与检测方面仍有一定的技术困难，可以通过现场试验确定水平腔形状控制方法，并与专业公司合作对声呐参数进行调整，从而精确检测腔体形状。

参 考 文 献

[1] Liang W G, Yang C H, Zhao Y S, et al. Experimental investigation of mechanical properties of bedded salt rock[J]. International Journal of Rock Mechanics and Mining Sciences, 2007, 44(3): 400-411.

[2] 郭凯, 李建君, 郑贤斌. 盐穴储气库造腔过程夹层处理工艺——以西气东输金坛储气库为例[J]. 油气储运, 2015, 34(2): 162-166.

[3] 李建君, 王立东, 刘春, 等. 金坛盐穴储气库腔体畸变影响因素[J]. 油气储运, 2014, 33(3): 269-273.

[4] 梁卫国, 张传达, 高红波, 等. 盐水浸泡作用下石膏岩力学特性试验研究[J]. 岩石力学与工程学报, 2010, 29(6): 1156-1163.

[5] 王彬, 陈超, 李道清, 等. 新疆H型储气库注采气能力评价方法[J]. 特种油气藏, 2015, 22(5): 78-81.

[6] Benefield R K. Brine production in a thin salt bedin NY[C]. Chester: Solution Mining Research Institute, 2003: 1-8.

[7] Berest P, Brouard B, Durup J G. Tightness tests in salt- cavern wells[J]. Oil & Gas Science and Technology, 2006, 56(5): 451-469.

[8] Thoms R L, Gehle R M. Feasibility of controlled solution mining from horizontal wells[C]. Louisiana: Solution Mining Research Institute, 1993: 1-9.

[9] Rezunenko V, Smirnov V, Kazaryan V. Tunnel type underground reservoir construction project[C]. Orlando: Solution Mining Research Institute, 2001: 1-12.

[10] Andrzej S K, Kazimierz M U. Modelling of horizontal cavern leaching: main aspects and perspectives[C]. San Antonio: Solution Mining Research Institute, 1995: 1-18.

[11] 杨海军, 李建君, 王晓刚, 等. 盐穴储气库注采运行过程中的腔体形状检测[J]. 石油化工应用, 2014, 33(2): 22-25.

[12] Jaime D, Luciano P, Ricardo C, et al. Improving dual well horizontal cavern volume and shape predictions by post-processing single well sonar data[C]. Leipzig: SMRI Fall 2010 Technical Conference, 2010: 1-9.

[13] 杨海军, 霍永胜, 李光, 等. TJ盐穴地下储气库注采气运行动态分析[J]. 石油化工应用, 2010, 29(9): 61-65.

[14] 韩琳琳, 廖凤琴, 蒋小权, 等. 盐岩储气库适用性评价标准的研究[J]. 岩土力学, 2012, 33(2): 564-568.

[15] 吴文, 侯正猛, 杨春和. 盐岩中能源(石油和天然气)地下储存库稳定性评价标准研究[J]. 岩石力学与工程学报, 2005, 24(14): 2497-2505.

［16］梁卫国，杨春和，赵阳升．层状盐岩储气库物理力学特性与极限运行压力［J］．岩石力学与工程学报，2008，27(1)：22-27．

［17］John A．Determination of salt cavern operating pressures using rock mechanics and finite element analysis［C］．Cleveland：Solution Mining Research Institute，1996：1-23．

［18］李建君，陈加松，吴斌，等．盐穴地下储气库盐岩力学参数的校准方法［J］．天然气工业，2015，35(7)：96-102．

［19］杨海军，郭凯，李建君．盐穴储气库单腔长期注采运行分析及注采压力区间优化——以金坛盐穴储气库西2井腔体为例［J］．油气储运，2015，34(9)：945-950．

［20］丁国生，杨春和，张宝平，等．盐岩地下储库洞室收缩形变分析［J］．地下空间与工程学报，2008，4(1)：80-84．

［21］王贵君．一种盐岩流变损伤模型［J］．岩土力学，2003，24(增刊1)：81-84．

［22］Hou Z M，Wu W．A damage and creep model for rock salt as well as its validation［J］．Chinese Journal of Rock and Mechanics and Engineering，2002，21(9)：1797-1804．

［23］陈雨，李晓．盐岩储库区地面沉降预测与控制研究现状与展望［J］．工程地质学报，2010，18(2)：252-260．

［24］李银平，孔君凤，徐玉龙，等．利用Mogi模型预测盐岩储气库地表沉降［J］．岩石力学与工程学报，2012，31(9)：1737-1745．

［25］屈丹安，杨春和，任松．金坛盐穴地下储气库地表沉降预测研究［J］．岩石力学与工程学报，2010，29(增刊1)：2705-2711．

［26］任松，姜德义，杨春和，等．盐岩水溶开采沉陷新概率积分三维预测模型研究［J］．岩土力学，2007，28(1)：133-138．

本论文原发表于《油气储运》2017年第36卷第8期。

基于分形理论的盐岩储气库腔底堆积物粒度分布特征

任众鑫[1]　李建君[1]　汪会盟[1]　刘建仪[2]　范　舟[2]

（1. 中国石油西气东输管道公司；2. 西南石油大学）

【摘　要】 国内盐岩储气库由于建库盐层本身地质特点，在造腔结束后，底部一般会堆积大量的不溶物，既浪费了部分有效库容，又为整体建库施工带来诸多不利。为科学评价储气库腔底堆积物，合理采取后续作业处理措施，需要对其分布特性进行研究。首先对其形成过程和存在影响进行了分析阐述，并通过室内实验获取了堆积物岩样，然后利用实验测定了粒度分布数据，最后采用分形分布理论研究其粒度分布特征。结果表明：在双对数坐标系下，腔底堆积物颗粒尺寸与累积数量呈线性关系，即其粒度分布具有分形特点，且分布分形维数可作为描述其粒度分布的特征参量，用以反映颗粒的均匀程度与集中性。研究成果可为堆积物体积评价、后续作业处理措施布置以及腔体库容计算提供理论参考与借鉴。

【关键词】 盐岩；储气库；堆积物；粒度分布；分形理论

Granularity distribution characteristics of deposits at the bottom of salt rock gas storage based on fractal theory

Ren Zhongxin[1]　Li Jianjun[1]　Wang Huimeng[1]　Liu Jianyi[2]　Fan Zhou[2]

（1. PetroChina West-East Pipeline Company；2. Southwest Petroleum University）

Abstract At domestic salt rock gas storages, a large amount of insoluble substance is deposited at the bottom of storages after its solution mining is completed, due to the inherent geological characteristics of the salt rock which is used for gas storage building. The existence of deposits occupies parts of effective storage and brings a lot of adverse factors to the overall construction of the gas storage. In order to evaluate the deposits at the bottom of the cavern scientifically and accordingly take the treatment measures in the following operation rationally, it is necessary to investigate the distribution characteristics of the deposits. At first, the formation process and effects of deposits were analyzed and illustrated. Then, the insoluble deposits were sampled from laboratory experiment and its granularity distribution data were measured. Finally, its granularity distribution characteristics were analyzed based on the fractal distribu-

---

基金项目：中国石油天然气集团专业公司项目"盐穴储气库腔底堆积物注气排卤扩容研究"，2013B-3401-0503。

作者简介：任众鑫，男，1987年生，工程师，2013年硕士毕业于中国石油大学（北京）油气井工程专业，现主要从事储气库工艺技术研究工作。地址：江苏省镇江市南徐大道60号，212000。电话：0511-84523166，E-mail：zhongxren@petrochina.com.cn。

tion theory. It is indicated that the particle size of deposits at the cavern bottom is in linear relationship to the cumulative quantity in the double logarithm coordinate system. It means that its granularity distribution is characterized by a fractal function, and the fractal dimension, as the characteristic parameter for describing the granularity distribution, can be used to reflect the uniformity and centrality of particles. The research results provide the theoretical reference and basis for the deposit volume evaluation, the treatment measure arrangement in the following operation and the storage capacity calculation.

Keywords　salt rock；gas storage；deposits；granularity distribution；fractal theory

随着国内天然气能源利用的快速推进，调峰和应急用气量逐年增长。地下储气库是天然气集输储运的重要基础设施，其中盐穴储气库由于具备储气量大、利用率高、注采气灵活等优势，越来越受到重视[1-7]。而国内建库由于盐层大多存在夹层多、不溶杂质含量高等特点，导致腔体建成后底部会堆积大量的不溶物，既浪费库容，又延缓了建库进程[8-10]。实验测定了储气库腔底堆积物颗粒粒度，并基于分形分布理论研究其粒度分布特征，研究成果可为堆积物体积评价、腔体库容计算以及堆积物后续作业处理提供方法和技术支持。

1　堆积物形成及影响

盐岩夹层中分布有大量泥岩，其不溶杂质占比较高，在一些泥质层位还存在蒙脱石、伊利石等黏土矿物。溶腔过程中，盐岩被注入淡水溶解并以卤水形式返至地面，盐层中不溶杂质则被释放并沉降，进而覆盖在尚未溶解的盐岩表面或上阶段已溶解完成的腔底，加之黏土矿物遇水软化膨胀，最终在腔底形成了大量的堆积物。

腔底堆积物的存在首先会影响库容。溶腔时不溶杂质部分被释放堆积在腔底，自身占据了部分有效体积，加之不溶物遇水膨胀，该部分损失体积更大。同时，堆积物的存在大大增加了注气排卤过程中管柱堵塞的风险。上返卤水携带不溶物进入排卤管也有堵塞管柱的可能，加上其对卤水结晶的促进作用，加快了故障发生的进程。另外，腔底堆积物还使对溶腔过程的控制管理变得复杂[11-14]。不溶物沉积在底部，占用之前溶腔形成的体积，既降低了成腔效率，又使溶腔形状难以控制。因此合理处置不溶物或转变利用，以尽可能地增大库容体积很有实际意义。

2　堆积物粒度分布测定

2.1　堆积物岩样的获取

由于腔底堆积物蓬松且浸满卤水，加之腔体空间较大，对其取样在技术工艺上有一定难度，故利用腔体不同层位岩心分别进行室内水溶实验，以获取堆积物岩样。以金坛矿区某井为例，给出了该研究所使用的岩心层位及描述（表1）。

表1　岩心层位及描述

岩样编号	深度/m	岩性描述	长度/m
1	641.82~641.94	钙芒硝质盐岩	0.12
2	689.25~689.55	泥质钙芒硝岩	0.20

续表

岩样编号	深度/m	岩性描述	长度/m
3	708.80~708.89	盐岩	0.09
4	695.20~695.35	盐岩	0.15
5	712.95~913.30	泥质石膏	0.25

2.2 粒度分布测定

粒度测定的方法有很多,如筛分法、沉降法和吸附法等,其中利用筛分法不仅可测定粒度,还可以绘制累积粒度分布曲线,故采用筛分法进行测定[15]。以 3# 岩心获取的堆积物岩样为例,得到测定结果(表2,图1)。

表2　3#岩心堆积物岩样粒度分布

尺寸/mm	筛物质量/g	分级质量分数/%	筛上质量分数/%	筛下质量分数/%
40.000	10.9117	30.80	30.80	100.00
30.000	7.1643	20.22	51.03	69.20
10.000	5.5664	15.71	66.74	48.97
5.000	4.5304	12.79	79.53	33.26
2.500	3.0999	8.75	88.28	20.47
1.250	1.5356	4.33	92.62	11.72
0.630	1.2946	3.65	96.27	7.38
0.315	0.4326	1.22	97.49	3.73
0.160	0.4864	1.37	98.87	2.51
0.315	19.1176	4.50	93.42	11.08
0.160	21.8800	5.15	98.57	6.58

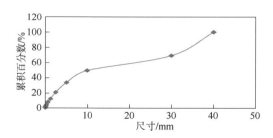

图1　3#岩心堆积物岩样粒度累积质量图

将各层位所得堆积物岩样累积叠加,得到堆积物岩样总体粒度分布(表3,图2)。

表3　堆积物岩样总体粒度分布

尺寸/mm	筛物质量/g	分级质量分数/%	筛上质量分数/%	筛下质量分数/%
50.000	6.3372	1.49	1.49	100.00
40.000	32.5595	7.66	9.15	98.51

续表

尺寸/mm	筛物质量/g	分级质量分数/%	筛上质量分数/%	筛下质量分数/%
30.000	111.5304	26.25	35.41	90.85
20.000	12.8405	3.02	38.43	64.59
10.000	68.3840	16.10	54.52	61.57
5.000	54.4206	12.81	67.33	45.48
2.500	42.2728	9.95	77.28	32.67
1.250	22.7324	5.35	82.63	22.72
0.630	26.7188	6.29	88.92	17.37
0.315	19.1176	4.50	93.42	11.08
0.160	21.8800	5.15	98.57	6.58

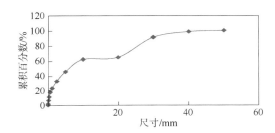

图 2　堆积物岩样总体粒度累积质量图

3　堆积物粒度分布模型

3.1　模型选定

对颗粒粒度分布特征进行分析,需要研究确定与其相符的函数表达式,以进行后续特征描述与评价。由岩样总体粒度分级质量数据可知,其分布特征符合双正态分布,但单个层位粒度分级却与之相差甚远。

鉴于分形理论近年来的拓展创新与独特视角,以及其在描述局部和整体的几何、数量规律及应用方面的优势[16-18],在此采用分形分布函数对堆积物粒度特征进行描述分析。根据 Mandelbrot 分形理论,多分散颗粒系统若符合分形分布,则应满足[19]:

$$n(>x) = Cx^{-D} \tag{1}$$

式中:$n(>x)$ 为系统中粒度大于 x 的颗粒数;x 为粒度,mm;C 为常数;D 为分形维数,用以表征颗粒组成集中性和均匀性的特征参数。

定义粒度 x 到 $x+\mathrm{d}x$ 间颗粒数 $m(x)$:

$$m(x) = \lim \frac{-\Delta n(>x)}{\Delta x} = -\frac{\mathrm{d}n(>x)}{\mathrm{d}x} \tag{2}$$

整理,得

$$m(x) = CDx^{-(D+1)} \tag{3}$$

根据分布矩概念,将累积分布与分形维数联系起来,累积表面积 $S(>x)$ 为

$$S(>x) = \int_x^{max} C_s x^2 m(x) dx = \frac{C_s CD}{2-D}(x_{max}^{2-D} - x^{2-D}) \quad (4)$$

累积质量 $W(>x)$ 为

$$W(>x) = \int_x^{max} \rho C_v x^3 m(x) dx = \rho C_v C \frac{D}{3-D}(x_{max}^{3-D} - x^{3-D}) \quad (5)$$

式中:C_s 为表面积形状系数;C_v 为体积形状系数;ρ 为颗粒密度,g/cm³。

则粒度上累积分布可用函数表示为

$$N(x) = \frac{W(>x)}{W_T} \times 100 = 100\left[1 - \left(\frac{x}{x_{max}}\right)^{3-D}\right] \quad (6)$$

若令 $m = 3-D$,转换可得以下累积分布函数:

$$M(x) = 100\left(\frac{x}{x_{max}}\right)^m \quad (7)$$

3.2 分布拟合及特征参数计算

将上述粒度值 X 和累积颗粒数 M 取对数,以对其粒度分布进行分形分布拟合(图3、图4)。

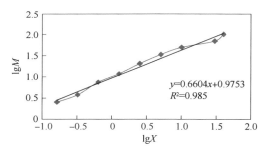
图3 3#岩心堆积物岩样分布拟合图

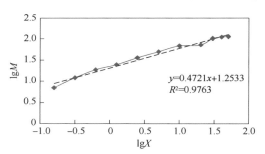
图4 堆积物岩样总体分布拟合图

由拟合结果可知,拟合偏差 R^2 在 0.97~0.99 之间,样品的粒度分布具有分形特点,即其粒度分布特征可用分形函数描述。根据分形维数计算结果(表4),对比堆积物岩样粒度数据可知,分形维数越大,颗粒的集中性和均匀程度越差。相反,分形维数越小,颗粒均匀性越好,颗粒尺寸越集中。

表4 各组堆积物岩样拟合分形维数计算数据表

岩心堆积编号	拟合斜率	分形维数
1#	0.8827	2.1173
2#	0.6627	2.3373
3#	0.6604	2.3396
4#	0.4173	2.5827

续表

岩心堆积物编号	拟合斜率	分形维数
5#	0.3051	2.6949
总体	0.4721	2.5279

3.3 特征参数与溶解度关系

由于腔底堆积物是各岩层段不溶物沉积累加的结果，为避免单一采用某特征参数表征其分布特性，并考虑颗粒级配产生的协同效应[20]，对特征参数即分形维数与溶解度数据之间关系进行分析。并根据岩心各层分形维数和溶解度数据（表5），对分形维数与溶解度进行拟合（图5）。

表5 各组堆积物岩样分形维数及溶解度数据表

岩心堆积物编号	分形维数	溶解百分数/%
1#	2.1173	43.52
2#	2.3373	55.31
3#	2.3396	62.74
4#	2.5827	85.21
5#	2.6949	91.34
总体	2.5279	81.44

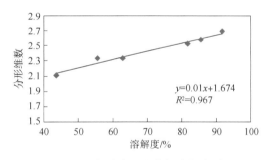

图5 堆积物岩样分形维数与溶解度关系图

对分形维数与溶解度关系做进一步分析验证，将各组堆积物岩样进行质量组合累加形成新的粒度分布，并采用相同处理方法进行分组拟合，计算得到分形维数及溶解度数据（表6）。

表6 堆积物岩样累加组合分组拟合数据表

岩样累加组合	溶解百分数/%	拟合斜率	分形维数
7#	49.55	0.7042	2.2958
8#	51.56	0.7289	2.2711
9#	78.84	0.4773	2.5227
10#	85.99	0.3873	2.6127
11#	58.34	0.6514	2.3486

续表

岩样累加组合	溶解百分数/%	拟合斜率	分形维数
12#	80.47	0.4689	2.5311
13#	87.15	0.3853	2.6147
14#	82.63	0.4500	2.5500
15#	88.96	0.3548	2.6452
16#	88.82	0.3605	2.6395

采用相同处理方法，对分布维数与溶解度进行拟合（图6）。堆积物分形维数与溶解度呈线性关系，相关度在0.98以上，即不溶物溶解度越大，形成的堆积物分形维数越大，亦即颗粒尺寸越不均匀。因此，可以用分形维数来表征腔底堆积物的粒度分布特征，其大小对应描述了颗粒粒度分布的均匀程度和集中性。

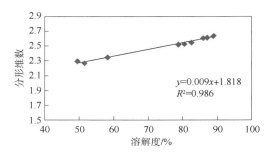

图6 堆积物岩样累加组合分形维数与溶解度关系图

4 结论

（1）腔底堆积物由岩层不溶物在溶腔过程中沉落堆积形成，其存在会对溶腔过程控制、后续注气排卤及库容体积建设等方面产生影响，从而延缓建库进程。

（2）通过室内实验获取了堆积物岩样，并采用筛分法对其粒度分布进行了实验测定，进而得到了样品粒度分级质量数据和总体累积质量数据。

（3）基于分形分布理论，研究分析了堆积物粒度分布特征，通过对粒度数据进行拟合并分析分形维数与粒度尺寸间关系，结合分形维数与溶解度关系，最后得出：腔底堆积物粒度分布可采用分形分布函数进行表征，且其分形维数能很好地描述颗粒分布的集中性与均匀程度。

参 考 文 献

[1] 宋桂华,李国韬,温庆河,等.世界盐穴应用历史回顾与展望[J].天然气工业,2004,24(9)：116-118.

[2] 王清明.盐类矿床水溶开采[M].北京：化学工业出版社,2003：18-21.

[3] Andrzej K D. Solution mining in salt deposits[M]. Poland：AGH University of Science and Technology Press，2007：35-46.

[4] Thoms R L, Gehle R M. A brief history of salt cavern use[C]. Amsterdam：Elsevier，2000：207-214.

[5] Yuan G J, Shen R C, Tian Z L. Review of underground gas storage in the bedded salt deposit in China[C]. Calgary：SPE Gas Technology Symposium, 2006：SPE-100385-MS.

[6] 郑雅丽, 赵艳杰. 盐穴储气库国内外发展概况[J]. 油气储运, 2010, 29(9)：652-655.

[7] 尹虎琛, 陈军斌, 兰义飞, 等. 北美典型储气库的技术发展现状与启示[J]. 油气储运, 2013, 32(8)：814-817.

[8] 班凡生, 耿晶, 高树生, 等. 岩盐储气库水溶建腔的基本原理及影响因素研究[J]. 天然气地球科学, 2006, 17(2)：261-266.

[9] 施锡林, 李银平, 杨春和, 等. 盐穴储气库水溶造腔夹层垮塌力学机制研究[J]. 岩土力学, 2009, 30(12)：3615-3620.

[10] 郭凯, 李建君, 郑贤斌. 盐穴储气库造腔过程夹层处理工艺——以西气东输金坛储气库为例[J]. 油气储运, 2015, 34(2)：162-166.

[11] 李建君, 王立东, 刘春, 等. 金坛盐穴储气库腔体畸变影响因素[J]. 油气储运, 2014, 33(3)：269-273.

[12] 高树生, 班凡生, 单文文, 等. 岩盐中不溶物含量对储气库腔体有效空间大小的影响[J]. 岩土力学, 2005, 26(增刊2)：179-183.

[13] 姜德义, 张军伟, 陈结, 等. 岩盐储库建腔期难溶夹层的软化规律研究[J]. 岩石力学与工程学报, 2014, 33(5)：865-873.

[14] 戴鑫, 马建杰, 丁双龙, 等. 金坛盐穴储气库JT1井造腔异常情况分析[J]. 中国井矿盐, 2015, 46(1)：26-29.

[15] 王元磊. 粒度趋势分析方法的研究进展[J]. 山东师范大学学报(自然科学版), 2008, 23(2)：81-84.

[16] 蔡建超, 胡祥云. 多孔介质分形理论与应用[M]. 北京：科学出版社, 2016：32-68.

[17] 朱华, 姬翠翠. 分形理论及其应用[M]. 北京：科学出版社, 2011：20-150.

[18] 舒志乐, 刘新荣, 刘保县, 等. 基于分形理论的土石混合体强度特征研究[J]. 岩石力学与工程学报, 2009, 28(增刊1)：2651-2656.

[19] Mandelbrot B. The fractal geometry of nature[M]. San Francisco：Freeman W H, 1982：15-127.

[20] 韩涛. 矿渣粉粒度分布特征及其对水泥强度的影响[D]. 西安：西安建筑科技大学, 2004：10-55.

本论文原发表于《油气储运》2017年第36卷第8期。

盐穴储气库造腔巨厚隔层处理的新思路

何 俊　赵 岩　井 岗　李建君

(中国石油西气东输管道公司储气库管理处工艺技术研究所)

【摘　要】 目前，在盐穴储气库造腔过程中，处理不溶物夹层的方式多为使其充分浸泡至弱化后自然垮塌，但对于厚度超过 10m 的巨厚不溶物隔层，使其垮塌十分困难，巨厚隔层垮塌的极限跨度可达 60m 左右，通过浸泡使其垮塌需要的时间长，造腔效率低，还可能损坏厚隔层下部的造腔管柱。以淮安储气库某井地质参数为例，提出一种在允许巨厚隔层存在的基础上进行造腔的新思路，即在隔层上下分别造腔，上部常规造腔、下部造水平腔，上下部的腔体在水平方向错开分布，以保证巨厚隔层作为下部水平腔顶板的稳定性。模拟结果表明：这种造腔方式比同等盐层条件下可多造约 $5×10^4 m^3$ 腔体，并且避免了下部造腔管柱被垮塌隔层损坏的风险。此外，提出在地面采用丛式井技术布井，以降低征地和钻机搬家安装费用，并使用移动式钻机优化钻井程序，改善造腔过程中钻井作业的经济性。

【关键词】 盐穴储气库；巨厚隔层；水平造腔；丛式井造腔；经济性

An innovative method to deal with thick insoluble interlayers in the solution mining of salt-cavern gas storage

He Jun　Zhao Yan　Jing Gang　Li Jianjun

(Technology Research Institute, Gas Storage Project Department of PetroChina West-East Gas Pipeline Company)

Abstract　At present, the most popular method to deal with thick insoluble interlayers in the solution mining of salt-cavern gas storage is to weaken the strength of the insoluble interlayers by immersing them sufficiently until they collapse naturally. For thick insoluble interlayers whose thickness is more than 10m, however, the collapse is quite difficult. The critical collapse span for the thick insoluble interlayers could be as much as 60m, and its collapse by full immersion takes long time, so that the solu-

---

基金项目：中国石油储气库重大专项"地下储气库关键技术研究与应用"子课题"盐穴储气库加快建产工程试验研究"，2015E-4008。

作者简介：何俊，男，1987年生，工程师，2012年硕士毕业于中国石油大学(北京)石油与天然气工程专业，现主要从事地下储气库造腔与综合评价的研究工作。地址：江苏省镇江市润州区南徐大道商务A区D座中国石油西气东输管道公司储气库管理处，212000。电话：18651287636。E-mail：xqdshejun@petrochina.com.cn。

tion mining efficiency is decreased and the leaching strings below thick interlayers may be damaged. In this paper, an innovative solution mining method in consideration of thick interlayers was proposed with the geological data of one well in Huai´an Gas Storage as an example. In this innovative method, solution mining is carried out separately above and under the interlayer. A conventional vertical cavern is leached above the interlayer, while a horizontal cavern is leached below the interlayer. The cavern above the interlayer is horizontally staggered from the one below the interlayer to maintain the stability of the thick insoluble interlayer as the roof of the horizontal cavern. It is indicated from the simulation results that by virtue of this innovative method, a cavern volume of 50000m^3 more can be created in the same conditions of salt layer, and the damage risk to the leaching string below the interlayer can be avoided. Finally, it was recommended to arrange wells by the cluster well technology so as to reduce land charges and rig moving and installation expenses while optimizing the drilling procedure by adopting the mobile drilling unit so as to improve the economical efficiency of drilling operation in the process of solution mining.

Keywords salt-cavern gas storage; thick insoluble interlayer; horizontal solution mining; cluster well solution mining; economical efficiency

在盐岩洞穴中储存油气资源，是能源地下战略储备的主要形式。与国外巨厚盐丘储库相比，中国盐岩地层埋深浅、成层分布、夹层多，地质条件复杂[1]。对于厚度较小的夹层，国内学者提出许多处理方法和夹层控制理论，并在工程现场开展了试验应用[2-9]。郑雅丽等[10]针对厚隔层提出"充分浸泡夹层、二次建槽"的设计思路：在厚隔层上下分别建槽，促进隔层充分浸泡后弱化垮塌，以达到扩大储气空间的目的。但是，这种促使厚隔层失稳的处理思路在实际造腔中会对造腔管柱造成危险，且造腔效率低。基于此，提出一种不破坏厚隔层的造腔新思路，以厚隔层为分界，分别对其上部和下部盐层进行开发，不仅可以简化工序，减小造腔风险，还能显著提高造腔效率。

1 厚隔层常规处理方式

目前，在储气库建设过程中，处理厚隔层的常规方法是在隔层上下分别造腔，同时采用各种措施破坏其稳定性，最终使其自然垮塌，以达到连通上下储气空间的目的。

以淮安储气库造腔为例，根据淮安储气库A井测井和岩心数据，该井1569~1581m处为12m厚隔层，平均不溶物质量分数为78%；隔层以上的纯盐段为104m，隔层以下的纯盐段为24m。根据同一库区的B井测井结果，造腔岩盐段内也有厚度大于10m的不溶夹层，给造腔工作带来许多挑战。淮安储气库A井造腔过程中处理这种巨厚隔层的方式是：在厚隔层上部注淡水，下部采卤水，通过裸眼同时溶蚀隔层的上下两侧，等待厚隔层悬空跨度达到一定程度后自行垮塌(图1)。

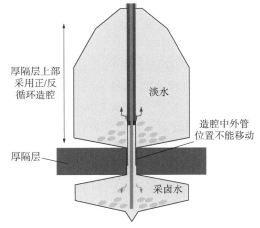

图1 淮安储气库造腔过程中厚隔层处理方式示意图

这种造腔方式的缺点：(1)夹层沟通上下腔体的通道较窄，易被上部不溶物堵塞，阻碍造腔；(2)隔层下部因卤水浓度较大，故造腔缓慢；(3)夹层垮塌后不溶物沉入腔底发生膨胀，占据腔体空间，缩小腔体的有效体积；(4)夹层下部造腔管柱在腔内暴露长度较大，易被垮塌的隔层损坏。根据模拟计算结果，使用该造腔方式，以腔体最大直径80m计算，最大只能达到$1.47×10^5 m^3$，造腔效率极低。

2 厚隔层处理新思路

邓祖佑等[11]通过对500多块不同年代的岩心进行测试分析，得出如下结论：泥岩对空气的渗透率约为$5.50×10^{-19} \sim 6.96×10^{-17} m^2$，可以作为腔体顶板。理论分析和现场试验发现，巨厚隔层作为直接顶板时的失稳垮塌概率较小[12]。根据淮安储气库A井的地层和钻井资料，可利用厚隔层上部盐层1496~1565m造腔，腔体高度69m，腔体体积估算为$1.2×10^5 m^3$，可作为独立的小腔体用于储气。隔层下部造腔段高度较小，可采用水平井造腔。为了保持厚隔层的稳定，隔层下部的水平腔和上部的常规腔在水平方向要间隔开来。按照这种思路造腔(图2)，可以将腔体分组进行布井。每一腔体组包括两个常规腔体和一个水平腔体，常规腔体位于厚隔层上方，水平腔体位于厚隔层下方，厚隔层上下方的腔体在水平方向上不可重叠。

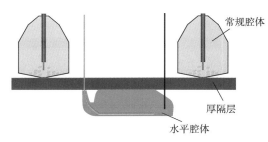

图2 淮安储气库厚隔层作为水平腔顶板造腔示意图

2.1 造腔方式

2.1.1 厚隔层上部开发

厚隔层上部采用常规造腔方式开发。一个腔体组内需要钻3口直井和1口水平井，为了避免地面条件的影响，采用丛式井布井。因井口位置未必在所建腔的盐层正上方，故部分厚隔层上部的造腔井采用定向井钻井。因下套管和造腔时需要满足移动造腔管柱的要求，故钻井井斜不可过大。根据国外经验，造腔定向井的井身剖面类型通常为"直—增—稳—降—直"的S形剖面。在国外造腔建库中，"S"形井的钻井工艺已形成较为固定的模式：(1)直井段打入盐层，固井水泥返至地面；(2)造斜，造斜点应距离管鞋50~100m；(3)进入增斜和稳斜段，当狗腿度小于4.6°/30m时，可下入473.08mm套管，当狗腿度小于8.6°/30m时，可下入339.73mm套管；(4)降斜至直井，直井段的井斜应控制在1.5°以内，进入直井段到最后一层套管鞋位置的长度应保持在50~100m，随后钻至目的井深。

采用常规的分段造腔方式进行造腔，通过使阻溶垫层逐渐上移来控制腔体形状，根据淮安储气库A井的实际数据，确定腔顶和腔底分别在1496m和1565m地层处，腔高为69m。以腔体最大直径80m计算，通过模拟计算得到腔体体积为$1.23×10^5 m^3$。

2.1.2 厚隔层下部开发

厚隔层下部采用水平井造腔技术[13-17]造腔。1993年，法国在东部Mulhouse附近25m厚的盐层中分5个阶段进行水平多步法造腔[18]，其盐层厚度与淮安厚隔层下部盐层相近，具有很高的参考价值。该项目在人工挖掘的巷道中进行钻井(图3)。CD和DE长度均为

15m，可以建造 30m 长的水平腔。水平井总长 44m，井眼直径 57mm，前 6m 进行扩眼并使用 127mm 套管固井，水平井裸眼段总长 38m。直井 CC′ 比水平井深 1m，总长 21m；中间井 DD′ 比水平井浅 1m，总长 19m，主要用以声呐测量。两口直井上部 10m 均采用 101.6mm 套管固井。

两口直井上部使用柴油阻溶，CC′ 井装有界面检测装置。为了使水平井与垂直造腔井连通，在 CC′ 井中下入双层管柱造腔，注

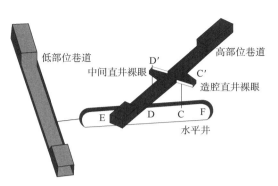

图 3　法国 Mulhouse 水平腔造腔示意图

水口在水平井以下 1m，排卤口在水平井以上 0.5m，以 10m³/h 排量注水、排卤一天时间。两井连通后，通过直井注水、水平井采卤方式进行造腔，注水、排卤排量范围 4~15m³/h（表 1）。该井实际使用的是双井单步法造腔，造腔过程中两口井距未发生变化。该项目最终实现整个水平段腔体形状基本相同；使用垂直井注水、水平井采卤模式，水平段腔体形状未出现锥形，均匀发展，这可能是因为水平段不够长；水平段截面形状是稳定的穹窿形，高宽比约为 0.5；夹层有 8° 的倾斜角，但最终腔体形状未发生倾斜。

表 1　法国 Mulhouse 水平腔造腔工艺参数

阶段	天数/d	平均流量/(m³/h)	最终体积/m³
1	2	10	22
2	10	5	130
3	8	8	300
4	10	15	540
5	7	15	750

杨海军等[19]曾对层状盐岩中的水平井造腔可行性进行探讨，在此对其提出的水平井造腔工艺进行优化，将水平井造腔分为以下步骤：钻井、直井造腔连通水平井、水平井多步造腔和直井反向造腔。

（1）钻井。先钻直井，再钻水平井。水平井靶点与直井距离不能太远，否则会增大两井连通难度（图 4）。水平井的水平段是水平腔造腔的基础，应该细致计算水平段的长度，确保达到水平腔体积要求。

（2）直井造腔连通水平井。直井常规造腔，直至连通水平井，通过直井压力变化判断连通情况（图 5）。

图 4　水平井造腔水平井 M 和直井 N 示意图　　图 5　水平井造腔直井 N 造腔连通水平井 M 示意图

（3）水平井多步造腔。双井连通后，起出 N 井造腔内管。改由 M 井注淡水、N 井采卤

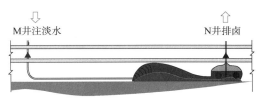

图 6 水平井 M 多步造腔示意图

水的模式造腔,并通过移动 M 井水平注入点深度进行逐段造腔(图6)。为避免腔体出现锥形顶,每次移动长度应细致计算,该阶段完成后腔体形状比较理想(图7)。

(4)直井反向造腔。在水平井多步造腔阶段,N 井作为采卤井,卤水含量较高,造成 N 井周围盐层溶蚀较慢。为了充分利用 N 井周围盐层,在水平井多步造腔阶段结束后,改变地面流程,从 N 井注入淡水造腔,M 井排卤(图8)。在直井反向造腔的过程中,N 井仍需用阻溶垫层保护腔顶,分阶段进行造腔。在造腔过程中通过化验卤水成分和声呐测腔来监测腔体体积和形状的发展。通过检测采出卤水各成分的质量分数可以计算得到地下已造腔体积,当每个阶段计算出的腔体体积达到设计要求时,通过 N 井进行声呐测量。声呐在卤水中能够探测 200m 以内的范围,当腔体水平段超过 200m 时,则需分别通过 M 井、N 井进行声呐检测,进而得出整个腔体形状。为了保证腔体形状的可确定性,建议水平段长度最大不超过 400m[20]。

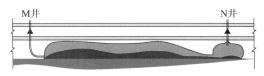

图 7 水平井 M 多步造腔完成图

图 8 水平井造腔直井反向造腔示意图

采用水平井造腔开发厚隔层下部盐层,不仅可以高效利用盐矿资源,同时将厚隔层由造腔障碍变为腔体顶板,节省了大量造腔时间和处理厚隔层的成本,也规避了夹层垮塌对管柱等造成的危险。

2.2 钻井工艺优化

2.2.1 布井

每个井组内需要钻 3 口直井和一口水平井,这对于地面条件的要求非常高。如果地面有湖泊、村庄等不宜钻井的地方,会对整个区块的开发产生影响。即使地面条件符合要求,3 口腔需要征用 4 块井场,征地费用将会使造腔成本大大增加。在油气井钻井中,该类问题通常采用丛式井技术解决。

丛式井是指在一个井场上有计划地钻出两口及以上的定向井组的新型钻井工艺。采用丛式井钻井技术面临的关键问题之一是丛式井组井口位置的优化设计。丛式井的布井方式有 4 边形布井、6 边形布井等。针对淮安地区,建议采取 10 边形的布井方式进行部署,每个丛式井平台包括两个井排,每个井排有 5 口井,共计 10 口井,其中 5 口常规腔,5 口水平腔(图9),这样布井可以大大提高盐层和地面的利用率。根据《丛式井平台布置及井眼防碰技术要求》(SY/T 6396—2014)的有关规定,井间距离不小于 2.5m,根据《密集丛式井井眼轨道设计与轨迹控制技术规范》(Q/SY 1639—2013)的有关规定,排间距离一般为 20~40m。国外用于造腔丛式井井间距为 15m,排间距为 15m。由于储气库钻井井深较浅,大多在 2000m 以内,钻机较油气钻井小,因此,可以适当降低距离要求。

2.2.2 钻井程序

考虑到储气库钻井多在同一地区或同一构造上进行，利用丛式井技术，可以采用移动钻机依次钻多口不同井相似层段的方式实施钻井作业。表层作业：集中打表层，表层固井不占用井口，将水泥头接在表层套管上，采用固井管道固井。固井后的候凝也不占用钻机时间，钻机可移至另外的槽口进行其他井的表层作业。当然，也可以采用多钻机同时进场实施钻井作业。使用移动式钻机组合各开次批量钻进，减少钻机搬家、候凝时间，简化不必要工序，形成流水线化作业流程。移动式钻机主要有液压轨道式和步进式两种，使用此类钻机能够实现当天平移当天开钻。钻机前后平移的动力来源主要是地面棘爪式液缸发动机，通过发动机带动钻机滑动，在钻机的底座前部和后部装设耳板和滑动轨道相连；如果钻机向前滑动一节，后面空出来的一节轨道应随之移动到前方，依次不断循环重复向前滑动(图10)。采用程序化作业方式省时，便于积累作业经验和优化程序，因而提高了钻井作业水平和钻井时效。

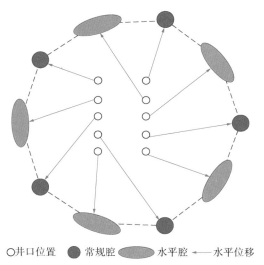

图9 淮安地区10边形丛式井布井示意图

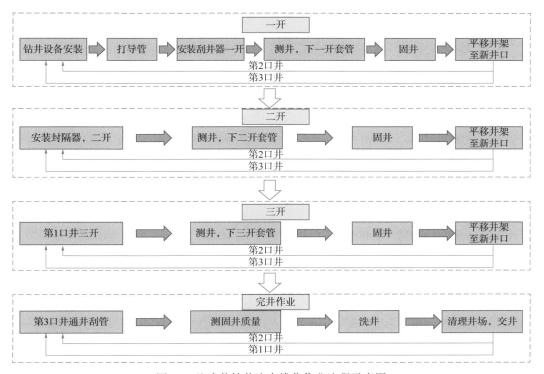

图10 丛式井钻井流水线化作业流程示意图

2.3 优势对比

根据淮安储气库 A 井测井数据，夹层上部盐层平均不溶物质量分数 27.2%，腔体高度 69m，以最大直径 80m 进行模拟，最终腔体体积约为 $1.23×10^5 m^3$；夹层下部盐层平均不溶物质量分数 34%，腔体高度 24m，以水平段 400m 进行计算，最终腔体体积约为 $9.7×10^4 m^3$。一组腔体由 2 个常规腔和 1 个水平腔组成，其总体积为 $3.43×10^5 m^3$。而采用常规方法处理厚隔层，相同的区域范围内，最多只能建造两个腔体，由造腔模拟可知两个腔体总体积最大为 $2.94×10^5 m^3$。此外，考虑到处理厚隔层的巨大资金和风险成本，新思路可以更高效地处理厚隔层问题。

同时，使用丛式井技术可以有效减少征地和井架搬家安装费用。油气井钻井实践表明：含两口井的丛式井比一般直井节约场地约 30%，含 3 口井的丛式井比一般直井节约场地约 53%。根据单口直井的投资费用，计算得出 10 口直井的总投资费用为 $1.01×10^7$ 元。以 10 边形布井方式为例，一个丛式井平台（10 口井）的投资成本共计 $9.58×10^6$ 元。丛式井的钻井成本按 10 口定向井的实际进尺以 550 元/m 的单价计算，并加上 10 口定向井的定向费用 $5×10^5$ 元。实际操作中，定向费用可以协商免付。采用丛式井的总投资费用比采用直井的投资费用节省 $5.2×10^5$ 元，如果再除去定向费，可以节省 $1.02×10^6$ 元。因此，采用丛式井造腔可以取得较好的经济效益和环境效益。

3 结束语

盐穴储气库巨厚隔层的处理新思路是保留巨厚隔层，在其上下分别造腔，因隔层不会垮塌，故避免了不溶物膨胀占据下部腔体体积，提高了腔体的利用率。通过模拟计算，采用这种新思路造腔最终形成的有效腔体比采用传统的巨厚隔层垮塌后形成的有效腔体约增大 $5×10^4 m^3$，并且规避了下部造腔管柱损坏的风险。此外，地面采用丛式井技术进行布井和钻井优化，可以大大提高建库的经济性和效率。

参 考 文 献

[1] 刘艳辉，李晓，李守定，等．盐岩地下储气库泥岩夹层分布与组构特性研究[J]．岩土力学，2009，30（12）：3627-3632．

[2] 徐孜俊，班凡生．多夹层盐穴储气库造腔技术问题及对策[J]．现代盐化工，2015(2)：10-14．

[3] 高树生，杨广雷，班凡生，等．岩盐储气库溶腔内夹层数值模拟研究[C]//沈阳：全国岩石力学与工程学术大会，2006：223-228．

[4] 马旭强．多夹层盐矿储气库稳定性和造腔效率研究[D]．北京：中国科学院大学，2016：8-21．

[5] 王同涛，闫相祯，杨恒林，等．多夹层盐穴储气库群间矿柱稳定性研究[J]．煤炭学报，2011，36（5）：790-795．

[6] 施锡林，李银平，杨春和，等．盐穴储气库水溶造腔夹层垮塌力学机制研究[J]．岩土力学，2009，30（12）：3615-3620．

[7] 郭凯，李建君，郑贤斌．盐穴储气库造腔过程夹层处理工艺——以西气东输金坛储气库为例[J]．油气储运，2015，34(2)：162-166．

[8] 孟涛，梁卫国，陈跃都，等．层状盐岩溶腔建造过程中石膏夹层周期性垮塌理论分析[J]．岩石力学与工程学报，2015，34（增刊1）：3267-3273．

[9] 袁炽. 层状盐岩储库夹层垮塌的理论分析[J]. 科学技术与工程, 2015, 15(2): 14-20.

[10] 郑雅丽, 赵艳杰, 丁国生, 等. 厚夹层盐穴储气库扩大储气空间造腔技术[J]. 石油勘探与开发, 2017, 44(1): 1-7.

[11] 邓祖佑, 王少昌, 姜正龙, 等. 天然气封盖层的突破压力[J]. 石油与天然气地质, 2000, 21(2): 136-138.

[12] 王元刚, 陈加松, 刘春, 等. 盐穴储气库巨厚夹层垮塌控制工艺[J]. 油气储运, 2017, 36(9): 1035-1041.

[13] Gomm H, Peters L. Cluster cavern well drilling-advantages and limitations[C]. Hannover: SMRI Fall 1994 Meeting, 1994: MP1994F_Gomm.

[14] Thoms R L, Gehle R M. Feasibility of controlled solution mining from horizontal wells[C]. Louisiana: SMRI Fall 1993 Technical Meeting, 1993: 1-9.

[15] Jaime D, Luciano P, Ricardo C, et al. Improving dual well horizontal cavern volume and shape predictions by post-processing single well sonar data[C]. Leipzig: SMRI Fall 2010 Technical Conference, 2010: 1-9.

[16] Andrzej S K, Kazimierz M U. Modelling of horizontal cavern leaching: main aspects and perspectives[C]. San Antonio: SMRI, Fall 1995 Technical Meeting, 1995: 1-18.

[17] Kazaryan V A, Pozdnyakov A G, Malyukov V P, et al. The use of solvent recirculation at Target Cavern I, leaching for horizontal cavity creation and quality brine output[C]. Basel: SMRI Fall 2007 Technical Conference, 2007: MP2007S_Kazaryan_2.

[18] Rezunenko V, Smirnov V, Kazaryan V. Tunnel type underground reservoir construction project[C]. Orlando: SMRI, Fall 2001 Technical Meeting, 2001: 1-12.

[19] 杨海军, 王元刚, 李建君, 等. 层状盐层中水平腔建库及运行的可行性[J]. 油气储运, 2017, 36(8): 867-874.

[20] Jaime D, Luciano P, Ricardo C, et al. Improving dual well horizontal cavern volume and shape predictions by post-processing single well sonar data[C]. Leipzig: SMRI Fall 2010 Technical Conference, 2010: 1-9.

本论文原发表于《油气储运》2019年第38卷第6期。

盐穴储气库表征渗透率研究

王元刚[1]　薛　雨[2]　李心凯[2]

(1. 中国石油管道有限责任公司西气东输分公司合肥管理处；
2. 中国石油管道有限责任公司西气东输分公司广东管理处)

【摘　要】 以整个盐腔为研究对象，进行盐腔注水试验，观察升压后的压力变化数据。应用达西公式建立了考虑腔体静溶因素的渗透率计算模型，选取3口盐穴储气库造腔井进行注水试压试验，计算出试验井的渗透率。通过相关试验数据分析，验证计算模型的准确性。

【关键词】 表征渗透率；盐穴储气库；腔体静溶；注水升压

Research on the characterization permeability of salt cavern gas storage

Wang Yuangang[1]　Xue Yu[2]　Li Xinkai[2]

(1. Hefei Region Department, PetroChina West-East Gas Pipeline Company;
2. Guangdong Region Department, PetroChina West-East Gas Pipeline Company)

Abstract　Taking the salt cavern as a whole unit, the pressure change data is recorded after water injection and boosting test by carrying press test to the salt cavern. A model for salt cavern permeability considering the static solution is established based on Darcy law. Three salt caverns are selected for the water pressure experiment and the permeability of the salt cavern is calculated. The accuracy of the model is verified combined with other experiment data.

Keywords　characterization permeability; salt cavern gas storage; static solution; water injection and boosting

盐岩因其低渗透率、低孔隙度及损伤自愈性等特点，成为地下储气库建设的首选岩体。金坛盐穴储气库的成功建设，标志着我国盐穴储气库建设技术的成熟。我国盐岩具有单层厚度薄、含夹层较多以及建库层段裂缝系统不确定等特点，层状盐岩层中储库的渗透性和安全性等问题引起了广泛重视，腔体的密封性是储气成功与否的决定性因素，因此对盐穴储气库的渗透率进行计算尤为重要。

基金项目：中国石油天然气集团公司科技重大专项"地下储气库关键技术研究与应用"(2015E-40)。
作者简介：王元刚(1986—)，男，硕士，工程师，研究方向为盐穴储气库造腔技术与仪电自动化技术。

通过盐穴储气库渗透性室内试验，研究人员获得了一定成果[1-5]。在循环及卸载压力条件下，盐岩中裂纹的产生情况表明，盐穴储气库渗透性是夹层系统与裂缝系统共同作用的结果；在水溶造腔过程中，泥岩与盐岩在高压条件下交界面处易产生微裂纹，裂缝表面有可能形成新的渗流通道；夹层位置是盐岩地下储库群最显著的影响因素，泥岩、盐岩夹层面的渗透率大于岩体本身，因而此位置所处区域是渗流最可能发生的区域。针对储气库注采运行的渗透率以及泥岩夹层与盐岩界面处的渗透率，国内学者通过模型在渗流特征、裂缝开裂扩展特征、气体渗透特征等方面进行了研究，发现渗透率与体积应变的关系可以用线性函数来近似描述[6-10]。

虽然目前关于盐穴储气库渗透率的试验方法已经非常成熟，但是大部分仅限于室内研究和模型推导，且研究目的单一，研究结果缺乏现场试验数据的验证支撑。在本次研究中，我们进行了现场盐穴储气库水试压试验，以整个盐腔为研究单位，观察盐腔注水升压后的压力数据变化。在此基础上，建立了考虑盐穴储气库注水净溶的表征渗透率计算模型，并结合现场试验结果，对盐穴储气库的渗透性进行了定性分析。

1 计算模型

1.1 计算原理

在保证套管密封性良好的情况下，进行盐穴储气库渗透性试验。目前，采用的渗透性试验介质主要有氮气（N_2）和水（H_2O）。在气体密封试验中，需要投入大量的高精度检测设备，操作流程较复杂，设备安全等级要求较高。现场操作中需要进行多方面协调，气测渗透率的计算方法也较为复杂，需要处理大量数据[11-13]。相对于氮气，水在地层中的渗流规律能够满足达西公式，计算方法简单。现场注水升压作业简单，见效快，可将注水升压数据用于储气库渗透率的快速计算。渗透率测试过程的时间较短，而岩盐蠕变、卤水热膨胀对腔体的影响需要经过数年到数十年的较长时间才能显现出来，因此，对这些影响予以忽略。

通过注水升压试验确定渗透率的原理是：当盐穴储气库达到稳定状态后，在短时间内注水时，由于水的压缩性较小，腔体会处于弹性变形状态；后续试验中，由于注入淡水溶盐以及储气库存在夹层、微裂缝等原因，腔体压力下降，若能计算出腔体渗漏量，则可根据达西公式计算出腔体的渗透率。其计算步骤如下：(1)注入淡水升压后，由于静溶及渗流作用的影响，腔体压力降低，根据压力变化计算出腔体的体积变化量；(2)根据注水量，计算腔体溶盐增加的体积；(3)根据腔体的体积变化量及溶盐体积，计算渗漏量；(4)根据达西公式计算腔体的渗透率。

1.2 渗透率计算模型

计算注水升压后腔体的总体积：

$$V_1 = V_{in}\left(1-\frac{V_{in}p_1}{K_v}\right) + V_0\left(1-\frac{p_1-p_0}{K_v}\right) \tag{1}$$

式中：V_1 为腔体总体积，cm^3；V_0 为腔体初始体积，cm^3；p_0 为升压前压力，MPa；p_1 为升压后压力，MPa；K_v 为水的体积模量，取356MPa；V_{in} 为注水量，cm^3。

计算降压后腔体的体积变化量：

$$\Delta V = V_1 \frac{p_1 - p_2}{K_v} \tag{2}$$

式中：p_2 为漏失后压力，MPa；ΔV 为腔体体积变化量，cm^3。

腔体体积变化受注水溶盐与储气库渗漏的相互作用。由于升压前腔内卤水达到饱和，且注水量相对较小，假设注入水溶盐在试验时间内可达到饱和，则注入水溶盐后腔体体积增量通过式(3)计算得出

$$\Delta V' = \frac{V_{in}s}{\rho_s} - V_{in}\left(\frac{1+s}{\rho} - 1\right) \tag{3}$$

式中：$\Delta V'$ 为腔体体积增量，cm^3；s 为氯化钠(NaCl)溶液在60℃下的饱和溶解度，取36%；ρ_s 为固体氯化钠密度，取 2.16g/cm^3；ρ 为注入水与氯化钠混合后的溶液密度，1.235g/cm^3。

在盐穴储气库水溶造腔过程中，由于夹层及微裂缝的存在，注入水发生一定程度的渗漏。实验过程中的渗漏量通过式(4)计算得出

$$q = (\Delta V - \Delta V')/t \tag{4}$$

式中：q 为腔体渗漏量，cm^3/s；t 为试验时间，s。

假设渗漏符合达西定律，沿腔体壁面均匀渗漏，则可根据球面流达西公式，计算盐穴储气库的表征渗透率：

$$K = \frac{q\mu\left(\frac{1}{R_w} - \frac{1}{R_e}\right)}{4\pi(p_1 - p_2)} \tag{5}$$

$$R_e = R_w + 1000 \tag{6}$$

式中：K 为储气库渗透率，mD；μ 为流体黏度，mPa·s；R_w 为腔体半径，cm；R_e 为升压/降压过程中注水压力波及的半径，cm。

2 现场试验

在盐穴储气库水溶造腔过程中，通过造腔管柱往盐层中注入淡水溶解盐岩，形成的饱和卤水排出至地面，最终使地下形成洞穴。夹层与微裂缝的存在，使得盐穴储气库会发生渗漏。当盐腔达到稳定状态后，往盐腔内注水使腔体压力升高，在静溶及渗漏过程中压力逐渐降低；继续注水，使腔体压力升高至设计值，并观察压力变化情况。多次重复以上过程后，盐腔会重新达到平衡。此时，可根据压力变化及注水量计算腔体的渗透率，选取3口盐腔进行注水试压试验。下面为W1井、W2井、W3井2017年的现场注水试压结果。

（1）W1井：腔体体积约 171602m^3；注水试压过程中腔内最高压力设计值为 16.400MPa；当腔内压力升至最高压力设计值后，共进行3次注水试压试验。其中，最后一次注水 2.800m^3，注水过程中压力升高了 0.014MPa，升压结束后21h内压力下降了

0.004MPa。图1所示为W1井注水试压结果。

（2）W2井：腔体体积为600696m³；注水试压腔内压力设计值为20.900MPa；注水使腔内压力达到设计值后进行3次升压试压。其中，最后一次注水1.286m³，注水期间压力升高0.230MPa，观察41h后压力回落0.066MPa。图2所示为W2井注水试压结果。

（3）W3井：腔体体积为61381m³；注水试压腔内压力设计值为19.500MPa；注水使腔内压力达到设计值后进行3次升压试压，其中最后一次注水5.000m³，注水期间压力升高0.129MPa，观察48h后压力回落0.051MPa。图3所示为W3井注水试压结果。

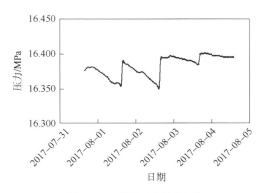

图1　W1井注水试压结果

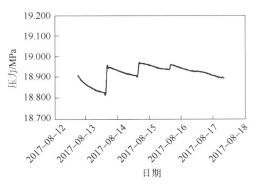

图2　W2井注水试压结果

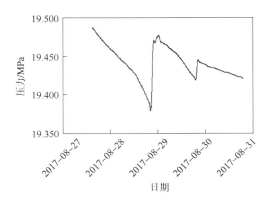

图3　W3井注水试压结果

以上3口井最后一次注水试验数据见表1。

表1　3口井最后一次注水试验数据

井名	升压前压力/MPa	升压后压力/MPa	注水量/m³	降压时间/h	漏失后压力/MPa	升压前腔体体积/m³
W1	16.385	16.399	2.800	21	16.395	171062
W2	18.937	18.960	1.286	41	18.894	600696
W3	18.970	19.099	5.000	48	19.048	61381

3　渗透率计算结果分析

结合试验数据，运用上述模型计算不同盐腔的渗透率。以W1井为例，由于每次升压的注水量相对于腔体体积而言都比较小，因此，可忽略最后一次升压前的腔体体积变化。

W1井最后一次注水升压结束后的腔体体积为

$$V_1 = 2.800 \times \left(1 - \frac{164}{3560}\right) + 171602 \times \left(1 - \frac{16.399 - 16.385}{356}\right) \approx 171064 \,(\mathrm{m}^3)$$

降压过程中的腔体体积变化量为

$$\Delta V = 171064 \times 10^6 \times \frac{16.399 - 16.395}{356} \approx 192207 (\text{cm}^3)$$

淡水溶盐后，溶解的氯化钠与卤水充分混合，腔体的体积增量为

$$\Delta V' = \frac{2.800 \times 36\%}{2.16} - 2.800 \times \left(\frac{1+36\%}{1.235} - 1\right) \approx 183266 (\text{cm}^3)$$

腔体的渗漏量为

$$q = (\Delta V - \Delta V')/h \approx 0.118 (\text{cm}^3/\text{s})$$

根据腔体体积计算出球面渗流界面的等效半径为3445cm，卤水的黏度取0.85mPa·s，代入式(5)得到的渗透率为

$$K = \frac{0.118 \times 0.85 \times \left(\frac{1}{3445} - \frac{1}{3445+1000}\right)}{4\pi \times 0.4} \approx 0.007 \text{mD}$$

同样计算出，W2井的渗透率为0.050mD，W3井的渗透率0.029mD。
3口盐腔井渗透率计算结果见表2。

表2　3口盐腔渗透率计算结果

井名	腔体等效半径/cm	远端半径/cm	溶盐体积/cm³	升压后腔体体积/m³	降压后腔体体积变化量/cm³	腔体的体积增量/cm³	渗漏量/(cm³/s)	渗透率/mD
W1	3444	4444	466667	171064	192207	183266	0.118	0.007
W2	2430	3430	214333	60070	1113666	84171	6.975	0.050
W3	2447	3447	833333	61384	879375	327260	3.195	0.029

通过分析数据可以看出，盐穴储气库渗透率极低。其中，W1井的渗透率为0.007mD，W3井的渗透率为0.029mD，而W2井的渗透率相对较高，但是也低至0.050mD。W2井的渗透率相对较高，其原因可能是井口微量泄漏等原因导致泄漏量增加，造成表征渗透率计算值偏大。即便如此，模型计算得到的渗透率与其他学者通过室内试验得到的结果相差不大，这就验证了所建立模型的准确性。

4　结语

根据水溶造腔基本原理和渗流力学基本理论，以整个盐腔为研究单元，考虑盐腔静溶的影响，推导出了盐穴储气库表征渗透率的计算方法。现场作业中通过注水升压的方法来计算渗透率，操作工艺较简单，计算量小，可在盐穴储气库水溶造腔过程中随时进行，实用性较强。采用现场试验数据，计算出的盐穴储气库渗透率极低，为0.007~0.050mD。通过与其他学者的室内试验研究结果相对比[14-17]，验证了所建立模型的准确性。所建立的模型适用于盐穴储气库渗透率的快速计算，在判断盐穴储气库造腔层段存在微裂缝及夹层等

条件下的渗透性时可予以参考。

参 考 文 献

[1] 李浩然，杨春和，刘玉刚，等．单轴荷载作用下盐岩声波与声发射特征试验研究[J]．岩石力学与工程学报，2014，33(10)：2108-2116．

[2] 武志德，周宏伟，丁靖洋，等．不同渗透压力下盐岩的渗透率测试研究[J]．岩石力学与工程学报，2012，31(增刊2)：3740-3746．

[3] 贾超，张凯，张强勇，等．基于正交试验设计的层状盐岩地下储库群多因素优化研究[J]．岩土力学，2014，35(6)：1718-1726．

[4] 陈卫忠，谭贤君，伍国军，等．含夹层盐岩储气库气体渗透规律研究[J]．岩石力学与工程学报，2009，28(9)：1297-1304．

[5] 纪文栋，杨春和，刘伟，等．层状盐岩细观孔隙特性试验研究[J]．岩石力学与工程学报，2013，32(10)：2036-2044．

[6] 王保辉，闫相祯，杨秀娟，等．地下储气库天然气运移的等效渗流模型[J]．中国石油大学学报(自然科学版)，2011，35(6)：128-130．

[7] 贾善坡，杨建平，王越之，等．含夹层盐岩双重介质耦合损伤模型研究[J]．岩石力学与工程学报，2012，31(12)：2548-2555．

[8] 王贵君，刘朝鹏，介少龙，等．考虑流—固耦合的天然气盐岩储库渗流特性研究[J]．地下空间与工程学报，2016，12(增刊2)：470-474．

[9] 王伟，徐卫亚，王如宾，等．低渗透岩石三轴压缩过程中的渗透性研究[J]．岩石力学与工程学报，2015，34(1)：40-47．

[10] 谭贤君，陈卫忠，杨建平，等．盐岩储气库温度—渗流—应力—损伤耦合模型研究[J]．岩土力学，2009，12(30)：3633-3641．

[11] 林勇，薛伟，李治，等．气密封检测技术在储气库注采井中的应用[J]．天然气与石油，2012，30(1)：55-58．

[12] 袁光杰，申瑞臣，袁进平，等．盐穴储气库密封测试技术的研究及应用[J]．石油学报，2007，28(4)：119-121．

[13] 杨石刚，方秦，张亚栋，等．盐岩地下储库气体泄漏量的计算方法[J]．岩石力学与工程学报，2012，31(增刊2)：3710-3715．

[14] 刘伟，李银平，杨春和，等．层状盐岩能源储库典型夹层渗透特性及其密闭性能研究[J]．岩石力学与工程学报，2014，33(3)：500-506．

[15] 周宏伟，何金明，武志德．含夹层盐岩渗透特性及其细观结构特征[J]．岩石力学与工程学报，2008，28(9)：2068-2073．

[16] 刘伟，Nawaz M，李银平，等．盐岩渗透特性的试验研究及其在深部储气库中的应用[J]．岩石力学与工程学报，2014，33(10)：1953-1961．

[17] 王新志，汪稔，杨春和，等．盐岩渗透性影响因素研究综述[J]．岩石力学与工程学报，2007，26(增刊1)：2678-2686．

本论文原发表于《重庆科技学院学报(自然科学版)》2019年第21卷第6期。

盐岩地层地应力测试方法

周冬林　杨海军　李建君　王晓刚　汪会盟　薛　雨

（中国石油西气东输管道公司储气库管理处工艺技术研究所）

【摘　要】　盐岩地层地应力是储气库形态设计、运行压力区间确定的重要影响因素，为了获得盐岩地层地应力参数，采用水压致裂法和多极子阵列声波测井计算盐岩地层地应力。针对盐岩的非渗透性特点，水压致裂法增加了回流工艺，在多个操作参数下进行多期裂缝重张测试，利用不同数据分析方法计算盐岩最小主应力得到了统一结果，误差低于5%。结合多极子阵列声波测井，分析了盐岩地层各向异性，依据密度测井资料和弹性波动理论，计算了地层水平主应力。阵列声波测井计算结果与水压致裂测试结果一致，测试盐岩地层中3个主应力值接近，为该地区盐穴储气库建设和优化设计提供了可靠的依据。

【关键词】　储气库；水压致裂；阵列声波；盐岩；地应力

A test method for the in-situ stress of salt rock

Zhou Donglin　Yang Haijun　Li Jianjun　Wang Xiaogang　Wang Huimeng　Xue Yu

(Institute of Process Technology, Gas Storage Project Department,
PetroChina West-East Gas Pipeline Company)

Abstract　The in-situ stress of salt rock is an important influencing factor for the design of gas storage shape and the determination of operation pressure interval. To obtain the in-situ stress parameters of salt rock, the in-situ stress of salt rock was calculated by means of hydraulic fracturing method and multipole array acoustic logging. Considering the impermeability of salt rock, the hydraulic fracturing method was added with thereflow process. Multistage fracture reopening tests were carried out on various operation parameters, and the minimum principal stress of salt rock was calculated by using various data analysis methods. And thus, the results were consistent with the error lower than 5%. Then, the anisotropy of salt rock was analyzed based on multipole array acoustic logging, and horizontal principal stress of the formation was calculated on the basis of density logging data and elastic wave theory. It is shown that the in-situ stress calculated from the array acoustic logging data is consistent with the result of hydraulic fracturing test and the three principal stresses of the salt rock are close. The research results pro-

---

基金项目：中国石油天然气集团公司重大科技专项"地下储气库关键技术研究与应用"，2015E-400801。

作者简介：周冬林，男，1987年生，工程师，2012年硕士毕业于中国石油大学（北京）地质工程专业，现主要从事盐穴储气库地质评价工作。地址：江苏省镇江市润州区南徐大道A区商务楼D座，212000。电话：0511-84522031，E-mail：zhoudonglin@petrochina.com.cn。

vide the reliable basis for the construction and optimal design of salt-cavern gas storage in this area.

Keywords gas storage; hydraulic fracturing; array acoustic wave; salt rock; in-situ stress

在含盐地层中建设储气库必须评价盐岩地应力场，地应力是储气库形态设计、运行压力区间确定的重要影响因素。长期以来中国对于盐岩地层的地应力特征缺乏研究，对于盐岩地层的原位地应力测试尚属空白。在众多地应力获取方法中，水压致裂法是测试深部原岩地应力的有效方法，被国际岩石力学学会试验方法委员会（International Society of Reconstructive Microsurgery，ISRM）和美国标准测试委员会（American Society for Testing and Materials，ASTM）推荐为地应力测试的主要方法之一，在各行各业得到了广泛应用。

该方法对测试设备和现场条件要求低，测试过程简单，操作性好，数据分析简便易行，测试的裂缝闭合压力可靠性高[1-5]。但该方法无法确定裂缝走向，因此无法判断主应力的方位，这在应力状态比较均一的地区更显著。因此在水压致裂测试地应力的基础上，实施多极子阵列声波测井，利用快横波方位指示岩层的最大水平主应力方向，同时该测井能够得到广泛的数据信息，可以计算岩石力学特征参数，结合密度测井资料，可以计算岩石强度，并估算岩层水平最大和最小主应力值[6-9]，其结果可与水压致裂法测试值进行对比。

1 测试地层概况

测试地层位于金坛盆地直溪凹陷，区域构造条件简单，无强烈挤压，无大型断层发育。含盐地层埋深800~1200m，主要由层状盐岩和泥岩组成，含少量钙芒硝和硬石膏。岩层处于弱构造应力区，垂向应力即为该地区主应力之一，并与井轴平行，适于水压致裂法测试地应力。由于盐岩有蠕变性和裂隙自愈性，地层裂缝不发育，同时岩层几乎不渗透，测试中增加回流工艺，采用人工方法将压裂液从裂缝中导出，使裂缝闭合[10-13]，完成测试。

2 测试方法

2.1 水压致裂法

水压致裂法是利用封隔器在井眼中将目标层段隔离，利用高压流体将目标地层压裂，在地层中产生张性裂缝并扩散到远离井筒的原始地层中，然后停注等待裂缝闭合，非渗透地层需要人工导出压裂液（即回流），测试过程中记录压力、流量随时间的变化曲线。通过地质力学和渗流力学理论分析压力变化曲线得到地层的原地应力状态[14-15]。根据测试数据计算得到裂缝闭合压力，其等效于地层最小主应力。

2.1.1 不同操作参数下的多期裂缝重张

每个测试层段都必须进行多期裂缝重张测试，每个测试期中使用不同的排量进行测试，并根据测试情况选择是否进行回流，以获得多个裂缝闭合压力的数据。裂缝闭合压力是最小主应力的真实反映，属于岩层固有属性，理论上不受测试操作参数变化的影响。多期裂缝重张应得到比较一致的结果，每个测试层段都重复了4~5个周期（图1，垂向应力通过密度测井计算得出；最小主应力为各个周期计算的裂缝闭合压力的算数平均值；流量由记录仪记录，负值代表回流。多期裂缝重张，共5个测试周期）。

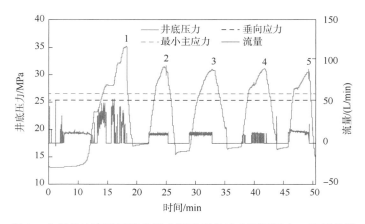

图 1　金坛盆地直溪凹陷盐岩水压致裂地应力测试压力—流量曲线

2.1.2　数据分析方法

利用均方根法、回流刚度法及 G 函数法计算测试数据。压力随时间变化的导数反映了线性流的状态，流态改变的拐点即为裂缝闭合的时间点，此时的压力即裂缝闭合压力。回流刚度是回流流量与压力的函数，曲线斜率的倒数即系统的刚度，回流过程开始与结束两部分曲线斜率的交点即为刚度变化点，对应裂缝闭合压力点。另外，G 函数曲线斜率的变化也可以指示裂缝闭合压力。对测试层位多个测试周期的结果进行分析计算，得到的裂缝闭合压力一致性较高，平均为 25.1MPa，方差为 0.504MPa，具有很高的可信度。

2.1.3　测试结果

水压致裂法共有 5 个测试层段，包括 3 个盐岩层段和 2 个泥岩夹层段，测试深度为 1040～1120m。每个层段的测试中，裂缝的压开与闭合至少需要重复 4~5 次，使得裂缝经过多次重张和扩展，多期裂缝重张获得多个裂缝闭合压降曲线[16-18]，分别采用均方根法、回流刚度法及 G 函数法计算裂缝闭合压力，将每种方法计算结果的平均值作为层位最小主应力值，并用数据标准差除以平均值，统计得出误差低于 5%，在可控范围内，表明水压致裂法测试的数据具有较好的一致性，可信度较高。测试 5 个层位的最小主应力介于 24.496～25.529MPa 之间（表 1）。

表 1　金坛盆地直溪凹陷水压致裂法测试各层段裂缝闭合压力

岩性	测井深度/m	真实垂深/m	裂缝闭合压力（平均值）/MPa	标准方差/MPa	统计误差/%
盐岩	1048.0	1042.0	24.496	0.782	3.2
泥岩	1080.5	1074.5	24.529	1.167	4.8
泥岩	1091.5	1085.5	25.127	0.174	0.7
盐岩	1110.0	1104.0	25.528	0.323	1.3
盐岩	1127.0	1121.0	25.309	0.798	3.2

2.2　多极子阵列声波测井

阵列声波测井是利用声波发射器发射声波，使地层产生轻微的变形，利用多个接收器阵列组合接收返回波，反映地层信息丰富，应用价值高。采用的交叉式多极子阵列声波测井仪（XMAC）包括 2 个声源和 8 个接收器，发射器正交摆放，可得到 12 个线性偶极波形，

12个交叉偶极波形，经过数据处理，提取纵波、横波及斯通利波等多种信息。

2.2.1 盐岩地层各向异性

地层产生裂缝后，定向分布的裂缝对声波的传播能力在各个方向上有不同程度的干扰：(1)裂缝的存在使得纵横波幅度下降，但不同的裂缝角度对纵横波衰减程度的影响不同，低角度裂缝或水平裂缝引起纵向上波阻抗增大，对纵波能量的衰减作用明显大于横波，而高角度裂缝或垂直缝使得地层横向上波阻抗增大，对横波能量的衰减作用明显大于纵波，因此纵波与横波的衰减情况反映了地层中裂缝的角度。(2)横波经过地层裂缝层等不连续面后，分裂为沿不连续面传播的横波和垂直于不连续面传播的横波，前者的传播速度比后者快，分别为快横波和慢横波[16-20]，利用横波比值(二者速度之差与二者速度之和的比值)来评价地层各向异性的大小，快横波沿不连续面传播，其方位指示了地层中水平最大主应力的方位。

水压致裂后实施多极子阵列声波测井，计算了地层中的纵波时差和横波时差，并分析了声波的慢度，结果表明：纵波和横波均存在一定程度的衰减，反映了地层中压裂缝角度的不规则性，横波比值介于1.78~1.84之间，表明了测试地层的各向异性特征较弱，在各向异性相对较强的井段，水平最大主应力总体为北东—南西向(图2，道1为自然伽马和声波时差曲线；道2为地层深度；道3为斯通利波变密度；道4为瑞利波振幅；道5为快慢横波时差和各向异性；道6为快横波方位图；道7为快慢横波开窗时间；道8为裂缝方位)。

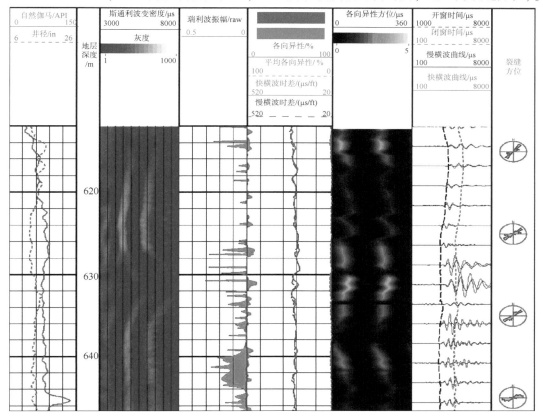

图2 多极子阵列声波测井法计算得到的金坛盆地直溪凹陷盐岩
地层各向异性及地应力评价结果(1in=2.54cm，1ft=30.48cm)

2.2.2 水平主地应力计算

利用阵列声波测井得到的纵波时差、横波时差，结合密度测井资料、泥质含量分析资料，建立模型来计算泊松比、杨氏模量、体积模量及剪切模量等参数(表2)。

表2 利用多极子阵列声波计算金坛盆地直溪凹陷岩石力学参数

井段/m	纵波时差/(μs/m)	横波时差/(μs/m)	体积密度/(g/cm³)	泊松比	杨氏模量/(10^4MPa)	体积模量/(10^4MPa)	剪切模量/(10^4MPa)	岩性
1039.2~1051.6	223.1	399.6	2.01	0.27	3.21	2.36	1.26	盐岩
1079.5~1082.4	270.1	494.6	2.20	0.29	2.44	1.88	0.95	泥岩
1090.7~1092.9	248.9	454.7	2.17	0.29	2.73	2.13	1.06	泥岩
1100.6~1123.6	222.6	393.6	2.05	0.26	3.35	2.37	1.32	盐岩
1125.2~1134.0	222.7	394.4	2.04	0.27	3.32	2.36	1.31	盐岩

利用计算得到的岩石力学参数，结合垂向应力、地层孔隙流体压力，双井径测井得到的构造不平衡系数等，计算了水平主应力[21-23]，并在此基础上计算了井周的剪切应力(表3)。

$$p_0 = \int_h^0 \rho g \mathrm{d}h \quad (1)$$

$$\sigma_H = \frac{\gamma p_0}{1-\gamma} + \alpha p_p [1-\gamma(1-\gamma)] F \quad (2)$$

$$\sigma_h = \frac{\gamma p_0}{1-\gamma} + \alpha p_p [1-\gamma(1-\gamma)] \quad (3)$$

$$\sigma_{r\theta} = \frac{\sigma_H - \sigma_h}{2} \left(1 - 3\frac{R^4}{r^4} + 2\frac{R^2}{r^2}\right) \sin 2\theta \quad (4)$$

式中：p_0 为垂向应力，MPa；ρ 为岩层密度，g/cm³；h 为深度，m；g 为重力加速度，m/s²；σ_H 为最大水平主应力，MPa；σ_h 为最小水平主应力，MPa；γ 为泊松比；α 为有效应力系数；p_p 为地层孔隙流体压力，MPa；F 为构造不平衡系数；$\sigma_{r\theta}$ 为井周剪切应力，MPa；R 为井周剪切质点外径，m；r 为井周剪切质点内径，m；θ 为地缝角度，(°)。

表3 金坛盆地直溪凹陷地层地应力计算结果

井段/m	剪切应力/MPa	垂向应力/MPa	水平主应力/MPa		岩性
			最大	最小	
1039.2~1051.6	2.212	23.26	26.54	24.31	盐岩
1079.5~1082.4	2.844	23.98	27.22	25.26	泥岩
1090.7~1092.9	2.663	24.21	27.79	25.65	泥岩
1100.6~1123.6	2.057	24.62	27.76	25.41	盐岩
1125.2~1134.0	1.011	24.98	28.18	25.81	盐岩

3 地应力状态

通过多极子阵列声波测井计算岩层各向异性表明测试地层各向异性较弱，同时确定了最大水平主应力方向为北东—南西。测试地层在弱构造挤压环境中，垂向应力即为该地区竖直方向的主应力之一，因此与之相垂直的最大水平主应力、最小水平主应力的方向分别为北东—南西、北西—南东。

水压致裂法利用多期裂缝重张和多种方法计算得到的数据结果一致性好，相对误差小，可靠性高，确定了岩层最小主应力。利用岩石力学参数计算得到了全井段的水平主应力，结合密度测井资料确定了垂向应力，计算结果与水压致裂测试结果吻合，表明了阵列声波计算结果的可靠性。从数值上看，3个主应力的值十分接近，通过参考计算得到的井周剪切应力发现，岩层中剪切应力较小，为1.0~2.8MPa，属于弱剪切应力环境。因此测试含盐地层中3个主应力数值趋于相等，与弱剪切力结果一致，反映了盐岩蠕变地层随时间推移，岩层中的剪切应力趋于0，地应力趋于各向同性的典型特征。

4 结论

增加回流工艺改进水压致裂法测试盐岩地层地应力有效可行，测试结果可信度高，为中国含盐地层地应力测试提供了借鉴。多极子阵列声波测井分析岩层各向异性，计算水平最大主应力的方向总体为北东—南西。综合水压致裂法测试结果和多极子阵列声波测井计算结果，分析盐层地应力特征，认为该地区属于弱剪切应力状态，3个方向主应力值相近，地应力趋于各向同性状态。

参 考 文 献

[1] Haimson B C, Cornet F H. ISRM suggested methods for rock stress estimation-Part 3：hydraulic fracturing (HF) and/or hydraulic testing of pre-existing fractures(HTPF)[J]. International Journal of Rock Mechanics and Mining Sciences, 2003, 40：1011-1020.

[2] 王成虎. 地应力主要测试和估算方法回顾与展望[J]. 地质评论, 2014, 60(5)：980-981.

[3] 景锋, 盛谦, 张勇慧, 等. 我国原位地应力测量与地应力分析研究进展[J]. 岩土力学, 2011, 32(2)：51-54.

[4] 董光, 邓金根, 朱海燕, 等. 重复压裂前的地应力场分析[J]. 断块油气田, 2012, 19(4)：485-488.

[5] 赵刚, 董事尔. 水压致裂法测量地应力理论与应用[J]. 山西建筑, 2009, 35(36)：77-78.

[6] 苏大明. XMAC测井资料评价水力压裂效果的应用分析[J]. 国外测井技术, 2011, 6(3)：41-43.

[7] 程道解, 孙宝佃, 程志刚, 等. 基于测井资料的地应力评价现状及前景展望[J]. 测井技术, 2014, 38(4)：379-381.

[8] 王军, Zhu Z Y, 郑晓波. 多极源随钻声波测井实验分析[J]. 地球物理学报, 2016, 59(5)：1909-1912.

[9] 严成增, 郑宏, 孙冠华, 等. 基于FDEM-Flow研究地应力对水力压裂的影响[J]. 岩土力学, 2016, 37(1)：237-243.

[10] Yuan Y, Xu B, Palmgren C. Design of caprock integrity in thermal stimulation of shallow oil-sands reservoirs[J]. Journal of Canadian Petroleum Technology, 2013, 52(4)：266-278.

[11] 王建军. 应用水压致裂法测量三维地应力的几个问题[J]. 岩石力学与工程学报, 2000, 19(2)：229-233.

[12] 马秀敏, 彭华, 李振, 等. 深孔水压致裂地应力测量[J]. 地质评论, 2013, 59(增刊1)：1016-1017.

[13] 张重远, 吴满璐, 陈群策, 等. 地应力测量方法综述[J]. 河南理工大学学报(自然科学版), 2012,

31(3):305-310.
[14] Bell J S, Bachu S. In-situ stress magnitudes in the Alberta Basin-regional coverage for petroleum engineers[C]. Alberta:SPE Canadian International Petroleum Conference,2004:155.
[15] Yuan Y, Xu B, Yang B. Geomechanics for the thermal stimulation of heavy oil reservoirs[C]. Kuwait City:Heavy Oil Conference and Exhibition,2011:150293.
[16] Settari A, Raisbeck J M. Fracture mechanism analysis in in-situ oil sands recovery[J]. Journal of Canadian Petroleum Technology,1979,18(2):252-254.
[17] 王晓杰,彭仕宓,吕本勋,等.用正交偶极阵列声波测井研究地层地应力场[J].中国石油大学学报(自然科学版),2008,3(4):43-45.
[18] 尹帅,丁文龙,赵威,等.基于阵列声波测井的海陆过度相碎屑岩地层裂缝识别方法[J].石油钻探技术,2015,43(5):75-79.
[19] 吴晓光,季凤玲,李德才.偶极声波测井技术应用现状及研究进展[J].地球物理学进展,2016,31(1):380-386.
[20] 杨红,许亮,何衡,等.利用测井压裂资料求取储层地应力的方法[J].断块油气田,2014,21(4):509-512.
[21] 赖富强,孙建孟,苏远大,等.利用多极子阵列声波测井预测地层破裂压力[J].地球物理勘探进展,2007,30(1):39-42.
[22] 高坤,陶果,仵岳奇,等.利用多极子阵列声波测井资料计算横向各向同性地层破裂压力[J].中国石油大学学报(自然科学版),2007,31(1):35-39.
[23] 董经利.多极子阵列声波测井资料处理及应用[J].测井技术,2009,33(2):115-119.

本论文原发表于《油气储运》2017年第36卷第12期。

Thermal analysis for gas storage in salt cavern based on an improved heat transfer model

Youqiang Liao[1,2]　Tongtao Wang[1,2]　Long Li[3]
Zhongxin Ren[4]　Dongzhou Xie[1,2]　Tao He[1,2]

(1. State Key Laboratory of Geomechanics and Geotechnical Engineering, Institute of Rock and Soil Mechanics, Chinese Academy of Sciences;
2. Hubei Key Laboratory of Geo-Environmental Engineering, Institute of Rock and Soil Mechanics, Chinese Academy of Sciences;
3. Jiangsu Gas Storage Branch Company, PipeChina West East Gas Pipeline Company;
4. PipeChina West East Gas Pipeline Company)

Abstract　Oversimplification of the heat transfer mechanism and assumption of uniform temperature and pressure in the salt cavern could cause significant errors when modeling temperature fields. This work proposed a coupled transient flow and heat transfer model for gas storage in the underground salt cavern to reflect the thermal behaviors between gas and surrounding rock, particularly the gas Joule–Thomson effect during gas operation. Then, a fully coupled numerical solution method based on a unified matrix is presented. An average error of 3.26% was observed between the model and field data quoted in literature. The case study indicates that during the gas withdrawal period, the maximum temperature difference in the salt cavern can reach 14.94℃, while the temperature drop of the cavity wall is only about 4.07℃, which is quite different from the traditional assumption. A special focus was given to the gas withdrawal rate, which seemed to have the most significant influence on the evolution of the temperature field in the salt cavern. Compared with the gas withdrawal rate of 35m^3/s, reducing the gas withdrawal rate to 20m^3/s can increase the minimum temperature by approximately 23.41%. This study could add further insights into the thermal performance during gas operation in the salt cavern and help to reveal the evolution of the temperature field in the cavity and surrounding rock.

Keywords　underground salt cavern; heat transfer model; joule–thomson effect; temperature field; gas withdrawal

1 Introduction

Owing to the typical regionality and uneven consumption of fossil fuels such as oil and natural

---

Corresponding author: Tongtao Wang, State Key Laboratory of Geomechanics and Geotechnical Engineering, Institute of Rock and Soil Mechanics, Chinese Academy of Sciences, Wuhan 430071, China, ttwang@whrsm.ac.cn.

gas, building a large-scale oil and gas storage system is an inevitable choice to cope with energy supply and peak regulation[1,2]. Energy storage in underground salt caverns (USCs) is one of the most potential methods[3], which has many advantages, such as fast injection-withdrawal rate, high safety, and low cost[4]. It has been widely applied in the United States, Europe, China, and other primary energy-consuming countries[3,5].

In natural gas peak regulation using USCs, injecting and withdrawing gas frequently in/from USC is regular, and the temperature field of the salt cavern and surrounding rock will change periodically with gas operation[6]. The mechanical properties, particularly salt rock's creep performance, are sensitive to temperature[7]. Therefore, periodic temperature changes will change the thermal stress distribution in the surrounding rock, resulting in irreversible thermal damage to the surrounding rock and then increasing the risk of salt cavern instability[8,9]. Additionally, because the physical and mechanical properties of the salt rock and the interlayer are quite different[10,11], the different geomechanical responses between them under the alternating temperature disturbance will lead to the significant development of microcracks in the surrounding rock and affect the tightness of the gas storage salt cavern[12,13]. Moreover, it was found from the simulation conducted by Li et al.[14] and Meng et al.[15] that when the temperature effect was incorporated into the thermo-hydro-mechanical model, the results for the evolution of cavity stability could better match the experimental data. Therefore, a further investigation on the heat transfer characteristics in the salt cavern was highly recommended, which is the premise for evaluating the mechanical stability[8,16,17] and tightness of underground gas storage in salt rock[18,19], and will also help optimize the gas injection and withdrawal parameters.

To our best knowledge, the research on salt cavern gas storage mainly focuses on the evolution of mechanical properties[20,21] and permeability evaluation[22,23]. In earlier studies, it was generally assumed that the gas temperature in the salt cavern was constant[24] or that the wall of the salt cavern was adiabatic[25]. Subsequently, Bérest et al.[26], and Li et al.[14], assuming uniform temperature and pressure in the salt cavern, based on the thermodynamic gas theory, established temperature prediction models of the salt cavern during gas injection and withdrawal. Based on the same assumption, Han et al.[27] further considered the gas flow and heat transfer in the wellbore in studying compressed air energy storage in salt caverns. Researchers tend to simplify the cavern thermodynamic problem by neglecting the spatial variations of pressure and temperature, which leads to a cavern uniform state. Although those models can still match the experimental or field temperature and pressure profiles quite well by tuning some thermophysical parameters for history-matching, the assumption of uniform temperature and pressure for a large-scale salt cavern may still lead to non-negligible errors[28].

Nomenclature

A	wellbore cross-sectional area, m²	r_{ci}	inner diameter of casing, m
b_{ol}	equal to 0 and 1 for formation and salt cavern	r_{co}	outer diameter of casing, m
C_e	specific heat capacity of gas, J/(kg·℃)	r_{to}	outer diameter of inner pipe, m
$C_{V,g}$	specific heat capacity of gas, J/(kg·℃)	t	simulation time, s
F_w	friction, N	T_e	formation temperature, ℃
g	gravitational acceleration, m/s²	T_g	gas temperature, ℃
h_t	convection heat transfer coefficient of inner wall of inner pipe, W/(m²·℃)	U	specific internal energy, J/kg
		U_{we}	comprehensive heat transfer coefficient between the gas and the surrounding environment, W/(m²·℃)
H	specific enthalpy, J/kg		
k_{cas}	heat conductivity of casing, W/(m·℃)		
k_{ces}	heat conductivity of cement, W/(m·℃)	v	gas flow velocity, m/s
k_e	heat conductivity of formation, W/(m·℃)	z	depth, m
p	gas pressure, Pa	ρ_e	density of formation, kg/m³
Q_e	heat flow rate from surrounding environment to wellbore, W/m	ρ_g	density of gas, kg/m³
		θ	inclination angle of wellbore, rad
r	distance from the wellbore, m	μ_{jT}	coefficient of the Joule-Thomson effect, K/Pa
r_{ces}	outer diameter of cement, m		

Taking the gas Joule-Thomson effect into account, Merey[29] studied the thermal behaviors in the salt cavern at different gas withdrawal and injection rates using TOUGH + RealGasBrine simulator. Leontidis et al.[30] and Niknam et al.[31] proposed a dynamic model for different gas injection scenarios to investigate the unsteady behavior of the geothermal power system. Considering the transient heat exchange between gas, wellbore, and formation, Bérest[6], Barajas and Civan[32] proposed a comprehensive model to analyze the spatial variations of temperature, pressure, and velocity in salt cavern. AbuAisha and Rouabhi[28] compared the temperature field prediction results of the simplified and complete models. It was found that under high-speed gas injection and production, the difference between the results of the two models can reach 7%. However, the complete simulation that addressed gas velocity field, turbulent flow model, and convective heat transfer took approximately 60 days. High simulation cost may limit the widespread application of the complete model. In addition, poor consideration of the conversion of internal energy to mechanical energy during the transient gas flow may cause potential challenges for temperature and pressure predictions without sufficient filed data.

Fig. 1 shows the mechanism of heat transfer during gas flow. The interactions between different factors during gas storage in the underground salt cavern can be summarized as follows:

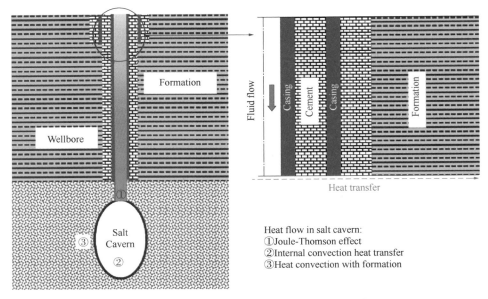

Fig. 1　The mechanism of heat transfer during gas flow

(1) The gas flow behaviors affect the rate and time of heat exchange with surrounding rock, which determines the gas's temperature distribution. The gas temperature affects the gas density, which affects the transient flow behaviors of the gas accordingly[33];

(2) In the gas flow process, a complex conversion between the internal energy, kinetic energy, and potential energy of the gas occurs with the change of gas pressure due to the expansion of the gas to work externally, which is known as the gas Joule–Thomson effect, macroscopically showing transient changes in gas temperature and flow velocity[34,35];

(3) There is continuous heat transfer between the gas in the salt cavern and the surrounding rock; thus, the inter-coupling between them should be fully considered. Otherwise, it will lead to specific numerical solution errors if the heat transfer in the wellbore, salt cavern, and surrounding rock is studied and solved independently[36].

In this work, acoupled transient flow and heat transfer model for gas storage in the underground salt cavern was proposed, in which the transient flow, heat transfer between the working gas and surrounding rock, and gas Joule–Thomson effect were both considered. Second, the heat transfer equations of the wellbore, salt cavern, and surrounding formation were constructed in the same matrix to reduce the numerical solution error. The fully implicit, synchronous solution of the heat transfer model for the underground salt cavern system was realized. Then, a thorough thermal and flow performance analysis was presented during gas withdrawal from the salt cavern. Special focus was given to the effects of gas withdrawal rate, time, and depth of salt cavern on the temperature evolution performance. This work could add further insights into the thermal and flow performance and provide a more accurate estimation of the temperature and pressure distribution in the salt cavern. It could become a powerful tool to optimize the gas injection and withdrawal parameters and also help to evaluate the mechanical stability of salt caverns.

2 Modeling of the transient flow and heat transfer

To model the complex transient flow and heat transfer coupling process, the following assumptions are made:

(1) To reduce the complexity and computation of the flow and heat transfer model while obtaining prediction results with acceptance accuracy, it's assumed that the gas flow in the wellbore is onedimensional;

(2) During the short gas injection period, the creep of the salt rock is ignored, and the volume of the salt cavern is assumed to be constant[20];

(3) Since the radius of the salt cavern is much larger than that of the wellbore, it is assumed that the gas flow velocity in the salt cavern is approximately equal to 0[6]. The gas natural convection simplified as heat conduction, gas compression/expansion work, and the heat exchange with the surrounding rock were all considered in this work.

2.1 Gas flow

Considering the density change along the wellbore with the pressure, the mass conservation equation of gas can be expressed as:

$$\frac{\partial(A\rho_g)}{\partial t}+\frac{\partial(A\rho_g v)}{\partial z}=0 \tag{1}$$

where, A is the wellbore cross-sectional area, m^2; ρ_g is the density of gas, kg/m^3; v is the gas flow velocity, m/s; t is the time, s; z is the depth, m.

When the gas flows in the wellbore, the quantitative relationship between its pressure and the hydrostatic column pressure, frictional pressure drop, and the momentum change can be expressed as follows based on the conservation of momentum.

$$\frac{\partial(A\rho_g v)}{\partial t}=-\frac{\partial}{\partial z}(Ap)-\frac{\partial(A\rho_g v^2)}{\partial z}-A\frac{f\rho_g v^2}{2D}+A\rho_g g\cos\theta \tag{2}$$

where, p is the gas pressure, Pa; f is the friction coefficient; D is the tubing inside diameter, m; g is the gravitational acceleration, m/s^2; θ is the inclination angle of the wellbore, rad.

The left-hand side of Eq. (2) represents the change in momentum of the gas. The right-hand side terms represent the momentum transfer due to pressure change, momentum transfer due to gas flow, friction pressure drop, and static head of the high-pressure gas, respectively.

The key to determining the pressure distribution along the wellbore lies in calculating the friction coefficient, which is related to the gas's flow state (laminar flow or turbulent flow) and the roughness of the pipe wall. The model proposed by Wang et al.[37] was used to calculate the friction coefficient.

$$\begin{cases} f = \dfrac{64}{Re} & (Re<2300) \\ f = 0.06539\exp\left[-\left(\dfrac{Re-3516}{1248}\right)^2\right] & (2300 \leqslant Re \leqslant 2300) \\ \dfrac{1}{\sqrt{f}} = -2.34\lg\left\{\dfrac{\varepsilon}{1.72D} - \dfrac{9.26}{Re}\lg\left[\left(\dfrac{\varepsilon}{29.36D}\right)^{0.95} + \left(\dfrac{18.35}{Re}\right)^{1.108}\right]\right\} & (Re>3400) \end{cases} \quad (3)$$

where, Re is the Reynolds number; ε is the mean roughness of the tubing, m.

2.2 Heat transfer

2.2.1 Heat transfer in wellbore

The transient gas flow in the wellbore has an essential effect on heat transfer. On the one hand, the gas flow process involves the conversion of internal, kinetic, and potential energy; on the other hand, it is also accompanied by a complex heat exchange with the surrounding environment. Taking the specific enthalpy of the fluid as the research object, the energy conservation equation can be expressed as[38]:

$$\dfrac{\partial}{\partial t}\left[A\rho_g\left(U+\dfrac{1}{2}v^2\right)\right] + \dfrac{\partial}{\partial z}\left[A\rho_g v\left(H+\dfrac{1}{2}v^2\right)\right] = -A\rho_g vg + Q_e \quad (4)$$

where, U is the specific internal energy, J/kg; H is the specific enthalpy, J/kg; Q_e is the heat flow rate from the surrounding environment to the wellbore, W/m.

The first term on the left-hand side of Eq. (3) represents the change in the internal energy and kinetic energy storage with time, and the second term represents the change in specific enthalpy and kinetic energy with depth. The terms on the right-hand side represent the gravitational potential energy and heat exchange between the wellbore and surrounding rock, respectively.

According to gas thermodynamic theory, the relationship between gas internal energy and specific enthalpy can be expressed as[39]:

$$H = U + p/\rho \quad (5)$$

Substitution of Eq. (4) into Eq. (3) and the gas energy conservation equation can be written as:

$$A\left(\rho_g\dfrac{\partial H}{\partial t} - \dfrac{\partial p}{\partial t} + \rho_g v\dfrac{\partial v}{\partial t}\right) + A\rho_g v\left(\dfrac{\partial H}{\partial z} + v\dfrac{\partial v}{\partial z}\right) = -A\rho_g vg + Q_e \quad (6)$$

The total derivative of enthalpy can be derived by using thermodynamic equilibrium relationships[35]:

$$\begin{cases} dH = C_{V,g}dT_g - \mu_{jT}C_{V,g}dp \\ \dfrac{\partial H}{\partial t} = C_{V,g}\dfrac{\partial T_g}{\partial t} - \mu_{jT}C_{V,g}\dfrac{\partial p}{\partial t} \\ \nabla H = C_{V,g}\nabla T_g - \mu_{jT}C_{V,g}\nabla p \end{cases} \quad (7)$$

where, $C_{V,g}$ is the specific heat capacity of gas, J/(kg·℃); T_g is the gas temperature, ℃; μ_{jT} is the coefficient of the Joule-Thomson effect, K/Pa.

Substitution of Eq. (6) into Eq. (5) and the gas energy conservation equation can be written as:

$$A\rho_g C_{V,g}\frac{\partial T_f}{\partial t} - A(\rho_g \mu_{jT} C_{V,g}+1)\frac{\partial p}{\partial t} + A\rho_g v\left(\frac{\partial v}{\partial t} + C_{V,g}\frac{\partial T_f}{\partial z} - \mu_{jT} C_{V,g}\frac{\partial p}{\partial z} + v\frac{\partial v}{\partial z}\right) = -A\rho_g vg + Q_e \quad (8)$$

the heat flow rate from the surrounding environment to the wellbore can be calculated by[40]:

$$Q_e = 2\pi r U_{we}(T_{e,1} - T_f) \quad (9)$$

where, U_{we} is the comprehensive heat transfer coefficient between the gas and the surrounding environment, W/(m²·℃). According to the research of Sun et al.[41] and Hasan et al.[42], the comprehensive heat transfer coefficient is related to the casing and cement sheath structures and their thermophysical parameters. It can be expressed as:

$$U_{we}^{-1} = \frac{1}{h_t} + \frac{r_{to}\ln(r_{co}/r_{ci})}{k_{cas}} + \frac{r_{to}\ln(r_{ces}/r_{co})}{k_{ces}} \quad (10)$$

where, h_t is the convection heat transfer coefficient of the inner wall of the inner pipe, W/(m²·℃); r_{to} is the outer diameter of the inner pipe, m; r_{ci} and r_{co} are the inner and outer diameter of the casing, respectively, m; r_{ces} is the outer diameter of cement, m; k_{cas} and k_{ces} are the heat conductivity of casing and cement, respectively, W/(m·℃).

It's noteworthy that the items on the right side of Eq. (10) represent the comprehensive heat transfer coefficient components of the inner pipe, casing, and cement sheath, respectively. The corresponding items can be added or subtracted according to the above equations for multi-layer casing or open-hole section.

2.2.2 Heat transfer in salt cavern and surrounding rock

The rules of convection heat transfer between the gas and surrounding environment are needed to solve the energy conservation equation of gas. The heat transfer mode for surrounding rocks is heat conduction, and the energy conservation equation can be expressed as follows.

$$\rho_e C_e \frac{\partial T_e}{\partial t} = \frac{1}{r}\frac{\partial}{\partial r}\left(k_e r \frac{\partial T_e}{\partial r}\right) \quad (11)$$

where, ρ_e is the density of formation, kg/m³; C_e is the specific heat capacity of gas, J/(kg·℃); T_e is the temperature of surrounding rock, ℃; r is the distance from the wellbore, m; k_e is the heat conductivity of surrounding rock, W/(m·℃).

While the heat transfer mode in the salt cavern is convective heat transfer and compression/expansion work. In this work, it's assumed that the gas flow velocity in the salt cavern is approximately equal to 0[6]; thus, to reduce the simulation cost, the convective heat transfer in the salt cavern can be simplified as heat conduction. Therefore, the energy conservation equation can be written as follows.

$$\rho_g C_g \frac{\partial T_g}{\partial t} = \frac{1}{r}\frac{\partial}{\partial r}\left(k_g r \frac{\partial T_g}{\partial r}\right) + b_{ol}(\rho\mu_{jT}C_{V,g}+1)\frac{\partial p}{\partial t} \qquad (12)$$

where, k_g is the heat conductivity of gas, W/(m·°C); b_{ol} is equal to 0 and 1 for the formation and salt cavern, respectively.

3 Numerical solution and verification

3.1 Wellbore-rock coupling solution

If the temperature field is solved by taking the wellbore or formation as the boundary condition, the full coupling between the wellbore and formation is not realized[36]. For numerical solutions with long simulation time and time steps, non-negligible errors may occur. The key to realizing the full coupling between the wellbore and formation lies in accurately characterizing the heat exchange mechanism at the boundary of those two physical bodies and realizing the full implicit solution of the boundary variables. In this section, the wellbore and the surrounding rock are regarded as a whole, and the two energy equations are constructed in a unified matrix for the coupled solution. The specific solution flow chart is shown in Fig. 2. The whole calculation process consists of two iterations: iteratively solve the pressure profile and temperature field, respectively. The specific steps are as follows:

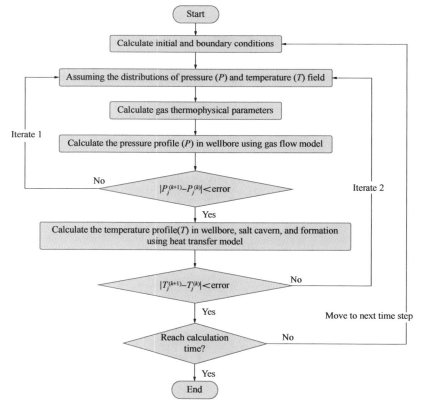

Fig. 2 Flow chart of the calculation process

(1) Assuming the wellbore temperature and pressure field distribution at $t = k$, the temperature and pressure distribution at $t = k-1$ are usually taken for the first iteration, and the calculation results of the previous iteration are used for subsequent iterations;

(2) Calculate the thermophysical parameters of the gas according to the assumed temperature and pressure distributions;

(3) Using the gas flow equation to solve the pressure profile in the wellbore;

(4) Determine whether the pressure profile satisfies the convergence condition; if not, go back to step(1);

(5) Coupling solution of the heat transfer model. Detailed discrete process and numerical solution algorithm can be found in Appendix A;

(6) Determine whether the temperature field distribution satisfies the convergence condition; if not, go back to step(1);

(7) Go to the next time step.

3.2 Model verification

The temperature and pressure data collected from a field gas withdrawal test presented by Crossley[43] and a mathematical model proposed by Bérest et al.[26] were used to verify the accuracy of this model. In the field experiment, a salt cavern in Melville, Canada, with a volume of 46153 m^3 was applied, and the initial temperature and pressure were approximately 32.0℃ and 13.4MPa, respectively. The detailed experimental process and parameters were described in their published paper. Therefore, the pressure and temperature evolution behaviors can be calculated by the model proposed by us and Bérest et al.[26], respectively. It is noteworthy that since the public literature does not clearly indicate the layout positions of temperature and pressure sensors during the field experiment, we thus assume that the sensor is in the center of the salt cavern to verify the improved thermal model presented in this work.

Fig. 3 compares the pressure and temperature profiles among the experimental results, the model proposed by Bérest et al.[26], and the model presented in this work. It can be clearly found that the trend of the model calculation results is basically the same as that of the measured data. It can accurately predict the rapid pressure drop due to gas withdrawal (before the 8th day) and the slow pressure increase due to the temperature increase after gas withdrawal stopped (after 8th day), the pressure prediction error in the whole stage is only 1.75%. For the prediction of the temperature field, the average error of the model proposed in this work is only 3.26% in the whole stage, and the maximum error occurs around the 6.5th day, which is about 6.56%. The main source of the error is a certain difference in temperature recovery characteristics in the salt cavern after gas withdrawal was stopped. Compared with the measured results, the temperature rises rate calculated by the model proposed in this paper is slower, while the Bérest model is faster. Nevertheless, a model with a maximum error of only 6.56% is still generally acceptable in engineering.

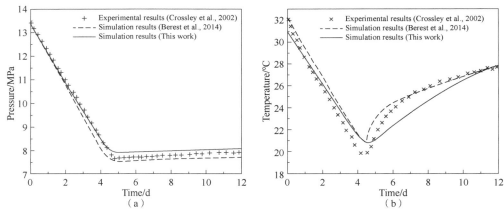

Fig. 3 Comparison of (a) pressure, and (b) temperature profiles among the experimental data, the model proposed by Bérest et al.[26], and the model presented in this work

4 Model performance

To further study the temperature and pressure field response characteristics of the gas storage

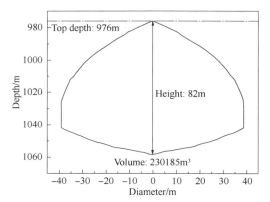

Fig. 4 The shape of the cavern obtained by Sonar measurement

salt cavern and surrounding rock during gas operation, this work takes a gas storage salt cavern in Jintan, Jiangsu, China, as an example. The basic parameters obtained by Sonar measurement are shown in Fig. 4. The depth of the salt cavern gas storage is 976 m, with a maximum diameter of 78 m, a total height of 82 m, and an effective cavity volume of 230185m^3.

Compared with the gas injection process, the pressure in the cavity gradually decreases during gas withdrawal. Generally, the risk of mechanical instability in the salt cavern is greater during this period[44]. Therefore, this section takes the gas withdrawal process as a case study, and the basic parameters are shown in Table 1.

Table 1 Summary of the basic parameters used for model performance

Parameter	Value	Parameter	Value
Diameter of wellbore/m	0.254	Ground temperature/℃	20
Gas withdrawal rate/(m^3/s)	35	Geothermal gradient/(℃/m)	0.018
Gas withdrawal duration/d	8	Simulation time/d	16
Initial pressure of salt cavern/MPa	16.5	Density of formation/(g/cm^3)	2.65
Specific heat of gas/[J/(kg·K)]	2347	Heat conductivity of gas/[W/(m·K)]	0.15
Specific heat of formation/[J/(kg·K)]	999	Heat conductivity of formation/[W/(m·K)]	2.09

4.1 Pressure distribution

Fig. 5 shows the evolution of the central pressure of salt cavern with time. It can be clearly found that the pressure gradually decreases during the gas withdrawal process, and it drops to a minimum value of approximately 7.24MPa at the end of gas withdrawal. After that time, due to the gradual recovery of the temperature in the salt cavern, the gas expansion will cause the pressure of the salt cavern to rise continuously, reaching 7.51MPa on the 16th day. About 22 days later, the pressure in the cavity tends to be stable, and the final pressure is 7.70MPa. The pressure increases by approximately 6.35% compared to the end of gas withdrawal.

Fig. 6 shows the pressure profile along the wellbore and salt cavern at $t = 0$d, 4d, 8d, and 16d. Due to the gas hydrostatic column pressure and friction pressure drop, there is an obvious pressure gradient from the wellhead to the bottom of the salt cavern, and the greater the wellhead pressure, the greater the pressure gradient. Taking the initial moment (0d) as an example, the wellhead pressure is 15.50MPa, and the bottom pressure of the salt cavern is 16.82MPa, with an increase of about 8.52%. For a salt cavern with a height of 82 m, the upper and lower pressure difference is only about 0.1MPa. Therefore, it is indicated that assuming the pressure in the salt cavern is a uniform value can obtain results with acceptable errors. However, the pressure distribution in the long wellbore must be considered.

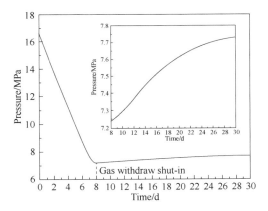

Fig. 5 Evolution of central pressure of salt cavern with time

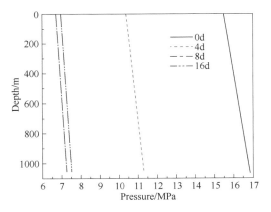

Fig. 6 Pressure profile along the wellbore and salt cavern at $t = 0$d, 4d, 8d, and 16d

4.2 Temperature distribution

4.2.1 Evolution of temperature with time

Fig. 7 shows the evolution of the central temperature of salt cavern with time. During the gas withdrawal process, the temperature in the gas storage salt cavern gradually decreases due to the expansion of the gas for external work (Joule-Thomson effect). At 7.45th days, the temperature of the gas storage salt cavern dropped to a minimum of approximately 23.12℃. Interestingly, as the temperature in the salt cavern gradually decreases, the heat transfer rate from the surrounding rock to the salt cavern gradually increases. Therefore, the temperature in the gas storage will increase before

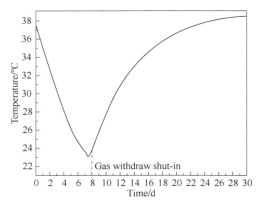

Fig. 7 Evolution of central temperature of salt cavern over time

the gas withdrawal is completed. The temperature in the salt cavern tends to be stable after about 30d, which is the same as the changing trend of the pressure.

4.2.2 Temperature profile in wellbore

Fig. 8 shows the temperature profiles along the wellbore at t = 0d, 4d, 8d, and 16d. It can be clearly seen that with the progress of gas withdrawal, the temperature in the wellbore shows an overall downward trend and reaches the lowest near the last moment of gas withdrawal (~ 8th d). From the evolution of wellhead temperature (as shown in Fig. 9), it's obvious that the gas temperature at the wellhead increases first and then decreases during the gas withdrawal stage and reaches a maximum temperature at 5.33th h. This is because the gas temperature in the deep salt cavern is relatively high at the beginning of gas withdrawal, and the high-temperature gas flows out to the wellhead, while as the gas pressure in the salt cavern decreases, the temperature gradually decreases correspondingly, and the gas temperature at the wellhead also decreases rapidly.

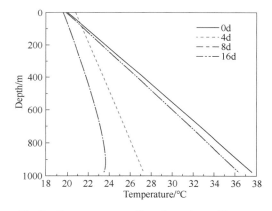

Fig. 8 Temperature profiles along the wellbore at t = 0d, 4d, 8d, and 16d

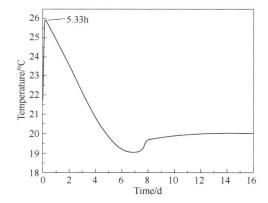

Fig. 9 Evolution of wellhead temperature over time

4.2.3 Temperature spatial distribution of salt cavern and surrounding formation

Fig. 10 shows the spatial distribution of temperature field of salt cavern and surrounding rock at t = 0d, 2d, 4d, 8d, 12d, and 16d. It can be found that the temperature in the salt cavern decreases significantly as gas withdrawal progresses. Concerning spatial distribution, the temperature at the center of the salt cavern is the lowest, and the temperature is higher as it is closer to the surrounding rock. At the end of the gas withdrawal process, the maximum temperature difference in the salt cavern can reach 14.94℃. It is because the higher-temperature surrounding rock continuously transfers heat to the salt cavern, and the heat transfer flux increases as it is closer to the cavity wall. It also indirectly confirms that the assumption that the gas temperature in the salt cavern is uniform could cause significant errors. As shown in Fig. 11, during the gas withdrawal period, the depth of the disturbance region of the temperature field is about 5.28 m. Compared with the original temperature, the temperature drop of the cavity wall is only about 4.07℃. Therefore, the influence of temperature disturbance on the evolution of the mechanical properties of the salt cavern needs to be further revealed in the follow-up research[45].

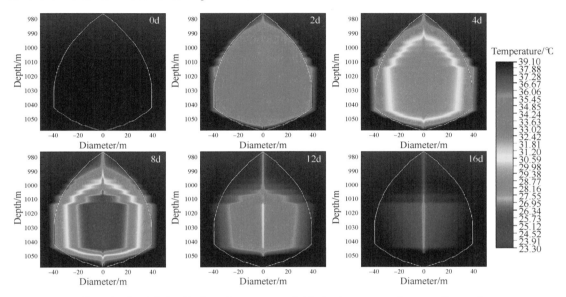

Fig. 10 Spatial distribution of temperature field of salt cavern and surrounding rock at $t=0$d, 2d, 4d, 8d, 12d, and 16d

In addition, the model established in this work can accurately simulate the spatial distribution of cavity temperature, while calculating the entire gas withdrawal process (12 days) in less than 1 min, which is significantly faster than that of the complete model proposed by AbuAisha and Rouabhi[28].

4.3 Sensitivity analysis

In this section, the influence of gas withdrawal rate, gas withdrawal duration, and salt cavern depth on the temperature evolution of the salt cavern will be analyzed to provide a theoretical basis for the optimal design of gas injection and withdrawal parameters.

4.3.1 Gas withdrawal rate

Fig. 12 shows the evolution of central temperature of salt cavern over time under different gas

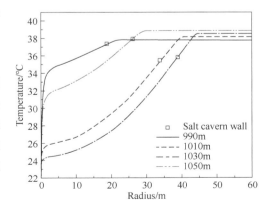

Fig. 11 Radial distribution of temperature profiles at different depths at the end of gas withdrawal

withdrawal rates. It can be found that the higher the gas production rate is, the greater the temperature drop rate in the salt cavern is. It is mainly because the temperature drop in the salt cavern gas storage is due to the pressure drop caused by the gas withdrawal. Specifically, the external work from gas expansion leads to a decrease in temperature. Meanwhile, detailed quantitative relationships between the temperature and gas withdrawal rate can be found in the energy conservation equation [Eq. (12)]. Compared with the gas withdrawal rate of $35m^3/s$, reducing the gas withdrawal rate to $20m^3/s$ can increase the minimum temperature by approximately 23.41%.

4.3.2 Gas withdrawal duration

Fig. 13 shows the evolution of the central temperature of salt cavern over time under different gas withdrawal duration. It can be found that with the continuous gas withdrawal, the temperature in the salt cavern gradually decreases due to the continuous decrease in pressure, and the heat transfer flux from the surrounding rock to the salt cavern also increases gradually. Therefore, at the gas withdrawal rate of $20m^3/s$, the lowest temperature in the salt cavern appears at the 10.5th d, which is about 27.50℃. After that, even if the gas withdrawal continues, the temperature in the cavity will still rise slowly. Predictably, when the heat flux transferred from the surrounding rock is equal to the gas expansion work, the temperature in the cavity can reach a dynamic equilibrium state.

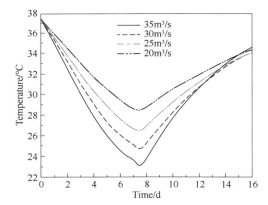

Fig. 12 Evolution of central temperature of salt cavern over time under different gas withdrawal rates

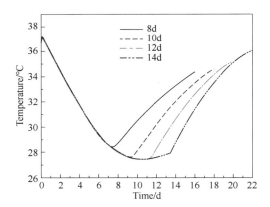

Fig. 13 Evolution of central temperature of salt cavern over time under different gas withdrawal duration

4.3.3 Depth of salt cavern

Fig. 14 shows the evolution of central temperature of salt cavern over time under different salt cavern depths. The greater the burial depth of the salt cavern, the higher the ambient temperature it is located, and the temperature in the cavity during the entire gas withdrawal process is basically at a higher level. Interestingly, under the same gas production conditions, although the difference in the initial temperature in the salt cavern, the temperature drop in the salt cavern is basically the same, about 14.71℃. It is mainly because the magnitude of temperature drop is related to the rate and duration of pressure drop. In other words, the gas withdrawal rate and duration are the key factors determining temperature drop.

To sum up, during the gas withdrawal period, the gas withdrawal rate has the most significant influence on the evolution of the temperature field in the salt cavern, which determines the temperature change range during the whole gas operation process. Therefore, to ensure the stability of the gas storage salt cavern, the gas withdrawal rate should be reduced concerning both the pressure field and the thermodynamic viewpoint, and an optimal design of the gas withdrawal rate should be carried out in combination with the economy.

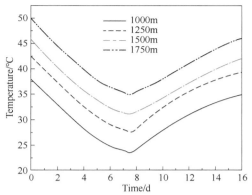

Fig. 14 Evolution of central temperature of salt cavern over time under different salt cavern depths

4.4 Outlook and future perspectives

In this study, a coupled transient flow and heat transfer model for gas storage in the underground salt cavern was proposed, in which the transient flow, heat transfer between gas and surrounding rock, and gas Joule-Thomson effect were both considered. The model established in this work can accurately simulate the spatial distribution of cavity temperature, while calculating the entire gas production process (12 days) in less than 1 min, which is significantly faster than that of the complete model proposed by AbuAisha and Rouabhi[28].

The mechanical properties of salt rock, particularly the creep performance, will change significantly with temperature, and the thermal stress in the salt rock caused by the temperature change will also significantly affect the permeability characteristics of the salt rock. Therefore, this model could become a helpful tool to predict the mechanical properties of the salt rock, and evaluate the tightness of the salt cavern of various gas injection and withdrawal parameters. Moreover, the research object composed of the wellbore and salt cavern has rich application background. In addition to the energy storage mentioned in this work, it, s also can be applied in energy efficiency evolution in compression air energy storage (CAES) and geothermal energy utilization, as well as heat/mass transfer simulation for CO_2 salt cavern storage.

5 Conclusions

Thermal behaviors in the salt cavern during gas operation are the premise for evaluating the mechanical stability and tightness of underground gas storage salt cavern. In this work, a coupled transient flow and heat transfer model for gas storage in the underground salt cavern was proposed to reflect the heat transfer behaviors between the working gas and surrounding rock, particularly the gas Joule-Thomson effect during gas operation. The following conclusions can be drawn from this study:

(1) For a salt cavern with a height of 82 m, the upper and lower pressure difference is only about 0.1MPa, while in the whole wellbore section, the pressure difference can exceed 8.5%.

(2) After gas withdrawal is stopped, due to the gradual recovery of the temperature in the salt

cavern, the gas expansion will cause the pressure of the salt cavern to rise continuously. About 22 days later, the pressure in the cavity tends to be stable. Compared with the pressure at the end of gas withdrawal, the final pressure increases by approximately 6.35%.

(3) During the gas withdrawal period, the maximum temperature difference in the salt cavern can reach 14.94℃, while the temperature drop of the cavity wall is only about 4.07℃, and the depth of the disturbance region of the temperature field is about 5.28 m. Therefore, assuming the uniform gas temperature in the salt cavern could cause significant errors.

(4) The gas withdrawal rate has the most significant influence on the evolution of the temperature field in the salt cavern. Compared with the gas withdrawal rate of 35m³/s, reducing the gas withdrawal rate to 20m³/s can increase the minimum temperature by approximately 23.41%.

Declaration of Competing Interest

The authors declare that they have no known competing financial interests or personal relationships that could have appeared to influence the work reported in this paper.

Data availability

Data will be made available on request.

Acknowledgments

The authors wish to acknowledge the financial support of the National Natural Science Foundation of China(Grant No. 42072307), Hubei Province Outstanding Youth Fund (2021CFA095), Strategic Research and Consulting Project of Chinese Academy of Engineering (Grant No. HB2022B08), and Strategic Priority Research Program of the Chinese Academy of Sciences (Grant No. XDC10020300).

Appendix A. Coupling solution method of temperature field

Implicit schemes are used to discretize the equations in the proposed model.

For the mass and momentum conservation equations, the detailed discretization scheme is as follows:

$$A_j \frac{\rho_j^k - \rho_j^{k-1}}{\Delta t} + A_j \rho_j^k \frac{v_j^k - v_{j-1}^k}{\Delta z} + A_j v_j^k \frac{\rho_j^k - \rho_{j-1}^k}{\Delta z} + \rho_j^k v_j^k \frac{A_j - A_{j-1}}{\Delta z} = 0 \quad (A.1)$$

$$\frac{A_j p_j^k - A_{j-1} p_{j-1}^k}{\Delta z} + A_j \frac{\rho_j^k v_j^k - \rho_j^{k-1} v_j^{k-1}}{\Delta t} + \frac{A_j \rho_j^k (v_j^k)^2 - A_{j-1} \rho_{j-1}^k (v_{j-1}^k)^2}{\Delta z} = -A_j F_{w,j}^k + A_j \rho_j^k g \cos\theta \quad (A.2)$$

For the temperature field model of the working gas and surrounding rock, the detailed discretization scheme is as follows:

Energy conservation equation of gas in wellbore and cavity:

$$\rho_j^k C_{V,g} \frac{T_{f,j}^k - T_{f,j}^{k-1}}{\Delta t} - (\rho_j^k \mu_{jT} C_{V,g} + 1) \frac{p_j^k - p_j^{k-1}}{\Delta t} + \rho_j^k v_j^k \left(\frac{v_j^k - v_j^{k-1}}{\Delta t} + C_{V,g} \frac{T_{f,j}^k - T_{f,j-1}^k}{\Delta z} - \mu_{jT} C_{V,g} \frac{p_j^k - p_{j-1}^k}{\Delta z} + v_j^k \frac{v_j^k - v_{j-1}^k}{\Delta z} \right)$$

$$= -\rho_j^k v_j^k g + \frac{2U_{hf}}{r} (T_{e,1,j}^k - T_{f,j}^k) \tag{A.3}$$

Eq. (A.3) can be simplified as follows:

$$\alpha T_{f,j}^k - \frac{2U_{hf}}{r} T_{e,1,j}^k = \beta \tag{A.4}$$

where,

$$\alpha = \frac{\rho_j^k C_{V,g}}{\Delta t} + \frac{\rho_j^k v_j^k C_{V,g}}{\Delta z} + \frac{2U_{hf}}{r} \tag{A.5}$$

$$\beta = \frac{\rho_j^k C_{V,g}}{\Delta t} T_{f,j}^{k-1} + (\rho_j^k \mu_{jT} C_{V,g} + 1) \frac{p_j^k - p_j^{k-1}}{\Delta t} - \rho_j^k v_j^k \left(\frac{v_j^k - v_j^{k-1}}{\Delta t} - C_{V,g} \frac{T_{f,j-1}^k}{\Delta z} - \mu_{jT} C_{V,g} \frac{p_j^k - p_{j-1}^k}{\Delta z} + v_j^k \frac{v_j^k - v_{j-1}^k}{\Delta z} \right) - \rho_j^k v_j^k g \tag{A.6}$$

Energy conservation equation of surrounding rock:

$$\rho_e C_e \frac{T_{e,i,j}^k - T_{e,i,j}^{k-1}}{\Delta t} = \lambda_e \left[\frac{1}{r_i} \frac{T_{e,i+1,j}^k - T_{e,i,j}^k}{r_{i+1} - r_i} + \frac{T_{e,i+1,j}^k - 2T_{e,i,j}^k + T_{e,i-1,j}^k}{(r_{i+1} - r_i)(r_i - r_{i-1})} \right] + b_{ol} (\rho_j^k \mu_{jT} C_{V,g} + 1) \frac{p_j^k - p_j^{k-1}}{\Delta t} \tag{A.7}$$

Eq. (A.7) can be simplified as follows:

$$(-\xi - \psi) T_{e,i+1,j}^k + (\xi + 2\psi + 1) T_{e,i,j}^k - \psi T_{e,i-1,j}^k = T_{e,i,j}^{k-1} + \omega \tag{A.8}$$

where,

$$\begin{cases} \xi = \dfrac{\lambda_e \Delta t}{\rho_e C_e r_i (r_{i+1} - r_i)} \\ \psi = \dfrac{\lambda_e \Delta t}{\rho_e C_e (r_{i+1} - r_i)(r_i - r_{i-1})} \\ \omega = \dfrac{\lambda_e b_{ol} (\rho_j^k \mu_{jT} C_{V,g} + 1)(p_j^k - p_j^{k-1})}{\rho_e C_e} \end{cases} \tag{A.9}$$

The heat transfer equation for the interface of wellbore and surrounding rock can be expressed as:

$$(1 + \eta + \gamma) T_{e,1,j}^k - \eta T_{f,j}^k - \gamma T_{e,2,j}^k = T_{e,1,j}^{k-1} + \omega \tag{A.10}$$

where,

$$\eta = \frac{U_{hf}\Delta t}{\rho_e C_e (r_1 - r_0)} \tag{A.11}$$

$$\gamma = \frac{\lambda_e \Delta t}{\rho_e C_e 2\pi r_0 (r_1 - r_0)} \tag{A.12}$$

Combined Eq. (A.4), Eq. (A.8), and Eq. (A.10), with initial conditions and boundary conditions, the heat transfer models of the wellbore, cavity, and surrounding rock were established in a unified tridiagonal matrix, then realizing the coupled solution of the temperature field. The relational expression of matrix operation is as follows:

$$A \times B = C \tag{A.13}$$

where,

$$A = \begin{bmatrix} \alpha & -2U_{hf}/r_0 & & & & & & \\ -\eta & 1+\eta+\gamma & -\gamma & & & & & \\ & -\psi_2 & (\xi+2\psi+1)_2 & -(\xi+\psi)_2 & & & & \\ & & \ddots & \ddots & \ddots & & & \\ & & & -\psi_i & (\xi+2\psi+1)_i & -(\xi+\psi)_i & & \\ & & & & \ddots & \ddots & \ddots & \\ & & & & & -\psi_{N-2} & (\xi+2\psi+1)_{N-2} & -(\xi+\psi)_{N-2} \\ & & & & & & -\psi_{N-1} & (\xi+2\psi+1)_{N-1} \end{bmatrix} \tag{A.14}$$

$$B = [T_{f,j}^k \quad T_{e,1,j}^k \quad T_{e,2,j}^k \quad \cdots \quad T_{e,i,j}^k \quad \cdots \quad T_{e,N-2,j}^k \quad T_{e,N-1,j}^k]^T \tag{A.15}$$

$$C = [\beta \quad T_{e,1,j}^{k-1}+\omega_{1,j}^k \quad T_{e,2,j}^{k-1}+\omega_{2,j}^k \quad \cdots \quad T_{e,i,j}^{k-1}+\omega_{i,j}^k \quad \cdots \quad T_{e,N-2,j}^{k-1}+\omega_{N-1,j}^k \quad T_{e,N-1,j}^{k-1}+(\xi_{N-1}+\psi_{N-1})T_{e,N,j}^k+\omega_{N,j}^k]^T \tag{A.16}$$

References

[1] J. F. Carneiro, C. R. Matos, S. van Gessel, Opportunities for large-scale energy storage in geological formations in mainland Portugal, Renew. Sustain. Energy Rev. 99(2019)201–211.

[2] R. Tarkowski, Underground hydrogen storage: Characteristics and prospects, Renew. Sustain. Energy Rev. 105 (2019)86–94.

[3] C. Yang, T. Wang, Y. Li, H. Yang, J. Li, D. Qu, B. Xu, Y. Yang, J. J. K. Daemen, Feasibility analysis of using abandoned salt caverns for large-scale underground energy storage in China, Appl. Energy 137(2015) 467–481.

[4] C. R. Matos, J. F. Carneiro, P. P. Silva, Overview of Large-Scale Underground Energy Storage Technologies for Integration of Renewable Energies and Criteria for Reservoir Identification, J. Storage Mater. 21 (2019)

241-258.

[5] P. Bérest, J. Bergues, B. Brouard, Review of static and dynamic compressibility issues relating to deep underground salt caverns, Int. J. Rock Mech. Min. Sci. 36(8)(1999)1031-1049.

[6] P. Bérest, Heat transfer in salt caverns, Int. J. Rock Mech. Min. Sci. 120(2019)82-95.

[7] W. Liang, C. Yang, Y. Zhao, M. B. Dusseault, J. Liu, Experimental investigation of mechanical properties of bedded salt rock, Int. J. Rock Mech. Min. Sci. 44(3)(2007)400-411.

[8] N. Böttcher, U. -J. Görke, O. Kolditz, T. Nagel, Thermo-mechanical investigation of salt caverns for short-term hydrogen storage, Environ. Earth Sci. 76(3)(2017).

[9] W. Li, X. Nan, J. Chen, C. Yang, Investigation of thermal-mechanical effects on salt cavern during cycling loading, Energy 232(2021)120969.

[10] Y. Li, W. Liu, C. Yang, J. J. K. Daemen, Experimental investigation of mechanical behavior of bedded rock salt containing inclined interlayer, Int. J. Rock Mech. Min. Sci. 69(2014)39-49.

[11] D. Li, W. Liu, D. Jiang, J. Chen, J. Fan, W. Qiao, Quantitative investigation on the stability of salt cavity gas storage with multiple interlayers above the cavity roof, J. Storage Mater. 44(2021)103298.

[12] P. Sicsic, P. Bérest, Thermal cracking following a blowout in a gas-storage cavern, Int. J. Rock Mech. Min. Sci. 71(2014)320-329.

[13] X. Shen, C. Arson, J. Ding, F. M. Chester, J. S. Chester, Mechanisms of Anisotropy in Salt Rock Upon Microcrack Propagation, Rock Mech. Rock Eng. 53(7)(2020)3185-3205.

[14] W. Li, C. Zhu, J. Han, C. Yang, Thermodynamic response of gas injection-and-withdrawal process in salt cavern for underground gas storage, Appl. Therm. Eng. 163(2019)114380.

[15] T. Meng, P. Jianliang, G. Feng, Y. Hu, Z. Zhang, D. Zhang, Permeability and porosity in damaged salt interlayers under coupled THMC conditions, J. Pet. Sci. Eng. 211(2022)110218.

[16] S. Na, W. Sun, Computational thermomechanics of crystalline rock, Part I: A combined multi-phase-field/crystal plasticity approach for single crystal simulations, Comput. Methods Appl. Mech. Eng. 338 (2018) 657-691.

[17] K. Serbin, J. Slizowski, K. Urbańczyk, S. Nagy, The influence of thermodynamic effects on gas storage cavern convergence, Int. J. Rock Mech. Min. Sci. 79(2015)166-171.

[18] T. Wang, L. Ao, B. Wang, S. Ding, K. Wang, F. Yao, J. J. K. Daemen, Tightness of an underground energy storage salt cavern with adverse geological conditions, Energy 238(2022)121906.

[19] L. Wei, C. Jie, J. Deyi, S. Xilin, L. i. Yinping, J. J. K. Daemen, Y. Chunhe, Tightness and suitability evaluation of abandoned salt caverns served as hydrocarbon energies storage under adverse geological conditions (AGC), Appl. Energy 178(2016)703-720.

[20] H. Mansouri, R. Ajalloeian, Mechanical behavior of salt rock under uniaxial compression and creep tests, Int. J. Rock Mech. Min. Sci. 110(2018)19-27.

[21] K. Fuenkajorn, D. Phueakphum, Effects of cyclic loading on mechanical properties of Maha Sarakham salt, Eng. Geol. 112(1-4)(2010)43-52.

[22] M. Tao, Y. Yechao, C. Jie, H. Yaoqing, Investigation on the Permeability Evolution of Gypsum Interlayer Under High Temperature and Triaxial Pressure, Rock Mech. Rock Eng. 50(8)(2017)2059-2069.

[23] M. Firouzi, K. Alnoaimi, A. Kovscek, J. Wilcox, Klinkenberg effect on predicting and measuring helium permeability in gas shales, Int. J. Coal Geol. 123(2014)62-68.

[24] G. Grazzini, A. Milazzo, Thermodynamic analysis of CAES/TES systems for renewable energy plants, Renew. Energy 33(9)(2008)1998-2006.

[25] N. Hartmann, O. Vöhringer, C. Kruck, L. Eltrop, Simulation and analysis of different adiabatic Compressed Air Energy Storage plant configurations, Appl. Energy 93(2012)541-548.

[26] P. Bérest, B. Brouard, H. Djakeun-Djizanne, G. H vin, Thermomechanical effects of a rapid depressurization in a gas cavern, Acta Geotech. 9(1)(2013)181-186.

[27] Y. Han, H. Cui, H. Ma, J. Chen, N. Liu, Temperature and pressure variations in salt compressed air energy storage (CAES) caverns considering the air flow in the underground wellbore, J. Storage Mater. 52 (2022)104846.

[28] M. AbuAisha, A. Rouabhi, On the validity of the uniform thermodynamic state approach for underground caverns during fast and slow cycling, Int. J. Heat Mass Transf. 142(2019)118424.

[29] S. Merey, Prediction of pressure and temperature changes in the salt caverns of Tuz Golu underground natural gas storage site while withdrawing or injecting natural gas by numerical simulations, Arab. J. Geosci. 12(6) (2019).

[30] V. Leontidis, P. H. Niknam, I. Durgut, L. Talluri, G. Manfrida, D. Fiaschi, S. Akin, M. Gainville, Modelling reinjection of two-phase non-condensable gases and water in geothermal wells, Appl. Therm. Eng. 223 (2023)120018.

[31] P. H. Niknam, L. Talluri, D. Fiaschi, G. Manfrida, Sensitivity analysis and dynamic modelling of the reinjection process in a binary cycle geothermal power plant of Larderello area, Energy 214(2021)118869.

[32] P. Barajas, F. Civan, Effective Modeling and Analysis of Salt - Cavern Natural - Gas Storage, SPE Prod. Oper. 29(01)(2014)51-60.

[33] Y. Liao, X. Sun, B. Sun, J. Xu, D. a. Huang, Z. Wang, Coupled thermal model for geothermal exploitation via recycling of supercritical CO_2 in a fracture-wells system, Appl. Therm. Eng. 159(2019)113890.

[34] Y. Liao, J. Zheng, Z. Wang, B. Sun, X. Sun, P. Linga, Modeling and characterizing the thermal and kinetic behavior of methane hydrate dissociation in sandy porous media, Appl. Energy 312(2022)118804.

[35] Z. Ziabakhsh-Ganji, H. Kooi, Sensitivity of Joule-Thomson cooling to impure CO_2 injection in depleted gas reservoirs, Appl. Energy 113(2014)434-451.

[36] L. Pan, C. M. Oldenburg, T2Well—An integrated wellbore-reservoir simulator, Comput. Geosci. 65(2014) 46-55.

[37] Z. Wang, B. Sun, J. Wang, L. Hou, Experimental study on the friction coefficient of supercritical carbon dioxide in pipes, Int. J. Greenhouse Gas Control 25(2014)151-161.

[38] B. Sun, X. Sun, Z. Wang, Y. Chen, Effects of phase transition on gas kick migration in deepwater horizontal drilling, J. Nat. Gas Sci. Eng. 46(2017)710-729.

[39] B. Liao, J. Wang, J. Sun, K. lv, L. Liu, Q. Wang, R. Wang, X. Lv, Y. Wang, Z. Chen, Microscopic insights into synergism effect of different hydrate inhibitors on methane hydrate formation: Experiments and molecular dynamics simulations, Fuel 340(2023)127488.

[40] W. Fu, J. Yu, Y. Xiao, C. Wang, B. Huang, B. Sun, A pressure drop prediction model for hydrate slurry based on energy dissipation under turbulent flow condition, Fuel 311(2022)122188.

[41] X. Sun, Y. Liao, Z. Wang, B. Sun, Geothermal exploitation by circulating supercritical CO_2 in a closed horizontal wellbore, Fuel 254(2019)115566.

[42] A. Hasan, C. Kabir, Wellbore heat-transfer modeling and applications, J. Pet. Sci. Eng. 86(2012)127–136.

[43] N. Crossley, Salt cavern integrity evaluation using downhole probes. A transgas perspective, SMRI Fall Meeting. (1996)21–54.

[44] P. Bérest, B. Brouard, Safety of Salt Caverns Used for Underground Storage Blow Out; Mechanical Instability; Seepage; Cavern Abandonment, Oil Gas Sci. Technol. 58(3)(2006)361–384.

[45] W. Li, C. Zhu, C. Yang, K. Duan, W. Hu, Experimental and DEM investigations of temperature effect on pure and interbedded rock salt, J. Nat. Gas Sci. Eng. 56(2018)29–41.

本论文原发表于《Applied Thermal Engineering》2023年第232期。

老腔改建

采盐井腔改建储气库和声呐测量技术的应用

屈丹安[1,2]　杨海军[1]　徐宝财[1]

(1. 中国石油西气东输管道公司储气库项目部；
2. 重庆大学西南资源开发及环境灾害控制工程教育部重点实验室)

【摘　要】 为确保西气东输管道的季节调峰和安全稳定运行，需要建设大型地下储气库，利用江苏金坛盐盆内的采盐井腔作为储气库是一条快速经济的捷径。文章论述了利用声呐测量技术对采盐井腔进行腔体形态测量、初选、稳定性评价，制订井筒修复改建的技术方案，确定注气排卤参数和现场施工技术方案的过程。

【关键词】 天然气；储气库；采盐井腔；声呐测量技术

Converting salt cavern into underground gas storage and applying sonar for survey

Qu Dan'an[1,2]　Yang Haijun[1]　Xu Baocai[1]

(1. Gas Stroage Project Department, PetroChina West-East Gas Pipeline Company;
2. Key Laboratory for the Exploitation of Southwestern Resources and the Environmental Disaster Control Engineering, Ministry of Education, Chongqing University)

Abstract　In order to ensure seasonal gas supply peak regulation and safe operation of West-East Gas Pipeline, large scale underground gas storages are needed. Utilizing the salt cavern in Jintan salt basin as the gas storage is a quick and economical way. The sonar technique is used for survey, preliminary selection and stability assessment of the cavern configuration. The technical scheme of brine well rehabilitation and recon struction is worked out, and the parameters of natural gas injection and bittern ejection as well as the technical scheme of field construction are determined.

Keywords　gas storage; salt cavern; sonar survey technique; reconstruction; natural gas

西气东输管道横贯中国东西，管道西起新疆塔里木盆地，东至上海白鹤镇，干线全长4000km，途经9个省(区)市，穿越沙漠、戈壁、黄土高原、山脉和江河等复杂地形地貌，是中国第一条大口径、长距离、高压力和采用先进钢材的现代化天然气管道，设计年输气量$120×10^8m^3$，为中国东部长江三角洲地区约8500万户居民提供生活用气。

作者简介：屈丹安(1965—)，男，陕西洛川人，高级工程师，中国地质大学(北京)石油与天然气工程专业硕士，重庆大学采矿专业在读博士，现从事储气库相关工作。

根据国外数十年长输管道的成功运行经验，为确保西气东输管道的季节调峰和安全稳定运行，需要在长江三角洲的天然气主要消费地区建设容量占年输气量15%的大型地下储气库。根据三维地震解释资料和前期地质综合评价，项目可行性研究确定了江苏省金坛盐盆为西气东输首批建库地址，储气库设计总容量为 $26.38\times10^8m^3$，有效工作气量为 $17.14\times10^8m^3$，日注气规模为 $900\times10^4m^3$，日采气规模为 $1500\times10^4m^3$。

在我国建设盐穴地下储气库是首次，采用水溶法采矿工艺，受单井注水量和卤水消化能力的制约，新的盐穴腔体的建造一般需要数年时间。2003年10月1日，随着西气东输管道东段的投产运行，急需一定的应急保安气量以保证管道的安全运行，利用金坛盐盆内的采盐井腔作为储气库是一条快速经济的捷径。

在金坛盐穴地下储气库可行性研究期间初选的建库区域内有一批小盐矿，在10多年的采卤生产过程中，形成了40多个采盐井腔，为尽快获得第一批盐穴储气腔体，向管道提供应急保安用气，中国石油西气东输管道公司从2003年开始进行了采盐井腔改建储气库的探索。

1 采盐井腔的选择

1.1 采盐井腔的开采历史和分布

金坛盐盆位于江苏省金坛市西北30km处，含盐面积约 $60.5km^2$，矿石储量为 162.42×10^8t，埋藏深度为900~1150m，平均氯化钠含量达85%，是我国东部地区综合指标最佳的大型盐矿，地下盐矿开发于20世纪80年代末，采用水溶法采矿工艺，生产的卤水主要作为真空制盐、烧碱和聚氯乙烯等盐化工产业的原料。

金坛盐盆地处经济发达的长江三角洲，东临上海，西靠南京，地理位置十分优越，区内有丹金漕河连接长江和太湖，水运交通十分便利，依托丰富的岩盐资源，生产的卤水可通过船运辐射宁(南京)、沪(上海)、杭(杭州)等盐化工产业发达地区，盐矿开发具有明显的独特优势。

在西气东输金坛盐穴地下储气库可行性研究期间，金坛盐盆内共有5个小盐矿，分属于两个行政县市管辖，年生产和销售卤水折合盐 165×10^4t 左右，共形成43个采盐井腔，其中金坛市盐业公司有26口，主要分布于金坛市茅兴盐矿、西阳盐矿、岗龙盐矿和新金冠盐矿，采盐井开采历史最长的15年左右，根据采盐量估算井下腔体的体积约 $21.79\times10^4m^3$；镇江市丹徒县荣炳盐矿有17口，采盐井开采历史最长的12年左右，估算井下腔体的体积约 $18.79\times10^4m^3$。

金坛盐盆内采盐井腔的分布如图1所示，其中A区21口、B区3口、C区4口、D区15口。

1.2 采盐井腔的预选

盐矿在10多年的采卤生产过程中，从未进行过盐穴腔体的形态测量。因此，根据盐穴储气库对单个腔体储气量和稳定性的基本要求，在进行声呐测量前，对40多口采盐井腔进行了预选，预选条件为：

(1) 根据采盐量计算每一口采盐井腔的腔体体积，要求腔体体积在 $10\times10^4m^3$ 以上；

（2）根据盐矿生产情况进行判断，要求各采盐井腔不能连通；
（3）要求各采盐井腔之间有一定的距离；
（4）要求各采盐井腔的地质、钻井、测井资料、生产日报等比较齐全。

根据上述条件，预选出15个采盐井腔，分别为：A区的A1井、A2井、A3井、A4井、A5井；B区的B1井、B2井；D区的D1井、D2井、D3井、D4井、D5井、D6井、D7井、D8井。

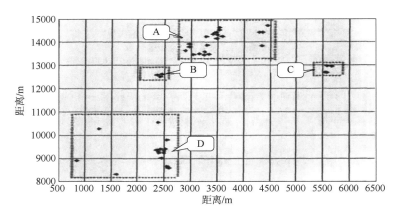

图1　金坛盐盆采盐井腔分布

1.3 首次声呐测量

1.3.1 盐穴声呐测量技术

盐穴声呐测量技术的工作原理为：沿采盐井井筒下放声呐测量井下仪器，井下仪器的声呐探头进入盐穴腔体后，在某一深度进行360°水平旋转，同时按设定的角度间隔向盐穴腔体壁发射声脉冲，检测回波信号，信号经井下仪器的连接电缆传回地面中心处理机，得到某一深度上的腔体水平剖面图；在盐穴腔体内不断改变检测深度，则可获得腔体不同深度上的水平剖面图；对于盐穴腔体的顶部、底部和异常部分，使用倾斜测量功能，可得到不同倾斜角度下的测量距离。两种原始检测数据经中心处理机软件处理后，最终可得到整个腔体的体积和三维形态图像，图2为声呐测量施工工艺图。

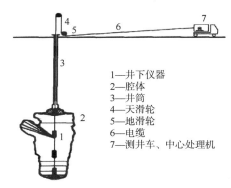

1—井下仪器　2—腔体　3—井筒　4—天滑轮　5—地滑轮　6—电缆　7—测井车、中心处理机

图2　声呐测量施工工艺示意

1.3.2 声呐测量技术与施工队伍的选择

盐穴地下储气库建设是一个较为特殊的领域，目前国际上只有德国、美国等国家的少数几家公司拥有此项技术。经过国内外调研比选，根据金坛盐矿采盐井腔的实际情况，确定了使用德国SOCON公司制造的声呐测量仪器和测量技术，其仪器组成为：陀螺稳定系统、方位仪（罗盘）、声速测量系统、超声波探头倾斜机构、超声波探头旋转驱动机构、压力平衡装置、S-CCL、超声波探头、压力测量、温度测量、计算机及系统软件等，德国SOCON公司的声呐测量仪器技术参数见表1。

表1 德国 SOCON 公司声呐测量仪器参数

设备参量	指标	设备参量	指标
使用范围	液体、天然气介质	深度定位精度/m	≤0.1
电源	100~240V，50Hz，5000W	罗盘	分辨率≤1°，精度≤+/-1.5°
探头水平旋转/(°)	360	工作频率	70kHz~1.8MHz
探头垂直倾斜/(°)	0~90	井下仪器耐温/℃	10~75
盐水介质测量距离	≤250m；隔一层套管≤70m；隔两层套管≤50m	井下仪器耐压/MPa	30
盐水介质体积测量精度	≤1%；隔一层套管≤4%；隔两层套管≤8%	井下仪器外径/mm	$D70$
天然气介质测量距离	≤80m（≥15MPa）	长度/mm	4000
天然气介质体积测量精度	≤1%（≥15MPa）；≤8%（≤10MPa）	质量/kg	75

1.3.3 声呐测量前的准备工作

根据腔体稳定性评价、井筒修复改造工艺等要求，设计的测井、声呐测量参数和软件计算结果有：井眼轨迹、套管接箍、腔体半径和直径、腔体温度和压力分布、腔体水平和垂直剖面图、腔体三维图、腔体体积、腔群投影图。

在研究分析了15个采盐井腔的地质、钻井、测井和采卤工艺等资料后，发现各采盐井的直径为139.7mm（5½in）的生产套管均下入盐层底部较深位置，由于盐矿各采盐井开采时间较长，其中最短的采卤时间为8年，最长的采卤时间有15年，估计直径139.7mm的生产套管腐蚀和结垢等情况比较严重，将会影响到声呐测量结果的精度，因此，声呐测量施工前做好了在直径139.7mm的生产套管内进行腔体形态测量的准备工作。

1.3.4 声呐测量现场施工

（1）井下作业。使用修井机进行井筒的清理工作，有13口井通井施工顺利到达设计深度；另外D1井由于在采盐过程中下入了永久分隔器，通井过程中使用了螺杆钻，最后也顺利到达设计深度；A1井由于采盐过程中出现井下事故，造腔内管遇卡，不得不放弃处理。最后，可用于声呐测量的采盐井腔为14口。

（2）测井。为做好14口采盐井腔的解释工作，在进行声呐测量前，利用国内测井仪器进行了井斜、方位、磁定位和伽马测井。

（3）声呐测量。根据设计的声呐测量程序，完成了13口采盐井腔的测量，得到了设计所需要的声呐测量参数和软件计算结果。D4井由于套管腐蚀严重，只得到了部分腔体的声呐测量数据。下面是14口采盐井腔的三维图，如图3所示。

(a) A区4口

(b) B区2口

(c) D区8口

图3 14口采盐井腔的三维图

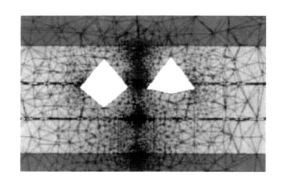

图4 FLAC3D网格剖分纵剖面示意

1.4 利用声呐测量数据对采盐井腔进行选择

通过对地质、钻井、测井资料的综合分析，利用声呐测量数据及相关地质资料，采用FLAC3D软件建立了三维数值计算模型，对14口采盐井腔进行了力学稳定性计算、分析和评价工作，最后选定A区4口、B区2口采盐井腔作为储气腔体的改造目标，并确定了每口井腔的注采气压力区间。图4为B区2口采盐井腔的FLAC3D网格剖分纵剖面示意。

1.5 利用声呐测量数据确定采盐井腔的改造技术方案

盐矿钻井的目的是溶盐采卤，使用直径139.7mm的非气密性生产套管，固井质量差，尺寸较小，加上套管长期浸泡在卤水中，腐蚀较为严重，不能满足储气腔体生产运行的需要，因此，需要对原有井身结构进行改造。经过多种方案论证，确定了全井筒套铣、扩眼后下入气密封套管的技术方案，在此技术方案中，利用声呐测量数据确定了井筒套铣深度、气密封套管的下入深度。

2 二次声呐测量

2.1 二次声呐测量前的井腔情况

在6口采盐井腔的井筒修复改造过程中，使用各种套铣工具进行了长井段套铣修井作业，连续套铣作业井段最长达959.5m，并进行了无井底岩盐固井、气密封试压等工作，其中 1 口井腔发现渗漏，二次声呐测量前有 5 口井腔具备注气排卤条件。

2.2 二次声呐测量

由于第一次测量是通过采盐井的套管进行的，套管本体和套管腐蚀、结垢严重，影响到声呐测量精度。在 6 口采盐井腔改造完成后，为掌握注气排卤施工作业中注气量、排卤量和气水界面在腔体内的位置，保证施工作业安全，在腔体内卤水介质中进行了第二次声呐测量，获得了精确的腔体形状和注气排卤施工设计所需要的声呐测量参数。图 5 是 6 口已改造采盐井腔的三维图（后面 2 口为观察井腔）。

图 5　完成改造后 6 口井腔三维图（后面 2 口为观察井腔）

2.3 利用声呐测量数据确定注气排卤施工技术方案

在对 5 口井腔的地质、钻井、测井和地质力学稳定性评价等资料分析的基础上，利用声呐测量的腔体温度和压力分布曲线、腔体测量体积等参数，计算出每个腔体的天然气注入量；利用声呐测量的腔体顶板和底板深度、垂直剖面图等参数，确定每个腔体注气排卤管柱的下入深度；利用声呐测量的腔体半径和直径、水平和垂直剖面图、腔体体积分布曲线等参数，根据注气排卤阶段的天然气注入速度和注入量，计算排出卤水的体积，掌握气水界面在腔体中的位置，保证注气排卤的安全施工。

在整个注气排卤施工过程中，利用声呐测量所提供的数据顺利完成了 5 口井腔的现场施工，注气量达到了设计要求，5 个盐穴腔体可为管道运行提供近 $5000 \times 10^4 m^3$ 应急工作气量。目前，建成的西注采气站已进行了多次安全注采气生产运行，为西气东输管道运行提供了调峰和应急保安用气。

3　结束语

目前在国际上，声呐测量技术是盐穴地下储气库腔体形态测量的唯一方法。

在采盐井腔改建储气库的过程中，声呐测量技术应用于现场施工的各个阶段，尤其在采盐井腔的选择、力学稳定性评价、井筒修复改造技术方案和注气排卤施工工艺等方面起到了非常重要的作用。

参　考　文　献

[1] 王希勇，罗雨香，龙刚，等．中国天然气供应安全战略研究[J]．中国能源，2006，28(2)：23-25.

[2] 丁国生,谢萍. 中国地下储气库现状与发展展望[J]. 天然气工业, 2006, 26(6): 111-113.
[3] James E B, Andreas R. Execution and results of a sonar survey in a horizontal leached cavern[A]. Technical Class of the Spring 2001Meeting[C]. Orlando, Florida, U.S.A., April 22-25, 2001.
[4] CSA Z341, Storage of Hydrocarbons in Underground Formations[S].
[5] Hartmut von Tryller, Andreas Reitze. Influence of the physical conditions in a cavern on the execution and results of sonar surveys [A]. Spring 2005 Conference[C]. Syracuse, New York, USA. 17-20 April, 2005.

本论文原发表于《石油工程建设》2009年第35卷第6期。

利用现有采卤溶腔改建地下储气库技术

丁建林

（中国石油西气东输管道公司）

【摘　要】 为确保西气东输的安全平稳供气，需尽快建设与管道工程相配套的储气库。与采用钻新井、溶新腔的建设方式相比，利用现有采卤溶腔改建地下储气库，可以大幅度缩短建设时间，节约建设投资。本文介绍了利用现有采卤溶腔改建地下储气库的技术流程和主要配套技术，详细阐述了全井长井段套铣、腔体气密封检测等工艺技术难点的解决方案，为我国后续盐穴储气库的建设提供了技术储备。

【关键词】 采卤溶腔；改建；地下储气库；技术

Reconstruction of existing salt-caverns into underground gas storage

Ding Jianlin

(PetroChina West-East Gas Pipeline Company)

Abstract The author considers that with the commissioning of the West to East Gas Pipeline Project, to insure of fering gas to consumers in a continuous and safe way, the gas storage must be constructed, and existing salt-caverns reconstructed into gas storage will speed up the development of underground gas storage and reduce construction investment compared with the construction of new wells and caverns. The process during the reconstruction is expounded thoroughly, and several crucial technologies are resolved and applied in practice, such as long casing milling, cavern tightness test, which can be taken as an important solution for other salt-caverns gas storage in construction. The technology provides a technical bank for our future construction of salt-caverns gas storages.

Keywords salt-caverns; reconstruction; underground gas storage; technology

随着西气东输管道工程的全线贯通，为确保向沿线用户安全平稳供气，需要建设与西气东输管道工程相配套的地下储气库。建设的主要目的，一是解决下游用户因用气不均衡所引起的季节调峰问题，部分满足用户的日调峰需要；二是解决由于长输管道意外故障不能正常供气时的应急供气需求问题，确保西气东输长输管道向下游用户安全平稳供气。

作者简介：丁建林，高级工程师，1964年生，1986年毕业于中国石油大学（华东）油气储运专业，1999年硕士毕业于清华大学工商管理专业，2007年博士毕业于中国石油大学油气储运工程专业，现任西气东输管道公司副总经理。

金坛盐穴地下储气库是国家重点工程西气东输管道工程的重要配套工程,是国内首次建设的盐穴型地下储气库。储气库构造位置位于江苏省金坛盆地直溪桥凹陷,根据三维地震资料和钻井资料,凹陷中古近系阜宁组四段沉积了厚度为60~250m的盐岩地层,含盐面积60.5km²,盐岩层埋深800~1200m。

盐穴地下储气库的建设有两种技术方式,一种是钻探新井,采用水溶法溶解盐岩层,在地下溶成形状较稳定的类似梨形的人工溶腔腔体,该种建库技术方式在西方发达国家是一项成熟的技术;另一种是利用现有采卤井溶成的地下溶腔腔体,经过严格筛选和稳定性评价后,按储气库的技术要求改造井身结构来储存天然气,此种建库技术方式在国际上尚无先例。采用第一种技术方式建设储气库,耗时最多的是溶腔阶段,据初步测算,用合理的井身结构、采卤管柱,配合相应的技术措施溶腔,建成设计所需的腔体容积,使其具有相当规模的储气能力,需要3~5年的时间;而采用第二种技术方式改建地下储气库,可大幅度缩短建设时间,能尽快满足西气东输长输管道意外故障的应急和部分调峰供气需求。

1 储气库改建的技术流程

中国石油西气东输管道公司在国内外没有成熟的技术经验可以借鉴的条件下,利用现有采卤溶腔改建地下储气库,结合金坛盐穴储气库的地质特点和实际情况,对该储气库进行了改造建设,建设的技术流程如图1所示。

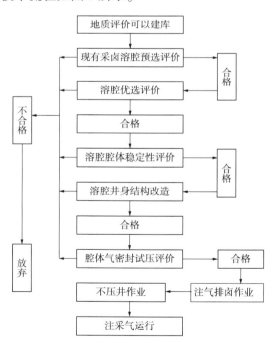

图1 采卤溶腔改建储气库技术流程

改建的工艺技术难点主要有五个方面,全井长井段套铣;特殊岩性盐膏岩地层与大井眼、大直径套管固井;腔体、生产套管鞋及生产管套整体气密封试压;腔体注气排卤过程中气水界面的准确判断；不压井起出大直径 114.3mm 排卤管柱作业。

2 储气库改建的主要配套技术

2.1 建库区地质评价

建库区内地质条件是储气库建设的基础。地质评价以测井、钻井资料以及三维地震构造解释成果为基础，采用室内和现场试验相结合的方式，对盐岩层及泥岩夹层的平面展布与纵向分布特征、断层平面及纵向延伸、上覆盖层物性及密封性、盐岩层及泥岩夹层的物性、盐岩层 NaCl 和不溶物含量、地层破裂压力、上覆盖层及泥岩夹层突破压力等做出了科学评价。评价结论认为，金坛盆地直溪桥凹陷陈家庄地区的盐层厚度大于 150m，岩层面积为 11.26km^2，是建设盐穴储气库最有利的区域。现有采卤溶腔的地质入选原则，一是远离断层，尤其要远离断穿盐岩层顶和底的大断层；二是腔体盐岩顶板层和底板层要有一定的厚度，以确保溶腔腔体的密封性。

2.2 溶腔预选评价

在金坛盐穴地下储气库库区附近共有采卤井腔 43 个，井距一般在 80~150m 之间。按图 2 的预选评价流程对 43 个采卤溶腔腔体进行筛选，共预选出独立腔体 14 个，单腔采卤容积在 $13×10^4$ ~ $33×10^4 m^3$ 之间，最大单腔容积约 $33×10^4 m^3$，最小单腔容积约 $12×10^4 m^3$，腔体总容积达到 $263.18×10^4 m^3$。

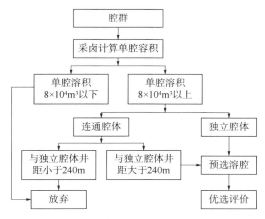

图 2 溶腔预选评价流程

2.3 溶腔优选评价

按照图 3 给出的优选评价流程，对通过预选的 14 个独立腔体进行优选，优选出了顶部盐层厚度大于 15m，底部盐层厚度大于 5m，溶腔形状类似于稳定梨形的 6 个溶腔腔体。6 个腔体单腔容积在 $15×10^4 m^3$ 左右，腔体总容积近 $98.52×10^4 m^3$，储气库容量达到 $14114.25×10^4 m^3$，调峰工作气量达到 $7826.35×10^4 m^3$，能够满足金坛储气库初期供气的基本需求。

2.4 溶腔腔体稳定性评价

蠕变是盐岩典型的力学特性之一。盐岩蠕变本构简化方程为：

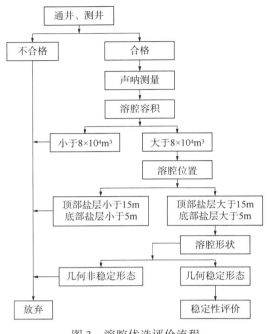

图 3 溶腔优选评价流程

$$\varepsilon = \beta(\sigma_1-\sigma_2)^n t + A[1-\exp(-\alpha t)]$$

式中：ε 为蠕变速率；β、n、A、α 为材料常数；$\sigma_1-\sigma_2$ 为应力差；t 为时间。

利用蠕变本构方程和拉格朗日分析程序 FLAC3D 的流变计算原理，对优选出的 6 个腔体进行静力及蠕变分析和稳定性评价。经过评价，确定了 6 个溶腔腔体的最高注气压力、最低采气压力、采气压降速率、溶腔腔顶距生产套管鞋的安全距离及 20 年腔体周边最大流变位移等。评价结果认为，在上述技术条件下，6 个溶腔腔体均可用作储气腔体。

2.5 溶腔井身结构改造

6 个通过稳定性评价的采卤溶腔的井身结构不能满足储气库对安全密封和注采气量的要求，主要存在的问题有，生产套管直径小，注采天然气的吞吐能力不能满足应急供气的需求；生产套管质量差，钢级低，壁薄，套管连接螺纹及螺纹密封脂不符合注采气井要求；生产套管水泥返高未全部返至地面，固井质量不能达到注采井要求；采卤井生产时间长，套管变形、腐蚀严重。

因此，遵循"弃井用腔"的原则，对 6 个采卤溶腔的井身结构进行了改造，即全井段套铣、扩眼、下入 244.5mm 生产套管及固井等。改造前的井身结构为，244.5mm 的表层套管下入深度为 35～50m，水泥返至地面；139.7mm 的生产套管下入深度为 1000～1050m，水泥未返至地面，固井质量差。改造后井身结构变为：339.7mm 的表层套管下入深度为 45～55m，水泥返至地面；244.5mm 的生产套管下入深度为 930～958m，水泥返至地面，固井质量合格。

在 6 个采卤溶腔井身结构的改造过程中，形成了两项改造工艺技术。

（1）全井长井段套铣技术。

在全井长井段套铣施工过程中，发现的技术难点有：原井套管、扶正器不规范，腐蚀严重；套铣过程蹩跳严重；铣管内环空易蹩泵、卡钻；易发生铣管断落事故；因盐腔具有蠕变特性，套铣最后阶段一旦连通盐腔，将发生井漏和井涌等复杂情况。为确保井身质量和固井质量达到储气对密封性的安全要求，需要预防井径扩大，保持盐层段井径规则。

针对全井套铣的技术难点，采取的主要技术措施有，在原生产套管中下入可回收式桥塞，保证套铣作业全过程井筒与腔体隔绝；根据地层岩性的不同和固井的质量程度选择合适的铣鞋，加快套铣进度，保证施工安全；采用合理的套铣参数，尽可能控制蹩钻；采取少套铣、勤切割的方法，根据进尺及泵压的变化情况决定起钻时机，尽可能避免复杂情况的发生，确保施工安全；套铣过程中，要平稳操作，及时切割，适时调整钻井液性能和排量，防止水泥环脱落卡死铣管；加大铣鞋内出刃，每套铣 4～5m 上下反复划眼两次，使套管外水泥环尽可能破碎,为外割刀切割创造条件;配备水力锻铣工具,特殊作业时代替内割刀,对套管实施切割,保证施工顺利进行;套铣最后阶段,在盐腔顶部以上预留 2~3m 的套管,不套通腔体,待下套管固井后,再套通腔体;采用饱和盐水钻井液保护盐层,防止盐层溶解造成井径扩大。通过采取上述技术措施,施工周期由第一口井的 108 天缩短到 54 天,全井套铣最长井段达到 959.5m,井身质量合格。

（2）特殊岩性盐膏岩地层与大井眼、大直径套管固井技术。

固井施工中存在的技术难点有,盐岩主要成分除氯化钠以外,还含有 Ca^{2+}、Mg^{2+}、SO_4^{2-} 等离子,其盖层含钙芒硝泥岩及含膏泥岩同样含有 Ca^{2+}、Mg^{2+}、SO_4^{2-} 等离子,这些离子会对固井水泥浆性能产生复杂的影响;井的深度浅,井底温度低;国内适用于浅井盐水水泥浆的外加剂少,配方筛选困难;盐水水泥浆密度高,起泡性强,密度不易控制,施工难度大;上覆玄武岩地层气孔发育、孔隙度高、易漏失;大井眼、大套管、井眼不规则,井壁稳定性差;储气库运行压力周期性变化,水泥石长期承受交变应力,对水泥环长期胶结性能及密封性能要求高。

为确保固井质量,针对上述技术难点,开发了 JSS 低温抗盐水泥浆体系,优化了固井施工工艺,逐步提高了改造井的固井质量。声幅测井资料显示,固井质量全部合格,部分井的固井质量达到优质水平。

2.6 溶腔气密封试压评价

改造后溶腔气密封性的优劣直接影响到储气库的安全可靠运行,因此,储气前必须对溶腔进行气密封试压评价。评价的主要目的是检测生产套管和腔体的技术状况,特别是检测生产套管鞋附近的固井质量是否满足高压储气对密封性的安全要求。

在充分吸收 API 与 Geostock 评价方法和标准优点的基础上,制定了适合于金坛盐穴地下储气库实际情况和特点的腔体气密封性评价方法和评价标准。腔体气密封性的检测方法如图 4 所示。向井腔下入单套试压管柱后,向环空中注入空气或氮气,使气水界面深度达到生产套管鞋以下 5~10m 的位置,在测试时间 24h 内利用气水界面测量仪检测气水界面的深度变化,根据气水界面深度的变化量来判定腔体的密封性是否合格。判定腔体密封性是否合格,需同时满足两个条件,即泄漏率随时间的变化逐渐减小,并最终达到一个稳定的水平;在测试时间 24h 内,气水界面的深度变化小于 1m。如果检测结果不满足第一个条件,则直接判定腔体的密封性不合格;如果检测结果满足第一个条件,但气水界面的深度变化大于 1m,可根据现场具体情况,通过延长测试时间来确定腔体的密封性是否合格。

利用上述评价方法和评价标准,对改造后的 6 个井腔进行了气密封试压评价,其中 5 个井腔的气密封性合格,可以作为储气腔体。

2.7 注气排卤作业

根据气密封试压评价结果,对 5 个气密封试压合格的井腔进行了注气排卤完井作业。其管柱组合自下而上为：177.8mm 的注采气管柱的下入深度为 925~1045m,永久性封隔器的下入深度为 900~1000m.安全阀的下入深度为 50~200m;114.3mm 的排卤管柱管鞋的下入深度为 1050~1100m,坐落接头的下入深度为 1040~1090m,如图 5 所示。

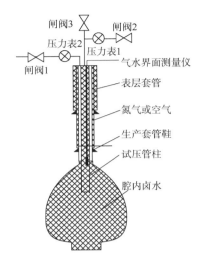 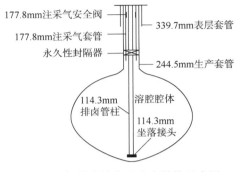

图 4　腔体气密封性检测方法　　　图 5　注气排卤管柱及井身结构示意图

在对 5 个井腔注气排卤作业期间，采用了计算与测井相结合的方法，对腔体内的气水界面进行了分析与判断，并逐渐摸索出气水界面随排卤量变化的规律。

2.8　不压井作业

由于注气排卤井腔具有较高的井口压力和独特的腔体、井身和管柱结构，其注气排卤结束后，需采用不压井作业装置将 114.3mm 的排卤管柱一次性起出，为后续的 177.8mm 管柱注采气运行和井下安全控制提供畅通通道。

不压井作业是在带压环境中采用不压井作业设备起出管柱的一种作业方法。金坛盐穴地下储气库采用的不压井作业设备是一种实用的、具有液压举升力和下推力的钻机辅助设备。其工作原理是靠修井机、不压井作业辅助机和桥塞(堵塞器)的相互配合来实现带压环境下起出管柱作业。不压井作业施工步骤为：拆井口→安装井口防喷器组→安装不压井作业设备→井口防喷器及不压井作业设备试压→起出 114.3mm 油管挂→起出 114.3mm 排卤管柱→拆除设备→装井口。共对 5 口井开展了不压井作业，井内 114.3mm 排卤管柱全部一次性起出。

2.9　注采气运行

目前，5 口井腔已累计注气四次，采气三次，总注气量为 $1.0441×10^8 m^3$，工作气量为 $0.5743×10^8 m^3$，已初步制定了较为合理的注采气工作制度。经关井期间长时间观察，244.5mm 生产套管与 177.8mm 注采气管柱环空压力及 177.8mm 注采气管柱压力稳定，注采气运行安全、平稳。

3　结束语

（1）利用现有采卤溶腔改建储气库，缩短了建设时间，节约了建设投资，为西气东输管道的安全平稳运行提供了保障。

（2）全井长井段套铣修井作业，是国内修井作业的难题。6 口采卤溶腔套铣改造作业，最长井段达959.5m，在国内尚属首次。经过不断总结、摸索、改进与创新，此项工艺技术已趋于成熟，大大缩短了改造作业周期。

（3）特殊岩性盐膏岩地层和大井眼、大直径套管固井，是国内钻井界的难题之一。经过对水泥浆体系和施工工艺的不断优化和完善，满足了储气库对密封性的安全要求。

（4）溶腔气密封试压评价技术综合了国外两种检测方法的优点，并具有方便、快捷、精度高、成本低和实用性强的特点。该项技术填补了国内空白，为后续盐穴储气库的大规模建设提供了技术储备。

（5）注气排卤作业和不压井起出大直径114.3mm排卤管柱作业风险高，难度大，在国内尚无先例。此项工艺技术的应用成功，为后续盐穴储气库的大规模建设积累了宝贵的经验。

本论文原发表于《油气储运》2008年第27卷第12期。

复杂对流井连通老腔改建储气库技术

刘继芹[1]　乔　欣[2]　李建君[1]　王文权[1]　周冬林[1]　肖恩山[1]　井　岗[1]　敖海兵[1]

（1. 中国石油管道有限责任公司西气东输分公司储气库项目部；
2. 中国石油北京油气调控中心）

【摘　要】 盐化公司拥有丰富的对流井老腔资源，如果能够有效利用此类腔体，不但可以缩短储气库建库时间，而且可以大幅降低工程投资。本文先介绍了盐化溶盐采卤生产模式，然后基于数值模拟与声呐检测数据，探讨了对流井采卤老腔改造过程中存在的形状检测、注气排卤模式等难点，论证了对流井老腔改造的可行性，提出了对流井改建储气库的方法。盐化公司对流井老腔改造技术拓宽了盐穴储气库建设的新思路，尤其是针对中国普遍夹层多、纯度低的盐层改建储气库具有重要意义。

【关键词】 盐穴储气库；对流井；仿真模拟；注气排卤；老腔改造

A technology for the reconstruction of existing convection connected salt caverns into underground gas storage

Liu Jiqin[1]　Qiao Xin[2]　Li Jianjun[1]　Wang Wenquan[1]　Zhou Donglin[1]
Xiao Enshan[1]　Jing Gang[1]　Ao Haibing[1]

(1. Gas Storage Project Department, West-East Gas Pipeline Branch of China Petroleum Pipeline Co. Ltd.; 2. PetroChina Oil & Gas Pipeline Control Center)

Abstract　Salt & Chemical Company has abundant resources of convection connected salt caverns. And if this type of salt cavern is utilized effectively, not only the reconstruction time of gas storage is shortened. but also the engineering investment can be reduced significantly. In this paper. the salt brine production model of Salt & Chemical Company was firstly reviewed. Then, based on numerical simulation and sonar detection data. the difficulties in the reconstruction process of existing convection connected salt caverns were investigated, such as shape detection and de-brining mode. Finally, the reconstruction fea-

---

基金项目：中国石油天然气集团有限公司重大科技专项"盐穴储气库加快建产工程试验研究"，2015E-4008。

作者简介：刘继芹，男，1988年生，工程师，2013年硕士毕业于中国石油大学（华东）油气田开发工程专业，现主要从事地下储气库造腔工艺技术研究及相关管理工作。地址：江苏省镇江市润州区南徐大道60号商务A区D座1801，222000。电话：15050863636。E-mail：liujiqin@petrochina.com.cn。

sibility of existing convection connected salt caverns into underground gas storage was discussed, and the reconstruction method was put forward. In conclusion, Salt & Chemical Company's technologies for the reconstruction of existing convection connected salt caverns into underground gas storage broaden the construction ideas of salt-cavem gas storage, and especially are of great significance to the construction of underground gas storage in salt layers with multiple interlayers and low purity in China.

Keywords salt-cavern gas storage; convection well; simulation; de-brining; reconstruction of existing salt cavem

随着国家能源结构调整,天然气消费量逐年增长,天然气应急调峰需求也不断提高,盐穴储气库因具有注采灵活、吞吐量大的特点而备受青睐[1-6]。中国拥有丰富的盐矿资源,但多存在盐层薄、夹层多、纯度低的特点,较厚纯盐层多已被选用或开采,并形成了大量的腔体资源[7-10]。建设金坛储气库时,提出应用盐化公司采盐形成的腔体资源改建储气库以加快储气库建设,金坛储气库成功改造5个单井老腔,形成的腔体体积为 $70×10^4 m^3$,库容为 $1×10^8 m^3$,工作气为 $5000×10^4 m^3$,5个老腔已安全运营10年[11-13],为了提高采卤效率,降低能耗,多采用双井对流连通采卤形式,并形成了丰富的对流井腔体资源。如果能够有效利用对流井连通老腔,可以加快盐穴储气库建设进度,从而尽快发挥调峰作用。目前,常规建库技术改建对流井腔体仍存在难点,在此,针对复杂对流井连通老腔改建储气库腔体相关技术进行研究,提出了对流井老腔改建储气库新思路,以期提高夹层多、纯度低盐层改建老腔或新建储气库的经济性。

1 盐化对流老腔基本概况

常规盐穴储气库建设一般采用单井垂直建腔方式(图1),该方式下腔体形状规则且易于控制,但存在摩阻大、能耗高、排卤质量浓度低等问题。盐化公司主要目的是获取盐卤资源,对形成的腔体形状要求较低,因此,多采用双井对流连通或三井对流连通形式生产[14-17]。由于三井连通对流生产形成腔体复杂,目前尚未考虑改建为储气库。盐化公司双井对流生产(图2)主要有两种模式:(1)两直井对流,两井压裂连通;(2)直井与水平对接井连通。因第一种采卤模式盐层中存在人为裂缝,且裂缝扩展难预测,故一般不改建为储气库。

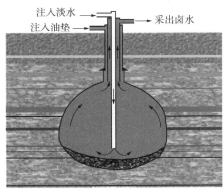 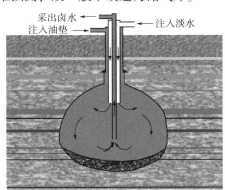

(a) 正循环　　　　　　　　　　　(b) 反循环

图1　盐穴储气库常规垂直造腔管柱组合方式示意图

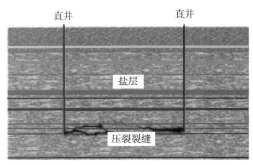
（a）压裂对流井

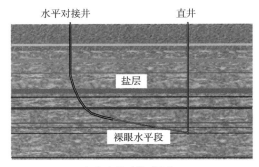
（b）水平对接对流井

图 2　盐化公司双井对流生产模式示意图

常规垂直造腔采用油垫保护腔顶，中心管与中间管控制淡水注入、卤水采出及循环模式，中间管与技术套管环空为保护液柴油，技术套管不长时间接触淡水与卤水。盐化对流井一般直接采用技术套管生产，一口井注淡水，另一口井采卤，定期转换注采方向。盐化公司应用已固井的技术套管直接注采生产，通过割管方式控制采盐层段，技术套管可能存在一定程度的腐蚀损伤。由于盐层含较多不溶物质，因此，在对流井生产过程中，技术套管容易被沉积的不溶物埋没。

2　改造难点

2.1　腔体形状检测

目前，盐穴腔体形状主要靠声呐仪器检测获取，采用电缆将声呐仪器下放至腔体内。检测方式包括水平测量和倾斜测量：水平测量一般用于腔体主体部位检测，这些部位形状比较规则，水平测量即可很好地描述腔体形状；倾斜测量则用于腔顶、腔底以及不规则部位的测量。开始测量时，一般先水平测量，对腔体形状进行整体把握，然后观察水平测量后腔体的垂直剖面图，查找需要加密测量和倾斜测量的部位(图3)。

目前，盐化公司对流井两井距为200~400m，其中水平段100~200m，采用常规声呐检测方法检测对流井腔体形状仍存在难点：（1）水平对接井端，声呐仪器下入困难，管柱端口处存在一定角度，仪器下入后回收存在风险；（2）声呐仪器无法检测到两井间水平段。随着对流井采卤生产，对接井采用割管形式控制上部盐层开采；采卤后期，对接井造斜段基本被割断掉入腔内，常规声呐仪器可以正常下入获取腔体形状，割管前采卤生产早期仍不能获取对接井端腔体形状。

以某盐矿对流井 M-N 井腔体形状(图4)为例，水

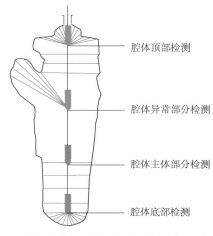

图 3　常规腔体形状声呐检测示意图

平对接井端与直井端腔体体积约为 $12×10^4 m^3$、$8×10^4 m^3$，常规声呐仪器不能检测到两井间是否存在水平通道。

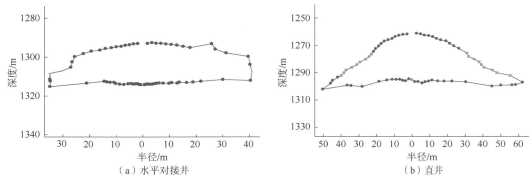

（a）水平对接井　　　　　（b）直井

图 4　某盐矿对流井 M-N 井腔体形状图

根据检测得到的对流井 M-N 井声呐形状与井身轨迹剖面图（图5）可见：由于该地区盐层不溶物含量高，成腔率较低，溶盐形成体积大部分被沉渣埋没，腔体水平段是否有被封堵的腔体空间目前仍无法检测到。

2.2　盐层动用体积预测

对流井造腔生产时，注水端腔体内流体流动模式等效于常规垂直腔体反循环造腔。假设对流井注水端腔体底部为常规垂直造腔排卤管，根据盐穴储气库成腔机理与金坛盐穴储气库成腔经验，当反循环造腔时，淡水注入腔体

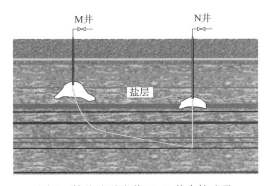

图 5　某盐矿对流井 M-N 井身轨迹及腔体形状剖面图

后，淡水由于密度差在浮力作用下首先向上运动，高质量浓度卤水向下运动，下部排卤管柱排出卤水几乎饱和（图6）。根据金坛储气库经验，在反循环造腔过程中，当两管距大于15m时，内管排出卤水为近饱和卤水。类比对流井造腔模式，水平通道和采卤端运移流体为饱和卤水，判断水平通道盐层溶蚀体积较少。依据对流井井射结构和生产模式，两井通道为生产初期溶蚀形成，此时整个腔体体积较小，注入淡水较快进入通道流向另一端，此时淡卤水与盐岩接触面有限且相对流速较快，进入水平通道中卤水质量浓度可能未饱和，对水平通道起到溶蚀作用，但采卤后期水平通道腔体多被不溶沉渣填充。

根据常规造腔成腔机理分析与数值模拟结果，可以推测对流井 M-N 井包括腔体体积与沉渣体积的动用体积（图7）。

综上，考虑对流井成腔模式与腔体形状的特殊性，对流井改建储气库注气排卤存在以下难点：（1）两腔 U 形连通，无法两腔同时排卤；（2）腔体低点不位于井眼垂线，有效体积利用率低；（3）水平段形状未知，注气排卤风险大。

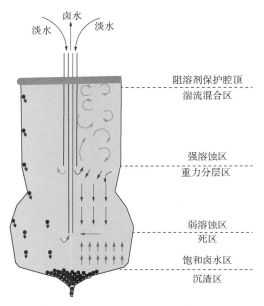

图6 常规单井单腔反循环造腔腔体流态示意图

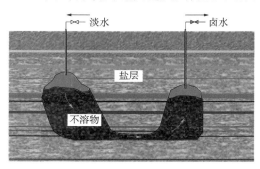

图7 某盐矿对流井 M-N 井腔体动用体积及沉渣剖面图

3 改造方法

盐化对流井老腔改造储气库不应局限于"改",更应强调"造"。通过对盐化对流井老腔有针对性地继续造腔,形成符合盐穴储气库需求的腔体。

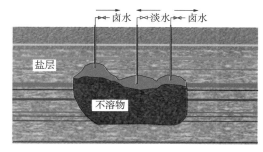

图8 小井距对流井改造示意图

3.1 小井距对流井改造

针对两井距较小(约200m)的对流井老腔,不溶物沉渣体积与有效空腔为整个采卤过程中盐层已动用体积,两井间未动用盐层宽度较小,两腔体边缘相距约100m。考虑在两井间新钻一口井,中间新钻井注淡水,两口老井采卤,随着中间井注水造腔进程的推进,形成新腔体(图8);新腔体将两老腔连通,利用常规声呐检测整个连通后的腔体形状。

新建腔体将两老腔连通为一体,可选取 3 口井中对应腔底最深的一口井为后期注气排卤井,也可有目标性地将中间腔体连通两腔后较早结束造腔,使该井腔底为最低点,选取该井为注气排卤井(图 9)。高于最终气卤界面的沉渣中的卤水也可排出,空隙体积可用于存储天然气,整个腔体损失较小。完成注气排卤后,可将其中两口井封堵,仅用一口井进行注采气作业;如果两口对流老井早期采卤过程中腐蚀较严重,可将该两口井完全封堵,采用新钻井作为后期注采气井。

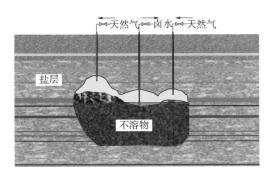

图 9　小井距对流井改造后注气排卤示意图

3.2　大井距对流井改造

盐化公司多采用大井距对流井,两井之间的距离为 300~400m。适当的大井距可以提高单井控制的盐层体积,延长单井采盐时间,两井间水平通道更长、体积更小。采用在两井中间钻一口新井的方法加以改造,新井钻至两井连通处进行循环造腔,采用新钻井注淡水造腔、两老井采卤。新钻井在底部形成较小腔体,主要应用此腔体进行排卤作业,防止沉渣堵塞排卤管柱。中间新建腔体满足排卤需求即可,新建腔体有效高度为 15~20m、直径约 30m 即可,新形成排卤腔体约为 $1.2×10^4 m^3$(图 10)。

采用中间新钻井排卤,为了防止排卤管柱被沉渣堵塞,排卤管柱置于腔顶位置,两老井在地面连通后同时注天然气,保持两腔内压力一致。腔体内卤水排空后,继续注天然气进行沉渣空隙中卤水采排,以天然气注入量来控制沉渣中气卤界面深度,最终沉渣中气卤界面深度仍高于 A、B 两点,天然气不会进入中间排卤腔体,以免高压天然气冲击排卤地面管道(图 11)。由于 A、B 两点深度不能具体检测,且沉渣空隙体积仅可基于盐岩不溶物分布及采卤数据计算获取,空隙体积计算存在较大误差,因此,注气排卤过程中排卤管柱应做好安全防护,以免突然涌入排卤腔的高压天然气冲击地面设备和管道。

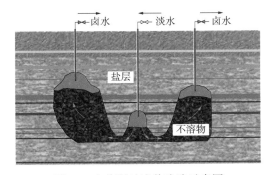

图 10　大井距对流井改造示意图

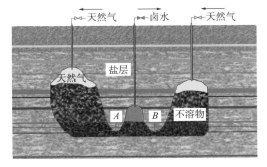

图 11　大井距对流井改造后注气排卤示意图

3.3　沉渣空隙体积计算

盐穴储气库盐岩层与泥夹层的组成有盐岩、泥质、钙芒硝及硬石膏等。盐矿地质资料显示,盐岩的主要组成成分为 NaCl,同时还有硫酸钠、钙芒硝等不溶物。采卤或造腔形成的腔底不溶物的堆积过程受多种因素影响,如溶腔方式、盐岩及夹层特性等[18-21]。通过盐岩水溶实验发现,盐岩中的盐被溶解后,其不溶物分散成粒径不同的颗粒。为了对生成不溶物的粒径进行分析,选取某盐矿目标岩层各个深度1~5号岩样进行溶蚀实验,获取不溶物沉渣,对不溶物粒径进行分级,发现单个岩样粒径的分布不符合正态分布(图12)。

地层中腔底堆积物是各个层位的累积叠加,研究各组堆积物叠加后的粒径分布(图13)表明:叠加后不溶物粒径分布特征符合双正态分布特征。因此,在后续空隙率的测定中可以使用双正态分布来模拟实验总粒径的分布情况。

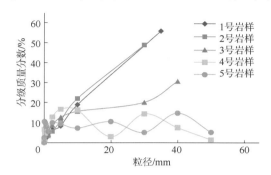

图12 某盐矿1~5号岩样不溶物粒径分级质量分数曲线

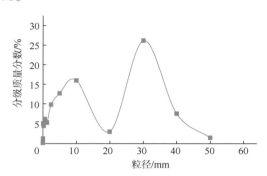

图13 某盐矿1~5号岩样不溶物粒径分级叠加质量分数曲线

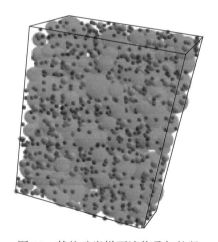

图14 某盐矿岩样不溶物叠加粒径分布下的仿真模拟图

根据岩样叠加粒径分布拟合的参数生成颗粒,应用仿真模拟软件模拟堆积过程(图14,蓝色、绿色颗粒均代表不溶物),模拟选用800mm×800mm×1200mm的长方体。在相同条件下进行多次模拟(表1),结果表明:堆积体空隙率变化不大,模拟方法可行,结果可靠,岩样不溶物粒径分布下模拟的空隙率$\phi=36.41\%$。

根据图11,进行大井距对接老腔改造时,为了保证排卤井安全,沉渣中仍有一定体积的水,以防止气水界面低于A、B两点。假设排卤沉渣体积占整个沉渣体积的70%,盐岩中原始不溶物质量分数$a=40\%$,不溶物综合膨胀系数(颗粒膨胀和堆积空隙)$b=1.6$,两腔体合计体积$V_c=20\times10^4\mathrm{m}^3$,则可计算M-N对流井老腔沉渣中可被利用的空隙体积V_z:

$$V_z = V_c \frac{ab}{1-ab}\phi = 12.95\times10^4\mathrm{m}^3 \quad (1)$$

表1 某盐矿岩样不溶物叠加粒径分布下仿真模拟总体粒径的空隙率

模拟次数	颗粒体积/($10^{11}\mathrm{mm}^3$)	总体积/($10^{11}\mathrm{mm}^3$)	空隙率/%
1	4.88	7.68	36.41
2	4.88	7.68	36.41
3	4.88	7.68	36.40

4 结论

（1）我国盐化公司采卤生产历史长，拥有丰富的采卤形成腔体资源，如果能对此类老腔筛选改造利用，可大大加快盐穴储气库建设，提高盐岩综合利用率。同时针对老腔改造，不仅要注重"改"，更要重视"造"。

（2）针对盐化丰富的盐化复杂对流井老腔资源，为了解决改建储气库中存在形状检测难度大、不溶物含量高的问题，提出两种改造方案，但小井距对流井改造后腔体仍需进行稳定性评价研究。

（3）大井距对流井老腔，通过在两井间钻新井、造排卤腔的方式，可以有效解决排卤低点难找、盐层成腔率低的问题。改造后的腔体在应急调峰时，可通过中间井注饱和卤水的方式来达到腔内天然气完全采空。

（4）对流井老腔改造后的最终气水界面低于沉渣深度100m左右，改造后腔体采气过程中气水界面以上沉渣可阻挡卤水在天然气中的挥发，降低地面天然气脱水压力，对水挥发起到"盖被"作用。

（5）针对对流井 M-N 井进行改造，考虑盐层不溶物质量分数为40%的情况下，不溶物沉渣中可挖掘 $12 \times 10^4 m^3$ 有效体积，大大提高腔体库容。

参 考 文 献

[1] 罗天宝，李强，许相戈．我国地下储气库垫底气经济评价方法探讨[J]．国际石油经济，2016，24（7）：103-106．

[2] 成金华，刘伦，王小林，等．天然气区域市场需求弹性差异性分析及价格规制影响研究[J]．中国人口资源与环境，2014，21(8)：131-140．

[3] 吴洪波，何洋，周勇，等．天然气调峰方式的对比与选择[J]．天然气与石油，2009，27(5)：5-9．

[4] 周学深．有效的天然气调峰储气技术——地下储气库[J]．天然气工业，2013，33(10)：95-99．

[5] 丁国生，梁婧，任永胜，等．建设中国天然气调峰储备与应急系统的建议[J]．天然气工业，2009，29(5)：98-100．

[6] 王峰，王东军．输气管道配套地下储气库调峰技术[J]．石油规划设计，2011，22(3)：28-30．

[7] 马旭强．多夹层盐矿气库稳定性和造腔效率研究[D]．北京：中国科学院大学，2016：5-34．

[8] 井岗，何俊，翁小红，等．CSAMT法在盐化老腔形状检测中的勘探试验[J]．石油化工应用，2016，35(7)：87-91．

[9] 周冬林，李建君，王晓刚，等．云应地区采盐老腔再利用的可行性[J]．油气储运，2017，36(8)：930-936．

[10] 杨海军，闫凤林．复杂老腔改建储气（油）库可行性分析[J]．石油化工应用，2015，34(11)：59-61，65．

[11] 杨海军．中国盐穴储气库建设关键技术及挑战[J]．油气储运，2017，36(7)：747-753．

[12] 田中兰，夏柏如，苟凤．采卤老腔改建储气库评价方法[J]．天然气工业，2007，27(3)：114-116，160．

[13] 丁建林，利用现有采卤溶腔改建地下储气库技术[J]．油气储运，2008，27(12)：42-46，74．

[14] 李君福，井矿盐开采中采卤对接井钻井技术的应用研究[J]．资源信息与工程，2017，32(3)：90-91．

[15] 夏筱红,杨伟峰,刘志强. 应城盐矿区采卤对接井施工技术[J]. 中国井矿盐,2002,33(3):22-24.

[16] 周铁芳,赵建亚,向军文. 采卤对接井钻井技术的研究[J]. 探矿工程(岩土钻掘工程),1995(1):6-10,17.

[17] 苗东涛. 盐岩地下储库不溶物排空方法实验研究[D]. 绵阳:西南科技大学,2016:15-43.

[18] 李银平,施锡林,刘伟,等. 盐穴水溶造腔建槽期不溶物运动性态及应用研究[J]. 岩石力学与工程学报,2016,35(1):23-31.

[19] 高树生,班凡生,单文文,等. 岩盐中不溶物含量对储气库腔体有效空间大小的影响[J]. 岩土力学,2005,26(增刊1):179-183.

[20] 韩杰鹏,盐穴储气库腔底堆积物空隙体积研究[D]. 成都:西南石油大学,2015:8-40.

[21] 垢艳侠,完颜祺琪,罗天宝,等. 层状盐岩建库技术难点与对策[J]. 盐科学与化工,2017,46(11):1-5.

本论文原发表于《油气储运》2019年第38卷第3期。

复杂老腔改建储气(油)库可行性分析

杨海军　闫凤林

(中国石油西气东输管道公司储气库项目部)

【摘　要】　我国地下存在大量溶盐老腔资源，科学合理地利用这些老腔进行储气(油)，对我国储气(油)库的发展有重要意义，但这些老腔形状很不规则而且存在多个井眼，利用难度很大。本文从复杂老腔的形状检测、老井封堵改造、老腔生产运行以及改建储气库与储油库的对比等方面进行论述，分析复杂老腔改建储气(油)库的可行性，为复杂老腔的改造利用提供一定的借鉴作用。

【关键词】　复杂老腔；储气(油)库；形状检测；老井改造

Feasibility analysis of gas (oil) storage constructions based on complex existing salt-caverns

Yang Haijun　Yan Fenglin

(Gas Storage Project Department, PetroChina West-East Pipeline Company)

Abstract　There are many existing salt-caverns resources in China. It is important signifi-cance for our country's gas (oil) storage to use these existing salt-caverns. But the shapes of these existing salt-caverns are irregular and more than one borehole in a existing salt-cavern, therefore. it is very difficult to use these complex existing salt-caverns. This paper dis cussed the shape measurement of complex existing salt-caverns, reconstruction of existing wells, production of reconstructed complex salt-caverns and the advantages of reconstruction oil storage, analyzed the feasibility of reconstruction gas (oil) storage, and supported references for the use of complex existing salt-caverns.

Keywords　complex existing salt-caverns; gas (oil) storage; shape measurement; reconstruction of existing well

　　盐岩具有非常低的渗透性且力学性能较为稳定，能够保证储库的密封性，所以盐岩溶解形成的腔体是能源(石油、天然气)理想的储存空间[1,4]；但我国盐穴储气(油)库建设起步较晚，盐岩层厚度薄且隔夹层多，所以储气(油)库建设进度较慢，如果将盐化企业老腔经过改造施工，达到储气(油)库的建设标准，用来做储气(油)库，将大幅增加我国储气(油)库的规模，以缓解我国储气(油)库建设严重滞后，储气(油)能力亟须提高的压力。

作者简介：杨海军，男(1961—)，高级工程师，大庆石油学院毕业，获得学士学位，现在中国石油西气东输管道公司储气库管理处从事储气库工程研究工作。E-mail：yhjun@petrochina.com.cn。

1 我国地下存在大量老腔

我国市场上供应的盐制品主要来自赋存于地下的盐岩矿藏,盐化企业通过在地面钻井,将淡水注入盐岩层,溶解盐岩形成卤水,再把卤水采至地面,通过地面的设备炼制成市场上所需的盐制品。盐化企业经过近三四十年的卤水开采,在地下形成了大量的老腔,据统计在我国金坛、云应、淮安、平顶山、楚州等地有300个以上的老腔存在(表1)。

表1 我国盐化老腔分布不完全统计表

地区	老腔数量
江苏金坛	43
江苏淮安	93
江苏楚州	26
湖北云应	159
河南平顶山	36

2 单腔改造获得成功经验

2007年中国石油西气东输管道公司储气库项目部对金坛地区43个老腔进行普查,经过初步筛选,对15个老腔进行了声呐检测工作(形状如图1),并应用全井套铣的方法将原采卤井的生产套管全部取出,用扩眼钻头扩眼后,重新下入新的生产套管固井,成功改造5个老腔用于储气生产。目前已经进行过几十轮注采运行周期,累计注气量为 $5.8 \times 10^8 m^3$,累计采气量为 $4.2 \times 10^8 m^3$,仍然保持安全平稳。

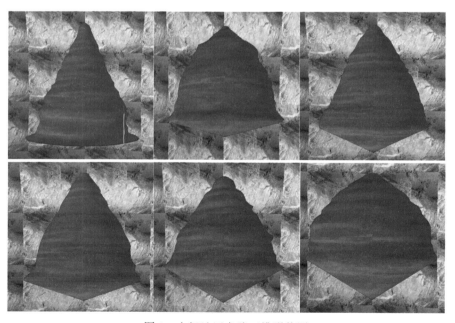

图1 金坛地区老腔三维形状图

在2013年通过带压声呐检测工作，监测投入生产运行后的腔体形状及净体积变化，测量结果显示腔体形状基本无变化，体积收缩率为1%～2%，处于合理范围之内（图2）。

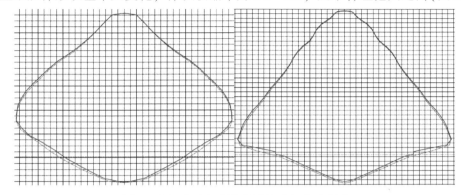

图2　金坛地区老腔投入运行后声呐对比图

3　复杂老腔改造难点

虽然在金坛对老腔改建储气库获得了成功，但仅限于单井单腔、形状规则的老腔进行改造，对多井单腔、形状复杂的老腔改造还未涉及，复杂老腔改造主要面临形状检测困难和改造施工困难两个难题。

目前盐化企业主要通过双井的方式进行采盐，分为钻水平定向井连通和水力压裂连通两种情况，双井连通后，不定期变换两井的注采关系，这样就会在采卤水的同时，在两井之间形成水平通道，但目前没有实测资料，无法准确判断；盐化企业在采盐过程没有控制腔体形状，而且有些采卤井进行过水力压裂施工，所以会造成腔与腔之间相互连通，这样就会出现多井单腔的情况（图3），腔体形状往往十分复杂，目前的声呐测量仪器无法完成复杂老腔的形状检测工作，需要对其进行改进。

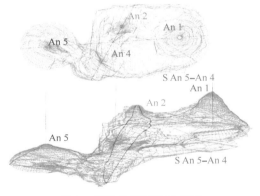

图3　复杂老腔形状示意图

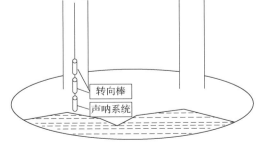

图4　复杂老腔声呐检测装置示意图

4　老腔利用的可行性

4.1　复杂老腔形状检测

通过对现有声呐测量仪器增加若干个转向棒（图4），当声呐仪器到达腔内不溶物顶面时，转向棒依靠自身重力与声呐仪器之间的电动机系统，使声呐仪器水平转向，并驱动其向前移动，这样就可以测量出两井之间的水平腔，进而弄清复杂老腔的具体形状，为复杂老腔后续改造施工提供了基础。

4.2 老井封堵改造

在金坛地区老腔改造的方式为对老井表层和生产套管进行套铣，然后重新下入套管固井，水泥返至地面，检查固井质量并进行气密封试压，如果均合格，则改造成功；考虑到后续的注气排卤作业需要将排卤管柱下入到腔内的最低点，这样才能最大限度地把卤水排出，充分利用腔内空间进行储气，但复杂老腔的不溶物分布情况复杂，很难确定老腔腔底，所以复杂老腔改造需要结合老腔的形状才能确定改造方案，如果老井的位置在腔内最低点，则该老井进行全井套铣、重新固井，将其他老井封堵；如果老井的位置都不在腔内最低点，则将老井全部封堵，然后在腔内最低点的位置重新钻一口新井，这样复杂老腔就改造成为单井单腔的简单形式，方便进行后续工作。

老井封堵的技术方案为锻铣部分下部套管，注水泥进行封堵，其工艺流程是：通井→测井→下部套管锻铣→扩铣水泥环至新地层→注水泥浆封堵→试压合格→锻铣其他部分套管→扩铣水泥环至新地层→注水泥浆封堵→在井筒内注一定高度的水泥塞[2,3]。

4.3 老腔运行压力确定

复杂老腔改造成功后将被用于储气(油)，腔体运行的上下限压力是确定工作气(油)量的重要参数，考虑到腔体的稳定性[4]，腔体运行的下限压力应不低于静水柱压力，腔体运行的上限压力则不高于地层破裂压力的80%。

4.4 老腔改建储油库的优势

如果改造成功后的老腔用于储气，由于气体容易泄漏，所以对腔体的密封性要求比用于储油更加严格，所以改造成功后的老腔用于储油有以下两点优势。

（1）油分子比气分子大得多，不易泄漏，对腔体的密封性要求相对较低，易于实现。

（2）如果单独建设储油库，需要在地面额外修建水池用于把腔内的油替换到地面，增加额外的建设成本；如果利用改造成功后的老腔进行储油，则不需要修建水池，可以利用其他腔内的卤水进行油水置换，工艺简单，建设成本较低。

5 结论与认识

（1）我国地下老腔资源丰富，科学合理地利用老腔进行储气(油)，对我国储气(油)库发展意义重大。

（2）形状检测与老井改造是利用复杂老腔改建储气(油)库面临的两个难点，需要对现有声呐仪器改进和老井封堵技术两个方面进行攻关。

（3）由于储存介质的性质不同，老腔改建储油库比改建储气库会更有优势。

参 考 文 献

[1] 杨春和，梁卫国，魏东吼，等. 中国盐岩能源地下储存可行性研究[J]. 岩石力学与工程学报，2005，

24(24)：4409-4417.

[2] 田中兰, 夏柏如, 苟凤. 采卤老腔改建储气库评价方法[J]. 天然气工业, 2007, 27(3)：114-116.

[3] 田中兰, 夏柏如, 申瑞臣, 等. 采卤盐矿老溶腔改建为地下储气库工程技术研究[J]. 石油学报, 2007, 28(5)：142-144.

[4] 尹雪英, 杨春和, 陈剑文. 金坛盐矿老腔储气库长期稳定性分析数值模拟[J]. 岩土力学, 2006, 27(6)：869-874.

本论文原发表于《石油化工应用》2015年第34卷第11期。

盐化对流井老腔改建水平腔储气难点分析

薛 雨[1]　王元刚[1]　周冬林[2]

(1. 中国石油西气东输管道公司
2. 中国石油盐穴储气库技术研究中心)

【摘　要】 盐穴储气库由于注采灵活，注采频次高，库容损失小等优势，近年来受到越来越多的关注，我国目前在大力开展盐穴储气库的建设，建库区域内存在大量对井连通老腔，如果能将盐化对井连通老腔改建为水平腔储气，就可以快速投入使用，缓解储气库建设的压力。盐化对流井老腔改建为储气库存在很多困难，从有效盐层损失，对井连通情况，阻溶剂界面控制以及对井井组之间的连通情况进行分析，归纳总结对流井老腔改建为水平腔过程中存在的技术难题，并指出对流井老腔改造面临的重要挑战。

【关键词】 有效盐层损失；对井连通空间；水平腔阻溶剂界面；对井井组

Analysis of the difficulties on rebuilding the old twin-wells of salt mine into strata aligned salt cavern gas storage

Xue Yu[1]　Wang Yuangang[1]　Zhou Donglin[2]

(1. PetroChina West-East Gas Pipeline Company；
2. PetroChina Salt Cavern Gas Storage Technology Research Center)

Abstract Salt cavern gas storage has been paid more and more attention in recent years as its small volume loss and flexible injection-production with high operation frequency. China is sparing no effort on the construction of salt cavern gas storage, there is a lot of old twin-wells in the salt mine, if the old twin-wells can be rebuilt into gas cavern storage, the construction period can be cut down to alleviate the pressure from the gas market. There is a lot of difficulties on rebuilding of the old twin-wells, this article stated the difficulties and challenges of the twin-well rebuilding from the point on the loss of effective salt rock, the connected situation of the twin-well, the controlling of the blanket level and the connection between the well groups, bring the challenge on the rebuild of salt cavern gas storage with twin-wells.

基金项目：中国石油天然气集团公司储气库重大专项"地下储气库关键技术研究与应用"（2015E-40）之课题八"盐穴储气库加快建产工程试验研究"（2015E-4008）。

作者简介：薛雨（1987—），男，工程师，硕士，现主要从事盐穴储气库技术研究工作。联系方式：17768775355。

Keywords loss of effective salt layer; connected space of twin-well; blanket level of strata aligned salt cavern gas storage; well group of twin-well

储气库在天然气行业中所占比重越来越大，作为储气库的重要组成部分，盐穴储气库经过近15年的发展，积累了宝贵的建设经验[1-6]。在金坛储气库已经成功改造5口盐化单井老腔并注气运行[7]的基础上，形成了成熟的直井单腔改造技术。但通过调研发现，盐化企业存在大量采用对井生产的老腔，部分对井老腔开采年限达到10年以上，在地下形成较大的腔体体积，随着天然气行业对储气库的需求日益增加，将对流井老腔改建为储气库从而加快储气库建设已经成为一种趋势[8-10]。

国内盐化老腔改建为储气库主要包括改造前评价工作，改造工程施工以及改造后的注采运行三大部分[11-13]，文章只针对改造前的评价部分遇到的难点进行分析。

1 盐化老腔改建水平腔存在问题

1.1 采卤过程中盐层损失较大

1.1.1 井眼轴线上盐层损失

盐化老腔在生产过程中，采用无阻溶剂采卤，且注水管与采卤管距离较大，采卤初期，因腔体体积较小，不溶物堆积较快，造成下部盐层得不到充分溶蚀就被不溶物填埋，盐层体积损失较大。以well-1井为例，如图1所示。该井内管初始下深1192m，由于采用无阻溶剂造腔，腔体体积只有6700m³时，阻溶剂界面上移约160m，不溶物在井底堆积，使腔体在短时间内上升25m，造成该段盐层无法继续造腔。

盐化老井采卤时间长，因此不溶物上升幅度更大，以某对井开采为例，直井造腔管柱经过一段时间开采后，测得不溶物顶面上升约150m；水平井不溶物顶面上升约180m，两口井盐层未得到充分利用，损失较大。

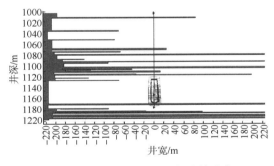

图1 well-1井管柱下深与腔体形状

1.1.2 对井连通部分盐岩损失

对井造腔建槽期首先在直井附近造腔，形成一定的腔体体积，然后钻水平井与直井对接，造腔过程中采用一注一采的方式进行造腔，由于水平段为裸眼完井，如果在两口井大面积连通前，进行水平井注、直井采的方式进行造腔，很容易造成井眼堵塞，连通中断，因此在直井注入水溶蚀到水平井管鞋之前，不能从水平井注水。注入水在直井段附近达到饱和，不再对水平段盐层进行溶蚀，水平段盐层利用率极低，盐岩浪费极大，如图2中A、B部分。

1.2 对井连通情况无法确定

如图3所示，对井造腔由于两口井之间的距离约300m左右，注入水在从注入井流向采卤井的过程中很快达到饱和，因此两井之间的盐层溶蚀程度较小，直径相对较小。声呐在卤水中探测半径在 200m 以内，腔体直径超过一定范围后无法测量腔体形状，因此声呐测腔只能对近井地带的腔体形状进行解释，连通区域形状以及溶蚀程度均无法判断。由于连通区域半径较小，在后期注采运行过程中，腔体稳定性存在较大风险，如果在注采气过程中该部分发生盐岩损伤变形脱落等情况，连通区间存在被隔断的风险。

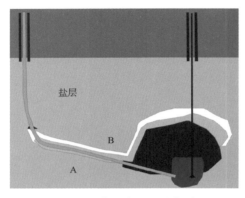

图 2　对井盐腔连通示意图

图 3　对井盐腔声呐测量形状

1.3　改造后继续造腔时阻溶剂界面控制困难

对于直井单腔，阻溶剂界面控制技术主要从环空中注入柴油，并定期检测阻溶剂界面位置，控制方法简单，易于实现。

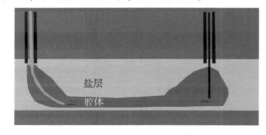

图 4　对井连通老腔示意图

大部分对井盐腔处于开采前期或中期，主力盐层段未进行开采，在改造为储气库之后需要继续进行造腔。不同于对井单腔，水平腔腔顶面积较大，阻溶剂界面无法控制，当盐岩上溶速度比侧溶速度快很多时，腔顶盐层迅速溶蚀，阻溶剂界面随时有失控的风险，对井连通老腔示意图如图 4 所示。

1.4　注采完井腔体有效体积损失较大

调研发现大部分水平腔均存在"A"形底问题，即腔底体积较大，注采完井结束后会有大量卤水残留在腔内，腔体体积不能最大化利用。

即使在两口井都是"V"形井的情况下，注采完井时腔体体积也会出现一定程度的损失，对井老腔注气排卤一般采用一注一采的形式，由于对井所在腔底不在同一水平面上，注采完井过程中排卤井位于腔底较深的井，如果注气井腔底呈"V"形，则造腔过程中"V"形底部分卤水不能排出腔底，该部分体积无法利用，因此在改造过程中需要考虑使腔体体积得到最大化利用的方案，对井老腔排卤过程中的有效体积损失示意图如图 5 所示。

1.5 对井井组的相互影响

1.5.1 安全矿柱距离

安全矿柱宽度与盐腔的稳定运行息息相关，它相当于地面建筑物的承重墙，用来支撑上部地层不发生塌陷，安全矿柱过窄，会导致邻近腔体运行时相互影响，在新腔建造的设计中一般要求矿柱直径比（P/D）在 2~3 的范围内。

老腔由于开采时未注意矿柱问题，因此对井井组距离往往较近，一些腔体的安全矿柱达不到腔体独立安全运行的标准，少数井由于距离较近很有可能在采卤过程中连通，无法改造为储气库。

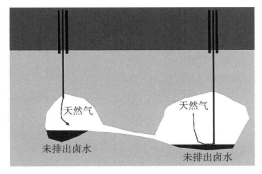

图 5 对井老腔排卤过程中的有效体积损失

1.5.2 老腔腔群连通情况

盐化老井在采卤过程中为保证两井连通并加快采卤速度，在采盐过程没有控制腔体形状，而且有些采卤井进行过水力压裂施工，压裂作业结束后，地层中出现裂缝条数以及裂缝方向、深度都无法估计，给对井老腔改造带来了极大困难，如果裂缝深度较大，储层渗透性增加，则无法改建成储气库。

2 对流井老腔改造存在的挑战

盐穴储气库以其独特优势受到越来越多的重视，盐化对井老腔改造在技术攻关方面存在以下挑战：

（1）建槽期以及双井连通之前存在大量的盐层浪费，对盐化采卤过程中损失的盐层进行再利用是一个挑战。

（2）双井连通情况无法确定，水平腔注采运行过程中的稳定性需要进现场的实践验证。

（3）对井老腔改造完成后继续造腔时，阻溶剂界面无法控制，腔顶有失控的风险，如何研究阻溶剂控制方法已经成为重要的攻关技术。

（4）盐化老腔对井间距较小，且部分井在采卤过程中采用压裂措施，对井井组有可能连通，无法改造成储气库，对连通老腔的筛选提出了更高要求。

参 考 文 献

[1] 田中兰,夏柏如,苟凤.采卤老腔改建储气库评价方法[J].天然气工业,2017,27(3):114-116.
[2] 李玥洋,田园媛,曹鹏,等.储气库建设条件筛选与优化[J].西南石油大学学报(自然科学版),2013,35(5):123-129.
[3] 田中兰,夏柏如,申瑞臣.采卤盐矿老溶腔改建为地下储气库工程技术研究[J].石油学报,2007,28(5):142-145.
[4] 丁建林.利用现有采卤溶腔改建地下储气库技术[J].油气储存,2008,27(12):42-46.
[5] 杨海军,闫凤林.复杂老腔改建储气(油)库可行性分析[J].石油化工应用,2015,34(11):59-61.
[6] 周冬林,李建君,王晓刚,等.云应地区老腔再利用可行性分析[J].油气储运,2017,36(8):

930-936.

[7] 田中兰,张芳.金坛盐矿采卤溶腔利用与改造技术[J].中国井盐矿,2005,36(2):17-20.

[8] 周俊驰,黄孟云,班凡生,等.盐穴储气库双井造腔技术现状及难点分析[J].重庆科技学院学报(自然科学版),2016,18(1):63-67.

[9] 易胜利.岩盐钻井水溶双井连通开采工艺的研究与推广[J].中国井矿盐,2003,34(6):20-22.

[10] 余勇进,陈千汉,周普松,等.薄层复层状盐矿对井连通开采溶腔形成机理研究[J].中国矿业,2010,19(3):82-85.

[11] 尹雪英,杨春和,陈剑文.金坛盐矿老腔储气库长期稳定性分析数值模拟[J].岩土力学,2006,27(6):869-874.

[12] 陈卫忠,伍国军,戴永浩,等.废弃盐穴地下储气库稳定性研究[J].岩石力学与工程学报,2006,25(4):848-854.

[13] 杨炳鑫.浅析昆明盐矿油垫对流井相互串通原因[J].中国井盐矿,2000,31(1):26-28.

本论文原发表于《盐科学与化工》2020年第49卷第1期。

盐穴地下储气库对流井老腔改造工艺技术

薛 雨 王元刚 张新悦

(中国石油西气东输管道公司)

【摘 要】 国内盐矿存在的老腔主要为对流井老腔,但目前国内外暂无对流井老腔改建为盐穴地下储气库的案例。为了充分增加盐腔的有效体积,在对比不同改造技术的基础上,提出封堵原有两口老井并钻新井,在通道中间打一口排卤井与原水平通道对接的对流井老腔改造工艺技术,并采用该工艺技术对国内某对流井老腔进行了改造试验。研究结果表明:(1)在水平通道中间打一口连接井的技术切实可行,可最大限度地利用盐腔下部残渣中的孔隙体积,并且改造后的盐腔能够满足储气库注采运行的要求;(2)在不溶物残渣中卤水排出 20%~40% 的情况下,试验井可增加腔体体积 13.94×10^4~27.88×10^4 m³,增加工作气量 1763×10^4~3526×10^4 m³。结论认为,该对流井老腔改造技术可充分排出对流井老腔不溶物残渣中的卤水,大幅度提高盐穴腔体的有效体积,可推广到类似对流井盐腔的改造实践中,具有广阔的应用前景。

【关键词】 盐穴;地下储气库;对流井老腔;排卤井;水平通道;连接井;老腔改造;孔隙体积;腔体体积

A technology of reconstructing salt cavern underground gas storages by use of the old chambers of those existing convection wells

Xue Yu Wang Yuangang Zhang Xinyue

(PetroChina West-East Gas Pipeline Company)

Abstract The existing caverns in salt mines in China are mainly convection wells. However, there is no case of using those existing convection wells to reconstruct UGS at home and abroad at present. In order to fully increase the effective volume of such salt chambers, we put forward the technology of plugging two old wells and drilling new wells, then drilling a brine drainage well in the middle of the channel to connect with the original horizontal channel, based upon a comparison analysis with other reconstruction technologies. Taking a convection well as an example, the test results show that the tech-

---

基金项目:中国石油天然气集团有限公司重大科研专项"盐穴储气库加快建产工程试验研究"(编号:2015E-4008)。

作者简介:薛雨,1987 年生,工程师,硕士;主要从事盐穴地下储气库现场施工管理及天然气管道输气技术研究工作。地址:(212000)江苏省镇江市润州区南徐大道 60 号商务 A 区 D 座 1813。电话:17768775355。ORCID:0000-0003-4393-7349。E-mail:xqdsxueyu@petrochina.com.cn。

nology of drilling a connecting well in the middle of the horizontal passage is feasible, and the pore volume in the lower residue of the salt chamber can be mostly utilized. The reformed chamber can also meet the injection-production operation requirement of a UGS. If 20%-40% brine in the insoluble residue can be discharged, the chamber volume can be increased by 139.4-278.8 thousand m³ and the working gas can be increased by 17630-35260 thousand m³. In conclusion, this technology can fully discharge the brine from the insoluble residue and increase the effective volume of the salt chamber. It shall be popularized and applied to similar convection wells, and has broad application prospects.

Keywords salt cavern; underground gas storage (UGS); old chamber of convection well; Brine well; horizontal channel; connecting well; reconstructing the old chamber; pore volume; chamber volume

盐穴储气库建设对于地下盐层的地质条件有很高的要求[1-4]。国内适合建设储气库的盐矿资源比较缺乏，在建或拟建的盐穴储气库只有江苏金坛、江苏淮安和河南等地[5-8]。而且在建的盐穴储气库中都存在盐层厚度较薄、夹层较多、夹层厚度大、不溶物含量高等不利条件，延缓了储气库建设的进度。我国是井矿盐生产大国，地下盐腔数量众多，仅湖北云应、河南平顶山等地就拥有老腔500多个，大部分充满卤水，处于废弃闲置状态[9]。为加快储气库建设，尽快形成储气能力，改造盐化老腔建设储气库是行之有效的方法，这种盐化采卤形成盐腔建设储气库的方法已经在金坛储气库6口井获得成功应用[10-12]。金坛盐矿采用单井采卤方式，而湖北云应、河南平顶山以及江苏淮安等地盐矿企业通常采用对流井采卤，为加快盐穴储气库建设，改造对流井老腔已经成为一种趋势。当前国内外没有对流井老腔改建为盐穴储气库的案例，笔者提出了一种封堵原有两口老井钻新井，在通道中间打一口排卤井与原水平通道对接的改造工艺技术并成功应用于国内某对流井老腔，在大幅度增加该对流井盐腔有效体积的同时，也为其他对流井老腔的改造提供了一定的借鉴。

1 对流井老腔改造必要性

国内绝大部分盐矿企业都采用对流井采卤，先钻一口直井well-1井，再钻一口定向井well-2井与之连通，两口井交替注采生产（图1）。盐矿采卤井基本都采用ϕ177.8mm生产套管的井身结构，如果直接下入ϕ89mm油管改为储气库注采井后，油管和套管同采最大采气量分别为$40×10^4 m^3/d$和$80×10^4 m^3/d$，只能满足正常调峰的需要，不具备应急供气的能力。

通过筛选盐化老腔发现，可用于进行改造的老腔采卤年限都较长，有些井甚至达到20年，井筒腐蚀、老化严重，并且盐化采卤井固井要求远低于储气库井对固井的要求，固井水泥未返至井口或者部分井段固井质量差，生

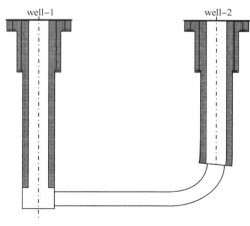

图1 盐矿对流井示意图

套管在井内长时间腐蚀后，其密封性能和剩余强度不能满足注采气长时间工作的要求。

同时对流井老腔底部及通道中存在大量的溶蚀空间，但大部分被不溶物所填埋，若能通过采取与单腔不同的注气排卤方式，排出不溶物之间的卤水将有效增大实际储气空间，增加库容。

2 对流井老腔改造工艺技术优选

对老腔进行改造必须使腔体、井筒、地层三者之间能够有效密封，同时井筒有足够的天然气注入和采出能力。成熟的单井老腔改造技术有以下3种[13-15]。

(1) 将原老井眼直接进行封堵，在边上按盐穴储气库井标准重新钻新井。

(2) 锻铣原老井部分下部套管和水泥，封堵后再打大直径井眼新井。

(3) 全井套铣，扩眼后下入大直径套管。

根据单井老腔改造经验，并结合对流井特点，对流井老腔改造工艺技术有以下6种。

(1) 利用原有两个老井筒，中间下入小尺寸注采管柱作为注采井。

(2) 全井套铣，将原有两口采卤井的 ϕ177.8mm 生产套管通过套铣方式取出，重新用 ϕ311.2mm 钻头扩眼，下入 ϕ244.5mm 生产套管固井。

(3) 封堵原有两口采卤井，在采卤溶腔上方按照储气库钻井标准重新钻井，下入 ϕ244.5mm 生产套管固井。

(4) 利用原有两个老井筒，中间下入小尺寸注采管柱作为注采井，然后在通道中在打一口新井作为排卤井。

(5) 封堵原有两口采卤井，在采卤溶腔上方按照储气库钻井标准重新钻井，下入 ϕ244.5mm 生产套管固井，然后在通道中再打一口新井作为排卤井。

(6) 全井套铣，即将原有两口采卤井的 ϕ177.8mm 生产套管通过套铣方式取出，重新用 ϕ311.2mm 钻头扩眼，下入 ϕ244.5mm 生产套管固井，然后在通道中在打一口新井作为排卤井。

对上述6种工艺进行比较，工艺5既能满足地下储气库最根本的密封性和可靠性要求，又能最大程度利用盐腔下部残渣中的孔隙体积，是最佳的改造工艺技术(表1)。

表1 改造工艺技术对比表

工艺代号	密封可靠性	调峰能力	储气能力	施工难度	投资
1	差	差	小	易	少
2	较好	优	一般	较大	一般
3	较好	优	一般	一般	一般
4	差	差	较大	一般	较小
5	较好	优	大	较大	较大
6	较好	优	大	大	大

3 工艺技术难点及对策

优选的工艺技术需要将原有的 well-1 井、well-2 井封堵，新钻 well-1′井、well-2′井和

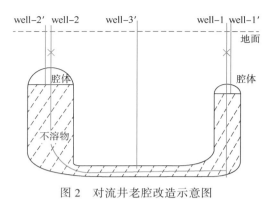

图 2 对流井老腔改造示意图

well-3′井这 3 口井(图 2)。后期投产注气排卤时，well-1′井和 well-2′井作为注气井，well-3′井作为排卤井。

3 口新井均按盐穴储气库常规钻井方式施工，其中 well-1′井和 well-2′井口距离老井井口中心线 10m 左右，井口正对盐腔底部最深位置以利于后期注气排卤，生产套管与腔体顶部保留一定的距离，以保证固井作业的安全性。盐腔上部的形状以及位置可以通过声呐技术准确测量，但腔体下部被不溶物填埋，现有的技术无法测出腔体下部以及水平连通通道的形状及位置，因此，well-1′井和 well-2′井钻井相对简单，well-3′井钻井难度较大。well-3′井在钻井施工时，需要考虑以下因素。

3.1 井口位置及井斜控制

well-3′井必须钻入水平通道，否则腔体下部无法利用，失去改造意义。由于水平通道的位置及形状均未知，因此 well-3′井口位置应尽量选在原水平段井眼轨迹上方。

盐穴储气库对井身质量要求较高，二开井段最大井斜不超过 1.5°，考虑到地层偏溶，为确保 well-3′井钻入通道，二开井段井斜可适当放宽至 3°。

3.2 井身结构选择

目前盐穴储气库生产井通常采用二开井身结构，若 well-3′井采用二开井身结构，二开直接钻入水平通道，钻入水平通道瞬间，由于井筒钻井液与通道内卤水存在压力差，可能会形成井喷现象。并且井眼与原水平通道对接，生产套管固井时需要先下入桥塞，满足承压条件后才能进行下套管及固井施工，工艺流程较为复杂。从工艺及安全角度考虑，well-3′井宜采用三开井身结构，待生产套管固井结束后，再三开钻入水平通道。

3.3 生产套管下深及固井

为避免二开钻井时提前进入水平通道，在施工前需要对通道位置进行预测，以确定生产套管下深。但由于此数值仅为预测值，实际深度可能比预测值浅，因此二开完钻时，可不钻测井口袋。

在固井时，若下部地层承压能力不够，可能发生将下部地层与水平通道压穿的情况，因此需要在固井前对老腔进行地层承压试验，盐穴储气库生产套管固井水泥浆密度介于 1.90~1.95g/cm³，承压试验时钻井液密度可达到 2.00g/cm³。

3.4 三开钻入水平通道

盐穴储气库井二开进入盐层后及三开井段通常采用饱和盐水聚合物钻井液钻进，考虑到避免污染盐腔中原有卤水，well-3′井三开采用饱和盐水钻进，同时尽量避免添加处理剂，在三开层段存在多个泥岩夹层的情况下，用高黏度盐水清洗井内岩屑。

三开进入水平通道瞬间，可能会形成先漏后喷的现象。通道内也有可能存在气体，连通瞬间上串造成气顶现场。三开施工时应加强观察，做好应对措施，防止发生井下复杂情况。一旦发现钻井液返出减少或者增多、钻时突然变快或者放空以及井漏等异常情况,说明已钻至水平通道。

well-3′井在钻井时也存在无法钻遇水平通道的可能,若钻至设计位置仍未钻遇通道,则继续钻进至盐底,通过溶腔连通水平通道。

3.5 水平腔体整体试压

为确保腔体整体密封性,除新钻井需单独每口井进行井筒气密封试压外,还需对整个水平腔体进行密封性测试。试验时 well-1′井和 well-2′井关井憋压,从 well-3′井环空持续注入氮气,观察压力变化情况,连续 24h 内压力降在 0.5MPa 以内为合格。

4 对流井老腔改造试验

4.1 对流老腔基本情况

选取我国某盐穴储气库一组对流井老腔进行改造试验,该对流井井身结构如图 3 所示,两口井井距为 330m,对接点深度为 1490m。

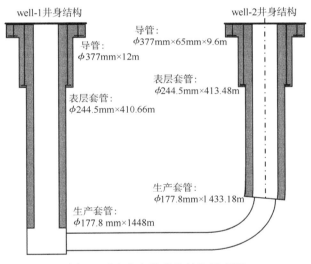

图 3 对流井老腔井身结构示意图

4.2 对流井老腔形态预测

对流老腔形态如图 4 所示,图中 A、C 可通过声呐实际测量得到(表 2),两口井刚投入采卤后,腔体主要向上扩展,由于测溶,水平方向也会向四周扩展,当腔体下部残渣填埋到一定程度后,不再向四周扩展,只向上发展,所以 well-1 井和 well-2 井下部残渣 B 和 D 部分的形状预测为圆柱体。水平通道 E 由于开采过程注入水很快达到饱和,不再溶蚀盐层,因此水平段截

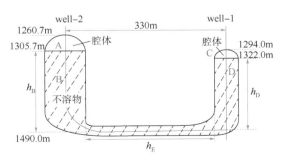

图 4 对流井老腔形态预测

面在水平段中部最小，越靠近 well-1 井和 well-2 井，截面越大，但此处也预测为理想状态下的圆柱体。

表2 well-1 井和 well-2 井声呐测腔结果表

井号	腔顶深度/m	腔低深度/m	腔体高度/m	腔底平均直径/m	腔体体积/m³
well-1	1294.0	1322.0	28	31.16	49904
well-2	1260.7	1305.7	45	41.39	121158

将 B、D 部分腔体预测为圆柱体的话，则 B 部分柱体半径为 41.39m，D 部分柱体半径为 31.16m。根据半径及高度可计算出 B 和 D 部分体积分别为 $74.93\times10^4m^3$ 和 $48.17\times10^4m^3$。再根据两井距离 330m 可计算出图 4 所示 h_E 为 257.45m。

该对流井组共采卤体积为 $678.1\times10^4m^3$，卤水质量浓度为 300g/L 左右，估算动用含盐地层地下空间体积为 $171.2\times10^4m^3$（表3）。

表3 well-1 井—well-2 井盐腔总体积估算表

采卤量/ 10^4m^3	采卤质量浓度/ (g/L)	盐密度/ (kg/m³)	采卤段 NaCl 含量/%	折算采盐体积/ 10^4m^3	盐腔总体积/ 10^4m^3
678.1	300	2160	55	94.2	171.2

根据盐层动用体积、ABCD 部分体积及 h_E，可计算出水平通道腔体 E 部分体积及水平通道平均半径（表4）。

表4 对流井老腔各部分体积预测表

部分	半径/m	高度/m	体积/10^4m^3	备注
A	—	—	12.12	实际测腔体积
B	41.39	184.30	99.14	预测体积
C	—	—	4.99	实际测腔体积
D	31.16	158.00	48.17	预测体积
E	9.16	257.45	6.78	预测体积
总计			171.20	

根据计算 E 部分柱体半径为 9.16m，因此水平通道顶面位置预测介于 1480.84~1490.84m。

4.3 生产套管下深

为保证二开固井时下部地层有足够的承压能力，同时还需留出 100m 左右的溶腔建槽深度，以防止注气排卤时不溶物堵塞排卤管，根据预测结果，综合考虑盐层分布情况，二开生产套管下深定为 1385m。

4.4 对流井老腔改造现场施工

well-3′井采用盐穴储气库常规钻井施工方式，实钻井身结构数据见表5。二开最大井斜 1.49°，井身质量合格，井筒气密封试压满足盐穴储气库密封性要求。

表5 well-3′井实钻井身结构数据表

序号	钻头直径/mm	井深/m	套管外径/mm	类型	套管下深/m
一开	660.4	20	508.0	导管	20
	444.5	516	339.7	表层套管	515
二开	311.2	1386	244.5	生产套管	1385
三开	215.9	1477	—	裸眼	—

生产套管固井后，关闭well-1井和well-2井，采用饱和卤水进行三开作业，当钻至1477m时进入水平通道，此时井筒中卤水发生漏失，补充注入60m³卤水后，卤水从井口返出，说明已钻至水平通道。使用钻头探底，水平通道底部位置在1486m，水平通道位置与预测值基本一致。

整体测试时向well-1井内注入清水，使well-1井口压力升至5.1MPa，此时well-2井压力为3.4MPa。关闭well-1井和well-2井，从well-3′井注入氮气，环空注气压力17.00MPa，气液界面深度为1396.81m。经过24h观察，气液界面降低0.08m，其余压力均无变化，整体试压合格。

由于盐化企业需要继续采卤生产，此次改造试验并未对原well-1井和well-2井进行封堵和钻新井，该两井生产已超过10年，若对该两口老井进行改造，则腔体的密封性和承压能力将进一步提高。

4.5 改造后注气排卤潜力分析

注气排卤时well-3′井作为排卤井，将排卤管柱下入水平通道即1477.0m的深度，1477.0m深度以上的卤水均可有效排出（图5）。

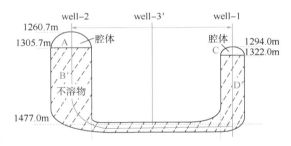

图5 对流井老腔注气排卤示意图

根据图4、表4可计算出1477m以上B′部分圆柱体积为92.15×10⁴m³，D′部分圆柱体积为47.26×10⁴m³，不溶物残渣膨胀后的总体积为139.41×10⁴m³。

根据表3，不溶物原始体积为77.0×10⁴m³（总体积171.2×10⁴m³减去采盐体积94.2×10⁴m³）。从图4可知，实际不溶物体积为154.09×10⁴m³，则不溶物膨胀率=实际不溶物体积/不溶物体积，即154.09×10⁴m³/77.0×10⁴m³=2.0。

根据上面的计算可知，不溶物遇水后膨胀后体积变为原来的两倍，则不溶物残渣中的卤水占比50%，1477m以上B′和D′部分不溶物残渣中的卤水体积为69.71×10⁴m³。假如能排出膨胀体积中20%~40%的卤水，则可增加腔体体积13.94×10⁴~27.88×10⁴m³，根据该储气库运行压力，通过改造该对流井老腔可增加库容3108×10⁴~6216×10⁴m³，增加工作气量1763×10⁴~3526×10⁴m³。

5 结论

（1）利用对流井老腔改建储气库可充分利用底部不溶物中的孔隙空间，增加腔体体积，缩短盐穴储气库建库周期。

（2）结合单腔改造经验和对流井特点，在综合考虑密封性，调峰能力，腔体利用程度等因素的基础上，提出封堵原有两口老井并钻新井、在通道中间再打一口排卤井的对流井老腔改造工艺。

（3）中间排卤井钻遇水平通道是改造成功的关键，从简化工艺及安全考虑，推荐采用三开井身结构，且需要留有足够的三开井段，保证下部地层的承压能力。

（4）我国某储气库的对流老腔改造试验的结果表明，本文提出的改造技术是可行的，试验井可增加腔体体积为 $13.94×10^4 \sim 27.88×10^4 m^3$，增加工作气量为 $1763×10^4 \sim 3526×10^4 m^3$。该试验井的成功改造可推广应用于其他对流老腔的改造。

参 考 文 献

[1] 完颜祺琪，丁国生，赵岩，等．盐穴型地下储气库建库评价关键技术及其应用[J]．天然气工业，2018，38（5）：111-117.

[2] 完颜祺琪，冉莉娜，韩冰洁，等．盐穴地下储气库库址地质评价与建库区优选[J]．西南石油大学学报（自然科学版），2015，37（1）：57-64.

[3] 李玥洋，田园媛，曹鹏，等．储气库建设条件筛选与优化[J]．西南石油大学学报（自然科学版），2013，35（5）：123-129.

[4] 井文君，杨春和，李银平，等．基于层次分析法的盐穴储气库选址评价方法研究[J]．岩土力学，2012，33（9）：2683-2690.

[5] 罩毅，党冬红，刘晓贵，等．盐穴储气库楚资1井 $\phi 244.5mm$ 套管固井技术[J]．石油钻采工艺，2015，37（2）：51-53.

[6] 陈波，沈雪明，完颜祺琪，等．平顶山盐穴储气库建库地质条件评价[J]．科技导报，2016.34（2）：135-141.

[7] 龙雪莲，贺德军．衡阳盐矿区建立地下盐穴储气库的地质条件分析[J]．中国井矿盐，2016，47（1）：19-21.

[8] 垢艳侠，武志德，祁红林，等．安宁盐穴储气库盖层密封性评价[J]．中国井矿盐，2013，44（4）：16-19.

[9] 周冬林，李建君，王晓刚，等，云应地区采盐老腔再利用的可行性[J]．油气储运，2017，36（8）：930-936.

[10] 张艺．金坛盐矿老腔改建储气库可行性研究[D]．重庆：重庆大学，2011.

[11] 田中兰，夏柏如，苟风．采卤老腔改建储气库评价方法[J]．天然气工业，2007，27（3）：114-116.

[12] 齐奉忠．金坛地区盐穴储气库固井技术研究与应用[D]．北京：中国石油勘探开发研究院，2005.

[13] 田中兰，张芳．金坛盐矿采卤溶腔利用与改造技术[J]．中国井矿盐，2005.36（2）：17-20.

[14] 田中兰，夏柏如，申瑞臣，路立君，袁光杰．采卤盐矿老溶腔改建为地下储气库工程技术研究[J]．石油学报，2007，28（5）：142-145.

[15] 丁建林．利用现有采卤溶腔改建地下储气库技术[J]．油气储运，2008，27（12）：42-46.

本论文原发表于《天然气工业》2019年第39卷第6期。

云应地区采盐老腔再利用的可行性

周冬林　李建君　王晓刚　刘继芹　井　岗

（中国石油西气东输管道公司储气库项目部）

【摘　要】　地下盐岩由于良好的物理力学特性，被公认为是存储石油、天然气最理想的介质，中国利用地下盐岩建设储气库起步较晚，已建成的储气库产能亟须提高。与此同时，中国储气库建设面临建库优质盐矿资源短缺、建库成本高、投产速度慢等问题，而各地盐矿企业采盐形成的地下老腔数量众多，因此利用老腔改建储气库是加快储气库建设投产的有效途径。通过分析云应地区盐层特征和采卤井资料，从盐腔的密封性、稳定性及经济性方面对该地区的老腔进行评价。结果表明：云应地区的盐层埋深较浅，构造平缓，无断层破坏，盖层密封性好。地下老腔数量众多，体积大，顶板厚度大，矿柱宽度大，部分老腔由于井距较小且采用压裂连通开采，存在盐腔连通串层等问题。相比之下，采用水平井对接方式采卤的老腔井距较大，矿柱宽度较为理想且未受压裂破坏，是该地区老腔改造的首选。

【关键词】　云应地区；储气库；老腔；盐岩

Feasibility of the reuse of existing salt mining caverns in Yunying area

Zhou Donglin　Li Jianjun　Wang Xiaogang　Liu Jiqin　Jing Gang

(Gas Storage Project Department, PetroChina West-East Gas Pipeline Company)

Abstract　Underground salt rock is globally recognized as the most ideal medium for oil and natural gas storage because of its good physical and mechanical properties. In China, however, the construction of underground gas storage (UGS) in salt rocks starts later, and the productivity of existing UGSs needs improving urgently. Furthermore, Chinese UGS construction is faced with the problems of deficient good salt mine resources, high construction cost and slow commissioning speed. Whereas there are a great number of old underground caverns that are formed by salt mine enterprises by means of the salt mining all around the country, the effective way to speed up UGS construction and commissioning is to rebuild UGS from existing caverns. In this paper, the existing salt caverns in Yunying area were evaluated from the aspects of sealing capacity, stability and economy by analyzing the characteristics of salt layers and the data of brine producers. It is indicated that the salt layers in Yunying area are characterized by shallo-

---

基金项目：国家重大科技专项资助项目"地下储气库关键技术研究与应用"，2015E-400801。

作者简介：周冬林，男，1987年生，工程师，2012年硕士毕业于中国石油大学（北京）地质工程专业，现主要从事盐穴储气库相关的地质评价工作。地址：江苏省镇江市润州区南徐大道A区商务楼D座，212000。电话：0511-84522031，Email：zhoudonglin@petrochina.com.cn。

wer burial depth, gentle structure, good sealing capacity of cap rocks and no faulting damage. There are considerable underground caverns and they are large with thick roofs and large pillar width. Some caverns are connected between layers for they are produced by means of fracturing connection with short well spacing. By comparison, those caverns where brine is produced in the mode of horizontal well butt with larger well spacing are not damaged by fracturing and their pillar width is relatively satisfactory, so they are taken as the preferred caverns for reconstruction.

Keywords Yunying area; underground gas storage; existing salt caverns; salt rock

盐岩是能源存储最理想的介质，世界上90%的能源都存储在盐岩介质或报废的盐矿井中[1]。中国是井矿盐生产大国，地下盐腔数量众多，仅湖北云应、河南平顶山等地有老腔500多个，大部分充满卤水，处于废弃闲置状态。中国石油及天然气等战略物资的储备所需地下空间缺口巨大，而可利用的建库资源非常短缺[2-5]，因此利用盐矿采卤形成的老腔改建储气库符合实际情况，前景广阔。尽管层状盐岩采卤形成的老腔存在夹层多、腔体形态不规则、井筒固井质量差、套管损坏等问题[6-11]，但在充分调查和评估的基础上，仍可以找到经济性好、安全性高、有改造前景的老腔，综合考虑建库时间和环境保护等因素，利用老腔改建储气库是加快储气库建设切实可行的方法。

20世纪50年代，荷兰、加拿大以及美国等国家已经使用废弃盐腔储存石油和天然气，20世纪80年代进一步使用盐腔储存氢气、废弃残渣以及核废料。中国盐穴储气库建设起步较晚，2005年，金坛储气库首次使用6个盐矿老腔改建储气库并获得成功。近10年来，随着中国石油天然气战略存储及废弃残渣处置问题日渐突出，废弃盐腔再利用研究越来越受重视。江苏淮安井神盐矿公司早在2011年已经利用自有盐腔处理废弃碱渣，陈结等[12-18]提出利用水平盐腔建立国家战略石油储备的技术思路，并探讨了相关技术问题。当前中国公开的利用废弃盐腔改建储气库或用作存储其他介质的案例屈指可数，储气库建设大部分通过新井溶盐造腔实施，与此相反，国外专门钻井溶盐建库的占比不到10%，绝大部分使用废弃盐腔或矿坑建库。无论从国际经验还是国内实际情况来看，老腔的再利用评价必然是未来盐穴储气库建设的重要研究方向。

1 云应地区特征

1.1 构造特征

云应地区位于湖北省云梦县和应城市交界处，是中国井矿盐主要生产区之一，地下盐矿资源丰富，且邻近西气东输二线，是中国中部地区储气库建设优选区域。含盐地层构造位于江汉盆地云应凹陷内，为白垩纪—新近纪的一个内陆盐湖断陷，北界不整合超覆于震旦系—寒武系及元古宇之上，构成向南西倾斜的斜坡，南部与龙赛湖低凸起毗连，西以皂市断裂为界，东至汉川凸起，是一个由断裂控制、北缓南陡的箕状凹陷盆地，面积2700km^2（图1）。

1.2 地层特征

含盐层赋存于古近系云应群膏盐组，矿床面积为188km^2，探明盐岩矿石储量为357×10^{8}t，折算成NaCl工业储量为250×10^{8}t，埋深500~1000m，含盐地层平均厚度为477m，为

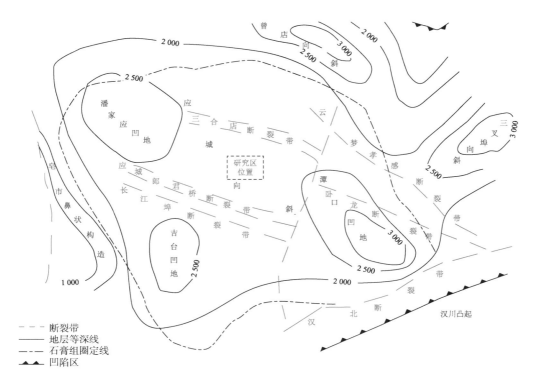

图 1 云应地区凹陷地质构造示意图(单位:m)

大型隐伏式盐岩矿床。岩性为灰—灰白色盐岩、泥质钙芒硝岩、泥质硬石膏岩、钙芒硝质泥岩、硬石膏质泥岩夹赭色泥岩、粉砂质泥岩、硬石膏质泥岩。从下而上可划分为下硬石膏段、下钙芒硝段、盐岩段、上钙芒硝段、上硬石膏段。盐岩段为盐岩主力层段,划分盐群50个,盐层累计厚度为190~240m,盐岩段内含矿率为37.42%~55.49%,盐岩组内含矿率为65.47%~84.13%。盐层品位低,以盐岩、泥岩、硬石膏和钙芒硝互层方式产出,盐群厚度为1.2~16m,夹层厚度为0.2~0.5m,整体上由边部向盆地中心盐层数增多、盐层总厚度加大,而夹层数则由边部向盆地中心减少,夹层总厚度由边部向盆地中心减小。

2 采卤井生产情况

云应地区共有7家盐矿企业,拥有盐井300余口,总制盐规模为$610×10^4$t/a,经过长期采卤生产,在地下形成的盐腔体积为$2726.15×10^4 m^3$。采卤利用双井对流方式开采,一口井注淡水,另一口井生产卤水。早期的对井都是两口直井生产,井间利用压裂造缝连通,2000年以后采用水平井加直井方式开采,井间利用水平井裸眼井段连通(图2)。该地区根据湖北省盐业开采规程和技术规范,在50个盐群中划分并确定K_2、K_5和K_8 3个采层作为开采对象,其他各层没有开采。其中K_2采层对应01盐群和1盐群,K_5采层对应13~15盐群,K_8采层对应29~30盐群,每对井都只开采一个层位。通过资料调研和收集,共明确盐矿井位87口,除了19口废弃井已无井组对应关系,明确对井34对,其中压裂井组29对,水平对接井组5对。

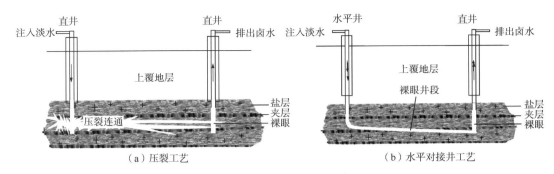

图2 对井采卤生产工艺示意图

3 老腔再利用的可行性

根据收集的盐矿采卤井测井解释资料和区域三维地震解释成果，云应地区断裂不发育，含盐地层构造平缓，整体呈现出北高南低的特点。基于此，结合盐层埋深、盐层岩性、上部盖层、井间距离以及生产资料，分析老腔的密封性、稳定性及经济性，评价该地区老腔的可利用性。

3.1 密封性

采卤层位由上至下分为 K_2、K_5 及 K_8 采层，因此 K_2 采层以上至盐顶之间的盐层段以及上硝段和上膏段都是腔体的盖层。K_2 采层至盐顶的盐层厚度为 25.1~75.2m，为盐岩和泥岩、钙芒硝和硬石膏互层，上覆钙芒硝和硬石膏段厚度为 140~300m，主要为钙芒硝、硬石膏、泥岩和粉砂岩互层。因此，从宏观来看，腔体上部盖层分布稳定，厚度较大，岩性致密，无断层破坏，密封性良好；从微观来看，盖层的密封性主要考虑上部岩层的渗透性，根据云应储气库 YZ1 井岩心样品的分析结果，上部盖层主要是盐岩和泥岩、粉砂岩和硬石膏，岩性致密，平均孔隙度为10%，平均渗透率为 0.13mD。06 盐群岩样突破压力为 3.5MPa，封闭气柱高度大于 300m，超过 06 盐群上部含盐岩地层总厚度 246m，且 K_2 采层与 06 盐群之间仍有 02~05 盐群，因此，盖层的密封性是有保障的。

3.2 稳定性

K_2 采层深度为 360~480m，开采时间较早，均采用压裂连通方式开采，对井距为 110~260m，井组距为 120~170m，井组距较小，由于各井组之间的布置缺乏规范，井组之间距离更小，甚至出现井组交叉的情况，因此 K_2 采层的盐腔已经成片连通，地下情况十分复杂，盐腔之间没有足够宽的矿柱支撑，不利于腔体的稳定，因此，这类成片连通的盐腔改建储气库的可能性很低。

K_5 采层深度为 470~510m，采卤井 17 对，其中仍在生产的井 8 对，均采用压裂连通方式开采，对井距为 180~250m，井组距为 250m，呈南北向排列分布；K_8 采层深度为 570~640m，采卤井 10 对，全部在生产，5 对井采用压裂连通方式开采，另外 5 对井采用水平对接方式开采，对井距为 250~280m，井组距为 265m，呈南北向排列分布。K_5 和 K_8 采层对井距和井组距较大，生产年限相对较短，腔体之间矿柱宽度相对较大（图3）。根据各井生产资料，利用采盐体积估算地下盐腔体积，参照金坛储气库实际溶腔情况，以盐腔半径

40m计算,最小矿柱宽度为120m,最大矿柱宽度为215m,平均为177.4m。矿柱宽度与直径之比平均值为2.1,在储气库建设中,为了保障盐腔稳定,一般要求盐腔之间的矿柱宽度与盐腔直径(80m)之比不小于2.0,以此判断老腔的安全矿柱条件是否符合稳定性标准。

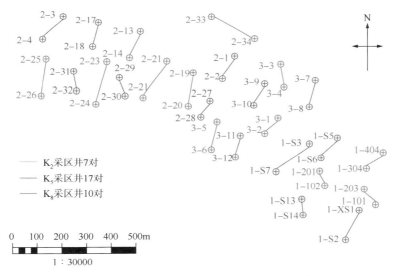

图3 云应地区K_2、K_5、K_8采区部分对井分布图

由于盐腔长期溶盐采卤,地下盐腔的腔顶可能已经上溶,甚至突破所在采层的顶板[19-22],因此无法明确各盐腔的腔顶深度。但盐腔顶板的稳定性仍然可以通过盐层厚度和井数据进行保守估计,较浅的K_2采层上部盐层厚度为25.1~75.2m,K_5采层上部盐层厚度为150~240m,K_8米层上部盐层厚度为250~380m,同时由各井套管鞋的下入深度来看,各井套管鞋以上盐层厚度在40~300m之前,虽然腔体上溶可能会造成套管鞋的裸露,但这一厚度也可以作为参考。由于盐岩具有较好的密封性,因此,一般要求盐腔顶部以上拥有15m盐层作为顶板以保障盐腔的稳定。综合各井数据来看,即使考虑腔体可能上溶后,各腔体的顶板厚度仍是有保证的。

盐矿早期的采卤井广泛使用压裂连通方式采卤,87口井中77口实施过压裂。由于早期压裂技术相对初级和简单,造缝技术无法做到定向和可控,必然会造成井周围地层应力状态的破坏,压裂将在地层中造成许多不定向延伸的小裂隙,形成高渗透区,不仅易导致腔体之间串层或者连通,也会破坏腔体的密封性,且盐腔的形态难以控制,不利于盐腔的稳定,因此,压裂对于老腔有破坏性影响,不利于老腔的再利用[23-24]。企业修井记录也证明部分井组如K_8层的2-13/2-14井组已经上溶,与K_5层的2-15/2-16井组串层连通,不排除其他压裂井组也存在连通的可能。

3.3 经济性

老腔的体积越大,改造的经济性越好。对比建造金坛等地新腔的建设成本,结合云应地区实际情况,改造的老腔体积应在$10×10^4 m^3$以上。根据云应地区采卤井的累积采盐体积和盐层不溶物含量,按泥岩膨胀系数1.5计算各腔体体积(表1)。结果表明:K_5和K_8采层的盐腔体积较大,平均约为$30×10^4 m^3$,改造经济性较好。

表 1 按泥岩膨胀系数 1.5 计算各生产井腔体体积

井号	采层	连通方式	套管鞋以上盐层厚度/m	盐顶深度/m	最小井组距/m	最小矿柱宽度/m	矿柱与直径比	腔体净体积/m³	备注
3-1	K₈	压裂	314	293	290	210	1.6	229524	—
3-2			322	300					
3-3	K₈	压裂	293	286	260	180	2.3	226517	—
3-4			111	487					
3-5	K₈	压裂	79	530	240	160	2.0	526564	—
3-6			187	280					
3-7	K₅	压裂	190	298	295	215	2.7	348765	—
3-8			187	280					
3-9	K₅	压裂	186	280	220	140	1.8	348765	—
3-10			203	286					
3-11	K₅	压裂	192	295	280	200	2.5	420281	—
3-12			—	—					
2-11	K₈	压裂	102	489	260	180	2.3	393231	—
2-12			311	296					
2-13	K₈	压裂	79	489	220	140	1.8	357437	上溶蚀与K₅层串层，排除串层连通，排除
2-14			323	274					
2-15	K₅	压裂	32	428	230	150	1.9	284978	
2-16			48	435					
2-17	K₅	压裂	38	423	200	120	1.5	429897	—
2-18			66	420					
2-19	K₈	水平对接	—	—	230	150	1.9	421428	
2-20									
2-21	K₈	水平对接	—	—	230	150	1.9	191178	
2-22			62	555					
2-23	K₈	水平对接	—	—	280	200	2.5	198015	
2-24			58	568					
2-25	K₈	水平对接	—	—	295	215	2.7	162018	
2-26			61	564					
2-27	K₅	压裂	32	448	210	130	1.6	262782	
2-28			59	441					
2-29	K₅	压裂	48	455	275	195	2.4	262782	
2-30			39	441					

续表

井号	采层	连通方式	套管鞋以上盐层厚度/m	盐顶深度/m	最小井组距/m	最小矿柱宽度/m	矿柱与直径比	腔体净体积/m³	备注
2-31	K_5	压裂	68	438	275	195	2.4	303529	—
2-32			—	—					
2-33	K_8	水平对接	—	—	290	210	2.6	135940	—
2-34			61	520					

4 结论

通过分析云应地区含盐地层构造和地层特征，结合采卤井生产资料，分析了老腔盖层的密封性，利用井资料计算了老腔的体积和井距，并估算了盐腔最小矿柱宽度和顶板厚度，分析了压裂对盐腔的破坏性影响。

（1）云应地区盐层构造平缓，无断层破坏，腔体埋深较浅，腔体顶板和盖层的稳定性和密封性好，适宜建设储气库。

（2）盐腔顶板厚度大，腔体体积大，矿柱宽度符合稳定性标准，改造经济性好，但部分老腔存在距离过近，受压裂连通方式采卤影响，容易出现连通串层等问题，不利于盐腔的稳定。

（3）采用水平对接井生产的5个老腔未受压裂破坏，是该地区老腔改建的优选对象，下一步应对老腔进行试压和声呐检测，进一步落实腔体的密封性和形态体积，为该地区老腔改造储气库提供依据。

参 考 文 献

[1] 杨海军. 中国盐穴储气库建设关键技术及挑战[J]. 油气储运，2017，36(7)：747-753.
[2] 杨春和，梁卫国，魏东吼，等. 中国盐岩能源地下储气储存可行性研究[J]. 岩石力学与工程学报，2005，24(24)：4409-4417.
[3] 王清明. 石盐矿床与勘查[M]. 北京：化学工业出版社，2007，12-14.
[4] 丁国生，李春，王皆明，等. 中国地下储气库现状及技术发展方向[J]. 天然气工业，2015，35(11)：107-112.
[5] Lina R. The development and new challenge of UGS in China[C]. Novy Urengoy：First Working Committee Meeting of WOC Storage 2015-2018 Triennium，2015：1-7.
[6] 杨海军，郭凯，李建君. 盐穴储气库单腔长期注采运行分析及注采压力区间优化——以金坛盐穴储气库西2井腔体为例[J]. 油气储运，2015，34(9)：940-950.
[7] 杨海军，于胜男. 金坛地下储气库盐腔偏溶与井斜的关系[J]. 油气储运，2015，34(5)：145-149.
[8] 李建君，王立东，刘春，等. 金坛盐穴储气库腔体畸变影响因素研究[J]. 油气储运，2014，33(3)：269-273.
[9] 郭凯，李建君，郑贤斌. 盐穴储气库造腔过程夹层处理工艺——以西气东输金坛储气库为例[J]. 油气储运，2015，34(2)：162-166.
[10] 杨海军，闫凤林. 复杂老腔改建储气(油)库可行性分析[J]. 石油化工应用，2015，34(11)：59-61.
[11] 田中兰，夏柏如，苟凤. 采卤老腔改建储气库评价方法[J]. 天然气工业，2007，27(3)：114-116.

［12］陈结，姜德义，刘春，等．盐穴建造期夹层与卤水运移互相作用机理分析［J］．重庆大学学报（自然科学版），2012，35(7)：107-113．

［13］Tian Z L, Yuan G J, Shen R C, et al. Appraisal method for gas storage construction based on exiting salt cavern［C］. Porto：Solution Mining Research Institute Spring Technical Conference, 2008：1-10.

［14］Hill L, Encinitas. Cavern well completion techniques including conversion of old cavern wells to alternate uses ［C］. Berlin：Solution Mining Research Institute Fall Technical Conference, 2004：1-6.

［15］Liu W, Chen J, Jiang D, et al. Tightness and suitability evaluation of abandoned salt caverns served as hydrocarbon energies storage under adverse geological conditions (AGC)［J］. Applied Energy, 2016, 178: 703-720.

［16］杨长来，孔君凤，刘伟．盐矿水溶开采地表塌陷发生机理及防治措施［J］．土工基础，2014，28(3)：128-131．

［17］任松，李小勇，姜德义，等．盐岩储气库运营期稳定性评价研究［J］．岩土力学，2011，32(5)：1465-1472．

［18］陈结，刘伟，任松，等．利用双井盐穴溶腔建立国家战略石油储备体系［J］．大科技，2016(16)：285-286．

［19］夏筱红，杨伟峰，刘志强．应城盐矿区采卤对接井施工技术［J］．中国井矿盐，2002，33(3)：22-23．

［20］李文魁．平顶山盐田现有盐穴老腔的再利用探讨［J］．中国井矿盐，2015，46(21)：20-22．

［21］Jun K, Yee P E. Solution mining plan for the enlargement of an existing Storage cavern capacity with additional leaching program［C］. Houston：Solution Mining Research Institute Spring Technical Conference, 2005：1-7.

［22］Rauche, Heidrum, Henry. Exploration and risk assessment of solution and caving caverns in the old mining of the Stassfurt area［C］. Nancy：Solution Mining Research Institute Fall Technical Conference, 2005：1-6.

［23］Dale S P, Jerry T F. Finite element analysis of salt cavern employed in the strategic petroleum reserve［C］. New Mexico：Solution Mining Research Institute Fall Technical Conference, 1982：1-6.

［24］Stefan W. Releaching and solution mining under gas (SMUG) of exiting caverns［C］. Basel：Solution Mining Research Institute Spring Technical Class Practical Aspect of Solution Mining, 2007：1-8.

本论文原发表于《油气储运》2017年第36卷第8期。

CSAMT 法在盐化老腔形状检测中的勘探试验

井 岗 何 俊 翁小红 王晓刚 薛 雨

(中国石油西气东输管道公司储气库管理处)

【摘 要】 盐化企业采卤形成的地下空间是一种宝贵的储气资源,盐化老腔直接改造成储气库可以缩减建库时间、迅速提高库容和节省资金。目前老腔形状检测是改造过程中的技术瓶颈,腔体的形状直接关系腔体稳定性。腔体卤水和围岩在电阻率上存在差异,这为 CSAMT 法提供了地球物理勘探前提条件。本次试验选择在淮安地区开展可控源音频大地电磁勘探法,采用合理的野外排列方式、数据处理、资料解释及腔体形态构建,确定了腔体的空间形态展布。

【关键词】 CSAMT;V8;盐腔;声呐

Exploration of CSAMT test method in shape detection of salt cavity

Jing Gang He Jun Weng Xiaohong Wang Xiaogang Xue Yu

(Gas Storage Department of PetroChina West East Gas Pipeline Company)

Abstract Underground space by salt company brined is a kind of precious natural gas resources which can directly transform into gas storage. It can reduce construction time, quickly increase capacity and save money. At present, the shape of salt cavity that is important to the stability of cavity detection is the technical bottleneck in the process of transformation. There are differences in the resistivity between the brine and surrounding rock, which provides a geophysical prerequisite for the CSAMT method. It carried out controlled source audio-frequency magnetotellurics method in Huai An for exploration test. By using reasonable field arrangement, data processing, data interpretation and construction of cavity shape, it determined the distribution of spatial morphology of the cavity.

Keywords CSAMT;V8;salt cavity;sonar

1 腔体检测技术现状

目前国内外成熟的地下腔体检测采用声呐测量技术,盐穴声呐测量技术的工作原理为:沿采盐井井筒下放声呐测量井下仪器,井下仪器的声呐探头进去盐穴腔体后,在某一深度

基金项目:中国石油储气库重大专项子课题"地下储气库关键技术研究与应用",项目编号:2015E-40。

作者简介:井岗,男(1986—),工程师,主要从事盐穴储气库地质研究工作。

进行360°水平旋转,同时按设定的角度间隔向盐穴腔体壁发射脉冲声波,接收回波信号,信号经井下仪器的连接电缆传回地面中心处理器,得到某一深度上的腔体水平剖面图;在盐穴腔体内不断改变检测深度,则可获得腔体不同深度的水平剖面,最终可得到整个腔体的体积和三维形态图像[1],声呐测量施工工艺图(图1),金坛地区A井的声呐三维图像(图2)。

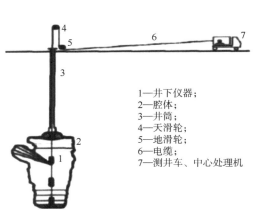

图1 声呐测量施工工艺示意图　　　　图2 金坛A井声呐三维图

声呐测量技术可以很好地测量单井单腔这种立式盐腔,盐化企业采卤一般采用直井+斜井水平对接的方式注水采卤,形成的腔体(图3),对于直井段的腔体形态可以通过声呐测量获取,对于水平段和斜井段的腔体形态目前还没有技术可以获取,这就导致无法深入评价水平盐腔的可用性,水平对接采卤井形状检测就成了老腔改造评价的技术瓶颈。

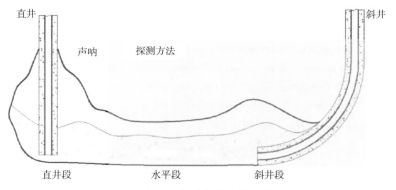

图3 水平对接采卤井腔体示意图

2 工区地质概况及勘探前提

2.1 工区地质概况

在大地构造上,淮安楚州盐矿赋存于苏北盆地盐阜坳陷淮安中断陷。淮安中断陷为中生代的断陷,新生代的凸起,南深北浅的箕状断陷。淮安盐矿内地层有第四系东台组、新近系盐城组上段、上白垩统赤山组和浦口组,期间缺失整个古近系和新近系盐城组下段。淮安盐矿含盐地层为上白垩统浦口组中部,为一套暗棕色、灰黑色泥岩、膏质泥岩、硬石膏、钙芒硝与盐岩组成不等厚互层。根据不同岩类的组合特点及剖面上的分布状况,可将含盐层剖面分为三个岩性亚段:上盐亚段、淡化段、下盐亚段。上盐亚段第四岩性组合底板埋深800~2900m,平均埋深1600m左右。上盐亚段第三岩性组合底板埋深1000~3300m,平均埋深2000m左右。上盐亚段底板埋深1300~3700m,平均埋深2200m左右。

2.2 地球物理勘探前提条件

根据区域地质资料显示:工区内主要被第四系覆盖,目标层段成岩性较好,溶腔内满含卤水。根据本次试验目的,结合钻探资料、区域地质资料,将区内主要岩性电阻率参数总结(表1)。

表1 工区地层电阻率参数

地层	组段	岩性	电阻率/($\Omega \cdot m$)
Q	顶部	黄白色夹灰色黏土亚黏土含少许钙质结核	5~22
Q	中部、底部	黄白色细砂和粉砂层	6~16
Ny		下段厚63.26~164m,岩性为灰色、灰绿色中粗砂岩与中细砂岩互层,局部夹泥岩,底部泥岩与细砂岩互层,局部含砾;上段厚125~197m,岩性为土黄、灰黄、灰绿色中细砂层、中粗砂层、黏土层互层	9~18
K_2r_p		砂质泥岩为主,富含硬石膏团块,局部夹钙芒硝条带,中下部产叶肢介化石,时代属晚白垩世;上部以棕红色泥岩及粉砂质为主夹细砂岩,普遍含有硬石膏团块和斑点	15~24
K_2r_p	上盐亚段	以灰色盐岩为主,夹钙芒硝岩、泥岩和粉砂岩	18~24

地层水电阻率与地下水矿化度的函数关系为:

$$\lg \rho_水 = a + b \lg C$$

式中:$\rho_水$为地层水的电阻率;C为矿化度;b为与矿化度有关的系数;a为与温度有关的系数。

为了更明确地说明地下水矿化度与电阻率的关系,下面给出苏联某地区松散类地下水矿化度与电阻率的关系曲线(图4)。盐卤水的电阻率约为$4\Omega \cdot m$,围岩电阻率为$18~24\Omega \cdot m$,两者之间电阻率差异较大[2],这就为CSAMT电磁勘探提供了地球物理前提。

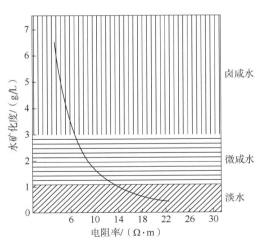

图4 电阻率与地下水矿化度的关系曲线

3 CSAMT 法野外数据采集

3.1 野外采集仪器设备

可控源音频大地电磁简称 CSAMT，是利用接地水平电偶源为信号源的一种频率域电磁测深法。CSAMT 采用了大功率的人工场源，具有信号稳定、信噪比高、穿透能力强、探测深度大等特点。近几年，CSAMT 作为研究深部矿成矿环境研究的有限手段，在研究盆地基底埋深起伏、盖层及成矿目的层展布规律、确定隐伏构造等方面得到广泛应用。探测深度可用公式 $H = 356\sqrt{\dfrac{\rho_s}{f}}$ 来计算，其中探测深度（H）与视电阻率 ρ_s 的平方根成正比，与频率（f）的平方根成反比[3]。本次 CSAMT 采用加拿大产 V8 电法系统，包括发电机组、发射机和接收机三个部分。发射机为 TXU-30(20kW)，发电机组是德国进口 30kW 的 24GF 型柴油发电机组，最高电压为 1000V（I=20A），最大电流为 40A（V=500V）；接收机系统由 V8 多功能电法仪和两个 RXU（辅助采集站）组成。V8 多功能电法仪具有以下特点：每道采用 A/D 转换器，低频 MT 弱信号采用 24 位，保持了最高的动态范围和分辨率；发射和接收为无线连接，始终采用 GPS 同步，避免了不断需要校对时钟同步的麻烦和出错的可能性；采用 TXU-30 发电机组，功率大、频率高，在提高观测信号的同时，可有效地避免我国 50Hz 的工业电干扰，有利于在矿区和城市附近开展工作。

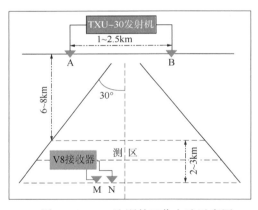

图 5 CSAMT 法野外工作方法示意图

3.2 观测系统及采集参数

CSAMT 野外观测系统示意图如图 5 所示。在 CSAMT 法野外工作中，观测区域布置在一个梯形范围内。发射偶极 AB 平行测线布置，AB 到测线的距离应大于 3 倍的趋肤深度，一般为 6~12km，测线的长度应保持在梯形范围内。

本次 CSAMT 试验共布置 9 条测线，近南北方向，分别为 20 线、30 线、35 线、40 线、50 线、55 线、60 线、70 线和 80 线，线距为 20m。

本次试验探测最大深度为 1800m，采集频率选择为 3072~0.44Hz，采用自动采集方式，共 38 个频点，深度在 1200~1500m 附近，设计加密，大大提高了分辨率；发射和接收采用 GPS 时钟对时，达到时间上的严格一致。本次 AB 偶极布设均平行测线，AB 长均大于 1.7km，收发距为 6.5km。接收测线信号均落在 AB 电极中间张角 60°范围内，发射最大电流达到 18A。

4 CSAMT 法室内数据处理

CSAMT 法数据处理的流程如图 6 所示。

（1）对采集数据进行预处理，转换成处理反演软件所要求的 SEC/PLT 文件，检查数据转换时给定的排列采集参数是否正确；

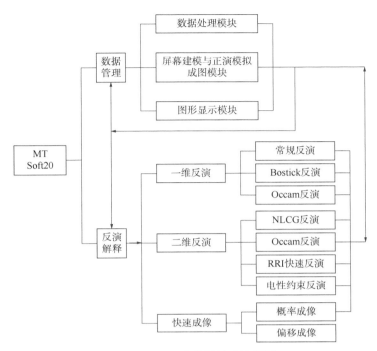

图 6 CSAMT 法数据处理流程图

（2）建立 L 记录文件，先后导入 SEC/PLT 文件及测线坐标 TXT 文件；

（3）人工剔除跳点、曲线平滑、静态校正及空间滤波。

由于干扰和观测误差的存在，在频率—视电阻率曲线上有时会出现非正常的跳跃。结合干扰记录，根据相邻测点的曲线特征，对野外采集的原始视电阻率和时间衰减曲线突变点进行平滑或丢弃[4]。

通过观测整条剖面上各测点电阻率—频率测深曲线的高频端曲线，结合野外地层的出露现状，采用电阻率曲线之间的以及相位曲线之间的相关系数识别 CSAMT 资料中的静态效应。

（4）建立相应的地电模型进行正反演计算，以确定最佳的反演方法；

（5）使用不同的方法包括近场 Bostick 反演以及二维带地形反演，输出反演结果绘制成电阻率等值线图。

5 CSAMT 法处理成果解释

5.1 电阻率剖面解释

反演处理的电阻率等值线剖面中，深约 1400m 处存在低阻异常，分析该异常由地下采卤形成的盐腔所致，该异常的边界电阻率为 $10^{0.85} \approx 7\Omega \cdot m$。盐卤水的电阻率约为 $4\Omega \cdot m$。本次试验深度较深，反演的电阻率结果是相对值，而不是真实的电阻率。通过建立模型进行正演并做二维反演，深度为 1450m 处电阻率为 $4\Omega \cdot m$ 的卤水腔反演出来的相对电阻率在 $7\Omega \cdot m$ 左右，故结合研究区地层电性特征，对反演剖面结果做出解释，用电阻率值为 $7\Omega \cdot m$

的等值线来圈定腔体的边界。

35线CSAMT反演电阻率解释剖面如图7所示，在测线平面位置1212～1270m，埋深1331～1490m处，存在$7\Omega \cdot m$的低阻异常，该异常为水平对接井采卤形成的腔体边界，该剖面上盐腔高差159m，宽度为58m。55线CSAMT反演电阻率解释剖面（图8），在测线平面位置1123～1173m，埋深1334～1490m处，存在$7\Omega \cdot m$的低阻异常，该剖面上盐腔高差为156m，宽度为50m。

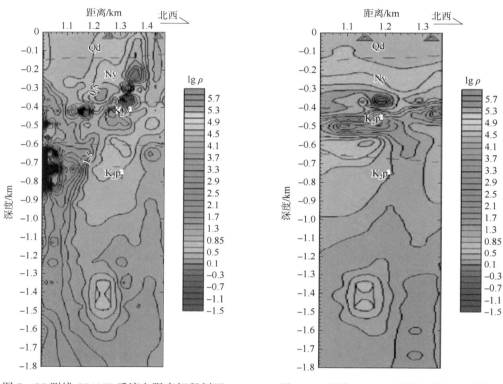

图7　35测线CSAMT反演电阻率解释剖面　　　图8　55测线CSAMT反演电阻率解释剖面

5.2　腔体三维形态构建

根据各条剖面刻画的低阻异常等值线（图9深色部分所示）进行空间差值，可构建探测的腔体三维形态（图10），计算其体积约为$180\times10^4 m^3$。根据实际采出卤水约$500\times10^4 m^3$，平均出卤质量浓度约300g/L，结合该矿区采卤成腔经验。估算实际形成的盐腔体积（含沉渣）应在$150\times10^4 m^3$左右，两者计算的体积差异分析可能原因见结论部分。

6　结论

本次试验的主要目的是在淮安地区开展可控源音频大地电磁勘探试验，通过野外数据采集、室内处理反演及解释电阻率剖面来获取地下腔体尺寸规模，探讨CSAMT法刻画腔体空间展布的能力。

（1）通过野外试验选取合理的观测系统参数，获取质量较好的原始采集数据；

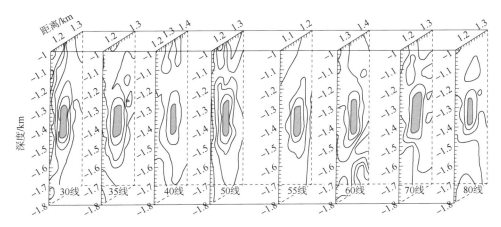

图9 CSAMT剖面叠加图

（2）通过V8现场测试及数据处理，基本掌握了由V8探测数据获取水平盐腔的空间形态方法；

（3）CSAMT法信号稳定，探测频率可根据需要加密，纵向分辨率高，对地下低电阻率腔体异常有较好的响应，可用于腔体尺寸定量—半定量的刻画；

（4）水溶采卤后腔体顶板可能会在一定范围内产生裂缝，高压卤水侵入顶板围岩，导致其电阻率大大降低，进而引起探测体积大于采卤计算体积；

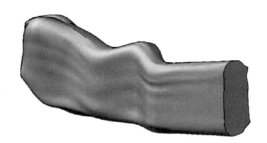

图10 腔体三维形态

（5）CSAMT法刻画的腔体形态属于间接数据获取的结论，建议在直井段进行声呐测量，对V8解释结果进行校正，以获得最可靠的测试结果。

参 考 文 献

[1] 屈丹安，杨海军，徐宝财．采盐井腔改建储气库和声呐测量技术的应用[J]．石油工程建设，2009，35(6)：25-28.
[2] 武毅，郭建强．用电阻率评价孔隙类地下水矿化度的方法技术[C]．地下水勘察与监测技术方法经验交流会，2003.
[3] 顾勇．V8多功能电法仪及其方法技术[J]．新疆有色金属，2009，(增刊)：20-22.
[4] 王大勇，李桐林，高远．CASMT法和TEM法在铜陵龙虎山地区隐伏矿勘探用的应用[J]．吉林大学学报(地球科学版)，2009，39(6)：1135-1139.

本论文原发表于《石油化工应用》2016年第35卷第7期。

Feasibility analysis of using abandoned salt caverns for large-scale underground energy storage in China

Chunhe Yang[1]　Tongtao Wang[1]　Yinping Li[2]　Haijun Yang[2]　Jianjun Li[2]
Dan'an Qu[2]　Baocai Xu[2]　Yun Yangc[3]　J. J. K. Daemen[4]

(1. State Key Laboratory of Geomechanics and Geotechnical Engineering, Institute of Rock and Soil Mechanics, Chinese Academy of Sciences; 2. West-to-East Gas Pipeline Company Gas Storage Project Department, PetroChina Company Limited; 3. McDougall School of Petroleum Engineering, University of Tulsa; 4. Mackay School of Earth Sciences and Engineering, University of Nevada)

Abstract　Rock salt in China is primarily bedded salt, usually composed of many thin salt layers and interlayers (e.g. anhydrite, mudstone, and glauberite). Thus, the feasibility analysis of abandoned salt caverns located in salt beds to be used as Underground Gas Storage (UGS) facilities is full of challenges. In this paper, we introduce the feasibility analysis of China's first salt cavern gas storage facility using an abandoned salt cavern. The cavern is located in Jintan city, Jiangsu province, China. The mechanical properties and permeability of the bedded salts are obtained by experiments. Based on the results of the analyses, it appears to be quite feasible to convert the abandoned salt caverns of Jintan city to UGS facilities. The stability of the cavern is evaluated by the 3D geomechanical numerical simulations, and the operating parameters are proposed accordingly. Results indicate that the maximum volume shrinkage of the cavern is less than 25% and the maximum deformations are less than 2% of the caverns' maximum diameters after operating for 20 years. It is recommended that the weighted average internal gas pressure be maintained as 11MPa to control the extent of the plastic zones to a safe level. Safety factors decrease with operating time, especially those of the interface between rock salt and mudstone layers decrease significantly. Effective strain is generally greater than 2%, and locally is greater than 3% after operating 20 years. The maximum pressure drop rate should be kept to less than 0.55MPa/day. Based on above proposed parameters, China's first salt cavern gas storage facilities were completed, and gas was first injected, in 2007. To check the status of the caverns after operating for 6 years, the volumes of the caverns were measured in 2013 by Sonar under working conditions. Measurement results show that the cavern shapes did not change much, and that volume shrinkages were less than 2%. Comprehensive results show that the feasibility analysis method proposed in this paper is reliable.

Keywords　energy storage; underground gas storage; abandoned salt cavern; numerical simulation stability; field monitoring

---

Corresponding author: Chunhe Yang, chyang@whrsm.ac.cn.

1 Introduction

Large-scale energy storage systems are used widely in the major industrial countries to reduce the disadvantages of energy demand fluctuations in electricity power grids[1-4]. Pumped hydropower, compressed air and natural gas energy storage are the main methods. The consumption demand for natural gas has similar fluctuations. Consumers' demand for gas changes daily and seasonally. During peak times, the largest amount of gas is needed (peak load), but a "base load" of gas is needed year-round. However, the supply from gas fields and the transport capacity of pipeline systems are basically constant. Therefore, the balancing of supply and demand is required in order to maintain a reliable gas supply system without interruptions or reductions. Large scale UGS systems can effectively overcome the disadvantages associated with the gas demand fluctuation, as well as some other emergencies. They can also greatly reduce the cost in long-distance gas transmission and can provide a steady gas source[5-10]. Because of these advantages, salt cavern UGS is widely used in many countries. About 554 salt cavern UGSs were in operation around the world by the end of 2012.

Salt caverns serving to store natural gas and compressed air, etc., has a long history[10]. As early as the Second World War, salt caverns were first used in Canada to store liquid and gaseous hydrocarbons. Britain used salt caverns to store crude oil during the Suez Crisis. Ten years later, Americans and Canadians began to use salt cavern for gas storage. Since then, salt cavern UGS had been widely used and developed. By the end of 2005, the number of salt cavern UGSs and working gas capacity were 15 and $100\times10^8 m^3$ in France, 4 and $32\times10^8 m^3$ in England, 41 and $198\times10^8 m^3$ in Germany, and 8 and $112\times10^8 m^3$ in Italy. Fig. 1 shows that the proportion of energy stored in salt caverns (mainly for natural gas) in the total reserved energy of America became bigger and bigger, and showed an accelerated trend in 1998-2008, according to information from the U.S. Energy Information Administration[11]. It increased from 11% in 1998, to 16% in 2005, and to 25% in 2008. Meanwhile, 31 salt caverns will be rebuilt and enlarged in the next few years to increase their reserves in the US[11]. By the end of 2012, there were 40 salt cavern gas storages in America. The British had invested 9.3 billion dollars to construct more than 20 UGSs in salt formations by 2010 and this would increase natural gas storage capacity by about 30%[12]. In conclusion, salt cavern storage has an important position in the international energy reserves, and will be constructed on a large scale in the future, for a long period of time.

With sustained and rapid development of its economy, China's external energy dependency increases. The net imports of petroleum products, including crude oil, oil, liquefied petroleum gas, etc., have reached 2.931×10^8 tons. The import of liquefied natural gas (LNG) and pipeline gas is $425\times10^8 m^3$, accounting for about 28.9% of China's natural gas consumption in 2012[13]. According to previous experience[14], the ratio of working gas volume in UGS to gas consumption should exceed 12% to assure the safety of the gas supply if the ratio of imported gas volume to gas consumption exceeds 30%. In China it was just 1.7% in 2012. It is much lower than the three main storage markets' 19%[14], and has caused many potential safety problems for China's gas supply, e.g., nationwide gas shortages causing many very negative impacts on China's economy and people's life in 2009. The Chinese government noticed the seriousness of the problem, and actively carried out research and construction of UGS. The Ministry of Science and Technology of China launched the National Basic Research Program of China "the Disaster Mechanism and Protection of Energy Reserves in Underground Storages ("973"Program)", which was mainly organized and implemented by the authors group. Research results have been used in the site selection and construction of salt cavern gas storages in Jintan and Huaian city in Jiangsu province, Qianjiang and Yingcheng city in Hubei province, and Pingdingshan city in Henan province, China. These engineering applications have demonstrated the research results with good practical implementation results[8,15-20]. Meanwhile, the China National Energy Administration formulated the "twelfth five-year" natural gas development plan in 2012, and invested 81.1 billion yuan to construct 24 UGSs to increase the ratio of working gas volume to gas consumption to 5%[21].

Salt caverns are one of the major UGSs in China. They have the following advantages compared with other types of gas storage[5,9-10]. (i) Flexibility. Salt cavern facilities operate under very high pressure, and can quickly inject or deliver large amounts of gas to the pipeline grid. As a result, these facilities are well-suited to meeting short-term changes in demand or supply. (ii) Cycling. Salt cavern operators can move natural gas in and out of these facilities more frequently, usually up to a maximum of 6-12 cycles per year, compared to traditional, seasonal gas storage where customers inject in the summer and withdraw in the winter. (iii) Requires less base natural gas. Salt cavern storage needs less base gas and has a higher proportion of working gas. (iv) Moderate distance to the consumer market. The rock salt mines of China are usually in the range of about 200 km to the consumers, which is an economical distance for the construction of UGS.

However, the disadvantages of the salt cavern construction are also very significant, as follows: (i) Construction time is long. (ii) Requires much more investment per cubic meter working gas than reservoirs in depleted oil and/or gas fields. (iii) Needs the supporting facility, e.g., brine treatment plant. For example, a cavern with a volume of $25 \times 10^4 m^3$ will take four years to finish the construction, will require more than twenty million yuan, and will produce more than $180 \times 10^4 m^3$ of brine. Therefore, converting caverns formed by brine production into gas storage has a good economic benefit. Jintan city of Jiangsu province, China, is a traditional brine production field, and has many caverns. By preliminary screening, 30 caverns with a total volume of $207.05 \times 10^4 m^3$ are deemed suitable to convert to UGS. Although a single cavern has a small volume and an irregular shape, and a well with a poor cementing quality, etc., there are still lots of attractions to the engineers to convert these caverns into UGSs. Main advantages include: (i) shortens construction time, (ii) reduces investment, and (iii) decreases negative influences on the environment. Based on the above background, the Chinese government decided to carry out the project "Convert the brine production cavern into UGS". To assure the safety and efficiency of the project implementation, six caverns were selected for the early experiments. The UGS facilities were completed and gas injected in 2007. By the end of 2013, the caverns had a good stability and low volume shrinkage, proved by Sonar technology, and they can satisfy the demand of the UGS.

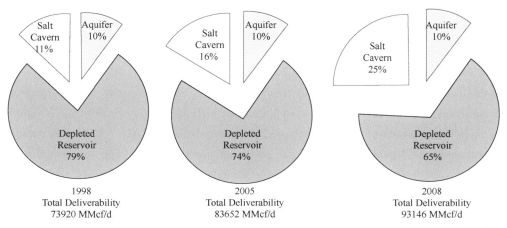

Fig. 1 Proportion of the natural gas stored in salt caverns as a fraction of the total USA stored gas[11]

Using abandoned underground spaces, such as coal and iron ore mines, to store natural gas or CO_2 has a long history. In 1961, Leyden mine was used to store natural gas for Colorado peak-day demands for almost 20 years[22]. Piessens and Dusar[23] studied the CO_2 sequestration in deep underground coal mine reservoirs, and pointed out that only few mines can meet the safety demands. Jalili et al.[24] reviewed the previous studies completed on CO_2 sequestration in abandoned coal mines around the world and presented preliminary assessments of the potential in Australia. Evans and Chadwick[12] introduced the current situation and perspective of UGSs in UK, and they stated that although hydrocarbons had been stored in abandoned mines, most of the facilities had closed due to leakage. However, literatures about converting abandoned salt cavern into UGS are rare for many critical conditions should be satisfied in the actual implementation of such conversions. Due to the creep and fracture self-healing characteristics of rock salt, abandoned salt caverns have significant advantages in technology and economy when used for gas storage compared to abandoned underground spaces in hard rocks.

Numerical simulation, one of the most powerful and popular tools in engineering, is widely used in the design and safety evaluations of underground structures. Its reliability is proved by many monitoring data. Hoffman[25] proposed a 3D finite model using JAC^{3D} software to evaluate the stability of UGS in salt domes. Van Sambeek[26] and Frayne and Van Sambeek[27] established a UGS pillar design equation and validated it by comparing the result of the equation with numerical results. Sobolik and Ehgartner[28] presented 3D models based on a close-packed arrangement of 19 caverns, and analyzed these models using a simplified symmetry involving a 30° wedge portion of the model. Kim et al.[2] presented a numerical modeling study of underground compressed air energy storage (CAES) in lined rock caverns. Their researches were used in a proposed underground CAES. Raju and Khaitan[3] developed a numerical model to simulate the mass and energy balance inside a cavern for CAES, and it was validated by the data from the Huntorf cavern operation. Wang et al.[5] used 3D geomechanical simulation to validate the analytical model of UGS salt cavern shape optimizing and pointed out that numerical and analytical solutions both were useful in the designs of salt cavern. As a UGS cavern has a large depth, the field monitoring and tests become difficult, especially for the abandoned caverns. Therefore, numerical simulation is preferentially used in the feasibility evaluation of abandon caverns to be converted to UGS.

In this paper, a systematic feasibility analysis of China's first salt cavern gas storage facility (using an abandoned salt cavern, located in Jintan city of Jiangsu province, China) is introduced. Content includes: analysis of geologic features, experiments on rock salt, numerical simulations, and assessment of field monitoring data. The operating parameters are proposed accordingly. Research results can provide references for the implementation of similar projects elsewhere in the world.

2 Geologic features of Jintan salt mine

2.1 Geologic structure

Jintan salt mine is located in a small basin with North-East orientation. Its northwest boundary is the junction of plains and Maoshan low hills, where the terrain has an elevation from 10 to 30 m. Its northern boundary is a relatively flat terrain with an elevation from 2.5 to 10 m. Total area of the Jintan salt mine basin is about 526 km^2, with lengths in North-East and North-West orientations of about 33 and 22 km respectively. Zhixiqiao sag is the deepest valley of Jintan basin, at the west of the basin. It has an east-north orientation with an area of about 265 km^2. Rock salt mines are located at the center of the Zhixiqiao sag, where the UGS is located. Its total area is 60.5 km^2.

Mesozoic, Upper Paleozoic, and Proterozoic strata of Jintan basin and surrounding areas outcrop in the north-east Ningzheng mountains and western Maoshan mountains. The well drilling in Jintan basin shows that the Quaternary loosely covers the whole basin. Cenozoic Paleogene is present only in the foothills of Maoshan mountain, and Neogene is not present. Cenozoic Paleogene is composed of Funing, Dainan, and Sanduo formations. Rock salt is present at the top of the Funing formation.

2.2 Distribution of rock salt layers

Rock salt of the Jintan basin developed from the late deposition of Zhixiqiao sag in Funing formation. According to the interpretation of seismic data and drilling data, the salt layers of Jintan salt mine are continuous in both horizontal and vertical directions. The long and short axes of the salt beds are 12 and 5.6 km respectively, and their total area is 60.5 km^2 with a thickness ranging from 67.85 to 230.95 m. The rock salt layer horizontal distribution shows ring-shaped thinning all around the edges. The layer tends to thin out in the north and west directions, and is cut off by the Zhixiqiao fault in the east. There is no information about the southwest boundary. According to the laws governing the Zhixiqiao concave deposit, the rock salt layer is speculated to gradually thin out in the southwest. Most importantly, the rock salt layer of the Jintan salt mine has a north-west slope with an inclination angle less than 10° on a macro level, and its thickness changes gently.

The rock salt formation in the Jintan basin forms a complete sedimentary cycle from the bottom to top. This means brine dilution, concentration, and then again dilution have occurred during the diagenetic development of rock the salt, therefore the section structure of the rock salt formation is relatively simple. The salt formation is divided by a continuous palm red and grey mudstone layer and a palm red mudstone layer into three main layers named Member 1, 2, and 3, from bottom to top.

Two main interlayers in the rock salt formation, between Members 1 and 2 and between Members 2 and 3, are present over the entire Jintan basin. The interlayer between Member 1 and 2 (ZY1) has a thickness ranging from 0.6 to 4.91 m, averaging 3.02 m. It is a secondary saline mudstone with fractures in the Maoxing area. The minimum, maximum, and average NaCl contents are 2.29%, 55.30% and 7.45%–13.89%, respectively. The interlayer between Member 2 and 3 (ZY2) has a thickness ranging from 0.28 to 4.84 m, with an average of 2.5 m over the entire Jintan basin. Its average thickness is 0.28–3.30 m in the Maoxing area. Its main component is saline mudstone with fractures, and its NaCl content ranges from 8.44% to 13.95%.

ZY1 and ZY2 interlayers are composed mainly of saline mudstone, mudstone, and calcium mirabilite mudstone, which have small fractures filled locally by the secondary salt. Thickness of a single interlayer generally ranges from 1 to 3 m, and the maximum thickness is no more than 5 m locally. Therefore, theses interlayers have little influence on the cavern shapes, and will eventually break and become insoluble precipitation in the cavern bottom.

From the above geologic feature analysis, Jintan salt mine has a good sealing potential, and the thicknesses of cavern roof and floor are sufficient to be stable. The thickness and distribution of the salt formation can satisfy the demands of converting abandoned salt caverns into UGSs.

3 Mechanical characterization tests

Compression tests and confined pressure rheological tests have been carried out to obtain the mechanical properties and creep parameters of Jintan rock salt. The damage characteristics of bedded rock salt, cavern stability, cover breakthrough pressure, and cavern sealing have close relations with the mechanical properties of interfaces between rock salt and interlayers. Therefore, direct shear tests, Brazilian splitting tests, interface microscopic inspections, and permeability tests have been carried out to determine the parameters of the interfaces. These test results provide the parameters and basis to evaluate the feasibility of China's first salt cavern gas storage facility.

3.1 Compression tests

Uniaxial and triaxial compression tests have been carried out to obtain the deformation and damage characteristics of Jintan bedded rock salt samples using the high temperature and high pressure rock rheological instrument, developed and made by the authors group. The samples are prepared from the rock salt from a depth ranging from 800 to 1200 m, which can be divided into three types, including rock salt, bedded rock salt, and mudstone samples. Rock salt samples are extracted from the rock salt, which contains a small amount of calcium mirabilite, and are beige and coarse-grained. Its main chemical components are NaCl, Na_2SO_4, $CaSO_4$, and some insoluble minerals. Bedded rock salt samples are extracted from the pultaceous anhydrite interlayer. Its average NaCl and average insoluble material contents range from 17.50% to 23.62% and from 36.24% to 42.62% respectively. Mudstone samples are extracted from the interlayer between two adjacent salt layers. Photographs of three typical samples are shown in Fig. 2.

(a) Rock salt (45-1)   (b) Bedded rock salt (85)   (c) Mudstone (137-1)

Fig. 2 Photographs of three typical samples before testing. The dimensions of the samples are listed in Table 1

Table 1 lists the results of uniaxial compression tests on three different types of Jintan rock samples. The average peak stresses of bedded rock salt samples are basically equal to those of the rock salt and mudstone samples, showing that the interfaces between rock salt and mudstone layers in the bedded rock salt formation have a high uniaxial strength.

Table 1 Results of uniaxial compression tests

Lithology	Sample number	Length/mm	Diameter/mm	Peak stress/MPa	Elastic modulus/GPa	Poisson's ratio
Rock salt	45-1	200.20	99.82	18.76	2.59	0.206
	84	201.04	101.78	17.75	5.39	0.274
	Average	—	—	18.26	3.99	0.240
Bedded rock salt	34-2	199.22	98.98	20.16	3.26	0.364
	41	194.92	101.12	19.12	3.51	0.190
	85	200.18	101.6	19.06	4.64	0.278
	Average	—	—	19.46	3.80	0.277

Lithology	Sample number	Length/ mm	Diameter/ mm	Peak stress/MPa	Elastic modulus/ GPa	Poisson's ratio
Mudstone	9	204.46	99.72	39.32	4.93	0.175
	135-2	201.26	101.68	36.05	5.68	0.290
	137-1	203.58	101.46	27.23	3.56	0.089
	Average	—	—	34.20	4.72	0.185

It is important to note that a stress drop takes place several times in the stress-strain curve of bedded rock salt under uniaxial compression or low confining pressure (<5 MPa) triaxial compression[16], which indicates that the mudstone has great influence on the damage of rock salt. This is because the axial peak strength of mudstone is larger than that of rock salt, and splitting failure in the mudstone layers of the bedded rock salt samples takes place first. Once the splitting fractures are formed, the deformation constraints on the rock salt layer caused by the mudstone layer weaken. The compression modulus of bedded rock salt decreases suddenly. This is similar with the process of material softening and causes the stress drop. As the confining pressure increases, the effects of the material difference on the stress concentration become smaller and smaller. When the confining pressure reaches a limit (5 MPa), the stress drop is not observed in the triaxial tests.

Tensile failure takes place in all three types of rock samples, and multiple splitting fractures form on the axial surface under uniaxial compression. Fig. 3 shows a bedded rock salt sample after failure in uniaxial compression. In mudstone, tensile failure occurs first, and then the fracture propagates to the rock salt and causes the failure of the rock salt. The unique failure characteristic of bedded rock salt samples differs from that of the other two kinds of samples. Experimental results also show that the damage characteristics of the three kinds of samples show obvious differences under confining pressures smaller than 5 MPa. In mudstone samples shear failure takes place, and the failure plane has an angle of about 10° with the sample axial direction. Dilatancy takes place in the rock salt samples on the lateral surface of the samples, without obvious fracture surfaces. Its failure is not a pure shear failure. When the sample reaches the peak stress, it still has large bearing capacity. The failure of bedded rock salt is very complicated. Shear failure takes place at the interlayers and then the fracture propagates to the rock salt, while the radial dilatancy becomes very obvious at the rock salt layer, while no sliding failure is observed at the interface between mudstone and salt layers.

3.2 Creep tests

Creep characteristics of rock salt have a decisive influence on the long-term stability and operating life of salt cavern UGS. The steady-state creep properties and corresponding constitutive relations of rock salt have been studied by many scholars[7,29-31]. Jintan bedded rock salt has low porosity, fracture content and permeability in the initial condition, but contains joints, anhydrite and other impurities. Under certain loads, the dislocations slide and rock salt turns into the steady-state creep stage after a very short time of initial creep. When the time or load is increased further, rock salt creep accumulates, and then progresses into the acceleration phase. For practical engineering, the initial creep is very small, and the accelerated creep stage is not reached. To obtain the creep constitutive relations, creep tests under different confining pressures and deviatoric stresses have been carried out. The results are shown in Fig. 4.

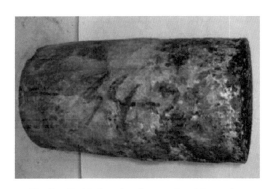

Fig. 3 Bedded rock salt sample with number of 34-2 after uniaxial compression test

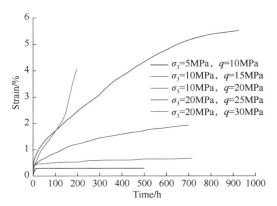

Fig. 4 Creep curves of Jintan rock salt under different confining pressures (σ_3) and deviatoric stress (q). q is deviatoric stress, $q = \sigma_1 - \sigma_3$, σ_1 is axial stress, and σ_3 is confining pressure

Fig. 4 shows that all the curves have reached the stable creep stage. Under a confining pressure of 20 MPa and a deviatoric stress of 30 MPa, the rock salt reaches the accelerating creep after about 180 h. Table 2 lists the creep test results of Jintan rock salt samples.

As shown in Table 2, the steady-state creep rate increases with the increase of the deviator stress, and its proportion of the total strain also increases. For example, the steady-state creep accounting for the proportion of the total strain reaches 90.63% when the confining pressure and deviatoric stress are 20 and 25 MPa respectively. This indicates that creep strain is the controlling factor of the deformation of rock masses surrounding a salt cavern. Therefore, it is important to precisely obtain the rock salt creep rule for the salt cavern UGS design and safety assessment.

Table 2 Creep test results of rock salt samples

Specimen number	Loading step	Confining pressure/ MPa	Deviatoric stress/ MPa	Time/ d	Instantaneous strain/%	Steady creep rate/s^{-1}	Creep strain/%	Creep strain/ total strain/%
52-1	1st	5	10	21	0.23	2.399×10^{-5}	0.06	20.79
52-1	2nd	10	15	30	0.05	1.646×10^{-4}	0.35	87.5
52-1	3rd	20	25	38.58	0.09	6.285×10^{-3}	5.49	98.387
52-1	4th	20	30	8.125	0.185	1.224×10^{-2}	4.04	95.621
53	1st	10	20	29.17	0.18	1.214×10^{-3}	1.74	90.625

According to our previous studies[15,16], the creep constitutive relation of rock salt in China generally follows the power function, and its standard form is

$$\dot{\varepsilon}_{cr} = A\bar{\sigma}^n \quad (1)$$

where $\dot{\varepsilon}_{cr}$ is the creep rate; A and n are the material property parameters; $\bar{\sigma} = \left(\dfrac{3}{2}\right)^{1/2}(\sigma_{ij}^d \sigma_{ij}^d)^{1/2}$, σ_{ij}^d is the deviator of σ_{ij}.

Based on Eq. (1) and the experimental data in Table 2, A and n of Jintan rock salt in steady state creep stage are 2.996×10^{-9} MPa/year and 4.480 respectively.

3.3 Direct shear tests

The main purpose of the direct shear test is to determine the shear strength parameters, i.e., cohesive strength (C) and internal friction angle (φ). The formulas of normal stress and shear stress are expressed as

$$\sigma_n = P/A \quad (2)$$
$$\tau = Q/A \quad (3)$$

where σ_n is the normal stress, Pa; P is the normal force, N; τ is the shear stress, Pa; Q is the maximum shear force that the sample can bear, N; A is the effective area along the shear failure direction, m^2.

On the basis of Coulomb's law[32], shear stress has an approximately linear relationship with normal stress when the normal force is relatively small, and its expression is as follows

$$\tau = \sigma_n \tan\varphi + C \quad (4)$$

where C is cohesive strength, Pa; φ is internal friction angle, degrees.

Direct shear tests have been carried out on mudstone, rock salt, and bedded rock salt samples extracted from different locations of the Jintan salt cavern. Based on the experimental results and Eq. (4), the shear strength parameters are obtained as shown in Table 3.

Table 3 Shear test results of samples extracted from different locations of Jintan salt cavern

Lithology	Location	C/MPa	φ/(°)
Mudstone	Cavern roof I	3.20	29.02
	Cavern roof II	4.59	26.08
	Cavern floor	2.33	33.35
	Average	3.37	29.48
Rock salt	Cavern roof	3.68	39.41
	Cavern I	3.84	42.03
	Cavern II	2.87	48.15
	Cavern III	2.21	48.43
	Average	3.15	44.50

continued

Lithology	Location	C/MPa	$\varphi/(°)$
Bedded rock salt	Cavern roof	1.26	52.28
	Cavern	3.97	38.07
	Average	2.62	45.18

As shown in Table 3, the strength of bedded rock salt samples has the same order of magnitude as that of rock salt and mudstone samples. This indicates the interfaces between rock salt and mudstone in bedded rock salt formation have a high shear strength.

3.4 Brazilian splitting tests

The purpose of Brazilian splitting tests is to indirectly determine the tensile strength of the rock. The formula used to calculate tensile strength based on the Brazilian splitting test results is[33]

$$\sigma_t = -\frac{2P}{\pi DL} \tag{5}$$

where σ_t is the tensile strength of the rock sample, Pa; P is the maximum load when the sample splits, N; D is the disk sample diameter, m; L is the disk length, m.

Brazilian splitting tests have been carried out to obtain the tensile strength of different lithology samples extracted from the Jintan salt mine. According to the experimental results, the average tensile strengths of Jintan rock salt, bedded rock salt, and mudstone are 1.04, 1.08 and 1.67 MPa respectively.

3.5 Interface microscopic observations

There always are some doubts about the sealing of UGS located in bedded rock salt in China. Generally, the interface between rock salt and mudstone layers is regarded as a material with low strength and high permeability, which should be considered separately in the design and numerical simulation. To obtain the microscopic characteristics of the interface between rock salt and mudstone layers in Jintan salt mine, electron microscopic observations have been carried out.

Fig. 5 shows photographs of Jintan rock salt core as well as Scanning Electron Microscope (SEM) photographs of the interface between rock salt and mudstone layers. As shown in Fig. 5, the interface between rock salt and mudstone layers is relatively clear, the salt particle size is large, and the particles are plate-shaped. The space between particles is large, and is filled with mineral filler (mainly mudstone). The particle size of the mudstone is smaller than that of rock salt. The mudstone particles are block-shaped and belt-shaped with a small space between adjacent particles. In the interface, the mudstone and salt particles are embedded in each other. Salt particles with large size directly embed small particles of argillaceous anhydrite, while small argillaceous anhydrite particles fill in the gap between the salt particles. Two kinds of particles cement together to form a jagged interlocking structure. Salt and mudstone particles change the cementation and arrangement mode outside the interfaces, and have no obvious boundaries. The spaces between salt particles are full of mudstone mineral aggregate. This indicates that the interface between rock salt and mudstone layers is not a weak structure plane but instead is a plane along which salt rock and shale particles are intermixed and cemented. This is the reason, from the micro-physics viewpoint, why the interface has a high strength. Moreover, testing results in Sections 3.1–3.4 show the interface has a good bearing capability. Therefore, the interface between rock salt and mudstone layers does not need to be considered separately in the numerical simulation and design.

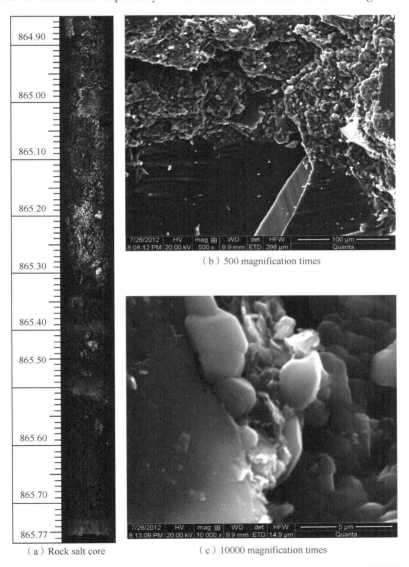

Fig. 5 Photos of Jintan rock salt core and Scanning Electron Microscope (SEM)

3.6 Permeability tests

To evaluate the sealing of caprock, interlayer, and rock salt of Jintan salt mine, permeability tests have been carried out on the CMS − 300 overburden pressure automatic permeability determination system. Through the pressure control system of the CMS - 300, confining pressure and axial compression are applied to the samples. The injection pressure is applied on one end of the sample, and the soap bubble flow meter is fixed at the other end of the sample to measure the gas flow rate. When the gas seepage reaches a stable stage, the gas volume flowing through the sample per minute is recorded. The permeability of the sample is calculated based on the testing results. Considering the confining pressure limit of the CMS - 300 and the gas pressure in the cavern, hydrostatic pressures are set as 7.5, 10, 12.5, 15, 17.5, 20, 25 and 30 MPa, and injection gas pressures as 0.4, 0.6, 0.8 and 1.0 MPa respectively. The disk sample diameter and length are about 25 and 50 mm respectively. Samples have been extracted from different locations of the target formation. Experimental results are shown in Fig. 6.

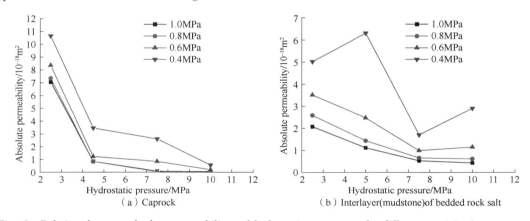

Fig. 6 Relations between absolute permeability and hydrostatic pressure under different gas injection pressures

Fig. 6 presents the relations between absolute permeability and hydrostatic pressure of caprock and interlayer (mudstone) of bedded rock salt under different injection gas pressures. As shown in Fig. 6(a), the absolute permeability of caprock decreases as the confining pressure increases when the confining pressure ranges from 2.5 to 10.0 MPa, and changes from 1.05×10^{-17} to 3.62×10^{-20} m^2. Therefore, the caprock can fully meet the sealing requirements of a salt cavern UGS.

Fig. 6(b) shows that the absolute permeability of the interlayer of bedded rock salt decreases as the confining pressure increases from 2.5 to 10.0 MPa, and changes from 4.95×10^{-18} to 4.37×10^{-19} m^2. This indicates that the interlayer of the bedded rock salt has a good sealing capacity.

During the tests of rock salt and the interface between rock salt and mudstone layers, there is no gas escape after 4 h. This shows that the absolute permeability of Jintan rock salt and interface is smaller than 10^{-20} m^2. Above experimental results show that the rock salt, mudstone interlayer, interface, and caprock in Jintan salt mine have low permeability and good sealing capacities, which can satisfy the sealing requirements of UGS.

4 Feasibility analysis

According to the basic requirements of cavern shape and volume, fifteen brine production caverns are preliminarily screened from thirty caverns as the candidates for converting into UGS in the first stage construction. They are located in the Maoxing, Maoxi, Zhixiqiao, Donggang, and Rongbin salt mine areas. The fifteen caverns all have separate stable structures, and each single cavern volume is larger than $8 \times 10^4 m^3$.

To reduce the risk associated with brine caverns being converted to UGS, the Dong-1 and 2, Gang-1 and 2, and West-1 and 2 caverns are selected for the preliminary experiments. Subsequently, the engineers can determine whether the rest of the caverns can be transformed into UGS according to the test results. The caverns selected in the preliminary experiments all have a stable pear-shaped or cone-shaped structure. Their ratios of height to diameter range from 1.0 to 1.6. Single cavern volumes are larger than $10 \times 10^4 m^3$. Their total volumes are about $100 \times 10^4 m^3$ with total storage capacity of $140 \times 10^6 m^3$ and working gas volume of $80 \times 10^6 m^3$, which can satisfy the basic balancing of supply and demand of the Jintan area in the early stage. Numerical models are built up based on the above experimental results and the strata parameters to evaluate the feasibility of converting the six brine caverns to UGS. Volume shrinkage rate, deformation, plastic zone, safety factor, effective strain, and delivery rate of the six caverns are given, and the operating parameters are proposed accordingly. As an example, the feasibility analysis of converting the West-1 and -2 caverns to UGS is discussed in detail. Fig. 7 shows the cavern location plan map of Zhixiqiao (West-1 and -2 caverns), where the West-3 cavern does not fit into the demand of UGS and is filled with brine to serve for observation. Fig. 8 gives the 3D shapes of West-1 and -2 caverns, obtained by Sonar. Other parameters of West-1 and -2 caverns and formation parameters used in the numerical simulations are listed in Table 4.

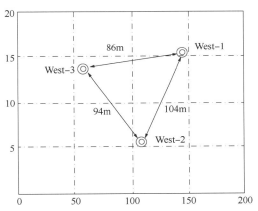

Fig. 7 Cavern location plan map

Table 4 Geometrical parameters of W-1 and -2 caverns

Number	Depth/m		Cavern height/m	Max. radius/m	Cavern volume/m³	Cavern shape
	Cavern top	Cavern bottom				
West-1	959.5	1013.4	53.9	52.6	104,923	Pear-shaped
West-2	937.4	1007.4	70	44.4	128,413	Cone-shaped

According to the geological conditions of Jintan salt mine, a 3D geomechanical model of West-1 and -2 caverns is established using the software ANSYS, as shown in Fig. 9. Then, the model is introduced into the software FLAC^{3D} for the calculations. The model is a cuboid including a bedded rock salt layer with a thickness of 146.67 m, and two mudstone layers above and below the bedded rock salt layer with thicknesses of about 300 m. The height, length and thickness of the model are

800, 900 and 400 m respectively. Considering the model symmetry, 1/2 model is shown in Fig. 9. Simply supported constraints are applied to the bottom of the model in the vertical direction, and to the four vertical boundary planes. Overburden pressure is calculated, based on the depth and formation average density, to have a value of 14.18 MPa (σ_v), and is applied on the top surface of the model. Given the large thickness and high salt content of bedded rock salt in Jintan salt mine, the in-situ stress is assumed to meet the hydrostatic pressure distribution, viz., in-situ stresses along three directions are equal at the same depth. The model includes 31565 nodes and 173130 elements. Tetrahedron, hexahedral, and pyramid transition elements are used in the model. The mesh quality is evaluated by the ''Mesh'' tool embedded in ANSYS software before calculation. To ensure the mesh independence, we changed the mesh sizes several times according to the monitoring results of points at different locations of the model. The residual error values are defined as 10^{-5}, and the deformations of the points located at the tops of West-1 and -2 caverns and domain imbalances are monitored in the numerical simulation. Monitoring results show the simulation converges, and the deformations of the points at the tops of West-1 and -2 caverns reach a stable state, no longer influenced by further decreases in mesh sizes. The domain has imbalances of less than 1% with the mesh size used. Other parameters used in the simulations are based on the above experimental results.

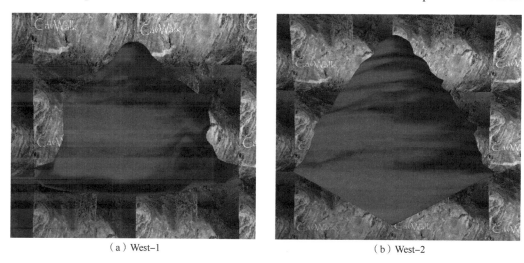

(a) West-1　　　　　　　　　　　　(b) West-2

Fig. 8　3D shapes of caverns obtained by Sonar in 2005

4.1　Cavern volume shrinkage

According to the results of geological exploration, internal gas pressure of Jintan salt cavern UGS is proposed to range from 6 to 14.5 MPa. Therefore, constant internal gas pressures with values of 6, 7, 9, 11, 13, and 14.5 MPa are simulated to study their effects on the volume shrinkages of West-1 and -2 caverns. The calculated results are shown in Fig. 10.

Fig. 10 shows the volume shrinkages of West-1 and -2 caverns over time for different internal gas pressures. Volume shrinkage is defined as the volume reduction as a proportion of the original volume of the cavern. As shown in Fig. 10, the volume shrinkages of West-1 and -2 caverns all

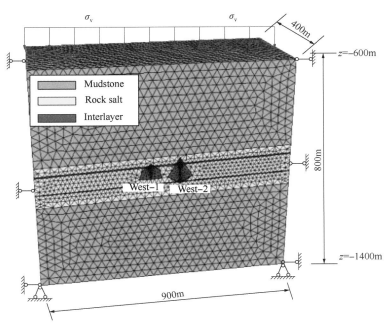

Fig. 9 Geomechanical model and boundary conditions of West-1 and -2 caverns.
σ_v is the overburden pressure with a value of 14.18 MPa

decrease with increasing internal gas pressure, and increase with increasing time. The reason is that increasing internal gas pressure counteracts some in-situ stresses causing the decrease of deviatoric stress to which the rock salt is subjected. Ultimately it decreases the salt creep rate. The volume shrinkages of West-1 and -2 are 22.24% and 14.32% respectively after operating times of 20 years under internal gas pressure of 6 MPa, while that decreases to 2.68% and 1.78% respectively under internal gas pressure of 14.5 MPa. The average internal gas pressure of Jintan UGS is about 10 MPa, and then volume shrinkages of West-1 and -2 caverns are all less than 10% after operating for 20 years. With reference to the allowable volume shrinkage of previous salt cavern UGS[31], West-1 and -2 caverns can be used as UGS from the viewpoint of volume shrinkage.

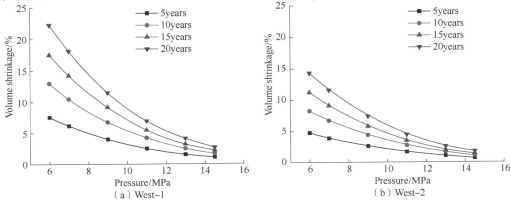

Fig. 10 Volume shrinkage of West-1 and -2 caverns as a function of internal gas pressure for different operating times

It can be seen in Fig. 10 that the volume shrinkage reduces as the internal gas pressure increases, and decreases only slowly once the internal gas pressure exceeds 11 MPa. Increasing the minimum internal gas pressure and reducing operating time with the minimum pressure can effectively reduce the volume shrinkage and increase the working life of salt cavern UGS. By comparison, West-2 has a more reasonable shape and better capability of withstanding volume shrinkage than West-1. This indicates that the cone-like shape of West-2 can be recommended in the new salt cavern UGS design.

4.2 Deformation

According to our previous studies[5,6], the deformations of rock masses surrounding caverns increase as the internal gas pressure decreases. Therefore, the minimum internal pressure with a value of 6 MPa is used in this section to study the effect of time on the deformation of West-1 and -2 caverns. Times with values of 0.1, 1, 5, 10, 15, and 20 years are simulated, and results are shown in Fig. 11.

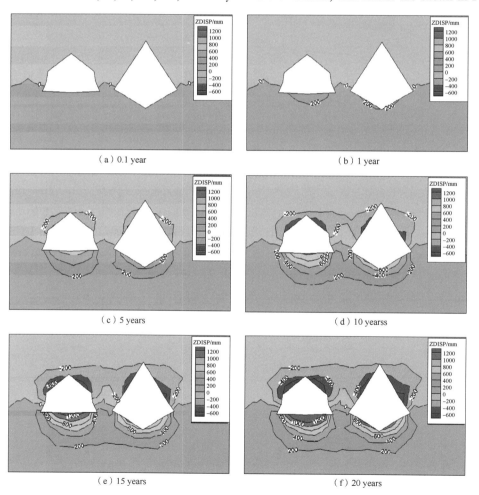

Fig. 11 Vertical deformation contours around West-1 and -2 caverns with a 6 MPa internal gas pressure after different operating times

Fig. 11 shows the vertical deformation contours of West-1 and -2 caverns after operating 0.1, 1, 5, 10, 15, and 20 years with internal gas pressure of 6 MPa. Upward and downward deformations are defined as positive and negative respectively. The vertical deformation of the cavern top is mainly downward, while that of the cavern bottom is mainly upward. The maximum deformation of the cavern bottom is much larger than that of the other locations. Therefore, cavern bottom deformation should be monitored carefully by Sonar. The maximum deformation of West-1 is much larger than that of West-2, especially after a long operating time. This shows that a cavern with a plane bottom is less able to withstand deformation, and should be avoided in the new salt cavern UGS design and construction. After operating for 20 years, the maximum deformations of West-1 and -2 caverns are about 1400 and 1200 mm, and account for 1.3% and 1.4% respectively of their maximum diameters. According to previous engineering experiences[17], these values are still in the allowable deformation range and can satisfy the safety demands.

4.3 Plastic zone

The Mohr-Coulomb criterion, one of the most extensively used criteria in geotechnical engineering, is used to predict whether rock masses undergo plastic failure. Plastic failure occurs when the shear stress on a certain plane reaches a limit called the shear yield stress. The general expression for the Mohr-Coulomb criterion is

$$\frac{1}{2}(\sigma_1 - \sigma_3) = c \cos \varphi - \frac{1}{2}(\sigma_1 + \sigma_3)\sin \varphi \tag{6}$$

where σ_1, σ_2, σ_3 are the maximum, intermediate, and minimum principal stress, respectively.

Fig. 12 presents the relation between the ratio of plastic zone volumes obtained using Eq. (6) to two cavern volumes and time. The ratio increases with time and decreases with an increase of internal gas pressure. For example, the ratio is 1.42 and 1.83 after operating 5 and 10 years respectively under the internal pressure of 6 MPa, while it is 0.21 and 0.37 under the internal pressure of 11 MPa. This is because the creep deformation increases the shear stress in the rock salt surrounding the cavern, and when the shear stress exceeds the rock salt strength, dilatancy takes place and plastic zones form. Internal gas pressure balances a part of the shear stress in the rock mass causing the decrease of plastic volume. Fig. 12 also shows that the increase of plastic volume becomes less when the internal gas pressure exceeds 11 MPa, and it nearly remains constant after operating 10 years. To improve the working life, the average internal gas pressure of West-1 and -2 is proposed to be 11 MPa.

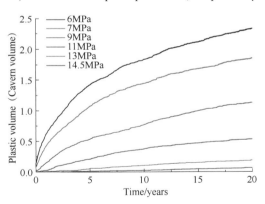

Fig. 12 Ratio of plastic volumes to volumes of West-1 and -2 caverns as a function of operating time under different internal gas pressures

4.4 Safety factors

To evaluate the safety of West-1 and -2 caverns, a rock salt damage criterion is introduced in this section. Based on previous test results of bedded rock salt of Jintan city, Jiangsu province, China[17], the Van Sambeek criterion best matches the experimental results among available criteria. Therefore, it is used in the numerical simulations. It is expressed as[34-36]

$$SF_{VS} = \frac{FI_1}{\sqrt{J_2}} \qquad (7)$$

where SF_{VS} is the safety factor for dilatancy; F is a material constant with a value of -0.27 (Tension is defined to be positive)[37]; I_1 is the first invariant of the stress tensor, $I_1 = \sigma_1 + \sigma + \sigma_3$; J_2 is the second invariant of the deviatoric stress tensor, $J_2 = [(\sigma_1-\sigma_2)^2 + (\sigma_2-\sigma_3)^2 + (\sigma_3-\sigma_1)^2]/6$.

Earlier studies [5, 37-40] have suggested that the Van Sambeek safety factor (SF_{VS}) indicates local damage when $SF_{VS} < 1.5$, failure when $SF_{VS} < 1.0$, and collapse when $SF_{VS} < 0.6$. These damage thresholds are used in this section. Fig. 13 presents the safety factor contours of rock masses surrounding the two caverns for different operating times, obtained using Eq. (7) under the internal gas pressure of 6 MPa. Safety factors decrease with operating time, especially those of the interface between rock salt and mudstone layers decrease significantly. This confirms that the interface is the weak point of West-1 and -2 caverns. The results show that local damage develops after operating 20 years, but the damage area is small. Therefore, West-1 and -2 caverns can be used as UGS from the viewpoint of safety factor.

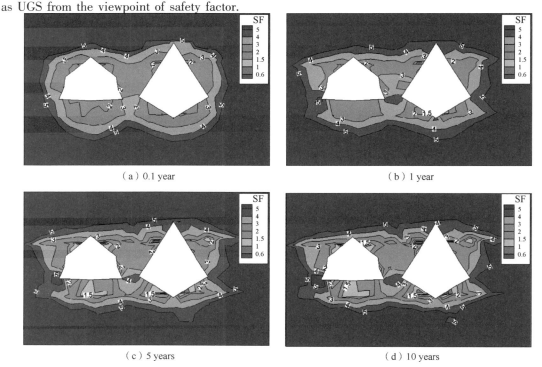

(a) 0.1 year

(b) 1 year

(c) 5 years

(d) 10 years

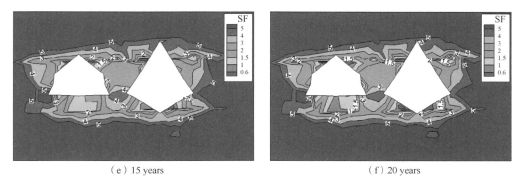

(e) 15 years (f) 20 years

Fig. 13 Safety factor contours around West-1 and -2 caverns with a 6 MPa internal gas pressure for different operating times

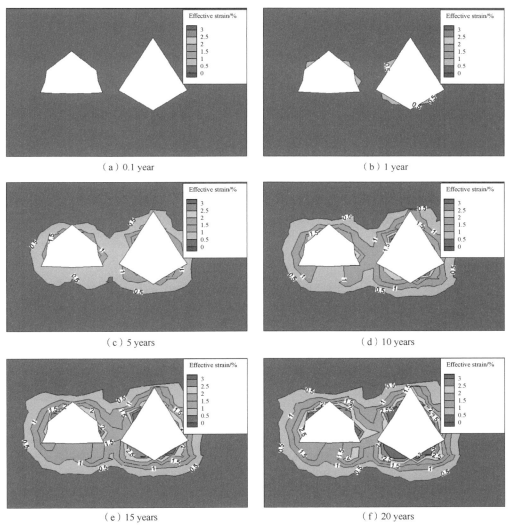

(a) 0.1 year (b) 1 year

(c) 5 years (d) 10 years

(e) 15 years (f) 20 years

Fig. 14 Effective strain contours around West-1 and -2 caverns with a 6 MPa internal gas pressure after different operating times

4.5 Effective strain

Taking into account that rock salt is subjected to three dimensional in-situ stresses and the damage accompanying plastic deformation, the effective strain is introduced in this paper to assess the damage of rock salt surrounding caverns. The effective strain is mainly used to define and measure the material damage by the change of modulus before and after damage, which has been widely applied in the damage evaluation of rock, coal, limestone, etc., for its simple definition, easy to calculate. A definition of effective strain is[41]

$$\varepsilon_{eq} = \sqrt{\frac{2}{3} \varepsilon^{dev} : \varepsilon^{dev}} \qquad (8)$$

where ε_{eq} is the effective strain; ε^{dev} is the deviatoric strain tensor.

According to available literature[42], the maximum effective strain of the rock mass surrounding salt caverns should be no more than 3% over the entire design service lifetime (usually 30 years) to avoid the creep damage of the rock salt. Fig. 14 presents the effective strain contours of West-1 and -2 caverns obtained using Eq. (8) under the internal gas pressure of 6 MPa for different operating times. The effective strain close to the caverns is much larger than that at other locations, and increases with increasing time. That is especially true of the interfaces between rock salt and mudstone layers, where it increases much more obviously. Effective strain in the pillar between West-1 and -2 caverns is generally greater than 2%, and locally is greater than 3% after operating 20 years. As West-1 and -2 caverns were formed for the brine production, without considering the working conditions of UGS, the working life is recommended to be no more than 20 years from the viewpoint of effective strain.

4.6 Pressure drop rate

A reasonable pressure drop rate is an important issue to ensure the cavern stability and to improve the operational economic benefits of UGS. It depends on the rate at which gas is withdrawn from the cavern. In this section, pressure drop rate is assigned as 0.3, 0.5, 0.6, 0.65, and 0.7 MPa/day respectively to study its effects on the shrinkage rate of West-1 and -2 caverns. In the numerical simulation, the internal gas pressure of the two caverns simultaneously decreases from 14.5 to 6.5 MPa. Results are shown in Fig. 15.

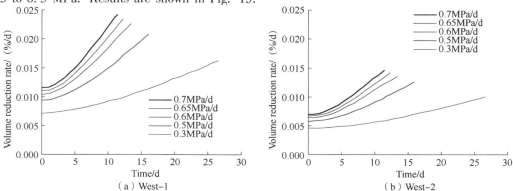

Fig. 15 Relation between volume reduction rate of West-1 and -2 caverns and time under different pressure drop rates

Fig. 15 presents the relation between volume shrinkages of West-1 and -2 caverns and time under different pressure drop rates. According to reference[43], volumetric strain of rock mass surrounding caverns should not exceed 3% in one cycle (usually one year). The stages of a cycle of West-1 and -2 caverns can be divided into: (i) storage with minimum internal pressure, (ii) injection, (iii) storage with maximum internal pressure, and (iv) production. The volume shrinkage is the largest during the production stage, followed by storage with minimum pressure. Based on our previous study results[17], the volume shrinkage rates of West-1 and -2 caverns with internal pressure of 6.5 MPa are both less than 0.0045%/d. In the numerical simulations, the production time is assumed to account for 1/3 of a cycle (actual production time is shorter than this), and the remaining time is calculated as storage with minimum pressure. Then, the volume shrinkage rates during the production stage should not be more than 0.015%/d to ensure the safety. Therefore, the volume shrinkage rate of 0.015%/d is used as the threshold to calculate the pressure drop rate of West-1 and -2 caverns during the production.

As shown in Fig. 15, the volume shrinkage of West-1 cavern is larger than that of West-2 cavern under the same condition, and approaches 0.0146%/d when the pressure drop rate reaches 0.7 MPa/d. From a safety viewpoint, it is recommended that the maximum pressure drop rates of West-1 and -2 caverns should be less than 0.55 MPa/d to keep some safety redundancy.

5 Operational testing

Based on above results of geologic exploration, testing, and numerical simulation, the abandoned salt caverns of Jintan city, Jiangsu province, China, have a good likelihood to be converted successfully to UGS facilities, and the corresponding operating parameters are proposed. The proposed parameters of West-1 and -2 caverns include the maximum, minimum, and average internal gas pressures of 14.5, 6, and 9-11 MPa respectively; the maximum pressure drop rate not greater than 0.55 MPa/day. In accordance with the conclusions and proposed parameters, the project of China's first salt cavern gas storage facilities (using abandoned salt cavern) was completed and gas was injected in 2007. To check the status of the caverns after running for 6 years, the volumes of the caverns were measured by Sonar, under working conditions, in 2013.

Fig. 16 presents the average radiuses of West-1 and -2 caverns obtained by Sonar in 2005 and 2013. As shown in Fig. 16, cavern shapes of West-1 and -2 caverns did not change much, the cavern walls did not collapse, nor were there large deformations. The volume of West-1 was reduced from 104,923 m^3 in 2005 to 103,360 m^3 in 2013 with a shrinkage of 1.49%, and that of West-2 was reduced from 128,413 m^3 in 2005 to 126,305 m^3 in 2013 with a shrinkage of 1.64%. This indicates that the method proposed in this paper can effectively evaluate the feasibility of salt cavern gas storage facilities in Jintan.

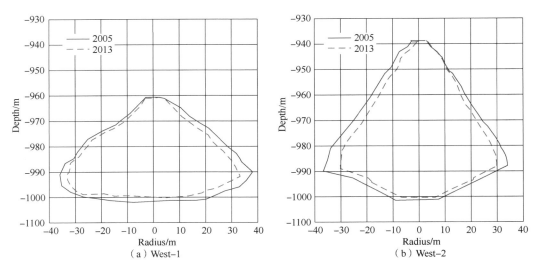

Fig. 16 Average radiuses of West-1 and -2 caverns obtained by Sonar in 2005 and 2013

6 Conclusions

(1) A feasibility analysis of China's first salt cavern gas storage facility (using abandoned salt caverns, located in Jintan city of Jiangsu province, China) has been performed. First, an overview is presented of the geology of the site. Second, the mechanical parameters of the host rock are determined using laboratory tests. Third, the interface between rock salt and interlayers is studied microscopically. Fourth, the permeability of the salt is determined experimentally. The feasibility of converting Jintan salt caverns for gas storage is evaluated by studying, using numerical simulations, the volume shrinkage, deformation, plastic zone, safety factor, effective strain, and pressure drop rate. The operating parameters of UGS converted from abandoned salt caverns are optimized based on the results of geological exploration, experimental tests, numerical simulations, and field observations.

(2) Results show that the Jintan bedded rock salt has good physical and mechanical performance, and the interfaces between rock salt and mudstone layers have high strength and sealing capability, which can satisfy the requirements for UGS. Indicators of UGS converted from Jintan abandoned salt caverns can meet the safety standard. The maximum, minimum, and average internal gas pressures, maximum pressure drop rate, and safe working life of West-1 and -2 caverns used as UGS are proposed.

(3) According to the proposed maximum, minimum, and average internal gas pressures, maximum pressure drop rate, and safe working life, the project of China's first salt cavern gas storage facilities (using abandoned salt caverns) was completed, and gas was injected in 2007. The volume of the cavern was measured by Sonar under working conditions in 2013. Results show that cavern shapes have not changed much, volume shrinkage was less than 2%, and cavern walls did not collapse, nor suffer large deformation.

(4) The method proposed in the paper can effectively evaluate the feasibility of salt cavern gas storage facilities in Jintan. Research results can provide a reference for the implementation of similar projects in other places.

Acknowledgments

The authors wish to acknowledge the financial supports of the National Basic Research Program of China (973 Program) (Grant No. 2009CB724600) and National Natural Science Foundation of China (Grant No. 41272391).

References

[1] Zafirakis D, Chalvatzis KJ, Baiocchi G, Daskalakis G. Modeling of financial incentives for investments in energy storage systems that promote the largescale integration of wind energy. Appl Energy 2013; 105: 138-154.
[2] Kim HM, Rutqvist J, Ryu DW, Choi BH, Sunwoo C, Song WK. Exploring the concept of compressed air energy storage (CAES) in lined rock caverns at shallow depth: a modeling study of air tightness and energy balance. Appl Energy 2012; 92: 653-667.
[3] Raju M, Khaitan SK. Modeling and simulation of compressed air storage in caverns: a case study of the Huntorf plant. Appl Energy 2011; 89: 474-481.
[4] Ozarslan Ahmet. Large-scale hydrogen energy storage in salt caverns. Int J Hydrogen Energy 2012; 37: 14265-14277.
[5] Wang TT, Yan XZ, Yang HL, Yang XJ, Jiang TT, Zhao S. A new shape design method of salt cavern used as underground gas storage. Appl Energy 2013; 104: 50-61.
[6] Wang TT, Yan XZ, Yang XJ, Jiang TT. Roof stability evaluation of bedded rock salt cavern used as underground gas storage. Res J Appl Sci Eng Technol 2012; 4: 4160-4170.
[7] Yang CH, Daemen JJK, Yin JH. Experimental investigation of creep behavior of salt rock. Int J Rock Mech Min Sci 1999; 36: 233-242.
[8] Yang CH, Jing WJ, Daemen JJK, Zhang GM, Du C. Analysis of major risks associated with hydrocarbon storage caverns in bedded salt rock. Reliab Eng Syst Saf 2013; 113: 94-111.
[9] Wang TT, Yan XZ, Yang HL, Yang XJ. Stability analysis of the pillars between bedded salt cavern groups by cusp catastrophe model. Sci China Tech Sci 2011; 54: 1615-1623.
[10] Wang TT. Study on deformation and safety of rock mass surrounding gas storage in bedded rock salt. Doctoral dissertation. Qingdao: China University of Petroleum; 2011 [in Chinese].
[11] Tobin James. U.S. Underground Natural Gas Storage Developments: 1998-2005. U.S. Energy Information Administration, Office of Oil and Gas; October 2006.
[12] Evans DJ, Chadwick RA. In: Underground Gas Storage: Worldwide experiences and future development in the UK and Europe, vol. 313. London: The Geological Society, Special Publications; 2009. p. 1-11.
[13] Tian CR. China's oil and gas imports and exports in 2012. Int Petro Econom 2013; 3: 44-55 [in Chinese].
[14] Sylvie CG. Underground gas storage in the world. France, Rueil Malmaison, CEDIGAZ; 2013.
[15] Guo YT, Yang CH, Mao HJ. Mechanical properties of Jintan mine rock salt under complex stress paths. Int J Rock Mech Min 2012; 56: 54-61.
[16] Li YP, Liu W, Yang CH, Daemen JJK. Experimental investigation of mechanical behavior of bedded rock salt containing inclined interlayer. Int J Rock Mech Min 2014; 69: 39-49.

[17] Yang CH, Li YP, Chen F. Bedded salt rock mechanics and engineering. Beijing: Science Press; 2009 [in Chinese].

[18] Du C, Yang CH, Yao YF, Li Z, Chen J. Mechanical behavior of deep rock salt under the operational conditions of gas storage. Int J Earth Sci Eng 2012; 5: 1670–1676.

[19] Li YP, Yang CH, Daemen JJK, Yin XY, Chen F. A new Cosserat-like constitutive model for bedded salt rocks. Int J Numer Anal Meth Geomech 2012; 33: 1691–1720.

[20] Li YP, Yang CH. On fracture saturation in layered rocks. Int J Rock Mech Min 2007; 44: 936–941.

[21] China National Development and Reform Commission. The twelfth five-year plan of natural gas development, no. 338; 2012. <http://www.sdpc.gov.cn/zcfb/zcfbtz/2012tz/W02012 1115581608265333.pdf> [in Chinese].

[22] Leroy WB. Abandoned coal mine stores gas for Colorado peak-day demands. Pipe Line Ind 1978: 55–58.

[23] Piessens K, Dusar M. CO_2 sequestration in abandoned coalmines. In: Proceedings of the international coal bed methane symposium, Tuscaloosa, Alabama. Paper no. 346; May 5–9, 2003.

[24] Jalili P, Saydam S, Cinar Y. CO_2 storage in abandoned coal mines. In: 11th Underground coal operators' conference. University of Wollongong & the Australasian Institute of Mining and Metallurgy; 2011. p. 355–360.

[25] Hoffman EL. Effects of cavern spacing on the performance and stability of gas filled storage caverns, SAND92-2545. Albuquerque, NM, US: Sandia National Laboratories; 1993.

[26] Van Sambeek LL. Salt pillar design equation. In: Peng SS, Holland CT, editors. Proceedings, 16th international conference on ground control in mining. Morgantown, WV: West Virginia University; August 5–7, 1997. p. 226–234.

[27] Frayne MA, Van Sambeek LL. Three-dimensional verification of salt pillar design equation. In: Cristescu ND, Hardy Jr JR, Simionescu RO, editors. Proceedings, fifth conference on the mechanical behavior of salt, university of bucharest, Bucharest, Romania. Netherlands: A. A. Balkema; August 9–11, 2002. p. 405–409.

[28] Sobolik SR, Ehgartner BL. Analysis of shapes for the strategic petroleum reserve. Albuquerue, USA: Sandia National Laboratories; 2006.

[29] Munson DE, DeVries KL, Fossum AF, Callahan GD. Extension of the M-D model for treating stress drops in salt. In: Proceedings of the 3rd conference on the mechanical behavior of salt. USDOE, Washington, DC (United States); 1993. p. 31–44.

[30] Chan KS, Bodner SR, Fossum AF, Munson DE. A damage mechanics treatment of creep failure in rock salt. Int J Damage Mech 1997; 6(1): 121–152.

[31] Bérest P, Minh DN. Deep underground storage cavities in rock salt: interpretation of in-situ data from French and foreign sites. In: Proceedings of the 1st conference on the mechanical behavior of salt. Clausthal-Zellerfeld: Trans. Tech. Publications; 1981. p. 555–572.

[32] Coulomb CA. Essai sur une application des règles des maximis et minimis à quelques problèmes de statique relatifs à l'architecture. Mém Acad Roy Div Sav 1973; 7: 343–387.

[33] ISRM. Suggested methods for determining tensile strength of rock materials. Int J Rock Mech Min Sci 1978; 15: 99–103.

[34] Van Sambeek LL, Ratigan JL, Hansen FD. Dilatancy of rock salt in laboratory test. In: Haimson BC, editor. Proceedings, 34th U. S. symposium on rock mechanics. University of Wisconsin-Madison, Madison, WI. Int J Mech Mining Sci & Geomech Abst, vol. 30. Pergamon Press; June 27–30, 1993. p. 735–738.

[35] Ratigan JL, Vogt TJ. A note on the use of precision level surveys to determine subsidence rates. Int J Mech Min Sci Geomech Abstr 1991; 28: 337-341.

[36] Park Byoung Yoon, Ehgartner Brian L. Allowable pillar to diameter ratio for strategic petroleum reserve caverns. Sandia report SAND 2011-2896. Albuqueraue, NM, USA: Sandia National Laboratories; 2011.

[37] DeVries KL. Geomechanical analyses to determine the onset of dilation around natural gas storage caverns in bedded salt. In: Spring conference SMRI, Brussels, Belgium; 2006.

[38] DeVries KL. Improved modeling increases salt cavern storage working gas, Winter Ed. GasTIPS; 2003.

[39] DeVries KL, Mellegard KD, Callahan GD. Proof-of-concept research on a salt damage criterion for cavern design: a project status report. In: Spring conference SMRI, Banff, Alberta; 2002.

[40] DeVries KL, Mellegard KD, Callahan GD. Laboratory testing in support of a bedded salt failure criterion. In: Fall conference SMRI, Chester, U.K., England; 2003.

[41] Im S, Atluri SN. A study of two finite strain plasticity models: an internal time theory using Mandel's director concept, and a general isotropic/kinematichardening theory. Int J Plast 1987; 3: 163-191.

[42] Wu W, Hou ZM, Yang CH. Investigations on evaluating criteria of stabilities for energy (petroleum and natural gas) storage caverns in rock salt. Chin J Rock Mech Eng 2005; 24: 2497-2505 [in Chinese].

[43] Hou ZM. Untersuchungen zum Nachweis der Stand sicherheitfür Untertagedeponien im Salzgebirge. Clausthal: Dissertation an der TU; 1997.

本论文原发表于《Applied Energy》2015年第137期。

钻采工程

中俄东线楚州盐穴储气库配套钻井液技术

薛 雨[1] 张新悦[1] 王立东[2]

(1. 中国石油西气东输管道公司；2. 中国石油盐穴储气库技术研究中心)

【摘 要】 楚州盐穴储气库是中俄东线配套储气库，具有重要建设意义。在楚州地区钻井施工时，上部东台组、盐城组易渗漏，黏土易造浆，下部浦口组含盐地层易溶蚀扩径和坍塌。针对楚州地区地层特点和施工难点提出采用分段钻井液技术：一开采用聚合物钻井液，防止表层漏失和坍塌；二开上部采用盐水聚合物钻井液，防止缩径和石膏侵；盐层段采用饱和盐水聚合物钻井液防止盐岩溶解。通过现场3口资料井对上述钻井液技术进行了验证，3口井钻井过程顺利，井径规则，未发生任何井下复杂情况，井身质量较高，盐层段岩心采取率均在99%以上。所形成的钻井液技术对建库钻井具有很好的指导意义。

【关键词】 钻井液；盐穴储气库；饱和盐水聚合物；楚州盐矿

Drilling fluid technology for Chuzhou salt-cavern gas storage for Sino-Russian east gas pipeline

Xue Yu[1] Zhang Xinyue[1] Wang Lidong[2]

(1. PetroChina West-East Gas Pipeline Company；
2. PetroChina Salt-cavern Gas Storage Technology Research Center)

Abstract The Chuzhou salt-cavern gas storage facilities is supporting gas storage for Sino Russian East Line；thus, it is of great construction significance. During the drilling operation in the Chuzhou area, the upper Dongtai and Yancheng formations are prone to leak, and the clay is prone to make slurry. The salt formations in the lower Pukou formation are prone to erode, expand and collapse. Drilling fluid technology for different well sections was used according to stratum characteristics and construction difficulties in the Chuzhou area. The polymer drilling fluid was used to prevent surface leakage and collapse in the first well section. The salt water polymer drilling fluid was used to prevent shrinkage and gypsum invasion in the upper part of the second section and the saturated brine polymer drilling fluid was used to prevent salt dissolving in the salt layer section. The above drilling fluid technology has been verified by 3wells in the field. The drilling process of the 3wells was smooth, the well diameter was regular, no downhole complications occurred, the quality of the well was high, and the coring rate of the salt layer was above 99%. The drilling fluid technology can provide good guidance for drilling of gas storage wells.

基金项目：中国石油天然气集团公司重大专项"地下储气库关键技术研究与应用"（编号：2015E-40）。
作者简介：薛雨，男，汉族，1987年生，工程师，油气田开发工程专业，硕士，从事盐穴储气库钻井技术研究工作。地址：江苏省镇江市南徐大道60号商务A区D座。E-mail：xqdsxueyu@petrochina.com.cn。

Keywords drilling fluid; salt-cavern gas storage; saturated brine polymer; Chuzhou Salt Mine

中国石油正在积极建设中俄东线配套楚州盐穴储气库来保障长三角地区的天然气供应。钻井是储气库建设过程中必不可少的环节，钻井的质量和效率决定着储气库的建设成本和注采开发进度[1]，而钻井液在确保安全、优质、快速钻井中起着越来越重要的作用。国内对于油气藏储气库钻井液做了大量研究[2-5]，但目前还未见到专门针对盐穴储气库的钻井液研究。穿越盐膏层时国内外主要采用高密度饱和盐水钻井液[6-10]，其密度基本都大于 $1.7g/cm^3$，盐膏层深度也都在3500m以深。相对而言盐穴储气库井深浅，一般不超过2000m，地层承压能力低，若采用高密度钻井液，容易发生井下漏失。盐矿钻井通常采用欠饱和或饱和盐水钻井液[11-15]，钻井时大部分存在井眼溶蚀严重，井径扩大率偏大以及卡钻或钻具脱落等复杂情况，井身质量远低于盐穴储气库井的要求。相比于土耳其盐穴储气库，楚州盐矿埋藏深，夹层多，地质条件更为复杂，土耳其盐穴储气库钻井液技术[16-19]不能完全适用于楚州。为此，根据楚州盐矿地层建立一套适用于楚州储气库的钻井液体系是楚州储气库钻井工作面临的一个重要难题，对加快储气库建设，提高经济效益，具有十分重要的意义。

1 地质及工程概况

楚州储气库位于江苏省淮安市淮安区（原楚州区），其地层从上至下分为东台组、盐城组、赤山组和浦口组，其中浦二段上盐亚段第三岩性组合为主力盐层，地质条件较好，是建库的目标层段，楚州储气库地层分布见表1。楚州储气库钻井一开为进入浦口组50m后完钻，二开为进入主力盐层20m后完钻，井身结构见表2，生产井采用二开井身结构，资料井出于取心及承压试验的需要采用三开井身结构。

表1 楚州储气库地层分布表

地层				主要岩性特征
组	段	亚段	岩性组合	
东台				灰色粉砂质黏土，下部含砾砂层
盐城	盐二			灰棕色砂、黏土，粉砂黏土互层
	盐一			缺失
赤山	赤二			剥蚀
	赤一			砖红色粉细砂岩
浦口	浦三			暗棕色粉砂质泥岩、粉砂岩、含石膏
	浦二	上盐	第四	灰色泥质岩和钙芒硝岩
			第三	灰白色盐岩
			第二	褐灰色钙芒硝岩
			第一	灰色盐岩和钙芒硝岩
		中淡化		泥质粉砂岩、细砂岩
		下盐		暗棕色灰黑色泥岩、膏质泥岩、粉细砂岩，硬石膏，钙芒硝与盐岩互层
	浦一			咖啡色泥岩夹砂岩、云质泥岩、泥质粉砂岩

楚州储气库在建库钻井时需要解决以下问题来保证安全钻进：（1）表层的不成岩黏土有

存在井口塌陷的风险；(2)东台组、盐城组上部成岩性差、胶结疏松、易渗漏和垮塌，盐城组中下部黏土易造浆；(3)盐城组与赤山组之间有断层且新老地层交界，易漏失和地层失稳；(4)浦三段中含有石膏，其钙离子将污染钻井液；(5)进入主力盐层，盐岩受钻井液溶蚀存在扩径现象，同时盐岩中夹杂的泥岩存在蠕变缩径及垮塌的风险；(6)浦二段含有芒硝层，其受温度影响溶解度变化大，下部可能欠饱和，但循环上部可能结晶，造成摩阻大甚至卡钻，容易发生井下复杂情况。

表2 楚州储气库钻井井身结构

序号	钻头直径/m	套管外径/mm	类型	水泥返高	备注
导管	660.4	508.0	导管		
一开	444.5	339.7	表层套管	地面	套管下入白垩系浦口组三段以下50m
二开	311.2	244.5	生产套管	地面	套管下入深度揭开主力盐群20m
三开	215.9		裸眼		资料井

2 钻井液技术

根据上述地层特点及施工难点，钻井液确定原则为：(1)上部东台组、盐城组和赤山组存在漏失风险，因此钻井液需具有防漏失的性能；(2)确保盐层段井径规则。根据以上原则，钻井时采用分井段钻井液方案：一开井段采用具有较好井眼净化能力的聚合物钻井液，防止上部地层发生漏失和水化膨胀；二开上部井段通过缓慢加入NaCl，从淡水钻井液逐步转换成饱和盐水钻井液；二开下部井段采用饱和盐水钻井液，来防止盐层的溶蚀及蠕变，钻井液密度 $1.3\sim1.35 g/cm^3$；三开井段采用的钻井液体系和性能指标与二开下部井段相同，钻井液设计方案见表3。

表3 楚州储气库钻井液设计方案

序号	钻井液体系	钻井液功能
导管	钠土浆	防止表层坍塌
一开	聚合物钻井液	抑制造浆、防止漏失、垮塌
二开	上部：盐水聚合物钻井液；下部：饱和盐水聚合物钻井液	防止盐岩溶蚀、蠕变及缩径、泥岩坍塌和石膏污染
三开	饱和盐水聚合物钻井液	防止盐岩溶蚀和石膏污染

2.1 导管 $\phi 660.4mm$ 井眼

开钻前以优质的膨润土加纯碱配制膨润土浆 $60m^3$，密度 $1.10g/cm^3$，黏度 $45\sim50s$，膨润土加量5t左右，纯碱的加量为膨润土的 $5\%\sim8\%$。由于井眼大，应随钻加入适量的提黏剂控制合理的钻井液黏切，使钻井液具有良好的悬浮携带能力，同时排量要达到 $45\sim50L/s$，以确保钻屑顺利返出。钻完进尺后，大排量充分洗井，循环干净井眼内的沉砂，起钻前配制稠浆保证下套管固井顺利。

2.2 一开 ϕ444.5mm 井眼

开钻前配制钻井液总量100m³，往配浆水中加入0.2%Na₂CO₃再用配浆泵加入5%的膨润土，水化时间>24h。使用膨润土浆开钻，钻井液中加入0.2%~0.3%PMHA-Ⅱ和0.4%~0.5%NH₄HPAN进行预处理，调整好钻井液性能后方可开钻。钻进中随时补充PMHA-Ⅱ、COATER、NH4HPAN，以提高钻井液的抑制能力和包被能力。

东台组、盐城组上部成岩性差，胶结疏松，易渗漏和垮塌，钻进过程中做好井漏预防工作，并储备足够数量的堵漏材料。钻进中加入零滤失井眼稳定剂（LXJ-1）、单向封闭剂（KD-23），配合使用超细碳酸钙（QS-4）随钻堵漏，保持良好的悬浮携带能力和造壁性能，确保井壁稳定。盐城组中下部含造浆黏土，钻进中加强监测，注意钻井液性能变化，加足抑制包被剂PMHA-Ⅱ（或COATER）抑制造浆。由于盐城组与赤山组之间新老交界地层和断层，需要避免产生漏失和地层失稳现象，若出现漏失按防漏堵漏措施执行；若出现地层剥垮等失稳情况，及时补充防塌剂（含量不小于2%），可配加1%~2%FT-1和2%~3%QS-4，改善滤饼质量，提高钻井液封堵防塌性能，并将钻井液密度提高至设计上限，防止井壁坍塌。在加防塌剂的同时，配合加入稀释剂，调整钻井液流变性能。

一开为ϕ444.5mm大井眼井段，钻井液要有较好的携砂及悬浮能力，合理使用固控设备，配合人工清砂，降低钻井液中的劣质固相。工程要保持双泵循环，保持泵排量>50L/s，确保钻屑及时从井底被带至地面。该井段应勤短起下钻，修复井眼，减少虚滤饼的形成，保持井壁干净。由于盐穴储气库井固井段井径扩大率要求小于10%，除钻井液适当提高黏切外，工程上不得定点循环，合理选择钻井参数。钻至下套管层位，要短起下钻并进行充分循环钻井液，修整井壁、清洁井眼，确保电测、下套管、固井作业顺利。

2.3 二开 ϕ311.2mm 井眼

二开前处理好水泥侵，调整好原浆性能钻进至进入浦口组二段含盐系地层200m之前将钻井液转换为饱和盐水钻井液。控制合适的膨润土含量，以利于钻井液流变性能的调整和固相含量的控制，在维护过程中采取等浓度等体积的处理方法。由于饱和盐水钻井液转化过程中盐的加入及地层中盐的溶入，在维护处理时要及时补充烧碱，保持体系的pH值[20]，以避免pH值降低，影响钻井液性能。

二开进入盐层后钻井液技术重点为确保Cl^-含量维持在$1.8×10^5$mg/L以上，需要定期对钻井液中的Cl^-含量进行监测，及时补充NaCl。同时加入抗盐结晶剂防止盐的重结晶而发生井下复杂情况。

为维持钻井液性能稳定，钻进中加入CMS、SMP和HV-CMC等护胶来调整流变性能；加大SMP、SPNH等抗温、抗盐材料的用量，来控制钻井液HTHP失水量，改善滤饼质量和增加抗温能力。同时，为了保证正常钻进，需要使钻井液维持较强的剪切稀释性，以提高悬浮和携带能力。加足防塌剂FT-1，钻进中继续补充并保持防塌剂的含量不小于2%。钻井液中防塌剂和高聚物、磺化处理剂可产生协同增效作用,有效地稳定井壁。将处理剂配成胶液,在充分溶解后再加入。避免处理剂干粉不能在饱和盐水钻井液中充分发挥效能。

在钻井液中可添加适量的防腐剂,来抑制盐水对管具材料及循环系统的腐蚀[21]。同时加强岩性分析判断和短起下钻,来给钻井液维护提供依据。

做好固控设备的维护与使用工作,有效平衡体系固相组分,以净化保优化,将无用固相含量降至最低,以细目振动筛为主,结合人工清砂,有效地控制钻井液中有害固相,降低滤饼表面的粗糙度,控制合理的膨润土含量,添加适量润滑剂,提高滤饼的润滑能力,来提高钻具在井眼内的安全性。

二开完钻后,应进行短起下钻,充分循环钻井液,性能均达设计要求后,方可起钻,确保电测、下套管顺利完成。

2.4 三开 ϕ215.9mm 井眼

虽然饱和盐水钻井液有很强的抗污染能力,但为了维持其原有性能,三开钻塞时,仍需做好抗钙、除钙处理,来降低扫塞过程中钙离子的污染。三开井段维护措施与二开下部井段相同。

2.5 钻井液性能参数

楚州储气库施工钻井液性能见表4,三开钻井液性能参数同二开下部井段钻井液性能参数。

3 应用效果评价

采用以上钻井液技术,在楚州储气库成功实施资料井3口,有效地解决了井眼净化、井壁稳定等问题,3口井均未发生任何井下复杂情况和事故。说明该钻井液体系与楚州地区地层特性配伍性较好,能够满足钻井工程的需要。

3.1 钻井施工顺利

上部地层采用聚合物钻井液,钻进过程中通过保持钻井液中大分子的有效含量,并随钻加入适量的封堵剂,控制滤失量小于8mL。施工时钻井液黏度变化不大,起下钻顺利,说明该钻井液体系能够有效抑制上部地层造浆,防止漏失,保证良好的滤饼质量和防塌性能。

进入盐层后转换为聚合物饱和盐水润滑钻井液体系,根据钻井液性能补充适当的提黏剂、润滑剂和抗盐结晶剂,整个二开三开过程施工顺利,盐层段井径规则(表5),未发生因为盐岩溶解而形成的"大肚子"井眼或者因为盐层塑性流动缩径而造成的卡钻现象,测井一次成功率100%。资料井工序繁多,包括取心、地应力测试、静置7d后井温测井、VSP测井、非造腔段回填、气密封试压等多道工序。通过上述钻井液措施的综合应用,3口井纯钻进时间仅占总施工时间的25%左右,缩短了钻井周期。

表 4 楚州储气库钻井液性能

序号	钻井液体系	常规性能						流变性能		MBT/(g/L)	固相含量/%	HTHP失水量/mL
		密度/(g/cm³)	马氏黏度/s	API失水量/mL	滤饼厚/mm	含砂量/%	pH值	塑性黏度/(mPa·s)	N值			
导管	膨润土	1.10	40~75									
一开	聚合物	1.10~1.15	40~55	<8	0.7	0.5	8~9	10~15	0.8			
二开上部	盐水聚合物	1.15~1.35	45~60	<5	0.5	0.4	9~11	15~20	0.4~0.6	55~60	<18	<13
二开下部	饱和盐水聚合物	1.30~1.36	45~75	<5	0.5	0.3	9~11	20~25	0.5~0.7	55~60	<18	<13

表 5 资料井钻井情况

井号	生产时间/h	纯钻进时间/h	时效/%	二开井深/m	二开井径扩大率/%
1井	3628	1003.17	27.65	1880	7.30
2井	4124	1004.67	24.35	1648	6.68
3井	3394	767.08	22.60	1796	7.22

3.2 取心收获率高

进入盐层段,为保证盐岩层不被溶解,形成规则的井径,必须确保钻井液中含盐量达到饱和。施工时将钻井液密度提高至设计上限1.35g/cm³,并不断补充NaCl,使Cl⁻含量始终大于1.8×10⁵ mg/L,同时加入足量的抗盐抗高温降滤失剂和羧甲基淀粉,控制API失水量<5mL,适当提高钻井液的黏度,以防止盐岩层的溶解。3口井盐层段岩心采取率均在99%以上(表6)。

表 6 资料井取心情况

井号	取心进尺/m	岩心长/m	岩心采取率/%
1井	483.83	480.11	99.20
2井	300.00	298.33	99.44
3井	50.00	49.78	99.56

3.3 钻井液费用占比低

3口井采用以上钻井液技术，并配合合理的钻井施工工艺，实现了优质、安全施工，整个施工过程中未发生任何因钻井液原因而导致的井下复杂情况。钻井液费用仅占钻井总费用的7%，降低了钻井成本。

4 结论

（1）采用分段钻井液技术，通过3口资料井的现场成功应用，证明制定的钻井液技术对策可行，能够保证楚州储气库钻井施工的质量和安全。

（2）楚州储气库一开地层采用高黏切、抑制性强的聚合物钻井液，工程上保持双泵循环，加强短起下，能够有效抑制上部地层造浆，确保井壁稳定。

（3）针对二开含盐层系，进入主力盐层前及时转换成饱和盐水聚合物钻井液，确保Cl^-含量维持在$1.8×10^5$ mg/L以上，提高钻井液黏度，防止盐层溶解，保证盐层段岩心采取率。

（4）提高各井段固控设备使用率，降低钻井液中无用的固相含量，保持钻井液性能稳定，能减少阻卡事故的发生。

参 考 文 献

[1] 代长灵,杨光,薛让平.长庆靖边储气库关键钻井技术[J].天然气勘探与开发,2016,39(1):65-69.

[2] 刘明峰,熊腊生,赵福祥,等.储气库救援井文23-6J井钻井液技术[J].钻井液与完井液,2014,31(5):43-45.

[3] 李称心,高飞,张茂林,等.新疆呼图壁储气库钻井液技术[J].钻井液与完井液,2012,29(4):45-48.

[4] 刘在桐,董德仁,王雷,等.大张坨储气库钻井液技术[J].天然气工业,2004,24(9):153-155.

[5] 南旭,杨勇,沈泉,等.双6储气库水平井钻井液技术[J].中国石油和化工标准与质量,2013,34(4):150.

[6] 路小帅,孙琳,王谱,等.塔河油田托甫台区复杂穿盐井快速钻井技术[J].石油机械,2014,42(12):1-5.

[7] 赵晖,王西峰,叶勇刚,等.高密度饱和盐水钻井液在胜利油区的应用探讨[J].中国石油大学胜利学院学报,2017,31(2):25-27.

[8] 刘伟,李华坤,徐先觉.土库曼斯坦阿姆河右岸气田复杂深井超高密度钻井液技术[J].石油钻探技术,2016,44(3):33-38.

[9] 孟庆生,江山红,石秉忠.塔河油田盐膏层钻井液技术[J].钻井液与完井液,2002,19(6):74-76.

[10] 陈红壮,黄河淳,高伟.TK1228井三开膏盐层钻井液技术[J].西部探矿工程,2010,22(9):29-32.

[11] 邝光升,孙宇,杨建军,等.云阳黄岭岩盐矿ZK0001深孔取心技术[J].探矿工程（岩土钻掘工程）,2018,45(3):53-56.

[12] 徐培远,袁志坚.青海盐溶地层钻探卤水泥浆配方研制及应用[J].探矿工程（岩土钻掘工程）,2017,44(6):41-44.

[13] 陈金照.宁夏固原硝口岩盐矿钻探施工技术[J].中国煤炭地质,2016,28(4):67-70.

[14] 王勇军,谭现锋,绍立宁,等.宁夏固原采卤井施工技术[J].探矿工程（岩土钻掘工程）,2016,43

(11):21-25.
- [15] 彭朝洪,肖长波,徐飞.叶舞凹陷ZKX井深部盐层钻井液技术研究[J].探矿工程(岩土钻掘工程),2015,42(2):28-32.
- [16] 刘在同,王建伟.土耳其盐穴地下储气库大井眼钻井技术[J].石油钻采工艺,2015,37(2):32-34.
- [17] 余广兴.土耳其盐穴储气库盐层长筒取芯技术[J].中国石油和化工标准与质量,2014,34(1):112.
- [18] 付洪涛.盐穴储气库井钻井关键技术研究与应用[D].西安:西安石油大学,2013.
- [19] 余广兴.土耳其盐穴储气库钻井表层失返性漏失处理[J].内江科技,2017,38(4):38-39.
- [20] 李轩,黄维安,贾江鸿,等.盐水钻井液pH值影响因素和缓冲方法研究[J].钻井液与完井液,2018,35(1):21-26.
- [21] 周永璋.盐井钻井的盐水泥浆液中钻具的腐蚀与防护研究[J].腐蚀科学与防护技术,2008,20(6):424-428.

本论文原发表于《探矿工程(岩土钻掘工程)》2019年第46卷第9期。

楚州盐穴储气库钻井工程难点及对策

薛雨[1]　王立东[1,2]　王元刚[1]　周冬林[1,2]　井岗[1]

(1. 中国石油西气东输管道公司；2. 中国石油储气库技术研究中心)

【摘　要】　中国石油天然气集团公司在楚州盐矿建设的中俄东线配套盐穴储气库存在地质条件复杂、井身质量要求高、钻井液维护困难、固井难度大和环保要求高等难题。为此，首先分析了楚州盐矿的地质特征，储气库钻进工程的难点，提出了适用于楚州地区的大井眼防斜优快钻井技术、分段钻井液技术、长封固段防漏、防憋堵固井技术和钻井不落地技术，并选取楚州储气库的3口井进行了试验。结果表明：(1)使用钟摆钻具组合，配合MWD随钻监测，能够实现高效钻进；(2)采用分段钻井液，严格控制滤失量，提高悬浮、携带能力，保证了正常钻进；(3)采用DRB-3S+JSS抗盐水泥浆体系可以有效提高顶替效率，保证固井质量；(4)采用钻井液不落地设备进行固液分离及无害化处理，达到了环保要求。该钻井技术为其他地区的盐穴储气库建库提供了技术支撑。

【关键词】　盐穴储气库；钻井；钻井液；固井；长封固段；防漏；防憋堵；技术

Drilling challenges and technical solutions for Chuzhou salt cavern gas storage

Xue Yu[1]　Wang Lidong[1,2]　Wang Yuangang[1]　Zhou Donglin[1,2]　Jing Gang[1]

(1. PetroChina West-East Gas Pipeline Company;
2. PetroChina Salt Cavern Gas storage Technology Research Center)

Abstract　One salt cavern gas storage constructed in Chuzhou salt mine and by CNPC, which backs the China-Russian Eastern Gas Pipeline, meets many challenges, such as complicated geological conditions, strict requirements to wellbore quality, difficult maintenance of drilling fluid, difficult cementing, and strict requirements to environmental protection. So, both geological characteristics of Chuzhou salt mine and drilling challenges of Chuzhou gas storage were firstly analyzed. Then, some solutions were put forward, e.g. large hole anti-deviation and fast drilling, subsection drilling fluid, long cementing section anti-leakage and anti-blocking cementing, and waste mud without landing. Finally, 3 wells in

---

基金项目：中国石油天然气集团公司重大科技专项"盐穴储气库加快建产工程试验研究"(编号：2015E-4008)。

作者简介：薛雨，1987年生，工程师；2012年毕业于中国石油大学(北京)油气田开发工程专业，主要从事盐穴储气库钻井管理工作。地址：(212000)江苏省镇江市润州区南徐大道60号商务A区D座。E-mail：xqdsx-ueyu@petrochina.com.cn。

Chuzhou gas storage were selected for testing these technologies. Results show that (1) the pendulum drill assembly combined with monitoring while drilling (MWD) can realize an efficient drilling; (2) the subsection drilling can control fluid loss strictly, and improve suspending and carrying capacity, so as to keep normal drilling; (3) an application of DRB-3S+JSS salt-resisting slurry system can increase displacement efficiency effectively and guarantee cementing quality; and (4) the waste mud without landing device is used for sol id-liquid separation and harmless treatment, so as to satisfy the requirements of environmental protection. In conclusion, these solutions may provide technical support for constructing salt cavern gas storages in other areas.

Keywords salt cavern gas storage; drilling; drilling fluid; cementing; long cementing section; anti-leakage; anti-blocking; technology

为积极应对气荒，落实国家能源发展战略，中国石油天然气集团公司正在建设中俄东线天然气管道来调节并保证东部地区的天然气供给。楚州盐穴储气库是中俄东线9座配套储气库中唯一的1座盐穴储气库，通过楚州储气库不但可以参与中俄东线和冀宁联络线的调峰和应急供气，更能保障长三角地区的天然气供应[1]，故而具有重要的建设意义。随着储气库的持续开发，钻井中的关键技术决定着储气库的钻井速度、成本消耗和注采开发进度[2]。针对盐穴储气库对钻井的要求，在分析楚州盐矿地质条件的基础上，提出了适用于楚州盐穴钻井的技术对策，为建库钻井施工提供指导。

1 楚州盐矿的地质特征

楚州盐矿位于江苏省淮安市淮安区，距在建的中俄东线天然气管道约61km。该区地层从上到下依次为第四系东台组、新近系盐层组、上白垩统赤山组和浦口组[3]。楚州盐穴储气库的建库层段为浦口组二段上盐亚段第三岩性组合，其埋深大于1800m，盐岩厚度大、泥质含量低，地质层位及岩性见表1。

表1 楚州盐矿地层表

地层					岩性特征
系	统	组	段	亚段	
第四系	全新统	东台组			灰色粉砂质黏土夹粉砂，下部含砾砂层、砂砾层
新近系	上新统	盐城组			灰棕色砂、砂砾与杂色黏土，含粉砂黏土互层
白垩系	上白垩统	浦口组	三段		暗棕色粉砂质泥岩、粉砂岩、泥质粉砂岩，泥岩含石膏芒硝
			二段	上盐亚段	以灰白色盐岩为主，夹钙芒硝岩、泥岩和粉砂岩

2 楚州储气库钻井工程难点

盐穴储气库井运行周期长，需超过50年[4-5]，储气库注采井要求强注强采，并且周期循环，井内压力交替变化，包括井内管柱、井口装置和固井质量等必须能承受交变应力的影响[6-7]。因此储气库对井身质量、固井质量和密封性的要求非常高。同时楚州盐矿本身的地质特点也给钻井施工带来了严峻挑战。

2.1 地层复杂，容易造成井下复杂情况

楚州储气库钻井时，表层为不成岩黏土，容易造成井口塌陷。第一次开钻的施工层段为东台组、盐城组泥岩，容易缩径，流沙层容易垮塌。盐城组与赤山组之间有断层且新老地层交界，容易产生漏失和地层失稳现象。赤山组、浦口组发育大段泥岩，地层较硬，机械钻速较低。第二次开钻的施工层段浦口组发育大段泥岩，机械钻速低。浦三段中下部含石膏团块或石膏线，钻井液易受钙污染。钻遇纯盐段时，盐层易溶蚀扩径，且中间夹大量泥岩，易蠕变缩径且泥岩容易因盐的溶蚀失去支撑而垮塌，造成井下复杂情况[8-9]。

2.2 井身质量要求高

盐穴储气库井身质量要求1000m内最大井斜不超过2°，1000m以后最大井斜不超过1.5°，井径扩大率不能超过10%。此外，楚州储气库受夹层多、新老地层交接面的影响，防斜打直难度大。直井全井段全角变化率要求小于1.25°/30m。若全角变化连续3个点超标，则判为不合格井。

2.3 钻井液维护困难

钻井时上部东台组、盐城组、赤山组成岩性差，胶结疏松，钻井液容易漏失。盐城组中下部的黏土容易造浆，浦三段中下部石膏团块中的钙也会污染钻井液。浦二段含有芒硝层，其受温度影响溶解度变化大，钻井液在下部层段可能欠饱和，但循环至上部层段又可能结晶，造成摩阻大甚至卡钻[10-11]。这些因素都增加了钻井液处理维护的难度。

2.4 一次封固段长，固井难度大

盐穴储气库固井时要求水泥返至地面[12]。楚州储气库生产套管平均下深在1800m左右，一次封固段长，上部承压能力低，在施工过程中容易产生漏失现象，难以保证水泥浆上返至地面。同时上部封固段盐层厚度超过600m，固井时盐层溶解，卤水进入水泥浆，造成水泥浆总体积减小，导致上部水泥浆回落，盖层和泥岩层的封隔质量受到影响。盐层部位的水泥浆也由于受到侵蚀，水泥浆稠化时间受盐层溶解的影响会大幅度延长，无法形成良好的水泥环，水泥与井壁间胶结质量差[13-16]。另外盐穴储气库井眼尺寸较大，由于地层复杂容易形成井径不规则井段，固井顶替效率也难以保证。

2.5 环保要求高

近年来，政府越来越重视工程建设中的环保工作，且楚州储气库地处经济发达的江苏地区，百姓的环保意识更为强烈，企业担负的环保责任也对本地区的钻井工作提出了更高的要求。

3 钻井技术对策

3.1 大井眼防斜优快钻井技术

与常规油气井相比，盐穴储气库井眼尺寸大，钻井液环空返速偏低，携砂困难。此外受夹层多、新老地层交接面和地层倾角的影响，防斜打直难度大。为此，根据地层岩性特点选用19mm复合片五刀翼PDC钻头施工，第一次开钻井段采用钟摆降斜钻具组合[17-18]，

同时保证钻井液排量不小于40L/s、钻进50m左右，单点测斜1次，保证井斜达标。第二次、第三次开钻下入"PDC+螺杆+稳定器"的塔式钟摆钻具组合，在施工中采用MWD仪器实时监测井身轨迹做好防斜打直工作，确保井身质量达标。

为减小井径扩大率，施工中做到均匀送钻，严禁定点循环，防止出现大井眼。针对盐城组与赤山组之间有断层且新老地层交界、井眼易漏失和失稳的特点，施工时严格控制钻压，处理好地层交界面，做好防斜打直工作。针对赤山组、浦口组发育大段泥岩、地层硬、机械钻速较低的特点，施工时以井下安全为主，不求速度，安全穿过了这一地层。

3.2 钻井液技术

钻进时采用分井段钻井液保证井壁稳定，确保井径规则，各井段钻井液方案见表2，钻井液设计参数见表3。

表2 不同井段钻井液方案表

井段	钻井液体系	作用
导管	膨润土浆	防表层坍塌
第一次开钻	聚合物钻井液	防井漏、缩径、垮塌
第二次开钻	盐水聚合物钻井液或饱和盐水聚合物钻井液	防井漏、缩径、垮塌、防膏侵、防盐溶蚀
第三次开钻	饱和盐水聚合物钻井液	防膏侵、防盐溶蚀

表3 钻井液性能参数表

钻井液		导管和第一次开钻		第二次开钻	
		导管 膨润土	第一次开钻 聚合物	非盐层 盐水聚合物	盐层 饱和盐水聚合物
常规性能	密度/(g/cm³)	1.10	1.10~1.15	1.12~1.35	1.30~1.35
	黏度/s	45~50	40~60	45~60	50~75
	API滤失量/mL		<8	<5	<5
	滤饼/mm		0.7	0.5	0.5
	含砂量/%		0.5	0.4	0.3
	pH值		8~9	9~11	9~11
流变性能	塑性黏度/(mPa·s)		10~15	15~20	20~25
	n值		0.8	0.4~0.6	0.5~0.7
固相含量/%				<18	<18
高温高压滤失量/mL				<13	<13
摩阻					0.09

第一次开钻钻进时针对上部地层成岩性差、易渗漏和垮塌的特点，钻井液中加入零滤失井眼稳定剂和单向压力封闭剂，并配合使用超细碳酸钙随钻堵漏，防止渗漏。盐城组中下部含造浆黏土，应加足抑制包被剂做好抑制造浆处理。盐城组与赤山组之间有断层且新老地层交界，若钻进中出现漏失和地层失稳情况，应及时补充防塌剂，并提高钻井液密度至设计上限，防止井壁坍塌。若钻井液流变性能发生变化，可通过稀释剂来调整。

第二次开钻钻至进入盐层前将钻井液转换为聚合物饱和盐水润滑钻井液体系[19-21]，由于盐的加入及地层中盐的溶解进入，钻井液pH值降低，应及时补充烧碱，保持钻井液体系的pH值。定期对钻井液中Cl^-含量进行监测，保证Cl^-含量维持在180000mg/L以上。加入抗盐结晶剂防止盐的重结晶而发生井下复杂情况。为抵抗盐水钻井液对管具材料及循环系统的腐蚀，在钻井液中添加适量的防腐剂。同时根据盐岩的蠕变情况，及时调整钻井液密度。

第三次开钻盐层段，尽管饱和盐水钻井液抗污染能力很强，但无法除去扫塞过程中钻井液中的钙离子，因此，钻进时仍按常规进行抗钙、除钙处理[22]，恢复其原有性能。

3.3 固井技术

楚州盐穴储气库固井难度主要体现在筛选稳定性好的抗盐水泥浆配方、提高大尺寸井眼段的顶替效率、防止上部地层的漏失以及防止环空憋堵等方面。

3.3.1 应用抗盐水泥浆体系

为保证楚州储气库盐层段的固井质量以及水泥环的长期密封性能，在借鉴国外盐穴储气库和其他盐穴储气库成功固井经验的基础上，确定楚州储气库固井时采用DRB-3S+JSS抗盐水泥浆体系，此水泥浆体系在金坛盐穴储气库、平顶山盐穴储气库和淮安盐穴储气库已经成功应用[23-24]。

3.3.2 提高顶替效率

楚州储气库生产套管固井时采用卤水作为冲洗液来保证冲洗液与钻井液的相容性及井壁稳定。同时为了增加钻井液与水泥浆的相容性，增加接触时间，有效驱替及携带环空中的稠化钻井液及浮滤饼，在注入要求密度的水泥浆前打入$1.70\sim1.75g/cm^3$的低密度水泥浆来隔离和缓冲钻井液和水泥浆。固井时中间浆密度介于$1.85\sim1.90g/cm^3$，尾浆密度介于$1.90\sim1.95g/cm^3$。

3.3.3 防漏、防憋堵

楚州盐穴储气库由于地层承压能力低等原因，在替浆过程中可能发生漏失现象。封固段长加上低压漏失层的存在，这给固井的替净工作造成了极大的困难。为此，固井前需对上部地层的承压能力进行测试，同时充分循环钻井液来调整钻井液性能，增加流动性。

施工时采用DRY-S增黏型隔离液，该隔离液在低温下增黏剂和悬浮剂易溶解、隔离液的黏度易调整，可防止不规则环空中钻屑沉积引起的环空憋堵问题。

3.4 钻井液不落地技术

钻井液是钻井过程中的主要污染物，对此楚州储气库钻井时采用钻井液不落地技术，对钻井作业产生的废弃钻井液进行固液分离，分离出的液相部分排到均质调节罐回用，部分转运；分离出的固相与钻屑一起转运，进行固化处理。处理工艺流程如图1所示。

4 应用实例

C1井、C2井、C3井是楚州储气库采用上述技术新完钻的3口井，其中C3井已投入造腔。从3口井实钻的井身结构、钻井方式、钻井液使用、固井等技术验证钻井技术方案的合理可行性，具体施工过程以C3井为例，其井身结构如图2所示。

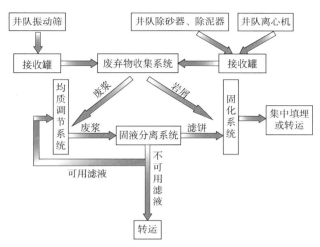

图1 钻井液不落地处理工艺流程图

C3井为楚州储气库张兴矿区的第1口资料井,设计井深2700m,为取全盐层资料,实际井深2753m。施工除正常第一、第二、第三次开钻外,还进行了取心、地应力测试、静置7d测井温、VSP测井、非造腔段封堵、造腔段扩眼、井筒气密封试压等工序,并进行了12次测井,所有工序及测井均一次成功,未发生任何井下复杂情况。全井施工周期162d,平均机械钻速3.59m/h。第一次开钻平均井径为487.04mm,第二次开钻平均井径为333.56mm,第三次开钻平均井径为235.53mm。

C3井施工中优选塔式钟摆钻具组合施工,第一次开钻钻具组合为:ϕ444.5mm 钻头+ϕ228.6mmDC×3根+ϕ203mmDC×3根+ϕ177.8mmDC×3根+ϕ158.8mmDC×15根+ϕ127mmDP(备注:DC表示钻铤,DP表示钻杆,以下相同);第二次开钻非盐层钻进:ϕ311.1mm钻头+ϕ203mmDC×3根+ϕ177.8mmDC×15根+ϕ127mm DP;盐层钻进:ϕ311.1mm钻头+ϕ177.8 mm DC×18根+ϕ127 mm DP;第三次开钻:ϕ215.9mm钻头+ϕ172mm螺杆1根+ϕ65 mm DC×5根+ϕ127mm DP(加重)×15根+ϕ127 mm DP。

在施工中严格按照钻井工程设计做好定点

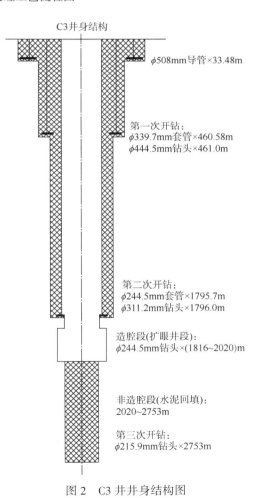

图2 C3井井身结构图

测斜工作,同时根据测斜数据及时调整钻井参数,井身质量符合设计要求见表4,井眼轨迹如图3所示。

表4 C3井井身质量表

项目	设计要求	实钻情况	结果
最大井斜	2°(0~1000m) 1.5°(1000m~完钻)	1.47°/700m	符合设计要求
全角变化率	1.25°	0.23°/25m	符合设计要求
水平位移	25m	24.06m	符合设计要求
第一次开钻平均井径扩大率	10%	9.57%	符合设计要求
第二次开钻平均井径扩大率	10%	7.22%	符合设计要求
第三次开钻平均井径扩大率	10%	9.09%	符合设计要求

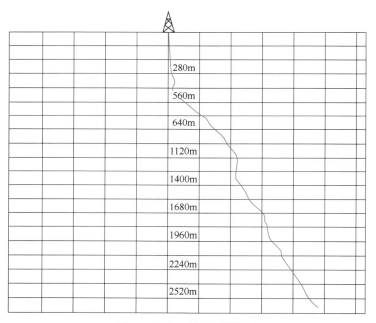

图3 井身垂直剖面投影图

为保证钻井液及时将岩屑携带出井外,防止沉砂卡钻,保持井壁稳定,降低摩阻,施工中采取了以下措施:(1)在大井眼段坚持双泵循环钻进,保证钻井液能及时携带出井底岩屑;(2)钻进时补充足量的提黏剂,增强钻井液的悬浮携带能力;(3)保持钻井液的强抑制性,加入足量的降滤失剂,提高黏度,从化学方面防塌;(4)充分利用固控设备,降低钻井液中有害固相和含砂,确保井眼干净清洁,降低摩阻。通过这些钻井液处理措施很好地配合了钻井施工,保证了安全高效钻进,钻井液实际性能参数见表5。

表5 钻井液实际性能参数表

钻井液		导管和第一次开钻		第二次开钻		第三次开钻
		0~20m	20~461m	461~640m	640~1796m	1796~2753m
		膨润土	聚合物	盐水聚合物	饱和盐水聚合物	饱和盐水聚合物
常规性能	密度/(g/cm³)	1.10	1.10~1.13	1.12~1.20	1.20~1.34	1.33~1.35
	黏度/s	45~50	45~55	48~53	38~45	40~53
	API滤失量/mL		5.6~7.8	5.2~5.8	4.8~6.2	4.2~5.5
	滤饼/mm		0.5~1.0	0.5	0.5	0.5
	含砂量/%		0.3~0.5	0.2	0.2	0.2
	pH值		9	9~11	11~12	10~11
流变性能	塑性黏度/(mPa·s)		15~18	12~17	12~17	12~26
	n值		0.48~0.65	0.60~0.61	0.54~0.77	0.53~0.77
固相含量/%				<18	<18	<18
高温高压滤失量/mL				<13	<13	<13
摩阻					0.09	0.09

C3井表层套管采用内插法固井,采用常规密度低温早强稳定性好的水泥浆,切实保证套管鞋处及上部井段的固井质量,第一次开钻固井难度相对较小。生产套管固井采用常规固井方法,减少套管内水泥浆与钻井液的混窜。采用高早强稳定性好的盐水水泥浆体系,以保证裸眼段及套管重合段的固井质量,固井施工过程见表6。

表6 C3固井施工流程表

顺序	操作内容	工作量/m³	密度/(g/cm³)	排量/(m³/min)	施工时间/min
1	循环处理钻井液	—	1.33	2.5~3.0	—
2	管汇试压	—	1.0	0.2	
3	注入隔离液(压下胶塞)	10	1.20	1.0	
4	双车注入低密度	5	1.59~1.54	2.1~1.8	3
5	双车注入领浆	20	1.94~1.87	2.1~1.8	11
6	双车注入中浆	20	1.94~1.89	2.1~1.8	10
7	双车注入尾浆	60	1.96~1.91	2.1~1.8	30
8	压上胶塞	2	1.00	0.8~1.0	3
9	替钻井液	60	1.33	1.0~1.5	50
10	碰压	5.6	1.0	0.5~1.0	10
11			检查回流		
12		注水泥替量施工总时间			117

候凝72h后进行固井质量测井,固井质量合格,并对井筒进行了气密封试压,24h后气液界面由1 804.16m降低至1803.89m,井口环空压力由21.29MPa降低至21.22MPa,井筒气密封试压合格,C3井能够满足盐穴储气库对密封性的要求。

C3 井在钻井时随钻收集钻井废水和岩屑，通过钻井液不落地设备实现固液分离，液相收集后回用或者拉运至有资质的污水处理厂进行深度处理达标排放，固相则通过加入水泥、固化剂、pH 值调节剂等后，在处理罐内实施无害化处理，拉送至指定地点填埋，实现了钻井废弃物与土壤隔离，未造成任何现场污染。

5 结论

（1）使用钟摆钻具组合，配合 MWD 随钻监测，能够有效应对楚州地区地层，控制井斜，保证井身质量，实现高效钻进。

（2）采用分段钻井液，在钻进过程中加强监测，及时调整钻井液性能，严格控制钻井液滤失量，改善滤饼质量，提高悬浮、携带能力，保证正常钻进。

（3）采用 DRB-3S+JSS 抗盐水泥浆体系，配合合适的顶替排量以及三段制水泥浆的液柱结构，可以有效提高顶替效率，保证固井质量。

（4）钻井时采用钻井液不落地设备，进行固液分离及无害化处理，能够保证钻井液、岩屑不落地，达到地方环保部门要求。

（5）通过 3 口井的现场成功应用，证明制定的技术对策可行，能够有效保证楚州盐穴储气库钻井施工的质量和安全。

参 考 文 献

[1] 垢艳侠，完颜祺琪，罗天宝，等．中俄东线楚州盐穴储气库建设的可行性[J]．盐业与化工，2017，46(11)：16-20.
[2] 代长灵，杨光，薛让平．长庆靖边储气库关键钻井技术[J]．天然气勘探与开发，2016，39(1)：65-69.
[3] 井岗，何俊，陈加松，等．淮安盐穴储油库潜力分析[J]．油气储运，2017，36(8)：875-882.
[4] 覃毅，连进报，肖龙雪，等．刘庄储气库注采气井 177.8mm 套管配套固井技术[J]．科学技术与工程，2012，12(8)：1892-1894.
[5] 侯树刚，胡建均，李勇军，等．低压低渗枯竭型砂岩储气库钻井工程方案设计[J]．石油钻采工艺，2013，35(5)：36-39.
[6] 孙海芳．相国寺地下储气库钻井难点及技术对策[J]．钻采工艺，2011，34(5)：1-5.
[7] 赵常青，曾凡坤，刘世彬，等．相国寺储气库注采井固井技术[J]．天然气勘探与开发，2012，35(2)：65-69.
[8] 吴虎，张克明，刘梅全，等．塔河油田深井穿盐膏层钻井液技术[J]．石油钻采工艺，2007，29(2)：86-90.
[9] 邢希金，赵欣，刘书杰，等．中东 Lower Fars 厚盐膏层钻井液室内研究[J]．石油钻采工艺，2014，36(4)：57-60.
[10] 蔺文洁，黄志宇，张远德．高密度饱和盐水钻井液在盐膏层钻进中的维护技术[J]．天然气勘探与开发，2011，34(1)：64-67.
[11] 刘政，黄平，刘静，等．川渝地区盐膏层钻井液技术对策[J]．天然气勘探与开发，2014，37(3)：75-77.
[12] 齐奉忠．金坛地区盐穴储气库固井技术研究与应用[D]．北京：中国石油勘探开发研究院，2005.
[13] 曾义金，王文立，石秉忠．深层盐膏岩蠕变特性研究及其在钻井中的应用[J]．石油钻探技术，2005，33(5)：48-51.

［14］郑建翔，周仕明，韩卫华．塔河油田石炭系盐层固井工艺技术［J］．西部探矿工程，2006，18（9）：59-63．
［15］齐奉忠，袁进平．盐层固井技术探讨［J］．西部探矿工程，2005，17（10）：58-61．
［16］张春涛．盐膏层固井技术及应用［J］．钻采工艺，2008，31（3）：146-148．
［17］张绍槐．深井、超深井和复杂结构井垂直钻井技术［J］．石油钻探技术，2005，33（5）：11-15．
［18］高德利．易斜地层防斜打快钻井理论与技术探讨［J］．石油钻探技术，2005，33（5）：16-19．
［19］刘在同，王建伟．土耳其盐穴地下储气库大井眼钻井技术［J］．石油钻采工艺，2015，37（2）：32-34．
［20］吉永忠，伍贤柱，邓仕奎，等．土库曼斯坦阿姆河右岸巨厚盐膏层钻井液技术研究与应用［J］．钻采工艺，2013，36（2）：6-8．
［21］左万里，田增艳，常峰，等．克深6井超常规大井眼穿盐膏层钻井液技术［J］．钻采工艺，2015，38（3）：115-117．
［22］唐军，聂福贵，卢虎，等．抑制性抗钙盐岩盐与石膏侵污钻井液工艺技术试验研究［J］．钻采工艺，2014，37（1）95-98．
［23］张幸，覃毅，李海伟，等．平顶山盐穴储气库固井技术［J］．石油钻采工艺，2017，39（1）：61-65．
［24］覃毅，党冬红，刘晓贵，等．盐穴储气库楚资1井ϕ244.5mm套管固井技术［J］．石油钻采工艺，2015，37（2）：51-53．

本论文原发表于《天然气勘探与开发》2019年第42卷第2期。

金坛地下储气库盐腔偏溶与井斜的关系

杨海军　于胜男

（中国石油西气东输管道公司储气库管理处）

【摘　要】 在金坛储气库造腔过程中，盐腔偏溶问题层出不穷，对腔体体积、造腔进度、注气排卤、运行安全等的不良影响不容小觑。为此，对金坛储气库28口造腔井的声呐数据、连斜数据、管柱提升记录、施工记录和原始溶腔设计等资料进行统计分析。结果表明：金坛储气库多口盐井偏移方向规律明显，大多偏北，少数偏南；多数盐腔偏溶方向与井轨迹偏移方向一致。由此提出盐腔偏溶与井斜有着密不可分的关系，并初步认定地应力对盐腔偏溶起到一定作用。

【关键词】 盐穴地下储气库；偏溶；声呐；连斜；井斜；地应力

Relationship between salt cavern partial melting and well deviation of Jintan underground gas storage

Yang Haijun　Yu Shengnan

(Gas Storage Project Department, PetroChina West-East Gas Pipeline Company)

Abstract In the cavity building process of Jintan Underground Gas Storage, salt cavern partial melting emerges frequently, which brings significantly adverse effects on the cavern volume, progress of cavity building, gas injection and brine discharge, and operational safety. Therefore, sonar data, continuous inclinometer data, string ascension records, as-built records, the original cavern design and other information of 28 cavity building wells in Jintan Underground Gas Storage are statistically analyzed. The results show that the offset directions of many wells of Jintan Underground Gas Storage are clear that are mostly northward and locally southward; the salt cavern partial melting direction is usually consistent with the well trajectory offset direction. Thus, it is proposed that salt cavern partial melting and well deviation have a close relationship, and it is initially determined that ground stress has a certain effect on salt cavern partial melting.

Keywords underground salt cavern gas storage; partial melting; sonar; continuous inclinometer; well deviation; ground stress

---

作者简介：杨海军，高级工程师，1961年生，1983年毕业于大庆石油学院开发系钻井工程专业，现主要从事地下储气库的运营管理工作。

金坛地下储气库位于江苏省金坛市直溪镇,毗邻镇江,属盐穴地下储气库,是中国盐穴第一库,规模为亚洲第一。它开创了中国利用深部洞穴实施能源储存的先河,在地址选区、区块评价、溶腔设计、造腔控制、稳定性分析、注采方案设计、钻完井工艺等方面积累了诸多研究成果和技术手段,为中国利用盐穴进行天然气储备提供了参考[1-5]。金坛地下储气库分西区与东区,建库盐层区域面积为 11.2km², 库深约 1000m。已经建成投产的西区利用采卤老腔6口,库容为 $1.21×10^8 m^3$。东区规划新建腔17口,库容为 $7.22×108m^3$。金坛矿盐层分布稳定,结构简单,封闭性好,易于水溶开采,在地下可以形成较大溶腔。每个盐穴的溶腔时间约3年。在建库过程中,由于溶盐的不对称性,往往造成盐腔形状不规则,甚至偏离中心管溶蚀而形成盐腔的偏溶。为此,以金坛28口造腔井的数据为基础,深入探讨了盐腔偏溶的诱因,并对其根源进行假设。

1 相关数据

收集了金坛储气库28口井的声呐数据、26口井的连斜数据、28口井的管柱提升记录、24口井的施工记录和24口井的原始溶腔设计资料。

1.1 声呐数据

声呐测腔数据包括了每口井每次声呐测腔的一系列数值。以 11# 井为例,每次声呐测腔的腔体形状包括各个横切面、纵切面的形状,对应的腔顶、腔底深度,各个深度对应的腔体半径,同时可见腔体3D立体图形。如需要,还可将前一次测腔或后一次测腔的腔体形状图叠加,以观察腔顶、腔底的变化,统计不溶物高度的变化情况(图1),并对应有文字报告。

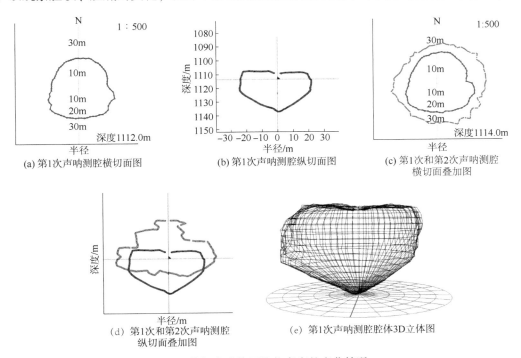

图 1 储气库腔体不溶物高度的变化情况

1.2 连斜数据

综合各口井的井斜角、方位角、最大位移等数据,分析可知每口井在全井段和溶腔井段深度范围内的井轨迹情况,以及各个深度内井口与井眼之间的偏移情况,并可与腔体最大半径 D_{max} 相对应,尝试寻找对应关系(表1)。

表1 24口井连斜数据统计表

井编号	全段井轨迹			溶腔段井轨迹			腔体情况	
	最大井斜角/(°)	对应方位角/(°)	最大位移/m	最大井斜角/(°)	最大方位角/(°)	D_{max}/m	D_{max}深度/m	D_{max}方位/(°)
1	1.620	103.450	12.090	1.370	45.000	71.6	1110	70~250
2	2.000	35.450	17.110	1.950	359.473	74.6	1082	155~335
4	3.350	359.650	8.270	3.100	358.066	76.6	1112	160~340
6	1.550	342.070	4.460	1.500	342.070	80.3	1070	175~355
7	1.340	10.020	7.820	1.300	9.500	72.8	1050	150~330
8	1.880	220.410	8.480	1.890	220.414	36.5	1170	0~180
9	2.470	329.000	16.900	2.470	329.000	81.6	1076	0~180
10	1.980	335.970	22.900	1.910	359.840	73.5	1096	165~345
11	1.600	338.380	6.490	1.580	338.380	81.1	1048	160~340
12	1.960	143.887	11.000	1.784	40.303	74.3	1078	160~340
13	1.520	335.000	12.910	1.165	329.766	79.0	1086	70~250
14	1.553	329.000	12.410	1.500	354.000	75.0	1082	175~355
15	1.400	13.000	7.130	0.530	324.000	74.7	1076	75~255
16	1.650	11.780	16.720	2.950	153.930	79.1	1150	179~359
17	0.940	156.000	—	0.940	156.000	73.6	1122	0~180
18	1.810	6.470	9.250	1.780	26.539	59.1	1130	165~345
19	1.950	336.280	14.100	1.850	302.824	68.0	1124	150~330
20	1.980	279.390	11.800	1.800	270.000	68.4	1158	35~315
21	1.600	271.400	—	1.600	271.421	62.2	1190	170~350
22	1.738	172.400	13.692	1.390	196.000	63.3	1142	110~290
23	1.890	131.780	13.900	1.894	131.778	57.9	1118	150~330
24	2.700	20.830	17.070	2.480	33.040	79.5	1164	160~340

1.3 管柱提升记录

统计每口井 GSDMAS 系统中的管柱提升记录,关注管柱提升数据和时间,对于探究管柱提升对偏溶的影响具有参考意义。管柱提升过快或提升距离过大,会造成溶蚀不充分;管柱提升过慢,会使相应深度段内的盐腔过度溶蚀,横向半径增大,影响腔体形状。

1.4 施工记录

统计每口井的每次作业时间,每次作业前17.78cm和11.43cm套管深度,中心管有无变形,腔底不溶物(顶面)深度。通过对比各个溶腔阶段内腔底不溶物高度和施工前中心管的深度,可知在溶腔过程中,中心管是否被不溶物掩埋而造成中心管弯曲。

1.5 原始溶腔设计

将原始溶腔设计中的理想模拟腔体形状和相关数值与实际溶腔结果数据对比,对实际溶腔情况进行评估,并探究相关影响因素。同时,溶腔设计中的油水界面深度也是重要的参考数据。

2 盐腔偏溶情况

2.1 总体偏溶情况

依据以上数据资料,统计金坛储气库 24 口井的偏溶情况,汇总在一张图上(图2):在金坛储气库偏溶的腔体中,绝大多数的偏溶方向(绿色宽箭头所示)偏北,个别偏南及东南向;井轨迹偏移方向(蓝色窄箭头所示)多偏北,个别偏南。

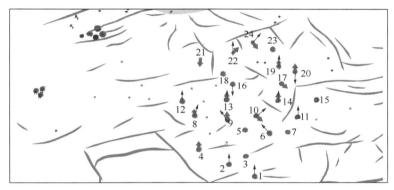

图 2　金坛储气库井轨迹及偏溶方向示意图

在 24 口井中,偏溶井 13 口,均有明显的井轨迹,其中 11 口井轨迹方向与腔体偏溶方向一致,9 口朝北,2 口朝南;24#井轨迹开始朝北,后朝南,腔体偏南;10#井轨迹方向与偏溶方向相反,并且井轨迹偏北,腔体偏南。不偏溶井 11 口,大多没有明显的井轨迹。同时,溶腔段最大方位角多朝北,腔体最大半径所处方位也多朝北。

2.2 单井偏溶情况

2.2.1 偏溶不明显的井

造腔段多井斜小,井轨迹偏移不明显(图3):偏溶不明显或基本未发生偏溶的井,横切面呈规则圆形,井轨迹与井口偏移不明显,3D 腔体立体图呈现对称特征,或长或扁,但大体以中心轴对称。

2.2.2 偏溶明显的井

造腔段井斜多与腔体偏溶方向一致,且大多偏北(图4):9#井造腔段井斜角 2.47°,方位角 329°,腔体朝北偏,偏移方向与井轨迹方向一致;17#井造腔段井斜角 0.94°,方位角 156°,倒数第 3 根轻微弯曲,腔体朝南偏,偏移方向与井轨迹方向一致;19#井造腔段井斜角 1.85°,方位角 302.8°,倒数第 2 根油管弯曲,笔尖弯曲严重,腔体朝北偏,偏移方向与井轨迹方向一致。

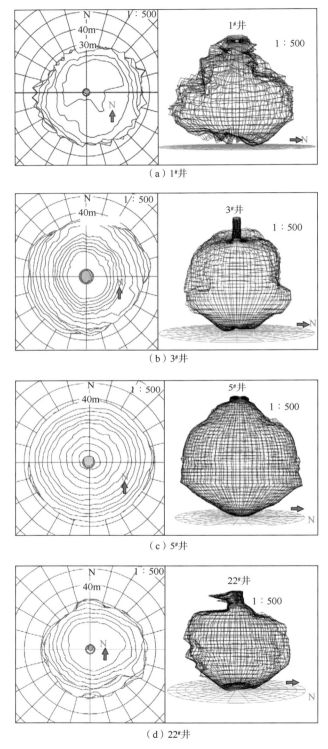

图 3　偏溶不明显的井的横切面和 3D 立体图

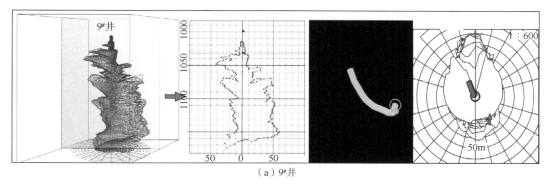

(a) 9#井

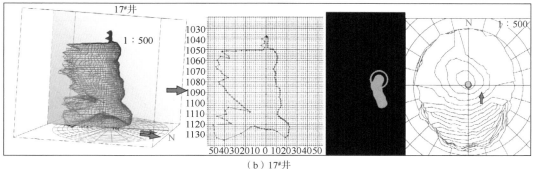

(b) 17#井

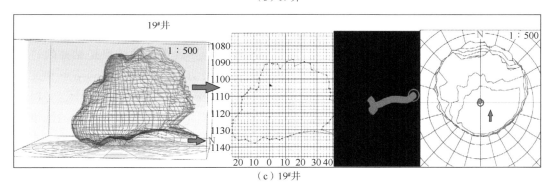

(c) 19#井

图 4 偏溶明显的井的横切面和 3D 立体图

综上，腔体偏溶方向多与井轨迹偏移方向一致，且偏溶越明显，井轨迹偏移越大；腔体偏溶不明显的井，井轨迹也不明显。但有几口井腔体偏移方向与井轨迹方向不一致。10#井造腔段井斜角 1.91°，方位角 359.84°；底部往上第 1 根油管严重弯曲，第 2 根油管轻微弯曲；腔体朝南偏，偏移方向与井轨迹相反。依据管柱提升记录，溶腔层段从 1145～1098m，恰为溶蚀不充分段，推测其可能为影响盐腔偏移的因素(图 5)。21#井造腔段井斜角 1.6°，方位角 271.42°；自下而上第 1 根、第 2 根油管严重弯曲，第 3 根、第 4 根油管轻微弯曲；井轨迹略朝东北方向偏，腔体偏南。管柱提升情况正常，推测腔体偏溶可能由地层非均质性引起(图 6)。

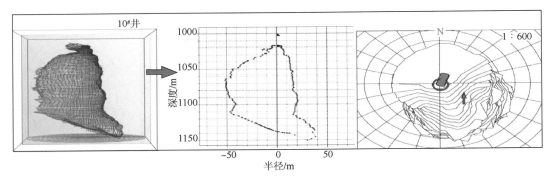

图 5　10#井腔体偏移情况

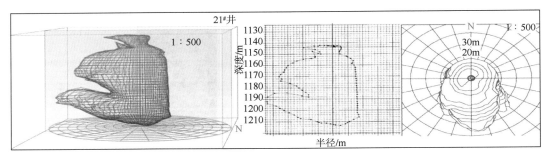

图 6　21#井腔体偏移情况

理论上，抛开其他影响因素，地应力越小，钻时越短，钻井速度越快。这是因为地应力小的区域对应较为疏松的地层，使得钻井更容易，因而导致直观的结果是钻井轨迹发生偏移，偏移方向指向地应力小的方向。此外，相关研究表明，应力对矿物的溶解和变质等具有重要影响。汤艳春等[6]通过建立盐岩溶蚀模型开展盐岩溶蚀试验，结果显示：（1）应力引起的应变能可以改变盐岩固体的物质活度，影响到盐岩固体表面溶解生成物的浓度，进而对盐岩的溶蚀速率产生直接影响；（2）应力的存在会改变盐岩固体外部水流状态、水流速度、固体表面边界层厚度及有效的水岩相互作用面积，从而间接对盐岩的溶解产生影响。

3　结论与认识

金坛储气库多口盐井偏移方向规律明显，大多偏北，少数偏南；多数盐腔偏溶方向与井轨迹偏移方向一致，说明地应力一定程度上间接影响盐腔的偏溶。对于井轨迹与腔体偏溶方向不一致的井，除地层的非均质性外，还可能是其他地质因素影响腔体偏溶，有待进一步考察。由于研究数据仅取自金坛库区，相关结论有待日后参考其他库区的实际情况加以验证。

参　考　文　献

[1] 纪文栋，杨春和，屈丹安，等．盐穴地下储气库注采方案[J]．油气储运，2012，31（2）：121-124.
[2] 班凡生，肖立志，袁光杰，等．地下盐穴储气库快速建槽技术及其应用[J]．天然气工业，2012，32（9）：77-79.

[3] 李丽锋,罗金恒,赵新伟.盐穴地下储气库井口破裂火灾事故危险分析[J].油气储运,2013,32(10):1054-1057.
[4] 赵艳杰,马纪伟,郑雅丽,等.淮安盐穴储气库注采循环运行压力限值确定[J].油气储运,2013,32(5):526-531.
[5] 陈雨,李晓,李守定,等.基于盐腔时效收敛特征的密集地下储气库群地面变形动态预测[J].天然气工业,2012,32(9):80-84.
[6] 汤艳春,房敬年.三轴应力作用下盐岩溶蚀特性试验研究[J].岩土力学,2012,33(6):1601-1607.

本论文原发表于《油气储运》2015年第34卷第2期。

平顶山盐穴储气库固井技术

张 幸[1,2] 覃 毅[3] 李海伟[2] 兰 天[2]

(1. 中国石油勘探开发研究院;2. 中国石油西气东输管道公司储气库项目部;
3. 渤海钻探工程有限公司第一固井分公司)

【摘 要】 平顶山盐穴储气库含盐地层厚度大,目的层埋藏深,泥岩夹层多,固井难度大、要求高。在分析前期固井施工过程中存在的易漏失、胶结质量差、未返出井口等难题的基础上,通过开展有针对性的盐水水泥浆配方优选、井眼准备、前置液优化、钻井和下套管施工等综合性技术措施的研究,保障了固井顺利施工和固井质量。PT1井是平顶山盐穴储气库的一口探井,该井生产套管一次封固段长,井眼尺寸大,井筒条件复杂,通过采用综合性能好的抗盐水泥浆体系,避免了固井期间漏失问题的发生,解决了平顶山盐穴储气库固井难题,为该地区后续的固井施工提供了宝贵经验。

【关键词】 固井;盐穴储气库;盐水水泥浆;氦气检测;顶替效率;现场试验

Cementing technology suitable for Pingdingshan salt cavern gas storage

Zhang Xing[1,2] Qin Yi[3] Li Haiwei[2] Lan Tian[2]

(1. PetroChina Research Institute of Petroleum Exploration & Development;
2. Gas Storage Project Department, PetroChina West-East Gas Pipeline Company;
3. No. 1 cementing branch, Bohai Drilling and
Exploration Engineering Co. Ltd.)

Abstract Pingdingshan salt cavern gas storage is characterized by thick salt formations, deep targets and multiple mudstone interbeds, so its cementing job is difficult and shall meet more requirements. The difficulties that occurred during the previous cementing were analyzed, such as circulation loss, poor cementation quality and no slurry returning to the wellhead. Then, a series of comprehensive measures were specifically studied to guarantee the smooth cementing and cementing quality, including brine cement slurry formula optimization, borehole preparation, prepad fluid optimization, well drilling and casing running. Well PT1 is an exploration well in Pingdingshan salt cavern gas storage. In this well, the single-stage cementing section of production casing is long, borehole size is large and wellbore condi-

---

作者简介:张幸(1986—),2010年毕业于中国石油大学(华东)石油工程专业,获学士学位,中国石油勘探开发研究院在读硕士研究生,主要从事盐穴储气库建设的相关工作,工程师。E-mail:cqkzhangxing@petrochina.com.cn。

tions are complex. Its circulation loss during cementing is avoided by adopting salt-resisting cement slurry system whose comprehensive performance is good and adjusting and circulating the performance of drilling fluids sufficiently. And consequently, the cementing difficulties of Pingdingshan salt cavern gas storage are solved. The research results provide the valuable experience for the subsequent cementing operation in this area.

Keywords cementing; salt cavern gas storage; saline cement slurry; helium detection; displacement efficiency; field test

平顶山盐穴储气库位于河南省平顶山市南部叶县境内，是目前国内计划建设埋藏最深的盐穴储气库，设计井深近 2000m，含盐层系赋存于舞阳箕状凹陷内，为一套中生界—新生界的碎屑化学岩系，属古近系核桃园组二段上部和核一段[1]。舞阳凹陷内核桃园组盐层厚、泥岩夹层多，地层倾角较大，夹层总厚度占 30%~40%，且各夹层厚度不等，最大可以达到 5m 以上，含盐层系最厚达 1000m，优选含盐质量较好的 14#~20# 盐群进行造腔。

盐穴储气库能否储存天然气，并在运行的过程中不发生泄漏主要取决于盐穴腔体及井筒的密封性能[2]，固井质量对于保障井筒密封最为重要。盐穴储气库注采气量大，并且井筒处于注和采的交变应力状态，在盐穴储气库的运行年限里，必须考虑交变应力，以保证储气库安全运行至少 30 年的目标[3]。因此为保证储气库长期安全运行，必须保证水泥环强度的稳定性和长期密封性能。盐穴储气库对固井质量的要求和普通油气井相比更加严格，如果固井质量差，则缺乏有效的补救措施。固井质量与注采气井的寿命及长期安全运行紧密相关，固井质量涉及钻井、固井施工、钻水泥塞、取心、扩眼等整个钻井施工过程，任何施工作业必须以保障固井质量为前提。同时，在盐穴储气库正常生产的过程中，仍需对套管及水泥环的胶结状态进行检测，确定水泥环与套管、地层之间的胶结情况[4]。在同区块其他井的固井施工中，多次出现易漏失、胶结质量差、未返出井口等复杂情况，造成固井质量差，为后续的溶腔、注气排卤等工作埋下了安全隐患。

1 固井技术难点

目前，盐穴储气库固井施工分为一开固井施工、二开固井施工，一开固井施工是为了封固上部疏松井段，防止垮塌，二开固井施工一般是进入设计盐群 14#~20#，下入气密封套管进行封固，为后续溶腔施工及注采气作业提供工程保障。

由于平顶山储气库地质情况复杂，目的层段埋藏深，特殊的地质条件给固井作业造成了诸多难点：其中一开井段由于地层疏松、井深浅、井眼尺寸大及井眼不规则等因素，在固井施工过程中，易发生漏失。采用内插法固井，容易发生固井附件密封失效，或造成插旗杆、灌香肠等复杂事故[5]，相对而言，一开固井技术难度较小。

二开固井施工较为复杂，其技术难点大，一旦失败将没有办法进行补救，目前平顶山储气库已发现存在的难点如下。

（1）由于造腔井段为 14#~20# 盐群，需封固盐岩层段较长，在钻进的过程中，易发生盐岩溶蚀、泥岩吸水膨胀等问题，造成井眼条件差，井眼不规则。

（2）二开井段较长，在表层套管鞋处，容易形成"大肚子"井段，虚滤饼附着，造成该处固井质量较差。

（3）井深较大，一次封固井段较长，且上部地层存在砂砾岩段，承压能力低，施工过程中易发生漏失，固井时不允许采用大排量顶替，提高顶替效率困难，保证水泥浆上返至地面困难。

（4）表层套管和技术套管的重合段，固井质量难以保证。

（5）下套管过程中，需进行氦气气密封检测，作业时间长，容易在井壁上形成较厚的虚滤饼，易发生缩径、遇卡等复杂情况，在下完套管顶通过程中以及注水泥过程中可能会因虚滤饼太多堵塞环空，造成憋堵，影响顶替效率。

（6）国内适合低温条件的抗盐水泥浆外加剂比较少，水泥浆体系必须满足抗盐要求，稠化性能必须满足施工要求，优选综合性能好的水泥浆体系较为困难[6]。

2 固井技术对策

根据近期盐穴储气库固井施工的情况及相应固井工艺措施的现场应用，采取以下针对性措施和施工方案可以有效提高固井质量，降低固井施工风险。

2.1 强化井眼准备

通井时，采用原钻具组合进行通井，到井底后充分循环钻井液，确保井壁稳定，井眼畅通，无沉砂、无坍塌。下完套管后，先小排量顶通，等返出正常、泵压稳定后，再以钻进时的排量充分循环洗井 2 周，同时调整钻井液性能，为后续固井施工准备良好的井眼条件[7]。

2.2 盐水隔离液体系

二开采用能够防止盐岩地层溶蚀及蠕变的饱和盐水钻井液钻进，为防止盐水钻井液和水泥浆接触污染，使用性能良好的盐水隔离液，在顶替钻井液的过程中起到冲洗、稀释、隔离、缓冲管壁和井壁滤饼的作用，提高水泥浆的顶替效率，同时可以有效清除井壁上附着的虚滤饼，保证水泥的良好胶结，确保固井质量[8]。饱和盐水隔离液占环空高度 200~300m。

2.3 优选水泥浆体系

对盐穴储气库的水泥浆要求和常规油气井及采盐井有很大不同，其技术难度较大[9]。目前，盐穴储气库二开固井施工均采用盐水水泥浆体系，为 DRB-3S+JSS 抗盐水泥浆体系，已在江苏金坛、江苏淮安盐穴储气库多口井进行了成功的应用。固井前，根据平顶山盐穴储气库的实钻情况、井下条件对水泥浆密度及配方进行具体调整。

2.3.1 水泥浆配方要求

（1）能配成设计密度的水泥浆，容易混合与泵送，具有良好的流动度、适宜的初始稠度，且均质、起泡性小，游离液为 0。

（2）水泥石早期强度发展快，并有长期的强度稳定性。

（3）外加剂配伍性好，敏感性低，稠化时间易调，对水泥水化、水泥内部结构、强度发展、长期胶结性能无不良影响。

（4）外加剂具有良好的抗盐性能。

2.3.2 水泥浆体系

盐穴储气库水泥石要承受长期交变应力的影响，对水泥环的胶结质量及长期密封性要求高。根据上述特点，满足生产套管固井要求的水泥浆体系，要求抗盐性好、低温快凝、高早强、微膨胀、浆体稳定性好、稠度适宜，和钻井液及隔离液具有良好的相容性。根据平顶山盐穴储气库的固井要求，综合考虑采用 JSS 抗盐降滤失剂，水泥浆稠化时间可用 ZH-2 型中温缓凝剂进行调节。

JSS 降滤失剂具有一定的分散性能，在正常水灰比情况下不用加入分散剂。但当水质中 Ca^{2+}、Mg^{2+} 较多时，需要加入适量的分散剂。FSS 分散剂可用于调节水泥浆的流动度，并可适当调节水泥浆的稠化时间。常规水泥浆中使用的消泡剂对于盐水水泥浆体系来说很难达到预期的消泡效果，D50 消泡剂能有效解决盐水水泥浆体系的消泡问题[10]。经过多次室内实验，筛选出了双密度双凝抗盐水泥浆体系，水泥浆性能见表1。水泥浆配方为：上部井段：G 级高抗硫油井水泥+3.0%~4.0%抗盐降滤失剂+8%~12%DRB-3S 增强材料+0.8%~1.5%分散剂+调凝剂+盐水；下部井段：G 级高抗硫油井水泥+3.0%~4.0%抗盐降滤失剂+8%~12% DRB-3S 增强材料+0.8%~1.5%分散剂+调凝剂+盐水。实验条件：52℃、20MPa。

表1 抗盐水泥浆配方及性能

序号	降滤失剂 JSS/%	分散剂 FSS/%	密度/ (g/cm³)	滤失量/ mL	稠化时间/ min	游离液/ %	24h 抗压强度/MPa	72h 抗压强度/MPa
1	3.0	0.8	1.95	108	142	0	18.2	24.9
2	3.5	0.8	1.95	96	155	0	17.6	24.0
3	4.0	1.2	1.95	84	185	0	17.0	24.5

2.4 提高顶替效率

提高平顶山盐穴储气库井顶替效率采取的主要技术措施如下。

（1）降低钻井液黏度与切力。下完套管后，充分循环钻井液，调整钻井液性能，保持密度不变，降低钻井液黏度与切力。隔离液对环空滞留的静止钻井液有一定的渗透力，降低黏土间的连接力，使钻井液或滤饼的结构松弛、拆散，易于顶替。

（2）提高套管居中度。根据实钻井眼状况与井径情况，合理加放套管扶正器，提高套管居中度，从而提高顶替效率。自由套管柱在井筒内受重力作用始终靠向铅直方向，因此任何一口井都存在套管不居中的问题。加放扶正器不仅可以提高套管的居中度，保持环空中的流速分布均匀，改善因环形空间间隙不均匀所造成的环空互窜现象，同样也可以减少水泥环厚薄不均的情况，并且可以降低替浆时的阻力，防止替浆时环空中钻井液的窜槽现象。

（3）使用饱和盐水隔离液。优选一种适用于平顶山盐穴储气库固井施工的隔离液，饱和盐水隔离液密度为 1.18~1.19g/cm³，稀释钻井液，冲洗净井壁和套管壁，提高顶替效率和水泥界面的胶结质量。

3 现场试验

以 PT1 井为例，该井为平顶山盐穴储气库的一口探井，井身结构见表 2，完钻井深为 2385m。一开 φ339.7mm 套管下深 384.66m 井段，二开 φ244.5mm 套管下深 1722.35m，水泥浆均要求返至地面。下入生产套管时要进行氦气气密封检测，下套管时间约为 48h，电测井底井温为 58℃左右。

表 2 PT1 井井身数据

参数	数值	参数	数值
一开井深/m	385.0	三开井深/m	2385
表层套管下深/m	384.66	注水泥塞井段/m	2004~2385
二开井深/m	1725	扩眼井段/m	1732~2004
生产套管下深/m	1722.35		

套管氦气检测近期在国内盐穴储气库开始应用，气密封检测技术可以检测出发生泄漏的套管，杜绝了不合格套管入井，可以降低后期生产中的天然气泄漏及环空带压的风险[11]。PT1 井是国内盐穴储气库首次应用该技术，为后续推广应用提供经验。

3.1 表层套管固井施工

表层套管固井采用内插法固井技术，采用稳定性好的常规密度低温早强水泥浆固井。下套管前充分循环钻井液 2 周以上，循环排量 3.0~3.5m^3/min，充分清洗井眼并处理好钻井液。下部 5 根套管每根安放 1 只扶正器，井口 2 根套管每根加 1 只扶正器，其余井段共安放 5 只扶正器，均加在套管接箍上。用清水作为隔离液并配合适当量的低密度水泥浆提高固井时的顶替效率，注水泥排量 1.0~1.2m^3/min，替浆排量 0.8m^3/min 左右，密度为 1.85g/cm^3 的水泥浆返出井口后停止注水泥。

固井施工过程中，密度为 1.85g/cm^3 的水泥浆返出地面后水泥浆发生回落，后采取回灌方法进行补救。固井质量检测，表层套管实现全井段封固，固井质量合格。

3.2 生产套管固井施工

生产套管固井套管串结构：φ244.5mm 气密扣浮鞋 + φ244.5mm 气密套管 2 根 + φ244.5mm 气密扣浮箍（带承托环）+ φ244.5 mm 气密扣套管串 + 变扣短节（气密公扣×LTC 母扣）+ φ244.5mmLTC 联顶节。扶正器安放方式为：下部 50m 井段每根套管安放 1 只扶正器，井底以上 50~400m 井段每 2 根套管安放 1 只扶正器，井底以上 400m 至井口段每 5 根套管安放 1 只扶正器，扶正器加在套管接箍上。

PT1 井生产套管一次封固井段长 1722.35m，且井眼尺寸大，井径不规则，最大井径 411.73mm，井径最小 318.897mm，平均井径扩大率 10.16%，上部井段地层承压能力低。现场施工时采取了以下措施。

（1）注水泥施工前充分循环钻井液，确保井眼畅通、无沉砂、无垮塌，为固井创造良好的井眼条件。通井完成后，在下部裸眼段打入黏度为 150s 以上的高黏钻井液，防止水泥浆下沉。

（2）由于该井是国内盐穴储气库第一次进行生产套管氦气气密封检测，下套管时间长。为了确保下套管施工顺利，对该井采取了裸眼静止72h，然后测量井径，在缩径严重井段，进行划眼，以改善井眼条件，再进行下套管作业。

（3）采用稳定性好的抗盐半饱和盐水水泥浆固井，优选水泥浆配方，使水泥浆达到高早强、稳定性好(游离液为0mL)，以保证裸眼段及套管重合段的固井质量。做好固井施工前水泥浆性能的复核检验，满足安全固井施工要求。

（4）充分调整钻井液性能，钻井液黏度控制在40~45s，采用饱和盐水冲洗液及低密度水泥浆作为隔离液、保证套管居中来提高固井时的顶替效率。

（5）顶替过程中为防止漏失，确保水泥全部返出地面，替浆排量控制在1.2~1.5m³/min，最后5m³采用水泥车顶替，排量控制在0.6~0.9m³/min。

（6）施工结束后泄压，确认浮箍回压阀密封良好后，采用套管内敞压候凝，候凝时间为72h。

PT1井生产套管固井72h后，水泥浆返至地面，固井质量合格率100%，优质率98%，创中国石油盐穴储气库固井质量最好指标，1600~1700m井段声幅曲线如图1所示。

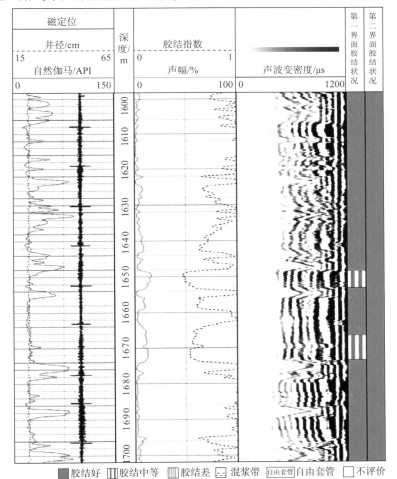

图1 平探1井井径及声幅曲线

4 结论

（1）平顶山盐穴储气库固井实践表明，采用的固井工艺、配套技术措施、盐水水泥浆可以有效保证施工安全及固井质量。

（2）针对上部地层承压能力低的问题，采用低排量顶替技术，保证了固井施工安全，有效预防了固井期间的漏失问题。

（3）固井前充分调整和循环钻井液，配合饱和盐水钻井液、套管居中等措施，保证了不规则大井眼条件下对钻井液的有效驱替。

参 考 文 献

[1] 张幸，李文魁，兰天，等.平顶山盐田采卤井生产套管固井技术研究[J].中国井矿盐，2015，46（5）：25-26.

[2] 袁光杰，田中兰，袁进平，等.盐穴储气库密封性能影响因素[J].天然气工业，2008，28(4)：105-106.

[3] 李国韬，郝国勇，朱广海，等.盐穴储气库完井设计考虑因素及技术发展[J].天然气与石油，2012，30(1)：52-53.

[4] 李丽峰，罗金恒，赵新伟，等.盐穴地下储气库风险评估技术与控制措施[J].油气储运，2010，29（9）：648-649.

[5] 覃毅.内插法固井工具失效典型案例及预防措施[J].石油钻采工艺，2015，37(6)：114-116.

[6] 尹学源.地下储气库盐穴及盐层固井技术[J].石油钻采工艺，2007，29(1)：92-93.

[7] 艾磊，刘子帅，张华，等.长庆陕224区块储气库井大尺寸套管固井技术[J].钻井液与完井液，2015，32(4)：63-66.

[8] 覃毅，党冬红，刘晓贵，等.盐穴储气库楚资1井ϕ244.5mm套管固井技术[J].石油钻采工艺，2015，37(2)：51-53.

[9] 丁国生，张昱文.盐穴地下储气库[M].4版.北京：石油工业出版社，2010：121-122.

[10] 朱海金，邹建龙，谭文礼，等.JSS低温抗盐水泥浆体系的研究及应用[J].钻井液与完井液，2006，23(4)：1-3.

[11] 林勇，薛伟，李治，等.气密封检测技术在储气库注采井中的应用[J].天然气与石油，2012，30（1）：55-56.

本论文原发表于《石油钻采工艺》2017年第39卷第1期。

盐腔声呐测量前的井下作业施工

杨清玉　李海伟　张　幸　马哲斌　闫凤林

(中国石油西气东输管道公司储气库项目部)

【摘　要】 为了解采卤溶腔现状及为下步溶腔方案提供依据，利用声呐进行盐腔形状的测量是目前最可靠的方法。由于溶腔盐岩井段往往存在难溶、不溶矿物质，甚至是不溶物地层，在采卤过程中不溶物质的塌落易造成管柱的掩埋和变形。因此在声呐测量前必须进行井下作业施工，确保声呐测量仪器顺利进出溶腔腔体内。

【关键词】　盐穴；溶腔；井下作业；声呐测量

Underground construction before sonar measurement in salt cavity

Yang Qingyu　Li Haiwei　Zhang Xing　Ma Zhebin　Yan Fenglin

(Gas Storage Project Department, PetroChina West-East Gas Pipeline Company)

Abstract　In order to understand the current situation of cavity and provide the basis for the next step of cavity scheme, sonar is the most reliable method to measure t he shape of cavity. Due to the poorly soluble and insoluble mineral, or even insoluble layer generally exists in the rock salt section of cavity, and the collapse of the insoluble mineral is easy to cause tubular column buried and deformation. Thus, underground construction must be completed before sonar measurement and guarantee the sonar devices can enter the cavity body successfully.

Keywords　salt carven; cavity; underground construction; sonar measurement

1　前言

目前国内盐矿企业主要采用单井对流、双井对流或多井对流方式进行采卤溶腔，近年来已有部分企业对采卤井腔体进行了声呐测量施工作业。声呐测量是通过超声波测距原理，实际测量出采卤溶腔形态和净容积，其主要目的是指导下步采卤溶腔作业施工，也为提高回采率、评价腔体的稳定性、评估腔体可利用性、废弃评估、预防地质灾害等提供可靠依据。

在采卤溶腔过程中，由于溶腔盐岩层段中不溶物及其隔夹层掉落、垮塌，会对采卤管

作者简介：杨清玉(1970—)，男，黑龙江五常人，高级工程师，主要从事盐穴储气库工程建设和技术管理。

柱造成挤压变形或掩埋，进行声呐测量时仪器不能通过管柱进入到自由腔体内。由于声呐仪器在管柱内无法实现倾斜测量，将不能测量出一些复杂溶腔的腔体形态，甚至导致声呐测量仪器卡在管柱里，造成井下复杂及重大经济损失。因此，在对盐腔进行声呐测量前必须进行井下作业施工，以满足声呐测量作业安全要求和对腔体的精确测量。

2 采卤溶腔工艺简介

2.1 采卤溶腔完井井身结构

目前，国内采卤溶腔井常见完井井身结构有三种方式：单管（生产套管），如图 1 所示；双管（生产套管和溶腔管柱），如图 2 所示；三管溶腔（生产套管和溶腔内、外管柱），如图 3 所示。

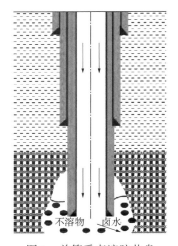
图 1 单管采卤溶腔井身结构示意图

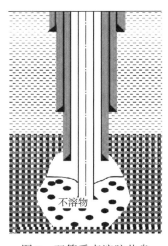
图 2 双管采卤溶腔井身结构示意图

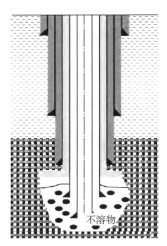
图 3 三管采卤溶腔井身结构示意图

2.2 常见采卤溶腔工艺

单管溶腔井身结构常见于双井对流或多井对流采卤工艺，即利用两口（或以上）直井和定向井组合通过盐岩地层建立循环进行采卤溶腔。

双管溶腔井身结构常见于单井无阻溶剂对流采卤工艺，即在生产套管内下入一个悬挂在井口处的溶腔管柱串并与生产套管环空建立循环进行采卤溶腔。双管溶腔井身结构也用于双井对流或多井对流采卤工艺。

三管溶腔井身结构常见于单井有阻溶剂对流采卤工艺，即在生产套管内下入两个悬挂在井口处的溶腔管柱串（溶腔内、外管），在溶腔内管和内、外管环空建立循环进行采卤溶腔。在溶腔外管与生产套管环空注入阻溶剂（一般为柴油或氮气）来控制腔体上溶。

3 井下作业施工

3.1 作业前施工准备

在井下作业施工前，收集作业井的地质、钻井、溶腔完井和采卤等资料，根据获取资料判断腔体形态和井下管柱工况，选取合适的修井设备(如修井机、泵车)、作业油管、井下作业工具(如通径规、刮削器、割刀等)。在施工前采取合适的循环采卤方式，尽量减少管柱内盐、石膏等结晶。

3.2 作业施工工序

准备道路、井场→拆卸地面工艺管线→拆卸井口→立井架

3.2.1 单管溶腔井施工工序

通井至溶腔顶部以下深度(必要时刮削、洗井)→测井(确定腔顶、腔底和管柱节箍深度)→确定管柱切割位置→选用机械式割刀→切割管柱(分两至三段)→判断切割效果→声呐测量→井口、地面恢复。

如果通井深度未达到预定深度处遇阻，遇阻处经刮削后通井仍然无法继续，可下入铅模判断井下复杂情况，如能确定生产套管发生了变形，可下入胀管器尝试恢复套管通径。

3.2.2 双管溶腔井施工工序

试提溶腔管柱，如果溶腔管柱成功提出，下步作业工序与单管溶腔井施工工序相同。

如果试提后仍不能提出溶腔管柱，按如下工序进行作业：

小油管及通径规对溶腔管柱进行通井(必要时刮削、洗井)→测井(确定腔顶、腔底和溶腔管柱节箍深度)→确定溶腔管柱切割位置(需在腔顶以下一定深度)→切割管柱→上提溶腔管柱。下步作业工序与单管溶腔井施工工序相同。

3.2.3 三管溶腔井施工工序

三管溶腔井中因有阻溶剂保护，生产套管周围盐岩不会被溶蚀，因而生产套管也不会被掩埋或变形(受地应力作用挤压变形除外)。

溶腔内外管试提、切割作业工序同双管溶腔井溶腔管柱作业工序。溶腔内外管切割位置必须在腔体里，并且切割位置尽可能靠近腔底。溶腔管柱是否需要从井内全部提出取决于下步溶腔方案或其他作业目的。

4 结论及认识

(1)通井及打铅模可判断管柱遇阻位置及状况，利用刮削、循环洗井及胀管技术是解除生产套管通径不足或局部变形的有效方法。

(2)利用地球物理测井(井温、自然伽马、磁定位)可以准确判断出管柱节箍、腔顶、腔底(取决于管柱变形情况)，为腔体尺寸的初步判断和管柱切割位置的选取提供依据。

(3)单管溶腔井作业时，由于不能对生产套管上提或者超拉，因此应选用机械割刀，切割点和切割次数应选择两次或两次以上。

(4)大多数情况下，处于自由悬挂状态下的溶腔管柱串可以利用修井机上提超拉，所以管柱切割方法的选择更加灵活，实际操作中可根据修井机悬重变化判断切割是否成功。

(5)利用本文中介绍的方法，在金坛盐矿、昆明盐矿、平顶山盐矿等盐矿对以上三种完井井身结构的采卤井都进行过声呐测量前的井下作业施工，并成功取得了腔体的声呐测量数据。

本论文原发表于《中国井矿盐》2015年第46卷第3期。

中俄东线楚州盐穴储气库固井难点及对策

薛 雨[1]　齐奉忠[2]　王元刚[1]　王立东[1]

（1. 中国石油西气东输管道公司；
2. 中国石油集团工程技术研究院有限公司）

【摘　要】　中国石油天然气集团公司正在建设楚州盐穴储气库来保障中俄东线的调峰和应急供气。楚州地区盐岩埋藏深、夹层多、地质条件复杂，相对于其他盐穴储气库，楚州储气库对固井技术提出了更高的要求。为保证楚州储气库固井质量，针对楚州储气库固井时存在的水泥浆配方筛选困难、提高顶替效率困难、易漏失及易憋堵的特点，提出了适合本地区的水泥浆体系和提高顶替效率、长封固段防漏以及防止环空憋堵等措施。通过现场 3 口井对固井措施进行了验证，经固井质量和气密封试压检测，3 口井固井质量全部合格，能够满足以后储气库生产的需要。提出的固井措施对以后该地区盐穴储气库固井具有很好的借鉴意义。

【关键词】　固井；盐穴储气库；顶替效率；盐水水泥浆；中俄东线

Cementing difficulties and solutions for Chuzhou salt cavern gas storage for sino-russian easternroute gas pipeline

Xue Yu[1]　Qi Fengzhong[2]　Wang Yuangang[1]　Wang Lidong[1]

(1. PetroChina West-East Pipeline Company;
2. CNPC Engineering Technology R&D Company Limited)

Abstract　CNPC is building Chuzhou salt cavern gas storage to ensure peak shaving and emergency gas supply for Sino-Russian Eastern Route Gas Pipeline. Chuzhou salt cavern gas storage is characterized by deep targets, multiple interbedded mudstone formations and complicated geological conditions. Compared with other salt cavern gas storage, Chuzhou gas storage puts forward higher requirements for cementing technology. To ensure the cementing quality, challenges like suitable slurry formulation selection, displacement efficiency improvement, lost circulation and pumping plugging are addressed by

---

基金项目：中国石油天然气集团公司储气库重大专项"地下储气库关键技术研究与应用"（2015E-40）之课题八"盐穴储气库加快建产工程试验研究"（2015E-4008）。

作者简介：薛雨，工程师，生于 1987 年，2012 年毕业于中国石油大学（北京）油气田开发工程专业，获硕士学位，现从事盐穴储气库钻井施工管理工作。地址：（212000）江苏省镇江市。E-mail：xqdsxueyu@petrochina.com.cn。

proposing proper measures. The cementing technology has been verified in three wells. The cementing quality and gas seal pressure test indicates qualified cementing qualities, which meet the requirements of later gas storage operation. The developed technology is of great significance for the future cementing of the salt cavern gas storage in this area.

Keywords cementing; salt cavern gas storage; displacement efficiency; brine cement slurry; Sino-Russian east route

江苏淮安楚州盐矿距在建的中俄东线天然气管道约61km，距冀宁联络线楚州分输站约7km，中国石油天然气集团公司正在此建设中俄东线楚州盐穴储气库。中俄东线规划的9座储气库中8座位于河北省以北，楚州储气库作为管线南段唯一的一座储气库，不但可以保证中俄东线和冀宁联络线的调峰和应急供气，更能保障长三角地区的天然气供应[1]，因此具有重要的建设意义。在盐穴储气库建设中，储气库的固井质量是钻完井工程各项作业中最为重要的作业之一，与储气库的寿命及长期运行安全紧密相关[2]。盐穴储气库与枯竭油气藏储气库相比，盐穴储气库井深浅、井眼尺寸大，对固井的要求更高，难度也更大。目前，国内唯一建成的盐穴储气库为金坛盐穴储气库，相比于金坛储气库，楚州盐矿虽然含盐层厚度大，但盐矿埋藏深，泥岩夹层多，建库条件更为复杂，金坛储气库的固井经验不能完全适用于楚州。同时，楚州库区周边盐化采卤井的固井要求也远低于储气库固井要求，邻井资料借鉴性不强。为此，笔者结合盐穴储气库对固井的要求，以及楚州地区特殊的地质条件对楚州盐穴固井存在的难点进行了分析，提出了适用于楚州盐穴的固井对策，以期为建库施工提供指导。

1 楚州盐穴储气库固井工程难点

1.1 地质条件复杂

楚州盐矿位于苏北盆地的淮安盐盆内，根据楚州地区二维、三维地震和钻井资料，楚州储气库选择浦口组二段上盐亚段第三岩性组合进行建库。地层主要由碎屑岩、硫酸盐岩和石盐岩组成，造腔段顶面埋深在1265～2120m。上部东台组和盐城组泥岩易缩径，流砂层易坍塌。盐城组与赤山组之间有断层，且新老地层交界，地层易漏失、易失稳。二开固井为进入Ⅲ-9盐层20m后下生产套管进行封固。根据钻井资料显示，上部封固段具有较多的石膏层、钙芒硝层以及难溶的泥岩夹层，单井夹层数都在40个以上，夹层厚度占到含盐地层厚度的35%，单个夹层厚度多在2m以上，最大可达8m。复杂的地质条件给楚州储气库固井带来了很大难度。

1.2 固井封固段长

盐穴储气库固井时要求水泥浆必须返至地面。楚州储气库生产套管平均下深在1800m，固井封固段长。上部地层受断层及地层交界的影响，施工过程中易发生漏失。通常使用低密度水泥浆防止漏失[3]，上部低温井段的固井质量难以保证。假如全井采用密度为1.86～1.95g/cm³的水泥浆封固，固井时容易发生井下漏失。如果水泥浆未返至井口，由于井口坐封有套管头，无法从环空灌注，不能满足盐穴储气库对固井的要求。同时上部封固段盐层厚度超过600m，水泥浆稠化时间受盐层溶解的影响会大幅度延长，无法形成良好的水泥环，盐层蠕变也会使套管发生变形或挤毁[4-6]。

1.3 优选综合性能好的抗盐水泥浆体系困难

楚州地区表层套管下入较深，约在650m，水泥浆体系不但要适应盐层、盖层、裸眼段固井的要求，还要保证ϕ339.7mm套管与ϕ244.5mm套管的固井质量。盐穴储气库固井时，通常采用盐水水泥浆体系[7]。但现有的低密度盐水水泥浆体系不够成熟，国内适合低温条件的抗盐水泥浆外加剂比较少，优选综合性能好的抗盐水泥浆体系困难。

1.4 复杂井眼条件下提高顶替效率困难

井径不规则是影响顶替效率的主要因素[8]。受断层及夹层影响，楚州储气库在钻井时容易造成井眼不规则，提高顶替效率困难。同时楚州储气库井较深，下套管时间也会相应增加，容易在井壁上形成较厚的虚滤饼，造成环空堵塞。固井时为避免压裂上部地层，只能采用小排量顶替，井径不规则井段的顶替效率难以保证。

1.5 存在憋堵现象

当钻井液黏度低、抑制性差，以及井下存在大肚子、糖葫芦井眼时容易发生环空憋堵现象[9-10]。根据老井资料，本地区已钻的9口井中有3口发生了憋堵现象，分别留有130m、126m和660m水泥塞，固井施工一次性成功率仅为66.7%，这严重影响了本地区固井施工的安全和质量。

2 楚州盐穴储气库固井对策

2.1 优选适应楚州储气库固井的水泥浆体系

根据盐穴储气库特点，水泥浆体系必须满足抗盐要求，稠化性能必须满足施工要求，需要具有良好的流变性能和早期强度来保证封固段胶结质量[11-12]。为保证楚州储气库盐层段的固井质量以及水泥环的长期密封性能，笔者在借鉴国外盐穴储气库和国内其他盐穴储气库成功固井经验的基础上[13-14]，结合楚州地区的地质情况，采集现场盐水、淡水和水泥在室内开展了配方筛选工作。通过大量室内试验分析不同含盐量对水泥浆稠化时间、水泥石抗压强度、水泥浆流动度、水泥石内部结构以及水泥浆冲蚀特性的影响，确定楚州储气库ϕ244.5 mm套管固井时采用DRB-3S+JSS抗盐水泥浆体系。

2.2 提高顶替效率的措施

2.2.1 卤水冲洗液的选择

楚州储气库二开钻井时采用饱和盐水钻井液，为保证冲洗液和钻井液有较好的相容性，并能对第一界面、第二界面进行良好的冲洗，同时考虑到冲洗液要能保证井壁的稳定，减少对盐层的冲蚀，结合现场的可操作性和成本，采用卤水作为楚州储气库固井的冲洗液。

2.2.2 注入低密度水泥浆

固井前钻井液性能受地层岩性、井眼状况及井下安全条件的限制，往往不能大幅度调整，且冲洗液在环空中只能占到约270m的环空高度，冲洗液量过大时容易引起井壁失稳，并且冲洗液携带性差，冲洗量过大时易引起岩屑在环空的堆积，造成固井事故。

为了增加钻井液与水泥浆的相容性,增加接触时间,有效驱替并携带环空中的稠化钻井液及浮滤饼,在注入要求密度的水泥浆前打入低密度水泥浆(密度为1.65~1.70g/cm³),低密度水泥浆可以很好地隔离并缓冲钻井液和水泥浆。由于该段水泥浆密度低、流动性好,所以易于达到紊流,提高顶替效率。表1为低密度水泥浆在不同井径下的临界返速及临界排量。由表1可以看出,在不同的井径扩大率下,低密度水泥浆均可以实现紊流顶替。

表1 低密度水泥浆在不同井径下的临界返速及临界排量

环空扩大率/%	0	5	10	15	20
井径/mm	311.20	326.70	342.30	357.80	373.40
环空临界返速/(m/s)	0.70	0.58	0.49	0.43	0.38
临界排量/(L/s)	20.70	21.50	22.30	23.00	23.80

2.2.3 其他配套措施

每口井都存在不同程度的套管不居中现象,为保证套管在固井过程中一直处于居中状态,需要合理使用套管扶正器[15]。

由于二开采用φ311.2mm钻头扩眼,井眼尺寸大,环空间隙大,还受井下状况、地层耐压强度、现场设备及水泥浆性能的限制,很难达到紊流顶替。由于注入水泥量大,为防止漏失,不能考虑采用大排量来提高顶替效率,在现有的施工条件下提出以下改进措施:

(1)确保井底没有沉砂并使钻井液保持清洁,保证钻井液性能在完钻前符合设计要求,裸眼段不存在漏失和阻卡。

(2)注水泥排量控制在0.9~1.1m³/min,初始替浆排量控制在1.0~1.2 m³/min,待平稳后替浆排量控制在1.5~1.8m³/min。采用水泥车碰压,碰压前5~8m³排量控制在0.5~0.7m³/min。

2.3 长封固段防漏措施

楚州盐穴储气库由于地层承压能力低,在替浆过程中可能发生漏失现象。封固段长加上低压漏失层的存在给固井的替净工作造成了极大困难,也影响了固井施工的安全和水泥浆返高。为此,在固井前和固井过程中提出以下措施:

(1)固井前对上部地层的承压能力进行验证,采用大排量洗井,通过测井井径确定具体排量[16],返速控制在1.0~1.2m/s;

(2)充分循环钻井液增加流动性,固井前将钻井液黏度调整至40~45s;

(3)井壁稳定性好的条件下进行井下清洁时,冲洗液采用饱和盐水体系,并将其密度调整至1.19g/cm³。

2.4 防环空憋堵措施

2.4.1 采用增黏隔离液防止环空憋堵

为防止盐岩地层溶蚀及蠕变,二开钻进采用饱和盐水钻井液[17]。为防止生产套管固井过程中环空憋堵,同时防止大肚子、不规则井眼中的钻屑在顶替过程中堵塞环空,在采用卤水冲洗液的基础上,新开发DRY-S增黏型隔离液,黏度60s左右。该隔离液在低温下使增黏剂和悬浮剂易溶解、黏度易调整,有利于提高不规则井眼条件下的顶替效率,防止不规则环空中钻屑沉积引起的环空憋堵问题。

2.4.2 防环空憋堵措施

（1）通井时在下部小井眼井段垫入稠浆（黏度120s以上），防止套管鞋处水泥浆下沉，同时对上部井段进行稠浆携砂，并记录携砂效果。

（2）下完套管采用单泵小排量顶通后，循环正常后配置20m³黏度120s以上稠浆携砂，携砂完后调整钻井液性能至密度1.35g/cm³、黏度45s左右，双泵以2.5~3.0m³/min的排量充分循环钻井液，确保井壁稳定、无沉砂、无垮塌，为固井作业创造一个良好的井眼条件。

3 现场应用

采用上述固井技术，选取楚州储气库C1井、C2井和C3井这3口井进行试验。3口井一开均采用φ444.5mm钻头，表层套管外径339.7mm；二开采用φ311.2mm钻头，生产套管外径244.5mm；三开采用φ215.9mm钻头，裸眼完井，固井时水泥浆均要求返至地面。

3.1 表层套管固井

表层套管采用内插法固井，为防止水泥浆和钻井液在套管内掺混缩短替浆时间，采用常规密度低温早强稳定性好的水泥浆固井。水泥浆配方为：G级高抗硫水泥+2%~3%早强剂+0.4%~0.6%分散剂+稳定剂+消泡剂+现场淡水。根据固井质量测井解释结果，3口井表层套管固井均合格。

3.2 生产套管固井

生产套管固井采用常规固井方式，采用高早强稳定性好的DRB-3S+JSS抗盐水泥浆体系。现场水泥浆性能见表2。

表2 水泥浆性能

水泥浆	密度/(g/cm³)	流动度/cm	每30min API失水/mL	游离液量/mL	稠化时间/min
领浆	1.90	23.5	103	0	248
尾浆	1.96	22.0	88	0	176

以新完钻的C2井为例来说明楚州盐穴储气库的固井措施。C2井施工流程见表3。下套管时，最底下5根套管每根套管加一只扶正器，井口附近，每5根套管加一只扶正器，其余井段每2根套管加一只扶正器。固井前以钻井时大排量循环钻井液2周以上，根据井眼情况调整钻井液性能，降低黏度和切力，增加流动性。注入饱和盐水冲洗液，冲洗液量控制在环空高度200~300m，采用60s以上的增黏型隔离液。

表3 C2井施工流程

操作内容	工作量/m³	密度/(g/cm³)	排量/(m³/min)	施工时间/min
循环钻井液	—	1.35	2.5~3.0	—
管汇试压	—	1.00	0.2	—
注入隔离液（压下胶塞）	10.0	1.05	1.0	—

续表

操作内容	工作量/m³	密度/(g/cm³)	排量/(m³/min)	施工时间/min
双车注入低密度	8.0	1.71~1.43	2.1~1.8	4
双车注入领浆	12.0	1.85~1.80	2.1~1.8	6
双车注入中浆	20.0	1.90~1.85	2.1~1.8	10
双车注入尾浆	42.0	1.96~1.90	2.1~1.8	21
压上胶塞	2.0	1.05	0.8~1.0	5
替浆	54.0	1.35	1.0~1.5	40
碰压	5.9	1.00	0.5~1.0	10

候凝72h后，固井质量测量结果显示全井固井质量合格。

综合采用上述技术对C1井、C2井和C3井进行固井，3口井生产套管固井质量均合格，对套管内和套管鞋处试压，保压30min压降均为0。完井后进行腔体密封性检测，3口井固井质量均能够满足盐穴储气库密封性要求。

4 结论

（1）通过C1井、C2井和C3井的现场成功应用，证明制定的技术对策可行，固井工艺、盐水水泥浆体系以及配套技术措施能够有效保证楚州盐穴储气库的固井施工质量和安全。

（2）采用DRB-3S+JSS抗盐水泥浆体系，配合合适的顶替排量以及饱和盐水前置液、低密度水泥浆、中间浆、尾浆的液柱结构，可以有效提高顶替效率，保证固井质量。

（3）采用DRY-S增黏型隔离液，黏度不小于60s，可以有效提高不规则井眼条件下的顶替效率，防止环空憋堵问题。

（4）采用综合配套固井技术，优选盐水水泥浆体系，提高顶替效率，降低上部地层漏失和环空憋堵对固井的影响是楚州储气库固井成功的关键。

参 考 文 献

[1] 垢艳侠,完颜祺琪,罗天宝,等.中俄东线楚州盐穴储气库建设的可行性[J].盐业与化工,2017(11):16-20.

[2] 齐奉忠.金坛地区盐穴储气库固井技术研究与应用[D].北京:中国石油勘探开发研究院,2005.

[3] 李桂平.深井低密度水泥浆体系的设计与应用[J].西安石油大学学报(自然科学版),2009,24(4):42-45.

[4] 马东华.土库曼亚速尔哲别油田盐层固井工艺技术[J].今日科苑,2010(8):127.

[5] 郑建翔,周仕明,韩卫华.塔河油田石炭系盐层固井工艺技术[J].西部探矿工程,2006,18(9):59-63.

[6] 齐奉忠,袁进平.盐层固井技术探讨[J].西部探矿工程,2005,17(10):58-61.

[7] 张弛,刘硕琼,徐明,等.低温胶乳盐水水泥浆体系研究与应用[J].钻井液与完井液,2013,30(2):63-65.

[8] 方春飞,周仕明,李根生,等.井径不规则性对固井顶替效率影响规律研究[J].石油机械,2016,44

（10）：1-5.

[9] 陈侃．关于S72-22井气层套管固井憋压漏失的原因分析[J]．中国科技博览，2014(7)：30-31.

[10] 王磊．8589-1井固井施工中发生憋泵事故的分析[J]．中国石油和化工标准与质量，2013(22)：100.

[11] 尹学源．地下储气库盐穴及盐层固井技术[J]．石油钻采工艺，2007，29(1)：92-94.

[12] 张幸，覃毅，李海伟，等．平顶山盐穴储气库固井技术[J]．石油钻采工艺，2017，39(1)：61-65.

[13] 张幸，李文魁，兰天，等．平顶山盐田采卤井生产套管固井技术研究[J]．中国井矿盐，2015(5)：25-26.

[14] 覃毅，党冬红，刘晓贵，等．盐穴储气库楚资1井244.5mm套管固井技术[J]．石油钻采工艺，2015，37(2)：51-53.

[15] 张晋凯，周仕明，陶谦，等．套管低偏心度下的水泥浆顶替界面特性研究[J]．石油机械，2016，44(7)：1-6.

[16] 石向前，蒋鸿，俞战山．吐哈油田盐膏层综合固井技术[J]．石油钻探技术，2002，30(3)：27-29.

[17] 班凡生，袁光杰，王思中，等．盐穴储气库井筒完整性风险分析与对策[C]//第四届中国管道完整性管理技术大会，2014：1135-1137.

本论文原发表于《石油机械》2019年第47卷第1期。

盐岩储气库堆积物注气排卤试验研究

任众鑫[1]　巴金红[1]　任宗孝[2]　张　幸[1]　张新悦[1]　王桂林[1]

（1. 中国石油西气东输管道公司；2. 西安石油大学）

【摘　要】 随着国内天然气工业的发展，地下储气库作为集输储运重要的基础设施，其必要性越加凸显。盐岩储气库是一种重要的地下储气库类型，国内建库盐层由于夹层多、品位低等诸多地质特点，建成后一般在腔内会堆积有大量不溶物。为加强对堆积物空隙体积的认识，明确堆积物空隙注气排卤效果，采用气体膨胀法对其空隙体积进行了试验测定，并基于物理模拟装置开展了堆积物空隙注气排卤试验研究与分析。研究结果表明，堆积物空隙注气排卤后，由于束缚水现象的存在预计有 19.83% 体积的卤水无法排出而剩余在空隙中；且排卤前空隙初始空间越大，排卤后剩余卤水体积越小，即注入气体所能驱替排出的卤水量越多。

【关键词】 盐岩储气库；堆积物；注气排卤；空隙体积

Experimental study on gas injection and brine ejection of deposits in rock salt gas storages

Ren Zhongxin　Ba Jinhong　Ren Zongxiao　Zhang Xing
Zhang Xinyue　Wang Guilin

（1. PetroChina West-East Gas Pipeline Company；2. Xi'an Shiyou University）

Abstract With the development of domestic natural gas industry, the necessity of underground gas storage as the important infrastructure of gas gathering and transportation has become increasingly obvious. Rock salt gas storage is an important type of underground gas storage. For domestic salt formation, there will be lots of deposits in cavity after completion because of the formation's geological characteristics such as many interlayers and low grade. In order to strengthen the understanding on void volume of deposits, and clarify the effect of injecting gas and ejecting brine (IGEB) in deposit void, the gas expansion method is adopted to test and measure the void volume, and based on physical simulation device study on IGEB in deposit void is carried out. According to the research results, after IGEB of the deposit void, there is 19.83% (volume fraction) of brine remains in void which cannot be ejected because of ex-

---

基金项目：中国石油集团专业公司项目"盐穴储气库腔底堆积物注气排卤扩容研究（2013B-3401-0503）"。

作者简介：任众鑫：工程师，硕士，2013 年毕业于中国石油大学（北京），从事储气库工艺、天然气管道与集输工艺技术研究。联系方式：15706171237，zhongxren@petrochina.com.cn，江苏省镇江市南徐大道 60 号，邮编：212000。

istence of irreducible water. What's more, the bigger the initial space of void before ejecting brine, the smaller the remaining brine volume after ejecting brine, namely the bigger the brine volume that injected gas can displace.

Keywords rock salt gas storage; deposits; gas injection and brine discharge; void volume

国家绿色低碳能源体系对天然气总量和实时供应的需求越来越大，地下储气库作为重要的应急调峰设施，所发挥的作用越来越明显[1-3]。盐岩储气库作为一种重要的储气库类型，具有注采气量大、垫底气可完全回收等特点[4-6]，其建设步伐逐渐加快。

国内盐层由于夹层多、品位低、泥质含量高等地质特点[7-9]，在建库溶腔过程中，随着盐岩被注入的淡水溶解并以卤水形式返至地面，其中不溶成分会被释放并沉降，进而覆盖在未溶解盐岩的表面形成空腔，加之黏土矿物遇水发生软化膨胀，从而在腔体下部形成了大量堆积物（图1），最终占用了相当一部分成腔体积。据现场数据，堆积物内部具有丰富的孔隙空间，造腔结束后该部分体积即被饱和卤水所填满。随后注气排卤进行到后期时，在高压气体驱动作用下，对于腔体下部或腔壁周边部分沉埋较突出的不溶堆积物逐渐脱离卤水环境而呈现在天然气中，其空隙中卤水逐渐被气体所充分驱替和置换，进而排至地面。

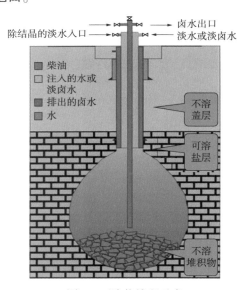

图1 不溶物堆积示意

国外由于盐岩纯度较高、泥质含量较低，建库完成后腔底堆积物一般较少，故研究多集中在水溶建腔、溶腔滤洗和形状控制等方面；国内盐穴储气库建设起步较晚，对于不溶堆积物尤其是堆积空隙体积和空隙空间内注气排卤问题的研究也较少提及。1988年，Yu等[10]研究了基于混合堆积颗粒情况下堆积体的空隙率计算模型及其影响因素；1995年，荷兰学者Hoffman等[11]研究了不同颗粒的物理特性及堆积空隙计算方法；2004年，法国路桥试验中心[12]基于可压缩堆积理论研究了预测混合堆积体系密实度的计算模型，可用于预测已知粒径分布的堆积密实度；2008年，吴成宝等[13]研究了中空玻璃微珠粒度分布特征与堆积空隙率的关系；2009年，陈锋等[14]模拟研究了不溶物影响下注气排卤后期的安全排卤速率。前述研究的着眼点多基于特殊用途材料如建材、混凝土等颗粒本身粒度特征，或考虑以工程安全性、连续性为前提的作业参数敏感性等方面的影响，针对盐层造腔后所形成的堆积物空隙情形，尤其在空隙中赋存有大量卤水情况下的具体排卤未有涉及。

因此，为加强对盐岩储气库腔底堆积物空隙部分体积的认识，明确其空隙内部进行注气排卤时所能取得的具体效果，基于相似性原理，利用设计试验装置开展了堆积物空隙率测定和注气排卤试验研究，可为相关专题探索研究和分析评价提供参考。

1 注气排卤原理

注气排卤,即造腔结束后,通过向盐腔注入天然气,将腔内先前采卤溶腔所形成的卤水驱替并返排至地面。注气排卤前进行注采完井,向井内下入注采和排卤两套管柱,天然气经压缩机增压后,经由注采管柱与排卤管柱间环空注入腔体,腔内卤水自排卤管柱通道排至地面,再由地面管道进行外输[15-16](图2)。排卤后期,随着气体的持续注入,气水界面稳步向下推进,在腔体下部或周边部分堆积较突出的不溶物逐渐脱离卤水环境而呈现在天然气中,其内部空隙中卤水也被气体所充分驱替和置换;但由于束缚水现象的存在,导致空隙内将有一定量的剩余卤水无法排出而残留在堆积物中[17-18]。

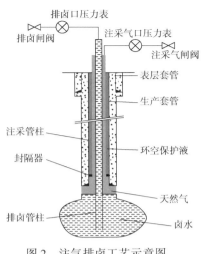

图 2 注气排卤工艺示意图

2 堆积物空隙率测定

2.1 堆积物样品准备

根据有关研究[19],堆积物颗粒总体特征符合双正态分布,其分布均匀程度和集中性可采用分形维数描述,且堆积物空隙率值取决于分形维数,而与颗粒粒度范围无关。基于相关研究成果,结合堆积物现场取样数量及试验设备条件限制,选用符合其分布特征的小颗粒砂石作为堆积物样品,进行空隙率测定和注气排卤试验研究。五组样品的数据参数见表1。

表 1 五组试验样品数据参数

粒径/mm	1#	2#	3#	4#	5#
1.250	0.31	0.26	0.23	0.28	0.21
0.630	33.68	33.91	25.62	33.80	25.70
0.315	25.07	12.27	23.39	18.67	18.59
0.160	39.22	43.22	45.62	41.22	47.12
0.080	1.48	9.49	4.66	5.48	7.67
0.050	0.23	0.85	0.48	0.54	0.71
分形维数	2.5495	2.6117	2.6604	2.6663	2.6992

2.2 测定方法和原理

根据盐腔堆积物的物性特征,结合相关空隙测定方法特点,研究采用气体膨胀法对堆积物样品进行空隙率测定。测定装置如图3所示。

试验装置主要由堆积物容器、氮气容器和高压自动泵等部分组成。在堆积物容器和氮气容器的外面使用恒温箱,以实现环境温度条件的控制;压力相关数据的变化可由压力传

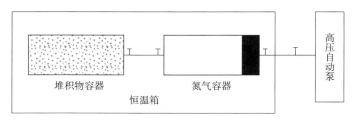

图 3 空隙率测定装置

感器监测得到；高压自动泵可为试验提供所需的稳定流速或驱替压力。测定原理即玻义尔定律，试验过程中已知体积 V_1 的氮气在一定压力 p_1 下向盛满堆积物样品（假定其空隙体积为 V_m）的容器 V_2 进行等温膨胀，最终平衡后系统压力为 p_2，则根据气体状态方程得出以下关系：

$$\frac{p_1 V_1}{Z_1} = \frac{p_2(V_1 + V_m)}{Z_2} \tag{1}$$

式中：Z_1、Z_2 分别为对应压力和温度条件下的气体压缩因子，可由相态软件计算得到，无量纲；V_m 为堆积物样品空隙体积，m³；V_1 为等温膨胀前氮气体积，m³；p_1 为等温膨胀前氮气压力，MPa；p_2 为等温膨胀后系统压力，MPa。

则堆积物样品空隙率 φ 可根据以上参数求得

$$\varphi = \frac{V_m}{V_2} = \frac{V_1(p_1 Z_2 - p_2 Z_1)}{V_2 p_2 Z_1} \tag{2}$$

式中：V_2 为堆积物容器体积，m³。

2.3 空隙率测定结果

根据上述方法对 5 组样品的空隙率值进行测定，数据结果见表 2。

表 2　5 组空隙率测定值

组别	1#	2#	3#	4#	5#
空隙率/%	47.45	46.82	46.53	45.90	44.08

由表 2 可知，各组样品空隙率值变化不大，分布范围在 44.08%~47.45%。同时，可以得知，随着各组样品颗粒分形维数的增加，其整体空隙率逐渐减小，分析原因是分形维数越大即颗粒粒度分布越分散时，均匀性和集中度越差，这是混合后出现颗粒间隙效应所致。

3 注气排卤试验

3.1 试验方法

试验装置如图 4 所示，按照相似性原理，在固定容器中设计安装内、外两套模拟管，并完成相关连接部位的密封。设计内管向下伸进容器底部，向外连接卤水收集装置；外管与氮气输送装置相连，安装连接至容器顶端。氮气生产完成后经由内外管环空向固定容器中输送。试验前将各组样品分别放进容器中，加入卤水至全部没过，从而模拟腔底堆积物中赋存饱和卤水的状态；然后从内外管环空向容器内注入氮气，并从内管驱替出卤水；待卤水尽可能排替完全后对容器抽真空处理，然后采用上述气体膨胀法测定其空隙率。

3.2 试验结果与分析

试验前堆积物样品的初始空隙率值 φ_1 已知，放入卤水后该部分空隙空间全部被卤水占据，在注气排卤过程中部分卤水被气体驱替置换出，试验完成后重新测定其空隙率值 φ_2，则 φ_2 即为注入气体所排替置换出的卤水体积，亦即注气排卤后气体组分所占体积，而 $\Delta\varphi = \varphi_1 - \varphi_2$ 为堆积物样品空隙中由于束缚水现象的存在所无法排出的剩余卤水体积。具体数据见表3。

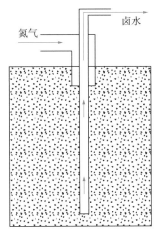

图 4 注气排卤试验装置示意图

表 3 试验前后五组空隙率数据

参　数	1#	2#	3#	4#	5#	平均值
初始空隙空间/%	47.45	46.82	46.53	45.90	44.08	46.15
排卤后空隙率/%	28.69	27.38	26.57	25.75	23.23	26.32
剩余卤水体积/%	18.76	19.44	19.96	20.15	20.85	19.83

由表3可知，5组堆积物样品进行试验后，即腔底堆积物在赋存饱和卤水的条件下进行注气排卤时，由于束缚水现象的存在，预计将有19.83%体积的卤水无法排出。

为探索排卤前空隙空间与排卤后空隙率值及剩余卤水体积值间关系，对各组数据进行对比分析。排卤后的空隙率值和剩余卤水体积分别随试验前初始空隙空间的变化关系如图5、图6所示。

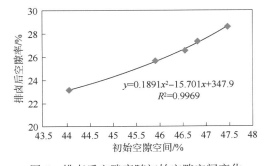

图 5 排卤后空隙率随初始空隙空间变化

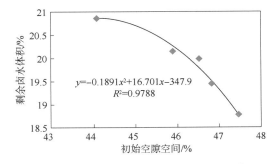

图 6 剩余卤水体积随初始空隙空间变化

综合图5、图6可知，排卤前初始空隙空间越大，排卤完成后堆积物的空隙率值也越大；相应地，剩余卤水体积值的变化与此相反，排卤前空隙空间越大，排卤后由于束缚水现象的存在而无法排出的剩余卤水体积越小，即注入气体所能驱替排出的卤水量越多；各项参数间变化关系的具体规律可分别参考图 5、图 6 中拟合得到的多项式。考虑到堆积物空隙的形成与赋存过程受多因素影响，在分析评价具体腔体的堆积物特征时需要与其相应地层成分及含量情况进行针对性结合。研究结果可为排卤体积分析和扩容效果评价及工艺经济性评价等提供借鉴与参考。

4 结论

（1）根据现场生产情况，结合盐岩储气库堆积物形成过程及注气排卤工艺原理，采用气体膨胀法对堆积物内部空隙体积进行了试验测定，并基于物理模拟装置开展了堆积物空隙注气排卤试验研究，最后完成了试验相关分析和评价。

（2）根据堆积物样品试验研究结果，由于束缚水现象的存在，注气排卤后预计有 19.83% 体积的卤水无法排出而剩余在空隙中；且排卤前空隙初始空间越大，排卤后剩余卤水体积越小，即注入气体所能驱替排出的卤水量越多。

（3）该研究对于加强堆积物空隙体积的认识、明确堆积物空隙中注气排卤效果，以及相关分析评价具有一定的指导意义。

参 考 文 献

[1] 刘炜，陈敏，吕振华，等. 地下储气库的分类及发展趋势[J]. 油气田地面工程，2011，30(12)：100-101.

[2] 秦云松，张吉军，郭帅. 天然气的战略储备及技术经济性分析[J]. 油气田地面工程，2014，33(9)：13-14.

[3] 任众鑫，李建君，巴金红，等. 盐岩储气库腔底堆积物扩容及工艺应用[J]. 天然气技术与经济，2017，11(5)：43-45.

[4] 崔红霞. 天然气地下储气库地面配套工艺技术研究[D]. 大庆：大庆石油学院，2003：10-33.

[5] 安丰春，涂彬. 我国天然气基础设施建设战略[J]. 油气田地面工程，2007，26(11)：6-7.

[6] 崔红霞，纪良才. 天然气地下储存的地面工艺技术[J]. 油气田地面工程，1999，18(4)：25-27.

[7] 任众鑫，李建君，朱俊卫，等. 盐岩储气库注气排卤工艺参数的数值模拟[J]. 油气储运，2018，37(4)：403-406.

[8] 王清明. 盐类矿床水溶开采[M]. 北京：化学工业出版社，2003：18-21.

[9] 张华宾，王芝银，赵艳杰，等. 盐岩全过程蠕变试验及模型参数辨识[J]. 石油学报，2012，33(5)：904-908.

[10] Yu A B, Standish N. An analytical-parametric theory of the random packing of particles[J]. Powder Technology, 1988, 55(3): 171-186.

[11] Hoffman A C, Finkers H J. A relation for the void fraction of randomly packed particle beds[J]. Powder Technology, 1995, 82(2): 197-203.

[12] 弗朗索瓦·德拉拉尔. 混凝土混合料的配合[M]. 廖欣，叶枝荣，译. 北京：化学工业出版社，2004.

[13] 吴成宝，段百涛. 中空玻璃微珠粒度分布分形特征及其与空隙率关系的研究[J]. 中国粉体技术，2008，14(1)：16-19.

[14] 陈锋，杨海军，杨春和. 盐岩储气库注气排卤期剩余可排卤水分析[J]. 岩土力学，2009，30(12)：3602-3606.

[15] 郝萍．金坛盐穴地下储气库地面工艺技术优化[J]．油气田地面工程，2011，30(6)：34-35．
[16] 庄清泉．注气排卤技术在盐穴造腔中的应用[J]．油气田地面工程，2010，29(12)：65-66．
[17] 任众鑫，杨海军，李建君，等．盐岩储库腔底堆积物空隙体积试验与计算[J]．西南石油大学学报（自然科学版），2018，40(2)：142-150．
[18] 王建俊，鞠斌山，罗二辉．低速非达西渗流动边界移动规律[J]．东北石油大学学报，2016，40(2)：71-77．
[19] 任众鑫，李建君，汪会盟，等．基于分形理论的盐岩储气库腔底堆积物粒度分布特征[J]．油气储运，2017，36(3)：279-283．

本论文原发表于《油气田地面工程》2018年第37卷第11期。

盐岩储气库注气排卤期剩余可排卤水分析

陈 锋[1] 杨海军[2] 杨春和[1]

(1. 中国科学院武汉岩土力学研究所 岩土力学与工程国家重点实验室;
2. 中国石油西气东输管道公司)

【摘 要】 地下盐穴溶腔注气排卤过程中天然气和卤水共存,排卤管进入到卤水的深度及排卤速率决定了天然气是否会被卤水带出地表,如何调整后期排卤速率,降低排卤过程中带出天然气的风险,是注气排卤过程中的施工技术难题。针对该问题,采用流场分析理论对注气排卤后期盐岩储气库腔底流场进行了模拟分析,得到排卤管外壁卤水向下流速;根据天然气气泡在卤水中的运移速率,分析确定了注气排卤后期不同剩余卤水深度下的安全排卤速率,为施工后期注气排卤速率调整提供了技术支撑。

【关键词】 盐岩;储气库;注气排卤;流场;许可速率

Analysis of residual brine of salt rock gas storage during injecting gas to eject brine

Chen Feng[1] Yang Haijun[2] Yang Chunhe[1]

(1. State Key Laboratory of Geomechanics and Geotechnical Engineering, Institute of Rock and Soil Mechanics, Chinese Academy of Sciences;
2. PetroChina West-East Gas Pipeline Company)

Abstract In the natural gas storage engineering, the casing depth in the residual brine and the velocity of ejecting brine decide if the natural gas overflows from the salt cavern during injecting gas to eject brine. It is a difficult problem to adjust the velocity of ejecting brine under different casing depth in the residual brine and decrease the danger of the natural gas from the cavern. This technical problem is researched. The flow field of gas storage in the process of injecting gas to eject brine is modeled by the flowing theory of fluid. The maximal velocity of brine in the outside of casing is obtained. Based on the theory of the rising velocity of natural gas bubbles in the brine, the permitted velocities of ejecting brine under the different depths of residual brine are determined. The research result gives the technical support in the gas storage engineering of salt rock.

基金项目:国家自然科学基金资助项目(No.50804045);国家重点基础研究发展规划(973)(No. 2009CB724602,No. 2009CB724603)。

作者简介:陈锋,男,1974年生,博士,主要从事盐岩地下溶腔稳定性方面研究。E-mail:fchen@whrsm.ac.cn。

Keywords salt rock; gas storage; injecting gas to eject brine; flow field; permitted velocity

由于盐岩具有良好的密封性、低渗透性和损伤自愈合性，使得深部地下盐矿成为世界各国地下储存的一种主要介质。由于盐矿地下开采后形成的地下空间，能够保持较长时间的稳定，这些地下空间能够为不溶解于盐的物质提供储存和处置场所。正是由于地下盐岩腔体优良的性能，其在地下天然气储存库中得到了广泛应用，1963年加拿大在Saskatchewan建成了世界上第一个天然气盐岩储存库，20世纪70年代美国在Mississppi的Eminence盐丘建成了2个天然气盐岩地下储存库，同年法国在Tersanne也开始建造盐岩地下储存库，1971年德国在Kiel附近的盐丘中也建造了天然气地下储存库。

在我国地下盐岩空间利用尚处于起步阶段，现仅有江苏金坛盐岩储气库建设处于前期建设阶段，针对相关的技术研究也处于空白阶段。盐岩储气库投入运行之前，盐岩腔体中存在着大量的卤水，需要使用天然气置换出卤水，图1为注气排卤示意图。在天然气置换卤水的过程中，当置换到后期时，由于卤水淹没排水管很浅，大量天然气可能会顺着管壁排出，形成不必要的安全事故。适当控制后期卤水排卤速率，减小排卤管口卤水速率，能有效降低排卤过程中带出天然气的风险。笔者结合我国天然气地下储存库中的排卤实际工况，采用流场分析理论对注气排卤后期盐岩储库腔底进行了流场模拟，根据天然气气泡在水中的上浮速率，分析确定了注气排卤后期不同剩余卤水深度下的安全排卤速率。

1 流场计算分析理论

FLUENT是用于计算复杂几何条件下流动和传热问题的程序。它提供了无结构网格生成程序，把计算相对复杂的几何结构问题变得容易和轻松。可以生成的网格包括二维的三角形和四边形网格；三维的四面体、六面体及混合网格。并且，可以根据计算结果调整网格。这种网格的自适应能力对于精度求解有较大梯度的流场如自由剪切流和边界层问题有很实际的作用。同时，网格自适应和调整只是在需要加密的流动区域里实施，而非整个流动场，可以节约计算时间。

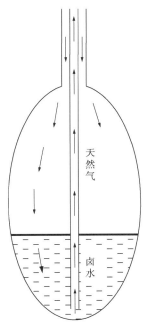

图1 储气库腔体注气排卤示意图

正是由于FLUENT强大的流场模拟功能，其可以求解下列问题：可压缩与不可压缩流动问题；稳态和瞬态流动问题；无黏流、层流及湍流问题；牛顿流体及非牛顿流体；对流换热问题(包括自然对流和混合对流)；导热与对流换热耦合问题；辐射换热；惯性坐标系和非惯性坐标系下的流动问题模拟；多运动坐标系下的流动问题；化学组分混合与反应；可以处理热量、质量、动量和化学组分的源项；用Lagrangian轨道模型模拟稀疏相(颗粒、水滴、气泡等)；多孔介质流动；一维风扇、热交换器性能计算；两相流问题；复杂表面形状下的自由面流动。

对于所有的流动，FLUENT都是解质量和动量守恒方程。对于包括热传导或可压性的流动，需要解能量守恒的附加方程。对于包括组分混合和反应的流动，需要解组分守恒方程或者使用 PDF 模型来解混合分数的守恒方程以及其方差。当流动是湍流时，还要解附加的输运方程。

流场求解过程中的质量守恒方程又称连续性方程，见式（1）：

$$\frac{\partial \rho}{\partial t}+\frac{\partial}{\partial x_i}(\rho u_i)=S_{\mathrm{m}} \tag{1}$$

该方程是质量守恒方程的一般形式，它适用于可压流动和不可压流动。源项 S_{m} 是从分散的二级相中加入连续相的质量（比方说由于液滴的蒸发），源项也可以是任何的自定义源项。

流场计算过程中的动量守恒方程，其惯性（非加速）坐标系中 i 方向上的动量守恒方程为

$$\frac{\partial}{\partial t}(\rho u_i)+\frac{\partial}{\partial x_j}(\rho u_i u_j)=-\frac{\partial p}{\partial x_i}+\frac{\partial \tau_{ij}}{\partial x_j}+\rho g_i+F_i \tag{2}$$

式中：p 为静压；τ_{ij} 为下面将会介绍的应力张量；ρg_i 和 F_i 分别为 i 方向上的重力体积力和外部体积力（如离散相相互作用产生的升力）；F_i 包含了其他的模型相关源项，如多孔介质和自定义源项。

应力张量由式（3）给出：

$$\tau_{ij}=\left[\mu\left(\frac{\partial u_i}{\partial x_j}+\frac{\partial u_j}{\partial x_i}\right)\right]-\frac{2}{3}\mu\frac{\partial \mu_l}{\partial x_l}\delta_{ij} \tag{3}$$

针对注气排卤的工程实际，流场分析中采用最常用的湍流模型—标准 k-ε。标准 k-ε 模型需要求解湍动能及其耗散率方程，湍动能输运方程是通过精确的方程推导得到，但耗散率方程是通过物理推理，数学上模拟相似原形方程得到的。该模型假设流运为完全湍流，分子黏性的影响可以忽略。

标准 k-ε 模型的湍动能 k 和耗散率 ε 方程形式如下：

$$\rho\frac{Dk}{Dt}=\frac{\partial}{\partial x_i}\left[\left(\mu+\frac{\mu_{\mathrm{t}}}{\sigma_k}\right)\frac{\partial k}{\partial x_i}\right]+G_k+G_{\mathrm{b}}-\rho\varepsilon-Y_{\mathrm{M}} \tag{4}$$

$$\rho\frac{D\varepsilon}{Dt}=\frac{\partial}{\partial x_i}\left[\left(\mu+\frac{\mu_{\mathrm{t}}}{\sigma_\varepsilon}\right)\frac{\partial \varepsilon}{\partial x_i}\right]+C_{1\varepsilon}\frac{\varepsilon}{k}(G_k+C_{3\varepsilon}G_{\mathrm{b}})-C_{2\varepsilon}\rho\frac{\varepsilon^2}{k} \tag{5}$$

式中：G_k 为平均速度梯度引起的湍动能；G_{b} 为用于浮力影响引起的湍动能；Y_{M} 为可压速湍流脉动膨胀对总的耗散率的影响；湍流黏性系数 $\mu_{\mathrm{t}}=\rho C_{\mu}\dfrac{k^2}{\varepsilon}$。

在注气排卤流场分析中，湍流模型的计算参数选取为：$C_{1\varepsilon}=1.44$，$C_{2\varepsilon}=1.92$，$C_{\mu}=0.09$，湍动能 k 与耗散率 ε 的湍流普朗特数分别为 $\sigma_k=1.0$，$\sigma_\varepsilon=1.3$。

2 腔体流场计算模型建立

在注气排卤流场分析中,由于各个腔体的腔体底部形状大体相似,以金坛某溶腔为例,对溶腔注气排卤过程中腔体底部流场进行了数值模拟研究。图2为盐岩储气库腔体排卤示意图,图3为腔体底部网格剖分图,分别建立了

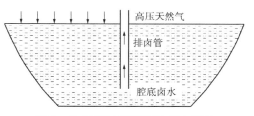

图2 盐岩储气库腔底排卤示意图

四个平面对称模型,分别考虑剩余可排卤水深度(即卤水水面至排卤管下部管口距离)为4m、3m、2m、1m时的情形。计算模型采用轴对称平面模型,套管位于模型中心,模型上表面为带压自由面,压力与实际工况相符,为注气排卤过程中的管口天然气压力与气柱压力之和13MPa。卤水为饱和卤水,相对密度为1.18,排卤管直径为11.4cm,排卤管管口离腔底为2m,腔体底面接近圆形,半径为3.5m。排卤管边界、腔底与腔周设为不透水边界。

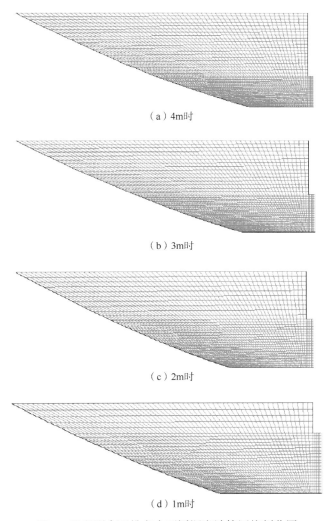

(a)4m时

(b)3m时

(c)2m时

(d)1m时

图3 腔底剩余可排卤水不同深度计算网格剖分图

3 腔底流场分布

利用所建不同剩余卤水深度所建的 4 种不同模型,对排卤速率分别为 $100m^3/h$、$80m^3/h$、$60m^3/h$、$40m^3/h$ 下,腔底形成的流场进行了分析,通过提取流场分析数据,获得了排卤气液界面处的最大流速。图 4 分别给出了不同模型一种排卤速率下的流场分布图。通过流场分布图可以看出,在剩余卤水较深(2m、3m、4m)时,卤水表面的流速比较均一,没有明显的流场梯度。当剩余卤水深度为 1m 时,流场对卤水表面具有明显的影响,在排卤管附近形成了明显的流场梯度,容易形成携带出高压天然气。

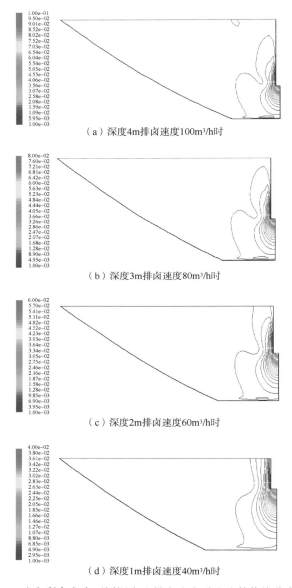

(a) 深度4m排卤速度100m³/h时

(b) 深度3m排卤速度80m³/h时

(c) 深度2m排卤速度60m³/h时

(d) 深度1m排卤速度40m³/h时

图 4　腔底剩余卤水不同深度和排卤速度时流速等值线分布图

表 1 为不同排卤速率、不同剩余卤水深度下的卤水表面流速分布。

表 1 100m³/h 排卤速度卤水表面速度分布

剩余排卤深度/m	排水速度 100m/h		排水速度 80m/h		排水速度 60 m/h	
	最大流速/(m/s)	截面平均流速/(m/h)	最大流速/(m/s)	截面平均流速/(m/h)	最大流速/(m/s)	截面平均流速/(m/h)
4	0.0008	0.11	0.0006	0.09	0.0005	0.07
3	0.0010	0.16	0.0008	0.13	0.0006	0.10
2	0.0050	0.23	0.0040	0.18	0.0030	0.14
1	0.0300	0.34	0.0240	0.27	0.0180	0.20

4 许可排卤速率分析

在注气排卤后期,只有腔体内卤水表面流速超过一定限值时,高速流动的卤水才能够携带出大量的天然气。通过不同排卤流量下的不同气液界面深度下的流场计算,获得当前工况下(排卤流量和剩余卤水深度)下的表面卤水的最大流速。通过对流速进行分析,当卤水表面最大流速超过一定限值时,就能够判断该种工况下的卤水是否能够携带出天然气,据此确定不同剩余卤水深度下安全排卤流量(不携带出天然气)。

在气浮理论中,液体中微气泡其受力平衡方程如下:

$$\rho_w V_b g - \rho V_b g = 1/2(C_D \rho_w U_b^2 A_b) \tag{6}$$

式中:ρ_b 为气体密度;d_b 为球形微气泡直径;V_b 为微气泡体积,$V_b = \pi d_b^3/6$;A_b 为微气泡垂直投影方向上面积,$A_b = \pi d_d^2/4$;U_b 为微气泡的上升速度:

$$U_b^2 = 4/3 \times d_b(\rho_w - \rho_b)g/(C_D \rho_w) \tag{7}$$

根据斯托克斯提出的绕圆球流动阻力系数的近似解为

$$C_D = \frac{24}{Re} \tag{8}$$

式中:Re 为水流流动雷诺数,$Re = U_b d_b/\nu$;ν 为流体的运动黏度。

由此可得

$$U_b^2 = \frac{\dfrac{4}{3}d_b \dfrac{\rho_w - \rho_b}{\rho_w}g}{\left(\dfrac{24\nu}{U_b d_b}\right)} \tag{9}$$

即

$$U_b = \frac{1}{18}d_b^2 \frac{\rho_w - \rho_b}{\rho_w} \frac{g}{\nu} \tag{10}$$

式中:ν 为液体的黏滞系数,取卤水的运动黏度 $1.01 \times 10^{-6} \mathrm{m^2/s}$ 取 13MPa 压力下 $\rho_b/\rho_w = 0.078$,则式(10)可变为

$$U_{b}=\frac{1}{18}d_{b}^{2}(1-0.078)\frac{9.8}{1.01\times10^{-6}} \qquad (11)$$

根据式(11)，表 2 给出了气体上浮速率与气泡直径的关系。

表 2 气泡上浮速率与气泡直径关系

气泡直径/mm	0.1	0.2	0.5	1	1.5	2
气泡上浮速率/(m/s)	0.005	0.02	0.12	0.50	1.12	1.95

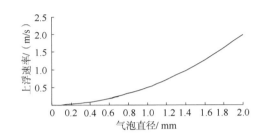

图 5 气泡上浮速率与气泡直径关系

水流以 U_k 速度向下流动，微气泡以 U_b 速度向上运动，则微气泡运动的表观速率为 $U=U_b-U_k$，只有当 $U_k>U_b$ 时，即水流向下流速大于气泡在水中的相对上浮速率时，气泡即能够随水体一起运动，水流能够携带出气体。

在排卤过程中，以注气过程中许可携带出直径为 1mm 的气泡，气液界面处的水体流速必须大于 0.5m/s；以可携带出 0.1mm 的气泡为例，气液界面处的水体流速必须大于 0.005m/s。

通过前述腔体底部流场分析，在腔体剩余 4m 卤水深度时，排卤速率为 100m³/h 时，其表面水流速为 0.0008m/s，远低于 0.005m/s，依据计算所得结果，其携带气泡的能力直径小于 0.1mm。在腔体剩余可排卤水 4m 深度时，以 100m³/h 的排卤速度，不会携带出天然气。

在腔体剩余 2m 卤水深度时，以 100m³/h 速率排卤时，其表面卤水最大流速为 0.005m/s，排卤速率为 60 m³/h 时，其表面水流速 0.003m/s，低于 0.005m/s。依据计算所得结果，若以可携带气泡的能力直径小于 0.1mm。在腔体剩余可排卤水 2m 深度时，以 60m³/h 的排卤速度，不会携带出天然气。

在腔体剩余 1m 卤水深度时，由以上流场等色线图看出，其为排卤流场急速变化区域，在以 40~100m³/h 的排卤速率，其靠近排卤管附近的水表面流速梯度大，容易形成气体带出。因此，在腔体剩余卤水 1m 深度时，易以小于 40m³/h 的排卤速率，缓慢排卤。

5　结语

地下盐穴溶腔注气排卤过程中，天然气和卤水共存，排卤管进入到卤水的深度及排卤速率决定了天然气是否会被卤水带出地表，如何调整后期排卤速率，降低排卤过程中带出天然气的风险，是注气排卤过程中的一个施工技术难题。针对该技术难题的研究，获得结论如下：

在储气库注气排卤过程中，在排卤后期，在腔体剩余可排卤水 4m 深度时，以 100 m³/h 的排卤速度，不会携带出天然气。在腔体剩余 2m 卤水深度时，以可携带气泡的能力直径小于 0.1mm 为限，以 60m³/h 的排卤速度，不会携带出天然气。

在腔体剩余 1m 卤水深度时，排卤管附近为排卤流场急速变化区域，在以 40~100m³/h 的排卤速率排卤时，其靠近排卤管附近的水表面流速梯度大，容易形成气体带出。在腔体剩余卤水 1m 深度时，易以小于 40m³/h 的排卤速率，缓慢排卤。

参 考 文 献

[1] Thoms R L, Gehle R M. Survey of existing caverns in U. S. salt domes[C]//The Mechanical Behavior of Salt Proceedings of the Second Conference. [S. l.]：[s. n.], 1984：703-717.

[2] Horseman S, Passaris E. Creep tests for storage cavity closure prediction. The mechanical behavior of salt[C]//Proceedings of the First Conference. Clausthal：Trans Tech Publications, 1981：119-157.

[3] Gerhard Staupendahl, Manfred W. Schmidt. Field investigation of the long term deformational behavior of a 10000m³ cavity at the asse salt mine. The mechanical behavior of salt [C]//Proceedings of the First Conference. Clausthal：Trans Tech Publications, 1981：511-525.

[4] Mirza U A. Prediction of creep deformations in rock salt pillars[C]//The Mechanical Behavior of Salt Proceedings of the First Conference. Clausthal：Trans Tech Publications, 1981：311-325.

[5] 吴秋丽, 王毅力. 溶气气浮过程动力学模型的分形观[J]. 环境污染治理技术与设备, 2005, 16(5)：29-34.

[6] 朱德祥, 马峥. 游泳水槽的气泡运动规律及去除方法研究[J]. 水动力学研究与进展, 2007, 22(3)：215-221.

[7] 张兆顺, 崔桂香. 流体力学[M]. 北京：清华大学出版, 2006：306-404.

本论文原发表于《岩土力学》2009 年第 30 卷第 12 期。

盐岩储气库注气排卤工艺参数的数值模拟

任众鑫[1]　李建君[1]　朱俊卫[2]　汪会盟[1]　王晓刚[1]

（1. 中国石油西气东输管道公司；2. 西南石油大学）

【摘　要】 盐岩储气库注气排卤后期，由于气水界面锥进形成气窜使得腔底大量卤水无法全部排出，从而导致腔体利用率无法最大化。为尽可能多地将卤水排出以实现腔体库容优质建设，以金坛库区 A 井为例，基于其声呐测腔数据，利用数值模拟软件建立腔体地质模型，进而模拟不同注气压力和注气量条件下注气排卤生产情况，并进行了相应的排卤效果对比分析，结果表明：注气压力或注气量越大，产生气窜现象时间越早，最终累计排出卤水也越少；随着注气参数的提高，日均排卤量增大，但增长幅度变小，并逐渐趋于平稳。

【关键词】 盐岩；储气库；注气排卤；工艺参数；数值研究

Numerical simulation on technological parameters of gas injection and brine discharge in salt-cavern gas storage

Ren Zhongxin[1]　Li Jianjun[1]　Zhu Junwei[2]　Wang Huimeng[1]　Wang Xiaogang[1]

(1. PetroChina West-East Gas Pipeline Company;
2. Southwest Petroleum University)

Abstract At the later stage of gas injection and brine discharge in salt-cavern gas storage, lots of brine at the bottom of the cavity cannot be discharged for gas channeling is generated due to the coning of gas-water interface, and consequently the utilization rate of cavity cannot be maximized. In order to eject brine as much as possible to achieve the goal of high quality storage building, the cavity geology model was established with Well A in Jintan gas storage as example. The model was based on the sonar cavity test data using numerical simulation software. Then, the gas injection and brine discharge situations under different gas injection pressures and rates were simulated. Finally, the brine discharge effects were compared and analyzed. It is indicated that the greater the injection pressure or injection rate is, the earlier the gas channeling arise and the less the ultimate cumulative brine discharge is. And with the increase of

---

基金项目：中国石油天然气股份公司科技攻关项目"盐穴储气库腔底堆积物注气排卤扩容研究"，2013B-3401-0503；中国石油天然气股份公司科技重大专项"地下储气库关键技术研究与应用"，2015E-40。

作者简介：任众鑫，男，1988 年生，工程师，2013 年硕士毕业于中国石油大学（北京）油气井工程专业，现主要从事盐穴储气库工艺技术研究。地址：江苏省镇江市南徐大道 60 号，212000。电话：0511-84523166，Email：zhongxren@petrochina.com.cn。

gas injection parameters, the daily average brine discharge increases, but its increasing amplitude declines gradually to the stable value.

Keywords salt rock; gas storage; gas injection and brine discharge; technological parameters; numerical simulation

随着经济的快速发展,中国对能源的需求持续增长,国家天然气配套设施建设也在不断完善与稳步推进[1-3]。地下储气库作为重要储集设施,其建设步伐正逐渐加快[4-5]。相比其他类型储气库,盐岩储气库具有储气量大、注采灵活、垫气量低并可完全回收等优点[6-7]。盐岩储气库建设过程较为复杂,自前期勘探确定库址至最后建成投产,其间会涉及钻井、采卤溶腔、注采完井、注气排卤及其他作业[8-9],其中注气排卤是非常重要的中心环节,工艺参数以及工程措施选择的合理与否决定了腔内卤水排出量,也直接影响着腔体最终建设质量的优劣[10-11]。因此,在注气排卤过程中有必要通过合理控制注气压力或注气量等工艺参数,以及优化排卤工艺等措施,尽可能多地将腔内卤水排出,以实现腔体库容的优质建设。基于此,采用数值软件,对该项工艺及主要相关参数进行模拟研究,为现场实践与施工提供理论参考与技术借鉴。

1 注气排卤工艺

注气排卤(图1)的目的是通过向盐腔注入天然气,将腔内先前采卤溶腔所形成的卤水驱替并返排至地面。卤水返出流经排卤管柱通道至地面,再由配套管道外输。天然气经压缩机增压后,通过注采管柱与排卤管柱间环空注入腔体[12-14]。排卤结束后,利用不压井作业将排卤管柱起出,并完成井口相关设备的安装。

在注气压力驱使下,作业后期排卤管下端气体会出现锥进突破,形成气窜现象,使得底部卤水无法继续排出。出现锥进后需要根据具体工况及生产条件确定下一步处理措施,一般情况下会通过不压井作业将排卤管柱带压起出,即注气排卤完成。卤水置换量直接关系着盐腔体积的利用率,在溶腔结束、腔体容积已确定的条件下,腔内卤水返排越多,则腔内可存储天然气量就越大。

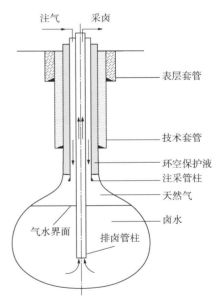

图1 盐岩储气库注气排卤工艺示意图

2 数值模拟

2.1 数学模型

对注气排卤过程做如下假设:(1)符合达西渗流定律;(2)等温过程;(3)气、水相态随压力变化;(4)流体可压缩。该过程中气水置换推进流动满足以下方程[15-17]。

质量守恒方程:

$$\sum_{j=1}^{2} \nabla \left[\frac{c_{ij}\rho_j K K_{rj}}{\mu_j} (\nabla p_j - \rho_j g \nabla D) \right] + q_i = \frac{\partial}{\partial t} \left(\phi \sum_{j=1}^{2} c_{ij}\rho_j S_j \right) \qquad (1)$$

式中：j 为相态，当 $j=1$ 为气相，$j=2$ 为液相；i 为组分；∇ 为汉密尔顿算子；c_{ij} 为 j 相 i 组分的质量分数；ρ_j 为 j 相密度，kg/m^3；K 为渗透率，μm^2；K_{rj} 为 j 相相对渗透率；μ_j 为 j 相黏度，$Pa \cdot s$；p_j 为 j 相分压，Pa；g 为重力加速度，m/s^2；D 为流体势能在坐标系 z 轴方向位移差，m；q_i 为 i 组分产量，m^3/d；ϕ 为孔隙度；S_j 为 j 相饱和度。

相渗透率方程：

$$K_{rj} = f(S_g, S_w) \qquad (2)$$

式中：S_g 为气相饱和度；S_w 为水相饱和度。

饱和度方程：

$$S_g + S_w = 1 \qquad (3)$$

2.2 模型建立

以金坛库区 A 井为例，根据声呐测腔数据，对盐岩储气库进行建模。储气库的声呐数据与传统油藏测井数据不同，其主要基于声呐技术逐层测定腔壁位置，进而确定腔体形状[18-20]。

采用数值模拟软件对声呐测腔结果进行数字化，并结合物性测井数据计算堆积物空隙率，模拟井深 $680 \sim 700m$，对腔体进行地质建模（图2）。为保证模拟预测结果的准确性，对所建模型与现场数据进行拟合。由于渗透率参数由实验测定，但该井模型中该参数不详，故通过修改不同参数进行模拟，对比遴选出最优结果。经分析发现，当空腔处渗透率为 $600\mu m^2$ 时模拟情况与现场数据接近，故选定此时模型参数作为基本参数，进行后续模拟预测。

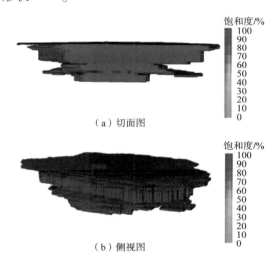

（a）切面图

（b）侧视图

图2 金坛库区 A 井腔体地质模型图

2.3 模拟结果

2.3.1 注气压力

在特定注气压力（7MPa、8MPa、9MPa、10MPa、11MPa、12MPa、13MPa）条件下，分别模拟其注气排卤情况，得到产生气窜现象的时间、气窜时的累计排卤量及日平均排卤量（图3）。对比可见：当注气压力越大时，出现气窜时间越短，相应的总排卤量越少；当注气压力较小时，不易出现气窜，最终累计排卤量较多，但所需排卤作业时间也更长；注气压力越大，日平均排卤量越大，但增长幅度逐渐变缓。

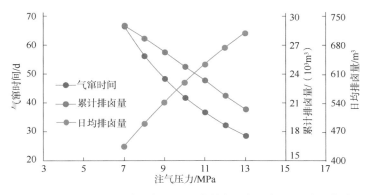

图 3　气窜时间、累计排卤量、日均排卤量与注气压力关系曲线

2.3.2　注气量

在特定的注气量（$10×10^4 m^3/d$、$30×10^4 m^3/d$、$50×10^4 m^3/d$、$60×10^4 m^3/d$、$70×10^4 m^3/d$、$80×10^4 m^3/d$、$100×10^4 m^3/d$）条件下，分别模拟其注气排卤情况，得到气窜时间、气窜时累计排卤量及日平均排卤量（图 4）。可见：注气量越大，气窜时间越短，累计排卤量也越少；注气量越小，累计排卤量越多，但排卤作业时间变长；随注气量增多，日均排卤量逐渐增大，但增幅缓慢且逐渐趋于平稳。

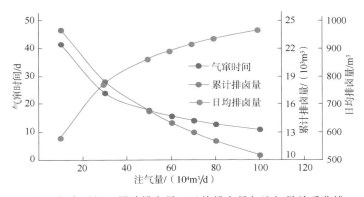

图 4　气窜时间、累计排卤量、日均排卤量与注气量关系曲线

3　结论

注气排卤工序是盐穴储气库建设过程中的重要环节。为准确了解排卤生产过程中注气工艺参数变化敏感性，为现场生产提供借鉴，针对金坛库区实例井某溶腔阶段所形成腔体进行注气排卤，通过模拟不同注气压力与不同注气量条件下气窜时间、气窜时累计排卤量及日平均排卤量的变化可知：注气压力或注气量越大，出现气窜越快，累计排卤量越少；随注气压力或注气量提高，日均排卤量增大即排卤速率加快，但增长幅度变小，并逐渐趋于平稳。因此注气排卤时可根据腔体实际情况进行相应工艺优化，如采用降低注气速率、间歇式排采等。

参 考 文 献

[1] 杨海军. 中国盐穴储气库建设关键技术及挑战[J]. 油气储运, 2017, 36(7): 747-753.

[2] 尹虎琛, 陈军斌, 兰义飞, 等. 北美典型储气库的技术发展现状与启示[J]. 油气储运, 2013, 32(8): 814-817.

[3] Haimson B C, Cornet F H. ISRM suggested methods for rock stress estimation-Part 3: hydraulic fracturing (HF) and/or hydraulic testing of pre-existing fractures (HTPF)[J]. International Journal of Rock Mechanics and Mining Sciences, 2003, 40: 1011-1020.

[4] 班凡生. 盐穴储气库造腔技术现状与发展趋势[J]. 油气储运, 2017, 36(7): 754-758.

[5] 杨海军, 郭凯, 李建君. 盐穴储气库单腔长期注采运行分析及注采压力区间优化——以金坛盐穴储气库西2井腔体为例[J]. 油气储运, 2015, 34(9): 945-950.

[6] 郭凯, 李建君, 郑贤斌. 盐穴储气库造腔过程夹层处理工艺——以西气东输金坛储气库为例[J]. 油气储运, 2015, 34(2): 162-166.

[7] 李建君, 王立东, 刘春, 等. 金坛盐穴储气库腔体畸变影响因素[J]. 油气储运, 2014, 33(3): 269-273.

[8] Bell J S, Bachu S. In-situ stress magnitudes in the Alberta Basin-regional coverage for petroleum engineers[C]. Calgary: Canadian International Petroleum Conference, 2004: 154-155.

[9] Berest P, Brouard B, Durup J G. Tightness tests in salt cavern wells[J]. Oil & Gas Science and Technology, 2001, 56(5): 451-469.

[10] 任众鑫, 李建君, 汪会盟, 等. 基于分形理论的盐岩储气库腔底堆积物粒度分布特征[J]. 油气储运, 2017, 36(3): 279-283.

[11] 杨春和, 梁卫国, 魏东吼, 等. 中国盐岩能源地下储存可行性研究[J]. 岩石力学与工程学报, 2005, 24(24): 4409-4417.

[12] 魏东吼. 金坛盐穴地下储气库造腔工程技术研究[D]. 北京: 中国石油大学(北京), 2008: 10-55.

[13] 袁进平, 李根生, 庄晓谦, 等. 地下盐穴储气库注气排卤及注采完井技术[J]. 天然气工业, 2009, 29(2): 76-78.

[14] 陈锋, 杨海军, 杨春和. 盐岩储气库注气排卤期剩余可排卤水分析[J]. 岩土力学, 2009, 30(12): 3602-3606.

[15] 赵斌, 李云鹏, 田静, 等. 含水层储气库注采效应的数值模拟[J]. 油气储运, 2012, 31(3): 211-214.

[16] 曹勋臣, 喻高明, 邓亚, 等. 基于地层压力恢复时间的注采比数值模拟优化方法——以肯基亚克某亏空油藏为例[J]. 断块油气田, 2016, 23(2): 193-196.

[17] 许珍萍, 蒋建勋, 葛静涛, 等. 枯竭气藏型地下储气库的注采动态数值模拟[J]. 天然气勘探与开发, 2011, 34(4): 53-55.

[18] 杨海军, 李建君, 王晓刚, 等. 盐穴储气库注采运行过程中的腔体形状检测[J]. 石油化工应用, 2014, 33(2): 22-25.

[19] 耿凌俊, 李淑平, 吴斌, 等. 盐穴储气库注水站整体造腔参数优化[J]. 油气储运, 2016, 35(7): 779-783.

[20] 魏东吼, 屈丹安. 盐穴型地下储气库建设与声呐测量技术[J]. 油气储运, 2007, 26(8): 58-61, 68.

本论文原发表于《油气储运》2018年第37卷第4期。

盐穴储气库气卤界面检测技术

付亚平　陈加松　李建君

(中国石油东部管道有限公司储气库项目部)

【摘　要】　盐穴储气库从建成到投产运行需要经历注气排卤阶段，气卤界面检测与控制是注气排卤的关键技术，现有检测方法检测成本高、不能连续检测，为此提出光纤式界面测试技术。该技术利用分布式光纤测试技术，通过井下光纤测量温度变化，根据温度曲线拐点判断气卤界面的位置，操作简单、探测范围广、监测成本低、能连续测量井下气卤界面位置。将该方法进行实际应用后，与现有检测方法进行对比和评估，效果良好，其不仅可应用于盐穴储气库注气排卤阶段，还可在氮气阻溶、天然气阻溶造腔过程中使用，具有广阔的应用前景。

【关键词】　盐穴储气库；注气排卤；光纤；气卤界面；连续检测

Measurement technique on gas-brine interface in salt cavern gas storage

Fu Yaping　Chen Jiasong　Li Jianjun

(Gas Storage Project Department, PetroChina East Gas Pipeline Co. Ltd.)

Abstract　Gas injection and brine discharge is an indispensable stage for the construction and operation of salt cavern gas storage, and the measurement and control of gas-brine interface is the key technology of gas injection and brine discharge. The existing measurement methods cannot do the measurement continuously and their cost is high. To solve these problems, an optical fiber contact measurement technique was developed in this paper. The distributed optical fiber measurement technique is adopted in this technology. The temperature is measured with the downhole optical fiber, and the location of the gas-brine interface is judged according to the inflection point of the temperature curve. This method is advantageous with simple operation, wide detection range, low monitoring cost and continuous measurement. In the practical application, this optical fiber interface measurement technique is proved reliable when compared and evaluated with the existing measurement methods. It can be used not only in the gas injection and brine discharge of salt cavern gas storage, but also in the process of solution mining under nitrogen and gas, so it has wied application prospect.

基金项目：中国石油重大科技专项"地下储气库关键技术研究与应用"，2015E-400301。

作者简介：付亚平，男，1985年生，工程师，2013年硕士毕业于中国石油大学(北京)地球探测与信息技术专业，现主要从事地下储气库方面的研究工作。地址：江苏省镇江市南徐大道60号商务A区D座1801室，212000。电话：15152909577。E-mail：fu_angle@163.com。

Keywords salt cavern gas storage; gas injection and brine discharge; optical fiber; gas-brine interface; continuous measurement

储气库具有实现常规调峰、应急调峰和能源战略储备等功能，是保障国家能源安全的一项重要工程。储气库建造完成后，腔体内填充着卤水，需要进行注气排卤，即通过注入天然气替换腔体内的卤水，在这一过程中，气卤界面的检测是关键技术，如果气卤界面失控，注气过多，天然气会进入排水管道，一旦发生泄漏，不仅会造成资源浪费，而且污染环境，甚至会发生爆炸；注气过少，腔体底部储集较多卤水，造成体积浪费[1-3]。

目前无可靠的连续气卤界面监测技术和方法，计算得出的气卤界面深度往往与真实气卤界面相差较大，容易导致气卤界面失控，存在井喷的风险。针对现有技术的不足，提出了一种新的气卤界面检测技术。该技术在现场进行了试验，试验效果良好，测量精度可达1~2m。

1 气卤界面检测方法

1.1 计算法

根据排出卤水的体积计算气卤界面的深度：

$$h = h_0 + \frac{V_b}{\sum V_i} \tag{1}$$

式中：h 为气卤界面的深度，m；h_0 为腔顶的深度，m；V_b 为排出卤水的体积，m^3；V_i 为腔体在每米深度处的体积，m^3。

该计算方法操作简单，可用于估算气卤界面深度，但实际生产中由于数据误差较大，计算出来的深度不够准确。

1.2 测井法

注气排卤进行一段时间后，在井口加设防喷设备，从排卤管柱中下入测井仪器，通过中子伽马测井的方法，检测气卤界面的深度。

中子伽马测井是采用同位素中子源发射的快中子连续照射井剖面，快中子被地层减速变成热中子，热中子继续在地层中扩散，并不断被吸收，有些核素能俘获热中子，并放出伽马射线，在离中子源一定距离处有一个伽马射线探测器，连续记录伽马射线。中子伽马测井值主要反映地层的含氢量，同时又与含氯量有关[4-5]。通过测量结果示意图(图1)，判断曲线的半幅点处为气卤界面的深度位置。

该方法测量结果相对准确，但是存在以下缺点：使用同位素中子源，存在辐射污染，对人体和环境有严重的危害；施工较为复杂，一次测量耗时一天左右，而且测量时需要停止生产；检测成本高，完成整个注气排卤阶段需要多次测井，并且无法

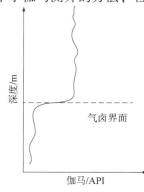

图1 中子测井法判断气卤界面深度示意图

实时检测。

1.3 光纤法

光纤法是利用光纤检测技术进行检测,其利用光的喇曼散射记录与温度相关的信号,再通过光时域反射技术确定温度信息对应的位置,得到温度与深度的剖面,从而判断气卤界面的深度位置。通过测量结果(图2)判断曲线的拐点处为气卤界面的深度位置。

目前主要利用光纤检测盐穴储气库造腔阶段油水界面,该方法测量前需要将光缆固定在造腔外管的外壁,在管柱的最底端缠绕光缆或者接箍保护器固定住光缆,在管柱中部利用钢带固定及接箍保护器保护光缆。井下安装完毕后从井口四通穿出光缆,将光纤连接好地面仪器,通电加热3000W左右,由于柴油和卤水比热容的不同,通过温度变化的快慢差异判断油水界面位置[6-9]。

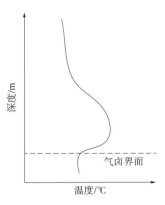

图2 光纤检测法判断气卤界面深度示意图

由于腔体内有易燃易爆的天然气,采用电阻加热可能出现电火花,也可能绝缘失效,存在比较严重的安全风险,因此测量油水界面的方法并不适用于注气排卤阶段气卤界面的检测。与光纤测量油水界面的技术相比,该方法无需加热单元,即可实现气卤界面的连续测量。该方法测量精度为1~2m,符合要求,具有广阔的应用前景[10-12]。

2 光纤检测原理

分布式光纤温度传感器系统,能在整个连续的光纤上,以距离的连续函数形式,测量出光纤上各点的温度值。分布式光纤温度传感器的工作机理是基于光纤内部光的散射现象的温度特性,利用光时域反射测试技术,在同步控制单元触发下,将较高功率的激光脉冲输入光纤,在高能脉冲向前传播的同时产生后向散射,然后将返回的散射光强随时间的变化探测下来。分布式光纤温度传感器基于背向散射或前向散射机理,其中背向散射具有温度测量的实际意义[13-14]。

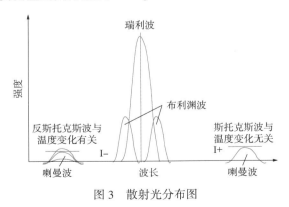

图3 散射光分布图

从光纤返回的散射光有3种成分:瑞利散射、布利渊散射和喇曼散射(图3)。喇曼散射光的强度与温度相关,其中反斯托克斯光信号的强度与温度有关,斯托克斯光信号与温度无关。在测得散射光的光强后,由反斯托克斯光信号的强度与斯托克斯光信号强度的比值可以推知相应的温度信息,再通过光时域反射技术(Optical Time Domain Reflectometer,OTDR),确定温度信息对应的位置,从而得到沿整条光纤的温度分布[15-17]。

2.1 光纤喇曼背向散射测温原理

光纤测温的原理是依据后向喇曼散射效应,激光脉冲与光纤分子相互作用发生能量交换产生散射。当光子被光纤分子吸收后会再次发射出来。如果有一部分光能转换为热能,那么将发出一个比原来波长长的光,$1/\lambda_s = 1/(\lambda-\Delta v)$,称为Stokes光;相反,如果一部分热能转换为光能,那么将发出一个比原来波长短的光,$1/\lambda_a = 1/(\lambda-\Delta v)$,称为Anti-Stokes光;其中$\lambda$为入射光波长,$\Delta v$为喇曼频移。喇曼散射光即由这两种不同波长的Stokes光和Anti-Stokes光组成,其波长的偏移是由光纤组成元素的固有属性决定的。

Stokes光强度:

$$I_s \propto \left[\frac{1}{\exp(hc\Delta\gamma/kT)-1}+1\right]\lambda_s^{-4} \tag{2}$$

Anti-Stokes光强度:

$$I_a \propto \left[\frac{1}{\exp(hc\Delta\gamma/kT)-1}\right]\lambda_a^{-4} \tag{3}$$

式中:λ_s、λ_a分别为Stokes和Anti-stokes光波长,m;h为普朗克常数,J·s;c为真空中光速,m/s;k为玻尔兹曼常数,J/K;$\Delta\gamma$为偏移波数,m^{-1};T为绝对温度,K。

为消除激光管输出不稳定、光纤弯曲、接头损耗等影响,提高测温准确度,温度信息的解调采用双通道双波长比较的方法,即对Anti-Stokes光和Stokes光分别进行采集,利用两者强度的比值解调温度信号。即从光波导内任何一点的反斯托克斯光信号和斯托克斯光信号强度的比例中,可以得到该点的温度[18-19]。由于Anti-Stokes光对温度更为灵敏,因此将Anti-Stokes光作为信号通道,Stokes光作为比较通道,则两者之间的强度比为

$$R(T) = \frac{I_a}{I_s} = \left(\frac{\lambda_a}{\lambda_s}\right)^{-4} \exp(-hc\Delta\gamma/kT) \tag{4}$$

得到温度的理论公式:

$$T = \frac{hc\Delta\gamma}{k}\left[\ln\left(\frac{I_s}{I_a}\right)+4\ln\left(\frac{\lambda_a}{\lambda_s}\right)\right]^{-1} \tag{5}$$

实际测量时,为了准确测量温度,需要进行温度标定,即在温度解调仪内设置恒温槽,根据恒温槽的温度来确定其他位置的温度值。因此将温度变为式(6):

$$\frac{1}{T} = -\frac{k}{hc\Delta\gamma}\left[\ln R(T)+4\ln\left(\frac{\lambda_a}{\lambda_s}\right)\right] \tag{6}$$

对于固定的温度T_0(恒温槽标定温度)有

$$\frac{1}{T_0} = \frac{k}{hc\Delta\gamma}\left[\ln R(T_0)+4\ln\left(\frac{\lambda_a}{\lambda_s}\right)\right] \tag{7}$$

得到温度值的测量公式:

$$\frac{1}{T} = \frac{1}{T_0} - \frac{k}{hc\Delta\gamma}\left[\ln R(T) - \ln R(T_0)\right] \tag{8}$$

在系统标定之后,通过式(8)测定 $R(T)$ 即可以确定沿光纤各点的温度值。

2.2 光纤光时域反射定位原理

对测量点的空间定位是通过光时域反射(Optical Time-domain Reflectometer,OTDR)技术实现的,当激光脉冲在光纤中传输时,与光纤中的分子、杂质等相互作用,发生散射(图4)。

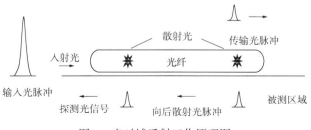

图 4　光时域反射工作原理图

光在光纤中走过的距离为

$$L = Vt \tag{9}$$

式中:L 为光在光纤中走过的距离,m;V 为光在光纤中的传输速度,m/s;t 为光在光纤中走过的时间,s。

光在光纤中的传输速度为

$$V = C/n_g \tag{10}$$

式中:C 为真空中的光速,m/s;n_g 为光纤实际折射率。

结合式(9)和式(10)得到散射光距离测量处的距离:

$$l = \frac{Ct}{2n_g} \tag{11}$$

式中:l 为散射光到测量处的距离,m。

因此利用光时域反射技术可以确定沿光纤温度场中每个温度采集点的距离,以及异常温度点、光纤断裂点的距离定位信息。

3　现场示意图及应用效果

3.1　示意图

将光纤检测技术应用到盐穴储气库的气卤界面检测,需要先将光纤设备安装于井下,安装方法有两种:将光缆安装于注采气管柱外壁,用扎带固定,通过地面加热的方法,判断出界面的深度;将光缆下端加一配重,类似于测井的方法下入排卤管柱中,不需要加热,利用光纤检测腔体内自身的热力场变化来判断界面的深度。

K井是一口需要进行注气排卤的盐穴储气库井,由于管外安装需要将注采气管柱提到地面,将光缆缓缓固定在外壁后再通过修井机下入井下,工艺复杂,且检测完成后无法回收井下设备。因此选取管柱内的安装方法(图5)[20]。

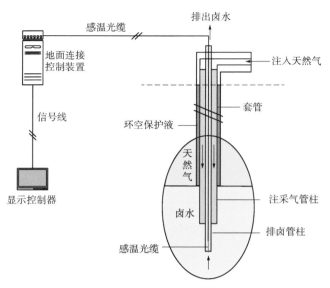

图 5 注气排卤现场检测示意图

3.2 应用效果

在 K 井注气排卤阶段,连续记录排出的卤水体积,根据声呐测量出的腔体体积估算出气卤界面的大致位置。同时利用光纤检测方法实时记录排卤管柱内的温度数据,并判断气卤界面深度,由光纤测量的排卤管柱内温度变化图(图6)可知:图 6(a)是刚排卤的温度曲线,图 6(b)是停止注气排卤后 1h 温度场相对稳定后的温度曲线,靠近底端温度的拐点 1116m 处为气卤界面的深度位置。

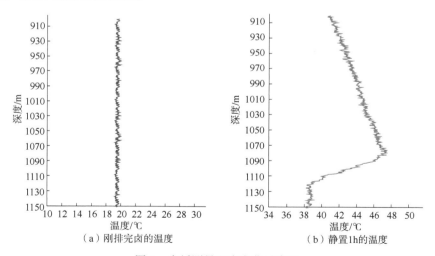

(a) 刚排完卤的温度   (b) 静置1h的温度

图 6 光纤测量温度变化示意图

选取该井某一个月内采用两种方法计算得到的气卤界面深度值,并将二者结果进行对比(表1)。其中第一次的气卤界面深度 1034m 是利用中子测井方法测量出的,作为参考深度。

表 1 两种方法测量 K 井气卤界面深度对比

光纤测量深度/m	计算深度/m	中子测井深度/m
1034.0	1034	1034
1036.5	1036	—
1038.5	1039	—
1041.0	1041	—
1042.5	1045	—
1045.0	1047	—
1046.5	1049	—
1049.0	1051	—
1051.5	1053	1052

由表 1 可知：二者的总体差距较小，但在注气排卤初期，二者的结果相差较小，随着注气排卤量的增大，二者差别逐渐加大，这是由于随着注气排卤量的增大，计量误差也增大，造成计算法估算的气卤界面深度误差也越来越大。通过 K 井最终测量结果可知，光纤检测气卤界面技术得到成功应用，气卤界面得到了很好的控制。

利用光纤检测的方法大大简化了注气排卤阶段气卤界面的控制，可以直观地显示气卤界面深度。该方法的应用能够加快注气排卤速度，通过设定一个深度报警值，在未达到此深度时，用于注气排卤的排量可以适当加大，接近或达到此深度时，放慢注气排卤速度，该技术的应用，能够减少中子测井次数，降低检测成本。

4 结论

采用光纤式气卤界面检测方法检测盐穴储气库注气排卤阶段的气卤界面，有以下几方面优势：测量深度范围大，操作简单，可实时连续检测气卤界面；节省作业时间，减少中子测井次数，节约作业成本；经过简单改造后可以同时测出井下温度分布和井下压力，有利于辅助判断井下管柱脱落、泄漏等故障；具有广阔的应用前景，可应用于注气排卤、天然气或氮气阻溶造腔等阶段。

该技术填补了国内外盐穴储气库气卤界面连续监测方法的空白，为进一步提高国内盐穴储气库储气能力奠定了坚实的基础，对国家天然气能源战略发展具有深远意义。

参 考 文 献

[1] 袁进平，李根生，庄晓谦，等．地下盐穴储气库注气排卤及注采完井技术[J]．天然气工业，2009，29（2）：76-78．

[2] 李龙，杨海军，刘玉刚，等．金坛盐穴储气库新溶腔井注气排卤情况分析[C]．银川：第七届宁夏青年科学家论坛，2011：452-454．

[3] 庄清泉．注气排卤技术在盐穴造腔中的应用[J]．油气田地面工程，2010，29（12）：65-66．

[4] 黄隆基．核测井原理[M]．青岛：中国石油大学出版社，2008：108-119．

[5] 孙林平，范宜仁，孙跃武，等．中子测井在储气库井腔体试压密封检测中的应用[J]．测井技术，2007，31（1）：35-38．

[6] 张建华,杨海军,李龙,等. 井下介质界面监测装置:201220371077.6[P]. 2013-02-27.
[7] 张建华,巴金红,李建君,等. 光纤式储油罐多相介质厚度测量装置:201520122874.4[P]. 2015-08-05.
[8] 彭勇,张映辉. 光纤油水界面监控仪研制与应用[J]. 仪器仪表学报,2005,26(8):857-859.
[9] Grosswig S, Vogel B. Permanent blanket-brine interface monitoring by temperature monitoring in salt caverns[C]. Porto:SMRI Spring 2008 Technical Conference,2008:2-6.
[10] 张晓威,刘锦昆,陈同彦,等. 基于分布式光纤传感器的管道泄漏监测试验研究[J]. 水利与建筑工程学报,2016,14(3):1-6.
[11] 史晓锋,蔡志权,李铮. 分布式光纤测温系统及其在石油测井中的应用[J]. 石油仪器,2002,16(2):20-23.
[12] 阎家光. 光纤传感器在煤矿井筒监测系统中的应用[J]. 自动化与仪器仪表,2016,206(12):63-64.
[13] Brentle J O, Möbius C. Fibre optic measurement system for the automatic and continues blanket level interface monitoring during the solution-mining process of salt caverns[C]. New York:SMRI Spring 2015 Technical Conference,2015:1-5.
[14] Grosswig S, Vogel B. Optic measurement system for temperature, automatic and continuous blanket interface monitoring in caverns[C]. Santander:SMRI Fall 2015 Technical Conference,2015:2-10.
[15] 刘德明,孙琪真. 分布式光纤传感技术及其应用[J]. 激光与光电子学进展,2009(1):29-33.
[16] 刘建霞. φ-OTDR 分布式光纤传感监测技术的研究进展[J]. 激光与光电子学进展,2013(8):199-204.
[17] 赵业卫,姜汉桥. 油井高温光纤监测新技术及应用[J]. 钻采工艺,2007(4):158-160.
[18] 王玉田. 光电子学与光纤传感器技术[M]. 北京:国防工业出版社,2003:106-109.
[19] 王剑锋,刘红林,张淑琴,等. 基于拉曼光谱散射的新型分布式光纤温度传感器及应用[J]. 光谱学与光谱分析,2013,33(4):865-868.
[20] José C. Pereira, common practices-gas cavern site characterization, design, construction, maintenance, and operation[C]. Houston:SMRI Spring 2012 Technical Conference,2012:18-19.

本论文原发表于《油气储运》2017年第36卷第7期。

金坛盐穴储气库新溶腔井注气排卤情况分析

李 龙 杨海军 刘玉刚 方 亮

(中国石油西气东输管道公司储气库管理处)

【摘 要】 金坛盐穴储气库经过多年发展，新溶腔井已相继完成或开始注气排卤。新溶腔井是我国首次完成水溶建库，所以需要充分收集、研究和分析新溶腔井注气排卤情况，为今后完腔井注气排卤施工提供充足的指导依据。

【关键词】 盐穴储气库；新溶腔井；注气排卤

1 先导试验井概况

先导试验井是我国专为盐穴储气库完成的一口资料井，同时也是我国第一个采用水溶建库的先导试验井。位于江苏省金坛市陈家庄盐矿西大伏庄西南0.42km，其构造位置为直溪桥凹陷紫阳桥—倪巷洼陷深凹部位，其所处盐岩层顶界井深945m，底界1186m，厚度241m。

先导试验井于2003年10月5日开钻，2003年11月29日完钻，完钻井深为1269m。2005年5月27日至2007年11月24日期间注水溶腔，累计排卤量为12319224m³，排盐量为269364.16L，造腔量为11.65m³，腔体设计运行压力为7~17MPa。库容量为1997m³、垫底气量为775m³、正常可采气量为980m³、应急可采气量为1222m³，先导试验井腔体形状如图1所示。

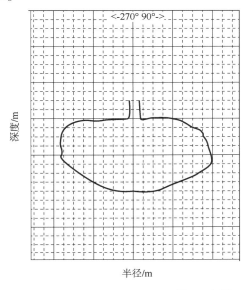 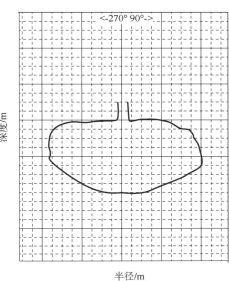

图1 金资一井声呐测腔示意图

先导试验井完钻井身程序数据：

(1) 表层套管：φ339.7mm×70.24m（钢级：J55，产地：宝钢，壁厚：9.65mm，内径：320.4mm）。

(2) 生产套管：φ244.5mm×1015.61m（钢级：N80 和 P110，产地：日本和天津，壁厚：10.03mm 和 11.99mm，内径：224.44mm 和 220.52mm）。

表1 先导试验井套管程序表

套管类别	钢级	套管直径/mm	壁厚/mm	下深/m	管鞋深度/m	水泥返高/m	固井质量	联入/m
表层	J55	339.7	9.65	70.24	70.24	地面	合格	
生产	N80		10.03	915.98				
	P110	244.5	11.99	1005.4	1015.61	150.5	合格	4.35
	N80		10.03	71015.61				

(3) 最大井斜、方位角及所在井深：井斜3.14°、方位47.11°、井深1268.5m，完钻井深1269.0m、总闭合方法69.78°、总水平位移10.06m。

(4) 盐层位置：盐顶978.41m，盐底1169.67m。

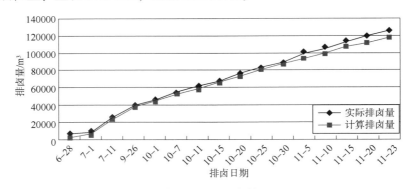

图2 先导试验井排卤情况图

2 新溶腔井概况

新溶腔井位于江苏省金坛市陈家庄盐矿大岸村附近，构造上位于金坛盆地直溪桥凹陷陈家庄次洼，盐层埋深977m，厚度为186.4m。2005年7月6日完钻，完钻井深1175m，套管下深999.77m。

2006年12月22日至2010年9月1日期间注水溶腔，累计排卤量为1901674m³，排盐量为424357t，造腔量为18.712m³，腔体设计运行压力为7~17MPa。库容量为3300m³、垫底气量为1300m³、正常可采气量为1450m³、应急可采气量为2000m³。新溶腔井腔体形状如图3所示。

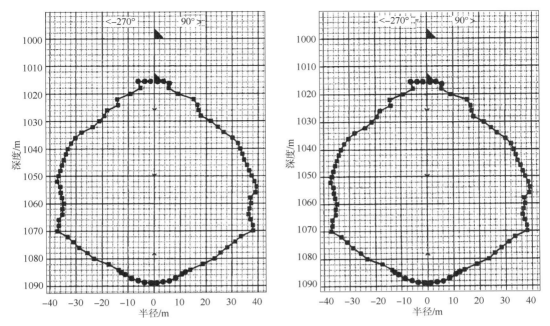

图 3 新溶腔井声呐测腔示意图

（1）新溶腔井钻井基本数据：

钻头程序：444mm×606.5m；

311mm×1002m；

215.9mm×1175m；

套管程序：339.7mm×606.5m（钢级：J55，壁厚：9.65mm）；

244.5mm×999.77m（钢级：P110，壁厚：11.99mm）。

阻位：975.85m。

人工井底：1154m。

水泥返高：地面。

固井质量：合格。

（2）完井井身程序数据：

ϕ444.5mm 钻头×606.5m+ϕ339.7mm 套管×606.5m；

ϕ311.15mm 钻头×1002m+ϕ244.5mm 套管×999.77m；

ϕ215.9mm 钻头×1175m。

套管类别	钢级	套管直径/mm	壁厚/mm	下深/m	管鞋深度/m	水泥返高/m	固井质量
表层	J55	339.7	9.65	606.5	606.5	地面	合格
生产	P110	244.5	11.99	999.77	999.77	地面	合格

（3）最大井斜、方位角及所在井深：

井斜 1.765°，方位为 200.215°，井深 425m；

完钻井深 1175m、总闭合方位为 97.1°、总水平位移为 7.35m。

（4）盐层位置：

盐顶为 977m，盐底为 1163.4m。

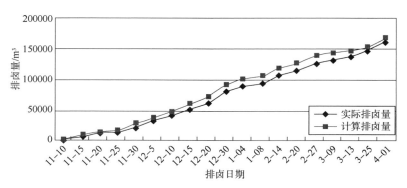

图 4　新溶腔井排卤情况图

3　结论与分析

通过注气过程中监测数据，对注气过程中注气量、排卤量及井口压力进行了记录，并根据腔体的声呐测腔数据将注气量换算成与其对应的计算排卤量，同时以排卤日期为横坐标绘制了注气过程中实际排卤量、计算排卤量曲线。在整个注气过程中，实际排出卤水量、注气量计算得出的理论排卤量结果十分接近，排除注气过程中各种异常工况引起的压力波动，排卤量的理论计算值准确度较好。

本论文原发表于《第七届宁夏青年科学家论坛论文集》2011 年。

Modeling debrining of an energy storage salt cavern considering the effects of temperature

Dongzhou Xie[1,2]　Tongtao Wang[1,3]　Long Li[4]　Kai Guo[5]　Jianhua Ben[4]
Duocai Wang[6]　Guoxing Chai[7]

(1. State Key Laboratory of Geomechanics and Geotechnical Engineering, Institute of Rock and Soil Mechanics, Chinese Academy of Sciences; 2. University of Chinese Academy of Sciences; 3. Hubei Key Laboratory of Geo-Environmental Engineering, Institute of Rock and Soil Mechanics, Chinese Academy of Sciences; 4. PipeChina West East Gas Pipeline Company Jiangsu Gas Storage Branch Company; 5. PetroChina Gas Storage Company; 6. PipeChina West East Gas Pipeline Company; 7. SINOPEC Petroleum Exploration and Production Research Institute)

Abstract　Debrining is one of the most important steps in the construction of a salt cavern. The debrining inner tubing may be blocked by salt deposits because the brine temperature drops during debrining. However, most previous studies have ignored the effects of salt deposits on debrining. In this paper, a novel debrining model of a gas storage salt cavern to predict the debrining parameters and salt deposit growth based on the crystallization kinetics and heat transfer principle is built. A finite difference program is developed to solve this model. The model is validated by debrining monitoring data. The effects of debrining rate on debrining parameters are analyzed. The results show that the brine and gas temperature increase gradually with debrining time and ultimately approaches a constant value. If the debrining rate is increased from 20 to 80 m^3/h, the salt deposits growth rate at the wellhead is decreased by 53.48%, and the total debrining time can be reduced by 77.3%, and the backflushing frequency can be decreased from 1.17 to 3.29 days/time. It is suggested that the debrining rate of Jintan UGS salt cavern increased to more than 80 m^3/h and the backflushing frequency decreased to 3.29 days/time.

Keywords　salt cavern gas storage; debrining; temperature; salt deposit; mathematical model

1　Introduction

Energy is an important material basis for human survival and the driving force of social development. With the development of the economy, global energy consumption has increased by 14.25% in the last decade. In 2021, global energy consumption reached 564.01 EJ, of which natural gas accounts for 24%[1]. Natural gas is a clean energy source that emits less carbon dioxide than coal and petroleum for producing the same amount of heat[2,3]. As the world population contin-

---

Corresponding author: Tongtao Wang, State Key Laboratory of Geomechanics and Geotechnical Engineering, Institute of Rock and Soil Mechanics, Chinese Academy of Sciences, Wuhan, 430071, China. ttwang@ whrsm. ac. cn.

ues to grow and industrialization accelerates, the problems of global warming and environmental pollution are getting worse[4,5]. In order to cope with global warming, many countries have proposed reducing greenhouse gas emissions and increasing the proportion of low-carbon energy and new energy in the energy consumption system [6-8]. In this context, the proportion of natural gas in global energy consumption is gradually increasing[9,10]. Fig. 1 presents global natural gas consumption from 2011 to 2021. Global energy consumption increases by 24.85% from 2011 to 2021, reaching $4037.5 \times 10^9 m^3$ by 2021. Such huge consumption stimulates the need for large-scale natural gas storage because natural gas consumption has significant seasonal fluctuations [11]. The construction of large-scale underground gas storage (UGS) is significant to the natural gas supply.

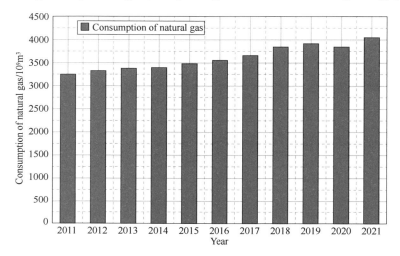

Fig. 1 The global natural gas consumption from 2011 to 2021[1]

There are four types of UGS: depleted oil and gas reservoirs, salt caverns, aquifers, and hard rock caverns[12,13]. Salt caverns are widely used for large-scale storage of natural gas and petroleum[14], because they have many advantages, including good tightness, high injection-production ratio, less cushion gas and low construction cost. The clean energy industry, such as photovoltaic and wind power, has been booming in recent years. Salt caverns also are used to store hydrogen[15] and compressed air[16]. The construction of a salt cavern includes two processes: leaching and debrining[17,18]. Debrining is a key step in converting the cavern into an effective gas storage space, and this step has high technical requirements and large safety risks. In particular, the debrining inner tubing (DIT) may be blocked by the salt deposits during debrining, as shown in Fig. 2. If the DIT is completely blocked by the salt deposits[19], the DIT needs to be replaced, which may lead to an explosion, fire disaster and other safety accidents. The accidents of salt deposits blocking the tubing have repeatedly occurred in salt mining[20-22]. In addition, the gas temperature changes dramatically during debrining. Complex conversion of heat energy, pressure energy and kinetic energy occurs during gas compression[23], which makes it difficult to predict the debrining parameters, such as gas injection pressure, gas velocity, injection gas volume, debrining rate, etc.

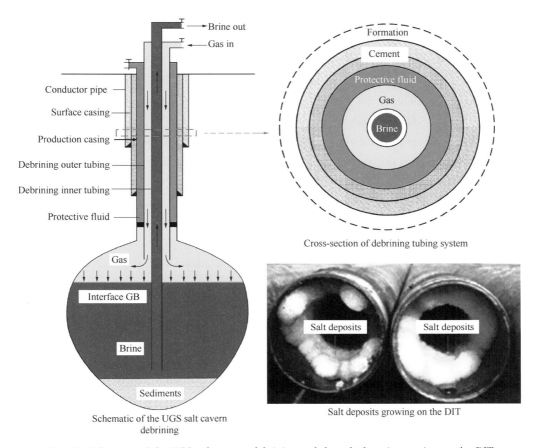

Fig. 2 Schematic of the UGS salt cavern debrining and the salt deposits growing on the DIT

Previous studies have mainly focused on the debrining scheme formulation and parameters optimization. Based on the implementation of the debrining project of Jintan UGS, the first debrining scheme was developed, and the debrining parameters were reasonably optimized by analyzing the debrining monitoring data[24-26]. One important finding from previous studies is that the gas injection pressure will gradually increase as the salt deposits grow when the debrining rate is constant. Therefore, the method of using the changes in gas injection pressure to determine whether to perform the backflushing operation was proposed[24]. Meanwhile, many mathematical models have been proposed to calculate the debrining parameters, such as gas injection pressure, debrining rate and the depth of the interface between gas and brine (GB interface), etc. Wang et al.[18,27] and Li et al.[28] proposed a mathematical model of debrining for UGS salt caverns and compressed air storage caverns and deduced equations for calculating debrining parameters. In addition, some researchers used numerical simulation to establish a geological model of UGS salt cavern and analyzed the quantitative relationship between gas injection pressure, debrining rate and GB interface depth[29-32]. Their research shows that the gas injection pressure is related to the GB interface depth in addition to salt deposit growth. The depth of the GB interface is dynamically changing during debrining.

Therefore, using the gas injection pressure to determine whether to backflush the DIT is unreliable. In addition, these mathematical models have several shortcomings: (1) The effects of temperature change on the physical properties of the gas have been ignored, (2) The salt deposit growth could not be predicted, and the influences on brine frictional resistance have not been considered, which makes it difficult to calculate the debrining parameters accurately. For salt deposit growth, Wei et al. [33] and Jing et al. [16] carried out a simulation experiment of salt deposit growth, and they found that the brine temperature variation significantly affected the salt deposit growth rate. Our previous studies demonstrated that the saturated brine temperature dropping in the DIT during debrining is the main reason for salt deposit growth[34,35]. However, the above methods cannot accurately predict the salt deposits growth and debrining parameters due to the lack of strict theoretical support of flow and heat transfer, which leads to the debrining programs in the field relying on the engineering experience and further leads to the waste of energy and freshwater resources. Therefore, it is urgent to establish a mathematical model for predicting debrining parameters and salt deposit growth to guide the construction of debrining.

The main objective of this study is to build a novel mathematical model of debrining for UGS salt caverns considering the effects of temperature, which is used to calculate the debrining parameters and predict the salt deposit growth. Firstly, a novel debrining mathematical model is built based on the crystallization kinetics and heat transfer principle. Secondly, a finite difference program is developed, and the model is validated by debrining monitoring data of the A-1 cavern of Jintan UGS. Finally, the effects of debrining rate on temperature field distribution, salt deposits growth and the GB interface depth are analyzed, and the debrining parameters are optimized. This study can provide a theoretical foundation for calculating the debrining parameters and predicting salt deposit growth during debrining.

2 Model development

Fig. 2 shows the schematic of debrining a UGS salt cavern. The gas is injected into the annulus between the DIT and debrining outer tubing (DOT), and the brine is discharged from the DIT during debrining. With the discharge of brine, the GB interface decreases continuously. Debrining is completed when the GB interface position decreases to the design depth.

2.1 Fluid flow and heat transfer model

2.1.1 Mass conservation equation

As shown in Fig. 2, the gas flow is a compressible transient onedimensional flow in the annulus. The gas flow needs to satisfy the mass conservation equation:

$$\frac{\partial(\rho_g A_g)}{\partial t}+\frac{\partial(\rho_g A_g v_g)}{\partial z}=0 \tag{1}$$

where ρ_g is the gas density, kg/m^3; A_g is the cross-sectional area of the annulus, m^2; v_g is the gas velocity, m/s; t is time, s; z is depth, m.

Generally, the brine is considered incompressible. The mass conservation equation of brine flow can be simplified:

$$\frac{\partial(A_b)}{\partial t}+\frac{\partial(A_b v_b)}{\partial z}=0 \qquad (2)$$

where A_b is the cross-sectional area of the brine flow channel, m^2; v_b is the brine velocity, m/s.

2.1.2 Momentum conservation equation

The gas flow in the annulus should satisfy the momentum conservation equation:

$$\frac{\partial(A_g\rho_g v_g)}{\partial t}+\frac{\partial(A_g\rho_g v_g^2)}{\partial z}+\frac{\partial(A_g p_g)}{\partial z}=A_g\rho_g g-\frac{A_g\rho_g \lambda_g v_g^2)}{2*2(r_{ai}-r_{to})} \qquad (3)$$

where p_g is the gas pressure, MPa. g is the acceleration of gravity, 9.81m/s^2; λ_g is the friction resistance coefficient of gas; r_{ai} is the inner radius of the DOT, m; r_{to} is the outside diameter of DIT, m. The effects of frictional resistance on gas pressure can be ignored because the gas density and viscosity are minimal.

The brine flow also needs to satisfy the momentum conservation equation:

$$\frac{\rho_b\partial(A_b v_b)}{\partial t}+\frac{\rho_b\partial(A_b v_b^2)}{\partial z}+\frac{\partial(A_b p_b)}{\partial z}=A_b\rho_b g-\frac{A_b\rho_b \lambda_b v_b^2}{4r_{tin}} \qquad (4)$$

where ρ_b is the brine density, kg/m^3; p_b is the brine pressure, MPa; λ_b is the friction resistance coefficient of brine; r_{tin} is the effective inner radius of the DIT, m.

2.1.3 Energy conservation equation

The gas compression leads to the conversion of pressure energy, potential energy and internal energy. The equation of gas energy conservation considering the enthalpy change of gas can be written as:

$$\frac{\partial\left[A_g\rho_g\left(u_g+\frac{v_g^2}{2}\right)\right]}{\partial t}+\frac{\partial\left[A_g\rho_g v_g\left(h_g+\frac{v_g^2}{2}\right)\right]}{\partial z}=-\rho_g A_g v_g g+2\pi r_{tin}U_{bg}(T_b-T_g)+2\pi r_{ai}U_{ge}(T_e-T_g) \qquad (5)$$

where u_g is the gas internal energy, J/kg; h_g is the gas enthalpy, J/kg; U_{bg} is the total heat transfer coefficient between brine and gas, W/(m^2·℃); U_{ge} is the total heat transfer coefficient between gas and formation, W/(m^2·℃); T_g is the gas temperature, ℃; T_b is the brine temperature, ℃; T_e is the formation temperature, ℃.

According to hydrodynamics, the relations between the gas enthalpy and gas internal energy and pressure can be expressed as[36]:

$$h_g=u_g+\frac{p_g}{\rho_g} \qquad (6)$$

$$dh_g=c_g dT_g-c_g\beta_g dp_g \qquad (7)$$

where c_g is the gas specific heat capacity, J/(kg·℃); β_g is the gas Joule-Thomson coefficient, ℃/MPa.

Substituting Eqs. (6) and (7) into Eq. (5), the equation of gas energy conservation in the form of temperature is obtained:

$$\frac{\partial(A_g\rho_g c_g T_g)}{\partial t} - \frac{\partial(A_g\rho_g c_g \beta_g p_g)}{\partial t} - \frac{\partial(A_g p_g)}{\partial t} + \frac{\partial(A_g\rho_g v_g^2/2)}{\partial t} + \frac{\partial(A_g\rho_g v_g c_g T_g)}{\partial z} - \frac{\partial(A_g\rho_g v_g c_g \beta_g p_g)}{\partial z} +$$

$$\frac{\partial(A_g\rho_g v_g^3/2)}{\partial z} = -\rho_g A_g v_g g + 2\pi r_{\text{tin}} U_{\text{bg}}(T_b - T_g) + 2\pi r_{\text{ai}} U_{\text{ge}}(T_e - T_g) \tag{8}$$

The thermophysical properties of brine are nearly constant, the Joule-Thomson cooling can be neglected. The energy conservation equation of brine can be expressed as:

$$\frac{\partial[A_b\rho_b c_b T_b]}{\partial t} - \frac{\partial[A_b\rho_b v_b c_b T_b]}{\partial z} = 2\pi r_{\text{tin}} U_{\text{bg}}(T_g - T_b) + \frac{A_b\rho_b \lambda_b v_b^3}{4 r_{\text{tin}}} \tag{9}$$

where c_b is the specific heat capacity of brine, J/(kg·℃).

2.1.4 Formation heat transfer model

The formation remains static during debrining. The modes of heat transfer between formation boundary and gas are heat conduction and heat convection. But the heat transfer mode inside the formation is mainly heat conduction. Assuming that the thermal physical parameters of the formation at different depths are the same, the energy conservation equation in the formation can be expressed as:

$$\rho_e c_e \frac{\partial T_e}{\partial t} = \frac{1}{r}\frac{\partial}{\partial r}\left(k_e r \frac{\partial T_e}{\partial r}\right) = k_e\left(\frac{\partial T_e}{r\partial r} + \frac{\partial^2 T_e}{\partial r^2}\right) \tag{10}$$

where ρ_e is the formation density, kg/m^3; c_e is the specific heat capacity of formation, J/(kg·℃); k_e is the heat conductivity of formation, W/(m·℃); r is the distance from the wellbore to formation, m.

2.2 Auxiliary equation

The gas density is related to pressure and temperature, it is calculated by the gas state equation:

$$p_g M_g = Z_g \rho_g R T_g \tag{11}$$

where M_g is the molar mass of gas, kg/mol; R is the ideal gas constant, $R = 8.31$ J/(mol·K); Z_g is the compressibility factor of gas, m.

The friction resistance coefficient of brine is related to the flow state and the pipe roughness. In this paper, an empirical formula is used to calculate the friction resistance coefficient[37]:

$$\lambda_b = 0.11\left[\frac{68}{Re} + \frac{\varepsilon}{2r_{\text{tin}}}\right]^{0.25} \tag{12}$$

where Re is the Reynolds number; ε is the roughness of the DIT, m.

The GB interface depth decreases as the brine is discharged during debrining. The GB interface depth affects the pressure of gas and brine. Therefore, the changes in the GB interface depth must be considered in the model. Because the shape of the cavern is irregular, it is difficult to determine the position of the GB interface directly. Therefore, a numerical calculation is used to predict the GB interface depth. According to the geometric relationship, the changes in GB interface depth can be calculated by the debrining brine volume at a time step, and it is written as follows:

$$dh_{\text{gb}} = \frac{dV_b}{A_g(z = H_{\text{gb}})} \tag{13}$$

where dh_{gb} is the change of the GB interface depth at a time step, m; dV_b is the debrining brine volume at a time step, m^3.

The GB interface depth can then be calculated as follows:

$$H_{gb} = H_0 + \sum dh_{gb} \tag{14}$$

where H_{gb} is the GB interface depth, m; H_0 is the initial GB interface depth, m.

2.3 Salt deposits growth

The salt deposits form on the DIT when the brine temperature decreases. Generally, the brine is saturated. The salt deposit growth rate depends on the supersaturation of the brine. According to the crystallization kinetics, the linear growth rate of salt deposits is written as[38-40]:

$$G = \begin{cases} 0 & c<c_s \\ \dfrac{K^g(c-c_s)^d}{\rho_s} & c<c_s \end{cases} \tag{15}$$

where G is the linear growth rate of the salt deposits, m/s; K_g is the coefficient of overall growth rate of salt deposit, (m/s)$^{-d}$; d the growth order of salt deposit; ρ_s is the density of salt deposits, kg/m^3; c is the brine concentration, kg/m^3; c_s is the brine saturation concentration, kg/m^3.

The main component of salt deposit is sodium chloride. The relationship between the saturation concentration of sodium chloride solution and temperature can be expressed by a quadratic polynomial[41]:

$$c_s(T_b) = 357.12 + 0.085T_b + 0.00317T_b^2 \tag{16}$$

If the brine reaches saturation in the cavern, substituting Eq. (14) into Eq. (13), the equation for the linear growth rate of the salt deposits is obtained:

$$G = K^g[0.085 \cdot (T_t - T_b) + 0.00317(T_t^2 - T_b^2)]^d \tag{17}$$

where T_t is the brine temperature in the cavern, measured by dieselbrine interface monitoring or sonar survey, ℃.

The salt deposits thicknesses and effective inner radius of DIT are calculated by:

$$e = G \cdot t \tag{18}$$
$$r_{tin} = r_{ti} - e \tag{19}$$

where e is the salt deposits thicknesses, m; r_{ti} is the inner radius of the DIT, m.

2.4 Total heat transfer coefficient

Fig. 3 presents a schematic diagram of heat transfer between the gas, brine and formation during debrining. The total heat transfer coefficient between brine and gas consists of four parts: the heat convection from brine to salt deposits, the heat conduction between salt deposits and DIT, and the heat convection from DIT to gas. The total heat transfer coefficient between brine and gas is written as:

$$\frac{1}{U_{bg}} = \frac{1}{h_b} + \frac{r_{tin}\ln(r_{ti}/r_{tin})}{k_c} + \frac{r_{tin}\ln(r_{to}/r_{ti})}{k_t} + \frac{r_{tin}}{r_{to}h_g} \tag{20}$$

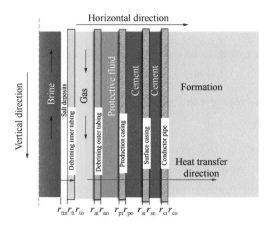

Fig. 3 Schematic diagram of heat transfer between the gas, brine and formation during debrining

where h_b is the convective heat transfer coefficient of brine in the DIT, W/(m² · ℃); k_c is the heat conductivity of salt deposits, W/(m · ℃); k_t is the heat conductivity of the DIT, W/(m · ℃); h_g is the convective heat transfer coefficient of the gas in the annulus, W/(m² · ℃); r_{to} is the outer radius of the DIT, m.

The total heat transfer coefficient between gas and formation consists of eight parts: the heat convection from gas to DOT, the heat conduction of DOT, protective fluid, PC, cement, surface casing and conductor pipe. The total heat transfer coefficient between gas and formation is written as:

$$\frac{1}{U_{ge}} = \frac{1}{h_g} + \frac{r_{ai}\ln(r_{ao}/r_{ai})}{k_a} + \frac{r_{ai}\ln(r_{pi}/r_{ao})}{k_{pf}} + \frac{r_{ai}\ln(r_{po}/r_{pi})}{k_p} + \frac{r_{ai}\ln(r_{si}/r_{po})}{k_{cem}} + \frac{r_{ai}\ln(r_{so}/r_{si})}{k_t} + \frac{r_{ai}\ln(r_{ci}/r_{so})}{k_{cem}} + \frac{r_{ai}\ln(r_{co}/r_{ci})}{k_t} \quad (21)$$

where k_a is the heat conductivity of the DOT, W/(m · ℃); k_{pf} is the heat conductivity of the protective fluid, W/(m · ℃); k_p is the heat conductivity of the PC, W/(m · ℃); k_t is the heat conductivity of the surface casing and conductor pipe, W/(m · ℃); k_{cem} is the heat conductivity of the cement, W/(m · ℃); r_{ao} is the outer radius of the DOT, m; r_{pi} is the inner radius of the PC, m; r_{po} is the outer radius of the PC, m; r_{si} is the inner radius of the surface casing, m; r_{so} is the outer radius of the surface casing, m. r_{ci} is the inner radius of the conductor pipe, m; r_{co} is the outer radius of the conductor pipe, m. Each term on the right side of Eq. (21) represents the heat transfer efficiency in the DOT, protective fluid, PC, cement, surface casing and conductor pipe, respectively. As shown in Fig. 2, the well structure is different at different depths. Therefore, when calculating the heat transfer coefficient at different depths, the partial terms in Eq. (21) should be added and deleted according to the well structure.

The convective heat transfer coefficient of fluid is related to the fluid's flow state and heat conductivity. Generally, the gas and brine flow states are turbulent during debrining. The convective heat transfer coefficient can be calculated by:

$$h = \frac{k}{R} Nu_d \tag{22}$$

where h is the convective heat transfer coefficient, W/(m² · ℃); k is the heat conductivity of fluid, W/(m · ℃); R is the hydraulic diameter, m; Nu_d is the Nusselt number.

An empirical model is used to calculate the Nusselt number[42], it is written as:

(1) The fluid is heated:
$$Nu_d = 0.023 Re^{0.80} Pr^{0.40} \tag{23a}$$

(2) The fluid is cooled:
$$Nu_d = 0.023 Re^{0.80} Pr^{0.40} \tag{23b}$$

where Pr is the Prandtl number.

3 Model solution and verification

3.1 Numerical solution method

In this paper, the finite difference method is used to solve the model. The detailed discrete methods for all equations in the model are presented in Appendix A. Because the shape of the cavern is irregular and the lengths of the DIT, DOT and PC are different, the cross-sectional area of the annular in the wellbore is the same but in the cavern it changes at different depths. Considering that the length of the wellbore is much greater than the height of the cavern, we only calculate the temperature field of gas, brine and formation in the wellbore. However, the diameter of the cavern is much larger than that of DOT, and the volume of the cavern has a great influence on the gas velocity field. Therefore, the pressure, velocity, and density fields of gas and brine are calculated at the GB interface position. Fig. 4 presents the diagram of spatial mesh division in the wellbore and cavern. The subscripts i and j represent node numbers in vertical and horizontal directions, respectively; the superscript k represents the time step.

In order to solve the model, a program has been developed using Visual Basic. Fig. 5 presents the flow chart of the numerical calculation. The main calculation steps are as follows:

(1) The temperature, pressure, density, and velocity fields of gas, brine, and formation are known at time node k. At the same time, we assume that the position of the GB interface and the salt deposits thickness distribution at time node k have been obtained. When $k = 0$, their parameters are obtained according to the initial conditions.

(2) The gas temperature field is assumed at time node $k+1$.

(3) According to the boundary conditions, the pressure field and velocity field of brine and gas are calculated.

(4) The convective heat transfer coefficient and total heat transfer coefficient are calculated. We calculate the temperature field of gas, brine and formation at time node $k+1$, and compare them with the assumption in step 2. If the simulation error limits cannot reach, we return to step 2.

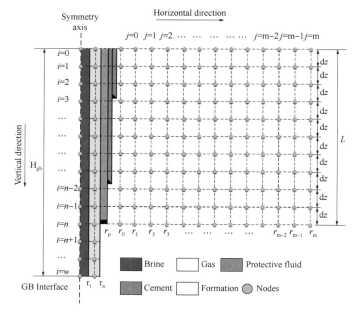

Fig. 4　Mesh of spatial in the wellbore and cavern

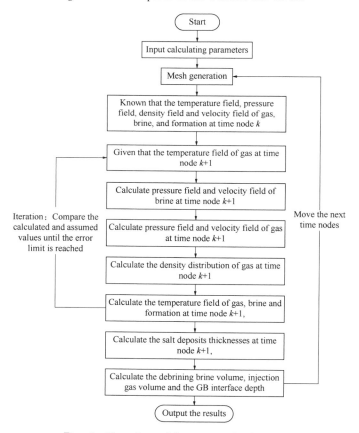

Fig. 5　Flow chart of the numerical calculation

(5) The salt deposits thicknesses are calculated at time node $k+1$ according to the distribution of the brine temperature.

(6) The debrining brine volume, injection gas volume and GB interface depth are calculated.

(7) Following the same sequence, the temperature, pressure, density, velocity and salt deposits thicknesses at all space nodes at different times are obtained.

3.2 Initial conditions and boundary conditions

(1) The initial temperature of the gas, brine and formation at different depths equals the geothermal gradient.

$$T_b(z, t=0) = T_g(z, t=0) = T_e(z, r, t=0) = T_0 + 0.03 * z \tag{24}$$

where T_0 is surface temperature, $T_0 = 20\text{℃}$.

(2) The initial state of gas and brine are motionless. The gas initial density distribution is affected by temperature and pressure. It is calculated by an iterative method. The brine initial pressure is calculated by the self-weight of the brine, written as:

$$p_b(z, t=0) = p_0 + \rho_b g z \tag{25}$$

where p_0 is the pressure caused by the back brine in the debrining tubing outlet, MPa.

(3) The gas injection temperature is equal to the surface temperature, and the brine temperature in the cavern and the formation temperature at the boundary remain unchanged at all time nodes:

$$T_b(z=L, t=k) = T_0 + 0.03 * L \tag{26}$$

$$T_g(z=0, t=k) = T_e(z=0, t=k) = T_0 \tag{27}$$

$$T_e(z, r=M, t=k) = T_0 + 0.03 * z \tag{28}$$

where L is the length of the DIT, m; M is the distance from the PC to the formation boundary, m.

(4) The gas pressure is equal to the brine pressure at the GB interface.

$$p_b(z=H_{gb}) = p_g(z=H_{gb}) \tag{29}$$

(5) When the gas reaches the GB interface, its velocity is equal to the descent velocity of the GB interface, which is related to the brine velocity. According to the geometric relation, the gas velocity at the GB interface can be calculated by the brine velocity in the DIT, written as:

$$v_g(z=H_{gb}) = \frac{v_b(z=H_{gb}) * A_b(z=H_{gb})}{A_g(z=H_{gb})} \tag{30}$$

3.3 Model verification

In section 2, a comprehensive mathematical model of debrining for salt cavern storage has been developed. The model should be validated to determine whether it accurately calculates the debrining parameters. Using the proposed model, the debrining process of the A-1 cavern of Jintan UGS is simulated. The model is validated by comparing with the monitoring and calculated results.

The A-1 cavern is located in Jintan City, Jiangsu Province, China. The A-1 cavern debrining started on July 25, 2022 and was completed on October 14, 2022. The total debrining time was 82 days. The interval between the end of leaching and the beginning of debrining is 748 days. In this long interval, the brine is heated by the formation and reaches saturation. After the completion of leaching, sonar surveys were carried out. Fig. 6 presents the shape of the cavern obtained by the

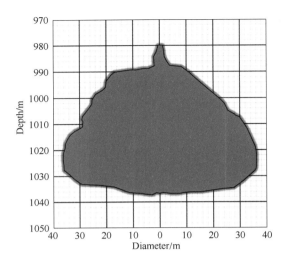

Fig. 6 The shape of the A-1 cavern obtained by sonar survey

sonar survey. The depth of the cavern roof and bottom are 980.0m and 1035.0m, respectively. The outer diameters and wall thicknesses of the DIT, DOT and PC used in the A-1 cavern are 244.5mm × 10.03mm, 177.8mm × 9.19mm and 114.3mm × 6.88mm, respectively. The maximum depth of the DIT is 1032.62 m. The debrining rate ranges from $20m^3/h$ to $120m^3/h$ with an average of $67m^3/h$. The actual time of debrining is 75 days. The continuous debrining and backflushing time daily are 22h and 2h, respectively. The average flushing rate is about $37m^3/h$. The parameters, such as gas injection pressure and GB interface depth, are monitored during debrining. A comparison of the monitored and calculated results is summarized in Table 1. As seen, the calculated results generally agree well with the monitored results.

The salt deposits growth increases the frictional resistance to brine flow. If the debrining rate remains constant, the gas injection pressure will increase with the debrining time. Fig. 7 shows the relation between the calculated and the monitored gas injection pressure on the 56th day of debrining. As seen in the figure, the gas injection pressure monitoring value is close to the calculated value. The gas calculated injection pressure increased from 15.14 MPa to 15.16 MPa, increasing by 0.02 MPa. The monitored gas injection pressure increased from 15.06 MPa to 15.14 MPa, increasing by 0.08 MPa. It demonstrates that the mathematical model has high accuracy.

Fig. 8 presents the relations between the calculated and the monitored debrining brine volume. The calculated value of debrining brine volume is very close to the monitored value. After 75 days of continuous debrining, the calculated and actual total debrining brine volume are 109350 and 109574 m^3, respectively. It also shows that the mathematical model has high accuracy.

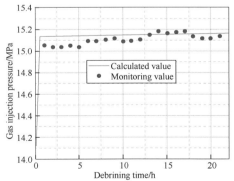

Fig. 7 Calculated and monitored gas injection pressure on the 56th day of debrining

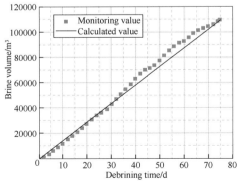

Fig. 8 Calculated and monitored debrining brine volumes

4 Results and discussion

Using the proposed model, the debrining process of the A-1 cavern is simulated under different debrining rates, and the effects of the debrining rate on the temperature field distribution, salt deposits growth, GB interface depth and total debrining time were analyzed. The main calculation parameters are shown in Table 2.

Table 1 Results of model verification

Scenario		This model	Monitored value
Parameters	Depth/m	980	
	Debrining rate/(m^3/h)	67	
	Injection gas temperature/℃	20	
	Time/d	75	
Results and Comparison	Initial injection pressure on the 56th day of debrining/MPa	15.14	15.06
	Ultimate GB interface depth/m	1029.1	1030.0
	Ultimate total debrining brine volume/m^3	109350	109574

Table 2 The main calculation parameters

Parameters	value	Parameters	value
Brine density/(kg/m^3)	1200	Cement heat conductivity coefficient/[W/(m·℃)]	0.65
Brine viscosity/(Pa·s)	0.0014	Coefficient of overall growth rate of salt deposits[40]	6.2×10^{-5} (m/s)$^{-1.25}$
Brine specific heat/[J/(kg·℃)]	3190	Growth order of salt deposits[40]	1.25
Brine heat conductivity coefficient/[W/(m·℃)]	0.6	Debrining tubing heat conductivity coefficient/[W/(m·℃)]	40
Formation density/(kg/m^3)	2600	Injection gas temperature/℃	20
Formation specific heat/[J/(kg·℃)]	2000	Debrining tubing roughness/mm	0.0004
Formation heat conductivity coefficient/[W/(m·℃)]	3.5	Geothermal gradient/(℃/m)	0.03
Salt deposits density/(kg/m^3)	2165	Vertical mesh spacing/m	2
Salt deposits heat conductivity coefficient/[W/(m·℃)]	3.5	Horizontal mesh spacing/m	0.5
Protective fluids heat conductivity coefficient/[W/(m·℃)]	1.2	Time step/s	300

4.1 Temperature distribution

Fig. 9 presents the temperature distribution of brine in the DIT at different times. The curves represent brine temperature at different depths. The variation trends of brine temperature under different debrining rates are similar. The brine temperature decreases with decreasing depth and its minimum at the wellhead. One important finding is that the brine cooling rate is larger in 0~100 m than in 100~980 m. This is mainly because the gas injection temperature is much lower than the brine temperature. When the gas reaches below 100 m, the gas temperature is similar to the brine temperature.

Fig. 10 presents the relationship between the brine temperature at the wellhead and debrining time. The brine temperature gradually increases with debrining time and ultimately approaches a constant value. The brine temperature at the same depth increases slightly with the increase of debrining rate. When the brine temperature approaches a constant value, the brine temperatures at the wellheads with the debrining rate of 20~100 m³/h are 45.1~46.1℃. It is worth noting that when the debrining rate is greater than 80 m³/h, the ultimate brine temperature at the wellhead under different debrining rates are similar.

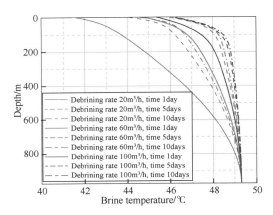

Fig. 9　The temperature distribution of brine in the DIT at different times

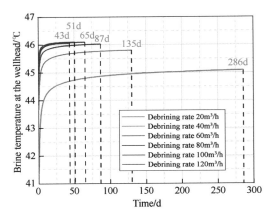

Fig. 10　The relationship between the brine temperature at the wellhead and debrining time

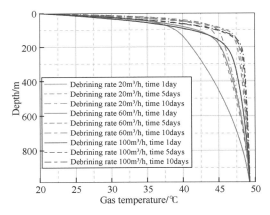

Fig. 11　The gas temperature distribution in the annulus at different times

Fig. 11 presents the gas temperature distribution in the annulus at different times. The variation trends of gas temperature under different debrining rates are also similar. The gas injection temperature is only 20℃. When the gas is injected into the annulus, its temperature increases rapidly in the 0~100 m depth range and slightly in the 100~980 m range. When the gas reaches the cavern, it is almost the same as that of the brine in the cavern. On the hand, the gas density is small, about a tenth of brine density, which causes the gas to be easily heated. On the other hand, the Joule-Thomson effect generated by gas compression during debrining also contributes to the temperature increase. Like the brine temperature variation law, the gas temperature increases gradually with debrining time and ultimately approaches a constant value.

Fig. 12 presents the temperature distribution contour in the formation at different times. In the figure, the horizontal axis represents the distance from the wellbore, and the vertical axis represents the depth. Before the debrining, the formation temperature equals the geothermal gradient. The temperature of the formation near the wellbore rapidly increases because the formation absorbs the heat from the gas. One important finding is that the increment of formation temperature increases in 0~100 m but decreases in 100~980 m with increasing depth. In the horizontal direction, the temperature decreases rapidly with the increase of the distance from the wellbore. This is because the thermal conductivity of the formation is low. After the completion of debrining, the formation temperature change exceeds 0.1 ℃ only within 8 m from the wellbore. This shows that the effect range of debrining on the formation temperature is small.

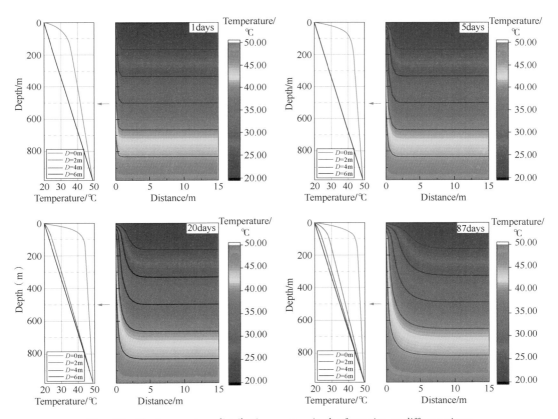

Fig. 12 The temperature distribution contour in the formation at different times

In order to analyze the effects of debrining rate on the formation temperature, Fig. 13 presents the formation temperature distribution after the completion of debrining under different debrining rates conditions. The influence range of debrining operation on formation temperature increases with debrining rate. Suppose the judgment standard of influence range is the change in formation temperature before and after debrining is less than 0.1℃. In that case, the influence range of debrining operation on formation temperature in horizontal direction is 13.5, 10, 8, 7 and 6.5 m for debrining rates of 20, 40, 60, 80 and 100 m³/h, respectively. This means that the effect of debrining on the formation temperature is less than 13.5 m.

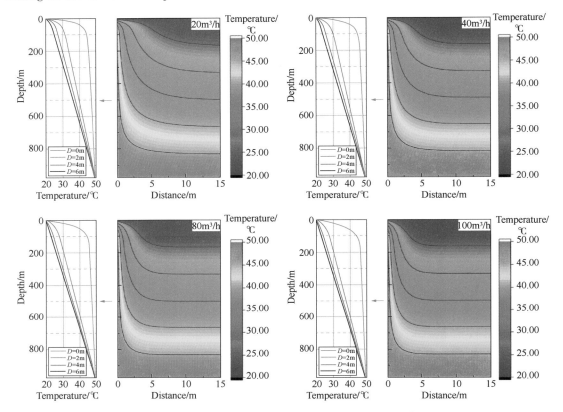

Fig. 13 The temperature distribution contour in the formation at different times

4.2 Salt deposits growth

In order to analyze the growth law of salt deposits on the DIT, Fig. 14 presents the distribution of salt deposits thicknesses on the DIT under different debrining rates. The salt deposits thicknesses gradually decrease with increasing depth. The salt deposits thicknesses reach their maximum at the wellhead. This is because the brine cooling at the wellhead is the largest and the salt deposits growth rate is the fastest. When the debrining rate is 20 m³/h, the salt deposits thicknesses at the wellhead after debrining for 4 days reached 33.7 mm, accounting for 67% of the inner radius of DIT. The greater the debrining rate, the smaller the salt deposits thicknesses at the same depth and debrining time. The salt deposits thicknesses at the wellhead are 13.6, 7.6, and 6.0 mm after debrining for 1 day when the debrining rate is 20, 60, and 100 m³/h, respectively. This indicates that the debrining rate has a significant effect on the salt deposits growth rate.

In order to quantitatively analyze the effects of debrining rate on the salt deposits growth rate

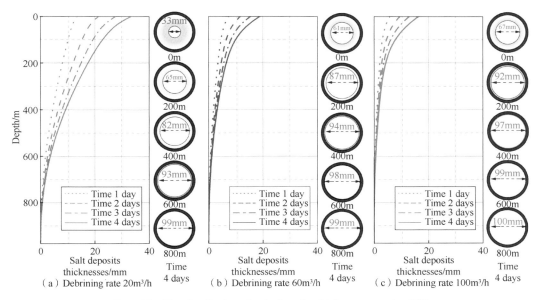

Fig. 14 The distribution of salt deposits thicknesses on the DIT

during debrining, Fig. 15 shows the relationship between salt deposits thicknesses and growth rate at the wellhead and the debrining time under different debrining rates. With the increase of debrining time, the salt deposits thicknesses increase, but the salt deposits growth rate gradually decreases. Increasing the debrining rate can significantly reduce the salt deposits growth rate. When the debrining rate was increased from 20 to 80 m^3/h, the salt deposits growth rate decreased by 53.48%. However, when the debrining rate was increased to 80 m^3/h, continuing to increase the debrining rate had little effect on the salt deposits growth rate. In order to inhibit the salt deposits growth, it is suggested that the debrining rate increased to 80 m^3/h.

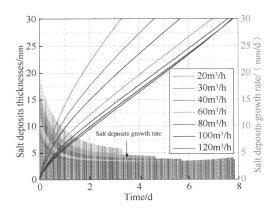

Fig. 15 The relationship between salt deposits thicknesses and growth rate at the wellhead and the debrining time under different debrining rates

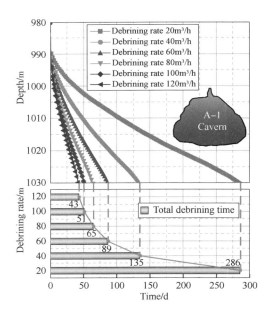

Fig. 16 The relationship between the GB interface depth and debrining time under different debrining rates and the relationship between the total debrining time and debrining rate

4.3 The GB interface depth and total debrining time

The GB interface depth is an important parameter for debrining. When the GB interface position is close to the mouth of DIT, the gas may be brought out by the brine, leading to explosion and fire. Therefore, it is important to accurately predict the GB interface depth to guide the debrining operation. Fig. 16 presents the relationship between the GB interface depth and debrining time under different debrining rates and shows the relationship between the total debrining time and debrining rate. With the increase of debrining time, the GB interface depth gradually decreases. In the initial stage of debrining, the decreasing speed of GB interface depth decreases rapidly. In the middle and late stages of debrining, the decreasing speed tends to become gentle. This is because the diameter of the cavern in the top is small, and in the middle and bottom is large. This indicates that the decreasing rate of GB interface depth depends on the shape of the cavern when the debrining rate is constant. The decreasing rate of GB interface increases and total debrining time with the increase of debrining rate. The debrining rate is increased from 20 to 120 m³/h, the total debrining time can be reduced by 77.3%.

4.4 Method for preventing the DIT blocking

Using fresh water to backflush the DIT after a period of continuous debrining is one of the effective methods for preventing the DIT is blocked by salt deposits. The backflushing frequency is an important parameter for debrining. Frequent backflushing significantly reduces the efficiency of debrining and increases the consumption of freshwater resources. However, if the backflushing frequency is too low may lead to the DIT being blocked. Therefore, a reasonable backflushing frequency is significant to reduce the risk of DIT blockage and increase the efficiency of debrining. Here, T_{30}, T_{50} and T_{70} are defined as the maximum continuous debrining

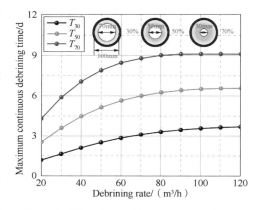

Fig. 17 The relationship between maximum continuous debrining time and debrining rate

time, representing the debrining time when the maximum salt deposits thicknesses reach 30%, 50% and 70% of the inner radius of DIT. We use the maximum continuous debrining time to determine backflushing frequency reasonably. Table 3 shows the maximum continuous debrining time value under different debrining rates. Fig. 17 presents the relationship between maximum continuous debrining time and debrining rate. As seen in the figure, when the debrining rate increased from 20 to 80 m^3/h, T_{30}, T_{50} and T_{70} increased significantly with increasing debrining rate. However, when the debrining rate reaches 80 m^3/h, T_{30}, T_{50} and T_{70} increase slightly with increasing debrining rate. The insoluble sediment at the bottom of the cavern can be sucked into the DIT during debrining[19], which may cause the DIT to be blocked by both the sediment and salt deposits. Therefore, we suggested that the salt deposits thicknesses should not exceed 30% of the inner radius of DIT, and the T_{30} is used as a reference to determine the backflushing frequency. In this case, the debrining rate is increased from 20 to 120 m^3/h, the backflushing frequency can be increased from 1.17 to 3.67 days/time. The actual debrining rate of A-1 cavern is about 70 m^3/h. The actual backflushing frequency of A-1 cavern is about 0.92 days/time, much less than the calculated value of 3.09 days/time. It is suggested that the debrining rate of Jintan UGS salt cavern increase to 80 m^3/h and backflushing frequency decrease to 3.29 days/time to improve the efficiency of debrining rate and save fresh water resources.

Table 3 Maximum continuous debrining time under different debrining rates

Maximum continuous debrining time/d	Debrining rate/(m^3/h)										
	20	30	40	50	60	70	80	90	100	110	120
T_{30}	1.17	1.63	2.09	2.49	2.83	3.09	3.29	3.44	3.55	3.63	3.67
T_{50}	2.54	3.59	4.47	5.13	5.63	5.99	6.24	6.40	6.50	6.54	6.55
T_{70}	4.28	5.87	7.05	7.88	8.44	8.77	9.00	9.08	9.09	9.10	9.10

5 Conclusions

In this paper, a novel mathematical model of debrining for salt cavern is established based on the crystallization kinetics and heat transfer principle. A finite difference program is developed to solve this model using Visual Basic. The A-1 cavern of Jintan UGS is taken as an example, the model is validated by debrining monitored results, and the effects of debrining rate on debrining parameters are analyzed. Through analysis of calculated results, the following are the main conclusions:

(1) The brine temperature decreases with the decrease of depth, and its minimum at the wellhead. On the contrary, the gas temperature increases significantly with increasing depth. The brine and gas temperature increases gradually with debrining time and ultimately approaches a constant value. The formation temperature increases significantly near the wellbore boundary. The influence range of debrining operation on formation temperature in the horizontal direction is less than 13.5 m.

(2) The salt deposits thicknesses decrease with the increase of depth, at the wellhead the salt deposits thickness is maximum. When the debrining rate was increased from 20 to 80 m³/h, the salt deposits growth rate decreased by 53.48%. However, when the debrining rate was increased to 80 m³/h, continuing to increase the debrining rate had little effect on the salt deposits growth rate.

(3) Increasing the debrining rate significantly reduces the total debrining time and backflushing frequency. If the debrining rate is increased from 20 to 80 m³/h, the total debrining time can be reduced by 77.3%, and the backflushing frequency can be decreased from 1.17 to 3.29 days/time. It is suggested that the debrining rate of Jintan UGS salt cavern increase to more than 80 m³/h and backflushing frequency decrease to 3.29 days/time.

Credit author statement

Dongzhou Xie: Methodology, Writing-original draft, Formal Analysis, Visualization. Tongtao Wang: Conceptualization, Resources, Supervision. Long Li: Validation, Data Curation. Kai Guo: Investigation, Formal analysis. Jianhua Ben: Writing – Review & Editing, Visualization. Duocai Wang: Project administration. Guoxing Chai: Investigation, Data Curation.

Declaration of competing interest

The authors declare that they have no known competing financial interests or personal relationships that could have appeared to influence the work reported in this paper.

Data availability

Data will be made available on request.

Acknowledgment

The authors wish to acknowledge the financial support of National Natural Science Foundation of China (Grant No. 42072307), Hubei ProvinceOutstandingYouth Fund(2021CFA095), andStrategicPriority Research Program of the Chinese Academy of Sciences (Grant No. XDC10020300). Thanks to Professor Jaak J. K. Daemen of Mackay School of Earth Sciences and Engineering, University of Nevada (Reno), USA, for his constructive suggestions and modifcations.

Appendix A. Equation discretization

(1) Gas flow and heat transfer

Gas state equation:

$$\rho_{g,i}^{k} = \frac{M p_{g,i}^{k}}{Z(p_{g,i}^{k}, T_{g,i}^{k}) R T_{g,i}^{k}} \quad (i = 0 \sim w) \tag{A.1}$$

Continuity equation:

$$\frac{A_{g,i}^{k} \rho_{g,i}^{k} - A_{g,i}^{k-1} \rho_{g,i}^{k-1}}{\Delta t} + \frac{A_{g,i}^{k} \rho_{g,i}^{k} v_{g,i}^{k} - A_{g,i-1}^{k} \rho_{g,i-1}^{k} v_{g,i-1}^{k}}{\Delta z} = 0 \quad (i = 1 \sim w) \tag{A.2}$$

Momentum conservation equation:

$$\frac{A_{g,i}{}^k \rho_{g,i}{}^k v_{g,i}{}^{k-1} - A_{g,i}{}^{k-1} \rho_{g,i}{}^{k-1} v_{g,i}{}^{k-1}}{\Delta t} + \frac{A_{g,i}{}^k \rho_{g,i}{}^k (v_{a,i}{}^k)^2 - A_{g,i-1}{}^k \rho_{g,i-1}{}^k (v_{g,i-1}{}^k)^2}{\Delta z}$$

$$= -\frac{A_{g,i}{}^k p_{g,i}{}^k - A_{g,i-1}{}^k p_{g,i-1}{}^k}{\Delta z} + A_{g,i-1}{}^k \rho_{g,i-1}{}^k (i = 1 \sim w) \qquad (A.3)$$

The momentum conservation equation is used to calculate the gas velocity field. The gas velocity has little effect on gas pressure because the gas density is small. To simplify the calculation, the effects of gas velocity on gas pressure are ignored. The momentum equation can be simplified as:

$$A_{g,i}{}^k p_{g,i}{}^k = A_{g,i-1}{}^k p_{g,i-1}{}^k + A_{g,i-1}{}^k g \rho_{g,i-1}{}^g \Delta z (i = 1 \sim w) \qquad (A.4)$$

Energy conservation equation:

$$\frac{A_{g,i}{}^k \rho_{g,i}{}^k c_{g,i}{}^k T_{g,i}{}^k - A_{g,i}{}^{k-1} \rho_{g,i}{}^{k-1} c_{g,i}{}^{k-1} T_{g,i}{}^{k-1}}{\Delta t}$$

$$\frac{A_{g,i}{}^k \rho_{g,i}{}^k c_{g,i}{}^k \beta_{g,i}{}^k p_{g,i}{}^k - A_{g,i}{}^{k-1} \rho_{g,i}{}^{k-1} c_{g,i}{}^{k-1} \beta_{g,i}{}^{k-1} p_{g,i}{}^{k-1}}{\Delta t} -$$

$$\frac{A_{g,i}{}^k p_{g,i}{}^k - A_{g,i}{}^{k-1} p_{g,i}{}^{k-1}}{\Delta t} + \frac{A_{g,i}{}^k \rho_{g,i}{}^k (v_{g,i}{}^k)^2/2 - A_{g,i}{}^{k-1} \rho_{g,i}{}^{k-1} (v_{g,i}{}^{k-1})^2/2}{\Delta t} +$$

$$\frac{A_{g,i}{}^k \rho_{g,i}{}^k v_{g,i}{}^k c_{g,i}{}^k T_{g,i}{}^k - A_{g,i-1}{}^k \rho_{g,i-1}{}^k v_{g,i-1}{}^k c_{g,i-1}{}^k T_{g,i-1}{}^k}{\Delta z} -$$

$$\frac{A_{g,i}{}^k \rho_{g,i}{}^k v_{g,i}{}^k c_{g,i}{}^k \beta_{g,i}{}^k p_{g,i}{}^k - A_{g,i-1}{}^k \rho_{g,i-1}{}^k v_{g,i-1}{}^k c_{g,i-1}{}^k \beta_{g,i-1}{}^k p_{g,i-1}{}^k}{\Delta z} +$$

$$\frac{A_{g,i}{}^k \rho_{g,i}{}^k (v_{g,i}{}^k)^3/2 - A_{g,i-1}{}^k \rho_{g,i-1}{}^k (v_{g,i-1}{}^k)^3/2}{\Delta z}$$

$$= -A_{g,i}{}^k \rho_{g,i}{}^k v_{g,i}{}^k g + 2\pi r_{\text{tin},i}{}^k U_{bg,i}{}^k (T_{b,i}{}^k - T_{g,i}{}^k) + 2\pi r_{ai,i}{}^k U_{ge,i}{}^k (T_{e,i,0}{}^k - T_{g,i}{}^k)(i = 1 \sim n) \quad (A.5)$$

(2) Brine flow and heat transfer:

$$\frac{\rho_b (A_{b,i}{}^k - A_{b,i}{}^{k-1})}{\Delta t} + \frac{\rho_b (A_{b,i}{}^k v_{b,i}{}^k - A_{b,i-1}{}^k v_{b,i-1}{}^k)}{\Delta z} = 0 (i = 1 \sim w) \qquad (A.6)$$

$$\rho_b \frac{A_{b,i}{}^k v_{b,i}{}^k - A_{b,i}{}^{k-1} v_{b,i}{}^{k-1}}{\Delta t} + \rho_t \frac{A_{b,i}{}^k (v_{b,i}{}^k)^2 - A_{b,i-1}{}^k (v_{b,i-1}{}^k)^2}{\Delta z} = -\frac{A_{b,i}{}^k p_{b,i}{}^k - A_{b,i-1}{}^k p_{b,i-1}{}^k}{\Delta z} +$$

$$\rho_b A_{b,i}{}^k g + \rho_b \frac{A_{b,i}{}^k \lambda_{b,i}{}^k (v_{b,i}{}^k)^2}{4 r_{\text{tin},i}{}^k} = (i = 1 \sim w) \qquad (A.7)$$

$$\frac{A_{b,i}{}^k \rho_b c_b T_{b,i}{}^k - A_{b,i}{}^{k-1} \rho_b c_b T_{b,i}{}^{k-1}}{\Delta t} - \frac{A_{b,i}{}^k \rho_b v_{b,i}{}^k c_b T_{b,i}{}^k - A_{b,i-1}{}^k \rho_b v_{b,i-1}{}^k c_b T_{b,i-1}{}^k}{\Delta z}$$

$$= 2\pi r_{\text{tin},i}{}^k U_{bg,i}{}^k (T_{g,i}{}^k - T_{b,i}{}^k) + \frac{A_{b,i}{}^k \rho \lambda_{b,i}{}^k (v_{b,i}{}^k)^3}{4 r_{\text{tin},i}{}^k} = (i = 1 \sim n) \qquad (A.8)$$

(3) Formation heat transfer.

On the border between the wellbore and formation:

$$\rho_e c_e (T_{e,i,0}{}^k - T_{e,i,0}{}^{k-1}) 2\pi r_{e,i,0} \Delta z (r_{e,i,1} - r_{e,i,0}) = 2\pi r_{ai}{}^k U_{ge,i}{}^k (T_{g,i}{}^k - T_{e,i,0}{}^k)$$

$$\Delta z \Delta t - k_e \frac{T_{e,i,0}{}^k - T_{e,i,1}{}^k}{r_{e,i,1} - r_{e,i,0}} r_{e,i,1} 2\pi \Delta z \Delta t \, (i=1 \sim n) \tag{A.9}$$

$$(f_1)_0 T_{e,i,0}{}^k + (f_2)_0 T_{e,i,1}{}^k = d_i \tag{A.10}$$

$$(f_1)_0 = \rho_e c_e r_{e,i,0}(r_{e,i,1} - r_{e,i,0}) + r_{ai}{}^k U_{ge,i}{}^k \Delta t + \frac{k_e r_{e,i,1} \Delta t}{r_{e,i,1} - r_{e,i,0}} \tag{A.11a}$$

$$(f_2)_0 = \frac{k_e r_{e,i,1} \Delta t}{r_{e,i,1} - r_{e,i,0}} \tag{A.11b}$$

$$d_i = \rho_e c_e r_{e,i,0}(r_{e,i,1} - r_{e,i,0}) T_{e,i,0}{}^{k-1} + r_{ai}{}^k U_{ge,i}{}^k T_{g,i}{}^k \Delta t \tag{A.11c}$$

in the formation:

$$\rho_c c_e \frac{T_{e,i,j}^k - T_{e,i,j}^{k-1}}{\Delta t} = k_e \left(\frac{1}{r_j} \frac{T_{e,i,j+1}^k - T_{e,i,j}^k}{r_{j+1} - r_j} + \frac{T_{e,i,j+1}^k - 2T_{e,i,j}^k + T_{e,i,j-1}^k}{(r_{j+1} - r_j)(r_j - r_{j-1})} \right) (i=1 \sim n, \; j=1 \sim m) \tag{A.12}$$

$$(f_0)_j T_{e,i,j-1}^k + (f_1)_j T_{e,i,j}^k + (f_2)_j T_{e,i,j+1}^k = T_{e,i,j}^{k-1} \tag{A.13}$$

$$(f_0)_j = -\frac{k_e \Delta t}{\rho_e c_e (r_{j+1} - r_j)(r_j - r_{j-1})} \tag{A.14a}$$

$$(f_1)_j = \frac{k_e \Delta t}{\rho_e c_e r_j (r_{j+1} - r_j)} + \frac{2 k_e \Delta t}{\rho_e c_e (r_{j+1} - r_j)(r_j - r_{j-1})} \tag{A.14b}$$

$$(f_2)_j = -\frac{k_e \Delta t}{\rho_e c_e (r_{j+1} - r_j)(r_j - r_{j-1})} - \frac{2 k_e \Delta t}{\rho_e c_e (r_{j+1} - r_j)(r_j - r_{j-1})} + 1 \tag{A.14c}$$

(4) Salt deposits growth:

$$G_i^k = \begin{cases} 0 & c < c_{s,i}{}^k \\ \dfrac{K^g [c - (c_{s,i}{}^k)^2]^d}{\rho_s} & c > c_{s,i}{}^k \end{cases} \tag{A.15}$$

$$c_{s,i}{}^k = 357.12 + 0.085 T_{b,i}^k + 0.00317 (T_{b,i}^k)^2 \tag{A.16}$$

$$e_i^k = G_i^k \cdot \Delta t \tag{A.17}$$

(5) GB interface depth:

$$H_{gb}{}^{k+1} = H_{gb}{}^k + \frac{v_{b,0}{}^k A_{b,0}{}^k \Delta t}{A_{g,w}^k} \tag{A.18}$$

References

[1] British petroleum. In. Bp Yearbook of world energy Statistics. 71st ed. 2022.

[2] Zhou H, Meng W, Wang D, Li G, Li H, Liu Z, et al. A novel coal chemical looping gasification scheme for synthetic natural gas with low energy consumption for CO_2 capture: modelling, parameters optimization, and performance analysis. Energy 2021; 225. https://doi.org/10.1016/j.energy.2021.120249.

[3] Al-Khori K, Bicer Y, Koç M. Comparative techno-economic assessment of integrated PV-SOFC and PV-Battery hybrid system for natural gas processing plants. Energy 2021; 222. https://doi.org/10.1016/j.energy.2021.119923.

[4] Kawakubo S, Ikaga T, Murakami S. Survey research on foreign urban assessment tool aimed at realization of sustainable cities. AIJ J Technol Des 2010; 16. https://doi.org/10.3130/aijt.16.601.

[5] Butt AA, Aslam HMU, Shabir H, Javed M, Hussain S, Nadeem S, et al. Climatic events and natural disasters

of 21st century: a perspective of Pakistan. Int J Econ Environ Geol 2020; 11. https://doi.org/10.46660/ijeeg.vol11.iss2.2020.445.

[6] Ghaemi Asl M, Rajabi S, Irfan M, Ranjbaran R, Doudkanlou MG. COVID-19 restrictions and greenhouse gas savings in selected Islamic and MENA countries: an environmental input–output approach for climate policies. Environ Dev Sustain 2022; 24. https://doi.org/10.1007/s10668-021-02018-3.

[7] Hamed TA, Alshare A. Environmental impact of solar and wind energy-A Review. J Sustain Dev Energy, Water Environ Syst 2022; 10. https://doi.org/10.13044/j.sdewes.d9.0387.

[8] Bjertnæs GHM. Efficient combination of Taxes on fuel and vehicles. Energy J 2019; 40. https://doi.org/10.5547/01956574.40.si1.gbje.

[9] Liao Y, Wang Z, Sun X, Lou W, Liu H, Sun B. Modeling of hydrate dissociation surface area in porous media considering arrangements of sand grains and morphologies of hydrates. Chem Eng J 2022; 433. https://doi.org/10.1016/j.cej.2021.133830.

[10] Wang T, Ao L, Wang B, Ding S, Wang K, Yao F, et al. Tightness of an underground energy storage salt cavern with adverse geological conditions. Energy 2022; 238. https://doi.org/10.1016/j.energy.2021.121906.

[11] Mikolajková M, Saxén H, Pettersson F. Linearization of an MINLP model and its application to gas distribution optimization. Energy 2018; 146. https://doi.org/10.1016/j.energy.2017.05.185.

[12] Lord AS, Kobos PH, Borns DJ. Geologic storage of hydrogen: scaling up to meet city transportation demands. Int J Hydrogen Energy 2014; 39. https://doi.org/10.1016/j.ijhydene.2014.07.121.

[13] Martínez Sánchez AM, Saldarriaga Cortés CA, Salazar H. An optimal coordination of seasonal energy storages: a holistic approach to ensure energy adequacy and cost efficiency. Appl Energy 2021; 290. https://doi.org/10.1016/j.apenergy.2021.116708.

[14] Zhang N, Yang C, Shi X, Wang T, Yin H, Daemen JJK. Analysis of mechanical and permeability properties of mudstone interlayers around a strategic petroleum reserve cavern in bedded rock salt. Int J Rock Mech Min Sci 2018; 112. https://doi.org/10.1016/j.ijrmms.2018.10.014.

[15] Liu W, Zhang Z, Chen J, Jiang D, Wu F, Fan J, et al. Feasibility evaluation of largescale underground hydrogen storage in bedded salt rocks of China: a case study in Jiangsu province. Energy 2020; 198. https://doi.org/10.1016/j.energy.2020.117348.

[16] Jin X, Xia Y, Yuan G, Zhuang X, Ban F, Dong A. An experimental study on the influencing factors of salt crystal in brine discharge strings of a salt-cavern underground gas storage (UGS). Nat Gas Ind 2017; 37: 130-4. https://doi.org/10.3787/j.issn.1000-0976.2017.04.016.

[17] Li J, Shi X, Yang C, Li Y, Wang T, Ma H. Mathematical model of salt cavern leaching for gas storage in high-insoluble salt formations. Sci Rep 2018; 8. https://doi.org/10.1038/s41598-017-18546-w.

[18] Wang T, Ding S, Wang H, Yang C, Shi X, Ma H, et al. Mathematic modelling of the debrining for a salt cavern gas storage. J Nat Gas Sci Eng 2018; 50. https://doi.org/10.1016/j.jngse.2017.12.006.

[19] Wang T, Chai G, Cen X, Yang J, Daemen JJK. Safe distance between debrining tubing inlet and sediment in a gas storage salt cavern. J Pet Sci Eng 2021; 196. https://doi.org/10.1016/j.petrol.2020.107707.

[20] Zhengkai Y, Weihua X, Xiaoliang H. Crystallization prevention measures and understanding of water soluble salt mining well production. CHINA WELL ROCK SALT 2015; 46: 29-31.

[21] Liu Y. Brief discussion on the cause and prevention of the pipe plugging caused by salt crystal and sand in docking connectivity ming. CHINA WELL ROCK SALT 2011; 42: 19-21.

[22] Zhou G, Cao S, Hu W. Reconsider ation on the laws to eliminate and Pr event plugging by salt crytallization in continental facies extr adinary deep salt well. CHINA WELL ROCK SALT 2007; 46.26-24.

[23] Liao Y, Zheng J, Wang Z, Sun B, Sun X, Linga P. Modeling and characterizing the thermal and kinetic behavior of methane hydrate dissociation in sandy porous media. Appl Energy 2022: 312. https://doi.org/10.1016/j.apenergy.2022.118804.

[24] Zhuang Q. Application of debrining technology in salt cavern construction. Oil-Gas F Surf Eng 2019; 29: 65-6.

[25] Wang C, Ba J, Guo K, Jiao Y. Treatment method of complicated situation of tubing for debrining of salt cavern gas

storage. Petrochemical Ind Appl 2015；34：10-2. https：//doi. org/10. 3969/j. issn. 1673-5285. 2015. 10. 014.

［26］Yuan J ping, sheng Li G, qian Zhuang X, wen Yu J, Du Y. Gas-injection production technology using underground gas-storage salt cavern to displace brine solution and contain injected gas. Nat Gas Ind 2009；29：76-9. https：//doi. org/10. 3787/j. issn. 1000-0976. 2009. 02. 020.

［27］Wang T, Yang C, Wang H, Ding S, Daemen JJK. Debrining prediction of a salt cavern used for compressed air energy storage. Energy 2018；147. https：//doi. org/ 10. 1016/j. energy. 2018. 01. 071.

［28］Li P, Li Y, Shi X, Xie D, Ma H, Yang C, et al. Experimental and theoretical research on the debrining process in sediments for a gas storage salt cavern. Geoenergy Sci Eng 2023；225. https：//doi. org/10. 1016/j. geoen. 2023. 211667.

［29］Ren Z, Li J, Zhu J, Wang H, Wang X. Numerical simulation on technological parameters of gas injection and brine discharge in salt－cavern gas storage. Oil Gas Storage Transp 2018；37：403-6. https：//doi. org/10. 6047/j. issn. 1000-8241. 2018. 04. 007.

［30］Ren Z, Ba J, Ren Z, Zhang X, Zhang X, Guilin W. Experimental study on gas injection and brine ejection of deposits in rock salt gas storages. Oil-Gas F Surf Eng 2018；37：50-3. https：//doi. org/10. 3969/j. issn. 1006-6896. 2018. 11. 012.

［31］Chen F, Yang HJ, Yang CH. Analysis of residual brine of salt rock gas storage during injecting gas to eject brine. Yantu Lixue/Rock Soil Mech 2009；30.

［32］Liu J, Wang Y, Xie K, Liu Y. Gas injection and brine discharge in rock salt gas storage studied via numerical simulation. PLoS One 2018；13. https：//doi. org/ 10. 1371/journal. pone. 0207058.

［33］Wei X, Liu Y, Shi X, Li Y, Ma H, Hou B, et al. Experimental research on brine crystallization mechanism in solution mining for salt cavern energy storage. J Energy Storage 2022；55. https：//doi. org/ 10. 1016/j. est. 2022. 105863.

［34］Xie D, Wang T, Li L, He T, Chai G, Wang D, et al. Temperature distribution of brine and gas in the tubing during debrining of a salt cavern gas storage. J Energy Storage 2022；50. https：//doi. org/ 10. 1016/j. est. 2022. 104236.

［35］Xie D, Wang T, Ben J, He T, Chai G, Wang D. Mathematic modeling of the salt deposits growing on the tubing during debrining for gas storage salt cavern. J Energy Storage 2022；55：105754. https：//doi. org/ 10. 1016/j. est. 2022. 105754.

［36］Sun X, Liao Y, Wang Z, Sun B. Geothermal exploitation by circulating supercritical CO_2 in a closed horizontal wellbore. Fuel 2019；254. https：//doi. org/10. 1016/j. fuel. 2019. 05. 149.

［37］Ав Черникин. Generalized equation to calculate the coefficient of hydraulic friction of pipeline. Oil Gas Storage Transp 1999；18：26-8.

［38］Bohnet M. Fouling of heat transfer surfaces. Chem Eng Technol 1987；10. https：// doi. org/10. 1002/ceat. 270100115.

［39］Pääkkönen TM, Riihimäki M, Simonson CJ, Muurinen E, Keiski RL. Modeling CaCO3 crystallization fouling on a heat exchanger surface – definition of fouling layer properties and model parameters. Int J Heat Mass Tran 2015；83. https：//doi. org/10. 1016/j. ijheatmasstransfer. 2014. 11. 073.

［40］Al-Jibbouri S, Ulrich J. The growth and dissolution of sodium chloride in a fluidized bed crystallizer. J Cryst Growth 2002；234. https：//doi. org/10. 1016/ S0022-0248(01)01656-6.

［41］Xie D, Wang T, Ben J, He T, Chai G, Wang D. Mathematic modeling of the salt deposits growing on the tubing during debrining for gas storage salt cavern. J Energy Storage 2022；55：105754. https：//doi. org/ 10. 1016/j. est. 2022. 105754.

［42］Dittus FW, Boelter LMK. Heat transfer in automobile radiators of the tubular type. Int Commun Heat Mass Tran 1985；12. https：//doi. org/10. 1016/0735-1933(85) 90003-X.

Temperature distribution of brine and gas in the tubing during debrining of a salt cavern gas storage

Dongzhou Xie[1,2]　Tongtao Wang[1,3]　Long Li[4]　Tao He[1,2]　Guoxing Chai[5]
Duocai Wang[6]　Hong Zhang[6]　Tieliang Ma[6]　Xin Zhang[7]

(1. State Key Laboratory of Geomechanics and Geotechnical Engineering, Institute of Rock and Soil Mechanics, Chinese Academy of Sciences; 2. University of Chinese Academy of Sciences; 3. Hubei Key Laboratory of Geo-Environmental Engineering, Institute of Rock and Soil Mechanics, Chinese Academy of Sciences; 4. Yinchuan Branch Office of WEGPC; 5. SINOPEC Petroleum Exploration and Production Research Institute; 6. PipeChina West East Gas Pipeline Company; 7. PipeChina West East Gas Pipeline Company Jiangsu Gas Storage Branch Company)

Abstract　During the debrining of an underground gas storage salt cavern (UGS), the decrease of brine temperature may cause the debrining inner tubing (DIT) to be blocked by salt crystal separating from brine. In this paper, a mathematical model used to calculate the temperature distribution of brine and gas in the tubing during the debrining is built based on the theories of heat transfer. A finite element iterative method is used to solve the mathematical model. A-1 cavern of Jintan UGS is taken as an example. The temperature distribution of brine and gas is calculated under different DIT sizes and debrining rates based on the mathematical model. The results show that the gas temperature increases rapidly in the 0~-30m and slowly in -30 ~ -925 m. Increasing DIT size and debrining rate have no significant effect on the gas temperature distribution. The brine temperature decreases non linearly with decreasing depth. The brine temperature increases with the increase of debrining rate and decrease of DIT size. Brine temperature falls slowly over the range from -925 m to -600 m, and significantly above -600 m. Increasing debrining rate and decreasing thermal conductivity of tubing can decrease the drop of brine temperature. Increasing gas injection temperature has a minor effect on the brine temperature distribution. In order to prevent the DIT blocking by salt crystal, we proposed that use a DIT with low thermal conductivity to debrine. The accuracy and reliability of the model are verified by comparing the calculated values of brine temperature at the wellhead with field measured values for an actual cavern debrining. This study provides a theoretical basis for evaluating the temperature distribution of brine and predicting salt crystal growth during debrining.

Keywords　salt cavern gas storage; debrining; mathematical model; temperature distribution; salt crystal

1 Introduction

Salt cavern underground gas storage (UGS), one of the main ways of energy storage, is widely

---

Abbreviations: DIT, debrining inner tubing; DOT, debrining outer tubing; GFRP, glass fiber reinforced plastics; PC, production casing; UGS, underground gas storage salt cavern.

Corresponding author: Tongtao Wang, ttwang@ whrsm. ac. cn.

used for energy storage, including petroleum [1], natural gas [2], hydrogen [3] and compressed air [4]. The energy consumption increases year by year in China, especially the natural gas consumption is growing rapidly, with an average annual growth rate of 14.1% in the past 20 years. In 2020, the natural gas consumption reached 320 billion m³ in China [5]. However, the natural gas storage total working capacity accounted for about 5% of total natural gas consumption of 2017, which means that the natural gas storage capacity is seriously inadequate [6]. Therefore, China is constructing UGS quickly to increase the capacity of natural gas seasonal peak regulation and emergency support [2,7,8]. At present, China mainly uses the single well oil pad method to construct UGS because this method is relatively mature and can deliver a cavern with ideal shape [9]. But, the construction speed of this method is slow. Especially in the stage of debrining, the construction speed will be slowed down greatly if the debrining inner tubing (DIT) becomes blocked by salt crystal [10].

Nomenclatures			
A_1	heat transfer area of a brine element (m²)	R	hydraulic diameter (m)
A_2	heat transfer area of a gas element (m²)	r_1	inside radius of DIT (m)
c	equilibrium concentration of the brine (g/100g H_2O)	r_2	outer radius of DIT (m)
c^*	actual concentration of the brine (g/100g H_2O)	r_3	inside radius of DOT (m)
Δc	supersaturation (g/100g H_2O)	r_4	outer radius of DOT (m)
c_b	brine specific heat [J/(kg·°C)]	T_b	brine temperature (°C)
c_g	gas specific heat [J/(kg·°C)]	T_g	gas temperature (°C)
c_p	specific heat at constant pressure of fluids [J/(kg·°C)]	T_p	protective fluid temperature (°C)
dq_b	transferred heat of a brine element per unit time (J/s)	t_b	brine temperature in the carven (°C)
dQ_b	heat transferred from a brine element (J)	t_g	gas injection temperature (°C)
$dQ_{b'}$	change in intrinsic energy of a brine element (J)	U_1	total heat transfer coefficient between brine and gas [W/(m·°C)]
dQ_g	heat transferred from brine to a gas element (J)	U_2	total heat transfer coefficient between gas and protective fluid [W/(m·°C)]
$dQ_{g'}$	change in intrinsic energy of a gas element (J)	v	flow velocity of the fluids (m/s)
dQ_p	heat transferred from the protective fluid to a gas element (J)	v_b	flow velocity of the brine (m/s)
G	crystal growth rate (mm/h)	v_g	flow velocity of the gas (m/s)
g	order of crystal growth	β	absolute error (°C)
K_g	crystal growth coefficient	λ	thermal conductivity of the fluids [W/(m·°C)]
h	length of each part (m)	μ	hydrodynamic viscosity of the fluids (Pa·s)
h_1	convective heat transfer coefficient from brine to inner wall of DIT [W/(m²·°C)]	ρ	density of the fluids (kg/m³)
h_2	convective heat transfer coefficient from outer wall of DIT to gas [W/(m²·°C)]	ρ_b	density of the brine (kg/m³)
h_3	convective heat transfer coefficient from gas to inner wall of DOT [W/(m²·°C)]	ρ_g	density of the gas (kg/m³)
k_1	thermal conductivity of DIT [W/(m·°C)]	Nu_d	Nusselt number
k_2	thermal conductivity of DOT [W/(m·°C)]	Re_d	Reynolds number
k_3	thermal conductivity of cement [W/(m·°C)]	Pr	Prandtl number
L	length of tubing (m)		

Debrining is one of the critical steps in the construction of UGS. Fig. 1 shows a schematic of the UGS debrining, where the yellow, blue and red represents the gas, brine and protective fluid, respectively. The primary steps of debrining consist of: the gas is injected into the annulus between debrining outer tubing (DOT) and DIT from the surface, which drives the brine into DIT from cavern to surface[8]. When most of the brine in the cavern is discharged, an effective gas storage space will be formed. The brine temperature in cavern is close to the formation temperature [11,12], but the gas temperature is roughly equal to the air temperature at ground level. Because the brine temperature is larger than the geothermal temperature and the gas temperature, the brine will transfer heat to the gas during the debrining. The brine temperature will gradually decrease. Meanwhile, the gas temperature will increase by absorbing heat from the brine. The concentration of brine nearly

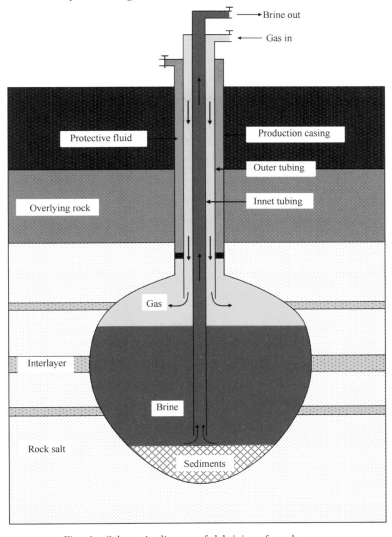

Fig. 1 Schematic diagram of debrining of a salt cavern

reaches saturation during debrining, and the brine will become supersaturated if its temperature falls [13]. The supersaturated brine may precipitate salt crystal under the induction of tiny insoluble particles [13]. The salt crystal reduces the debrining rate and increases the power and energy consumption of the compressor. Under some extreme conditions, the DIT is completely blocked by the salt crystal, and then the debrining is stopped. The main reason for salt crystal is the brine temperature decrease. Thus, it is important to study the temperature distribution of brine and gas in the tubing during debrining for preventing the salt crystal blocking the DIT.

The UGS is an important way of energy storage. Many scholars have carried out research related to debrining technologies and heat transfer problems. Chen et al. [14] built a mathematical model to predict the cavern temperature in the stage of gas injection and production and used FLAC^{3D} to calculate the changes of temperature of gas and cavern walls. They obtained the temperature variation in cavern walls. Bérest et al. [15] analyzed the temperature distribution of cavern walls during the gas injection and production and used a numerical simulation to research the impact of temperature change on the cavern walls. He believed that thermal stress could cause tensile cracks in cavity walls, but that the depth of these cracks is limited. Khaledi et al. [16] established a mathematical model to predict the temperature and pressure of air during gas injection and production for a compressed air energy storage, and used finite element software to simulate the mechanical behavior of salt rock, and analyzed the stability of a cavern under the two ultimate states of low pressure and high pressure. Wang et al. [8] built a mathematical model for calculating the debrining parameters, and deduced the equations for calculating the debrining time, gas injection pressure, daily gas injection volume and cumulative gas injection volume. In addition, they developed a calculation program of debrining parameters. Jin et al. [13] carried out a series of experiments regarding the salt crystal blocking the DIT, and analyzed the reasons for brine precipitates salt crystal. Liu [17] used Fluent software to simulate the wellbore temperature field in leaching stages for UGS and analyzed the effects of debrining rate, temperature and material of DIT on the wellbore temperature distribution. He proposed some measures to prevent salt crystal. Li et al. [18] based on the thermodynamic principle, proposed a temperature calculation model of gas injection and production and carried out numerical simulation of the fully coupled thermal−mechanical model of a salt cavern under different gas injection and production rates. They believed that the rapid gas recovery would increases the tensile stress, and a tensile stress zone may appear at the top and bottom of the cavern. Bérest [11] discussed the thermodynamic behavior of salt caverns in which is stored brine, petroleum, natural gas and hydrogen, respectively. He believed that the temperature of the liquid in cavern changes slowly in a static state, and the temperature change is small during rapid injection and production. Li et al. [19] proposed a method for discharge the brine from sediments in a salt cavern. The main principle is constructing a new cavern to connect with the original cavern at its bottom, and injecting the gas into the new cavern and displacing brine from original cavern. According to above literature reviews, we can conclude that the debrining is a key step for construction of UGS and the temperature distribution of brine during debrining has an important effect on salt crystal. There are few studies on the temperature distribution of brine and gas during debrining for UGS.

The main objective of this paper is to build a mathematical model to calculate the temperature distribution of brine and gas during the debrining for a UGS. The contents of this paper are: (1) a mathematical model is established to predict the temperature distribution of brine and gas based on the heat transfer of fluid in the tubing, and a finite element iterative method for solving the mathematical model is proposed; (2) taking A-1 cavern of Jintan UGS as example, the temperature distribution of brine and gas is calculated under different DIT sizes and debrining rates; (3) the effects of three factors on the brine temperature are analyzed, including the debrining rate, gas injection temperature and thermal conductivity of tubing. (4) the accuracy and reliability of the mathematical model are verified by comparing calculated results with field measured values of an actual cavern; (5) the method of salt crystal growth calculated is discussed. This study can provide a theoretical foundation for evaluating the temperature distribution of brine and gas and predicting salt crystal growth during debrining.

2 Mathematical model

A set of debrining tubing systems consisting of three coaxial tubings with different diameters is used to debrine. The structure of the tubing system and a schematic diagram of fluids motion and heat transfer in the tubing during debrining is shown in Fig. 2. The three coaxial tubing from inside to outside are respectively debrining inner tubing, debrining outer tubing and production casing (PC). Moreover, the salt crystal layer may form when the ions in brine deposit on the walls of DIT. The DIT is the channel in which the brine moves. The annulus between DIT and DOT is the channel for gas injection. The annulus between DOT and PC is filled with protective fluid that remains stationary during debrining. During debrining the temperature of brine and gas in the tubing constantly changes because the brine, gas and protective fluid are affected by heat conduction and heat convection. In order to accurately depict the temperature distribution of brine and gas in the tubing, a mathematical model is established in this section.

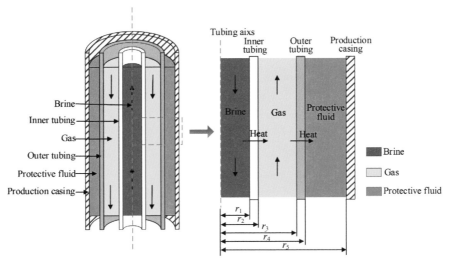

Fig. 2 Schematic diagram of fluids motion and heat transfer in the tubing during debrining

2.1 Brine temperature distribution

The brine in the tubing transfers heat to gas mainly by conduction and convection during debrining. The wellhead is taken as the coordinate origin, and the vertical downward is positive. The brine element of length dz is selected from the DIT to analyze. Fig. 3a presents a schematic diagram of the heat transfer of a brine element. The temperature difference between brine and gas in the radial direction is larger than in the axial direction. Thus, only the radial heat transfer is considered and the axial heat transfer is ignored. According to the theories of heat transfer, the transferred heat of a brine element per unit time is written as[20]:

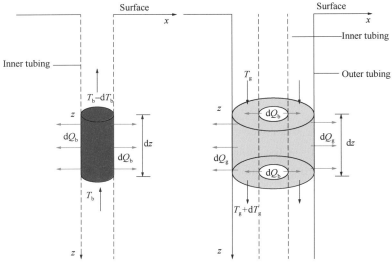

(a) Heat transfer of a brine element in DIT (b) Heat transfer of a gas element in the annulus between DIT and DOT

Fig. 3   Schematic diagram of heat transfer of brine and gas element

$$dq_b = -U_1 A_1 \Delta T = -U_1 A_1 (T_b - T_g) \tag{1}$$

$$A_1 = 2\pi r_1 dz \tag{2}$$

where dq_b is the transferred heat of a brine element per unit time, J/s; U_1 is the total heat transfer coefficient between brine and gas, W/(m·℃); A_1 is heat transfer area of a brine element, m²; T_b is brine temperature, ℃; T_g is gas temperature, ℃; r_1 is the inside radius of DIT, m.

The duration of a brine element flow through length dz of DIT is written as:

$$dt = \frac{dz}{v_b} \tag{3}$$

where v_b is the flow rate of brine, m/s.

The transferred heat of a brine element is written as [20]:

$$dQ_b = -U_1 A_1 (T_b - T_g) \frac{dz}{v_b} \tag{4}$$

where dQ_b is the heat transferred from a brine element, J.

The change in intrinsic energy of a brine element resulting from heat loss is written as:

$$dQ'_b = c_b \pi r_1^2 \rho_b dz dT_b \tag{5}$$

where dQ'_b is the change in intrinsic energy of a brine element, J; c_b is the specific heat of brine, J/(kg·℃); ρ_b is the density of brine, kg/m³.

Based on the law of conservation of energy, the change in intrinsic energy of a brine element is equal to the heat transferred by a brine element, written as:

$$dQ_b = dQ'_b \tag{6}$$

Substituting Eqs. (4) and (5) Into Eq. (6), the differential equation of the temperature distribution of brine can be expressed as:

$$\frac{dT_b}{(T_b - T_g)} = -\frac{2U_1}{c_b r_1 \rho_b v_b} dz \tag{7}$$

U_1 is the total heat transfer coefficient between brine and gas, consists of the heat conduction and convective heat transfer coefficient, written as [20]:

$$U_1 = \frac{1}{\frac{1}{h_1} + \frac{r_1 \ln(r_2/r_1)}{2k_1} + \frac{r_1}{h_2 r_2}} \tag{8}$$

where r_2 is the outer radius of DIT, m; h_1 is the convective heat transfer coefficient from brine to inner wall of DIT, W/(m²·℃); h_2 is the convective heat transfer coefficient from outer wall of DIT to gas, W/(m²·℃); k_1 is the thermal conductivity of DIT, W/(m·℃).

2.2 Gas temperature distribution

Fig. 3b presents a schematic diagram of the heat transfer of a gas element. Selecting a gas element of length dz to analyze, the heat transferred from brine to a gas element is written as [20]:

$$dQ_g = U_1 A_1 (T_b - T_g) \frac{dz}{v_g} \tag{9}$$

where dQ_g is the heat transferred from brine to a gas element, J; v_g is the flow speed of gas, m/s. The heat transferred from the protective fluid to a gas element can be expressed as:

$$dQ_p = U_2 A_2 (T_p - T_g) \frac{dz}{v_g} \tag{10}$$

$$A_2 = 2\pi r_3 dz \tag{11}$$

where dQ_p is the heat transferred from the protective fluid to a gas element, J; U_2 is the total heat transfer coefficient between gas and protective fluid, W/(m²·℃); T_p is protective fluid temperature, ℃. A_2 is the heat transfer area of gas element, m²; r_3 is the inside radius of DOT, m.

The change in intrinsic energy of a gas element due to heat loss is written as:

$$dQ'_g = c_g \pi (r_3^2 - r_2^2) \rho_g dz dT_g \tag{12}$$

where dQ'_g is the change in intrinsic energy of a gas element, J; c_g is the specific heat of gas, J/(kg·℃); ρ_g is the density of gas, kg/m³.

According to the law of conservation of energy, the change in intrinsic energy of a gas element is equal to the heat transferred by a gas element, written as:

$$dQ'_g = dQ_g + dQ_p \tag{13}$$

Substituting Eqs. (9), (10) and (12) into Eq. (13), the differential equation of temperature distribution of gas can be expressed as:

$$\frac{\mathrm{d}T_g}{[U_2 r_3 (T_p - T_g) + U_1 r_1 (T_b - T_g)]} = \frac{2}{c_g (r_3^2 - r_2^2) \rho_g v_g} \mathrm{d}z \tag{14}$$

The heat transfer between gas and protective fluid consists of two parts: the heat convection from gas to inner wall of DOT and the heat conduction in the DOT. The total heat transfer coefficient from gas to protective fluid is written as [20]:

$$U_2 = \frac{1}{\dfrac{1}{h_3} + \dfrac{r_3 \ln(r_4/r_3)}{2k_2}} \tag{15}$$

where r_3 is the inside radius of DOT, m; r_4 is the outer radius of DOT, m; h_3 is the convective heat transfer coefficient from gas to inner wall of DOT, W/(m² · ℃); k_2 is the thermal conductivity of DOT, W/(m · ℃).

The thermal conductivity k_1 and k_2 are determined by their materials. In general, the DIT is made of steel J55, and the DOT is made of steel J55 or P110. Their thermal conductivity is about 40W/(m · ℃). The convective heat transfer coefficients h_1, h_2 and h_3 are related to the physical properties and flow conditions of fluids. The convective heat transfer coefficient of the forced convective heat transfer fluid in the tubing is the function of the thermal conductivity, hydraulic diameter and Nusselt number [20]. It is written as:

$$h_i = \frac{\lambda_i}{T} Nu_d \quad (i = 1, 2, 3) \tag{16}$$

where λ_i is the thermal conductivity of the fluids, W/(m · ℃); Nu_d is the Nusselt number; R is the hydraulic diameter, m.

The hydraulic diameter is valued as the actual size of channel for fluid flow. For brine flow, the hydraulic diameter is equal to inside diameter of DIT. For gas flow, the hydraulic diameter is equal to difference between the outer diameter of DIT and inner diameter of DOT. The Nusselt number depends on the flow condition of the fluids. If the fluid flow is laminar in the tubing, the Nusselt number is written as:

$$Nu_d = 3.66 + \frac{0.0668(R/L) Re_d Pr}{1 + 0.04[(R/L) Re_d Pr]^{2/3}} \tag{17}$$

where Re_d is the Reynolds number of the fluids; Pr is the Prandtl number of the fluids; L is the length of tubing, m.

If the fluid flow is fully-developed turbulence in the tubing and the wall of the tubing is smooth, the Nusselt number of the fluids in the coaxial tubing with different diameters is written as [21], respectively:

(i) in the inner wall of DIT

$$Nu_d = 0.023 Re_d^{0.80} Pr^{0.30} \tag{18a}$$

(ii) in the outer wall of DIT

$$Nu_d = 0.018Re_d^{0.82}Pr^{0.52} \tag{18b}$$

(iii) in the inner wall of DOT

$$Nu_d = 0.016Re_d^{0.82}Pr^{0.52} \tag{18c}$$

The Reynolds number depends on the flow condition of the fluids in the tubing. The computational equation of the Reynolds number is written as:

$$Re_d = \frac{\rho v R}{\mu} \tag{19}$$

(i) When $1200 \leqslant Re_d < 2000$, the fluids flow in the tubing is laminar;
(ii) When $2000 \leqslant Re_d < 2300$, the fluids flow in the tubing is unsteady;
(iii) When $Re_d \geqslant 2300$, the fluids flow in the tubing is turbulent.

where μ is the hydrodynamic viscosity of the fluids, Pa·s; ρ is the density of the fluids, kg/m³; v is the flow velocity of the fluids, m/s.

The Prandtl number represents the relationship of energy and momentum transfer processes in the fluid. The Prandtl number is the function of specific heat at constant pressure, hydrodynamic viscosity and thermal conductivity of the fluids. It is written as:

$$Pr = \frac{c_p \mu}{k} \tag{20}$$

where c_p is the specific heat at constant pressure of fluids, J/(kg·℃).

Based on Eq. (8) and Eqs. (15) ~ (20) the total heat transfer coefficient U_1 and U_1 can be determined.

2.3 Calculation method

Eqs. (7) and (14) are respectively the differential equation of temperature distribution of brine and gas. Finding their general solution is difficult because both T_b and T_g are unknown functions. Therefore, a finite element iterative method is used to calculate the temperature distribution of brine and gas in the tubing. Firstly, Dividing the tubing into n equal parts, we obtain $n+1$ nodes and the length of each part is equal to $h = L/n$. We use subscripts to distinguish brine, gas and protective fluid and superscripts to represent the nodes. Secondly, using the T_b^i, T_g^i and T_p^i to calculate the T_b^{i+1} and T_g^{i+1}. If the length h of each part is small, T_b, T_g and T_p in the tubing parts can be replaced by the T_b^i, T_g^i and T_p^i, respectively. Therefore, the Eqs. (7) and (14) can be found the general solution. The calculating equation of T_b^{i+1} and T_g^{i+1} are written as:

$$T_b^{i+1} = T_g^{i+1} + (T_b^i - T_g^{i+1})e^{\frac{2U_1 h}{c_1 \rho_b v_b r_1}} \quad (i = 0, 1\cdots, n) \tag{21}$$

$$T_g^{i+1} = \frac{U_2 r_3 T_p^i + U_1 r_1 T_b^i}{(U_2 r_3 + U_1 r_1)} + e^{-\frac{2(U_2 r_3 + U_1 r_1)z}{c_g(r_3^2 - r_2^2)\rho_g v_g}}\left[T_g^i - \frac{U_2 r_3 T_p^i + U_1 r_1 T_b^i}{(U_2 r_3 + U_1 r_1)}\right] \quad (i = 0, 1\cdots, n) \tag{22}$$

Thirdly, in order to accurately determine the temperature distribution of brine and gas, it is necessary to use the iteration method for calculation. The boundary conditions for temperature of brine, gas and protective fluid are written as:

$$z = L, \quad T_b(L) = t_b \tag{23a}$$
$$z = 0 \quad T_g(0) = t_g \tag{23b}$$

where L is the length of the tubing, m; t_b is the temperature of brine in the carven, measured by diesel-brine interface monitoring or sonar survey, ℃; t_g is the gas injection temperature, ℃.

The protective fluid is warmed by the host rock to the same temperature because the protective fluid remains motionless during debrining. Therefore, protective fluid temperature is equal to the geothermal temperature and remains constant during debrining. The geothermal gradient is 0.03℃/m. Therefore, the equation of temperature distribution of protective fluid is written as:

$$T_p(z) = T_0 + \frac{3z}{100} \tag{24}$$

There are three steps for iterative calculation: (1) Selecting T_b^0 as the iterative initial value; (2) Using Eqs. (21) and (22) to calculate the temperature of brine and gas at all nodes. (3) Comparing the calculated value of brine temperature T_b^{n+1} with actual value t_b using Eq. (25). If the error does not satisfy Eq. (25), it is necessary to re-select the initial value T_b^0 use the dichotomy and continue to calculate the temperature of all nodes according to Steps (1) and (2) until the error satisfies Eq. (25). If the error satisfies Eq. (25), the temperature of brine and gas at all nodes is output as the temperature distribution of brine and gas.

$$|T_b^{n+1} - t_b| \leq \beta \tag{25}$$

where β is a positive number, take $\beta = 0.01℃$.

So far, a mathematical model is established to calculate the temperature distribution of brine and gas in the tubing during debrining, and the calculation method is proposed base on the principle of finite elements. The flow chart of the calculation method is shown in Fig. 4.

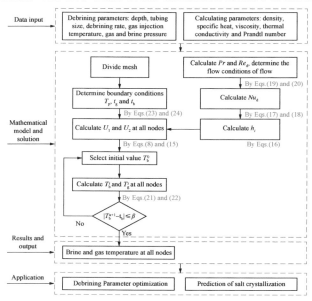

Fig. 4 Flowchart of calculation of the temperature distribution of brine and gas

3 Application and verification

In this section, the A-1 cavern of Jintan UGS is selected as an example [8]. By using the mathematical model, the temperature distribution of brine and gas in the tubing is calculated for different DIT sizes and debrining rates. The influences of the debrining rate, gas injection temperature and thermal conductivity of tubing on brine temperature are analyzed. By comparing the calculated values and measured values of brine temperature at the wellhead of the B cavern, the accuracy and reliability of this model are verified.

3.1 Background of A-1 cavern

A-1 cavern is located in the salt rock mining area, 800 m east of Maolu town, Jintan city, Jiangsu province, China. The shape of the A-1 cavern obtained by the sonar survey is shown in Fig. 5. The cavern has a relatively regular cone-like shape and is basically a rotational body about its axis. The depth of the A-1 cavern roof and bottom are -925.0 m and -985.5 m, respectively. The outer diameters and wall thicknesses of the DIT, DOT and PC used in the A-1 cavern are 244.5 mm × 10.03 mm, 177.8 mm × 9.19 mm and 114.3 mm × 6.88 mm, respectively. Jintan is in the north subtropical monsoon zone, with four distinct seasons. Its lowest air temperature is 3.2℃, its highest air temperature is 39.4℃, and its annual average temperature is 17.5℃.

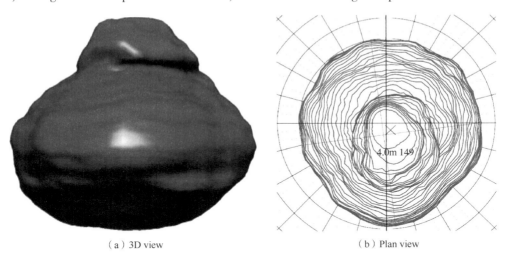

(a) 3D view (b) Plan view

Fig. 5 Shape of A-1 cavern

3.2 Temperature distribution of brine and gas

The temperature distribution of brine and gas during debrining is calculated under different debrining rates and DIT sizes. According to the debrining operation experiences of Jintan UGS, the possible sizes of DIT are 114.3 mm × 6.88 mm (outer diameter and wall thickness), 127 mm × 7.52 mm and 139 mm × 6.98 mm [6]. The maximum debrining rate is usually less than 160 m³/h and more than 40 m³/h. This is because the debrining rate is too large may lead to large insoluble particles at the bottom of the cavern becoming sucked into the DIT and significant increase the compressor's energy consumption, but the debrining rate is too low will increase the debrining time. Therefore, the debrining rates of 50, 100 and 150 m³/h have been selected for calculations.

The mathematical model cannot calculate the temperature of brine and gas in the cavern. Therefore, the maximum calculation depth is determined as −925m, the depth of the A−1 cavern roof. Usually, the tubing is made of steel, and its thermal conductivity is approximately 40 W/(m · ℃). The gas injection pressure slowly increases with time during debrining, and the trend of increase is slight[8]. Therefore, we assume that the gas injection pressure and brine pressure remain constant, and the physical properties of the brine and gas are constant during debrining. Table 1 presents the main properties used in the calculations. The gas injection temperature is generally 20~30℃.

Table 1 The main properties used in the calculations

	Density/(kg/m³)	Specific heat/ [J/(kg℃)]	Thermal conductivity/ [W/(m · ℃)]	Hydrodynamic viscosity/ (Pa · s)	Prandtl number
DIT	8030.00	502	40	—	—
DOT	8030.00	502	40	—	—
Gas	81.40	2266[22]	0.044	0.11×10^{-4}[23]	0.57
Brine	1200.00	3290[17]	0.60[11]	0.14×10^{-2}[24]	7.84

Fig. 6 presents the temperature distribution of brine and gas under different debrining rates when the DIT size is 114.3 mm × 6.88 mm. It can be seen from the figure that the gas temperature increases with depth. The gas temperature rises rapidly in the 0 ~ −30m and slowly in −30 ~ −925 m. When the gas reaches the roof of the cavern the gas temperature is basically equal to the protective fluid temperature. This is mainly because the temperature difference at the wellhead between brine and gas is larger than between gas and protective fluid, which leads to the gas absorbing heat more than released heat. The gas density and specific heat are relatively small, so that the gas temperature rapidly increases when it absorbs a little heat. But when the gas temperature reaches 25℃, the gas absorbing heat is slightly different from the released heat, which leads to the speed of the gas temperature rises to slow down. The data also shows that the temperature distribution of gas is similar under different debrining rates, which indicates that the effects of debrining rates on temperature distribution of gas are not significant.

The temperature distribution of brine is basically identical under different debrining rates. The brine temperature non linear decrease with the decrease of depth. The brine temperature slowly decreases within −925 ~ −600 m and significantly decreases above −600 m. This is caused mainly by the temperature difference between brine and gas being small in the bottom of DIT, so that the brine releases less heat. As the depth continues to decrease, the temperature difference between brine and gas gradually increases, and the heat released by brine consistently increases. The calculation results also show that the lower debrining rate results in faster brine temperature decreases, but the differences of brine temperature at the wellhead under different debrining rates are not large. For

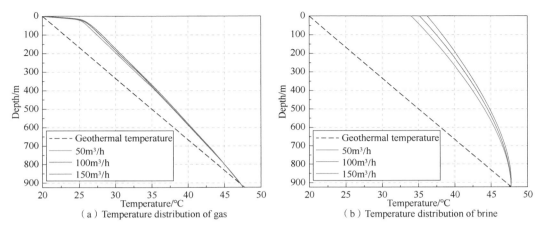

Fig. 6 Temperature distribution of brine and gas under three different debrining rates when the DIT size is 114.30 mm × 6.88 mm

example, when the debrining rates are 50, 100 and 150m³/h, the brine temperatures at the wellhead are 34.05, 35.36 and 36.18℃, respectively. This is mainly because the greater the debrining rate, the shorter the flow time of brine in DIT, and the smaller the range of brine temperature decreases. However, the convective heat transfer coefficient increases with the increasing debrining rate and results in total heat transfer coefficient increases. Fig. 7 presents the relations between the total heat transfer coefficient U_1 and U_2 and debrining rates using Eqs. (8) and (15). It shows that the total heat transfer coefficients U_1 and U_2 increase significantly with the increasing debrining rate. This indicates that the heat flux increases with the increasing debrining rate. That is, the heat released by brine per unit time increases.

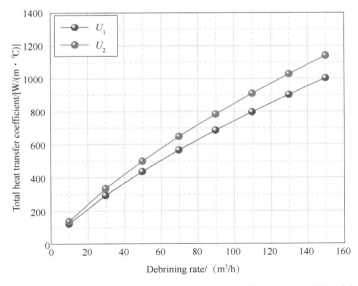

Fig. 7 Relations between the total heat transfer coefficient U_1 and U_2 and the debrining rate

Fig. 8 presents the temperature distribution of brine and gas under different debrining rates when the DIT size is 127.00 mm × 7.52 mm. Fig. 9 presents the temperature distribution of brine and gas under different debrining rates when the DIT size is 139.00 mm × 6.98 mm. By comparing Figs. 6, 8 and 9, it can be seen that the shapes of the temperature distributions of brine and gas under different DIT sizes are essentially the same. When the debrining rate is identical, the temperature drop range of brine increases with the increase of DIT size. For example, when the debrining rate is 100 m^3/h and the DIT size is respectively 114.30 mm × 6.88 mm, 127.00 mm × 7.52 mm and 139.00 mm × 6.98 mm, the brine temperature at the wellhead is 35.36, 33.10 and 30.23 ℃. This indicates that using a small DIT decreases the drop range of brine temperature and the speed of salt crystal. Considering that the DIT size too small will increase the energy of debrining, then the DIT can easily become blocked by salt crystal. Therefore, we proposed that the A−1 cavern use DIT with size is 127.00 mm × 7.52 mm to debrine.

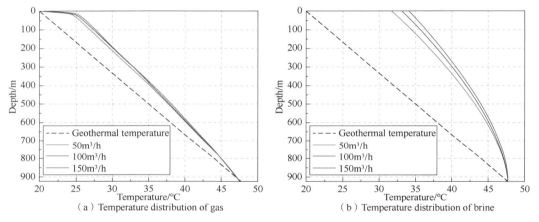

Fig. 8 Temperature distribution of brine and gas under three different debrining rates when the DIT size is 127.00 mm × 7.52 mm

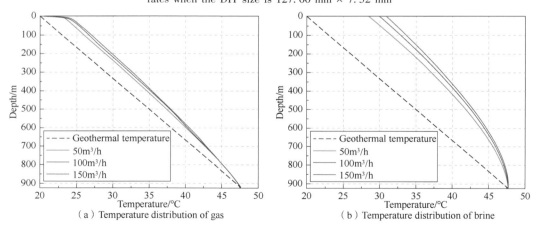

Fig. 9 Temperature distribution of brine and gas under three different debrining rates when the DIT size is 139.00 mm × 6.98 mm

3.3 Influencing factors of brine temperature

(1) Debrining rate.

Fig. 10 presents the relations between debrining rate and the temperature drop range of brine for different DIT sizes. The temperature drop range of brine decreases non linearly with the increase of debrining rate. When the debrining rate is low, increasing the debrining rate can significantly reduce the temperature drop range of brine. When the debrining rate increases to 100m³/h, continuing to increases the debrining rate cannot significantly reduce the drop range of brine temperature. It is indicated that the

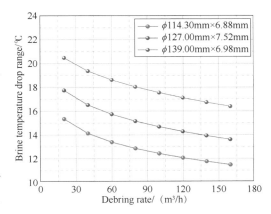

Fig. 10 Relations between debrining rate and drop range of brine temperature under different DIT sizes

debrining rate can be appropriately increased in order to prevent the salt crystal blocking the DIT. Increasing debrining rate can reduce the drop range of brine temperature, further reducing the growth rate of salt crystal. However, according to a previous study [6], the debrining rate too large may increase the compressor's energy consumption and the risk of large particles being sucked into the DIT. According to the above analysis, it is suggested that the A-1 cavern debrining rate should be 100m³/h.

(2) Gas injection temperature.

Some researchers proposed that increasing gas injection temperature can decrease the drop of brine temperature and then prevent the salt crystal blocking the DIT [13]. Fig. 11 presents the temperature distribution of brine and gas under different gas injection temperatures. The distributions of gas temperature are different within 0~-40 m, but beneath -40 m are largely identical. When the gas injection temperature is less than the brine temperature, the gas temperature rises rapidly, while the gas temperature rapidly reduces. This is mainly because the gas specific heat and density are low, which leads to the temperature sharply changing when gas absorbs or transfers a little of heat. The results in Fig. 11 also show that the temperature distributions of brine are almost the same under different gas injection temperatures, which means that increasing the gas injection temperature can not effectively slow down the decrease of brine temperature.

(3) Tubing thermal conductivity.

The thermal conductivity of DIT has a significant effect on the heat transfer of brine and gas. The main reason for the sharp brine temperature decrease is the large thermal conductivity of the tubing. If the tubing material is replaced by glass fiber reinforced plastics (GFRP), the drop of the brine temperature can be significantly reduced. GFRP is a composite material, with a density of 1600 kg/m³ and its thermal conductivity is 0.5 W/(m·℃) [17]. Fig. 12 presents the temperature distribution of brine and gas when the material of tubing is respectively J55 steel and GFRP. The smaller the thermal conductivity of the tubing, the smaller the drop of brine temperature. The brine

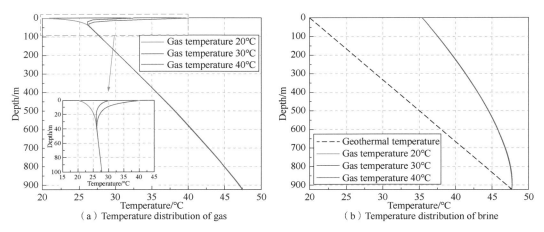

Fig. 11 Temperature distribution of brine and gas under three different gas injection temperatures, when debrining rate is 100 m³/h and DIT size is 114.30 mm ×6.88 mm

temperature at the wellhead is 35.36 and 44.72℃ when the tubing is made out of J55 steel or GFRP, respectively. This indicates that using tubing with low thermal conductivity can significantly slow the drop of brine temperature and effectively prevented the salt crystal blocking the DIT.

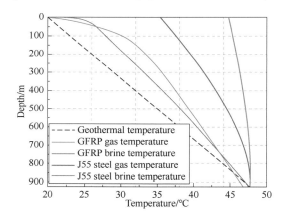

Fig. 12 Temperature distribution of brine and gas for different tubing materials, when the debrining rate is 100 m³/h and the DIT size is 114.30 mm × 6.88 mm

3.4 Field verification

Base on the mathematical model, the temperature distribution of brine during debrining of the B cavern is calculated. Fig. 13 shows a schematic of B cavern debrining. The B cavern uses two wells for debrining, the gas injection well B-1, and the debrining well B-2. B-1 is a vertical well, and B-2 is a horizontal well. The depth of vertical parts of B-2 is -940m. The wells B-1 and B-2 use debrining tubing and production casing to debrine. Both debrining tubing and production casing are made of J55 steel, and their sizes is 139.70 mm × 7.74 mm and 218.00 mm × 9.00 mm, respectively. The annulus between debrining tubing and production casing is filled by cement which the thermal conductivity of 1.8 W/(m·℃). The debrining rate of B cavern remains at 40 m³/h. The total heat transfer coefficient between brine and formation is written as:

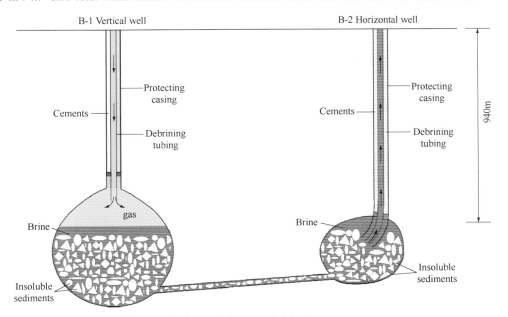

Fig. 13 Schematic diagram of debrining of B cavern

$$U_1 = \cfrac{1}{\cfrac{1}{h_1} + \cfrac{r_1 \ln(r_2/r_1)}{2k_1} + \cfrac{r_1 \ln(r_3/r_2)}{2k_3}} \quad (26)$$

where k_3 is the thermal conductivity of cement, W/(m·℃).

Although the well B-2 only has two-layer tubing and no protective fluid, the mathematical model still applies to the well B-2 debrining. Based on the mathematical model and Eq. (26), the temperature distribution of brine is calculated. The calculation depth is −940 m, the initial brine temperature is 48.2℃, other calculation parameters as shown in Table 1. Fig. 14 presents the temperature distribution of brine for the well B-2 debrining. Results show that the calculated brine temperature at the wellhead is 36.2℃. The brine and tubing wall temperature at the wellhead during the well B-2 debrining were measured, as shown in Fig. 15. According to field measured results, the brine and tubing wall temperature at the wellhead are 34.3 and 34.5℃. Because the brine released a little heat during measurement, the brine temperature is lower than the tubing wall temperature, which means that the tubing wall temperature is closer to the actual brine temperature in debrining tub-

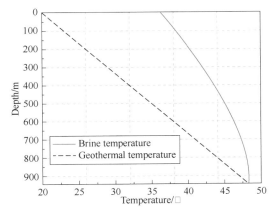

Fig. 14 Temperature distribution of brine for the well B-2 debrining

ing. The difference of brine temperature at the wellhead between calculated values and measured values is small, which indicates that the mathematical model established in this paper has a high calculation accuracy and reliability.

(a) Wall temperature (b) Brine temperature

Fig. 15 The brine and tubing wall temperature at the wellhead

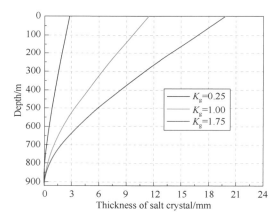

Fig. 16 A-1 cavern of crystal growth thickness at different depth, when debrining rate is 50 m³/h and DIT size is 114.30 mm × 6.88 mm

4 Discussion on calculated salt crystal growth

How to accurately calculate salt crystal growth on the tubing is a difficult problem. According to the calculated results in Section 3, the brine temperature decreases with the decrease of depth. The saturated brine will become supersaturated when its temperature decreases. The supersaturated brine may precipitate salt crystal under the induction of tiny insoluble particles on the tubing. Based on the above analysis, the brine temperature decreases is main reason for salt crystal growth. Therefore, calculating the distribution of brine temperature is the theoretical basis for predicting salt crystal growth.

According to crystal growth kinetics, supersaturation is the driving force of crystal growth, which has a great influence on crystal growth rate and quality [25]. The salt crystal growth has two processes: ion diffusion process and surface reaction process [26]. The ion diffusion rate and surface reaction rate are controlled by supersaturation. The crystal growth rate is the function of supersaturation [27], which can be written as:

$$G = K_g \Delta c^g \qquad (27)$$

where G is the crystal growth rate, mm/h; K_g is the crystal growth coefficient; g is the order of crystal growth; Δc is the supersaturation, g/100g H_2O.

Supersaturation is the difference between the actual and equilibrium concentration of the brine:

$$\Delta c = c^* - c \qquad (28)$$

where c^* is the actual concentration of the brine, g/100g H_2O; c is the equilibrium concentration of the brine, g/100g H_2O.

Because salt crystals are mainly composed of sodium chloride (NaCl), the equilibrium concentration of the brine can be expressed by the equilibrium concentration of NaCl, written as:

$$c = 0.0003 T_b^2 + 0.011 T_b + 35.691 \qquad (29)$$

where T_b is the brine temperature, ℃.

Substituting Eqs. (29) and (28) into Eq. (27), the salt crystal growth rate can be expressed as:

$$G = K_g [0.0003(t_b^2 - T_b^2) + 0.011(t_b - T_b)^g] \qquad (30)$$

where t_b is the brine temperature in cavern, ℃.

The NaCl crystal growth rate increases linearly with the increasing supersaturation [25-27], which means the g is equal to 1. Because the crystal growth coefficient is unknown, three cases are selected for calculating the salt crystal growth during debrining. Using the mathematical model and Eq. (30), the NaCl crystal growth thickness in the DIT is calculated under different crystal growth coefficients. Fig. 16 presents the A-1 cavern of crystal growth thickness at different depths after debrining 24h. It can be seen from the figure that the crystal thickness decreases with the depth increases, the crystal thickness is maximum at the wellhead. This is because the crystal growth rate decreases with the depth increase, and the crystal growth rate is maximum at the wellhead. When the crystal growth coefficients are 0.25, 1.00 and 1.75 mm/h, the crystal thickness at the wellhead are 2.85, 11.30 and 19.77 mm, respectively. The calculated results show that the mathematical model established in this paper can provide a theoretical basis for predicting the growth of salt crystals during debrining.

5 Conclusions

(1) According to the rule of heat transfer of brine and gas in the tubing during debrining of an UGS, a mathematical model used to calculate the temperature distribution of brine and gas is established, and a finite element iterative method is used to solve this mathematical model.

(2) Based on the established mathematical model, the A-1 cavern of Jintan UGS is selected as an example, the temperature distribution of brine and gas under different debrining rates and DIT sizes are calculated. The calculated results show that the gas temperature increases rapidly in the $0 \sim -30$m and slowly in $-30 \sim -925$m. When the gas reaches the roof of the cavern the gas temperature is basically equal to 47.6℃. The brine temperature decreases non linearly with the decrease of depth, and the temperature distribution of brine increases with the increases of debrining rate and the decreases of DIT size.

(3) Increasing the debrining rate and decreasing the thermal conductivity of tubing can reduce the drop of brine temperature, but increasing gas the injection temperature has a minor effect on the temperature distribution of brine. In order to prevent the salt crystal blocking the DIT, it is proposed to use a DIT with low thermal conductivity to debrine.

(4) By comparing the calculated values of brine temperature at the wellhead with field measured values for B cavern debrining, it is shown that the proposed model has a high calculation accuracy and reliability.

CRediT authorship contribution statement

Dongzhou Xie: Methodology, Writing - original draft, Visualization. Tongtao Wang: Conceptualization, Resources, Supervision. Long Li: Investigation, Formal analysis. Tao He: Writing - review & editing, Visualization. Guoxing Chai: Investigation, Data curation. Duocai Wang: Software, Project administration. Hong Zhang: Investigation, Project administration. Tieliang Ma: Supervision, Data curation. Xin Zhang: Investigation, Formal analysis.

Declaration of Competing Interest

The authors declare that they have no known competing financial interests or personal relationships that could have appeared to influence the work reported in this paper.

Acknowledgments

The authors wish to acknowledge the financial support of National Natural Science Foundation of China (Grant No. 42072307), Hubei Province Outstanding Youth Fund (2021CFA095), Special Fund for Strategic Pilot Technology of Chinese Academy of Sciences (Grant No. XDPB21), and Strategic Priority Research Program of the Chinese Academy of Sciences (Grant No. XDC10020300). Thanks to Professor Jaak J. K. Daemen of Mackay School of Earth Sciences and Engineering, University of Nevada (Reno), USA, for his constructive suggestions and modifications.

References

[1] N. Zhang, C. Yang, X. Shi, T. Wang, H. Yin, J. J. K. Daemen, Analysis of mechanical and permeability properties of mudstone interlayers around a strategic petroleum reserve cavern in bedded rock salt, Int. J. Rock

Mech. Min. Sci. 112 (2018), https: // doi. org/10. 1016/j. ijrmms. 2018. 10. 014.

[2] C. Yang, T. Wang, D. Qu, H. Ma, Y. Li, X. Shi, J. J. K. Daemen, Feasibility analysis of using horizontal caverns for underground gas storage: a case study of Yunying salt district, J. Nat. Gas Sci. Eng. 36 (2016), https: //doi. org/10. 1016/j. jngse. 2016. 10. 009.

[3] W. Liu, Z. Zhang, J. Chen, D. Jiang, F. Wu, J. Fan, Y. Li, Feasibility evaluation of large-scale underground hydrogen storage in bedded salt rocks of China: A case study in Jiangsu province, Energy 198 (2020), https: //doi. org/10. 1016/j. energy. 2020. 117348.

[4] T. Wang, C. Yang, H. Wang, S. Ding, J. J. K. Daemen, Debrining prediction of a salt cavern used for compressed air energy storage, Energy 147 (2018), https: //doi. org/10. 1016/j. energy. 2018. 01. 071.

[5] China's National Bureau of Statistics, China Statistical Yearbook, China Statistics Press, Beijing, 2020.

[6] T. Wang, G. Chai, X. Cen, J. Yang, J. J. K. Daemen, Safe distance between debrining tubing inlet and sediment in a gas storage salt cavern, J. Pet. Sci. Eng. 196 (2021), https: //doi. org/10. 1016/j. petrol. 2020. 107707.

[7] Jinlong Li, X. Shi, C. Yang, Y. Li, T. Wang, H. Ma, H. Shi, Jianjun Li, J. Liu, Repair of irregularly shaped salt cavern gas storage by re-leaching under gas blanket, J. Nat. Gas Sci. Eng. 45 (2017), https: //doi. org/10. 1016/j. jngse. 2017. 07. 004.

[8] T. Wang, S. Ding, H. Wang, C. Yang, X. Shi, H. Ma, J. J. K. Daemen, Mathematic modelling of the debrining for a salt cavern gas storage, J. Nat. Gas Sci. Eng. 50 (2018), https: //doi. org/10. 1016/j. jngse. 2017. 12. 006.

[9] T. Wang, C. Yang, X. Shi, H. Ma, Y. Li, Y. Yang, J. J. K. Daemen, Failure analysis of thick interlayer from leaching of bedded salt caverns, Int. J. Rock Mech. Min. Sci. 73 (2015), https: //doi. org/10. 1016/j. ijrmms. 2014. 11. 003.

[10] C. Wang, J. Ba, K. Guo, Y. Jiao, Treatment method of complicated situation of tubing for debrining of salt cavern gas storage, Petrochem. Ind. Appl. 34 (2015) 10-12, https: //doi. org/10. 3969/j. issn. 1673-5285. 2015. 10. 014.

[11] P. Bérest, Heat transfer in salt caverns, Int. J. Rock Mech. Min. Sci. 120 (2019), https: //doi. org/10. 1016/j. ijrmms. 2019. 06. 009.

[12] L. Haiwei, H. Jun, Z. Xing, L. Tian, L. Mengxue, Diesel-brine interface monitoring in the solution mining of salt-cavern gas storage based on optical fiber technology, Oil Gas Storage Transp. 37 (2018) 1037-1040.

[13] X. Jin, Y. Xia, G. Yuan, X. Zhuang, F. Ban, A. Dong, An experimental study on the influencing factors of salt crystal in brine discharge strings of a salt-cavern underground gas storage (UGS), Nat. Gas Ind. 37 (2017), https: //doi. org/10. 3787/ j. issn. 1000-0976. 2017. 04. 016.

[14] J. Chen, W. Jiang, C. Yang, X. Yin, S. Fu, K. Yu, Study on engineering thermal analysis of gas storage in salt formation during gas injection and production, Yanshilixue Yu Gongcheng Xuebao/Chinese J. Rock Mech. Eng. 26 (2007).

[15] P. Bérest, B. Brouard, H. Djakeun-Djizanne, G. Hévin, Thermomechanical effects of a rapid depressurization in a gas cavern, Acta Geotech. 9 (2014), https: //doi. org/ 10. 1007/s11440-013-0233-8.

[16] K. Khaledi, E. Mahmoudi, M. Datcheva, T. Schanz, Analysis of compressed air storage caverns in rock salt considering thermo-mechanical cyclic loading, Environ. Earth Sci. 75 (2016), https: //doi. org/10. 1007/s12665-016-5970-1.

[17] J. Liu, Effect of Wellbore Temperature Distribution on Brine Crystallization in Salt Cavern Storage Wellbore, China University of Petroleum, Beijing, 2018.

[18] W. Li, C. Zhu, J. Han, C. Yang, Thermodynamic response of gas injection-andwithdrawal process in salt cavern for underground gas storage, Appl. Therm. Eng. 163 (2019), https://doi.org/10.1016/j.applthermaleng.2019.114380.

[19] P. Li, Y. Li, X. Shi, K. Zhao, X. Liu, H. Ma, C. Yang, Prediction method for calculating the porosity of insoluble sediments for salt cavern gas storage applications, Energy 221 (2021), https://doi.org/10.1016/j.energy.2021.119815.

[20] J. P. Holman, Heat Transfer, version 10. ed, China Machine Press, BeiJing, 2011.

[21] P. Li, Research of Heat Transfer Characteristics of Casing Ground Heat Exchanger with Medium and Deep Ground Source Heat Pump, Harbin Institute of Technology, 2018.

[22] D. Zhu, H. He, Calculation of well bore temperature distribution in condensate gas well, Tianranqi Gongye/Natural Gas Ind. 18 (1998).

[23] A. L. Lee, M. H. Gonzalez, B. E. Eakin, The viscosity of natural gases, J. Pet. Technol. 18 (1966), https://doi.org/10.2118/1340-pa.

[24] X. Ma, Determination of Viscosity of Brine at Different Temperatures, China Well Rock Salt, 1979, pp. 33–36.

[25] J. Zhao, H. Miao, L. Duan, Q. Kang, L. He, The mass transfer process and the growth rate of NaCl crystal growth by evaporation based on temporal phase evaluation, Opt. Lasers Eng. (2012), https://doi.org/10.1016/j.optlaseng.2011.07.013.

[26] A. Naillon, P. Joseph, M. Prat, Sodium chloride precipitation reaction coefficient from crystallization experiment in a microfluidic device, J. Cryst. Growth 463 (2017) 201–210, https://doi.org/10.1016/j.jcrysgro.2017.01.058.

[27] M. Bohnet, Fouling of heat transfer surfaces, Chem. Eng. Technol. 10 (1987), https://doi.org/10.1002/ceat.270100115.

本论文原发表于《Journal of Energy Storage》2022年第50期。

造腔设计与分析

井间距对盐穴储气库小间距对井造腔的影响

李 龙 侯 磊 李建君 范丽林 李小明 李海伟 王桂林

(国家管网集团西气东输公司江苏储气库分公司)

【摘 要】 中国盐岩赋存的主要特点是夹层多、盐层薄,对于单直井造腔方法有诸多限制,为此提出采用小间距对井造腔方法克服不良地质条件。该方法采用大尺寸造腔套管对井直流排卤,具有造腔速度快、建库周期短、盐层利用率高等优点。开展不同双井间距以及不同腔体连通方式的物理模型试验,探究井间距对腔体最终形态的影响规律。结果表明:井间距较小时,腔体水平方向长度小,未较好利用水平空间;井间距较大时,水平段的腔体溶解速度明显降低,且腔体形态不规则;预制对接井的造腔方法具有更高的成腔速率。实际对井造腔工程中对井间距需依据地质条件与工程需求确定。研究结果为盐穴储气库小间距对井造腔腔体形态设计与工艺参数优化提供了借鉴与指导。

【关键词】 层状盐岩;储气库;对井间距;腔体形状;溶腔规律

盐岩因其孔隙度小、渗透率低、可塑性强等优点,被认为是优越的天然气储存介质[1-2]。目前,国内外普遍采用单直井水溶造腔法建造盐穴储气库。这种造腔方法对地质条件要求高,通常应用于厚度大、杂质少的盐岩地层中。而中国盐岩具有盐层薄、品位低、夹层多等劣势[3]。在薄盐层中,采用传统的单井对流法造腔,不仅成本高,而且造腔周期长。此外,受到盐层中杂质沉积以及薄夹层垮塌堆积的影响[4],腔体的利用率普遍偏低,使腔体单位体积造价上升[5]。基于中国盐岩赋存条件,亟需探究一种能够充分利用薄盐层空间、降低沉渣堆积影响的造腔方法。

小间距对井造腔方法的特点是同时采用 2 口井进行溶蚀作业,形成一水平放置的椭圆形腔体[6]。相比广泛应用的单直井水溶造腔,小间距对井造腔方法可以更好地利用水平方向的空间,提高水平薄盐层的空间利用率[7]。井底间距的设定受两方面因素的制约:(1)较大的井间距难以进行腔体形态控制,容易形成偏溶,不利于腔体稳定性;(2)较小的井间距则难以拓展腔体的水平空间。目前,已有众多学者对小间距对井造腔工艺进行了研究。班凡生[8]分析了采用对井造腔技术提高层状盐层造腔速度的可行性;郑雅丽等[9]提出采用自然溶通的方法进行双井造腔;姜德义等[10]通过数值方法获取小间距对井水溶造腔过程中中流场和浓度场的特征;任松等[11]通过开展对井造腔水溶试验,分析了注水流量对于腔体拓展的影响;易亮等[12-13]探究了不同提管方式对双井造腔腔体形态的作用。

基金项目:国家管网集团西气东输公司科技项目"楚州盐穴储气库小间距对井造腔模拟及实验研究",KJ202102。

作者简介:李龙,男,1966 年生,高级工程师,1991 年 7 月毕业于西南石油大学采油工程专业,现主要从事地下盐穴储气库工程建设及生产运营技术管理的研究工作。地址:江苏省镇江市南徐大道 60 号商务 A 区 D 座,212000。E-mail:lilong@pipechina.com.cn。

法国于 2008 年开始在 Geosel-Manosque 地区采用双井造腔法进行 TA&TB 储气库造腔，储气库体积可达 $50×10^4 m^3$。荷兰由双井建成的储气库位于格罗宁根 Zuidwending 村附近的盐丘，总计容积约为 620000m^3。中国目前尚无小间距对井造腔工程案例，目前只有湖北云应盐穴储气库采用双井造腔法建造。由于小间距对井造腔方法中双井间距、注水排卤方式等尚未得到充分研究，使得关键造腔参数与腔体形态发展之间的关系还未明确[14-15]，而腔体几何形态和体积直接关乎储气库的安全和经济性问题。为使小间距对井造腔技术能够尽快投入工程应用，亟须对小间距对井造腔方法的工程参数进行研究。

在此采用物理模型试验方法对井间距的选取和造腔效果进行探究。通过对比各组试验中的腔体形态与造腔速率的变化，探究小间距对井造腔规律，并基于地质条件与工程需求提出小间距对井造腔参数选取方法。

1 试验参数与装置

相似理论研究作为一种科学有效的研究手段，可用以指导建立与现场原位实物相似的实验室尺度模型，通过观测相似模型上的力学行为和现象，反推原位实物。对于盐岩室内水溶造腔模拟试验而言，由于室内模型试验的尺寸远小于现场盐腔尺寸，因此需要对试验过程中的其他参数(注水量、管柱尺寸、两口距、溶蚀时间等)进行调整，使其满足相似理论的要求。根据水溶造腔工艺的技术特点，影响盐岩水溶造腔的参数见表 1，其中有 4 个基本量纲(L，M，T，θ)，L 为长度量纲，M 为质量量纲，T 为时间量纲，θ 为温度量纲。

表 1 小间距对井造腔试验中各参数及量纲的统计表

参数	量纲	参数	量纲
几何尺寸 l	L	卤水浓度 c	ML^{-3}
溶解时间 t	T	温度 T_e	θ
盐岩密度 ρ	ML^{-3}	注水流量 q	$L^{-3}T^{-1}$
溶解速率 ω	$ML^{-2}T^{-1}$		

由于彼此相似的现象必定具有数值相同的相似准则，因此实验需测定各个相似准则中所包含的一切物理量。因此需要将试验结果整理成无量纲量的形式，才能与工程尺度的规律进行对比。选取 $l[L]$、$t[T]$、$\omega[ML^{-2}T^{-1}]$、$T_e[\theta]$ 为基本物理量，其他 3 个物理量可用基本物理量来表示，而本次试验需确定的物理量为注水流量，可表示为

$$q = l^\alpha t^\beta \omega^\lambda T_e^\gamma \tag{1}$$

由方程量纲齐次的原则，可得系数 $\alpha=3$，$\beta=-1$，$\lambda=0$，$\gamma=0$，则与 q 有关的 π 项为

$$\pi_q = \frac{qt}{l^3} \tag{2}$$

同理：

$$\pi_\rho = \frac{\rho l}{t\omega} \qquad \pi_c = \frac{lc}{t\omega} \tag{3}$$

参数的相似比用 K 表示，几何相似比可表示为

$$K_l = \frac{l_p}{l_m} \tag{4}$$

式中：l_p 为原型尺寸，l_m 为模型尺寸。

其他参量相似比类似，则式(3)中各参量相似比的关系可表示为

$$\frac{K_q K_t}{K_l^3} = 1 \quad \frac{K_\rho K_l}{K_t K_\omega} = 1 \quad \frac{K_l K_c}{K_t K_\omega} = 1 \tag{5}$$

同时根据相似原理，在相似准则相等的条件下，将试验得到的无量纲量推广到原型系统中，进而得到所需的有量纲量：(1)几何相似比为 1∶200。实验室中盐腔直径 30cm 对应实际盐腔直径 60m。(2)流量相似比为 $1/200^3$。实验室流量为 $0.0015\text{m}^3/\text{h}$，即 25mL/min，现场流量为 $60\text{m}^3/\text{h}$。(3)时间相似比为 1∶200。实验室为 3d，现场应为 600d。

按照相似理论搭建小间距对井造腔试验装置(图1)。高精度蠕动流量泵可将淡水注入

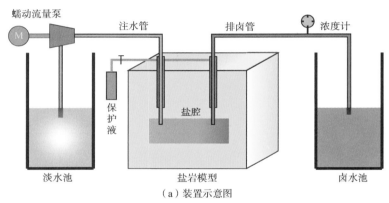

(a) 装置示意图

(b) 装置实物图

图 1　小间距对井溶腔试验装置示意图

盐岩空腔内，在注水压力下腔内卤水由右端排卤管排出，此过程模拟了盐穴储气库对井分别注水采卤的过程。每一套管柱系统包含造腔套管与生产套管两种结构：造腔套管用于注

水淡水和排出卤水,生产套管用于注入环空保护液以防卤水上溶过多。蠕动流量泵包含流量监测功能,可实时测量和更改注水流量。盐岩模型为1/2模型结构,其对称面被亚克力板覆盖并用环氧树脂进行密封,通过可视化模型观察溶腔形态拓展规律,并通过摄像机进行记录。改变溶腔入口流量、提管高度、两管间距,可以模拟不同工况下腔体形态的发展。该试验是对腔体形态变化规律进行探究,选取巴基斯坦盐矿中的方形盐砖作为试验模型,尺寸为30cm×20cm×20cm。

2 试验方法

井间距对腔体扩展的影响不可忽视,不同井间距下的淡水流分布区域差别很大。设计以下试验探究不同井间距下的腔体扩展过程。试验利用相同尺寸盐砖(30cm×20cm×20cm),实际工程盐砖尺寸为60m×40m×40m。根据几何相似原理,试验分别选用5cm、10cm、15cm的钢管间距表示10m、20m、30m的井底间距(图2)。此外,为了论证对接井对于小间距对井造腔工程的必要性,设置不含水平连接段的模型作为对照组[图2(d)]。实际工程的造腔管柱布置方案如下:(1)对井间距10m,同时最下端设计水平对接连通段;(2)对井间距20m,设置水平连接段;(3)对井间距30m,设置水平连接段;(4)双井间距10m,不设置水平连接段。

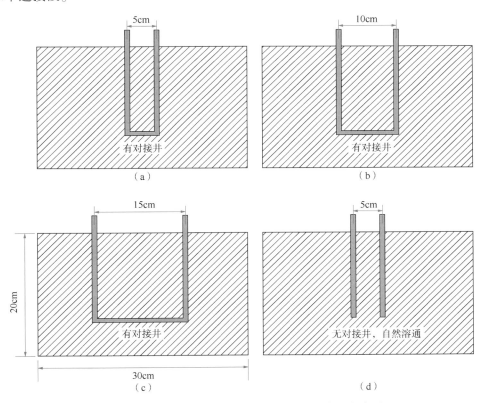

图2 井间距对腔体扩展的影响试验中管柱布设方案图

不同于采卤工程中的水平对接井溶腔工艺,小间距对井造腔需采用循环注水的方式来控制腔体的形态，即间隔一定时间切换注水端来使腔体两侧平衡发展。在造腔试验中开始选用一端注水一端排水的方式进行溶腔，随后每间隔2h，将注水管与排卤管对调，切换注水方向。而在对照组中，试验开始时两管柱分别进行单直井造腔。直到两侧腔体连通后，采用一端注水、一段排水的方式继续造腔，随后同样每间隔2h，将注水管与排卤管对调。试验共设置6个造腔阶段，每个阶段的造腔时间为2h，此外每间隔4h将管柱与保护液提升5cm，以使腔体向竖直方向拓展（表2）。

表2 对井造腔模型试验造腔参数

阶段序号	注水端	流量/(mL/min)	造腔时间/h	造腔套管提升高度/cm	保护液提升高度/cm
1	左	30	2	0	0
2	右	30	2	0	0
3	左	30	2	5	5
4	右	30	2	5	5
5	左	30	2	10	10
6	右	30	2	10	10

3 试验结果

3.1 腔体形态分析

盐腔的形态扩展过程即盐腔内壁盐岩持续的溶解过程，盐腔呈现出的各种形态则表明盐腔内部各处不同的溶解速率。试验过程中通过透明亚克力板可实时观测盐腔轮廓，每隔30min拍照获取盐腔形态，从而记录盐腔形态整个变化过程。

3.1.1 5cm对井间距有对接井

腔体的最初形态为人工切割的横槽（图3），对应于对井井眼。注入淡水之后，腔体逐渐开始向上和侧向发展，但发展的速度是不均匀的（一阶段）。左侧注水造腔2h之后，左侧腔体宽度明显大于右侧腔体宽度，腔体顶面同样呈轻微的左高右低的形态（二阶段）。随着造腔继续进行，切换到右侧管柱进行注水，使得右侧注水管口附近加快溶蚀（三阶段）。然后切换到左侧注水，左侧再次加快溶蚀，使得腔顶从出水管口至排水管口呈现一个降低的坡面形态（四阶段）。随着注水位置切换到右侧，右侧的溶解速度开始增加，使得原左高右低的腔顶斜率缓慢变小，左右侧腔体大小再次相近（五阶段）。在注水井与排卤井反复交替下，左右侧腔体交替快速扩展，最终腔形呈现左右近似相同的形状（六阶段）。可见，最终腔体直径约等于两倍井间距，腔形在水平方向上较紧凑。

分析腔体左右侧交替扩展的原因：淡水从注水管柱进入腔体之后，在较高质量浓度的卤水中向前运动的距离有限，相反极易受到浮力而向上运动，淡水首先在井口附近参与盐岩溶解，因此在距离注水口近的位置卤水浓度更低，腔壁溶解速度更快，注水井一侧的腔体发展更快。随着造腔继续进行，卤水向排水口运动的过程中继续参与溶解，质量浓度不断升高，到排水口时达到最高。由于整个水平通道质量浓度的不均匀性，注水口附近卤水质量浓度低，溶解快，腔顶抬升快；而排卤口附近卤水质量浓度高，腔顶抬升慢，因此最终形成左右侧腔体交替发展的过程。

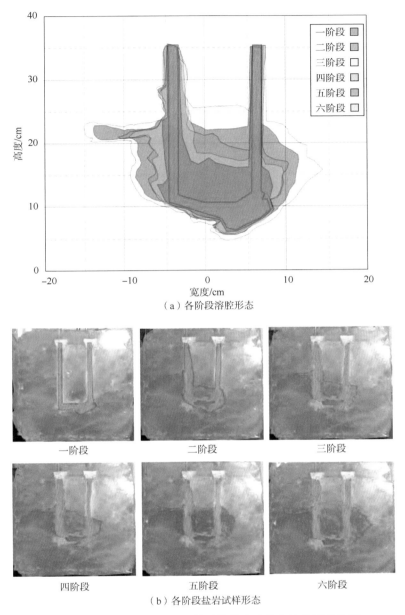

(a) 各阶段溶腔形态

(b) 各阶段盐岩试样形态

图 3 对井间距为 5cm 时预制对接井造腔的腔体轮廓变化图

3.1.2 10cm 对井间距有对接井

在 10cm 对井间距下腔顶亦呈现注水口侧高、排卤口侧低的发展趋势（图 4）。相比 5cm 对接井间距造腔试验，该试验左右侧腔体差距明显增大，左侧腔顶与右侧腔顶的高度差进一步增加。这是由于对井间距变大，淡水向排卤井运动的路径明显增加，腔体内溶质扩散程度加强，导致腔体内卤水沿水平方向运动时更容易达到高质量浓度，所形成的斜面更倾斜，腔体更不规则。但是最终形成的腔体在水平方向跨度更大，在同一高度的盐岩层中，形成的腔体体积更大。腔体最终形态长轴在水平方向呈橄榄形，更好地利用了薄盐层的空间。

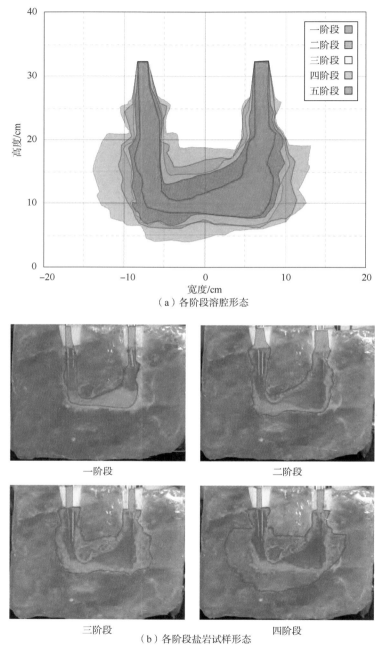

(a) 各阶段溶腔形态

一阶段　　　　　　　　　　　　　二阶段

三阶段　　　　　　　　　　　　　四阶段

(b) 各阶段盐岩试样形态

图 4　对井间距为 10cm 时预制对接井造腔的腔体轮廓变化图

3.1.3　15cm 对井间距有对接井

当对井间距为 15cm 时，腔体扩展过程与以上试验类似，但由于井距增大，腔体内水平段明显增长，在建槽期淡水可以在水平段内充分吸收盐溶质，因此水平段中腔顶较平整，未出现明显倾斜面(图5)。但是随着腔体直径逐渐增大,由于淡水注入速度较小,淡水从注水管柱进入腔体之后,在较高质量浓度的卤水中向前运动的距离有限,相反极易受到浮力而向上运动。在注水井与排卤井附近区域,腔体顶部迅速上溶,溶解速度远超过水平段。最终腔体两端溶解程度更高,而水平段溶解程度低,使得腔体最终呈U形。这种腔形虽然较好地利用了水平空间,但是管柱附近的腔体快速上溶难以控制,给腔体的稳定性和气密性带来风险。

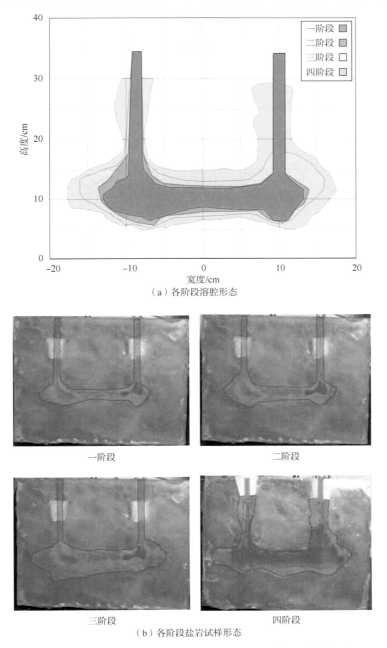

图5 对井间距为15cm时预制对接井造腔的腔体轮廓变化图

3.1.4　5cm 双井间距无对接井

将双井间距为 5cm 时无对接井造腔试验(图6)与 5cm 双井间距含对接井造腔(图3)对比发现，在初始单直井造腔阶段，无对接井的腔体拓展速度明显更低。但是这种造腔方法中腔体内排出的卤水质量浓度明显更高。这是由于流量降低使得卤水与腔体内壁接触时间更长，溶解的盐量增多，因此腔内卤水质量浓度增加。在后期腔体溶通后，注水流量增加，腔体的拓展速度明显增大。此时由于腔内卤水质量浓度低，腔体发展迅速，在很短时间内便达到一定高度。

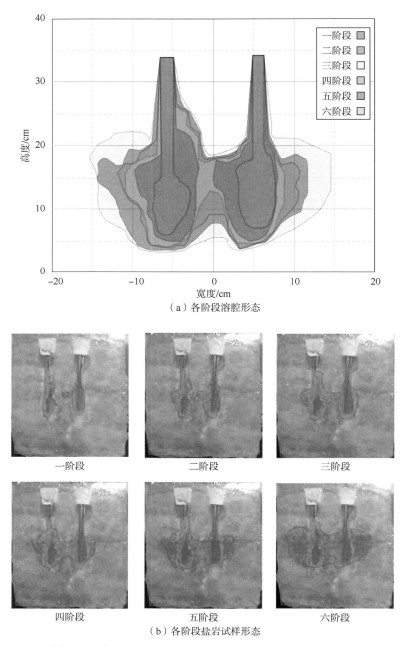

(a) 各阶段溶腔形态

(b) 各阶段盐岩试样形态

图 6　双井间距为 5cm 时无对接井造腔的腔体轮廓变化图

综上，较大和较小的对井间距都不利于溶腔工作：当井间距较小时，腔体水平方向长度小，未能较好地利用水平空间；当井间距较大时，水平段的腔体溶解速度明显降低，且腔体形态不规则。而采用不预制对接井进行造腔，在初期单直井分别造腔阶段，注水流量明显更小，腔体拓展速度更低，建腔效率低。因此，预制水平腔可缩短建槽期时间，提高造腔效率，同时合适的水平对接段能够大幅提高腔体的水平方向拓展长度。

3.2 腔体体积分析

将上述4种造腔方案各阶段腔体体积进行对比(表3)，并对不同造腔累计时间后的腔体体积变化曲线进行分析(图7)可见，10m对井间距无对接井的造腔方案中腔体体积增加速度明显低于另外3种方案。这是由造腔初期(0~6h)采用两口井分别造腔的方法导致的，这也表明单井造腔法的成腔效率远低于对井造腔。在造腔初期(0~6h)，双井间距为10m的造腔速率为683.33cm³/h，双井间距为20m的造腔速率为620.17cm³/h，双井间距为30m的造腔速率为494.17cm³/h，无对接井的造腔速率为260.17cm³/h，含对接井的造腔速度相比不含对接井的造腔速度增加了89%以上。而在整体造腔过程中，双井间距为10m的造腔速率为656.25cm³/h，双井间距为20m的造腔速率为577.58cm³/h，双井间距为30m的造腔速率为485.71cm³/h，无对接井的造腔速率为375cm³/h，含对接井的造腔速度相比不含对接井的造腔速度增加了29%以上。对比可知，采用双井造腔法比传统单直井造腔法成腔速度高20%以上。

表3 4种造腔方案各阶段腔体体积对比表

造腔方案	各阶段造腔体积/cm³					
	一阶段	二阶段	三阶段	四阶段	五阶段	六阶段
10m对井间距有对接井	474.7	1264.3	1927.2	2196.5	3167.0	4518.4
20m对井间距有对接井	986.8	1896.4	2488.8	3233.0	5147.3	—
30m对井间距有对接井	1344.6	2139.1	2837.9	5107.3	—	—
10m双井间距无对接井	831.7	1859.6	2712.3	3417.8	4556.9	6517.4

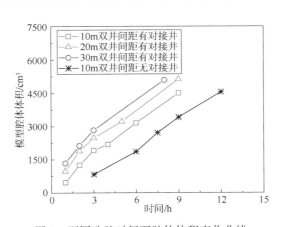

图7 不同造腔时间下腔体体积变化曲线

在同一个造腔方案中，随着造腔时间延长，成腔速率逐渐降低。这表明随着腔体体积增大，注入的淡水从管柱流向腔体侧壁的过程中，会与腔内卤水接触更长的时间，使得造腔后期腔体侧壁的溶蚀速度降低，成腔速率变慢，单位造腔成本增加。在对井造腔工程中，需要确定合适的预期腔体体积，以保持成腔效率并降低单位体积造腔成本。井间距越小的造腔方案，成腔速率越高。这是由于双井间距增大后，注入的淡水需流经更长的水平段才能被另一口井排出，即卤水将与盐岩腔壁接触更长的时间，使得腔体卤水吸收了更多的溶质质量浓度更高，最终使成腔速度降低。

4 双井间距的选取方法

小间距对井造腔方法与单直井造腔技术的溶蚀建腔原理相似，都是注入淡水对目标区域盐岩进行溶解，控制造腔参数形成稳定的溶腔[16-17]。但不同的是小间距对井造腔方法同时采用两个无中心管的井筒进行注水排卤作业，大大提升了排卤能力。依据目前已有双井建库工程案例，井底间距大多在20m左右。对井间距直接影响腔内卤水浓度分布，最终决定溶腔效率和腔体形态，需要依据地质条件和造腔方案确定最优的对井间距。

中国古盐湖矿床的赋存形式多为含夹层薄盐矿层[1]，盐卤开采方式多为水溶压裂对井连通开采技术。但是受测试方法所限，该技术形成的溶腔形态及机理研究尚基本处于空白状态[14]。对于溶腔形态的传统认识主要通过理论计算[18]，认为是在盐层中形成以两井为中心的近椭圆形水平溶腔，在薄层顶板垮塌后逐层上溶[图8(a)]。在建造水平井阶段，利用水平井钻井技术连通两个竖井的底端，可以更好地确保腔体围岩的完整性和密闭性。在排卤阶段，预制水平对接井后，建槽时即可采用一端注水另一端排卤的形式以提高造腔效率[19-20]，只需在竖井中插入生产套管及造腔套管。在注入保护液后，其中一口井注入淡水，另外一口井排出高质量浓度卤水进行造腔。由于注水端的卤水质量浓度更低，导致注水区域的溶蚀速度更快，而排卤区域的腔体较小。为了平衡两侧溶腔的大小，需要适时交替调整进水与出水的顺序。最后分步提升造腔套管及油垫层高度，达到要求的溶腔形状。最终溶腔的横截面近似为长轴为水平方向的椭圆形，并在腔顶形成了具有一定跨距的平台结构[12][图8(b)]。

对于自然溶通法，两口井的井间距不宜过大，如湖北云应储气库采用小间距双井水溶造腔现场的先导试验，得出两井间距约为18m。设定井间距需考虑腔体设计的最大直径。法国TA&TB储气库设计最大直径为90m，双井间距为10m。对于预制水平对接井的对井造腔法，从造腔工艺角度来看，井距偏小可能导致井间对流及涡流，将影响腔体顶部油垫的控制[9]。根据以往的室内对井采卤模拟试验与盐矿对井开采声呐检测结果，对井的井距较大时，两口井之间的通道水溶不发育，水平段盐腔的腔体形状控制成为难题[21]。参考金坛储气库造腔经验设计盐腔最大直径为80m，建槽阶段拟采用单井分别溶滴、自然溶通的方式，双井间距设置在30m以内；对于建造水平对接井以提高薄盐层水平空间利用率的造腔方案[22]，推荐采用井间距较大的对井。

5 结论

小间距对井溶腔技术更适用于中国薄盐层地质条件，在湖北云应储气库已经初步证明了其工程适用性。盐穴储气库溶腔阶段的形状控制是影响腔体质量的关键因素。在小间距对井造腔中，腔体形态不仅与油垫高度、造腔内外管高度等参数有关，还与两井间距有关。

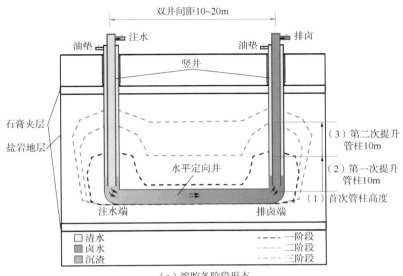

(a)溶腔各阶段形态

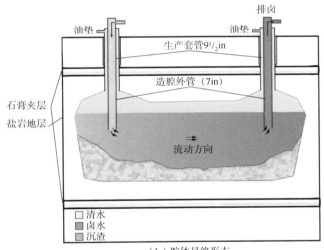

(b)腔体最终形态

图 8 两竖井连通时小间距对井溶腔示意图

在此开展不同井间距的对井造腔物理模型试验，得到以下规律：

(1)小间距对井造腔法可选用更大尺寸的造腔管柱且需更大的注水流量，可有效缩短溶腔时间，大幅提升建槽期工作效率，降低单位体积造腔成本。

(2)相比单井造腔方法，小间距对井造腔法可以在相同高度盐层中形成更大容积的腔体，提高盐矿地层空间的利用率。通过两竖井间水平定向井的布置，可更加灵活地控制腔体形态。

(3)对井间距直接影响腔体的最终形态。当井间距较小时，腔体水平方向长度小，未能较好地利用水平空间；当井间距较大时，水平段的腔体溶解速度明显降低，且腔体形态不规则。

（4）预制对接井的造腔方法拥有更高的成腔速率。随着井间距的增加，造腔速率会小幅增加。但是随着腔体体积的增加，成腔速度会降低。在对井造腔工程中，需确定合适的预期腔体体积以保持成腔效率。

参 考 文 献

[1] Chen J, Lu D, Liu W, et al. Stability study and optimization design of small-spacing two-well (SSTW) salt caverns for natural gas storages[J]. Journal of Energy Storage, 2020, 27: 101131.

[2] 杨春和，李银平，屈丹安，等．层状盐岩力学特性研究进展[J]．力学进展，2008，38(4)：484-494.

[3] 姜德义，李晓康，陈结，等．层状盐岩双井流场浓度场试验及数值计算[J]．岩土力学，2019，40(1)：165-172，182.

[4] Li J L, Yang C H, Shi X L, et al. Construction modeling and shape prediction of horizontal salt caverns for gas/oil storage in bedded salt[J]. Journal of Petroleum Science and Engineering, 2020, 190: 107058.

[5] Li J L, Shi X L, Zhang S. Construction modeling and parameter optimization of multi-step horizontal energy storage salt caverns[J]. Energy, 2020, 203: 117840.

[6] Jiang D Y, Wang Y F, Liu W, et al. Construction simulation of large-spacing-two-well salt cavern with gas blanket and stability evaluation of cavern for gas storage[J]. Journal of Energy Storage, 2022, 48: 103932.

[7] Liu W, Jiang D Y, Chen J, et al. Comprehensive feasibility study of two-well-horizontal caverns for natural gas storage in thinly-bedded salt rocks in China[J]. Energy, 2018, 143: 1006-1019.

[8] 班凡生．层状盐层造腔提速技术研究及应用[J]．中国井矿盐，2015，46(6)：16-18.

[9] 郑雅丽，赖欣，邱小松，等．盐穴地下储气库小井距双井自然溶通造腔工艺[J]．天然气工业，2018，38(3)：96-102.

[10] 姜德义，李晓康，李晓军，等．层状盐岩小井间距双井水溶造腔流场相似模拟研究[J]．工程科学与技术，2017，49(6)：65-72.

[11] 任松，唐康，易亮，等．小井间距双井水溶造腔腔体扩展规律研究[J]．地下空间与工程学报，2018，14(3)：805-812，858.

[12] 易亮，姜德义，陈结，等．小井间距双井水溶造腔模型试验研究[J]．地下空间与工程学报，2017，13(增刊1)：155-161，169.

[13] 易亮，邓清芮．小井间距双井溶腔长期稳定性研究[J]．重庆科技学院学报（自然科学版），2017，19(5)：25-29，51.

[14] 周俊驰，黄孟云，班凡生，等．盐穴储气库双井造腔技术现状及难点分析[J]．重庆科技学院学报（自然科学版），2016，18(1)：63-67.

[15] 何俊，井岗，赵岩，等．盐穴储气库快速造腔方案[J]．油气储运，2020，39(12)：1435-1440.

[16] Yang C H, Wang T T, Qu D A, et al. Feasibility analysis of using horizontal caverns for underground gas storage: A case study of Yunying salt district[J]. Journal of Natural Gas Science and Engineering, 2016, 36(Part A): 252-266.

[17] 完颜祺琪，李康，丁国生．复杂层状盐岩储气库造腔新工艺探讨[C]．银川：2016年全国天然气学术年会，2016：1285-1289.

[18] Li J L, Shi X L, Yang C H, et al. Mathematical model of salt cavern leaching for gas storage in high-insoluble salt formations[J]. Scientific Reports, 2018, 8(1): 372.

[19] 任松，唐康，易亮，等．小间距双井水溶造腔腔体扩展特性[J]．东北大学学报（自然科学版），2017，38(11)：1654-1658.

[20] Wan J F, Peng T J, Jurado M J, et al. The influence of the water injection method on two-well-horizontal salt cavern construction[J]. Journal of Petroleum Science and Engineering, 2020, 184: 106560.

[21] Yang J, Li H, Yang C H, et al. Physical simulation of flow field and construction process of horizontal salt cavern for natural gas storage[J]. Journal of Natural Gas Science and Engineering, 2020, 82: 103527.

[22] Li J L, Tang Y, Shi X L, et al. Modeling the construction of energy storage salt caverns in bedded salt[J]. Applied Energy, 2019, 255: 113866.

本论文原发表于《油气储运》2023年6月2日。

盐穴储气库水溶造腔工艺优化研究与现场应用

王元刚　周冬林　邓　琳　付亚平　管　笛

（中国石油西气东输管道公司储气库项目部）

【摘　要】 盐穴储气库在天然气行业所占比重越来越大，金坛储气库作为国内第一座投产的盐穴储气库，经过工程建设实践发现，很多溶腔工艺存在较大的优化空间。通过分析储气库水溶造腔流程，对造腔过程中涉及的提高造腔效率，减小造腔耗能以及盐层最大化利用进行分析，提出了水溶造腔的优化工艺：采用反循环造腔大幅度增加排卤盐度，减小井下作业次数缩短不必要的停井时间以加快水溶造强进程，安装光纤界面仪实现阻溶剂界面的实时监控，优化注水量大幅度减小能耗，处理厚夹层以及整体扩容腔体对盐层进行最大化利用。

【关键词】 反循环水溶造腔；光纤界面仪；注水量优化；厚夹层处理；腔体扩容

Study on Improving the Solution-mining Processes of Salt-cavern Gas Storage and Field Applications of the Improved Process Technology

Wang Yuangang　Zhou Donglin　Deng Lin　Fu Yaping　Guan Di

(Gas Storage Project Department, PetroChina West-East Gas Pipeline Company)

Abstract　While the salt-cavern gas storage solution is becoming increasingly popular in the natural gas industry, engineering practice shows significant improvement opportunities for many solution-mining processes of the Jintan gas storage, the first commissioned salt-cavern gas storage in China. The solution-mining processes of the gas storage were analyzed to identify opportunities to improve solution-mining efficiency, minimize solution-mining energy consumption, and maximize the utilization of salt layers. The improvement opportunities identified are: employing reverse-circulation solution mining to increase greatly the salinity of the discharged brine; minimizing the number of underground operations and eliminating unnecessary interruptions to mining operations, thereby reducing solution-mining cycle time; installing a fiber-optic interface meter to enable real-time monitoring of the solution-inhibitor interface; optimizing the quantity of water injected to reduce energy consumption; and treating thick intercalated beds and expanding the overall cavern capacity to maximize the utilization of salt layers.

基金项目：中国石油重大科技专项（2015E-40）。

作者简介：王元刚，1986年生，男，汉族，山东临沂人，工程师，硕士，主要从事盐穴储气库水溶造腔相关方面的研究工作。E-mail：515935924@qq.com。

Keywords　reverse-circulation solution mining; fiber-optic interface meter; water injection quantity optimization; thick intercalated bed treatment; cavity expansion

盐穴储气库以其吞吐量大，注采灵活等独特的优势在世界范围的石油工业领域中受到越来越多的重视，金坛储气库作为中国第一座盐穴储气库，经历了10多年的发展历程[1-4]。建库初期由于溶腔经验缺乏，建设进度缓慢，随着金坛盐穴储气库的建成以及投产运行，在原有溶腔工艺的基础上[5-7]，通过不断试验，对溶腔工艺进行优化，经济高效地加快了储气库建设进度。以金坛储气库建设为例，从提高效率、减小建设成本以及盐层最大化利用方面对水溶造腔过程中的工艺进行优化。

1　水溶造腔基本流程

盐穴储气库建设主要选取含盐量较高的层段，采用水溶盐的方式进行建腔。根据地层情况，先确定可用于建库的盐层层位，然后将注采管柱下入到该盐层的指定位置，从地面管线进行注水，淡水在盐腔内与盐层充分接触后，从采卤管返出高浓度卤水。

溶腔过程中，将阻溶剂注入设计位置，保证腔内注入水与阻溶剂上部隔离[8-9]，防止上部盐层过早溶蚀造成盐岩浪费。根据注入和排出卤水的方式分为正循环和反循环两种方式，其中正循环是内管注入淡水，环空排出卤水[图1(a)]，反循环注水排卤方式则相反：环空注入淡水，内管排出卤水[图1(b)]。

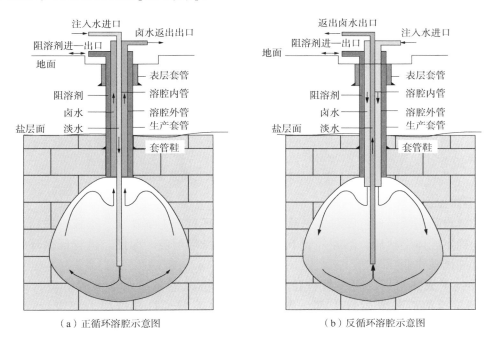

(a)正循环溶腔示意图　　(b)反循环溶腔示意图

图1　正反循环溶腔示意图

储气库建设初期大部分采用正循环过量注入阻溶剂的方式进行溶腔，即从环空中注入大量阻溶剂，过剩的阻溶剂通过油套管环空返至地面，该种方法操作简单，易于观察。

2 优化方案

水溶造腔涉及的工艺参数主要包括循环模式、井下作业、阻溶剂界面监控、注采量以及盐层利用率等，在保证溶腔效果的前提下，可对各工艺参数进行优化，在提高溶腔速度的同时降低了建设投资。

2.1 采用反循环技术加快溶腔速度

2.1.1 采用反循环溶腔

储气库建设初期，正循环溶腔由于操作简单、易于控制被广泛采用，反循环溶腔由于阻溶剂界面位于外管管鞋上方，精确控制难度大，建设初期现场采用较少。正循环溶腔从腔体底部进行溶蚀，由于注入淡水密度小，容易上浮，在未充分溶蚀盐层的情况下就从排卤管返至地面，卤水采出浓度较低，溶腔速度缓慢。反循环溶腔由于注入水在腔体上部与盐层充分溶蚀，溶解出的低浓度卤水，在重力差和浓度差的作用下向底部和四周充分进行扩散，排出卤水质量浓度可达到300g/L。选取well-1井进行试验，采用反循环溶腔时，相同排量时的排卤盐度数比正循环高25g/L，溶腔速度可加快10%（图2）。

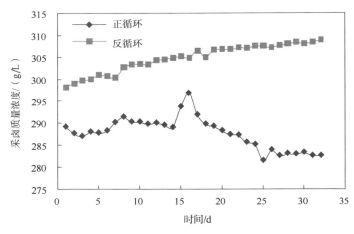

图2 well-1井不同循环模式下的采卤盐质量浓度对比

2.1.2 井下作业次数优化

储气库建设采用反循环溶腔时，一般先采用正循环打开上部盐层，形成一段窗口，进行声呐作业确定腔体形状后，溶腔外管下放，改用反循环溶腔。因此，进行反循环时需要进行两次管柱调整作业。

反循环技术成熟之后，可在管柱位置未调整的情况下完成两个溶腔阶段，即将阻溶剂界面调整至溶腔外管管鞋上部，先进行正循环溶腔，打开盐层上部窗口；然后，流程直接转为反循环继续溶腔，优化后减少了井下管柱作业次数，并节省了井下作业的时间，大幅度提高了溶腔效率。选取well-2井进行试验，在保持溶腔管柱位置以及阻溶剂界面不变的情况下，先进行正循环溶腔然后流程转为反循环。结果显示，腔体形状模拟结果与实际测量数据基本一致（图3），说明优化后的溶腔方案在实际溶腔过程中切实可行。

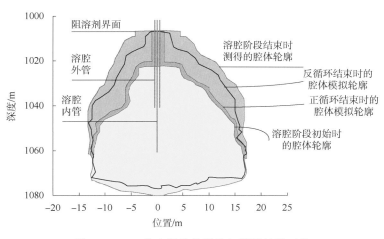

图 3 well-2 井实际腔体形状与模拟结果对比

2.2 光纤界面仪的使用以及排量优化降低溶腔成本

2.2.1 光纤界面仪的使用

溶腔过程中为严格控制阻溶剂界面，需定期注入阻溶剂，当阻溶剂界面在管鞋位置时，一般采用地面观察法，即从生产套管与溶腔外管环空套管注入过量阻溶剂，直至阻溶剂从环空返至地面，该种方法需要注入大量阻溶剂，尤其是在腔体直径很大的情况下，即使注入大量阻溶剂也不一定有阻溶剂返出地面，阻溶剂界面位置无法准确判断，造成资源浪费。well-3 井某溶腔阶段结束后腔体形状如图 4 所示，该阶段后期腔体直径约 35m，在采用地面观察法注入阻溶剂的过程中时，由于腔顶直径较大，为保证阻溶剂返至地面，每次阻溶剂注入量均超过 10m³，该阶段结束时，阻溶剂累计注入量约 100m³。

光纤是一种对温度特别特别敏感的介质，由于卤水和阻溶剂的比热容不同，卤水和阻溶剂从光缆处的热交换也不同，处在不同介质中的光缆升温速度或降温速度不同，在介质分界面位置产生温差梯度[10-11]。根据这一原理研发出一种光纤式阻溶剂界面仪，通过对光纤进行辅助加热，记录加热过程中不同深度处的温度分布，并通过软件计算和记录分界面的实际位置，实现全溶腔井段阻溶剂界面连续监测。

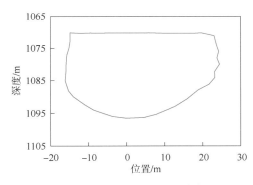

图 4 well-3 井某溶腔阶段结束时的腔体形状

well-4 井安装光纤界面仪后可在地面实时监控界面的微小变化，根据光纤界面仪的测量结果可明显判断出阻溶剂位于 1128.8m（图 5），根据测得的阻溶剂界面深度，确定是否要注入阻溶剂，不再需要定期注入过量阻溶剂，大幅度减小了阻溶剂的消耗量。在 well-4 井安装光纤界面仪后，即使在腔顶很大的情况下，阻溶剂累计注入量只有 30m³，消耗量减小了近 70%。

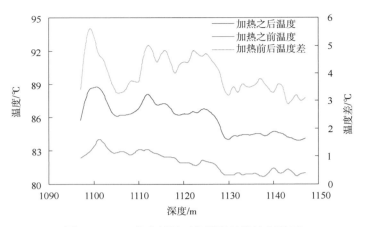

图 5 well-4 井光纤界面仪测得的阻溶剂界面

2.2.2 注水排量优化

储气库建设初期为保证快速溶腔,采用大排量方式溶腔,注入水由于排量较大与盐岩接触时间短,卤水质量浓度低,无法满足盐化企业需求,必须重新注入腔内再循环[12-16]。

现场水溶造腔数据显示,在注入量较高的状态下,注水泵压较高,耗能较大,但是卤水返出质量浓度较低,溶腔量较低,并且二次循环过程中卤水密度较大,需要更大的注入压力才能将卤水排出,储气库建设投资大幅度增加;随着排量逐渐降低,泵压逐渐降低,且注入水与盐岩充分溶蚀,返出卤水质量浓度较高,综合比较,存在一个临界值,既可以保证采出卤水能到达盐化要求,又能大幅度降低溶腔注水耗能。

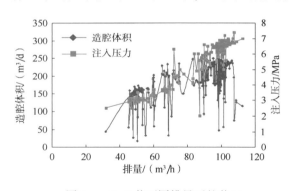

图 6 well-5 井不同排量下的井口注入压力与溶腔体积

well-5 井在不同排量下的泵压以及排卤盐度数据显示,排量较小时,随注入量的增加,泵压以及溶腔体积几乎成比例增加,当注水量超过一定值时,注入压力随注水量增加而增加,但是溶腔量基本不增加(图 6)。说明溶腔过程中,排量不是越大越好,而是存在一个临界值,在这一临界排量下,溶腔速度快且效率最高,统计结果显示,well-5 井最优排量为 80m³/h。

根据 well-5 井的生产状况,设定不同的注水量进行溶腔。统计表明,单位溶腔体积电费减少了 40%。

2.3 充分利用盐层体积

2.3.1 厚夹层处理工艺

中国适用于建造盐穴储气库的盐层普遍为近水平层状分布,建库层段存在大量的石膏层、钙芒硝层和泥岩层等难溶夹层,若干夹层厚度在 2m 以上,最大可达 20m。以往溶腔往往避开厚夹层[17-22],只在夹层下部溶腔,造成厚夹层下部盐层体积损失,最终完腔体积较小。

为了验证厚夹层的可垮塌性，选取溶腔层段含有厚夹层的 well-6 井与 well-7 井进行溶腔试验(图 7)[12-13]。

如图 7 所示，两口井厚夹层下部存在约 15m 纯盐层，通过对厚夹层下部的盐层进行水溶造腔使厚夹层下部形成一定的腔体体积后发现，厚夹层下部腔体体积足够大时厚夹层会大规模垮塌，垮塌之后形成的腔体体积分别为 8000m^3 与 12000m^3，且垮塌之后可继续在厚夹层上部溶腔，下部的盐层得到了充分利用。

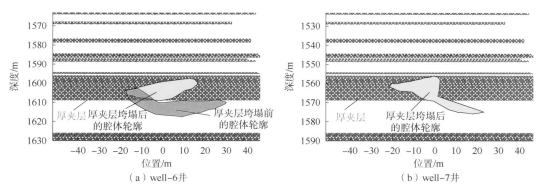

图 7　well-6 井与 well-7 井厚夹层处理结果

2.3.2　盐腔整体扩容技术

盐穴储气库建设初期，主要参照国外模式，矿柱距离较大，两个腔体之间的盐岩有一程度的损失。通过对建库层段地层盐性和腔体稳定性进行分析，腔体之间的矿柱距离在保证盐腔稳定性的基础上，可适当减小。因此，对将完腔的井可进行扩容，适当增大腔体直径，可大幅增加腔体体积。

well-8 井扩容结果显示，该井腔体直径在增加 4m 的情况下，腔体体积增加了约 $1.3×10^4 m^3$，稳定性评结果显示，腔体在扩容后能安全稳定运行(图 8)。

3　结论

(1) 采用反循环溶腔，大幅度提高了外输卤水质量浓度，相对于正循环使溶腔速度可提高 10%；减少井下作业次数，减少了不必要的停井时间，缩短了溶腔周期，降低了溶腔成本。

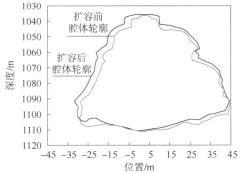

图 8　well-8 井腔体整理扩容效果

(2) 采用光纤界面仪实时监控阻溶剂位置，不需要注入过量阻溶剂，减少了阻溶剂损耗；在保证溶腔速度的同时，减小注水量，大幅度降低水溶造腔的能耗。

(3) 进行厚夹层处理以及腔体整体扩容，在合理范围内，最大限度增加腔体体积，保证盐层最大化利用。

参　考　文　献

[1] 李铁，张永强，刘广文. 地下储气库的建设与发展[J]. 油气储运，2000，19(3)：1-8.

[2] 潘楠.欧美俄乌地下储气库现状及前景[J].国际石油经济,2016,24(7):80-92.
[3] 张刚雄,李彬,郑得文,等.中国地下储气库业务面临的挑战及对策建议[J].天然气工业,2017,37(1):153-159.
[4] 金虓,夏焱,袁光杰,等.盐穴地下储气库排卤管柱盐结晶影响因素实验研究[J].天然气工业,2017,37(4):130-134.
[5] 班凡生,袁光杰,申瑞臣.多夹层盐穴腔体形态控制工艺研究[J].石油天然气学报(江汉石油学院学报),2010,32(1):362-364,390.
[6] 李晓鹏.盐穴储气库建腔技术[J].当代化工,2016,45(7):1460-1463.
[7] 赵志成,朱维耀,单文文,等.盐穴储气库水溶建腔机理研究[J].石油勘探与开发,2003,30(5):107-109.
[8] 李海伟,杨清玉.盐穴储气库溶腔过程中腔体净容积及油水界面计算实例[J].石油化工应用,2015,34(6):56-58.
[9] 胡开君,巴金红,王成林.盐穴储气库造腔井油水界面位置控制方法[J].石油化工应用,2014,33(5):59-61.
[10] 付亚平,陈加松,李建君.盐穴储气库气卤界面光纤式检测[J].油气储运,2017,36(7):769-774.
[11] 张晓威,刘锦昆,陈同彦,等.基于分布式光纤传感器的管道泄漏监测试验研究[J].水利与建筑工程学报,2016,14(3):1-16.
[12] 王文权,杨海军,刘继芹,等.盐穴储气库溶腔排量对排卤盐度及腔体形态的影响[J].油气储运,2015,34(2):175-179.
[13] 李龙,王文权.盐穴储气库造腔跟踪分析方法应用研究[J].石油化工应用,2015,34(7):37-40.
[14] 肖恩山,刘继芹,王晓刚,等.盐穴储气库造腔节能优化技术[J].石油化工应用,2017,36(4):77-88.
[15] 耿凌俊,李淑平,吴斌,等.盐穴储气库注水站整体造腔参数优化[J].油气储运,2016,35(7):779-783.
[16] 李建君,巴金红,刘春,等.金坛盐穴储气库现场问题及应对措施[J].油气储运,2017,36(8):982-986.
[17] 施锡林,李银平,杨春和,等.多夹层盐矿油气储库水溶建腔夹层垮塌控制技术[J].岩土工程学报,2011,33(12):1957-1963.
[18] 施锡林,李银平,杨春和,等.盐穴储气库水溶造腔夹层垮塌力学机制研究[J].岩土力学,2009,30(12):3615-3620.
[19] 郭凯,李建君,郑贤斌.盐穴储气库造腔过程夹层处理工艺——以西气东输金坛储气库为例[J].油气储运,2015,34(2):162-166.
[20] 郑雅丽,赵艳杰,丁国生,等.厚夹层盐穴储气库扩大储气空间造腔技术[J].石油勘探与开发,2017,44(1):137-143.
[21] 龙继勇,陆守权,谢雷,等.W盐穴地下储气库泥岩夹层垮塌试验研究[J].中国井矿盐,2016,47(4):24-29.
[22] 袁炽.层状盐岩储库夹层垮塌的理论分析[J].科学技术与工程,2015,15(2):14-20.

本论文原发表于西南石油大学学报(自然科学版)2018年7月4日。

光纤技术在盐穴储气库油水界面监测中的应用

付亚平[1] 吴 斌[2] 敖海兵[1] 陈加松[1] 管 笛[1] 井 岗[1]

(1. 中国石油西气东输管道公司;2. 中国石油西气东输管道公司南京应急抢修中心)

【摘 要】 盐穴储气库腔体形态控制是制约储气库库容和实现储气库稳定运行的重要因素,控制阻溶剂界面位置是控制腔体形态的关键。为了改进现有控制方法监控成本高、不能实时测量等缺点,提出光纤式界面监测方法:使用新型光纤界面测试仪进行测量,利用分布式光纤测试技术,通过光缆中的电缆加热,由于卤水和垫层介质(一般用柴油、氮气)的比热容不同,光缆与周围介质进行热交换发生温度变化,根据温差即可判断介质界面位置。基于光纤式界面监测方法,研制了新型的分布式光纤界面测试仪,其具有操作简单、探测范围广、监测成本低、实时测量井下介质界面的优点。利用该技术可实现盐穴储气库的大规模反循环造腔,加快造腔速度,保障腔体安全。

【关键词】 盐穴储气库;造腔;光纤;油卤界面;连续监测

Application of optical fiber technology to the oil/water interface monitoring of salt-cavern gas storage

Fu Yaping[1] Wu Bin[2] Ao Haibing[1] Chen Jiasong[1] Guan Di[1] Jing Gang[1]

(1. PetroChina West-East Gas Pipeline Company; 2. Nanjing Emergency Repair Center, PetroChina West-East Gas Pipeline Company)

Abstract The shape control of salt-cavern gas storage is the key to restrict the gas storage capacity and realize the stability of gas storage, and the key technology to control the cavity shape is to control the interface position of dissolution inhibitor. Existing control method is high in monitoring cost and cannot provide real-time measurement. In order to improve these shortcomings, an optical fiber interface monitoring method was proposed in this paper. In this method, a new type of optical fiber interface tester is used for measurement, and the distributed optical fiber testing technology is used and the cable in the optical cable is heated. Brine and cushion medium (generally diesel and nitrogen) are different in specific heat capacity, so heat exchange happens between the optical cable and the surrounding medium, which leads to temperature change. In this way, the position of the medium interface can be determined accord-

---

基金项目:中国石油重大科技专项"地下储气库关键技术研究与应用",2015E-400301。
作者简介:付亚平,男,1985年生,工程师,2013年硕士毕业于中国石油大学(北京)地球探测与信息技术专业,现主要从事油气储运和光纤技术相关的研究。地址:湖北省武汉市洪山区雄楚大道977号,430073。电话:021-50950395。Email:fu_angle@163.com。

ing to the temperature difference. This device is advantageous with simple operation, wide detection range, low monitoring cost and real-time measurement of downhole medium interface. The application of this technology can realize large-scale reverse-circulation solution mining of salt-cavern gas storage, increase solution mining speed and guarantee cavity safety.

Keywords　salt-cavern gas storage; solution mining; optical fiber; oil-brine interface; continuous monitoring

盐穴储气库腔体形态控制是制约储气库库容和实现储气库稳定的关键,控制阻溶剂界面位置可以有效地控制腔体形态。一般选用密度小于卤水的流体介质作为盐穴造腔阻溶剂,如柴油、机油、氮气或天然气等。阻溶剂的控制是造腔的关键,阻溶剂界面位置失控可能会造成腔体形态畸变,严重时可能会导致腔体形态失控,最终致使腔体垮塌甚至报废[1-2]。常用的阻溶剂控制方法有地面观察法、压力表监测法及中子测井法等[3-4]。其中,地面观察法判断方法简单、直接成本低,但只适用于正循环造腔;压力表监测法测量结果不准确;中子测井法测量准确,但成本高;井下电阻式传感器测量稳定性差且测量范围有限。综上,由于这些方法均无法实现大规模反循环造腔,因此提出光纤式介质界面监测方法。

1　光纤式介质界面监测原理

光纤式介质界面监测方法(图1)基于分布式光纤测试技术和光缆加热控制技术[5],首先测量井下各深度处光缆附近介质的初始温度,随后利用地面仪器对光缆进行加热并测得其加热后温度,通过光缆在加热前后探测得到的温差来判断界面的深度,并将参数实时传输到地面解调器进行显示及存储,根据设计要求实时调整界面位置。该方法需要将光缆固定在造腔外管并置于井下,对井下阻溶剂和卤水界面进行实时监控,操作简单,能够探测全部溶腔井段的深度[6-13]。

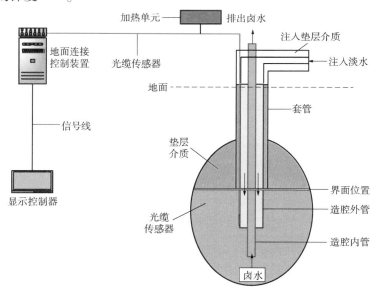

图1　光纤式介质界面监测方法示意图

1.1 光纤测试理论

分布式光纤温度传感系统(Distributed Temperature Sensing, DTS)能够以距离的连续函数形式在整个连续的光纤上测量出各点的温度。在同步控制单元触发下,将较高功率的激光脉冲输入光纤,激光脉冲在光纤中传输时与光纤中的分子、杂质等相互作用而发生散射。测量 Raman 散射产生的两种散射光的光强,根据其比值即可得到温度信息,通过光时域反射(Optical Time Domain Reflectometer, OTDR)技术,确定温度所对应的位置,从而得到沿整条光纤的温度分布情况[14-16]。

1.1.1 光时域反射定位

OTDR 技术可以实现测量点的空间定位(图2),确定沿光纤温度场中每个温度采集点、异常温度点以及光纤断裂点与 OTDR 光纤接头的距离[17-19]。

在时域里,入射光经过背向散射返回至光纤入射端所需时间为 t,激光脉冲在光纤中走过的路程为 $2L$,由此可以得到散射光距离测量处的距离:

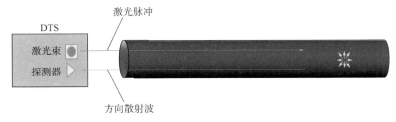

图 2　激光脉冲在光纤中传播示意图

$$L = \frac{ct}{2n_g} \tag{1}$$

式中:c 为真空中的光速,m/s;n_g 为光纤实际折射率。

1.1.2 温度测量

光纤测温原理:依据后向拉曼散射效应,激光脉冲与光纤分子相互作用发生能量交换产生散射。当光子被光纤分子吸收后会再次发射出来。如果部分光能转换为热能,将发出一个波长较长的 Stokes 光。如果部分热能转换为光能,将发出一个波长较短的 Anti-stokes 光。拉曼散射光则是由这两种不同波长的光组成,其波长的偏移由光纤组成元素的固有属性决定。

为了消除激光管输出不稳定、光纤弯曲以及接头损耗等的影响,提高测温准确度,温度信息的解调采用双通道双波长对比法。该方法分别采集两种光,利用两种光强的比值解调温度信号。由于 Anti-Stokes 光对温度更为灵敏,因此将 Anti-Stokes 光作为信号通道,而 Stokes 光作为对比通道。故从光波导内任何一点的 Anti-Stokes 光强与 Stokes 光强有关的比例关系中,可以得到该点的温度计算式[4]:

$$T = \frac{hc\Delta\gamma}{k}\left[\ln\left(\frac{I_s}{I_a}\right) + 4\ln\frac{\lambda_a}{\lambda_s}\right]^{-1} \tag{2}$$

式中:T 为绝对温度,K;h 为普朗克常数;$\Delta\gamma$ 为偏移波数,m^{-1};k 为玻尔兹曼常数;I_s、I_a 分别为 Stokes 和 Anti-stokes 光强,cd;λ_s、λ_a 分别为 Stokes 和 Anti-stokes 光波长,m。

1.2 光纤界面监测

光纤式介质界面监测方法能够实时、连续监测盐穴储气库造腔井段阻溶剂界面，保障大规模反循环造腔的安全实施。通过前期的试验，研制了分布式光纤界面测试仪。光纤特有的防燃、防爆、抗腐蚀、抗干扰等优点，使其能在有害环境中安全运行，有效解决了其他井下监测设备存在的易腐蚀、易损坏问题。

分布式光纤界面测试仪工作原理(图3)：当光缆与井下的流体达到温度平衡时，光纤界面测试仪记录当前光纤所在各深度处的温度分布。对光缆进行辅助加热，保证整条光缆上每个单元的发热量相同。由于不同介质的比热容不同[水为4200J/(kg·℃)，柴油为2100J/(kg·℃)，空气为1030J/(kg·℃)]，阻溶剂(油或气)和卤水在光缆上发生的热交换也不同，处于不同介质中的光缆升温速度或降温速度不同，会在介质分界面位置产生温差梯度。通过记录加热过程中温度的空间分布，处理温度数据，可以计算得出该分界面的深度位置，并借助软件显示和记录。

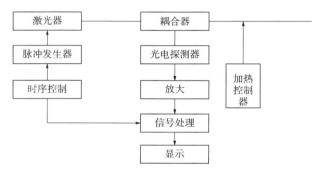

图3 分布式光纤界面测试仪工作原理示意图

光纤式介质界面监测法主要用到的设备：复合铠装光缆(包括感温多模光纤单元和加热单元)，分布温度解调仪、监测控制器及加热控制器。与传统的电缆监测油水界面不同，光纤式介质界面监测方法主要有以下优点[20-23]：(1)光纤具有抗腐蚀，抗射频和电磁干扰，防燃、防爆、耐高温、强电场、电离辐射，且稳定、耐用等特点，能在有害环境中安全运行，有效避免了其他井下监测设备存在的易腐蚀、易损坏问题；(2)监测范围广，可以实现全造腔井段油水界面监测；(3)安装后，能够实时、连续监测。

2 地面试验验证

为了验证该方法是否能够准确识别盐穴储气库溶腔过程中的油水界面位置，进行了地面试验。试验步骤：(1)将调制解调仪、加热单元及光缆连接好；(2)将具有加热功能的一段光缆(加热段为608～700m)置于加入柴油和自来水的水管中，待光缆稳定后测试光缆的初始温度即为标定原始温度，根据温度范围调整温度基准线；(3)用尺测量出油水界面所处的深度位置约614m；(4)利用地面加热单元对测试光缆进行加热，通过仪器记录加热后的温度变化曲线，并在显示控制器上显示温差变化曲线，通过温差曲线的半幅点法以识别出柴油和卤水界面的位置614m[图4(a)]；(5)改变油水界面的位置，用尺量出油水界面位置在615m处；(6)静置一段时间，待流体稳定后，记录测试光缆的初始温度；(7)重复步骤

(4),识别出油水界面的位置615m[图4(b)]。

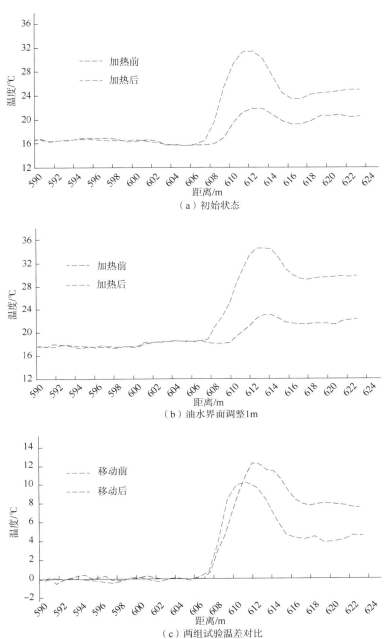

图4 不同界面位置下加热前后温度对比曲线

将油水界面移动前后的温差曲线[图4(c)]对比可见,移动前后的油水界面深度相差1m。由于油和水的比热容相差不大,如果能够准确识别出油水界面,则容易识别气水界面。

3 实际应用

光纤式介质界面监测仪在测量前需要将光缆固定在造腔外管外侧：在管柱最底端，利用缠绕光缆或其他物件焊接固定住光缆；在管柱中部，用钢带固定；在管柱的套管接箍处使用接箍保护器，防止管柱下井过程中损坏光缆。井下安装完毕后，从井口四通的旁侧穿出光缆，将光纤头熔接上光纤跳线，连接好地面设备，测试检查仪器的状况，完成仪器的安装准备工作后，即可进行介质界面测试。

利用光纤测试法测得某盐穴储气库 A 井的界面温度相关曲线[图5(a)]，利用中子寿命测井可得该井的界面示意图[图5(b)]。对比某盐穴储气库2个造腔井利用光纤式介质界面监测方法测量结果与中子测井方法测量结果(表1)可见，两种方法的测量差值均小于或等于0.6m。

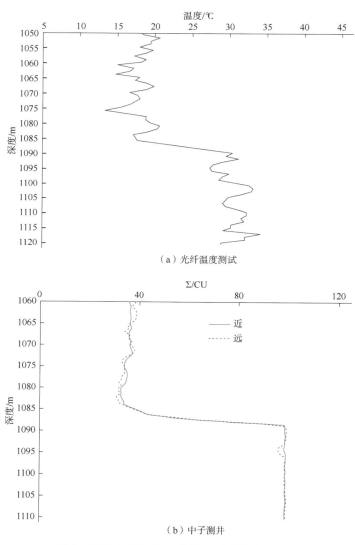

图 5 某盐穴储气库 A 井界面测试结果对比图

表 1 某储气库 A 井、B 井两口井界面测量结果对比

造腔井号	测试次数	注退油量/m³	界面位置/m 中子测量	界面位置/m 光纤测量	测量差值/m
A 井	1	0	1109.1	1 109.1	—
A 井	2	退 0.5	1105.3	1105.0	0.3
A 井	3	退 0.1	1104.3	1103.8	0.5
A 井	4	退 1.5	1086.5	1086.0	0.5
A 井	5	注 2.1	1109.8	1109.2	0.6
B 井	1	0	1069.0	1069.0	—
B 井	2	退 3.0	1053.5	1053.4	0.1
B 井	3	注 0.7	1066.4	1066.5	0.1
B 井	4	注 2.3	1068.7	1068.9	0.2

在实际应用中，通过光纤界面监测方法能够实时测量油水或气水界面，拓展了造腔工程的设计思路。在早期，由于造腔时界面控制较困难，主要利用柴油作为阻溶剂，且采用正循环造腔方法。目前，可改用氮气等气体作为阻溶剂，并大规模采用反循环造腔方法[24]，能够大大提高造腔速度。在一个造腔阶段结束后，可以只提升垫层与卤水界面的位置，而不用通过井下作业上提管柱，减少了造腔管柱调节次数、作业费用。根据界面测试结果，可适量减少阻溶剂的注入量，节省成本。此外，通过温度变化可以对井下造腔管柱损坏或造腔过程中腔顶塌落进行预警，避免造腔阶段风险事故的发生。

4 结论

为了监控盐穴腔体造腔过程中的阻溶剂界面，在盐穴储气库造腔阶段可以采用光纤式阻溶剂界面监测方法。

（1）该方法具有测量深度范围大、操作简单、实时连续监控介质界面等优点。

（2）若造腔阶段能更多地采用反循环造腔方法，可以大大缩短造腔时间。

（3）在造腔过程中使用该方法监测界面，在其稳定工况下，可以减少井下作业次数、中子测井次数、阻溶剂注入量，大大节省单井造腔成本。

（4）推广使用该方法，能够拓宽造腔工程设计思路，可以进一步推广使用天然气或氮气阻溶造腔。

（5）分布式光纤界面测试仪加入光纤测量压力的功能后，可以同时测出井下温度和压力，有助于提高判断井下管柱脱落故障的准确率。

参 考 文 献

[1] Pereira J C. Common practices – gas cavern site characterization, design, construction, maintenance, and operation[C]. Houston：SMRI Spring 2012 Technical Conference，2012：18-19.

[2] 李建君，王立东，刘春，等. 金坛盐穴储气库腔体畸变影响因素[J]. 油气储运，2014，33（3）：269-273.

[3] 胡开君，巴金红，王成林. 盐穴储气库造腔井油水界面位置控制方法[J]. 石油化工应用，2014，33（5）：59-61.

[4] 付亚平，陈加松，李建君．盐穴储气库气卤界面光纤式检测[J]．油气储运，2017，36(7)：769-774.

[5] 吴冰，朱鸿鹄，曹鼎峰，等．基于主动加热光纤法的冻土相变温度场特征分析[J]．工程地质学报，2019，27(5)：1093-1100.

[6] 西涛涛．分布式光纤温度传感器在油田高温测井中的应用[J]．石化技术，2018，25(8)：16-17.

[7] 饶云江．长距离分布式光纤传感技术研究进展[J]．物理学报，2017，66(7)：139-157.

[8] 张建华，巴金红，李建君，等．光纤式储油罐多相介质厚度测量装置：201520122874.4[P]．2015-08-05.

[9] Grosswig S, Vogel B. Permanent blanket-brine interface monitoring by temperature monitoring in salt caverns[C]．Porto：SMRI Spring 2008 Technical Conference，2008：2-6.

[10] 汪会盟，巴金红，程林，等．盐穴储气库新型光纤连续油水界面监测仪研究[C]．成都：第三届全国油气储运技术交流大会，2013：1303-1305.

[11] Brentle J O, Möbius C. Fibre optic measurement system for the automatic and continues blanket level interface monitoring during the solution-mining process of salt caverns[C]．New York：SMRI Spring 2015 Technical Conference，2015：1-5.

[12] 李海伟，何俊，张幸，等．基于光纤技术的盐穴储气库溶腔过程油水界面监测[J]．油气储运，2018，37(9)：1037-1040.

[13] 赵雪峰，张鑫旺．基于主动加热测温的分布式光纤传感技术在海底管道冲刷悬空监测中的应用[J]．中国海洋平台，2018，33(2)：22-29.

[14] 况洋，吴昊庭，张敬栋，等．分布式多参数光纤传感技术研究进展[J]．光电工程，2018，45(9)：62-77.

[15] 王玉田，郑龙江，侯培国，等．光电子学与光纤传感器技术[M]．北京：国防工业出版社，2003：106-109.

[16] 王传琦，伍历文，刘阳．分布式光纤温度和应变传感系统研究进展[J]．传感器与微系统，2017，36(4)：1-4，7.

[17] Bazzo J P, Pipa D R, Silva E V D, et al. Sparse reconstruction for temperature distribution using DTS fiber optic sensors with applications in electrical generator stator monitoring[J]．Sensors，2016，16(9)：1425.

[18] Nuñez-lopez V, Muñoz-torres J, Zeldounl M. Temperature monitoring using distributed temperature sensing (DTS) technology[J]．Energy Procedia，2014，63：3984-3991.

[19] Lu Y, Zhu T, Chen L, et al. Distributed vibration sensor based on coherent detection of phase-OTDR[J]．Journal of Lightwave Technology，2010，28(22)：3243-3248.

[20] 张晓威，刘锦昆，陈同彦，等．基于分布式光纤传感器的管道泄漏监测试验研究[J]．水利与建筑工程学报，2016，14(3)：1-6.

[21] 铁成军，燕云，张娟，等．检测水泥环密封性的光纤传感技术试验研究[J]．石油机械，2018，46(3)：22-28.

[22] 徐锲，许海燕，宋耀华，等．基于光纤分布式传感器的时频定位技术[J]．仪器仪表学报，2014，35(10)：2161-2169.

[23] 任利华，陈德飞，潘昭才，等．超深高温油气井永久式光纤监测新技术及应用[J]．石油机械，2019，47(3)：75-80.

[24] 李建君，陈加松，刘继芹，等．盐穴储气库天然气阻溶回溶造腔工艺[J]．油气储运，2017，36(7)：816-824.

本论文原发表于《油气储运》2020年第39卷第1期。

淮安盐穴储气库厚夹层盐层造腔工艺设计

齐得山 李建君 巴金红 王元刚 敖海兵 李淑平 王文权

（中国石油西气东输管道公司储气库项目部）

【摘 要】 为了充分利用已探明的优质盐矿资源建设盐穴储气库，尽量避免盐层中厚夹层对造腔的影响，需要开展厚夹层盐层条件下的造腔工艺研究。通过造腔试验，明确了厚夹层在造腔过程中是可以垮塌的，其下部盐岩可以造腔。基于盐层地质特征和中国同类项目造腔实践经验，将淮安储气库腔体的理想形态设计为顶底近似为圆锥体、主体为圆台，最大直径位于圆台底部。基于钻井资料，设计了造腔工艺方案并进行数值模拟，思路如下：首先采用正循环方式在厚夹层下部盐岩造腔，然后调整造腔管柱，在厚夹层上下同时造腔，促使厚夹层垮塌；进入造腔中后期，改为反循环造腔，可加快造腔进度。数值模拟与实际应用表明该造腔工艺方案可充分利用造腔段的盐岩，该设计思路可为造腔过程中厚夹层的处理工艺提供借鉴。

【关键词】 盐穴储气库；造腔工艺；厚夹层；腔体形态；盐岩

Design of solution mining technology for thick-interlayer salt beds in Huai'an salt cavern gas storage

Qi Deshan Li Jianjun Ba Jinhong Wang Yuangang
Ao Haibing Li Shuping Wang Wenquan

(Gas Storage Project Department, PetroChina West-East Gas Pipeline Company)

Abstract For the full use of proven high-quality salt resources to build salt cavern gas storage and minimizing the impact of thick interlayer in salt beds on solution mining, it is required to carry out research on solution mining technology under the condition of thick interlayer salt beds. It can be clearly learnt from the cavern building test that, the thick interlayer is likely to collapse during the solution mining process, and the lower salt rock is considered to be capable of cavern building. Based on the geological characteristics of salt beds and the practical experience of cavern building in similar projects in Chi-

---

基金项目：中国石油储气库重大专项"地下储气库关键技术研究与应用"子课题"盐穴储气库加快建产工程试验研究"，2015E-4008。

作者简介：齐得山，男，1987年生，工程师，2013年硕士毕业于中国石油大学（北京）地质工程专业，现主要从事盐穴储气库溶腔工艺优化设计及动态跟踪分析等研究工作。地址：江苏省镇江市润州区南徐大道商务A区D座中国石油西气东输管道公司储气库管理处，212000。电话：15862931327。E-mail：1427225650@qq.com。

na, the cavity of Huai'an gas storage should be designed in a desirable shape with the top and bottom as an approximately cone and the main body as frustum of a cone with the maximum diameter located at the bottom of frustum of a cone. Based on the drilling data, the cavern building plan was designed and numerically simulated. The idea is as follows: firstly, a positive circulation method should be adopted for cavern building in the salt rock at the lower part of the thick interlayer, and then the cavern building string should be adjusted to simultaneously build the cavern at the upper and lower part of the thick interlayer for accelerating the collapse of the thick interlayer. In the middle and late stage of cavern building, the reverse circulation cavern building method should be adopted to speed up the progress of cavern building. The numerical simulation and practical application show that the solution mining technology could take full advantage of salt rocks in the cavern building areas. The design idea is expected to provide a reference for the treatment of medium and thick interlayer in the cavern building process.

Keywords salt cavern gas storage; solution mining technology; thick interlayer; cavity shape; salt rock

盐穴储气库利用在地下盐层或盐丘中采用水溶方式开采盐岩形成的地下洞穴储存油气[1-4]。中国含盐盆地众多，岩盐资源丰富，但是盐矿普遍以层状为主，盐层内夹层多，夹层厚度从1~10m及以上不等，适合建设盐穴储气库的优质盐矿资源并不多[5-7]。经过多年的探索与实践，目前只有江苏金坛、江苏淮安、河南平顶山和湖北潜江等地的部分盐矿适合建设盐穴储气库[8]。为充分利用现有盐矿资源建设储气库，开展多夹层尤其是含厚夹层盐层条件下的造腔工艺研究具有重要意义。

在夹层垮塌造腔方面，班凡生等[9-14]以薄夹层盐层为对象，研究了造腔过程中夹层的垮塌机理及垮塌控制工艺，认为盐穴储气库建设应尽量避开夹层多且厚的盐层。李建君等[15-17]研究了夹层对盐腔形态体积的影响，总结了多夹层盐岩段造腔工艺，认为只要工艺方案合理，就可避免薄夹层对造腔的影响。在厚夹层盐层造腔方面，郑雅丽等[18]通过数值模拟获得了厚夹层的垮塌规律。王元刚等[19]基于造腔实践总结了厚夹层的垮塌控制工艺。

以下基于淮安盐穴储气库地质特征、造腔实践等数据，论述厚夹层盐层条件下的腔体形态设计、造腔方案设计及造腔注意事项等，以期为后续同类盐穴储气库的造腔设计和建设提供借鉴。

1 淮安储气库概况

1.1 地质特征

淮安储气库构造位于苏北盆地洪涟阜坳陷洪泽断陷中的赵集次凹内，研究区内断层不发育，构造形态比较简单，整体表现为一个向东南倾没的单斜形态，产状平缓。造腔目的层段为古近系阜四段上盐亚段，盐层以层状稳定分布，从东北向西南埋深逐渐增加，厚度逐渐增大，顶界埋深1000~2500m，厚度为37.5~169.5m，平均厚度为123.9m，盐岩主要由石盐组成，夹部分硬石膏、钙芒硝等，石盐体积分数为55.4%~86.8%。

造腔目的层段盐层自上而下依次发育6个盐层，各盐层在不同方向上有不同的变化趋势，除盐层5在库区西侧尖灭不发育外，其他盐层均在库区范围内普遍发育。库区西侧造腔段盐层 4 下发育一套 10m 左右的厚夹层，厚夹层下发育有厚度约 20m 的盐层 6，盐岩品位较好，不溶物含量低(图 1，其中 GR 和 DT 分别代表自然伽马测井曲线和声波测井曲线)。

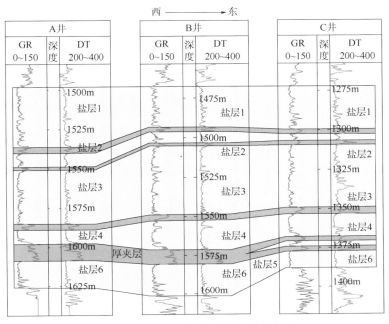

图 1　淮安储气库东西向连井剖面图

1.2　造腔试验

由于造腔过程中厚夹层下部的盐岩有可能溶解形成腔体，从而导致厚夹层出现垮塌，为此对淮安储气库 A 井进行了造腔验证试验。A 井位于库区西侧，钻遇盐顶深度为 1499m，盐底深度为 1626m，盐层下部泥质厚夹层顶深为 1596m、底深为 1608m、厚度为 12m。A 井造腔试验从开始到结束共经历 4 个阶段，测过 4 次腔体形态体积(图 2)。

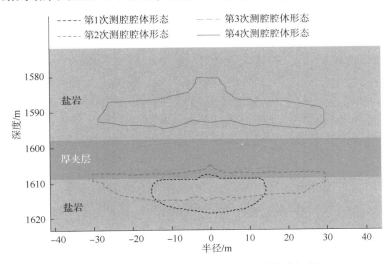

图 2　淮安储气库 A 井 4 次声呐测腔腔体剖面图

（1）厚夹层下正循环造腔。油垫位于厚夹层下，造腔外管深度与油垫位置一致，造腔内管深度位于盐底上部3m处，阶段目标主要是建立腔体底槽，以堆放盐层中的不溶物。阶段结束时，腔顶直径为30m，内管被埋，油垫控制较好，厚夹层未垮塌。

（2）厚夹层下反循环造腔。油垫和造腔外管深度不变，将造腔内管提至腔底以上2m处，阶段目标是加快造腔速度并最大限度地增加腔顶直径。阶段结束时，腔顶直径为60m，厚夹层未垮塌，但由于反循环造腔时油垫与造腔外管深度一致，油垫不好控制，厚夹层底部与卤水充分接触，导致厚夹层部分垮塌。

（3）厚夹层上下同时造腔。油垫深度位于厚夹层上，造腔外管与油垫深度一致，造腔内管深度位于腔底以上2m处，正循环造腔，阶段目标是在厚夹层上下同时造腔，确定厚夹层能否垮塌。该阶段造腔开始10天后，排卤质量浓度由240g/L突降至50g/L，随后进行的测井作业中，仪器下入遇阻，推断是厚夹层垮塌将管柱砸断变形导致。阶段结束时的腔体形态也表明厚夹层已经大面积垮塌。

（4）厚夹层上正循环造腔。油垫与造腔外管深度一致，位于厚夹层上部，造腔内管深度位于腔底以上2m处，阶段目标是充分利用厚夹层上部的盐岩造腔。阶段结束时，腔体已完全在厚夹层上部，最大直径达到60m，说明厚夹层已经完全垮塌。

造腔试验结果表明：造腔段下部的厚夹层是可以垮塌的，厚夹层下的盐岩可以用来造腔；在厚夹层下造腔，以厚夹层作为顶板时，最大跨度可以达到60m；反循环造腔速度比正循环快，且能够有效增大腔顶直径，但是应用在厚夹层下造腔时，油垫不好控制，卤水和厚夹层容易过早接触，导致厚夹层过早垮塌，厚夹层下的盐岩不能充分利用；新井造腔初期应采用正循环造腔，可以充分利用下部盐岩。

1.3 造腔注意事项

建槽期第1阶段应注意控制油垫深度，否则油垫上升，厚夹层过早与卤水接触，会导致厚夹层过早垮塌，下部盐岩不能被充分利用。建槽期第2阶段应注意监控排卤浓度，厚夹层垮塌时可能会砸断、砸弯造腔内管，导致排卤浓度急剧下降，出现这种情况后需要及时进行井下作业，更换造腔管柱。另外，造腔时尽量不要长时间停井，以免造腔管柱被不溶物或卤水中析出的盐堵塞。

2 造腔方案设计

淮安储气库盐层地质特征和造腔试验分析结果表明，厚夹层在造腔过程中可以垮塌，厚夹层下部盐岩可以利用。根据中国同类项目已完腔腔体形态，确定腔体形态的理想参数，并在此基础上设计造腔工艺方案，进行数值模拟，进一步预测腔体完腔后的形状。

2.1 腔体形态设计

稳定性评价结果表明，椭球形、球形及梨形的腔体稳定性最好，但从金坛盐穴储气库的资料来看，建成形态规则的腔体难度很大。金坛储气库多口完腔腔体不同造腔阶段的盐腔形态表明，盐腔的顶、底近似为锥体，主体形态近似为圆台（图3）。基于金坛储气库已完腔的腔体形态和淮安储气库造腔试验成果，将淮安储气库腔体的理想形态设计为顶底近似圆锥体，主体为圆台，主体最大直径80m，位于圆台底部；主体最小直径70m，位于圆台顶部(图4)。

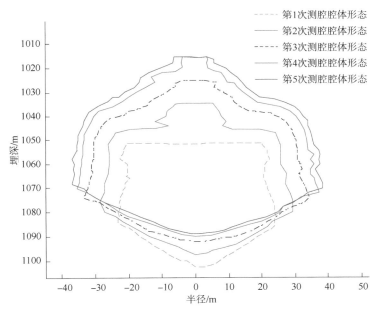

图 3　金坛储气库某完腔井历次测腔腔体剖面图

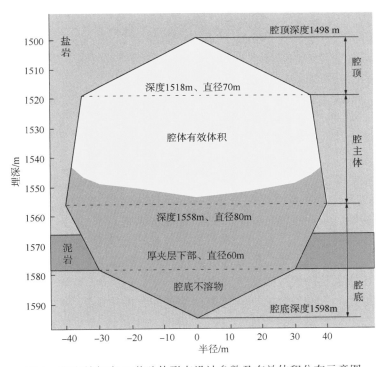

图 4　淮安储气库 B 井腔体形态设计参数及有效体积分布示意图

淮安储气库 B 井钻遇盐顶深度为 1468m，盐底深度为 1600m，盐层厚度为 132m；盐层下部泥质厚夹层顶深为1568m、底深为1578m，厚度为10m。腔体顶板预留30m，则腔顶深度为1498m；腔体底部预留2m，则腔底深度为1598m。剩余可造腔盐层厚度为100m，圆台最大直径位于厚夹层上部10m，最小直径位于腔顶以下20m处，则上部圆锥体高度为20m，主体圆台高度为40m，下部圆锥高度为40m，计算腔体的理论体积为27×10^{4}m^3。B井造腔段内水不溶物体积分数为26.5%，水不溶物膨胀系数为1.6。根据上述参数计算得到B井的单腔有效体积为15.5×10^{4}m^3，其计算公式为

$$V_1 = V_2(1-\alpha\beta) \tag{1}$$

式中：V_1 为腔体有效体积，m^3；V_2 为腔体理论体积，m^3；α 为造腔段不溶物体积分数；β 为不溶物蓬松系数，一般取1.2~1.9。

2.2 造腔工艺方案

根据金坛储气库建设经验，盐穴储气库造腔方案设计原则如下：(1)采用正循环建槽反循环造腔的方式，可加快造腔进度。(2)在保证卤水浓度饱和或近饱和的前提下设计注入排量，建槽期两管距较小，因此采用小排量造腔；建槽完毕进入造腔期后，逐渐增大排量[20]。(3)造腔时用淡水。

2.2.1 建槽期方案

建槽期是新井造腔的初始阶段，主要目的是在盐层底部溶解形成一个类似槽状的盐腔，用以堆放盐层中不能被卤水带出的不溶物[21]。B井主要利用厚夹层下部20m的盐岩建槽，分两个阶段进行。

第1阶段：在厚夹层下部盐岩造腔，采用正循环的方式，造腔外管和油垫深度设置在厚夹层以下2~3m处，造腔内管距离盐底2m，目的是形成一个直径约60m的腔体，以充分利用下部盐岩，并为垮塌的厚夹层提供一个堆积空间。

第2阶段：厚夹层上下盐岩同时造腔，采用正循环的方式，造腔外管和油垫深度设置在厚夹层上部约10m处，造腔内管设置在厚夹层下部2~3m处，目的是使厚夹层腾空，并使厚夹层浸泡在卤水中，加快厚夹层垮塌。

2.2.2 造腔期方案

B井造腔段盐层厚度较小，为使腔体的最大直径能达到80m，造腔时的两管距和管柱提升的高度不宜过大。经过多次优化管柱位置和溶解时间，造腔期可通过5个阶段完成。第1阶段为造腔初期，采用正循环造腔，油垫深度与造腔外管位置一致，有利于扩大腔体直径。1~4阶段为造腔中后期，采用反循环造腔，适当增大油垫与中间管的相对位置，使腔体上部形态逐渐变为锥形，最后形成穹形腔顶。第5阶段为造腔末期，采用正循环造腔，以控制腔顶直径在稳定性评价可接受范围内。

通过数值模拟预测，经过建槽期和造腔期的造腔，B井最终可形成14.7×10^{4}m^{3}的有效体积，与理想形态下有效体积的比值为95%。说明采用该设计方案造腔，可充分利用造腔段的盐岩。

2.3 实施效果

基于上述设计方案，B井采用正循环的方式利用厚夹层下部盐岩进行了第一阶段的建槽，因阶段末期油垫上升，厚夹层与卤水过早接触，致使该阶段建槽提前结束，阶段累计

造腔时间 115 天，累计产盐量为 3.35×10⁴t。声呐测腔结果表明，该井已初步完成建槽，厚夹层已经垮塌，腔体直径约 52m，腔体有效体积为 1.17×10⁴m³（图 5）。

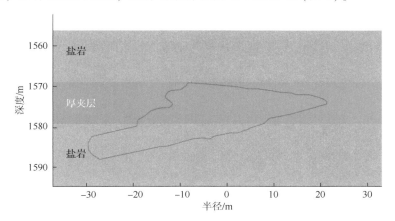

图 5　淮安储气库 B 井第 1 阶段末声呐测腔腔体剖面图

3　结论

（1）造腔试验结果表明：造腔段盐层中的厚夹层在造腔过程中是可以垮塌的，厚夹层下部的盐岩可以用来造腔；在厚夹层下造腔，以厚夹层作为顶板时，最大跨度可以达到 60m。

（2）在厚夹层下部造腔时，应注意控制好油垫深度，避免厚夹层过早垮塌；在厚夹层上下同时造腔时，应注意监控排卤浓度，当排卤浓度急剧降低时，应及时更换造腔管柱。

（3）基于造腔实践经验，将腔体的理想形态设计为顶底近似圆锥体，主体为圆台，最大直径位于圆台底部。

（4）设计造腔方案时，首先采用正循环方式在厚夹层下部盐岩造腔，然后调整造腔管柱，在厚夹层上下同时造腔，促使厚夹层垮塌；进入造腔中后期，改为反循环造腔，可加快造腔进度。

参 考 文 献

[1] Seto M, Nag D K, Vutukur V S. In-situ rock stress measurement from rock cores using the emission method and deformation rate analysis[J]. Geotechnical and Geological Engineering, 1999, 17：241-266.
[2] Bérest P, Brouard B, Durup J G. Tightness tests in salt-cavern wells[J]. Oil & Gas Science and Technology, 2001, 56(5)：451-469.
[3] Liang W G, Yang C H, Zhao Y S, et al. Experimental investigation of mechanical properties of bedded salt rock[J]. International Journal of Rock Mechanics and Mining Sciences, 2007, 44(3)：400-411.
[4] 周学深. 有效的天然气调峰储气技术——地下储气库[J]. 天然气工业, 2013, 33(10)：95-99.
[5] 杨春和, 梁卫国, 魏东吼, 等. 中国盐岩能源地下储存可行性研究[J]. 岩石力学与工程学报, 2005, 24(24)：4409-4417.
[6] 丁国生, 李春, 王皆明, 等. 中国地下储气库现状及技术发展方向[J]. 天然气工业, 2015, 35(11)：107-112.

［7］肖学兰．地下储气库建设技术研究现状及建议［J］．天然气工业，2012，32（2）：79-82．
［8］杨海军．中国盐穴储气库建设关键技术及挑战［J］．油气储运，2017，36（7）：747-753．
［9］班凡生，袁光杰，赵志成．盐穴储气库溶腔夹层应力分布规律［J］．科技导报，2014，32（16）：45-48．
［10］袁光杰，班凡生，赵志成．盐穴储气库夹层破坏机理研究［J］．地下空间与工程学报，2016，12（3）：675-679．
［11］班凡生，袁光杰，申瑞臣．多夹层盐穴腔体形态控制工艺研究［J］．石油天然气学报，2010，32（1）：362-364．
［12］施锡林，李银平，杨春和，等．多夹层盐矿油气储库水溶建腔夹层垮塌控制技术［J］．岩土工程学报，2011，33（12）：1957-1963．
［13］施锡林，李银平，杨春和，等．盐穴储气库水溶造腔夹层垮塌力学机制研究［J］．岩土力学，2009，30（12）：3615-3620．
［14］姜德义，张军伟，陈结，等．岩盐储库建腔期难溶夹层的软化规律研究［J］．岩石力学与工程学报，2014，33（5）：865-873．
［15］李建君，王立东，刘春，等．金坛盐穴储气库腔体畸变影响因素［J］．油气储运，2014，33（3）：269-273．
［16］郭凯，李建君，郑贤斌．盐穴储气库造腔过程夹层处理工艺——以西气东输金坛储气库为例［J］．油气储运，2015，34（2）：162-166．
［17］李建君，陈加松，吴斌，等．盐穴地下储气库盐岩力学参数的校准方法［J］．天然气工业，2015，35（7）：96-102．
［18］郑雅丽，赵艳杰，丁国生，等．厚夹层盐穴储气库扩大储气空间造腔技术［J］．石油勘探与开发，2017，44（1）：1-7．
［19］王元刚，陈加松，刘春，等．盐穴储气库巨厚夹层垮塌控制工艺［J］．油气储运，2017，36（9）：1035-1039．
［20］王文权，杨海军，刘继芹，等．盐穴储气库溶腔排量对排卤浓度及腔体形态的影响［J］．油气储运，2015，34（2）：175-179．
［21］丁国生，张昱文．盐穴地下储气库［M］．北京：石油工业出版社，2010：82-88．

本论文原发表于《油气储运》2020年第39卷第8期。

盐穴储气库水溶造腔的影响因素

王元刚[1]　高寒[2]　薛雨[1]

（1. 中国石油西气东输管道分公司；2. 华东石油技师学院）

【摘　要】 盐穴储气库由于其吞吐量大，注采灵活等优势受到越来越多的关注。大量腔体的完腔数据显示，盐穴储气库的完腔体积与初始设计体积存在一定差距，因此保证盐层利用率达到最大化尤为重要。结合金坛盐穴储气库水溶造腔实例，详细分析了井下异常情况以及地面临井的相互影响这两大因素在造腔过程中对腔体体积的影响，认为建槽期腔体直径、夹层的存在以及腔体偏溶、井眼轨迹偏离等因素对造腔有较大影响。提出造腔过程中应在建槽期充分扩容，制定有效的夹层处理方案并监控各动态参数，防止造腔过程中出现偏溶以及管柱脱落等异常情况，从而保证盐层利用率最大化。

【关键词】 盐穴储气库；体积损失；井下异常；临井影响；改进方案

Influencing factors of solution mining for salt cavern gas storage

Wang Yuangang[1]　Gao Han[2]　Xue Yu[1]

(1. PetroChina West-East Gas Pipeline Company; 2. East China Petroleum Technician College)

Abstract Salt cavern gas storage has attracted growing attention for its advantages of large throughput and flexible injection and production. A mass of cavity completion data shows that the final cavity volume of salt cavern gas storage deviates from the initial design volume to some extent, so it is particularly important to maximize the utilization of salt bed. Combined with the example of solution mining in Jintan Salt Cavern Gas Storage, through detailed analysis of the impact of two major factors, i.e. the underground anomalies and the mutual interaction of surface adjacent wells, on the cavity volume in the solution mining process, it is believed that the cavity diameter, the existence of interbed, partial dissolution of cavity, deviation of well trajectory, etc. in cavity construction stage have great effect on the solution mining. Therefore, in the process of solution mining, volume expansion should be fully carried out in cavity construction stage, effective interbed treatment schemes should be prepared, and various dynamic parameters should be carefully monitored to prevent anomalies such as partial dissolution and pipe string falling off in the solution mining process for the maximum utilization of salt bed.

Keywords　salt cavern gas storage; volume loss; underground anomalies; interaction of adjacent wells; improved scheme

目前世界各国都在大力开展盐穴储气库建设[1-7]。中国第一座盐穴储气库——金坛盐穴储气库于2007年开始注采运行，经过10多年的发展以及经验积累，目前投入注采运行的井已经有20余口，腔体平均完腔体积约20×10⁴m³，在天然气调峰方面发挥了重大作用。但是部分腔体的完腔体积与最初设计存在一定的差距，根据现场造腔经验，阻溶剂异常以及腔体偏溶等异常情况均会对完腔体积产生不同程度的影响。造腔过程中的阻溶剂失控[8-9]、夹层难溶[10-14]等异常情况经常发生，类似情况也会造成盐腔体积损失。将影响盐腔有效体积减小的因素分为井下异常情况和地面临井的相互影响两大类，并对影响因素进行了详细分析。

1　井下异常情况

1.1　建槽期腔体直径较小

金坛储气库建设初期由于缺乏造腔经验，只能借鉴国外技术，忽视了盐穴储气库建库层段不溶物含量高的特点，没有意识到建槽期扩容的重要性。造腔过程中，上部盐层释放的不溶物在底部堆积，在腔体直径较小的情况下，不溶物顶面快速上升，顶面以下盐层不再溶蚀(图1)。

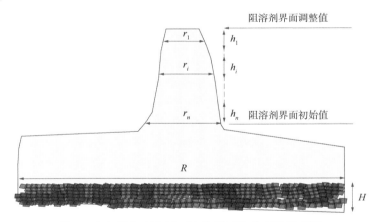

图1　金坛储气库盐层不溶物顶面抬升高度示意图

阻溶剂界面调整后，溶蚀半径达到r_i，该部分溶蚀产生的不溶物膨胀后的体积为

$$V_1 = \sum_{j=1}^{n} \pi r_i^2 h_i \xi \lambda \quad (i = 1, 2, \cdots, n) \tag{1}$$

式中：h_i为阻溶剂界面调整距离，m；r_i为腔体半径，m；ξ为不溶物质量分数；λ为不溶物遇水后的膨胀系数。

假设释放的不溶物在腔底均匀堆积，则该部分体积使不溶物顶面抬升的距离为

$$H=\frac{V_1}{\pi R^2} \quad (2)$$

式中：R 为腔体主体的半径，m。

通过式(1)、式(2)可见，不溶物顶面的堆积速度与腔体主体半径的平方成反比关系，若腔体半径较大，则不溶物堆积速度缓慢，但是在建槽期初始阶段，如果腔体半径较小，则不溶物顶面快速上移，下部盐层得不到充分利用。

以金坛储气库 well-1 井(图2)、well-2 井(图3)为例说明建槽期腔体直径较小对不溶物顶面抬升的影响。well-1 井建槽期造腔管柱下深1205m，腔体直径只有10m，上部盐层中释放的不溶物在底部堆积，腔体直径较小使得不溶物顶面上升很快，未对底部盐层进行充分溶蚀，该井在造腔仅 6766m³ 时，不溶物顶面已经上升约25m，该部分盐层体积损失；well-2 井建槽期腔体直径仅16m，上部盐层中释放的不溶物在腔底快速堆积，第一阶段造腔内管下深为1153m，在造腔仅 7749.4m³ 时，盐层损失已经达到31.5m。

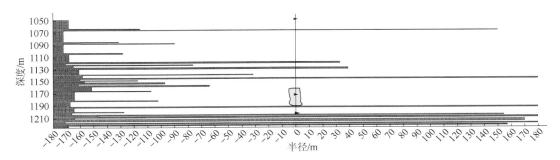

图2　金坛储气库 well-1 井建槽期腔体形状示意图

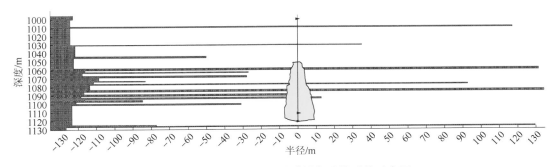

图3　金坛储气库 well-2 井建槽期腔体形状示意图

通过对比两口井的数据可知：建槽期腔体半径较小时，造腔段的不溶物在腔底大量堆积，填埋段的盐层不能被溶蚀，盐层大量损失，最终完腔体积减小。

1.2 夹层分布影响

夹层是影响造腔体积以及腔体形状的重要因素，夹层中不溶物的质量分数最高可达95%以上。由于夹层的难溶性[15-18]，造腔过程中需要谨慎处理，若处理不当，很可能在造腔夹层上部二次溶腔或者造成腔体形状严重不规则。

1.2.1 夹层上部二次造腔

造腔层段中存在夹层时,在夹层完成溶蚀或垮塌前,如果采用的造腔工艺不当,受夹层难溶性影响,夹层位置附近的盐层未被溶蚀,导致在夹层上部二次造腔,完腔体积产生较大损失。

在 well-3 井夹层段直径小于 10m 的情况下,调整管柱在夹层上部进行造腔,不溶物在脖颈段堆积,堵塞了夹层上下的通道,下部腔体体积损失,在夹层上部形成二次造腔(图 4,蓝色部分为二次造腔)。

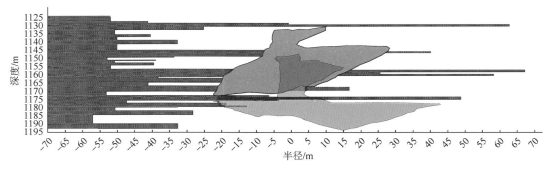

图 4 金坛储气库 well-3 井腔体形状示意图

well-4 井夹层下部有大量残余阻溶剂,淡水无法、上浮溶蚀夹层附近盐岩,不溶物在夹层上部堆积,形成二次造腔,虽然采取了一系列措施,但始终不能溶蚀,因此完腔体积变小(图 5)。

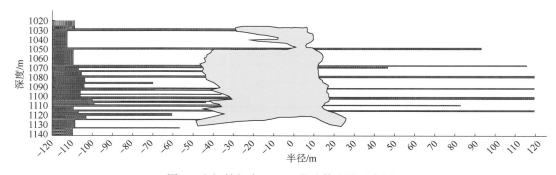

图 5 金坛储气库 well-4 井腔体形状示意图

1.2.2 夹层/岩脊悬空

well-5 井(图 6)造腔段夹层过多,在注水速度较快的情况下,含矿率较高的易溶盐层溶蚀速度较快,而含矿率较低的层段由于不溶物的难溶性,溶蚀速度较慢,容易出现盐层被溶蚀而夹层未被溶蚀的情况,形成岩脊悬在腔中,后续造腔过程中如果造腔管柱下入较深,很可能出现大块夹层因为受力不均脱落将造腔管柱管砸弯或砸断,引发安全事故风险。因此,基于安全考虑,腔体下部不能继续造腔,盐层有效利用率降低。

1.3 造腔过程中管柱脱落

如果管柱作业过程中丝扣未拧紧,其在造腔过程中受注入水连续冲击[19-20]作用或者被

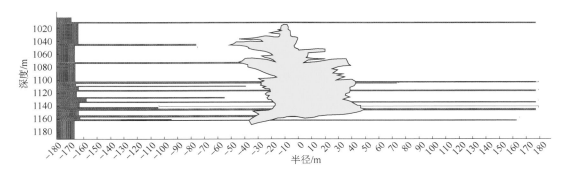

图 6　金坛储气库 well-5 井腔体形状示意图

掉落的大块夹层砸中,则造腔管柱容易从丝扣处脱落,脱落位置下部盐层无法溶蚀,下部盐层有一定体积损失,如 well-6 井(图 7)。

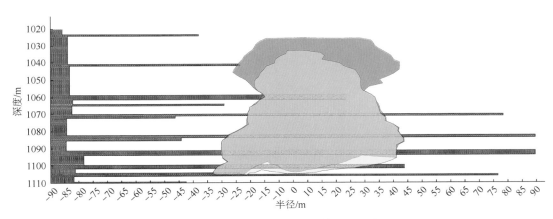

图 7　金坛储气库 well-6 井腔体形状示意图

2　地面临井影响

临井的影响主要体现在两个腔体的距离对腔体最大直径的影响,根据法国 Geostock 公司的建设经验,为保证腔体安全平稳运行,腔体之间必须存在一定的安全矿柱,在盐穴储气库注采运行过程中,矿柱距离(图 8)需要满足以下表达式:

$$\frac{L_P}{(D_1+D_2)/2} \geqslant S \quad (3)$$

$$L_P = L_W - R_1 - R_2 - L_d \quad (4)$$

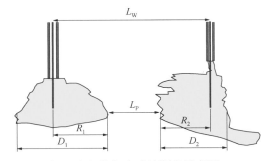

图 8　金坛储气库矿柱距离示意图

式中:L_P 为两个腔体之间的矿柱距离,m;D_1

为井眼连线上腔体 A 的最大设计直径，m；D_2 为井眼连线上腔体 B 的最大设计直径，m；S 为最小安全矿柱比；R_1、R_2 分别为两井连线上腔体相对方向最大的半径，m；L_w 为两井的井口距离，m；L_d 为两口井井底位移的总和，m。

将式（4）代入式（3）可变为

$$\frac{L_w - R_1 - R_2 - L_d}{(D_1 + D_2)/2} \geq S \qquad (5)$$

为保证盐层最大化利用，在腔体最大直径确定的情况下，井距一般按照满足最小安全矿柱比的距离进行设计。

2.1 腔体偏溶

现场造腔过程中，由于夹层的不均匀分布，地层倾角较大[21-24]，以及 KCl、$MgCl_2$ 等易溶组分质量分数较高等因素，偏溶现象经常发生，如果造腔过程中两个腔体向相对方向发生偏溶，即 R_1 或 R_2 增加，则矿柱距离将缩小，如果按照初始设计的腔体最大直径继续造腔，则矿柱比将不满足安全要求，这种情况下，需要缩小腔体的最大设计直径 D_2 或 D_1 来保证安全矿柱比满足安全要求，而当腔体最大直径减小时，腔体的完腔体积则会大幅度减小。

2.2 井眼轨迹

储气库钻井对井眼轨迹要求较高，必须严格控制井斜，但是在岩性复杂或倾角较大的特殊地层中进行储气库钻井时，井眼轨迹会偏离设计轨迹，根据现场数据统计，金坛储气库井底位移最大偏移量为 20m。井底位移偏移较大时，若腔体继续采用初始最大直径造腔，则矿柱距离不能满足安全要求，只能通过减小腔体最大直径的方式使后期运行满足安全要求，而完腔体积也会大幅度减小。

3 结束语

完腔体积是决定盐穴储气库建设的关键因素之一，通过金坛储气库的造腔经验认为，造腔过程中存在的井下异常情况以及临井影响是影响腔体体积的主要因素。针对可能导致盐腔体积损失的因素，提出了保证盐层得到最大化利用的措施：在建槽期尽量扩容，增加腔体半径，减缓不溶物堆积速度；制定合理的溶蚀方案，防止夹层上部二次造腔；钻井过程中严格控制井眼轨迹，并在造腔过程中实时监控各造腔参数，及时发现偏溶以及管柱脱落等异常情况。

参 考 文 献

[1] 宋桂华，李国韬，温庆河，等. 世界盐穴应用历史回顾与展望[J]. 天然气工业，2004，24(9)：116-118.

[2] 王希勇，熊继有，袁宗明，等. 国内外天然气地下储气库现状调研[J]. 天然气勘探与开发，2004，27(1)：49-51.

[3] 丁国生，李文阳. 国内外地下储气库发展现状与趋势[J]. 国际石油经济，2002，10(6)：23-26.

[4] 郑雅丽，赵艳杰. 盐穴储气库国内外发展概况[J]. 油气储运，2010，29(9)：652-655.

［5］杨伟，王雪亮，马成荣．国内外地下储气库现状及发展趋势［J］．油气储运，2007，26（6）：15-19.
［6］丁国生，谢萍．中国地下储气库现状与发展展望［J］．天然气工业，2006，26（6）：111-113.
［7］李铁，张永强，刘广文．地下储气库的建设与发展［J］．油气储运，2000，19（3）：1-8.
［8］胡开君，巴金红，王成林．盐穴储气库造腔井油水界面位置控制方法［J］．石油化工应用，2014，33（5）：59-61.
［9］付亚平，陈加松，李建君．盐穴储气库气卤界面光纤式检测［J］．油气储运，2017，36（7）：769-774.
［10］郭凯，李建君，郑贤斌．盐穴储气库造腔过程夹层处理工艺——以西气东输金坛储气库为例［J］．油气储运，2015，34（2）：162-166.
［11］袁光杰，班凡生，赵志成．盐穴储气库夹层破坏机理研究［J］．地下空间与工程学报，2016，12（3）：675-679.
［12］班凡生，袁光杰，申瑞臣．多夹层盐穴腔体形态控制工艺研究［J］．石油天然气学报（江汉石油学院学报），2010，32（1）：362-364.
［13］徐孜俊，班凡生．多夹层盐穴储气库造腔技术问题及对策［J］．现代盐化工，2015（2）：10-14.
［14］王元刚，陈加松，刘春，等．盐穴储气库巨厚夹层垮塌控制工艺［J］．油气储运，2017，39（9）：1041-1047.
［15］张光华．中石化地下储气库建设现状及发展建议［J］．天然气工业，2018，38（8）：112-118.
［16］李建君，王立东，刘春，等．金坛盐穴储气库腔体畸变影响因素［J］．油气储运，2014，33（3）：269-273.
［17］梁卫国，张传达，高红波，等．盐水浸泡作用下石膏岩力学特性试验研究［J］．岩石力学与工程学报，2010，29（6）：1156-1163.
［18］班凡生，袁光杰，赵志成．盐穴储气库溶腔夹层应力分布规律［J］．科技导报，2014，32（16）：45-48.
［19］李银平，杨春和，屈丹安，等．盐穴储油（气）库水溶造腔管柱动力特性初探［J］．岩土力学，2012，33（3）：681-686.
［20］郑雅丽，赖欣，邱小松，等．盐穴地下储气库小井距双井自然溶通造腔工艺［J］．天然气工业，2018，38（3）：96-102.
［21］杨海军，于胜男．金坛地下储气库盐腔偏溶与井斜的关系［J］．油气储运，2015，34（2）：145-149.
［22］齐得山，李淑平，王元刚．金坛盐穴储气库腔体偏溶特征分析［J］．西南石油大学学报（自然科学版），2019，41（2）：75-83.
［23］完颜祺琪，丁国生，赵岩，等．盐穴型地下储气库建库评价关键技术及其应用［J］．天然气工业，2018，38（5）：111-117.
［24］孙军昌，胥洪成，王皆明，等．气藏型地下储气库建库注采机理与评价关键技术［J］．天然气工业，2018，38（4）：138-144.

本论文原发表于《油气储运》2020年第39卷第6期。

金坛盐穴储气库精细造腔腔体体积优化

齐得山　李建君　赵　岩　刘　春　王元刚　李淑平　刘继芹

（中国石油西气东输管道公司储气库项目部）

【摘　要】 为了充分利用金坛盐矿区有限的盐矿资源，使整个库区盐腔的总体积最大化，需要从金坛储气库正在造腔的井中，筛选出直径可适当增大的盐腔，并优化腔体体积。首先排除即将完腔的、严重偏溶的、临腔最小矿柱比小于 2.3 的 3 类直径不可增大的腔体，然后通过临腔腔体评价，从金坛库区 26 口正在造腔的井中筛选出 10 口腔体，确定最大直径后，重新对腔体体积进行模拟。优化后，10 口腔体的总体积约增加 $26×10^4 m^3$，不仅充分利用了盐岩资源，而且降低了造腔成本，可为今后其他盐穴储气库的建设提供参考。

【关键词】 盐穴储气库；天然气调峰；盐岩；矿柱比；水溶造腔

The cavity volume optimization of fine cavity building in Jintan Salt-Cavern Gas Storage

Qi Deshan　Li Jianjun　Zhao Yan　Liu Chun
Wang Yuangang　Li Shuping　Liu Jiqin

(Gas Storage Project Department, PetroChina West-East Gas Pipeline Company)

Abstract To make full use of the limited salt resources in the area of Jintan salt mine and maximize the total cavity volume of the whole Jintan gas storage, it is necessary to select the cavities whose diameter can be increased appropriately in the cavity building wells, and then optimize their volumes. In this paper, three types of cavities whose diameter cannot be increased were firstly excluded, including the one to be completed soon, suffered from serious partial solution, or with minimum pillar/diameter ratio (P/D) less than 2.3 in its adjacent cavity. Then, based on the evaluation of its adjacent cavity, 10 cavities were selected from 26 cavity building wells. Finally, the cavity volume was simulated again after its maximum diameter was determined. After the optimization, the total volume of 10 cavities is increased by $26×10^4 m^3$. Thus, not only the limited salt resources are utilized fully, but also the solution mining

---

基金项目：中国石油储气库重大专项"地下储气库关键技术研究与应用"子课题"盐穴储气库加快建产工程试验研究"，2015E-4008。

作者简介：齐得山，男，1987 年生，工程师，2013 年硕士毕业于中国石油大学（北京）地质工程专业，现主要从事盐穴储气库溶腔工艺优化设计及动态跟踪分析等工作。地址：江苏省镇江市润州区南徐大道商务 A 区 D 座中国石油西气东输管道公司储气库管理处，212000。电话：15862931327，Email：xqdsqideshan@petrochina.com.cn。

cost is cut down. The research result provides the theoretical guidance for the construction of other salt-cavern gas storages in future.

Keywords salt-cavern gas storage; natural gas peak-shaving; salt rock; pillar/diameter ratio; solution mining

金坛储气库是中国第一座盐穴地下储气库，位于江苏省金坛市，距西气东输一线镇江分输站34.8km，是西气东输管道的重要配套设施[1-5]。盐穴地下储气库是利用水溶方式在地下盐层或盐丘中开采盐岩形成的用于储存油气的地下洞穴，具有操作灵活、工作气量比例高的优点，适用于调峰[6-12]。

盐穴储气库建设对库区盐层地质特征和地面条件有很高要求[13-15]，目前，国内仅筛选出江苏金坛、江苏淮安和河南平顶山等几个可用于建设盐穴地下储气库的库区，除金坛储气库在建之外，其他几个库区均处于先导性试验阶段。鉴于中国可用于建设盐穴地下储气库的盐矿资源严重匮乏[16-17]，充分利用金坛盐矿区有限的盐矿资源，并在保证储气库运行安全的前提下，建设较大储气规模的盐穴储气库群尤为重要。

1 可行性

1.1 造腔设计现状

金坛储气库建设前期，在考虑地面条件的前提下，腔体大小按照矿柱比不小于2.5、腔体最大直径不大于80m进行设计（图1）。矿柱比是相邻两腔间矿柱宽度P与两个腔体平均直径$(D_1+D_2)/2$的比值。随着造腔实践经验的积累，设计腔体时所采用的最小矿柱比虽有所降低，但仍介于2.3～2.4之间，腔体设计的最大直径不超过83m。已完腔的和部分正在造腔的腔体与临腔的矿柱比均较大，腔体的最终直径相对较小，个别完腔腔体的直径仅为70m，致使金坛库区可用于造腔的盐矿资源并未被充分利用。

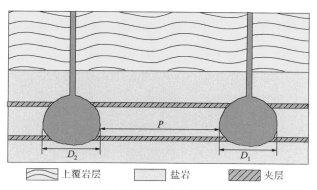

图1 金坛盐穴地下储气库盐腔示意图

1.2 稳定性评价

为了确定单腔直径能否再增大，临腔矿柱比能否再降低，基于金坛储气库地层数据，应用FLAC³ᴰ软件，分别建立了单腔及双腔的三维数值模型并进行稳定性评价，模型能够较好地反映含夹层盐穴储气库在运营期内的长期蠕变变形特征[18-21]。在模拟计算中，考虑造腔和稳定性等因素，溶腔形状最终确定为上部为半椭球、下部为半球的组合图形（图2），球体半径为r_2，椭球长半轴为r_1，短半轴为r_2，腔体高度为r_1+r_2，腔体直径为$2r_2$。

为了分析直径扩大对单腔稳定性的影响，在保持腔体高度不变的情况下，分别建立直径80m、90m的两个单腔模型进行稳定性评价。结果表明：当直径为80m时，单腔蠕变30年的体积收缩率为8.04%；当直径为90m时，单腔蠕变30年的体积收缩率为8.19%，可见溶腔直径扩大至90m时，腔体体积收缩率变化不大，仅为80m时的1.019倍（图3）。

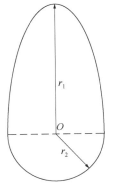

图2 金坛盐穴储气库腔体形状模拟图　　图3 金坛盐穴储气库不同直径下单腔的蠕变曲线

此外，对不同直径、不同矿柱比条件下的多个双腔模型进行稳定性评价。对双腔进行数值模拟时，不仅要考虑腔体直径增加对体积收缩率的影响，还要考虑矿柱比降低对双腔间无损距离的影响。腔体在长期蠕变过程中会在周围形成塑性区，双腔塑性区边界之间的距离，即双腔之间非塑性区单元的宽度为"无损宽度"。研究结果表明：（1）当双腔腔体直径均为80m时，蠕变30年后的最大体积收缩率为8.8%；当双腔腔体直径均为90m时，蠕变30年后的最大体积收缩率为9.81%，增大腔体直径对体积收缩率影响不大。（2）当矿柱比一定时，无损宽度随蠕变时间的延长而减小；若蠕变时间相同，随矿柱比增大，无损宽度也逐步增大（图4）；当矿柱比为2.2时，蠕变30年后腔体间的无损宽度约为80m，足以满足储气库安全要求（图5）。

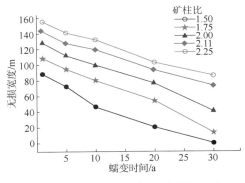 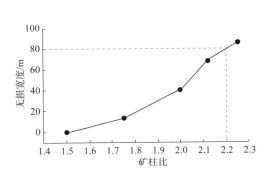

图4 金坛盐穴储气库不同矿柱比下　　图5 金坛盐穴储气库蠕变30年后
　　无损宽度与蠕变时间关系曲线　　　　　无损宽度与矿柱比关系曲线

金坛储气库部分腔体的体积还有进一步增大的空间,因此如何从正在造腔的井中筛选出合适的腔体,是精细造腔腔体体积优化的关键。

2 腔体优化

2.1 腔体筛选

基于盐腔腔体稳定性的考虑,腔体设计体积不仅受临腔最小矿柱比的限制,而且受腔体偏溶程度等多种因素的影响。虽然造腔设计时所采用的最小矿柱比有所降低,但并非所有大于最小矿柱比的腔体的设计直径及体积均可增大。

从正在造腔的腔体中筛选可优化的腔体,首先应该排除直径不可增大的腔体,主要包括3类:即将完腔的腔体、严重偏溶的腔体、临腔最小矿柱比小于2.3的腔体。如果临近两腔的直径均可增大,但是受矿柱宽度和矿柱比的限制,只能选择一口腔优化时,则需要对两口腔进行全腔模拟,评价增大哪一口腔的直径可以获得更大的腔体体积。对优选出的腔体进行重新模拟,并基于模拟结果重新计算临腔最小矿柱比。腔体优化后需要满足两个条件:与临腔的最小矿柱比介于2.2~2.3之间;腔体的最大直径不超过90m。

2.1.1 即将完腔的腔体

造腔进入封顶期,主要对盐腔进行溶滤以形成良好的穹形顶部。当腔体即将完腔,尤其当进入最后一个造腔阶段后,腔体的穹形顶部已经形成,若此时增加腔体直径,则不利于腔顶直径控制,容易形成大平顶,影响腔体的稳定性。参照国内造腔经验,在穹形腔顶已经形成的前提下,腔顶最大直径不得超过18m。

2015年底声呐数据显示,金坛储气库腔体B的穹形腔顶已经形成,但距离设计腔顶深度还有7m,油水界面还需再上提一次,该腔体才能完腔。基于声呐更新造腔方案时,经方案优选,确定JK6-7腔体的最大设计直径为82m,腔顶直径为16m,与临腔最小矿柱比为2.42。若将该井腔体最大直径增至87m,则需延长最后一阶段的造腔时间,设计腔顶直径会达到30m,腔顶不再呈穹形,而是一个大平顶,极不利于腔体的稳定(图6),所以该井不再考虑优化。

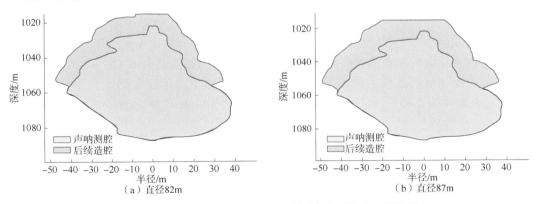

图6 金坛盐穴储气库腔体B声呐测腔及后续造腔模拟图

2.1.2 严重偏溶的腔体

受盐层非均质性的影响,腔体在淋滤过程中,不同方向的溶蚀速率不同。有的方向上溶快，有的方向上溶慢，会造成腔体某一方向半径大于其他方向半径，即偏溶。偏溶在造腔过程中普遍存在，但偏溶程度较轻，腔体在不同方向的半径差别较小，对腔体稳定性的影响也很小。

若腔顶偏溶严重，最大半径是最小半径的数倍，腔顶已不再是近似对称的穹形，所能承受的压力降低，则会对腔体的稳定性产生较大影响。虽然腔体偏溶程度对腔体稳定性的影响还有待进一步深入研究，但是考虑到腔体的稳定性，腔体体积优化时还需排除严重偏溶的腔体。

经筛选，从金坛库区正在造腔的腔体中找出3口严重偏溶的腔体，其最大半径均超过最小半径的3倍。典型实例为腔体C，最近一次声呐数据表明腔体向东南方向严重偏溶，腔体最大直径为79.1m，深为1132m，110°方向的半径已达到61.7m，其反方向的半径仅为13.6m。后续造腔模拟结果表明：腔体C的偏溶现象一直持续到腔顶，致使腔顶110°方向的半径远大于其他方向(图7)。

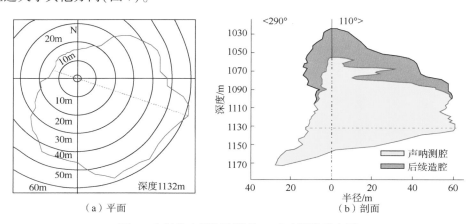

图7 金坛盐穴储气库腔体C声呐测腔形态图

2.1.3 矿柱比小于2.3的腔体

在实际造腔过程中，受盐层中不溶物夹层和造腔工艺参数的影响，个别腔体的边界发生畸变[15-16]，因而造成腔体的实际直径略大于设计直径，腔体间的实际矿柱比略小于设计值。由此可见，当相邻两腔的矿柱比小于2.3时，若某腔体直径再增加，则优化后两腔的实际矿柱比可能会小于2.2。为此，腔体优化不再考虑临腔矿柱比小于2.3的腔体。

经筛选，金坛库区正在造腔的井中矿柱比小于2.3的腔体共有6对，12口，腔体体积优化时，首先排除这些腔体。典型实例是腔体E与腔体F，其最大设计直径分别为79m、75.8m，两个腔体的直径平均值为77.4m，矿柱宽度为174.6m，矿柱比为2.25。

2.2 临腔腔体评价

排除不可优化的腔体后，如果相邻的两个腔体体积都存在优化空间，则需对两口腔进行评价，优先增大一口腔的直径，然后再考虑另一口腔的直径能否增大以及增大到多少。评价思路：在矿柱比的限制下，分别增大两口腔的直径，比较哪一口腔体直径增大时，两口腔体积和最大，则首先选取那一口腔进行优化。

典型实例是腔体G和腔体H。腔体G原设计直径为80.0m，腔体H原设计直径为77.7m，两腔间的矿柱宽度为182.2m，矿柱比为2.31。如果将其中一口腔的直径增大3m，则两腔间的理论矿柱比降至 2.24，尚存在可优化空间。为了选出最适合优化的腔体，设计了两种方案，将其中一口腔的直径增大约 3m 进行全腔模拟，并基于模拟结果计算两腔的实际矿柱比(表1)。

表1 腔体 G 与腔体 H 在两种优化方案下的参数表

造腔方案	腔体	直径/m	体积/m³	体积之和/m³	矿柱宽度/m	矿柱比
方案1	G	82.7	173084	312454	178.7	2.23
	H	77.7	139370			
方案2	G	80.0	163775	307833	179.1	2.23
	H	80.3	144058			

模拟结果表明：当腔体 G 的直径从 80m 增至 82.7m 时，腔体体积增大了 9309m³，腔体直径平均每增加 1m，腔体体积增大 3448m³；当腔体 H 直径从 77.7m 增至 80.3m 时，腔体体积增大了 4688 m³，腔体直径平均每增加 1m，腔体体积增大 1803m³。对比可知：当腔体 G 直径增大时，两个腔体的体积和更大，因此选择腔体 G 进行优化。

2.3 优化结果

经过排除不可优化腔体和临腔腔体评价，从金坛储气库正在造腔的井中筛选出 10 个可用于优化的腔体。腔体优化思路：首先根据腔体与临腔之间矿柱宽度，以临腔最小矿柱比不小于 2.2、腔体最大直径不超过 87m 为标准，计算出腔体直径增大后的最大理论值；其次对筛选出的腔体进行全腔模拟，直径不能超过理论最大值；最后基于模拟结果计算与临腔的最小矿柱比，若矿柱比小于 2.2，则重新进行全腔模拟，直至矿柱比大于 2.2。

经优化，10 口腔体的总体积共增大了 261459m³，平均单腔体积增大了 26146m³(表2)。金坛库区单腔腔体体积原设计的平均值为 250000m³。10 口腔优化后，总体积增加了约一口腔的体积。对于筛选出的单腔，优化后单腔体积增加了 10.5%，表明在前期经济投入相同的情况下，腔体体积优化增大了整个库区的总体积，这不仅充分利用了盐层资源，而且降低了造腔成本。

表2 金坛盐穴储气库区腔体优化结果

可优化腔体	原造腔方案		优化后造腔方案		优化后体积增大值/m³
	最大直径/m	最小矿柱比	最大直径/m	最小矿柱比	
A	81.0	2.40	86.6	2.27	25508
G	80.0	2.30	82.7	2.23	9309
J	80.0	2.33	83.3	2.22	7348
K	80.0	2.40	86.5	2.28	31057
L	70.0	2.30	72.0	2.23	10558
M	80.0	3.90	86.3	3.50	19743
N	79.0	3.60	86.9	3.20	44363
P	81.4	5.40	86.8	4.90	27287
R	80.6	3.16	86.7	2.80	50895
T	81.6	3.00	86.2	2.70	35391

3　结论

（1）稳定性评价分析结果表明，金坛储气库单腔的理论最大直径可增大至90m，双腔间的理论矿柱比可降至2.2，库区内的腔体体积还存在优化空间。

（2）从正在造腔的井中筛选出合适的腔体增大直径，是精细造腔腔体体积优化的关键，通过排除不可优化腔体和临腔腔体评价，金坛库区共筛选出10口腔。

（3）优化后，10口腔的总体积共增大了261459m^3，相当于增加了一口腔的体积，充分利用了金坛库区有限的盐矿资源，降低了造腔成本。

参 考 文 献

[1]　杨海军，郭凯，李建君．盐穴储气库单腔长期注采运行分析及注采压力区间优化——以金坛盐穴储气库西2井腔体为例[J]．油气储运，2015，34（9）：945-950．

[2]　李建君，王立东，刘春，等．金坛盐穴储气库腔体畸变影响因素[J]．油气储运，2014，33（3）：269-273．

[3]　杨海军，于胜男．金坛地下储气库盐腔偏溶与井斜的关系[J]．油气储运，2015，34（2）：145-149．

[4]　李建君，陈加松，吴斌，等．盐穴地下储气库盐岩力学参数的校准方法[J]．天然气工业，2015，35（7）：96-101．

[5]　周学深．有效的天然气调峰储气技术——地下储气库[J]．天然气工业，2013，33（10）：95-99．

[6]　赵志成，朱维耀，单文文，等．盐穴储气库水溶建腔机理研究[J]．石油勘探与开发，2003，30（5）：107-109．

[7]　班凡生，耿晶，高树生，等．岩盐储气库水溶建腔的基本原理及影响因素研究[J]．天然气地球科学，2006，17（2）：261-266．

[8]　王文权，杨海军，刘继芹，等．盐穴储气库溶腔排量对排卤浓度及腔体形态的影响[J]．油气储运，2015，34（2）：175-179．

[9]　刘继芹，焦雨佳，李建君，等．盐穴储气库回溶造腔技术研究[J]．西南石油大学学报（自然科学版），2016，38（5）：122-128．

[10]　郑雅丽，赵艳杰，丁国生，等．厚夹层盐穴储气库扩大储气空间造腔技术[J]．石油勘探与开发，2017，44（1）：137-143．

[11]　郭凯，李建君，郑贤斌．盐穴储气库造腔过程夹层处理工艺——以西气东输金坛储气库为例[J]．油气储运，2015，34（2）：162-166．

[12]　齐得山，巴金红，刘春，等．盐穴储气库造腔过程动态监控数据分析方法[J]．油气储运，2017，36（9）：1078-1082．

[13]　丁国生，张昱文．盐穴地下储气库[M]．北京：石油工业出版社，2010：6-46．

[14]　杨海军．中国盐穴储气库建设关键技术及挑战[J]．油气储运，2017，36（7）：747-753．

[15]　李建君，巴金红，刘春，等．金坛盐穴储气库现场问题及应对措施[J]．油气储运，2017，36（8）：982-985．

[16]　丁国生，李春，王皆明，等．中国地下储气库现状及技术发展趋势[J]．天然气工业，2015，35（11）：107-111．

[17]　肖学兰．地下储气库建设技术研究现状及建议[J]．天然气工业，2012，32（2）：79-82．

[18]　刘建平，姜德义，陈结，等．盐岩水平储气库的相似模拟建腔和长期稳定性分析[J]．重庆大学学报，2017，40（2）：45-50．

［19］王同涛，闫相祯，杨恒林，等．多夹层盐穴储气库群间矿柱稳定性研究［J］．煤炭学报，2011，36（5）：790-795.

［20］李占金，杨美宏，孙文诚，等．基于FLAC^{3D}的深部大规模开采围岩稳定性分析［J］．矿业研究与开发，2017，37（6）：90-93.

［21］陈结，刘剑兴，姜德义，等．围压作用下盐岩应变与损伤恢复试验研究［J］．岩土力学，2016，37（1）：105-112.

本论文原发表于《油气储运》2018年第37卷第6期。

金坛盐穴储气库腔体偏溶特征分析

齐得山　李淑平　王元刚

(中国石油西气东输管道公司储气库项目部)

【摘　要】 国内可用于盐穴储气库建设的盐矿以层状盐岩为主，造腔过程中普遍存在腔体偏溶现象，研究其特征及成因对国内以后盐穴储气库的建设具有一定的借鉴作用。以国内第一个盐穴储气库金坛储气库为研究对象，基于声呐测腔数据，提出了以偏溶系数，即腔体最大半径与同一平面最小半径的比值，来定量表征腔体的偏溶程度，最大半径方向即为腔体偏溶方向。统计结果表明，金坛储气库腔体偏溶系数1.13~11.88，偏溶方向以北东南西向为主。结合夹层、可造腔盐层厚度和地应力数据，分析了腔体偏溶发生的原因，认为造腔过程中夹层的不均匀垮塌可促使腔体发生偏溶；可造腔盐层厚度越大，腔体发生偏溶的可能性就越大，偏溶程度就越严重；地应力方向对腔体的偏溶方向具有重要影响。

【关键词】 盐穴储气库；金坛；腔体形态；偏溶成因；地应力；水溶造腔

Characteristics of cavity differential dissolution of Jintan salt cave gas reservoir

Qi Deshan　Li Shuping　Wang Yuangang

(Gas Storage Project Department, PetroChina West-East Gas Pipeline Company)

Abstract In China, salt mines that can be used for salt cave gas reservoir construction are mostly composed of layered salt rocks, and differential dissolution of the cavity often occurs during solution mining. Research into the characteristics and causes of such phenomenon can provide references for future construction of salt cave gas reservoirs in China. This work investigates the Jintan Salt Cave Gas Reservoir, which is the first salt cave gas reservoir in China. Based on sonar cavity data, differential dissolution in the cavity can be quantitatively analyzed using the differential dissolution coefficient, which is the ratio of the maximum cavity radius to the minimum radius in the same plane. The direction of the maximum radius is the direction of differential dissolution in the cavity. The statistical results reveal that, for the Jintan Gas Reservoir, the differential dissolution coefficient in the cavity is 1.13~11.88, and differential dissolution occurs primarily along the northeast-southwest direction. The causes of differential dis-

---

基金项目：中国石油储气库重大专项(2015E-4008)。

作者简介：齐得山，1987年生，男，汉族，河南周口人，工程师，硕士，主要从事盐穴储气库溶腔工艺优化设计及动态跟踪分析方面的研究。E-mail：xqdsqideshan@petrochina.com.cn。

solution in the cavity are analyzed by integrating the thickness and ground stress data of interlayers and salt layers that can be used for mining. It is believed that non-uniform collapses of interlayers during solution mining can lead to differential dissolution in the cavity. Thicker salt layers that are more suitable for mining result in greater likelihood and severity of differential dissolution in the cavity. The ground stress directions significantly influence partial melting in the cavity.

Keywords salt cave gas reservoir; Jintan; cavity morphology; causes of differential dissolution; ground stress; solution mining

盐穴储气库具备工作气量比例高、注采气操作灵活等优势，最适合用于天然气管道调峰，在国内具有很好的发展前景。国内盐穴储气库的建设已有十多年的历史，积累了丰富的造腔实践经验，在水溶造腔机理及工艺技术等方面取得了一些研究成果[1-11]。金坛储气库是国内第一个盐穴储气库，研究分析其建设过程存在的问题，可为国内其他盐穴储气库的建设提供借鉴。

有关金坛储气库的研究，概括起来主要有以下几个方面：（1）关于溶腔工艺技术及腔体形态体积等造腔方案参数的优化分析[12-14]；（2）总结分析了造腔生产过程中可能存在的问题，并提出了相应的动态监控分析方法[15-16]；（3）在国内首次开展了以天然气阻溶为核心的腔体扩容及修复试验[17-18]；（4）通过地质力学参数校正及腔体运行损伤评价分析等来优化腔体的运行压力区间[19-21]。但是关于腔体形态异常，尤其是腔体偏溶方面的研究成果还很少。

基于声呐测腔数据，统计了金坛储气库盐腔的偏溶系数和偏溶方向，结合库区的地质和地应力等数据分析了腔体发生偏溶的原因，对金坛及国内其他盐穴储气库以后的造腔设计与施工具有一定的指导意义。

1　金坛储气库简介

金坛储气库地理位置位于江苏省常州市金坛区直溪镇，构造位置处于金坛盆地直溪桥凹陷内，盐岩层发育于古近系阜宁组上部，埋深 975~1200m，厚度为 150~230m，不溶物含量为 14.95%~24.48%[10]。金坛储气库设计库容为 $26.00\times10^8m^3$，工作气量为 $17.00\times10^8m^3$，目前已形成库容为 $9.00\times10^8m^3$、工作气量为 $5.40\times10^8m^3$。从 2007 年开始注采运行，截止到 2017 年底，金坛储气库累计注气量为 $25.40\times10^8m^3$，累计采气量为 $19.05\times10^8m^3$，注采气量呈逐年递增趋势，在西气东输管道调峰中发挥了重要作用（图 1）。其中，2017 年共进行了 4 轮注气、6 轮采气，累计采气量为 $7.40\times10^8m^3$，是库区已形成工作气量的 1.4 倍，凸显了盐穴储气库注采气运行的灵活性。

2　声呐测腔技术

单个盐腔从开始溶腔至最终完腔，需经历数个溶腔阶段，每个阶段结束后需通过专门的声呐设备测量腔体的形态和体积，为后续溶腔方案的设计或调整提供依据。声呐测腔的基本原理就是声波测距原理[11]。声呐测量仪器下入盐腔中后，可以向卤水中发出超声波并在其中传播。超声波遇到盐腔壁后，产生反射波，一部分反射波会被声呐测量仪接收到，并记录下其从发射到返回的时间，这个时间是超声波从声呐测量仪到腔壁的双程传播时间。

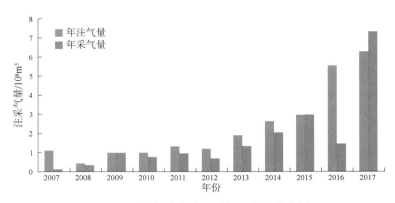

图 1 金坛储气库年度注采气量统计柱状图

另外,声呐测量仪还可以测量超声波在卤水中的传播速度,乘以传播时间后就是声呐测量仪到腔壁的双程距离,除以 2 就是声呐测量仪到腔壁的距离。

声呐仪器下入到一定深度,首先,进行定位,从正北方向开始测量,然后顺时针旋转,每隔一定角度测量一个值,旋转一周后重新回到正北方向。然后,上提或下放仪器至另一深度,根据需要可进行水平或倾斜测量,如此重复,直至完成整个腔体形态的检测。测量结果就是不同深度上以井眼为圆心,以测量值为半径的一系列数据点,在专业软件上将各点连接起来就直观反映出了腔体的形态特征及体积大小(图 2)。

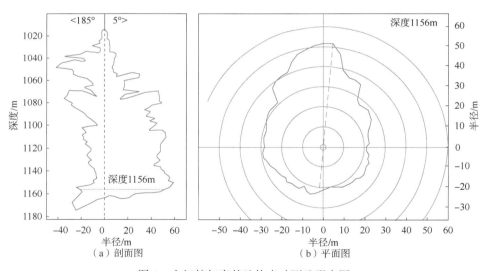

图 2 金坛储气库某腔体声呐测腔形态图

3 金坛储气库腔体偏溶特征

腔体的偏溶是指盐穴储气库在注水造腔过程中,由于同一深度不同方向上的盐岩溶解速率不同而产生的腔体半径大小不同的现象。偏溶是促使腔体形态不规则的重要因素,对腔体的体积也会产生一些不良影响[22]。为定量表征腔体的偏溶现象,基于声呐测腔数据

(图2)，提出了偏溶系数这个参数，即最大半径与同一深度最小半径的比值，来表征腔体的偏溶程度。最大半径方向即为腔体偏溶方向。基于金坛储气库36口腔的声呐数据，计算了每口腔的偏溶系数，并统计了每口腔的偏溶方向(表1)。

表1 金坛储气库各盐腔偏溶系数及偏溶方向统计表

井号	偏溶系数	偏溶方向/(°)	偏溶程度	井号	偏溶系数	偏溶方向/(°)	偏溶程度
1	1.13	100	正常腔体	19	1.85	20	轻微偏溶
2	1.16	10		20	1.88	10	
3	1.16	30		21	1.94	330	
4	1.30	50		22	2.06	190	中等偏溶
5	1.31	190		23	2.06	10	
6	1.33	340		24	2.26	300	
7	1.44	110		25	2.32	300	
8	1.48	100		26	2.34	210	
9	1.48	350		27	2.42	10	
10	1.52	160	轻微偏溶	28	2.48	10	
11	1.52	20		29	3.28	190	严重偏溶
12	1.53	350		30	3.29	0	
13	1.56	60		31	3.95	180	
14	1.63	40		32	4.98	120	
15	1.64	90		33	5.12	40	
16	1.66	170		34	5.93	170	
17	1.76	30		35	7.04	20	
18	1.82	20		36	11.88	200	

金坛储气库腔体偏溶系数最大值为11.88，最小值为1.13，说明腔体偏溶是一个普遍现象。根据偏溶系数的大小对金坛储气库的腔体进行了分类：1.00<偏溶系数≤1.50 的为正常腔体，共 9 口，占比 25%；1.50<偏溶系数≤2.00 的为轻微偏溶腔体，共 12 口，占比 33%；2.00<偏溶系数≤2.50 的为中等偏溶腔体，共 7 口，占比 20%；偏溶系数>2.50 的为严重偏溶腔体，共 8 口，占比 22%。腔体偏溶方向以北东南西向为主（图3）。

腔体的偏溶达到一定程度后，后期就很难通过造腔工艺参数的调整来改变这种现

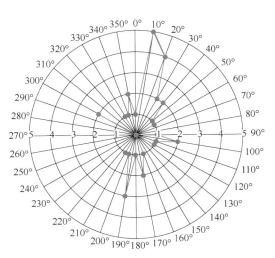

图3 金坛盐穴偏溶方向玫瑰花图

象。金坛储气库完腔井造腔过程中各阶段偏溶特征分析结果表明,造腔过程中腔体的偏溶程度有3种变化趋势:(1)造腔前期腔体形态正常,造腔中后期腔体出现偏溶现象[图4(a)];(2)造腔前期腔体轻微偏溶,造腔中后期偏溶系数增大,偏溶程度加重[图4(b)];(3)腔体形态正常,未出现偏溶[图4(c)]。

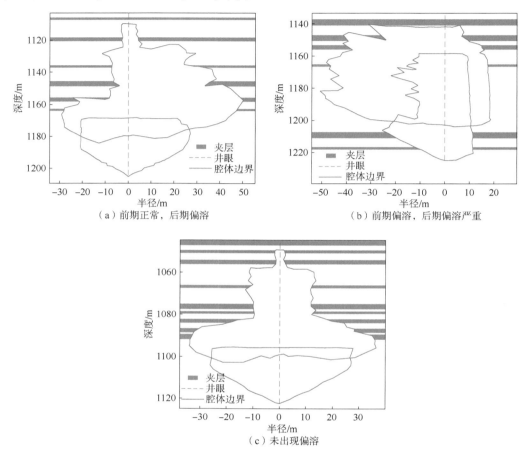

图4 部分腔体不同时期腔体剖面叠加图

4 偏溶原因分析

4.1 夹层不均匀垮塌

金坛储气库的造腔段是阜宁组四段的层状盐岩,含有一定数量的泥质夹层。因泥质不溶于水,夹层在水溶造腔过程中会随着腔体的扩大而逐渐垮塌,垮塌物堆积于腔体底部。受夹层地质特征及力学性质空间分布非均质性等多种因素的影响,不同夹层及同一夹层不同方向上的垮塌特征是不同的,这对腔体的形态有重要影响。

如果夹层垮塌的速度与盐岩水平方向上的溶解速度一致,腔体壁面就比较光滑[图4(c)]。如果夹层垮塌的速度与盐岩水平方向上的溶解速度不一致,腔体壁面就会出现岩脊或岩谷,形态就会呈现出一定程度的不规则。同一个夹层,如果不同方向上的垮塌速度不一致,就会造成腔体不同方向上的半径不一致,而不同方向的半径差距较大时就是偏溶现象。

典型实例是 A 井,该井可用于造腔的盐层中发育有 7 个泥质夹层,前两个阶段在比较纯的盐层段造腔,腔体形态比较规则,最大半径 26m,同一平面上的最小半径为 21m,计算出的腔体偏溶系数为 1.24,属正常腔体[图 5(a)]。第二阶段结束后,油垫由 1169m 提升到 1130m,造腔段内含有 4 个泥质夹层。第三阶段造腔过程中,1 号夹层均匀垮塌,其下部腔体形态比较规则,但是 2 号、3 号、4 号夹层不均匀垮塌,腔体一侧夹层垮塌速度与盐岩溶解速度一致,壁面比较光滑;腔体另一侧夹层垮塌速度快于盐岩溶解速度,壁面出现岩谷。腔体上部不同方向上的半径大小出现较大差别,其中 2 号夹层垮塌处偏溶系数达到了 2.40,3 号夹层垮塌处偏溶系数达到了 3.80,4 号夹层垮塌处偏溶系数达到了 3.00,腔体上部出现了严重偏溶[图 5(b)]。

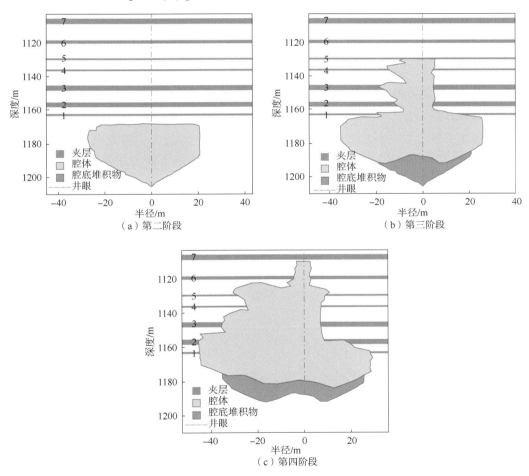

图 5 A 井造腔过程中不同阶段结束后腔体形态剖面图

腔体一侧岩谷的出现,增大了盐岩与水的接触面积,加快了这一侧盐岩的溶解,使得腔体的偏溶现象更加明显,第四阶段结束后腔体的最大偏溶系数已经达到了 5.00[图 5

（c）］。夹层垮塌后形成的不溶物堆积在腔体底部，因 2 号、3 号、4 号夹层在一侧的垮塌程度较另一侧严重，致使该方向上腔底的堆积物明显多于另一侧［图 5（b），图 5（c）］。

4.2 造腔段盐层厚度

统计了金坛储气库已完腔腔体的偏溶系数大小与其可造腔盐层厚度、盐层埋深、隔夹层数量等各项地质参数之间的关系。结果表明，腔体的偏溶系数有随着可造腔盐层厚度增大而增大的趋势（图 6），即可造腔盐层厚度越大，腔体发生偏溶的可能性也就越大、偏溶程度也可能就越严重。分析其原因，有两条：（1）可造腔盐层厚度越大，垂向上各小层及夹层间的非均质性差异越大，造腔过程中不同盐层的溶蚀速度及不同夹层的垮塌程度出现差异的可能性就越大，致使腔体不同深度、不同平面上的半径差异就越大；（2）

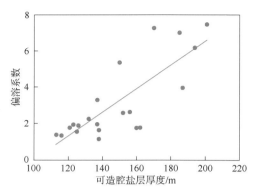

图 6 已完腔腔体偏溶系数与
可造腔盐层厚度交会图

可造腔盐层厚度越大，腔体完腔的时间就越长，即溶蚀时间也就越长。而腔体的偏溶可以加速偏溶方向上盐岩的溶蚀，因此，溶蚀时间越长，腔体的偏溶现象就越严重。

4.3 地应力方向

金坛储气库最大主应力方向为北东—南西向，最小主应力方向为北西—南东向[23]。金坛储气库 36 口腔的偏溶方向统计结果表明：腔体的主要偏溶方向是北东南西向，与最大主应力方向一致；另一偏溶方向为北西—南东向，与最小主应力方向一致（图 3）。说明盐穴储气库建库区的地应力方向对腔体的偏溶方向具有重要的影响。

钻井过程中形成的井眼，打破了地层中原有的地应力平衡状态，井壁上常常会产生应力释放缝，应力释放缝的走向与最大水平主应力方向一致［图 7（a）］。另外，钻井过程中，如果钻井液液柱压力过大，井壁上会形成压裂诱导缝，压裂诱导缝的走向与最大水平主应力方向一致［图 7（a）］；如果钻井液液柱压力过小，井壁岩石就会崩落，形成椭圆井眼，崩落方位与最小水平主应力方向一致［图 7（b）］[24-26]。

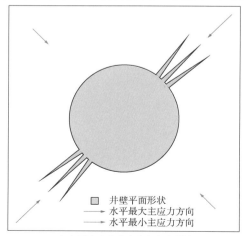

（a）最大主应力方向上产生裂缝

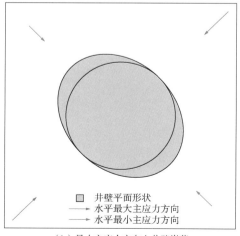

（b）最小主应力方向上井壁崩落

图 7 钻井过程中地应力影响井壁变化示意图

裂缝或椭圆井眼的形成，增大了盐岩与卤水的接触面积，加速了最大水平主应力或最小水平主应力方向上盐岩的溶蚀，致使这两个方向上的腔体半径大于其他方向，这是地应力方向影响腔体偏溶方向的根本原因。金坛储气库腔体偏溶方向与最大主应力方向一致，分析发现，是钻井过程中井壁上产生的裂缝主导了腔体偏溶方向的发展。

5 结论

（1）基于声呐测腔数据，提出了偏溶系数这个参数，即最大半径与同一深度最小半径的比值，来定量表征腔体的偏溶程度。最大半径方向即为腔体偏溶方向。

（2）金坛储气库腔体偏溶系数最小值1.13、最大值11.88，说明腔体偏溶是一个普遍现象，偏溶方向以北东—南西向为主。

（3）造腔过程中腔体的偏溶程度有3种变化趋势：造腔前期腔体形态正常，造腔中后期腔体出现偏溶现象；造腔前期腔体轻微偏溶，造腔中后期偏溶系数增大，偏溶程度加重；腔体形态正常，未出现偏溶。

（4）造腔过程中夹层的不均匀垮塌可促使腔体发生偏溶；可造腔盐层厚度越大，腔体发生偏溶的可能性就越大，偏溶程度越严重；地应力方向对腔体的偏溶方向具有重要的影响。

参 考 文 献

[1] 李建中．利用岩盐层建设盐穴地下储气库[J]．天然气工业，2004，24(9)：119-121.

[2] 班凡生，高树生．岩盐储气库水溶建腔优化设计研究[J]．天然气工业，2007，27(2)：114-116.

[3] 肖学兰．地下储气库建设技术研究现状及建议[J]．天然气工业，2012，32(2)：79-82.

[4] 丁国生，李春，王皆明，等．中国地下储气库现状及技术发展方向[J]．天然气工业，2015，35(11)：107-111.

[5] 郑雅丽，赵艳杰，丁国生，等．厚夹层盐穴储气库扩大储气空间造腔技术[J]．石油勘探与开发，2017，44(1)：1-7.

[6] 杨海军．中国盐穴储气库建设关键技术及挑战[J]．油气储运，2017，36(7)：747-753.

[7] 金犇，夏焱，袁光杰，等．盐穴地下储气库排卤管柱盐结晶影响因素实验研究[J]．天然气工业，2017，37(4)：130-134.

[8] 任众鑫，杨海军，李建君，等．盐穴储库腔底堆积物空隙体积试验与计算[J]．西南石油大学学报（自然科学版），2018，40(2)：142-150.

[9] 郑雅丽，赖欣，邱小松，等．盐穴地下储气库小井距双井自然溶通造腔工艺[J]．天然气工业，2018，38(3)：96-102.

[10] 丁国生，郑雅丽，李龙．层状盐岩储气库造腔设计与控制[M]．北京：石油工业出版社，2017.

[11] 杨海军，李龙，李建君．盐穴储气库造腔工程[M]．南京：南京大学出版社，2018.

[12] 王文权，杨海军，刘继芹，等．盐穴储气库溶腔排量对排卤浓度及腔体形态的影响[J]．油气储运，2015，34(2)：175-179.

[13] 郭凯，李建君，郑贤斌．盐穴储气库造腔过程夹层处理工艺——以西气东输金坛储气库为例[J]．油气储运，2015，34(2)：162-166.

[14] 李建君，王立东，刘春，等．金坛盐穴储气库腔体畸变影响因素[J]．油气储运，2014，33(3)：269-273.

[15] 李建君，巴金红，刘春，等．金坛盐穴储气库现场问题及应对措施[J]．油气储运，2017，36(8)：982-986．
[16] 齐得山，巴金红，刘春，等．盐穴储气库造腔过程动态监控数据分析方法[J]．油气储运，2017，36(9)：1078-1082．
[17] 刘继芹，焦雨佳，李建君，等．盐穴储气库回溶造腔技术研究[J]．西南石油大学学报(自然科学版)，2016，38(5)：122-128．
[18] 李建君，陈加松，刘继芹，等．盐穴储气库天然气阻溶回溶造腔工艺[J]．油气储运，2017，36(7)：816-824．
[19] 李建君，陈加松，吴斌，等．盐穴地下储气库盐岩力学参数的校准方法[J]．天然气工业，2015，35(7)：96-101．
[20] 杨海军，郭凯，李建君．盐穴储气库单腔长期注采运行分析及注采压力区间优化——以金坛盐穴储气库西2井腔体为例[J]．油气储运，2015，34(9)：945-950．
[21] 敖海兵，陈加松，胡志鹏，等．盐穴储气库运行损伤评价体系[J]．油气储运，2017，36(8)：910-917．
[22] 杨海军，于胜男．金坛地下储气库盐腔偏溶与井斜的关系[J]．油气储运，2015，34(2)：145-149．
[23] 周冬林，杨海军，李建君，等．盐岩地层地应力测试方法[J]．油气储运，2017，36(12)：1385-1390．
[24] 赵永强．成像测井综合分析地应力方向的方法[J]．石油钻探技术，2009，37(6)：39-43．
[25] 程道解，孙宝佃，成志刚，等．基于测井资料的地应力评价现状及前景展望[J]．测井技术，2014，38(4)：379-381．
[26] 苏大明．XMAC测井资料评价水力压裂效果的应用分析[J]．国外测井技术，2011，6(3)：41-43．

本论文原发表于《西南石油大学学报(自然科学版)》2019年第41卷第2期。

盐穴储气库造腔设计与跟踪

刘 春 高云杰 何邦玉 王元刚 周冬林 李建君

（中国石油西气东输管道公司）

【摘 要】 中国适合建库的地质条件为层状盐岩，造腔盐层段具有厚度小、夹层多、含盐品位低等特点，其地质结构决定了中国盐穴储气库造腔设计不能完全借鉴国外设计经验。根据近十年来盐穴腔体造腔设计和现场实践，首次提出了适合层状盐岩造腔设计关键参数的选取原则，并基于此原则编制了初步造腔设计与调整方案，以指导现场造腔实践，效果良好。在现场造腔跟踪期间，形成了造腔动态分析、阻溶剂界面监控、造腔进度跟踪及声呐测腔等一系列跟踪与监测技术，有效避免和减少了现场造腔异常并及时发现和解决造腔突发问题，保证了腔体在建造期间正常有序。研究成果可为其他盐穴储气库建设提供设计依据和技术支持。

【关键词】 盐穴储气库；造腔；设计；现场跟踪

Solution mining design and tracking of salt cavern gas storages

Liu Chun Gao Yunjie He Bangyu Wang Yuangang
Zhou Donglin Li Jianjun

(PetroChina West-East Gas Pipeline Company)

Abstract In China, the layered salt rocks are geologically suitable for the construction of salt cavern gas storages, and the leaching salt intervals are characterized by thin single layer, multiple interbed and low salt grade. Due to the geological structure, the solution mining design of salt cavern gas storages in China cannot learn from foreign design experience completely. Based on the design and site practice of solution mining in the past ten years, the principle to select the key parameters of solution mining design in layered salt rocks was put forward for the first time in this paper, and it fills the domestic blank. Based on this principle, the preliminary solution mining design and adjustment plan were prepared to guide the

---

基金项目：中国石油储气库重大专项子课题"地下储气库关键技术研究与应用"，2015E-40。
作者简介：刘春，男，1984年生，工程师，2011年硕士毕业于中国石油大学（北京）油气田开发专业，现主要从事盐穴储气库溶腔工艺优化设计及动态跟踪分析等技术的研究工作。地址：江苏省镇江市润州区南徐大道商务A区D座中国石油西气东输管道公司储气库管理处，212000。电话：13101922108。E-mail：cqkliuchun@petrochina.com.cn。

practical solution mining, and its application results are good. In addition, a series of tracking and monitoring techniques were developed during the on-site tracking of solution mining, such as dynamic analysis for solution mining, blanket interface monitoring, leaching progress tracking and sonar survey, so as to avoid and reduce the leaching abnormalities effectively, identify sudden problems timely and ensure the normality and order of solution mining. The research results can provide design basis and technical support for the construction of other salt cavern gas storages.

Keywords　salt cavern gas storages; solution mining; design; on-site tracking

地下盐穴腔体建造是一项复杂的系统工程，且单腔建造周期长[1-6]。中国与国外盐丘地质结构差异大，盐穴储气库造腔设计不能完全借鉴国外设计经验。近年来，国内学者对盐穴储气库建库造腔工艺技术[7-14]、隔夹层垮塌机理[15-18]、管柱损坏机理[19-21]、注采运行工艺技术[22-25]等进行了研究，但针对中国地质特征的盐穴储气库造腔设计和现场监控技术等研究较少。根据国内近十年来盐穴腔体设计和现场工程实践，提出了适合中国盐穴储气库造腔设计关键参数的选取原则，并建立了现场造腔监控体系。

1　造腔设计

1.1　单井双管造腔工艺原理

盐穴腔体建造是一个水溶盐过程，通过钻井将生产套管下入到盐顶以下 10～15m，生产套管鞋以下盐层采用裸眼完井。单井双管分步式造腔工艺是在生产套管内再下入两根同心管柱，其外径分别为 117.8mm、114.3mm，通过定期向设计腔顶注入阻溶剂控制腔体上溶速。盐穴腔体形状控制通过分阶段调整造腔管柱组合位置及阻溶剂界面深度，自下而上逐步揭开上部盐层段，最终设计理想盐穴腔体形状为梨形（图1）。

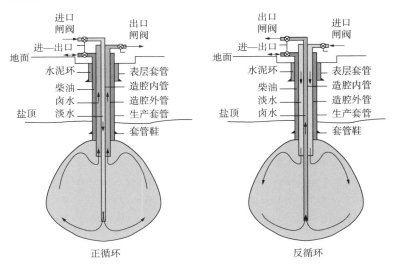

图 1　单井双管分步式造腔示意图

1.2　关键设计参数选取原则

国外盐穴储气库多为盐丘型盐穴储气库，造腔管柱一般采用大管串结构（两根同心管柱的外径分别为269.2mm、177.8mm），每个造腔阶段揭开盐层厚度大，盐层非均质性和不溶物隔夹层对腔体形状影响小，设计参数选取相对简单。中国适合建库的层状盐岩非常有限，且地质特征具有特殊性。造腔设计时参数选取尤为关键，有必要建立适合中国地质特征的造腔设计关键参数选取原则。

1.2.1 造腔循环模式

盐穴腔体建造周期可划分为建槽期、腔体主体期及封顶期3个阶段。建槽期一般采用正循环建槽，打开腔底盐层，建立不溶物口袋；腔体主体期采用正、反循环相结合的模式造腔，加快造腔速度，提高排卤质量浓度；封顶期采用正循环，阻溶剂界面和腔顶形状易于控制。该造腔循环模式已在国内盐穴储气库建造期间得到验证。

1.2.2 造腔管柱两管距

造腔管柱两管距直接影响排卤质量浓度高低，同时也影响腔体形状发展。造腔管柱两管距主要与可造腔盐层厚度、不溶物含量及隔夹层等有关。中国盐穴储气库可造腔盐层厚度主要分布在100~250m，平均不溶物含量为10%~40%。根据多年造腔经验，建槽期两管柱的合理间距为30~50m，建立不溶物底坑；腔体主体阶段正循环的两管距为40~80m，反循环两管距选择10m以上，排卤质量浓度即可达到290g/L以上；造腔末期采用正循环造腔，此时腔体净腔高较大，造腔两管距一般在50m以上。对于不溶物含量高、隔夹层多的盐层段，造腔管柱两管距不易过大，过大易导致腔体形状发展不规则，且腔底底坑抬升过快会导致局部盐层溶蚀不充分。

1.2.3 阻溶剂界面位置及上提高度

阻溶剂界面位置选择在品位较高的盐层上，距离隔夹层3~5m为宜。阻溶剂界面位置还与造腔模式有关，正循环造腔时，阻溶剂界面位置一般与造腔外管管鞋深度相同，也可位于造腔外管管鞋之上；反循环造腔时，阻溶剂界面位于造腔外管管鞋上方。每个造腔阶段阻溶剂界面上提高度受不溶物体积分数、隔夹层分布等影响。在此，综合考虑造腔盐层段平均不溶物体积分数 α 对阻溶剂上提高度的影响。

（1）当造腔段为纯盐层（$\alpha<10\%$）时，每个造腔阶段阻溶剂界面上提高度 H 可控制在30m以上。以金坛A1井、A2井为例，这两口井可造腔盐层段盐岩品位较高（图2）。

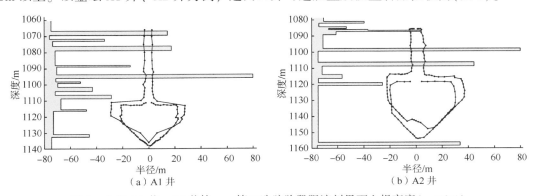

图2　金坛A1井、A2井第一、第二造腔阶段阻溶剂界面上提高度（$\alpha<10\%$）

（2）当造腔段为较纯盐层（$10\%<\alpha<30\%$）时，每个造腔阻溶剂界面上提高度可控制在15~30m之间。以金坛A3井、A4井为例，这两口井可造腔盐层段盐岩品位较高（图3）。

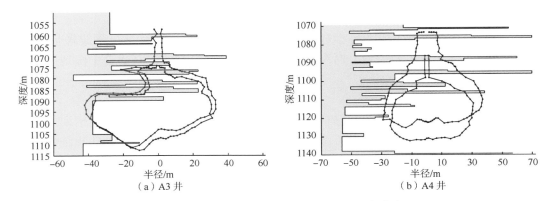

图 3　金坛 A3 井、A4 井第三、四造腔阶段阻溶剂界面上提高度($10\%<\alpha<30\%$)

（3）当造腔段为不纯盐层（$\alpha>30\%$）时，每个造腔阶段阻溶剂界面上提高度可控制在 5~15m 之间。以云应 A5 井为例，该井平均不溶物含量约 50%（图 4）。

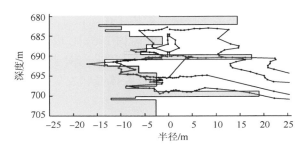

图 4　云应 A5 井阻溶剂界面上提高度（$\alpha>30\%$）

1.2.4　造腔阶段划分

造腔阶段的划分直接影响建造周期内造腔管柱调整作业和声呐测腔次数。根据金坛盐穴储气库工程实践经验，盐穴腔体建槽期 1~2 个阶段，造腔期 4~6 个阶段，封顶期 1~2 个阶段。整个造腔周期划分 6~10 个造腔阶段为宜，每个造腔阶段设计体积为 $3×10^4$~$5×10^4 m^3$。

1.2.5　腔体最大直径与矿柱比

为了保证注采运行期间腔体的安全性和稳定性，设计时应保证腔体与邻腔之间的矿柱比（相邻矿柱宽度与腔体最大直径之比）满足稳定性评价要求。在建造期间，腔体形状可能发展不规则导致腔体偏溶等，应不断根据声呐数据调整腔体最大直径。

1.3　单井造腔设计

单井造腔设计分为两种：一是基于新完钻井基础上新井初步造腔设计；二是基于已有声呐资料造腔井造腔方案调整设计。在进行造腔设计时，需要收集区域地质情况（构造、断层等），基本溶蚀速率，目的层段温度，设计井井身结构参数，可造腔盐层厚度，不溶物隔夹层分布，声呐测腔资料及邻腔分布等资料，并根据这些资料建立关键参数选取原则，设计新井初步造腔方案或造腔井造腔调整方案。

1.3.1　新井初步造腔方案

新井初步造腔设计方案主要是：确定初步腔体形状、设计腔体体积、腔顶底位置、设计腔体最大直径、初始造腔管柱下入深度、阻溶剂界面深度、设计阶段体积等。以金坛 A6

井为例，其造腔基本参数：盐层段顶底深度分布为929~1106m；可造腔盐层厚度为143m；泥质夹层共计8层，累计厚度为9m；平均不溶物体积分数约为12.1%；生产套管鞋深度为948m；完钻深度为1110m；造腔管柱为管串结构，套管内两根同心管柱外径分别为117.8mm、114.3mm；设计排量为100m³/h。A6井设计腔体总体积为278500m³，最大直径为80m，初步设计8个造腔阶段：1~2阶段为腔体建槽，3~6阶段为腔体主体，7~8阶段为腔体封顶(表1、图5)。

表1　金坛A6井新井造腔初步设计关键参数统计表

造腔阶段	循环方式	时间/d	排量/(m³/h)	油水界面/m	内管深度/m	外管深度/m	阶段体积/m³	累计体积/m³
1	正循环	150	100	1071	1106	1071	30000	30000
2	正循环	150	100	1056	1079	1056	40000	70000
3	正循环	90	100	1041	1075	1041	23500	93500
	正循环	30	100	1021	1075	1041	8000	101500
4	反循环	150	100	1021	1046	1070	35000	136500
	反循环	30	100	1001	1046	1070	8000	144500
5	反循环	180	100	1001	1030	1062	43000	187500
	反循环	30	100	981	1030	1062	8000	195500
6	反循环	90	100	981	998	1030	38000	233500
7	反循环	120	100	968	1030	1010	21000	256500
8	正循环	90	100	963	1040	963	22000	278500

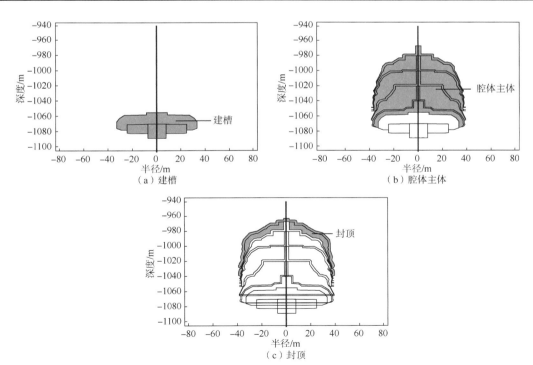

图5　金坛A6井初步设计全腔模拟结果图

1.3.2 造腔调整方案

盐穴建造时腔体形状发展受盐层非均质性、不溶物隔夹层及造腔工艺等因素影响，造腔期间腔体形状发展可能会偏离设计。为防止腔体形状偏离设计，造腔期间应定期安排声呐仪器检测腔体形状、腔体体积及腔体最大直径等，并在声呐检测腔体形状的基础上设计调整造腔方案，确定后续造腔管柱位置和阻溶剂界面深度、阶段设计体积等参数，优化腔体形状和体积，保证造腔井与邻腔之间的矿柱比在设计范围之内。基于已完成的 A6 井 6 次声呐检测结果调整设计造腔方案(图 6)，最终实际腔体形状与设计形状基本一致。造腔期间随着声呐测腔次数增多，声呐腔体形状越来越接近最终完腔腔体形状。

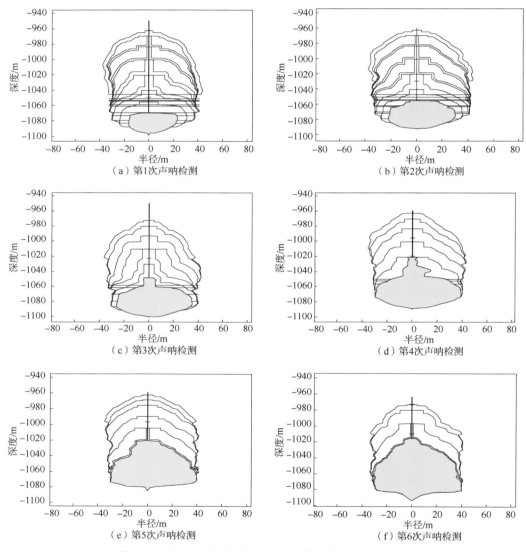

图 6　金坛 A6 井基于不同次声呐检测结果的全腔模拟图

2 造腔现场跟踪

中国盐穴单腔建造周期一般持续3~5年。盐穴建造期间经常出现阻溶剂界面失控、造腔管柱脱落及腔体偏溶等异常问题。若造腔时出现异常未能及时发现和解决，将会导致阻溶剂界面上部盐层过早打开、部分造腔盐层段溶蚀不充分或腔体间矿柱宽度缩短等问题，影响腔体完腔体积和后期注采运行安全。为了确保造腔井正常运行，井下出现异常问题时能够及时发现解决，腔体建造期间的现场跟踪至关重要。目前现场已建立了完善的造腔跟踪监控体系。

2.1 造腔动态参数监测

记录每天井口注入压力、油垫压力、排卤质量浓度、进出口排量等，并定期进行井口卤水取样。通过分析上述动态参数，可及时发现阻溶剂界面失控、井下管柱堵塞或损坏等异常。当阻溶剂井口压力突然明显降低或井口取样发现大量阻溶剂时，应及时确认阻溶界面位置；排卤质量浓度下降明显的原因可能是造腔管柱损坏或脱落；注入压力增大且进出口排量突然差异较大的原因可能是排卤管柱内盐结晶造成局部堵塞。金坛A7井在造腔期间排卤质量浓度突然明显下降（图7），推测是井下管柱部分脱落，经井下作业确认为底部脱扣。

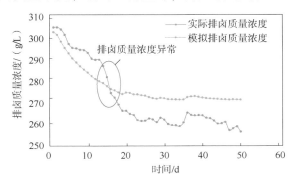

图7 金坛A7井实际排卤质量浓度与模拟排卤质量浓度对比曲线

2.2 阻溶剂界面监控

阻溶剂界面控制是造腔跟踪的关键因素之一。为了监测阻溶剂界面，现场定期安排向造腔井井下注阻溶剂，并先后发展了地面观察法、中子测井法及光纤界面仪法等多种有效的阻溶剂界面监测技术。地面观察法仅适用于正循环模式造腔，且要求阻溶剂界面深度与造腔外管管鞋一致；中子测井法针对未下入光纤界面仪且界面位于造腔外管管鞋以上的造腔井；已下入光纤界面仪的造腔井采用光纤界面仪监测阻溶剂界面位置，其不仅能够实时监测，而且监测范围广（图8）。目前，现场已能较好地实现造腔阻溶剂界面监控。

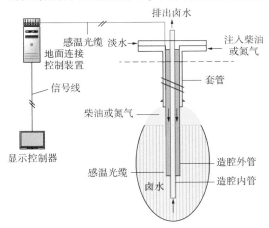

图8 光纤界面仪法监控阻溶剂界面工艺原理图

2.3 造腔进度跟踪

盐穴腔体分步建造，造腔进度跟踪指导现场井下作业和声呐测腔安排。造腔进度一般基于每天卤水化验结果来估算日累计造腔体积，由于化验数据仅计算地面采盐体积，因此与井下腔体净体积存在较大误差。为了更精确地安

排声呐检测作业，需要将地面采盐体积换算为井下净造腔体积，理论计算公式为

$$\Delta V_{\mathrm{f}} = \frac{\Delta V_{\mathrm{s}}(1-\alpha\beta)}{1-\alpha-\dfrac{C_{\mathrm{b}}}{\rho_{\mathrm{b}}}(1-\alpha\beta)} \quad (1)$$

式中：ΔV_{f} 为井下日累计造腔净体积，m^3；ΔV_{s} 为地面日累计采盐体积，m^3；C_{b} 为日排卤质量浓度，g/L；ρ_{b} 为纯盐密度，kg/m^3；α 为造腔层段平均不溶物体积分数；β 为不溶物膨松系数。

基于式(1)，当计算得到的阶段造腔体积达到设计体积时，即可进行声呐测腔作业（图9）。在造腔施工期间，还应使用声呐检测结果校核式(1)中的不溶物膨松系数和平均不溶物体积分数，采用校核过的参数结果指导下阶段的造腔施工。

2.4 声呐测腔

在造腔期间，为了监测腔体形态发展，防止腔体超出设计范围，监控其与邻腔之间的矿柱比，避免发生不可预见的意外情况，应定期检测腔体尺寸、形状、体积及最大直径是否与设计目标一致，是否有严重偏离。每个造腔阶段结束后都安排一次声呐测腔（图10）。

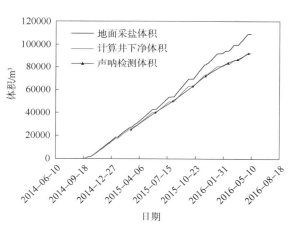

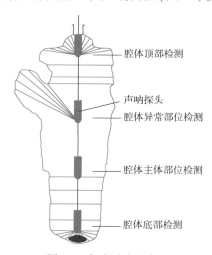

图9 造腔过程中采盐体积、净体积、声呐体积匹配曲线

图10 声呐测腔示意图

3 结论

（1）基于中国特殊的盐穴储气库建库地质条件，提出了适用于层状盐层造腔关键设计参数的选取原则，填补了国内空白。

（2）根据盐穴储气库造腔关键设计参数选取原则，设计了盐穴储气库单井初步造腔与调整方案，指导现场造腔取得良好效果。

（3）在现场造腔跟踪实践中，形成了造腔动态分析、阻溶剂界面监控、造腔进度跟踪及声呐测腔等跟踪与监测技术，可以有效避免和减少现场造腔异常并及时发现、解决造腔期间的突发问题，从而保证腔体在建造期间正常有序。

（4）造腔设计和现场跟踪是盐穴储气库腔体建造的关键技术。中国盐穴储气库发展历史短，造腔设计相关原则和跟踪技术还不够完善，绝大多数成果都是在金坛盐穴储气库建设期间获得的。随着中国其他盐穴储气库的投入兴建，盐穴储气库造腔设计与监控技术将日趋完善。

参 考 文 献

[1] 魏东吼．金坛盐穴地下储气库造腔工艺技术研究[D]．北京：中国石油大学（北京），2008：3-4.

[2] 丁国生，李春，王皆明，等．中国地下储气库现状及技术发展方向[J]．天然气工业，2015，35(11)：107-111.

[3] 丁国生．金坛盐穴地下储气库建库关键技术综述[J]．天然气工业，2007，27(3)：111-113.

[4] 李晓鹏．盐穴储气库建腔技术[J]．当代化工，2016，45(7)：1460-1463.

[5] 李铁，张永强，刘广文．地下储气库的建设与发展[J]．油气储运，2000，19(3)：1-8.

[6] 李建中，李奇，胥洪成．盐穴地下储气库气密封检测技术[J]．天然气工业，2011，31(5)：90-95.

[7] 胡开君，巴金红，王成林．盐穴储气库造腔井油水界面位置控制方法[J]．石油化工应用，2014，33(5)：59-61.

[8] 孔令峰，赵忠勋，赵炳刚，等．利用深层煤炭地下气化技术建设煤穴储气库的可行性研究[J]．天然气工业，2016，36(3)：99-107.

[9] 郭凯，李建君，郑贤斌．盐穴储气库造腔过程夹层处理工艺——以西气东输金坛储气库为例[J]．油气储运，2015，34(2)：162-166.

[10] 刘继芹，焦雨佳，李建君，等．盐穴储气库回溶造腔技术研究[J]．西南石油大学学报（自然科学版），2016，38(5)：122-128.

[11] 齐得山，李建君，赵岩，等．金坛盐穴储气库精细造腔腔体体积优化[J]．油气储运，2018，37(6)：633-638.

[12] 张刚雄，李彬，郑得文，等．中国地下储气库业务面临的挑战及对策建议[J]．天然气工业，2017，37(1)：153-159.

[13] 杨海军，于胜男．金坛地下储气库盐腔偏溶与井斜的关系[J]．油气储运，2015，34(2)：145-149.

[14] 李建君，王立东，刘春，等．金坛盐穴储气库腔体畸变影响因素[J]．油气储运，2014，33(3)：269-273.

[15] 杨春和，李银平，陈锋．层状盐岩力学理论与工程[M]．北京：科学出版社，2009：31-40.

[16] 袁光杰，班凡生，赵志成．盐穴储气库夹层破坏机理研究[J]．地下空间与工程学报，2016，12(3)：675-679.

[17] 施锡林，李银平，杨春和，等．卤水浸泡对泥质夹层抗拉强度影响的试验研究[J]．岩石力学与工程学报，2009，28(11)：2301-2308.

[18] 施锡林，李银平，杨春和，等．盐穴储气库水溶造腔夹层垮塌力学机制研究[J]．岩土力学，2009，30(12)：3615-3620.

[19] 李银平，杨春和，屈丹安，等．盐穴储油（气）库水溶造腔管柱动力特性初探[J]．岩土力学，2012，33(3)：681-686.

[20] 李银平，葛鑫博，王兵武，等．盐穴地下储气库水溶造腔管柱的动力稳定性试验研究[J]．天然气工业，2016，36(7)：81-87.

[21] 郑东波，黄孟云，夏焱，等．盐穴储气库造腔管柱损坏机理研究[J]．重庆科技学院学报（自然科学版），2016，18(2)：90-93.

[22] 杨海军，李建君，王晓刚，等. 盐穴储气库注采运行过程中的腔体形状检测[J]. 石油化工应用，2014，33(2)：23-25.
[23] 杨海军，霍永胜，李光，等. TJ 盐穴地下储气库注采气运行动态分析[J]. 石油化工应用，2010，29(9)：61-65.
[24] 杨海军，郭凯，李建君. 盐穴储气库单腔长期注采运行分析及注采压力区间优化——以金坛盐穴储气库西 2 井腔体为例[J]. 油气储运，2015，34(9)：945-950.
[25] 金虓，夏焱，袁光杰，等. 盐穴地下储气库排卤管柱盐结晶影响因素实验研究[J]. 天然气工业，2017，37(4)：130-134.

本论文原发表于《油气储运》2019 年第 38 卷第 2 期。

盐穴储气库反循环造腔的试验研究

李 龙　屈丹安　李建君　史 辉　肖恩山　杨海军

（中国石油西气东输管道公司储气库管理处）

【摘　要】 金坛盐穴储气库是我国第一个地下盐穴储气库，自开工建设以来，已累计造腔两百余万立方米，并积累了大量造腔经验。目前金坛盐穴储气库正处于造腔快速发展的阶段，以前使用的造腔模式已不能满足造腔需要，研究新的高效的造腔模式是大势所趋。反循环造腔模式是一种低成本、高效率的造腔模式，但现在国内没有实现反循环造腔的先例，必须加快反循环造腔研究，使国内盐穴储气库造腔工作取得新突破、造腔进度再上一个新台阶。

【关键词】 盐穴储气库；反循环；油水界面检测

金坛盐穴储气库使用双层管柱造腔，并使用柴油阻溶，这种配置下具有正循环和反循环两种造腔模式（图1、图2）。由于正循环容易掌握且易于控制，因此从2005年至今金坛盐穴储气库一直使用正循环造腔模式。经过五年多的学习和实践，目前国内已经完全掌握了正循环造腔，为了进一步提高技术水平，加快造腔进度，金坛盐穴储气库开始研究反循环造腔技术，进行了反循环造腔试验，这是国内第一次完整的反循环造腔现场试验。根据国外经验，反循环与正循环相比，其具有如下的优势和劣势。

优势：

（1）反循环造腔速度快。

根据国外盐穴储气库造腔经验，反循环造腔速度在相同排量和两口距下要比正循环快得多。使用业内知名的造腔软件WINUBRO模拟，在相同的注水量的条件下，反循环造腔速度是正循环造腔速度的1.2倍左右。

（2）返出卤水浓度高。

由于造腔过程中盐穴内卤水浓度分布是自上而下逐渐增加的，而反循环造腔是从较深的造腔内管排出卤水，所以排出卤水浓度就高，一般一次循环就能满足盐业公司的需要，这样就省去了正循环模式下的二次循环，降低了造腔成本。

劣势：

（1）造腔层段溶蚀速率不同，不易控制盐穴的形状；

由于反循环造腔模式下，盐穴内自上而下的壁面附近与主体部分的浓度梯度逐渐减小，越接近腔顶，浓度梯度越大，这就使得溶蚀速率越接近顶部越大。而溶蚀速率越不均衡，盐穴越容易出现不规则形状。

（2）油水界面不易检测。

由于反循环模式下，油水界面在造腔外管管鞋以上，必须使用测井的方法来控制油水界面，而由于受到现场条件和测井仪器自身条件的限制，之前无法精确地检测油水界面。

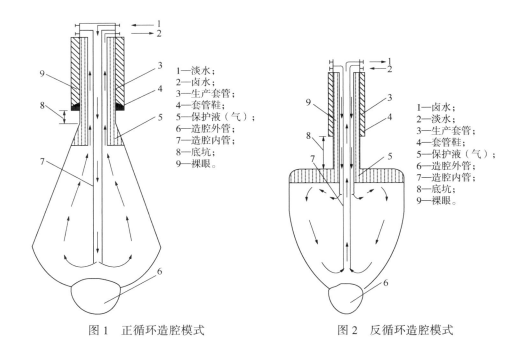

图 1　正循环造腔模式　　　　　　图 2　反循环造腔模式

1　反循环造腔试验过程

1.1　试验井选择及步骤

1.1.1　选择原则

由于仅仅是反循环试验，必须保证不会对盐穴带来不利影响，原则上有两点要求：一是要求造腔外管必须距离盐穴肩部较短，并且处于初期造腔阶段，二是为了防止出现异常，反循环试验时间不能太长，即盐穴即将完成阶段造腔。根据这些条件，从当前的造腔井中筛选出一口井进行试验。

1.1.2　试验步骤

（1）使用造腔设计软件进行反循环模拟确定造腔管柱位置。

（2）进行声呐测量，取得反循环开始时盐穴形状和体积。

（3）进行井下作业调节造腔管柱到设计位置，造腔内管注意避开腔底一定高度，防止造腔过程中堵管。

（4）注油至造腔外管管鞋。

（5）导通反循环流程并进行注水造腔，记录造腔参数。

（6）使用 WINUBRO 协助计算补油量，定期补充柴油。保证油垫厚度能够阻止盐穴上溶。

（7）完成反循环试验并进行声呐检测，取得反循环结束后的盐穴形状和体积。

1.1.3　模拟结果

由模拟的阶段造腔图可以看出（图3），在试验的造腔参数设置下，经过半个月的反循环造腔，盐穴顶部溶蚀较快，由顶之下逐渐减慢，试验结束后，盐穴顶部发展基本与中部持平。

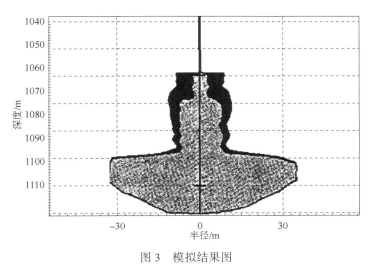

图3 模拟结果图

1.2 反循环试验结果

1.2.1 试验前后声呐检测结果比较

由声呐检测结果可知(图4)，反循环模式下溶蚀速率由上至下逐渐减小，造腔内管以下基本不溶。盐穴顶部已经出现轻微不规则性，这在正循环模式下是从来没有过的。盐穴顶部在造腔外管管鞋附近，没有发生上移，这说明在造腔过程中定期补充一定量的柴油可以满足保护腔顶的需要。

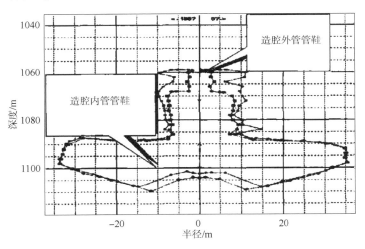

图4 反循环前后声呐检测结果比较

1.2.2 试验与模拟结果的比较分析

使用造腔设计软件预测在反循环造腔试验周期内增加自由体积为1546m³，稳定排卤质量浓度为300g/L；实际增加的自由体积为2642m³，稳定排卤质量浓度为296g/L(表1)。软件预测的造腔体积比实际的造腔体积小 41.5%，这个误差实际上是由于声呐仪器的测量误差导致的，反循环期间盐穴自由体积在 6m³ 左右，而声呐有 5% 的误差，因此测量的声呐体积与真实的体积有 3000m³ 左右的误差，所以产生预测与实测的误差是有可能的。实测的排卤质量浓度与预测的结果非常接近，这说明使用的造腔设计软件用于金坛储气库的层状盐层造腔设计还是可行的，图 5 的比较结果进一步支持了这个观点，软件预测的盐穴形状与声呐实际检测的形状吻合得很好。

表 1　反循环模拟预测与实测结果比较

	造腔时间/d	增加自由体积/m³	稳定排卤质量浓度/(g/L)
预测	15	1546	300
实测		2642(声呐体积差)	296

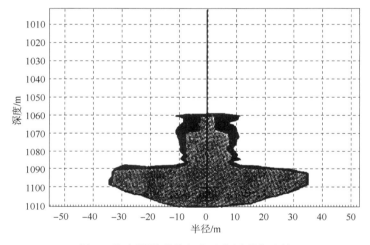

图 5　盐穴预测形状与声呐实测形状比较

1.2.3　反循环参数变化

如图 6 所示，反循环井口注水压力、油垫压力随注入量增加而增加。根据该井前期造腔数据，在造腔管柱位置相同、油垫位置相同的正循环模式下淡水注入量为 97m³/h，其注水压力为 7.8MPa、油垫压力为 58MPa，而反循环模式下淡水注入量为 90m³/h，注水压力为 73MPa，油垫压力为 6.6MPa。这说明相同造腔参数下，由单井来看两种造腔模式需要的压头相同，反循环造腔单井并不能节能。图 7 说明，反循环浓度变化受注入量影响较小，并且反循环模式在开井后很快就能达到很高的浓度，这一点可以用于造腔时需要的快速浓度调节。另外，根据该井前期造腔数据，在造腔管柱位置相同、油垫位置相同的正循环模式下淡水注入量为 97m³/h，排卤质量浓度为 243.1g/L，而相同条件下反循环排卤质量浓度为 296.9g/L，反循环的造腔速率约为同条件下正循环的 1.2 倍左右。

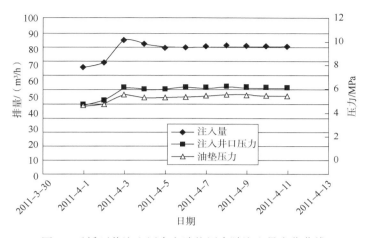

图 6　反循环井注入压力和油垫压力随注入量变化曲线

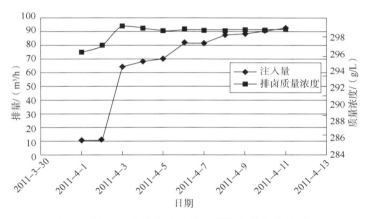

图 7　反循环井排卤质量浓度随注入量变化曲线

2　反循环油水界面控制

反循环模式造腔时，由于盐穴顶部卤水浓度很低，这就会使得腔顶上溶很快，特别是在裸眼里进行造腔时，不及时注油，油水界面很快就会上移，如果造腔外管再有泄漏的话这种情况会更严重，所以反循环造腔必须有精准的油水界面控制方法。

研究油水界面的控制方法就是反循环模式的关键所在。同时，油水界面的控制对正反循环造腔，提高造腔质量和造腔效率都有很大意义。

2.1　油水界面的控制方法

2.1.1　地面观察法

这种方法用在反循环模式中必须将反循环模式临时改成正循环模式，由于注油频率频繁，这样必然会导致生产套管鞋处压力频繁处于剧烈的交变应力下，这是不允许的，所以反循环中地面观察法不可取。图 8 是在实验过程使用地面观察法控制油水界面时，油垫压力随时间变化曲线，注入量保持不变。

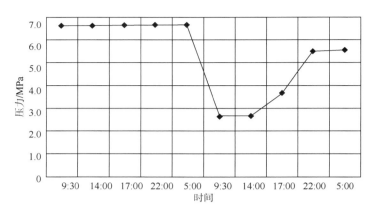

图 8　反循环过程中使用地面观察法注油时油垫压力变化曲线

2.1.2　油水界面尺检测法

这是一种永久井下仪器。优点是可以随时进行油水界面测量，使用柴油与卤水电导率不同来测量油水界面，实现在线测量。缺点是测量范围有限，不能在全部造腔井段范围内进行油水界面控制，目前该仪器还处于小范围试验阶段，有待进一步改进。

2.1.3　高能中子测井

该测井方法利用氯离子对热中子特定的俘获能力来区分卤水和柴油，氯离子含量高，俘获的热中子就多，反射的热中子数就少，探头测得的计数值就小，另外，氯离子俘获热中子后发生核反应，放出伽马射线，仪器探头探测的伽马射线能值就高，这样理论上可以综合使用中子计数值和伽马能值两条曲线来确认油水界面深度。在反循环试验中，试验了高能中子测井仪器，以确定仪器是否能分别出油水界面，并在此基础上调节参数使测量结果达到最佳。试验的结果是，这种仪器可以在 8½in 和 12¼in 井眼中使用，并且裸眼直径越大，效果越好，不但可以测量静止油水界面还可以测量移动的油水界面，这种仪器的试验成功，使反循环造腔的推广成为可能。

3　结论与建议

3.1　结论

反循环排卤质量浓度较高，一次循环基本可达到盐化的要求。反循环由腔顶之下溶蚀速率逐渐减小，内管以下基本不溶。

反循环造腔由于腔顶附近浓度低，溶蚀快，容易出现不规则性。

目前使用的造腔软件可以用于金坛储气库层状盐层的反循环造腔模拟。

反循环比正循环造腔快 1.2 倍左右，但是反循环造腔单井不能节能。

定期注入一定量的柴油可保证反循环模式下腔顶不上溶。

高能中子测井仪器不仅可以检测静止油水界面，还可以检测移动油水界面。因此这种测井仪器可以用于盐穴储气库的油水界面控制。

3.2　建议

（1）使用高能中子测井仪器来控制油水界面，减少作业次数。反循环模式必须使用该仪器，正循环模式下使用该仪器，可以在不作业情况下调整油水界面，同时由于可以将油水界面设定在造腔外管管鞋上方，可以减少柴油损失，降低补注频率，进一步降低油水界面检测的频率，减少作业次数，降低造腔成本。

（2）安装油垫压力高压和低压警报。反循环模式下，由于造腔内管离腔底较近，造腔过程中，腔底抬升会导致造腔内管被埋，从而容易发生堵塞，一旦造腔内管堵塞，腔内压力上升很快，会使得盐穴超压。另外，如果造腔外管由于破损获密封不严而发生严重泄漏，会使得油水界面迅速上移，从而使低浓度卤水接触到生产套管鞋附近盐层而将其溶蚀，这对今后盐穴的安全运行将造成严重地影响。上述的两种情况，压力变化很快，仅仅靠人工来监测油垫压力达不到要求，必须安装实时的高压和低压警报。

参 考 文 献

[1] Jug S Manocha, Terry Carter. Solution Mining and Cavern Storage in Bedded Salts of Ontario[C]//SMRI Meeting, 1993.
[2] Klafki M, Bannach A, Achtzehn L. ModelIling and Control of Cavern Development for Underground Gasstorage Stabfurt Germany[C]//SMRI Spring Meeting, 1998.
[3] Frank H, Hajo D, Tryller V. Hartmut. Computer Controlled Blanket Control System（BCS）4th New Generation-BCS Value[C]//SMRI Fall Conference, 2005.
[4] GEOSTOCK, UbroAsym Mannual under Windows System, 2009.

本论文原发表于《第七届宁夏青年科学家论坛论文集》。

盐穴储气库回溶造腔技术研究

刘继芹[1]　焦雨佳[1]　李建君[1]　施锡林[2]

(1. 中国石油东部管道有限公司储气库项目部；
2. 中国科学院武汉岩土力学研究所岩土力学与工程国家重点实验室)

【摘　要】　为提高地下盐穴储气库的经济性，在确保运行安全前提下，开发企业要求盐穴储气库地下净体积尽可能大，而在造腔过程中由于地质情况、工程故障以及认识不到位等原因导致腔体部分盐层未得到充分溶蚀，体积没有达到最优，为充分利用日渐匮乏的适合建库盐层，有必要进行回溶造腔，提高单腔有效体积。利用注气排卤管柱，在注气排卤后，回注淡水，驱顶天然气使卤水界面至设计深度，对界面下盐层进行回溶扩容。为此，建立了回溶造腔数学模型，并针对金坛储气库 L 井进行回溶造腔模拟研究。结果表明，L 井回溶造腔一个轮次可有效提高腔体有效体积 13.04%，效果显著。研究结果对提高盐层利用率和已有腔体有效体积等具有重要意义。

【关键词】　盐穴储气库；回溶造腔；腔穴修补；天然气阻溶；数值模拟

Back-leaching Technology in The Construction of Underground Salt Cavern Gas Storage

Liu Jiqin[1]*　Jiao Yujia[1]　Li Jianjun[1]　Shi Xilin[2]

(1. Gas Storage Project Department of PetroChina Eastern Gas Pipeline Co. Ltd.;
2. State Key Laboratory of Geomechanics and Geotechnical Engineering,
Institute of Rock and Soil Mechanics, CAS)

Abstract　In order to improve the economic efficiency, the developer always requires to maximize the net volume of the underground gas storage facilities on the premise of operation safety. However, due to geological reasons, engineering failures and inadequate comprehemsion, some salt layers often fail to dissolve during the storage-building process. To make the best use of the increasingly deficient salt formations suitable for underground gas storage facilities, back-leaching can be used to optimize the volume of the gas storage facility. After gas-injection and debrining, fresh water is injected by tubing strings to push the interface of the brine and gas at the designed depth. By doing that the salt formations below the interface can be leached again, which will enlarge the volume of the leached caverns. We established the

---

基金项目：中国石油科技重大专项(2015E 40)、国家自然科学基金(51404241)。
作者简介：刘继芹，1988年生，男，汉族，山东济宁人，助理工程师，硕士，主要从事盐穴储气库造腔工艺技术研究。E-mail：liujiqin@petrochina.com.cn。

back-leaching mathematic models and wrote calculation program. The pilot back-leaching in well-L in Jintan project shows the effective volume of the cavern can be raised by 13.04 percent. This research is important for enhancing the efficiency of the use of salt formations and will improve the effective volume of the built caverns.

Keywords salt cavern gas storage; back-leaching; cavity repair; SMUG; numerical simulation

随着天然气消费量的逐年增长，天然气应急调峰需求也在不断提高[1-2]，盐穴储气库作为一种安全、高效的地下存储方式越来越受到中国天然气企业的青睐[3-5]。盐穴地下储气库的建设投资成本较高，为了提高地下盐穴储气库的经济性，在确保运行安全前提下，开发企业要求盐穴地下净体积尽可能大，而在建库过程中由于地质情况、工程故障以及认识不到位等原因导致盐穴腔体部分盐层未得到充分溶蚀，体积没有达到最优[6-7]。同时，随着中国盐穴储气库的建设，综合考虑盐穴储气库选址的影响因素，中国可供盐穴储气库建腔的盐矿资源越来越缺乏，造腔过程中必须充分利用盐层，优化资源配置[8-9]。使用柴油作为造腔阻溶剂安全、可控性好，是目前盐穴储气库造腔过程中普遍的选择[10]。但是由于各种复杂情况导致部分层段未得到充分溶蚀的腔体，一般腔顶直径较大或出现严重偏溶，继续使用常规柴油阻溶方式进行修补，柴油需求量巨大，经济成本太高，这种情况在金坛储气库造腔过程中已经出现。

董建辉等[11]开展了造腔过程中使用氮气作为阻溶剂的造腔工艺理论研究，给出了腔体内气水界面稳定时的井口注气量计算模型。19世纪90年代，美国和德国等国已经开展利用天然气回溶工艺技术研究[12-13]，Staβfurt S 106使用回溶工艺，该井库容增加$7×10^4 m^3$，Victor7井使用回溶技术单腔库容增加了$11×10^4 m^3$，采用天然气作为阻溶剂回溶技术在德国普遍使用。虽然国外进行了相关回溶造腔矿场试验，但还没有成熟的商业模拟软件可以有效模拟回溶造腔过程，指导回溶工程试验。中国还没有进行过相关理论研究及矿场试验。

对存在由于各种复杂情况导致部分层段未得到充分溶蚀的腔体进行造腔扩容，要求尽量减少作业工序，管柱组合简单。此时造腔作业基本完成，要用大量阻溶剂控制卤水界面深度保护腔顶，可以考虑使用天然气作为阻溶剂，经济成本相对较低，天然气直接来源于管道，又可重新回采入管道[14]。本文考虑应用注气排卤管柱进行回注淡水造腔，通过控制腔内淡水的注入及天然气的排出而控制气卤界面，保护腔体上部偏溶部分或已达设计要求的部分，防止腔顶继续溶蚀，同时对下部未达溶部分进行扩容。

1 工艺设计

1.1 管柱结构

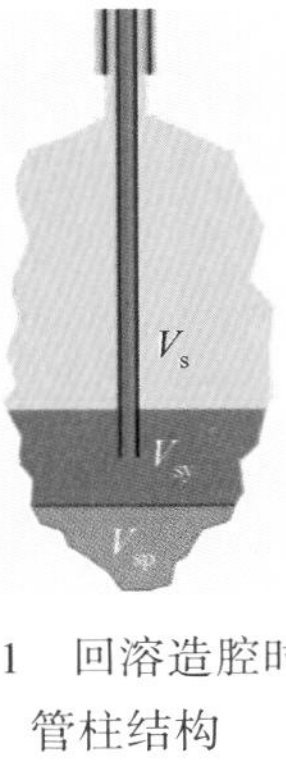

回溶造腔作业一般在注气排卤完成后，此时地面天然气管线已连接至井口，方便天然气的取用，114.3mm内管及177.8mm外管造腔管柱已更换为114.3mm单管（图1，V_s—腔体有效体积，m^3；V_{sy}—剩余液体积，m^3；V_{sp}—不溶物体积，m^3）。进行回溶造腔作业时，减少了地面管线及地下管柱改造，降低了作业成本，回溶造腔结束后，提出单管，即可进行注采气运

图1 回溶造腔时管柱结构

行[15]。注气排卤时留一定的剩余卤水可防止天然气漏出，引发安全问题，不溶物为造腔时所沉积。

1.2 回溶造腔

回溶造腔过程通过中心管注入淡水，天然气从环空管中排出[图2(a)]，气液界面下盐壁在淡卤水的作用下逐步溶解[图2(b)]，盐穴半径增大[图2(c)]，当淡水注入至设计深度时，停注静溶，待腔内卤水饱和后环空注天然气进行注气排卤作业。将气卤界面排至设定安全深度[图2(d)]。此过程为回溶造腔一个轮次，如腔体仍未达到要求，可再进行一个轮次回溶造腔。回溶造腔主要溶蚀腔体中下部盐岩，同时保护了腔体顶部。

回溶注淡水到腔顶一个轮次腔体溶盐为

$$C_s(V_s + \Delta V_{salt} - \Delta V_{salt} \cdot a \cdot b) = \Delta V_{salt}(1-a)\rho_{salt} \tag{1}$$

式中：C_s 为饱和卤水质量浓度，kg/m^3；ΔV_{salt} 为回溶掉的盐所占体积，m^3；a 为盐岩中不溶物含量，%；b 为不溶物膨胀系数，无量纲；ρ_{salt} 为纯盐密度，kg/m^3。

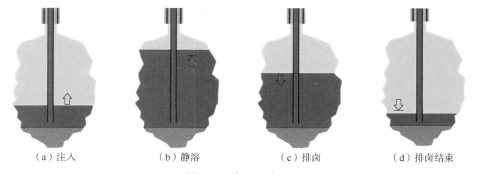

（a）注入　　（b）静溶　　（c）排卤　　（d）排卤结束

图 2　回溶造腔步骤

腔体有效体积增加量为 $\Delta V_s = \Delta V_{salt} - \Delta V_{salt} \cdot a \cdot b$，由式(1)，有

$$\frac{\Delta V_s}{V_s} = \frac{1}{\dfrac{1-a}{1-a \cdot b} \cdot \dfrac{\rho_{salt}}{C_s} - 1} \tag{2}$$

2　回溶造腔数学模型

将腔体纵向划分为等高度网格，横向划分为以井为中心的等角度扇形网格[16-19]。设网格高度足够小，为 ΔH（图3），从下往上对网格编号，设气卤界面所在网格编号为 N；腔体径向等角度划分为 K 等份，如图4所示；腔内不溶物界面所在网格编号为 M，则 t 时刻

$$\sum_{i=M}^{i=N}\sum_{j=1}^{j=K} \pi R_{ij}^2 \Delta H \bigg|_t = \frac{Q\rho_{注} t + \sum_{t=0}^{t} M_{salt\,t}}{\rho_t} \tag{3}$$

式中：R_{ij} 为纵向第 i 个，平面 j 方向上网格半径，m；ΔH 为网格高度，m；Q 为淡水注入量，m^3/s；$\rho_{注}$ 为注入淡水密度，kg/m^3；$M_{salt\,t}$ 为 t 时刻的溶盐量，kg；ρ_t 为 t 时刻腔内卤水

密度，kg/m³。

不溶物沉积所占体积为溶盐体积中不溶物膨胀后体积

$$\sum_{i=1}^{i=M}\sum_{j=1}^{j=K}\pi R_{ij}^2 \Delta H \Big|_t = ab\sum_{i=1}^{i=M}\sum_{j=1}^{j=K}\pi R_{ij}^2 \Delta H \Big|_t - ab\sum_{i=1}^{i=N}\sum_{j=1}^{j=K}\pi R_{INij}^2 \Delta H \Big|_t \tag{4}$$

式中：R_{INij} 为纵向第 i 个，平面 j 方向上网格初始半径，m。

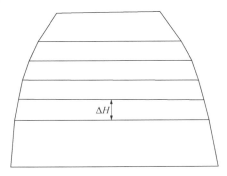

图 3　腔体纵向网格划分

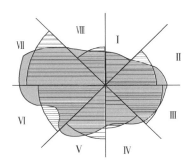

图 4　腔体径向网格划分

t 时刻腔内不同深度，不同方向的半径为

$$R_{ij}\big|_t = R_{ij}\big|_{t-\Delta t} + K_{ij} \cdot \Delta t \tag{5}$$

式中：K_{ij} 为盐岩溶蚀速率，m/s；Δt 为时间步长，s。

盐岩溶蚀速率计算公式为

$$K_{ij} = k(\psi)\left[1+0.0262(T-20)\right] \cdot (C_s - C_t)^{\frac{3}{2}} \sqrt{C_s} \tag{6}$$

式中：$k(\psi)$ 为盐壁角度为 ψ 时的盐岩溶蚀系数，$\dfrac{\text{mm}}{\text{h}(\text{°Kap})^2}$（1°Kap = 10 g/L）；$T$ 为腔内温度，℃；C_t 为 t 时刻腔内平均卤水质量浓度，kg/m³。

盐岩溶蚀系数主要受侧溶角（图5）的影响，其计算公式为：

当 $0 \leqslant \psi \leqslant \psi_g$ 时

$$k(\psi) = 0 \tag{7}$$

当 $\psi_g < \psi < 90°$ 时

$$k(\psi) = \left[k_w - (k_r - k_w)\sqrt{\cos\psi}\right]\left\{1-\left(\dfrac{\tan\psi_g}{\tan\psi}\right)^2\right\} \tag{8}$$

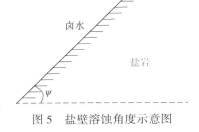

图 5　盐壁溶蚀角度示意图

当 $90° \leqslant \psi \leqslant 180°$ 时

$$k(\psi) = k_w - (k_r - k_w)\sqrt{-\cos\psi} \tag{9}$$

式中：k_w 为水平方向的盐岩溶蚀系数，$\dfrac{\text{mm}}{\text{h}(°\text{Kap})^2}$；$k_r$ 为垂直方向的盐岩溶蚀系数，$\dfrac{\text{mm}}{\text{h}(°\text{Kap})^2}$；$\psi_g$ 为侧溶角，(°)。

腔内卤水在 t 时刻的平均质量浓度为

$$C_t = \dfrac{\sum\limits_{t=0}^{t} M_{\text{salt}t}}{\sum\limits_{i=M}^{i=N}\sum\limits_{j=1}^{j=K}\pi R_{ij}^2 \Delta H\big|_t} \tag{10}$$

根据数学模型，回溶造腔模拟计算流程如图6所示，具体计算步骤为：

（1）根据注入淡水量、溶盐量结合卤水质量浓度计算卤水量，求取 t 时刻腔内卤水与天然气界面位置，根据溶盐量体积计算不溶物生成量计算不溶物界面位置；

（2）根据卤水与天然气界面位置判断是否到达设计深度，如到达，停注静溶，否则保持原注入速度注淡水；

（3）计算 $t \sim t+\Delta t$ 期间腔体半径增加量，及 $t+\Delta t$ 时刻的腔体半径；

（4）计算 $t+\Delta t$ 时刻腔内卤水的浓度，并判断卤水是否饱和，如饱和计算结束，否则继续步骤(1)。

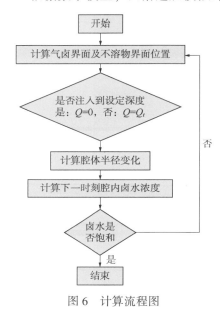

图6　计算流程图

3　实例应用

金坛储气库L井腔体主体部分平均直径基本都在60m以内，还远小于初设最大直径80m，且L井与周围邻井安全矿柱值都大于200m，$P/D \geq 2.5$，但由于考虑到腔顶明显向南侧溶，腔顶最大偏溶半径达约40m，如图7和图8所示。若该井要继续溶腔，需防止腔顶继续溶蚀，柴油量需求量将巨大，经济成本太高；若使用回溶造腔，经济成本将大大降低，扩容潜力巨大，将大大增加单腔库容。

L井轴线腔底深度1122m，考虑网络划分方法局限，1122m以下兜状体积不考虑入模型，轴线腔顶深度为1026m，假设管柱深度为1118m，气卤界面位置为1117m，将腔体垂直划分网格间隔为0.25m，径向每5°划分一个网格，考虑盐岩中不溶物垂向非均质性，回溶淡水注入速率为150m³/h，为保护腔顶，设定1035m为回溶气卤液面上限。

回溶造腔一个轮次后，腔体最大直径由70.90m增加到76.94m，符合安全标准，腔体

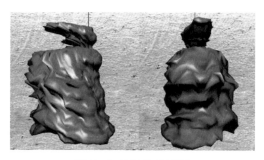

图7　L井三维显示图

体积由 172070m³ 增大到 194500m³，增加腔体体积 22430m³，扩容 13.04%，效果显著，与理论计算吻合。回溶后腔体形状与初始形状截面对比如图 9 和图 10 所示。由图 9(蓝色虚线为溶蚀后的腔体，下同)可知，腔底由于不溶物的堆积深度抬升到 1118m，回溶造腔很好地保护了腔顶。由图 11 可知，腔体下部平均半径增加较上部显著，最大值为 4.5m。

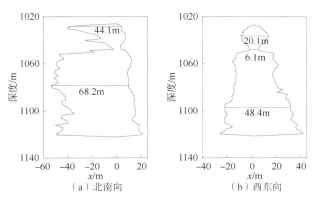

图 8　L 井北南、西东方向截面图

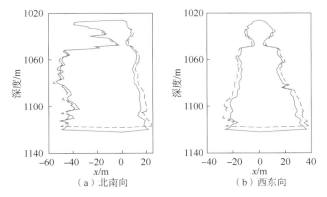

图 9　L 井回溶一个轮次后纵向截面对比图

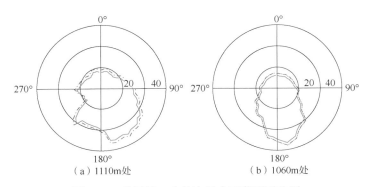

图 10　L 井回溶一个轮次后水平截面对比图

初始管柱深度决定着腔内剩余饱和卤水的量，管柱下放太深，注气排卤时不溶物容易堵塞管柱，管柱下放太浅，剩余液太多，不能充分利用腔体。一般情况下，至少预留液面超过管柱底部1m左右以保护注气排卤过程。同时，管柱深度决定的初始饱和卤水深度也影响着回溶造腔的效率及腔体形状的变化，假设初始气卤液面深度分别在1117m，1110m和1100m，完成一个回溶造腔轮次后，腔体半径平均增加量如图12所示，初始液面越低，腔内剩余饱和卤水越低，腔体下部半径较上部增加越显著。这主要是初始液面越高，腔内饱和卤水越多，回溶初期，虽然注入了淡水，但腔内卤水质量浓度仍较高，由式(6)知，液面以下盐壁的溶蚀速率仍较低，降低了腔体下部和上部半径的变化差距。

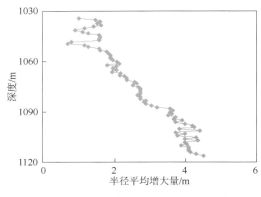

图11　L井不同深度半径平均增加量　　　图12　管柱位置对半径平均增加量影响

表1为不同初始液面位置时进行回溶造腔一个轮次造腔增加的腔体体积，初始液面位置越高，虽然可以降低腔体底部半径增加量，但同时也降低了腔体体积的增加，适合针对底部半径扩容潜力较小的腔体。

表1　不同初始液面下回溶效果

初始液面位置/m	腔体增加体积/m³	增加比例/%
1117	22430	13.04
1110	21847	12.70
1100	16948	9.85

4　结论

（1）回溶造腔时，可直接利用注气排卤管柱，管柱结构简单，可有效对中下部未达容腔体进行扩容，管护腔顶，充分利用有限可建库盐层，利用管道天然气作为阻溶剂，节约工程成本。

（2）建立了回溶造腔数学模型，根据数学模型，编制回溶造腔软件，考虑盐岩不溶物纵向非均质性及腔壁角度对溶蚀速率的影响。

（3）回溶造腔模拟结果显示，经过一个轮次回溶造腔后，有效扩容22430m³，充分利用L井下部盐层，同时研究了不同初始管柱深度对扩容体积及腔体形状发展趋势的影响。

参　考　文　献

[1] 成金华，刘伦，王小林，等. 天然气区域市场需求弹性差异性分析及价格规制影响研究[J]. 中国人口

资源与环境，2014，24（8）：131-140.
［2］吴洪波，何洋，周勇，等．天然气调峰方式的对比与选择［J］．天然气与石油，2009，27（5）：5-9.
［3］周学深．有效的天然气调峰储气技术——地下储气库［J］．天然气工业，2013，33（10）：95-99.
［4］丁国生，梁婧，任永胜，等．建设中国天然气调峰储备与应急系统的建议［J］．天然气工业，2009，29（5）：98-100.
［5］王峰，王东军．输气管道配套地下储气库调峰技术［J］．石油规划设计，2011，22（3）：28-30.
［6］李建君，王立东，刘春，等．金坛盐穴储气库腔体畸变影响因素［J］．油气储运，2014，33（3）：269-273.
［7］王汉鹏，李建中，冉莉娜，等．盐穴造腔模拟与形态控制试验装置研制［J］．岩石力学与工程学报，2014，33（5）：921-928.
［8］井文君，杨春和，李银平，等．基于层次分析法的盐穴储气库选址评价方法研究［J］．岩土力学，2012，33（9）：2683-2690.
［9］谭羽非，宋传亮．利用盐穴储备战略石油的技术要点分析［J］．油气储运，2008，27（8）：1-4.
［10］班凡生，熊伟，高树生，等．岩盐储气库水溶建腔影响因素综合分析研究［C］．第九届全国岩石力学与工程学术大会，2006，754-760.
［11］董建辉，袁光杰．盐穴储气库腔体形态控制新方法［J］．油气储运，2009，28（12）：35-37.
［12］Walden S. Releaching and solution mining under gas（SMUG）of existing caverns［C］．Basel：SMRI 2007 Spring Meeting，2007.
［13］Kunstman A，Urbańczyk K. Designing of the stor age caverns for liquid products，anticipating its size and-shape changes during withdrawal operations with use of unsaturated brine［C］．Porto：SMRI 2008 Spring Meeting，2008.
［14］Edler D，Abdel-Haq A. Interpretation of a tightness test of a gas storage cavern leached with a gaseous blanket［C］．Nova Scotia：Solution Mining Research Institute Fall Conference，2007.
［15］袁进平，李根生，庄晓谦，等．地下盐穴储气库注气排卤及注采完井技术［J］．天然气工业，2009，29（2）：76-78.
［16］Edler D. Numerical simulation of cavern shape de velopment during product withdrawal with fresh water theory and case study［C］．The 9th International Sympo-sium on salt，Beijing，China，2009.
［17］Chemkop. Basis of asymmetrical model of Win-Ubro［EB/OL］．http：//www.chemkop.pl/winubro/index.php?view=theory，2015-10-26.
［18］Saberian A. Numerical simulation of development of solution-mined storage cavities［D］．Austin：Texas Univ.，1974.
［19］Kunstman A，Urbanczyk K，Mlynarska K P，Solution mining in salt deposits-outline of recent development trends［M］．Krakow，Poland：AGH University of Science and Technology，2007：106-118.

本论文原发表于《西南石油大学学报（自然科学版）》2016年第38卷第5期。

盐穴储气库快速造腔方案

何 俊　井 岗　赵 岩　王成林　成 凡

（国家管网集团西气东输管道公司）

【摘　要】　储气库作为天然气管网的配套设施，对于维持天然气供应平稳有重要作用。中国储气库建设近年来发展飞速，但由于其建设相对较晚，尚无法满足日益增加的天然气需求，冬季"气荒"现象时有发生。储气库建设投入大、周期长，一个储气库从造腔到投产约需要4年时间。在此提出一种快速造腔投产方案，即在造腔的同时注气，利用注入的天然气作为保护腔顶的阻溶剂。这种快速造腔方式不仅降低了阻溶剂的成本，同时还将腔体投产时间大大提前。模拟分析表明，边建边储的盐穴地下储气库造腔方式是可行的。

【关键词】　盐穴储气库；快速造腔；天然气阻溶；方案；数据模拟

Accelerated mining method for salt-cavern gas storage

He Jun　Jing Gang　Zhao Yan　Wang Chenglin　Cheng Fan

(PipeChina West East Gas Pipeline Company)

Abstract　Underground gas storage, which serves as the auxiliary facility of the nature gas pipeline network, is very important to maintain the stability of gas supply. In recent years, the construction of underground gas storage develops rapidly in China. However, due to the late construction, the underground gas storage cannot satisfy the growing demand of nature gas, and gas shortage has happened occasionally in winter. In view of the large investment and long construction time required for gas storage construction, generally 4 years required from mining to put into service, an accelerated mining method is put forward herein. Definitely, it is to inject gas while mining and to take the injected gas as the protective blanket of the cavity roof. In this way, not only the cost of blanket material can be reduced, but also the construction time can be shortened significantly. As shown in the simulation analysis, it is feasible to store gas while the underground storage is constructed.

Keywords　salt-cavern gas storage; accelerated mining; gas blanket; plan; data simulation

随着天然气管网的逐渐普及，天然气需求量越来越大。尤其在冬季，天然气供给不足的情况时有发生[1]。地下储气库作为天然气管网的配套设施，其调峰和应急供气功能[2-10]有利于保障供气平稳。中国储气库建设起步较晚，虽然经历了10余年的快速发展，但目前

作者简介：何俊，男，1987年生，工程师，2012年硕士毕业于中国石油大学（北京）石油与天然气工程专业，现主要从事地下储气库的造腔与综合评价工作。地址：上海市浦东新区世纪大道1200号，200122。电话：18651287636。E-mail: xqdshejun@petrochina.com.cn。

的库容仍无法完全满足需求增长[4]。中国盐穴储气库大多为层状盐层建库，盐岩品位低，夹层数量多且厚[5]，采用常规的注淡水排卤水造腔工艺，建造 $20×10^4 m^3$ 的腔体约需 4 年时间[11]。如何缩短建腔周期，实现快速建腔是储气库建设的难点。

目前，最普遍的盐穴造腔方式是注入阻溶剂对腔体进行保护，并且控制腔体形状[12]。阻溶剂多采用柴油，但是柴油阻溶剂成本较高，气体阻溶剂成本较低。国外采用气体作为阻溶剂的盐穴造腔技术发展较早。20 世纪 90 年代，德国 Neuehuntorf NK Ⅲ 盐穴储气库成功采用氮气作为阻溶剂进行造腔[13]；美国和德国也早已成功开展天然气阻溶回溶造腔工艺技术研究[14-16]。中国在金坛储气库分别进行了氮气和天然气阻溶造腔试验，取得了良好的效果[17]。

考虑到使用天然气阻溶造腔技术是可行的，当地下腔体达到一定的体积后，可以利用上部腔体储气，将所储存的天然气作为阻溶剂控制腔体形状，并进一步造腔，从而缩短建腔周期，使腔体尽快投入使用。

1 造腔方式

常规的造腔方式是从下向上分阶段造腔，每个阶段结束后向上移动阻溶剂，进行下一阶段的造腔。而采用新工艺的快速造腔方式是先从下往上造腔，再从上往下造腔。这种造腔方式必须满足两个前提条件：(1) 在建槽阶段建好足够容纳全腔不溶物（考虑蓬松系数）的底坑；(2) 腔体裸眼已经打开，有足够的通道使后续阶段造腔过程中产生的不溶物掉入底坑，确保不会因不溶物落下而堵塞通道(图1)。

采用这种新工艺进行造腔，主要包括以下 3 个阶段：

(1) 建槽阶段。首先，根据目标体积和该地区的不溶物含量，确定需要建造底坑的最小体积。底坑为存放造腔过程中落入底部不溶物的空间，由于底坑最终会被不溶物掩埋，因此底坑的体积不能作为腔体有效存储体积的一部分。但是如果底坑体积不足，不溶物会占用一部分用来存储天然气的腔体体积，从而减少腔体的有效存储体积。在实际造腔过程中发现，不溶物从腔壁掉落入腔底会不断膨胀，计算所需底坑体积时，必须考虑不溶物的膨胀，不同地区的蓬松系数不同，该系数可以通过试验取得。底坑建造完毕后，进行声呐测腔，对底坑形状、体积等参数加以分析。

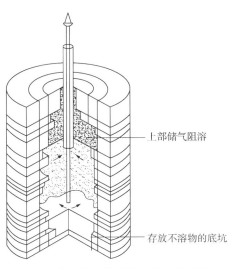

图 1 采用新工艺的快速造腔示意图

(2) 建造初始腔体。对于常规的地下储库，盐腔用于储存的主体部分一般为圆柱体，这种形状较为稳定而且可以保证盐层使用的最大化。但对于快速成腔的盐穴地下储气库，如果第二阶段完成后的初始腔体为圆柱体，那么后续建造将会引起下部直径超过设计直径(图2)。即使控制初始腔体直径，在后续建设中使下部腔体直径在允许范围之内，对于上部腔体而言也将损失大量盐层，降低了盐层的利用效率。因此，圆柱形的初始腔体并不适合这种类型的储气库造腔方式。

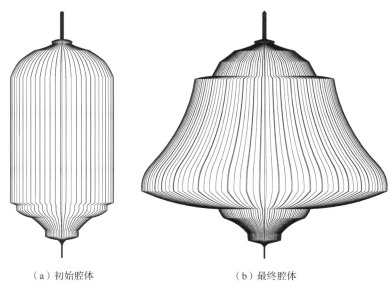

（a）初始腔体　　　　　　　　　　（b）最终腔体

图 2　初始腔体为圆柱体时天然气阻溶造腔形态图

为此，建造初始腔体时，应尽量缩小下部腔体直径，保持上部直径，形成上面直径较大、下面直径较小的花盆状腔体形态（图3）。使腔体的最终形态保持为圆柱体，直径控制在允许直径范围内。

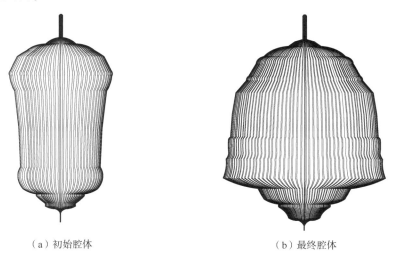

（a）初始腔体　　　　　　　　　　（b）最终腔体

图 3　初始腔体为花盆状时天然气阻溶造腔形态图

（3）完善下部腔体。当初始腔体建造完成后，可利用上部腔体储气，将所储存的天然气作为阻溶剂，在下部继续造腔。利用正反循环快速造腔（图4），由中心管或中间管注入淡水及排出卤水，环空注采天然气阻溶剂，天然气与卤水界面应始终高于中心管管口一定距离，保证天然气不进入中心管对地面管道造成压力冲击[18]。这一阶段的造腔从上到下进行，随着下部腔体扩大，腔体储气量也不断增大，同时将气水界面不断下压，直至达到最终的目标体积和腔体形态。

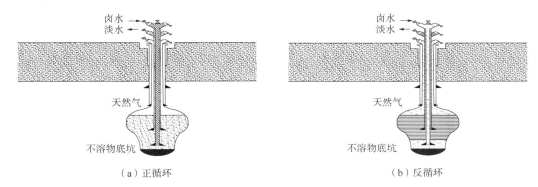

图 4 正反循环快速造腔示意图

2 模拟分析

通过软件模拟采用新工艺的快速造腔方法，所用的地质资料均为金坛储气库某井的实际数据。造腔层段深为 942～1107m，厚度为 165m；造腔层段平均不溶物质量分数为 19.67%，蓬松系数为 1.9；腔体的目标体积为 $25×10^4 m^3$；保证矿柱比大于 2.5，最大直径不超过 80m[19-20]；计算得出造腔层段的不溶物体积大约为 $5×10^4 m^3$，考虑蓬松系数，底坑的体积应大于 $9.5×10^4 m^3$。

2.1 建槽

将初始参数输入软件(表 1)，进行建槽阶段的腔体模拟(图 5，其中不同颜色表示不同的造腔阶段)。造腔共进行 500 天，底坑模拟体积为 $111841.2 m^3$，不溶物埋深为 1067.3m，最大直径为 80m。底坑体积完全可以储存膨胀后落入的不溶物。

根据建槽阶段排出卤水的质量浓度变化曲线(图 6)可见：建槽阶段排卤质量浓度逐步上升，最终稳定在 290g/L 左右。建槽阶段共进行了 4 次气水界面操作，每次气水界面操作之前均需声呐测腔，根据声呐测腔结果对下一阶段的造腔设计进行调整。

表 1 建槽阶段造腔初始参数

阶段	循环模式	阶段天数	注入体积/m^3	油水界面/m	注入深度/m	排卤深度/m	不溶物埋深/m	阶段体积/m^3	总体积/m^3
1	正循环	30	72000	1074	1107	1074	1089.5	3082.2	3120.3
2	正循环	120	288001	1074	1107	1074	1080.7	14712.0	17832.3
3	正循环	30	71992	1064	1107	1074	1079.5	7668.6	25500.9
4	反循环	80	191985	1064	1067	1074	1076.4	19480.6	44981.5
5	正循环	30	72017	1049	1074	1067	1075.3	8489.6	53471.1
6	反循环	60	144020	1049	1054	1064	1072.8	16292.7	69763.8
7	反循环	30	72017	1034	1054	1064	1071.8	9086.4	78850.2
8	反循环	120	288072	1034	1053	1063	1067.3	32991.0	111841.2

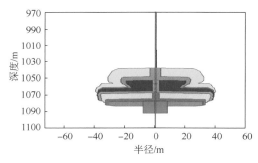
图5 建槽阶段腔体发展情况模拟图

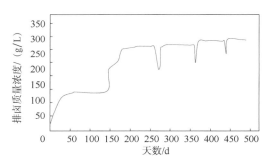
图6 建槽阶段排卤质量浓度的变化曲线

2.2 初始腔体建造

建槽阶段完成后，开始对初始腔体主体储存部分进行建造。金坛储气库对于最大直径的要求是不超过80m，腔顶直径约为15m。为此，该阶段控制下部腔体直径在65m左右，上部腔体直径在75m左右，腔顶直径在15m左右。

对初始腔体建造阶段进行模拟（表2、图7），该阶段造腔持续570天，模拟总体积为269939m³，不溶物埋深1050.2m。下部腔体最大直径为66m，上部腔体最大直径为71m。

表2 初始腔体建造阶段造腔参数

阶段	循环模式	阶段天数	注入体积/m³	油水界面/m	注入深度/m	排卤深度/m	不溶物埋深/m	阶段体积/m³	总体积/m³
1	正循环	30	71987	1024	1063	1053	1066.3	8165.5	120006.7
2	反循环	80	191910	1024	1049	1056	1064.0	23201.9	143208.6
3	反循环	60	143944	1012	1023	1033	1061.9	17664.2	160872.8
4	反循环	60	143932	997	1021	1031	1059.7	15184.6	176057.4
5	反循环	60	143932	985	1006	1016	1057.6	16136.1	192193.5
6	反循环	60	143932	975	1004	1014	1056.0	16610.3	208803.8
7	反循环	60	143928	960	1002	1012	1054.5	16669.3	225473.1
8	反循环	80	191900	950	991	1010	1052.2	23493.1	248966.2
9	正循环	80	192061	942	1000	942	1050.2	20972.8	269939.0

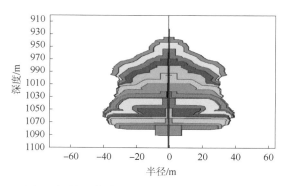
图7 初始腔体建造阶段腔体发展情况模拟图

根据该阶段的排卤质量浓度变化曲线(图8)可知：排出卤水的质量浓度保持在280g/L左右。该阶段进行了8次气水界面操作，每次油水界面操作后进行声呐测腔，根据声呐测腔结果对下一阶段的造腔设计进行调整。

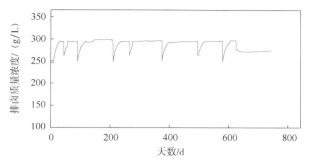

图8 初始腔体建造阶段排卤质量浓度的变化曲线

2.3 下部腔体完善

完成第2阶段后，腔体上部则可以储气，并将所储天然气作为下部继续造腔的垫层。该阶段设计进行4次气水界面操作，气水界面由腔顶942m逐步下移至1016m，通过声呐测腔分析腔体形态等参数(表3)。最终的腔体体积为326903.1m³，不溶物埋深为1043m，造腔240天。经过该阶段造腔，下部腔体直径逐步增大至最大直径，主腔体形状也逐步转变为圆柱体(图9)。

表3 下部腔体完善阶段造腔参数

阶段	循环模式	阶段天数	注入体积/m³	油水界面/m	注入深度/m	排卤深度/m	不溶物埋深/m	阶段体积/m³	总体积/m³
1	正循环	60	956	990	956	1048.5	15150.9	285089.9	3253.1
2	反循环	60	991	1008	1018	1047.0	13183.1	298273.0	89157.4
3	反循环	50	998	1015	1025	1044.9	11242.5	309515.5	122847.7
4	反循环	70	1016	1033	1043	1043.0	17387.6	326903.1	207255.5

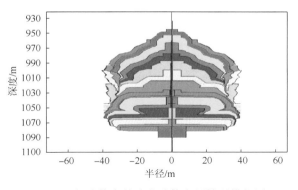

图9 下部腔体完善阶段腔体发展情况模拟图

根据以上造腔设计的模拟结果可知,使用边建边储的盐穴地下储气库造腔方式,造腔1070天即可储气,而该井如果按照常规方式进行造腔初步设计,待完腔后进行储气,则需1310天,储气可以提前240天。常规方式造腔的完腔体积约为$33×10^4 m^3$,使用边建边储造腔设计的完腔体积为$32.6×10^4 m^3$,两者基本相同(表4)。

表4 边建边储造腔设计各阶段储气空间数据

阶段	造腔层段/m	储气层段/m	储气空间/m^3	造腔段腔体初始直径/m	造腔段腔体最终直径/m
1	956~990	942~956	3253.1	68	74
2	991~1018	942~991	89157.4	70	80
3	998~1025	942~998	122847.7	73	80
4	1016~1043	942~1016	207255.5	69	78

3 结论

(1)金坛储气库的现场试验证明利用天然气作为阻溶剂在中国造腔是可行的,天然气作为阻溶剂不仅成本低,且不会对地层造成污染,具有良好的应用前景。

(2)在造腔过程中,有意控制腔体形态,使初步建造的腔体形成上面直径较大、下面直径较小的花盆状。在造腔体积达到设计总体积的80%左右时储气,利用储存的天然气作为阻溶剂,继续造腔,完善腔体形状,提高盐层利用率。模拟分析表明腔体投入注气时间可以提前240天。

(3)老腔改建储气库时,建议关注符合该形态的老腔,进行现场试验。

(4)金坛储气库在现场试验中,造腔管柱采用普通管柱,并未出现天然气泄漏情况。但若推广使用,出于安全考虑,建议采用气密性管柱。使用气密管柱无疑会增加造腔成本,建议进行进一步的经济性评价。

参 考 文 献

[1] 董秀成,李君臣.我国"气荒"的原因及对策[J].天然气工业,2010,30(1):116-118.

[2] 李铁,张永强,刘广文.地下储气库的建设与发展[J].油气储运,2000,19(3):1-8.

[3] 杨春和,梁卫国,魏东吼,等.中国盐岩能源地下储存可行性研究[J].岩石力学与工程学报,2005,24(24):4410-4417.

[4] 丁国生,梁婧,任永胜,等.建设中国天然气调峰储备与应急系统的建议[J].天然气工业,2009,29(5):98-100.

[5] 赵锐,赵腾,李慧莉,等.塔里木盆地顺北油气田断控缝洞型储层特征与主控因素[J].特种油气藏,2019,26(5):8-13.

[6] 王峰,王东军.输气管道配套地下储气库调峰技术[J].石油规划设计,2011,22(3):28-30.

[7] 罗天宝,李强,许相戈.我国地下储气库垫底气经济评价方法探讨[J].国际石油经济,2016,24(7):103-106.

[8] 成金华,刘伦,王小林,等.天然气区域市场需求弹性差异性分析及价格规制影响研究[J].中国人口资源与环境,2014,24(8):131-140.

[9] 吴洪波,何洋,周勇,等.天然气调峰方式的对比与选择[J].天然气与石油,2009,27(5):5-9.

[10] 周学深. 有效的天然气调峰储气技术——地下储气库[J]. 天然气工业, 2013, 33(10): 95-99.
[11] 班凡生. 层状盐层造腔提速技术研究及应用[J]. 中国井矿盐, 2015(6): 16-18.
[12] 李建君, 陈加松, 刘继芹, 等. 盐穴储气库天然气阻溶回溶造腔工艺[J]. 油气储运, 2017, 36(7): 816-824.
[13] Holschumacher F, Saalbach B, Mohmeyer K U. Nitrogen blanket initial operational experience[C]. Hannover: Solution Mining Research Institute Fall 1997 Meeting, 1997: 1-18.
[14] Stefan W. Re-leaching and solution mining under gas(SMUG) of existing caverns[C]. Basel: Solution Mining Research Institute Spring 2007 Technical Class, 2007: 102-110.
[15] Andrzej K, Kazimierz U. Designing of the storage cavern for liquid products, anticipating its size and shape changes during withdrawal operations with use of unsaturated brine[C]. Porto: Solution Mining Research Institute Spring 2008 Technical Conference, 2008, 1-15.
[16] Jack W, Gatewood, Market H. Solution mining and storing natural gas simultaneously-operational experience[C]. Cracow: Solution Mining Research Institute Spring 1997 Meeting, 1997: 1-21.
[17] 陈加松. 金坛储气库完成国内首次天然气阻溶造腔试验[EB/OL]. (2016-08-30)[2020-05-18]. http://news.cnpc.com.cn/system/2016/08/30/001608402.shtml.
[18] 刘继芹, 焦雨佳, 李建君, 等. 盐穴储气库回溶造腔技术研究[J]. 西南石油大学学报(自然科学版), 2016, 38(5): 122-128.
[19] 张强勇, 王保群, 向文. 盐岩地下储气库风险评价层次分析模型及应用[J]. 岩土力学, 2014, 35(8): 2299-2306.
[20] 杨海军, 闫凤林. 复杂老腔改建储气(油)库可行性分析[J]. 石油化工应用, 2015, 34(11): 59-61.

本论文原发表于《油气储运》2020年第39卷第12期。

盐穴储气库天然气阻溶回溶造腔工艺

李建君 陈加松 刘继芹 付亚平 汪会盟 王成林

（中国石油西气东输管道公司储气库项目部）

【摘　要】 盐穴储气库完腔后无法继续使用柴油阻溶剂对不规则腔体进行回溶造腔，因此迫切需要发展天然气阻溶回溶造腔技术。研究了该技术的工艺流程、工艺参数、造腔模拟技术，并在西气东输金坛储气库开展现场试验，效果超出预期：A 腔体增加自由体积 14000m^3，腔体上部突出的岩脊发生自然垮塌，排除了后期注气排卤过程中盐层垮塌砸坏排卤管柱的风险。该试验为国内首次腔体扩容及修复试验，对今后盐穴储气库建腔技术的推广及削减造腔成本具有重要的工程指导意义。

【关键词】 盐穴储气库；天然气阻溶剂；回溶造腔技术；腔体扩容；腔体修复

Re-leaching solution mining technology under natural gas for salt-cavern gas storage

Li Jianjun Chen Jiasong Liu Jiqin Fu Yaping
Wang Huimeng Wang Chenglin

(Gas Storage Project Department, PetroChina West-East Gas Pipeline Company)

Abstract After salt-cavern gas storages are completed, diesel oil can't be used as the blanket for re-leaching solution mining of deformed caverns, so it is in urgent need to develop the re-leaching solution mining technology under natural gas. In this paper, the process flow, process parameters and cavern building simulation technology of this new technology were investigated and tested on site in Jintan Gas Storage of West-to-East Gas Pipeline. The test results are better than the expectations. The free volume of the cavern A increases by 14000m^3. And besides, the prominent rock ridge at the upper part of cavern A collapsed naturally, so the damage risk of the falling salt rocks on brine displacement strings in the later gas injection de-brining stage was eliminated. This test is the first salt cavern expansion and repair test in China. It is of great engineering guiding significance to the popularization of salt cavern building technologies and the reduction of cavern building cost in the future.

基金项目：中国石油储气库重大专项"地下储气库关键技术研究与应用"子课题"盐穴储气库加快建产工程试验研究"，2015E-4008。

作者简介：李建君，男，1982 年生，工程师，2008 年硕士毕业于清华大学动力机械及工程专业，现主要从事盐穴储气库建设技术研究。地址：江苏省镇江市南徐大道 60 号商务 A 区 D 座 1801 室，212000。电话：13814795191，E-mail：cqklijianjun@petrochina.com.cn。

Keywords salt-cavern gas storage; gas blanket; re-leaching solution mining technology; cavern expansion; cavern repair

近些年,天然气消费需求逐年攀升,消费比例大幅增加[1],西气东输销售气量年均增速约16%。地下盐穴储气库作为油气管道的配套设施,在保障市场安全平稳用气和季节性应急调峰方面发挥着重要作用[2-5],其具有注采灵活、吞吐量大、垫层气量少等优势,因而国内外都在加快盐穴储气库的建设。

地下盐穴储气库主要采用水溶造腔法进行建造[6],造腔过程中需要注入阻溶剂(柴油、氮气、天然气等)来控制腔体的形状[7]。国内在层状盐岩中建腔,地质条件较复杂,盐岩夹层多、品位差、造腔难度大,为了更好地控制腔体形状,目前普遍使用柴油阻溶剂进行造腔,但是受限于复杂地质条件以及造腔工艺的发展,仍有部分腔体在造腔阶段未溶蚀完全,腔体畸形,损失了大量库容。为了充分利用现有盐岩资源,必须对部分未完全溶蚀的畸形腔体进行扩容和修复,若再次使用柴油阻溶剂成本太高,若使用其他气体阻溶剂则需要增建相关配套设施,成本也较高,因此天然气成为阻溶剂的最佳选择。欧美国家对盐穴储气库的研究较早,使用各种阻溶剂造腔技术较为成熟,德国 Neuehuntorf NK Ⅲ 盐穴储气库成功使用氮气作为阻溶剂进行造腔,平均每立方米腔体消耗 $0.4m^3$ 氮气[8]。美国和德国于20世纪90年代开展天然气阻溶回溶造腔工艺研究[9-10,13],其中德国利用 Staβfurt S 106 技术增加库容 $7×10^4m^3$,Victor7 井也利用该技术增容 $11×10^4m^3$。德国 Peckensen 储气库甚至利用天然气阻溶技术清除了腔内堵塞物并处理了故障[11-12]。但是使用天然气作为阻溶剂也存在一定的安全风险,美国 Moss Bluff 盐穴储气库使用该技术在造腔过程中发生了井喷并导致大火[14]。

使用天然气作为阻溶剂对现有腔体进行扩容和修复主要有两种回溶造腔方法:双管回溶造腔模式和单管回溶造腔模式。双管回溶造腔模式与柴油阻溶剂相似,但是需要将造腔井口和造腔管柱更换成气密封井口和管柱组合,腔体扩容及修复完毕后还需将现有井口和井下设备更换成注气排卤井口和注采完井装置,成本巨大;单管回溶造腔模式较为简单,在注气排卤阶段便可实现,利用现有注气排卤井口和管柱组合即可达到要求,成本较低。西气东输利用天然气阻溶回溶造腔技术率先在金坛储气库开展了工程应用试验,并取得了重大进展。

1 工程概况

西气东输金坛储气库作为国内第一座盐穴储气库,其盐岩层的地质条件在已发现的盐矿资源中最好,最适合建设盐穴储气库。但是由于地下盐层的不均质性和造腔工艺等原因,部分腔体在造腔过程中并未完全发展,形成偏溶,导致畸形腔体的形成,损失了大量库容(图1),部分畸形腔体突出的岩脊在后期注气排卤过程中发生垮塌则会砸坏、砸断排卤管柱,导致井喷,存在重大安全隐患[图1(b)]。因此需要充分利用金坛优质盐矿,对未充分溶蚀的腔体进行修补,继续进行回溶造腔,以增加库容量。

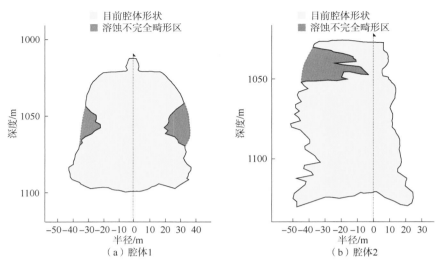

图 1 盐穴储气库畸形腔体形状示意图

2 天然气阻溶造腔工艺

天然气阻溶造腔有两种模式，双管模式和单管模式。由于开展天然气阻溶回溶造腔有一定的风险性，因此对井口采气树和管柱的气密性具有严格要求，而造腔过程中气水界面的控制是关键[15-16]。

2.1 双管回溶造腔模式

天然气阻溶回溶造腔双管模式与目前的柴油阻溶剂造腔模式相似，只是用天然气替代柴油作为阻溶剂[图 2(a)]，从生产套管与溶腔外管的环空注入天然气垫层至指定深度后实施造腔，可以进行正反循环造腔。

2.2 单管回溶造腔模式

天然气阻溶回溶造腔单管模式，目前注气排卤井口装置和管柱组合基本满足造腔要求，通过注淡水排气和注气排卤流程切换，气卤界面在安全范围内浮动变化，达到了对目标层段进行溶蚀的目的[图 2(b)]，注淡水排气由排卤管柱注淡水，从生产管柱和排卤管柱环空排出天然气，将气卤界面控制至设计深度的上部，注入淡水实施造腔；注气排卤是从生产管柱和排卤管柱环空注入天然气，将卤水从排卤管柱中排出，将气卤界面控制至设计深度的下部。

2.3 优缺点对比

2.3.1 双管模式

优点：注水量大，造腔速度快；灵活性好，可根据需要进行正反循环切换；天然气垫层达到设计深度后，根据需要进行增补，增补天然气量有限。

缺点：井口模式较为复杂，需要重新设计造腔井口，以达到气密封的要求；长期从生产套管和溶腔外管环空中注入和采出天然气，会对生产套管造成永久性损伤，导致今后长期注采运行过程中井筒的密封性存在一定的安全风险；回溶造腔结束后仍需要进行注采完井，安装注气排卤井口进行注气排卤。

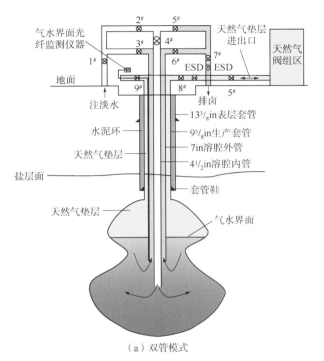

(a)双管模式

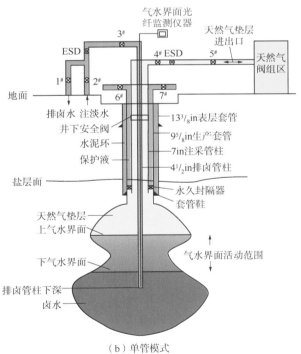

(b)单管模式

图2 天然气阻溶回溶造腔工艺流程图(1′=8in=2.54cm)

2.3.2 单管模式

优点：井口模式比较简单，目前注气排卤的井口即可满足要求，造腔结束后可以直接进行注气排卤；造腔前已完成注采管柱的座封，避免了造腔过程中对生产套管的腐蚀和损伤。

缺点：需要频繁切换流程，操作相对复杂，灵活性欠缺；单个轮次注水量有限，造腔周期相对较长。

3 工艺参数

天然气阻溶回溶造腔单管模式工艺技术参数的确定是造腔成功与否的关键，需要针对不同的腔体形状制定不同的技术参数。

3.1 主要目标溶蚀层段深度

将已经完腔的腔体形状与腔体设计形状进行对比，未充分溶蚀的盐岩层段即为主要目标溶蚀层段，通过对比确定腔体体积挖掘潜力，目标溶蚀段的上下深度定为 h_1 和 h_2。

3.2 排卤管柱下入深度

排卤管柱的下入深度根据目标溶蚀层段的位置确定，排卤管柱下入深度定为 h_3，$h_3 > h_2$。为了避免气水界面失控，气体从排卤管柱排出，导致井喷事故，通常需要排卤管柱的下入深度超过下气水界面 20m。

3.3 气水界面变化深度

气水界面变化深度即注水排气和注气排卤流程切换过程中的气水界面深度变化，气水界面间体积变化幅度不宜太小，否则需要频繁切换流程。上下气水界面变化深度定为 h_4 和 h_5，$h_4 \leqslant h_1$，$h_5 \approx h_2$。若气水界面变化幅度较大，则 $h_5 \leqslant h_2$；若气水界面变化幅度较小，则 $h_5 \geqslant h_2$。

3.4 保护层段深度

部分腔体由于地质因素或者造腔工艺原因会发生较严重的偏溶，特别是腔体顶部区域[图1(b)]，若在回溶造腔过程中进一步溶蚀，则会使得腔体进一步偏溶，腔体稳定性变差，进而导致腔体失稳、崩塌，因此在造腔过程中需要确保保护层段不受溶蚀。保护层段深度定为 h_6 和 h_7。若保护层段位于较高部位，则需要上气水界面位于保护层段下部，$h_7 < h_4$；若保护层段位于较低部位，则需要排卤管柱下深高于保护层段，$h_6 > h_3$。

4 数值模拟

由于盐穴储气库建设成本巨大，腔体的稳定性关系到储气库今后能否正常注采运行，为保证造腔成功率，需要开发回溶造腔模拟技术来指导造腔工程[17-20]。

4.1 物理模型

天然气阻溶回溶造腔模拟流程主要包括3个阶段（图3，此为一个轮次），分别为注水排气阶段、停井静溶阶段及注气排卤阶段，造腔过程中腔壁溶蚀盐岩层段主要为上气水界面与排卤管柱下深之间盐岩深度段。

（1）注水排气：从排卤管柱注入淡水，从注采管柱和排卤管柱环空采气，气水界面将由下往上推，注入的淡水可以进行溶腔[图3(a)]。

（2）停井静溶：当气水界面推至上气水界面位置时，停井静溶，注入的淡水不会立刻饱和，稀释腔内浓卤水时，通过对流扩散溶蚀腔壁后盐岩渐变饱和，具体停井时间视井口取样结果确定[图3(b)]。

（3）注气排卤：从注采管柱和排卤管柱环空注入天然气，将气水界面由上往下推至下气水界面，浓卤水从排卤管柱排出腔体[图3(c)]。

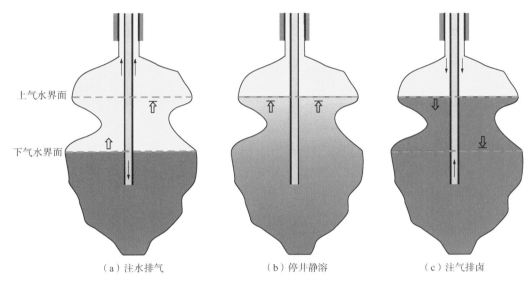

(a) 注水排气　　　　　(b) 停井静溶　　　　　(c) 注气排卤

图3　天然气阻溶回溶造腔模拟流程图

造腔过程中可根据需要进行多个轮次的造腔，造腔之前需要对腔体的发展趋势，尤其是腔壁半径变化情况进行模拟计算，确定腔体可进行造腔的轮次数，确保腔体形状发展的安全稳定。同时，在造腔过程中需要时刻关注腔体发展情况，以避免重大损失。

4.2　数学模型

4.2.1　造腔基本模型

（1）盐的物质平衡模型。

根据物质平衡理论，腔体内卤水中盐的增量来源于两部分：一部分是腔壁的溶盐，另一部分是注入的淡水夹带的盐分（通常为0）。卤水中盐的增量的计算式为

$$V_i C_c = M_s + V_i C_i \tag{1}$$

式中：V_i 为注入的淡水体积，m^3；C_c 为腔内饱和卤水质量浓度，kg/m^3；M_s 为腔体溶盐质量，kg；C_i 为注入淡水含盐质量浓度，kg/m^3。

（2）腔体体积增加计算模型。

腔体增加的自由体积主要为溶蚀体积扣除不溶物堆积体积，该模型认为注入的淡水为纯淡水，不含盐量。其计算式为

$$V_s = \frac{V_i C_c}{(1-a)\rho_s} \tag{2}$$

$$\Delta V = V_s(1-ab) \tag{3}$$

式中：V_s 为腔体溶蚀体积，m³；ΔV 为腔体增加自由体积，m³；a 为不溶物体积分数，%；b 为不溶物膨胀系数，无量纲；ρ_s 为纯盐岩密度，kg/m³。

4.2.2 腔体形状变化计算模型

腔体溶蚀层段主要集中在排卤管柱下入深度至上气水界面深度段，因此将该深度段腔体划分为等高度网格，从下向上将划分为 N 等份，网格高度（ΔH）足够小；腔体平面径向等角度划分为 K 等份，则1轮次造腔为

$$\sum_{j=1}^{j=N}\sum_{i=1}^{i=k} \pi R_{ij}^2 \Delta H = \frac{V_i\rho_w + V_0\rho_b + V\rho_{salt}(1-a)}{\rho_b} \tag{4}$$

式中：R_{ij} 为纵向第 i 个、横向第 j 个扇形网格半径，m；ΔH 为网格高度，m；ρ_w 为注入水的密度，kg/m³；V_0 为排卤管柱下深至下气水界面深度段腔体体积，该体积已知，m³；ρ_s 为纯盐岩密度，kg/m³；ρ_b 为饱和卤水密度，kg/m³。

造腔结束后腔体半径计算式为

$$R_{ij} = R_{ij}^0 + K_{ij}t \tag{5}$$

$$K_{ij} = k_\psi \left[1 + 0.0262(T-20)\right](C_s - C_t)^{\frac{3}{2}}\sqrt{C_s} \tag{6}$$

式中：R_{ij}^0 为纵向第 i 个、横向第 j 个扇形网格造腔前初始半径，m；K_{ij} 为网格盐岩溶蚀速率，m/s；t 为时间，h；k_ψ 为盐岩溶蚀系数；T 为腔内温度，℃；C_s 为饱和卤水体积浓度，kg/m³；C_t 为腔内平均卤水质量浓度，kg/m³。

5 现场试验

5.1 腔体形状

试验选取金坛储气库 A 腔体（图4）进行回溶造腔试验，该腔体最大直径 D 约为60m，远小于设计直径80 m，并且与周围临腔的矿柱宽度 P 大于200m，$P/D>2.5$，损失大量库容。腔体上部盐岩未溶蚀充分，呈葫芦口形，有一块突出的岩脊，葫芦口处直径仅为5~10m。腔体顶部偏溶严重，腔顶大平顶直径达到了40m，常规造腔方法无法将突出的岩脊溶蚀掉，对后期注气排卤阶段排卤管柱造成一定威胁。

造腔试验的目的主要有：单腔扩容10000m³；突出岩脊自然垮塌，消除日后注气排卤风险；防止腔体顶部继续发生溶蚀。

5.2 工艺参数

根据 A 腔体的形状，为了确保造腔试验安全，尽量使腔体上部突出岩脊溶蚀或者自然垮塌，下气水界面深度需要定于岩脊的下部，更有利于岩脊的垮塌。综合分析确定了以下造腔技术参数：目标溶蚀层段为1034~1050m；排卤管柱下入深度为1080m；气水界面深度

为 1034~1050m；保护层段深度为 1027~1032m。

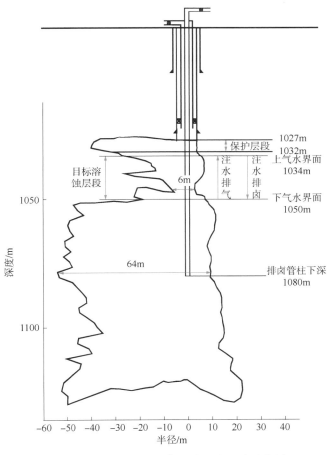

图 4 金坛储气库 A 腔体回溶造腔试验示意图

5.3 模拟计算

根据该腔体前期造腔声呐形状和造腔数据，模拟并拟合了相关参数（溶蚀速率、不溶物膨胀系数等），计算出了各个轮次的注水量和造腔量（表 1），用以指导现场造腔操作。根据腔体形状发展趋势（图 5），该腔体的体积增加段主要集中在 1050~1080m，目标层段溶蚀效果比较差，主要原因是目前该模拟技术仍存在缺陷，无法模拟盐岩上溶和盐层垮塌，模拟数据仅作参考，后期仍需要根据现场实际情况进行修正。但即使 A 腔体按照模拟的情况发展，在经历 20 个轮次的造腔之后，该腔体的最大直径仍远小于 80m，可以保证腔体的安全稳定性。

表 1 A 腔体回溶造腔 20 个轮次腔体增加体积　　　　单位：m³

轮次	阶段注水量	累计注水量	溶腔体积	累计溶腔体积	不溶物体积	不溶物累计体积	增加自由体积	累计增加自由体积
1	6217	6217	1104	1104	189	189	915	915
2	6330	12547	1124	2228	192	381	932	1847

续表

轮次	阶段注水量	累计注水量	溶腔体积	累计溶腔体积	不溶物体积	不溶物累计体积	增加自由体积	累计增加自由体积
3	6444	18991	1144	3372	196	577	949	2795
4	6561	25552	1165	4537	199	776	966	3761
5	6680	32232	1186	5723	203	979	983	4744
6	6801	39033	1208	6931	206	1185	1001	5746
7	6924	45958	1229	8160	210	1395	1019	6765
8	7050	53008	1252	9412	214	1609	1038	7803
9	7178	60185	1274	10686	218	1827	1057	8859
10	7308	67493	1298	11984	222	2049	1076	9935
11	7440	74933	1321	13305	226	2275	1095	11030
12	7575	82508	1345	14650	230	2505	1115	12145
13	7712	90221	1369	16019	234	2739	1135	13280
14	7852	98073	1394	17414	238	2978	1156	14436
15	7994	106067	1419	18833	243	3220	1177	15613
16	8139	114206	1445	20278	247	3468	1198	16811
17	8287	122493	1471	21750	252	3719	1220	18030
18	8437	130930	1498	23248	256	3975	1242	19272
19	8590	139520	1525	24773	261	4236	1264	20537
20	8746	148265	1553	26326	266	4502	1287	21824

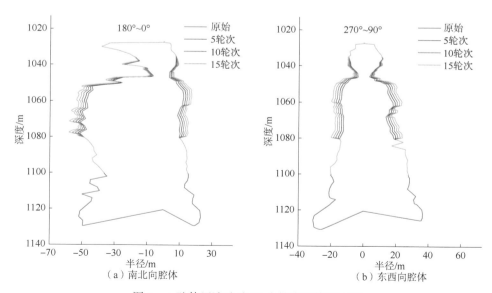

图 5　A 腔体回溶造腔 20 个轮次腔体剖面图

5.4 结果分析

实际试验过程中,以表1数据作为参考,并根据光纤监测的气水界面的变化,对每个轮次的注水量和排卤量进行调整,试验共进行了8个轮次,预计增加自由造腔体积11000m³。但是根据A腔体造腔前后声呐形状数据对比结果(图6),无论是造腔体积增加还是腔体形状变化,该造腔试验效果均远超出预期。

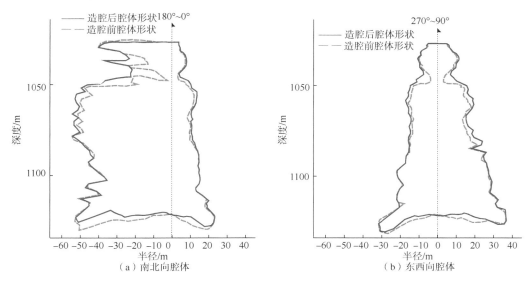

图6 A腔体回溶造腔前后腔体剖面对比图

5.4.1 溶腔体积

试验中腔体的主要溶蚀深度段为1034~1080m,1027~1032m保护层段基本未发生溶蚀,与预计一致。1050m处腔体的上溶效果尤其好,导致目标层段突出的岩脊底部失去支撑,形成较大范围的垮塌。根据A腔体回溶造腔体积的分析结果(表2),此次试验共增加自由体积约14000m³,比预期造腔目标超出近3000m³,主要体现在目标层段的溶蚀效果上,由于腔体上部岩块整体垮塌,导致腔体底部不溶物整体压实,增加腔体自由体积3140m³,与超出预期体积基本一致。

表2 A腔体回溶造腔体积分析结果

深度段/m	造腔前体积/m³	造腔后体积/m³	增加体积/m³	增加体积的模拟值/m³	备注
全腔体	185353	199622	14269	11400	超出预期3000m³
1034~1050	5052	10984	5932	1229	主要目标溶蚀层段超出预期4700m³
1050~1080	60668	72914	12246	10759	主要体积增加段
1118~1130	19118.2	15282.9	-3835.3	-2275	压实体积约3140m³

5.4.2 腔体形状

整体来看,A腔体回溶造腔试验后,腔体半径在1034~1080m深度段有不同程度的扩大,最大直径由74.5m增至75.0m,未超过80m上限,且腔体形状更加规则,稳定性更好。

葫芦口处突出岩脊垮塌使该处盐层打开，直径由 5～10m 增至 20～30m，基本排除了后期岩脊垮塌砸坏排卤管柱的风险。腔体底部由于腔体上部的岩脊垮塌和不溶物的沉淀形成不同程度的抬升。

突出岩脊垮塌主要有两方面原因：一是岩脊底部盐岩层受到溶蚀，失去支撑垮塌；二是当气水位于岩脊以下时，突出岩脊悬空，失去卤水浮力，经历多个轮次的气水界面变化时，岩脊由于周期性的加载、卸载作用，形成疲劳损伤最终垮塌。

6 结论

（1）在注气排卤阶段，利用天然气作为阻溶剂对腔体未溶蚀完全层段进行回溶造腔，该技术的研究与应用在国内尚属首次。

（2）造腔工艺流程可采用注气排卤井口装置及管柱组合，并在排卤管柱中加入光纤气水界面监测仪器，造腔结束后，取出监测仪器和排卤管柱，重新下入排卤管柱至指定深度即可进行注气排卤作业，可较大程度地削减成本。

（3）目前开发的模拟技术可以模拟腔体的变化趋势，总体趋势正确，可以指导现场操作，但是仍然存在缺陷，无法模拟腔体上溶和盐层垮塌，需要继续开展研究。

（4）提出了确定造腔工艺参数的方法，可以根据不同腔体形状，制定具体技术参数。

（5）在西气东输金坛储气库 A 腔体进行了造腔试验，效果超出预期，成功扩容约 14000m^3，腔体突出岩脊基本垮塌，排除了后期砸坏排卤管柱的风险。

（6）中国盐岩资源匮乏，可用于建造盐穴储气库的盐矿较少。盐岩层地质条件复杂、不均质性强、夹层较多，导致更多的腔体形状不规则，损失大量库容。在已建的盐穴腔体基础上，利用天然气阻溶回溶造腔工艺技术实现修复和扩容，大大增加了盐穴储气库的库容量和工作气量，降低了建造成本，增加了经济效益，具有重要的应用和指导意义。

参 考 文 献

［1］严铭卿，廉乐明，焦文玲，等. 21 世纪初我国城市燃气的转型［J］. 煤气与热力，2002，22（1）：12-14.

［2］李铁，张永强，刘广文. 地下储气库的建设与发展［J］. 油气储运，2000，19（3）：1-8.

［3］杨春和，梁卫国，魏东吼，等. 中国盐岩能源地下储存可行性研究［J］. 岩石力学与工程学报，2005，24（24）：4410-4417.

［4］丁国生，梁婧，任永胜，等. 建设中国天然气调峰储备与应急系统的建议［J］. 天然气工业，2009，29（5）：98-100.

［5］王峰，王东军. 输气管道配套地下储气库调峰技术［J］. 石油规划设计，2011，22（3）：28-30.

［6］班凡生，熊伟，高树生，等. 岩盐储气库水溶建腔影响因素综合分析研究［C］. 沈阳：第九届全国岩石力学与工程学术大会，2006：754-760.

［7］李建君，王立东，刘春，等. 金坛盐穴储气库腔体畸变影响因素［J］. 油气储运，2014，33（3）：269-273.

［8］Holschumacher F, Saalbach B, Mohmeyer K U. Nitrogen blanket initial operational experience［C］. Hannover：Solution Mining Research Institute Fall 1997 Conference，1997：1-18.

［9］Stefan W. Releaching and solution mining under gas（SMUG）of existing caverns［C］. Basel：Solution Mining

Research Institute Spring 2007 Technical Conference, 2007: 1-20.

[10] Andrzej K, Kzimierz U. Designing of the storage cavern for liquid products, anticipating its size and shape changes during withdrawal operations with use of unsaturated brine[C]. Porto: Solution Mining Research Institute Spring 2008 Technical Conference, 2008: 1-15.

[11] 袁进平, 李根生, 庄晓谦, 等. 地下盐穴储气库注气排卤及注采完井技术[J]. 天然气工业, 2009, 29(2): 76-78.

[12] Matthieu K, Yvan C, Helge T. Elimination of an obstruction blocking access to a cavernpart by drilling and solution mining[C]. Galveston: Solution Mining Research Institute Spring 2011 Technical Conference, 2011: 1-30.

[13] Jack W, Gatewood, Market H. Solution mining and storing natural gas simultaneously-operational experience[C]. Cracow: Solution Mining Research Institute Spring 1997 Technical Conference, 1997: 1-21.

[14] Benoît B, Joel N, Kerry D. Analysis of Moss Bluff cavern $1^{#}$ blowout data[R]. France: Solution Mining Research Institute 2013 Research Report, 2013: 1-188.

[15] Jan O B, Chrlstian M, Henning M, et al. Fibre optic measurement system for the automatic and continues blanket level interface monitoring during the solution-mining process of salt caverns[C]. New York: Solution Mining Research Institute Spring 2008 Technical Conference, 2015: 1-7.

[16] Stephan G, Bernhard V, Henning M, et al. Optic measurement system for temperature, automatic and continuous blanket interface monitoring in caverns[C]. Santander: Solution Mining Research Institute Spring 2008 Technical Conference, 2015: 1-10.

[17] 刘继芹, 焦雨佳, 李建君, 等. 盐穴储气库回溶造腔技术研究[J]. 西南石油大学学报(自然科学版), 2016, 38(5): 122-128.

[18] 董建辉, 袁光杰. 盐穴储气库腔体形态控制新方法[J]. 油气储运, 2009, 28(12): 35-37.

[19] 王汉鹏, 李建中, 冉莉娜, 等. 盐穴造腔模拟与形态控制试验装置研制[J]. 岩石力学与工程学报, 2014, 33(5): 921-928.

[20] 赵志成, 朱维耀, 单文文, 等. 盐穴储气库水溶建腔机理研究[J]. 石油勘探与开发, 2003, 30(5): 107-109.

本论文原发表于《油气储运》2017年第36卷第7期。

盐穴储气库溶腔参数优化方案的研究

李建君　李　龙　屈丹安

(西气东输管道公司储气库项目部)

【摘　要】 金坛盐穴储气库进行了工艺流程改造，各单井可以在注入卤水与淡水之间自由切换。2010 年金坛盐业公司将进一步提高卤水消化能力。在这两个前提下，盐穴储气库的溶腔的工程建设的优化问题日益突出。本文就金坛盐穴储气库的注水排卤优化问题进行了系统的研究。

【关键词】 地下储气库；盐岩矿床；精细三维地震；特殊观测系统

金坛盐穴储气库溶腔优化的目标是在盐业公司要求的卤水盐度、卤水处理量的限制条件下使得溶盐量最大化。目前，由于各个腔体属性不同，如泥岩含量深度分布、腔体体积、地层温度、注水排卤两口距等，各腔体在相同卤水或清水注入量下，排出卤水的盐度不同。因此，必须寻找各腔体注入量的最佳组合使得排出卤水混合后既能达到盐化的要求又能使溶盐量最大。

注入量理论上在一定范围内是可连续调节的，但是实际的注入量调节是离散的。因此各个腔体的注入量可以选择一定范围内的离散值。根据现场溶腔经验选择 40、45、50、55、60、65、70、75、80、85、90、95、100、105、110、115、120 这 17 个离散点。使用金坛盐穴储气库溶腔数值模拟软件发现，如果使用现场数据可以比较精确地模拟溶腔过程，即在一定注入量下，可以预测排出卤水的盐度。据此可以采取如下优化方案：

（1）根据所选井当前状态，使用 WINUBRO 软件模拟不同清水注入量下排出的卤水密度和盐度；

（2）由于每口井有 17 个数据点，假如有 13 口井，则就有 17^{13} 中组合方式，这种巨量计算不能用遍历的方法来编程，而遗传算法是一个比较好的编程选择；

（3）目标函数选取标准：使得在给定排量下的卤水混合后能达到盐化要求盐度下溶盐量最大。

1　腔体排量盐度预测

将盐度预测精确是优化成功与否的重要体现，必须根据以往现场溶腔数据以及声呐检测结果来调节模拟程序使得模拟结果接近实际溶腔结果，然后在此基础上再进行预测。

作者简介：李建君(1982—)，助理工程师，清华大学动力机械及工程专业硕士；现就职于西气东输管道公司储气库项目部，从事地下储气库地质评价工作。

2 算法设计

2.1 遗传算法介绍

由于要搜索表1中所有排量组合，总共有 $17^9 = 118587876497$ 种组合方式，每种组合需要计算 1ms 左右时间，所以不能用遍历的方法计算。而遗传算法（GA）是一种行之有效的全局搜索算法，目前在多个领域广泛应用，其基本计算流程如图1所示。

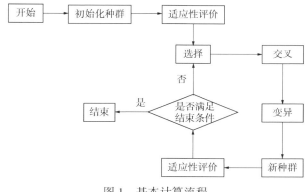

图1 基本计算流程

2.2 遗传编码

根据实际情形不太好用常用的二进制编码，采用自定义十进制数组编码，如下为一条染色体编码：

JK6-5	JK3-2	JK5-2	JK6-4	JK7-6	JK5-3	JK6-6	JK8-5	JK8-6
0	12	1	4	5	7	10	11	2

对于每个腔体将其注入量从 40~100 用 0~12 这 13 个数字表示（表1）。上边的染色体表示 JK6-5 注入量为 40，JK3-2 注入量为 100，JK5-2 注入量为 45，JK6-4 注入量为 60，JK7-6 注入量为 65，JK5-3 注入量为 75，JK6-6 注入量为 90，JK8-5 注入量为 95，JK8-6 注入量为 50。

表1 各腔体排量盐度关系

注入量\井号	JK6-5	JK3-2	JK5-2	JK6-4	JK7-6	JK5-3	JK6-6	JK8-5	JK8-6
40	291.2	285.7	278.3	294.5	279.5	282.0	277.7	282.3	279.6
45	289.0	283.4	275.5	292.6	277.0	279.5	274.9	280.0	277.1
50	287.0	281.2	272.8	290.7	274.6	277.1	272.4	277.8	274.8
55	285.1	279.2	270.3	288.9	272.4	274.9	270.0	275.8	272.7
60	283.2	277.2	267.8	287.8	270.3	272.5	267.7	273.9	270.8
65	281.3	275.4	265.5	286.0	268.4	270.4	265.5	272.1	268.9
70	279.3	273.7	263.4	284.4	266.6	268.5	263.4	270.4	267.1

续表

注入量＼井号	JK6-5	JK3-2	JK5-2	JK6-4	JK7-6	JK5-3	JK6-6	JK8-5	JK8-6
75	277.4	272.0	261.4	283.0	264.8	266.7	261.4	268.8	265.5
80	275.6	270.4	259.4	281.8	263.0	265.0	259.6	267.3	263.9
85	273.8	268.8	257.6	280.7	261.3	263.3	257.8	265.8	262.4
90	271.9	267.4	255.8	279.6	259.8	261.6	256.1	264.4	260.9
95	270.1	265.9	254.0	278.4	258.3	259.7	254.5	263.1	259.5
100	268.4	264.6	252.3	277.2	256.9	257.6	252.9	261.6	258.1
105	266.8	263.2	250.7	276.1	255.5	255.5	251.4	260.3	256.8
110	265.4	261.9	249.1	275.1	254.2	253.5	249.8	259.1	255.6
115	264.1	260.7	247.6	274.1	252.9	251.5	248.3	257.9	254.3
120	263	259.5	246.2	273.1	251.6	249.4	246.8	256.8	253.1

2.3 选择操作

种群中个体被选中的概率与其适应性有关，个体适应度使用如下公式计算：

$$f(n) = \begin{cases} 0, & Q_{inj} > Q_{inj\text{-}max} \\ 0, & S_{inj} < S_{inj\text{-}min} \\ V(n), & Q_{inj} \leq Q_{inj\text{-}max} \end{cases}$$

式中：Q_{inj} 为总注入量；$Q_{inj\text{-}max}$ 为最大总注入量；V 为种群中个体所有腔体的总造腔量；S_{inj} 为排出卤水混合后的盐度；$S_{inj\text{-}min}$ 为盐化可接受的卤水最小盐度。

个体选择使用轮盘赌进行，个体被选中的概率与其适应度相关，使用如下公式计算：

$$P(n) = \frac{V(n)}{\sum V(n)}$$

2.4 交叉操作

从种群中随机选择两个个体，将其染色体基因序列某 N 位交叉，N 根据基因序列长度随机选取。交叉按照一定的概率进行，这个概率事先给定。

如上是两个个体的染色体基因序列，交叉操作就是将两个体的红色部分分别与对方白色部分交叉。

JK6-5	JK3-2	JK5-2	JK6-4	JK7-6	JK5-3	JK6-6	JK8-5	JK8-6
0	12	1	4	5	7	10	11	2
5	6	7	4	5	3	8	0	2

2.5 变异操作

变异操作就是按一定概率选择个体基因序列上的某位，并按一定概率进行变异，就是此位的值换成其他注入量。

JK6-5	JK3-2	JK5-2	JK6-4	JK7-6	JK5-3	JK6-6	JK8-5	JK8-6
0	12	1	4	5	7	10	11	2

↓ 从 0~16 中随机选择一个数替换 4

JK6-5	JK3-2	JK5-2	JK6-4	JK7-6	JK5-3	JK6-6	JK8-5	JK8-6
0	12	1	9	5	7	10	11	2

2.6 适应性评价

对经过上述遗传操作的新种群中的个体进行适应性评价，还要根据优化的目标函数确定，优化的目标是在满足盐化盐度和卤水处理量的情况下造腔量最大。适应度计算公式已给出，这里再细化一下：

$$f(n) = \begin{cases} 0, & Q_{inj} > Q_{inj-max} \\ 0, & \dfrac{\sum_{k=0}^{M} Sal(k)Q(k)}{\sum_{k=0}^{M} Q(k)} < S_{inj-max} \\ \dfrac{\sum_{k=0}^{M} Sal(k)Q(k)}{\rho_{inj}}, & Q_{inj} \leqslant Q_{inj-max} \end{cases}$$

式中：$Sal(n)$ 为单个腔体卤水盐度；$Q(n)$ 为 ρ_{inj} 单个腔体注入量；M 为纯盐密度；M 为个体中单井数 -1。

在经过一定计算步数后，以满足盐化要求的造腔量最大的注入量组合为结果。

3 计算实例

3.1 算例 1

根据盐化要求盐度 270g/L 以及总注入量不超过 500，根据所编制的程序计算。

计算种群代数为 100000 代（图 2）。

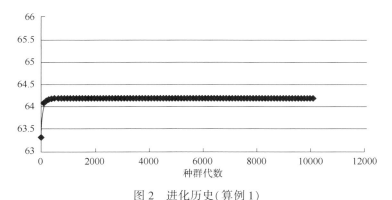

图 2 进化历史（算例 1）

计算出最佳注入量组合如下：
JK6-5：inj = 65　SalVol = 8.45331

JK3-2: inj = 60   SalVol = 7.68932
JK5-2: inj = 40   SalVol = 5.14656
JK6-4: inj = 100   SalVol = 12.8155
JK7-6: inj = 45   SalVol = 5.76283
JK5-3: inj = 50   SalVol = 6.40546
JK6-6: inj = 40   SalVol = 5.13546
JK8-5: inj = 55   SalVol = 7.01294
JK8-6: inj = 45   SalVol = 5.76491

以上 inj 表示注入量，单位是 m^3/h。SalVol 表示溶盐速度，单位是 m^3/h。

造腔量 TotSalVol = 64.1863m^3/h；

盐度 TotSal = 277.67gL；

适应性 Fittness = 64.1863；

每天造腔量为 64.1863×24 = 1540.4712m^3。

3.2 算例2

盐化要求盐度270g/L以及总注入量不超过500，9口井中由于要保证JK3-2、JK5-2的造腔速度，使其注入量保持100m^3/h。

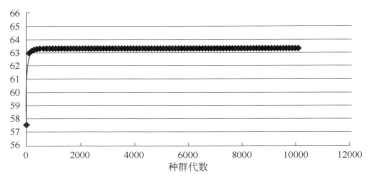

图3 进化历史(算例2)

计算出最佳注入量组合如下：

JK6-5: inj = 55   SalVol = 7.24942
JK3-2: inj = 100   SalVol = 12.233
JK5-2: inj = 100   SalVol = 11.6644
JK6-4: inj = 70   SalVol = 9.20388
JK7-6: inj = 35   SalVol = 4.56796
JK5-3: inj = 35   SalVol = 4.6068
JK6-6: inj = 30   SalVol = 3.93897
JK8-5: inj = 40   SalVol = 5.22053
JK8-6: inj = 35   SalVol = 4.56634

以上 inj 表示注入量，单位是 m^3/h。SalVol 表示溶盐速度，单位是 m^3/h。

造腔量 TotSalVol = 63.2513m³/h；

盐度 TotSal = 273.625；

适应性 Fittness = 63.2513；

每天造腔量为 63.2513×24 = 1518.0312m³。

金坛盐穴储气库实际有 13 口新腔在溶，盐化对盐度的需求也在 285 以上，这里取其中 9 个新腔优化，将剩余的 4 口腔用于 2 次循环，可以使盐度达标。以算例 2 为例，最优方案造腔量是 63.2m³/h，最差方案是 57.5m³/h，最优方案每天多造腔 137m³，目前每方腔体积造腔成本约 100 元，因此，每天最优方案比最差方案可增加效益 13700 元左右，而由于盐穴储气库造腔工程又是一项长期的工程，如此算来带来的效益是很可观的。

4 结论

对于金坛盐穴储气库的溶腔组合优化问题，遗传算法能快速找到满足限制的注入量组合。这将为以后更多腔体的同时溶腔提供一个优化方法，指导生产。这种优化在腔体数目比较少，卤水处理量不大时优势不明显，如果增加溶腔数目和卤水处理量则此优化就非常有意义。而金坛盐穴储气库即将增加 9 口新腔，并且盐化也将要增加卤水处理量到 800m³/h，因此优化注入量组合的需要在今后将日益突出。

参 考 文 献

[1] 陆金桂. 遗传算法原理及其工程应用[M]. 徐州：中国矿业大学出版社，1997.
[2] 段玉倩，贺家李. 遗传算法及其改进[J]. 电力系统及其自动化学报，1998，(1).

本论文原发表于《第六届宁夏青年科学家论坛论文集》。

盐穴储气库巨厚夹层垮塌控制工艺

王元刚　陈加松　刘　春　李建君　薛　雨　周冬林　王晓刚

（中国石油西气东输管道公司　储气库管理处工艺技术研究所）

【摘　要】 在盐穴储气库的建设过程中，巨厚夹层的存在不仅使造腔难度大幅增大，而且在不同程度上影响腔体形状。为了充分利用盐层，对巨厚夹层的垮塌过程及其控制工艺进行研究。以国内某储气库12m的厚夹层为例，选取2口井进行现场试验。结果表明：巨厚夹层作为直接腔体顶板时，出现整体失稳垮塌的概率较小；在巨厚夹层达到一定跨度后，上下同时造腔，井眼处易产生局部破坏，从而引起夹层的大面积垮塌。结合现场数据分析可得：通过控制井下工艺使厚夹层跨度达到60m时，既可以使厚夹层大面积垮塌，又可以减小井下工艺的复杂程度。研究结果对盐穴储气库厚夹层处理具有一定借鉴的意义。

【关键词】 盐穴储气库；巨厚夹层；局部破坏；腾空跨度

The technology for controlling the collapse of thick interlayer during the construction of salt-cavern gas storage

Wang Yuangang　Chen Jiasong　Liu Chun　Li Jianjun
Xue Yu　Zhou Donglin　Wang Xiaogang

(Gas Storage Technology Research Institute of Gas Storage Project Department, PetroChina West-East Gas Pipeline Company)

Abstract The existence of thick interlayer makes the salt-cavern leaching more difficult during the construction of salt-cavern gas storage, and impact the cavern shape in different degrees. In order to make full use of salt beds, therefore, it is necessary to study the collapse process of thick interlayer and its control technologies. In this paper, an interlayer of 12 m thick in one salt-cavern gas storage in China was taken as the example for study, and two wells were selected for field tests. It is indicated that the probability of integral instability and collapse is lower if the thick interlayer is taken as the direct roof of cavern. If the diameter of the thick interlayer is large enough, partial damage tends to occur along the

---

基金项目：中国石油天然气集团公司储气库重大专项"地下储气库关键技术研究与应用"，2015E-40。
作者简介：王元刚，男，1986年生，工程师，2012年毕业于中国石油大学（北京）油气田开发工程专业，现主要从事盐穴储气库造腔技术研究。地址：江苏省镇江市润州区南徐大道商务A区D座中石油西气东输管道公司储气库管理处，212000。电话：13218370363，E-mail：wangyuangang@petrochina.com.cn。

wellbore when solution mining is carried out simultaneously above and below the interlayer, and consequently the interlayer will be collapsed in the large area. It is shown from the analysis on the field data that the thick interlayer with diameter of 60m by controlling the downhole technologies can not only lead to the collapse of thick interlayer in the large area, but also reduce the complexity of the downhole process. The research results can be used as the reference for processing the thick interlayer of salt-cavern gas storage.

Keywords salt-cavern gas storage; thick interlayer; partial damage; collapse diameter

目前，我国盐穴储气库建设大多采用单井阻溶剂对流法水溶造腔[1]，从地面注入低浓度卤水，流动的卤水不断溶蚀盐岩，腔体体积不断扩大，同时通过注入阻溶剂阻止盐岩上溶控制腔体形状。我国适于建造盐穴储气库的盐层中含有较多的难溶夹层，在造腔过程中将逐渐成为腔体的直接顶板或悬挑于腔内[2]。

班凡生等[3-7]针对薄夹层建立了不同的理论模型，对薄夹层的垮塌机理及其垮塌控制技术进行了研究。西气东输外围储气库建库层段含有 10m 以上的巨厚夹层，且夹层中不溶物含量达到 95% 以上，其垮塌性质与薄夹层不同，因此有必要研究巨厚夹层的垮塌过程。若巨厚夹层垮塌跨度较小，则将在夹层上部形成二次造腔，造成盐岩体积大量损失，最终影响腔体库容。根据现场造腔经验，夹层垮塌后将造腔管柱砸弯或砸断的情况时有发生，使井下工艺难度增加。在此，通过建立理论模型研究巨厚夹层的垮塌方式，选取造腔段存在 12m 巨厚夹层的 2 口井分析其夹层垮塌的控制过程，并对类似厚夹层垮塌控制工艺提出建议，以期对后续盐穴储气库巨厚夹层的垮塌控制工艺提供有益参考。

1 巨厚夹层垮塌性分析

在卤水浸泡过程中，注入水从夹层表面浸入夹层，夹层所含盐岩颗粒遇水呈蜂窝状溶解，表面产生微小裂隙[8]，岩石力学强度降低，使得夹层剥落[9]，夹层中盐岩含量越高，越容易剥落。当夹层含盐量较低时，经过卤水浸泡后巨厚夹层仍能基本保持原有骨架，抗拉强度无明显变化，内部无明显溶蚀现象[10]，因此对于厚度超过 10m 的难溶巨厚夹层，在受卤水浸泡过程中只有与卤水直接接触的部分会受剥蚀作用而脱落，而大部分厚夹层未受卤水影响，后续理论模型可以近似认为巨厚夹层整体力学性质短时间内不受卤水影响。

1.1 巨厚夹层作为直接顶板

国内的盐腔稳定性分析方法已经成熟[11-15]。袁炽等[16-17]认为夹层破坏主要考虑局部破坏准则和整体失稳破坏准则，针对巨厚难溶夹层建立了稳定性评价模型，模拟发现厚夹层出现局部应力破坏点，则厚夹层垮塌形式符合局部破坏准则；反之，则厚夹层垮塌形式符合整体失稳破坏准则。

根据造腔经验，为了保证储气库长期安全运行，腔体最大直径的参考值为 80m。厚夹层下部腔体空间足够大，溶蚀上覆盐层时，造腔管柱需要下入夹层下部，当前腔体直径会继续扩大，为了保证腔体最大直径不超过参考值，厚夹层下部需要预留一定的溶蚀空间，因此，厚夹层的最大腾空跨度按 70m 考虑，采用 Abaqus 软件模拟厚夹层作为直接顶板时腾空跨度分别为20m、40m、60m和70m的稳定性。由于平面应变模型模拟的腔体形状是对称的，因此计算时只需计算一半腔体即可，为了得到腔体附近应力场的变化细节，在腔壁附近进行网格加密。

模拟结果显示即使厚夹层腾空跨度达到70m时，也未出现应力破坏点（图1）。初步判断在盐穴储气库造腔过程中，厚夹层作为直接顶板时存在整体失稳垮塌的可能性较小，并且夹层的厚度越大，出现整体失稳的概率越小。

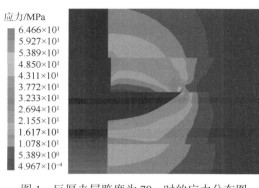

图1 巨厚夹层跨度为70m时的应力分布图

1.2 巨厚夹层悬空

厚夹层悬空时顶底面与卤水充分接触，且厚夹层上下同时参与造腔，使其在悬空状态下垮塌。施锡林等[18-20]提出的极限跨度概念目前仍被广泛应用，认为达到极限跨度后夹层是可以垮塌的。极限跨度的表达式为

$$D = 2h\sqrt{\frac{1.224E}{K\sigma_r(1-\mu^2)}} \quad (1)$$

式中：D为夹层垮塌的极限跨度，m；h为夹层厚度，m；E为夹层弹性模量，GPa；K为修正系数，$K \geqslant 1$；σ_r为夹层所在深度处的水平向地应力，MPa；μ为夹层泊松比，通过卤水浸泡条件下的室内岩心力学实验获得。

以某储气库为例，该储气库造腔层段含有巨厚夹层，将其地层参数（表1）代入式（1），可得该厚夹层在悬空状态下的极限跨度约为10m。

表1 某储气库主要地层参数

h/m	E/GPa	μ	K	σ_r/MPa
12	7.48	0.32	1.2	49.95

1.3 现场巨厚夹层处理

根据上述分析，巨厚夹层作为直接顶板时的失稳垮塌概率较小；悬挑于腔内的跨度约为10m时则很容易失稳引起大规模垮塌，若巨厚夹层垮塌时下部腔体体积较小，则会造成夹层下部盐岩体积大量浪费。为了充分利用夹层底部盐岩，现场巨厚夹层的处理（图2）可以分两步进行：

（1）厚夹层底部造腔。将阻溶剂控制在厚夹层以下进行造腔，充分利用厚夹层下部盐层，该阶段厚夹层为腔体直接顶板，由于厚夹层作为直接顶板时出现失稳垮塌的概率较小，因此该阶段结束时厚夹层的腾空跨度应该尽可能大。

（2）厚夹层顶、底同时造腔。将造腔外管调至夹层顶部以上3~5m溶蚀夹层顶部盐层，内管调至腔底不溶物以上2~3m进行造腔，进一步扩大厚夹层下部腔体体积的同时，使厚夹层处于悬空状态，促使厚夹层大规模垮塌。

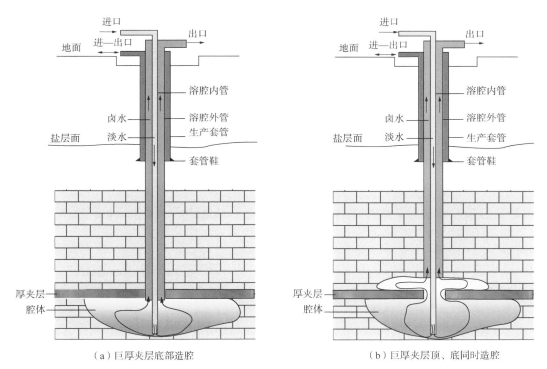

(a) 巨厚夹层底部造腔 (b) 巨厚夹层顶、底同时造腔

图 2　现场巨厚夹层处理示意图

2　实例分析

国内某盐穴储气库 A 井、B 井建库层段含有 12m 巨厚夹层，夹层段埋深约 1600m，平均不溶物含量在 97% 以上，属于难溶巨厚夹层。

2.1　A 井

A 井巨厚夹层垮塌前共有 3 个造腔阶段，各阶段的造腔方案及腔体形状如下：

（1）巨厚夹层下部正循环造腔。为了观察厚夹层作为直接顶板时造腔的腔体形状，在腔体体积较小时进行腔体形状测量。测量结果显示该阶段结束时厚夹层的腾空跨度约 30m，阻溶剂界面控制较好，厚夹层无垮塌迹象，初步证明厚夹层作为直接顶板时的稳定性较好，出现失稳垮塌的概率较小（图 3）。

（2）厚夹层下部反循环造腔，最大限度增加厚夹层腾空跨度。第 1 阶段厚夹层未出现垮塌，表明腾空跨度可以继续增加。声呐结果显示，厚夹层腾空跨度为 60m 时，未出现失稳垮塌。

该阶段由于反循环造腔，厚夹层底部与卤水充分接触，除夹层底部受到卤水浸泡作用溶解剥落外，厚夹层主体部分未与卤水接触，结构完好，表明难溶巨厚夹层受卤水浸泡发生失稳垮塌的概率较小（图 4）。

（3）巨厚夹层悬空垮塌。将内管调整至厚夹层底部注水，外管调整至厚夹层上部排卤，当阶段产盐量为 1400m³ 时，排卤质量浓度便下降至 50g/L（图 5），初步判断夹层大面积垮塌将管柱砸断，卤水循环出现短流程。声呐结果显示，井眼周围厚夹层大面积垮塌，但是腔体边缘处夹层未垮塌(图6)。

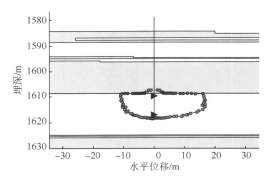

图3　A井第1造腔阶段结束时的腔体形状

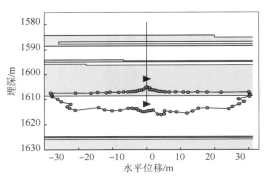

图4　A井第2造腔阶段结束时的腔体形状

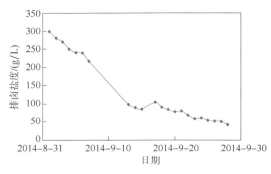

图5　A井第3造腔阶段的排卤盐度曲线

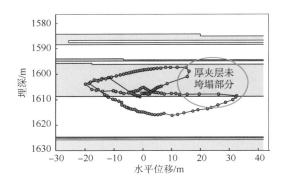

图6　A井巨厚夹层垮塌阶段的腔体形状

2.2　B井

B井地层条件与A井类似,在第1造腔阶段后期,阻溶剂界面失控从厚夹层底部上升至夹层顶部,继续产盐至1000 m³ 时进行声呐测腔。声呐结果显示,厚夹层发生大面积垮塌,腔体最大直径约51 m,由于夹层垮塌的不均匀性,腔体部分形状无法通过声呐检测到,因此,B井厚夹层垮塌前的腾空跨度很可能大于51 m(图7)。

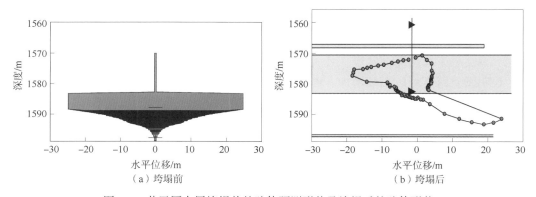

图7　B井巨厚夹层垮塌前的腔体预测形状及垮塌后的腔体形状

2.3 垮塌原因分析

对 A 井、B 井两井的造腔数据进行分析,厚夹层作为直接顶板时腾空跨度达到 50～60m,也未出现大规模垮塌,说明整体失稳垮塌的可能性较小;当阻溶剂界面上升至厚夹层顶部时,在造腔体积基本未增加的情况下,夹层大面积垮塌,因此厚夹层垮塌时夹层悬空跨度基本为 0,这是由于裸眼段卤水流速较大,受卤水流速影响产生局部应力破坏,导致厚夹层大面积垮塌。

当卤水排量 Q 为定值时,不同卤水截面处的流速为

$$v_R = Q/(2\pi R^2) \tag{2}$$

$$v_0 = Q/[\pi(R_0^2 - r_0^2)] \tag{3}$$

式中:v_R 为球面半径 R 处的卤水截面流速,m/s;v_0 为厚夹层中裸眼段卤水流速,m/s;R_0 为井眼裸眼半径,m;r_0 为造腔内管半径,m。

在现场实际工况中,卤水排量一般为 100m³/h,由流速公式可知,远离裸眼处的流速可以忽略不计。巨厚夹层不溶物含量达到 97%且不溶于水,经过卤水浸泡后厚夹层仍能基本保持原有骨架,抗拉强度无明显变化,内部无明显溶蚀现象。阻溶剂界面上移后,厚夹层不溶物含量较高,难溶于水,裸眼段井径未扩大,流速最大可以达到 1.524km/h,井眼附近由于卤水冲刷等原因造成应力集中而产生局部破坏,引起厚夹层大规模垮塌。

3 建议

3.1 作为直接顶板时选择最优腾空跨度

通过理论分析与现场试验发现,巨厚夹层出现整体失稳垮塌的概率较小,厚夹层达到极限跨度后,可以在夹层上下同时造腔使厚夹层悬空引起垮塌。若夹层腾空跨度较小时垮塌,残渣在井眼附近大量堆积,后续造腔管柱不能下入腔底,夹层下部盐岩未充分溶蚀,盐层利用率较低,并且厚夹层跨度较小,溶蚀上部盐层时,很容易形成二次造腔,夹层上部盐岩体积大量损失(图 8)。因此单纯从理论角度分析,厚夹层垮塌前的腾空跨度越大越好。

但通过 A 井、B 井现场试验可知,随着厚夹层腾空跨度增大,阻溶剂控制难度也增大,如果阻溶剂界面失控,厚夹层将很快大规模垮塌,导致造腔管柱被砸弯或砸断,增加井下工艺困难。

A 井夹层跨度达到 60m 后,裸眼段不溶

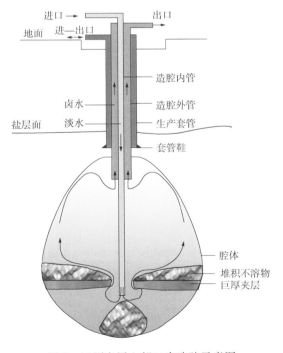

图 8 巨厚夹层上部二次造腔示意图

物小部分脱落，阻溶剂界面缓慢上升，若后续继续造腔，则阻溶剂界面很有可能失控，引起夹层提前垮塌。B井阻溶剂界面失控，厚夹层垮塌后的腔体最大直径为51m，由于厚夹层垮塌的不均匀性，可以判断厚夹层在垮塌前的跨度大于51m。根据A井造腔经验，厚夹层在跨度达到60m时是可以维持阻溶剂界面稳定的。综合分析A井、B井巨厚夹层的垮塌过程，当巨厚夹层的腾空跨度约为60m时为最优腾空跨度，既可以调整管柱在厚夹层上下同时造腔，保证厚夹层大面积垮塌，又降低了井下工艺的复杂程度。

3.2 严格控制阻溶剂界面

在现场造腔过程中，随厚夹层腾空跨度的增加，阻溶剂界面控制难度相应增加，随时有界面失控引起裸眼段局部应力破坏使厚夹层提前垮塌的风险。厚夹层作为直接顶板时，夹层底部裸眼段受卤水浸泡，小部分剥落，阻溶剂界面很容易上移，因此现场垮塌前，必须严格控制阻溶剂界面，定期注入阻溶剂使其位于厚夹层下部，若有必要应该检测阻溶剂的界面位置。

4 结论

（1）巨厚夹层作为腔体直接顶板时，稳定性较好，发生整体失稳垮塌的概率较小。

（2）巨厚夹层在卤水的充分浸泡过程中，夹层底部小部分与卤水直接接触部分易受到溶蚀作用而产生裂缝剥落，未与卤水直接接触部分受影响较小，夹层结构完整。

（3）厚夹层跨度达到一定值后，在厚夹层顶、底同时造腔，裸眼段由于卤水冲刷等原因会产生局部应力破坏，引起厚夹层大面积垮塌。

（4）以A井、B井处理造腔层段12m厚夹层为例，为了保证厚夹层垮塌前腔体体积足够大且减小井下工艺的复杂程度，可以在厚夹层腾空跨度约为60m时调整管柱，在厚夹层上下同时造腔使其大面积垮塌，研究成果对后续巨厚夹层的处理具有一定借鉴意义。

参 考 文 献

[1] 胡开君，巴金红，王成林．盐穴储气库造腔井油水界面位置控制方法[J]．石油化工应用，2014，33(5)：59-61．

[2] 李建君，王立东，刘春，等．金坛盐穴储气库腔体畸变影响因素[J]．油气储运，2014，33(3)：269-273．

[3] 班凡生，袁光杰，赵志成．盐穴储气库溶腔夹层应力分布规律[J]．科技导报，2014，32(16)：45-48．

[4] 袁光杰，班凡生，赵志成．盐穴储气库夹层破坏机理研究[J]．地下空间与工程学报，2016，12(3)：675-679．

[5] 郭凯，李建君，郑贤斌．盐穴储气库造腔过程夹层处理工艺——以西气东输金坛储气库为例[J]．油气储运，2015，34(2)：162-166．

[6] 班凡生，袁光杰，申瑞臣．多夹层盐穴腔体形态控制工艺研究[J]．石油天然气学报，2010，32(1)：362-364．

[7] 徐孜俊，班凡生．多夹层盐穴储气库造腔技术问题及对策[J]．现代盐化工，2015，4(2)：10-14．

[8] 任松，文永江，姜德义，等．泥岩夹层软化试验研究[J]．岩土力学，2013，34(11)：3110-3116．

[9] 姜德义，张军伟，陈结．岩盐储库建腔期难溶夹层的软化规律研究[J]．岩石力学与工程学报，2014，

33(5): 865-873.
[10] 施锡林, 李银平, 杨春和, 等. 卤水浸泡对泥质夹层抗拉强度影响的试验研究[J]. 岩石力学与工程学报, 2009, 28(11): 2301-2308.
[11] Kerry L, Vries D, Kirby D, et al. Salt damage criterion proof-of-concept research[R]. Pittsburgh: National Energy Technology Laboratory of United States Department of Energy, 2002: 40-58.
[12] 吴文, 侯正猛, 杨春和. 盐岩中能源(石油和天然气)地下储存库稳定性评价标准研究[J]. 岩石力学与工程学报, 2005, 24(14): 2497-2505.
[13] 梁卫国, 杨春和, 赵阳升. 层状盐岩储气库物理力学特性与极限运行压力[J]. 岩石力学与工程学报, 2008, 27(1): 22-27.
[14] 李建君, 陈加松, 吴斌, 等. 盐穴地下储气库盐岩力学参数的校准方法[J]. 天然气工业, 2015, 35(7): 96-102.
[15] 杨海军, 郭凯, 李建君. 盐穴储气库单腔长期注采运行分析及注采压力区间优化——以金坛盐穴储气库西2井腔体为例[J]. 油气储运, 2015, 34(9): 945-950.
[16] 袁炽. 层状盐岩储库夹层垮塌的理论分析[J]. 科学技术与工程, 2015, 15(2): 14-20.
[17] 施锡林, 李银平, 杨春和, 等. 多夹层盐矿油气储库水溶建腔夹层垮塌控制技术[J]. 岩土工程学报, 2011, 33(12): 1957-1963.
[18] 施锡林, 李银平, 杨春和, 等. 盐穴储气库水溶造腔夹层垮塌力学机制研究[J]. 岩土力学, 2009, 30(12): 3615-3620.
[19] 郑雅丽, 赵艳杰, 丁国生, 等. 厚夹层盐穴储气库扩大储气空间造腔技术[J]. 石油勘探与开发, 2016, 44(1): 1-7.
[20] 屈丹安, 施锡林, 李银平. 盐穴储气库建槽工程实践与顶板极限跨度分析[J]. 岩石力学与工程学报, 2012, 31(增刊2): 3703-3709.

本论文原发表于《油气储运》2017年第36卷第9期。

盐穴储气库溶腔排量对排卤浓度及腔体形态的影响

王文权　杨海军　刘继芹　齐得山　张新悦　胡志鹏　付亚平

（中国石油西气东输管道公司）

【摘　要】 为了研究盐穴储气库溶腔过程中不同排量对排卤质量浓度及腔体形态的影响规律，通过对溶蚀速率公式的推导及变形，分析了排量与排卤质量浓度及造腔体体积的相关性，而后利用 WinUbro 造腔数值模拟软件建立了理想的腔体模型并赋予其一定的假设条件进行对比分析。为了验证所得结论的可靠性，对金坛现场溶腔井进行数据统计与整理，对各个溶腔阶段进行对比与分析，并以部分井的典型阶段为例进行详细分析，所得结论与理论分析一致，即溶腔时注淡水排量越小，排卤质量浓度越高，腔体形态越规则。

【关键词】 盐穴储气库；溶腔排量；排卤质量浓度；腔体形态

Effect of cavity displacement of salt cavern gas storage on brine displacing concentration and cavity form

Wang Wenquan　Yang Haijun　Liu Jiqin　Qi Deshan
Zhang Xinyue　Hu Zhipeng　Fu Yaping

（PetroChina West-East Gas Pipeline Company）

Abstract　In order to identify the effect of different displacements on the brine displacing concentration and cavity form in the process of dissolving cavity, the dissolution velocity formula is deduced and transformed to analyze the correlation between the displacement and brine displacing concentration and the cavity volume. Then, using WinUbro cavern numerical simulation software, an ideal cavity model is established and used for comparative analysis under certain assumptions. To verify the reliability of the conclusions, the data of cavity wells in Jintan are collected, and compared and analyzed for each stage of dissolving cavity; detailed analysis is made for typical phase of some wells, deriving the conclusion which is consistent with results of theoretical analysis, namely, the smaller the freshwater displacement is, when dissolving the cavity, the higher the brine displacing concentration and the more regular the cavity form are.

Keywords　salt cavern gas storage; cavity displacement; brine displacing concentration; cavity form

---

作者简介：王文权，助理工程师，1987 年生，2013 年硕士毕业于中国石油大学（北京）石油工程学院石油与天然气工程专业，现主要从事盐穴储气库造腔设计及管理工作。电话：15050857898，E-mail：wenquan-wang@163.com。

金坛盐穴储气库[1]是我国首次利用地下盐层建设的天然气储库,因盐穴储气库注采频次高、反应快、不伤害环境等诸多优点,被视为西气东输工程不可或缺的配套工程,主要用于保障长江中下游用户的用气安全和季节调峰,解决管道供气不均衡引起的供气过剩或供气不足的问题,目前金坛盐穴储气库已进入大规模溶腔的阶段[2-3]。溶腔工程是一项复杂的系统工程,简单地说,溶腔过程就是将淡水以一定的速率注入地下盐层中,使盐岩溶解于淡水形成卤水并排出处理,使地下盐层形成一定形状的腔体或盐穴的过程。然而,该过程除了受到不溶物含量、盐溶解速率等盐层物性影响外,还受到诸多工艺参数的制约,如排量大小、循环方式、两管柱距离和溶腔时长等,这些参数直接影响腔体的形态和排卤质量浓度等,如果能有效利用这些参数的影响规律,制定合理的工作制度,可大大节省成本,增加经济效益。当前,在溶腔参数的相互影响及调整方面已有诸多研究成果[4-6],但在排量对腔体形态上的影响方面鲜见报道。基于此,开展溶腔过程中注淡水排量对排卤质量浓度及腔体形态的影响规律研究,以期对实际溶腔工作不同阶段的淡水排量选择提供指导。

1 数学模型与数值模拟

1.1 数学模型

盐穴储气库中水溶盐的溶蚀速率计算公式[7-9]为

$$\omega = \kappa(\psi)[1+\beta(T-T_0)](10^4 C_N - 10^4 C)^\alpha C_N^{2-\alpha} \tag{1}$$

式中:T 为腔体温度,℃;T_0 为参考温度,取 20℃。β 为温度系数,取 0.0262℃$^{-1}$;C_N 为饱和卤水质量浓度,g/L;C 为排卤质量浓度,g/L;α 为数值常数,取 1.5;$\kappa(\psi)$ 为溶蚀速率系数。

假设在某个溶腔阶段中,溶蚀速率系数、腔体温度及饱和卤水质量浓度均为常数,则 $\omega = \kappa(\psi)[1+\beta(T-T_0)]10^4 C_N^{2-\alpha}$ 为定值,设为 γ,则式(1)变换为

$$\omega = \gamma(10^4 C_N - 10^4 C)^\alpha \tag{2}$$

可见,此时溶蚀速率仅与排卤浓度有关,且当溶蚀速率变大时,排卤质量浓度变小。溶蚀速率指在某个溶腔阶段中,盐岩被淡水溶解的速度,可通过对比两次声呐测量期间腔体直径增加值、腔顶位移及溶腔时间计算溶蚀速率:

$$\omega = \Delta L/t \tag{3}$$

式中:ΔL 为腔体在水平或垂直方向上溶蚀的长度,m;t 为该阶段时长,h。

联立式(2)、式(3)可得

$$\Delta L = \gamma(10^4 C_N - 10^4 C)^\alpha t \tag{4}$$

可见,在一定时间内,当排卤质量浓度变小时,腔体在水平或垂直方向上溶蚀的长度变大,腔体体积亦相应变大。而卤水排量公式为

$$P = \Delta V/t \tag{5}$$

式中:ΔV 为溶腔所用淡水体积,m³。

对比式(4)、式(5)可知,在同一溶腔阶段内,溶腔时长相同,若形成的腔体体积越

大,则进出腔体的淡水体积越大,即排量越大。

综合以上公式分析可知,排卤质量浓度越小,造腔体积越大,排量越大,反之亦然。

1.2 数值模拟

WinUbro 造腔数值模拟软件是一个专门针对盐穴溶腔过程进行模拟分析的软件。目前,国际上绝大多数盐穴储气库均使用 WinUbro 进行数值模拟及设计分析[7]。利用该软件建立模型,假设某溶腔阶段采用 7in+4½in 油管的造腔管柱组合(1in=25.4mm),内外管相距 30m,利用正循环模式造腔 150d,分别以排量 50m³、100m³、200m³进行溶腔,不溶物含量及盐溶解速率均设为相等的常量,观察腔体形态变化,并对排卤质量浓度进行对比(图1、图2)。

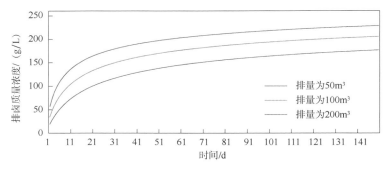

图1 不同排量下排卤质量浓度随时间变化曲线

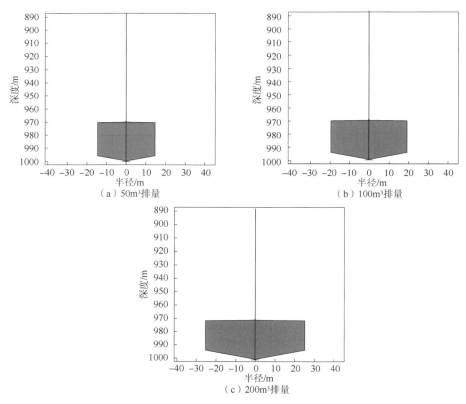

图2 不同排量下的腔体形态特征

显见，排量与排卤质量浓度之间形成了良好的相关性，排量越小，排卤质量浓度越高，而在腔体发育上，在相同时间内，排量越大，腔体半径增加越多，腔体形态在横向上发展越快，且腔体形态变化是规则的。然而，为保证腔体的稳定性，每口溶腔井会在初始设计中根据邻井矿柱比的计算结果设定一个最大半径值，大排量导致的腔体横向发展过快不利于腔体最大半径的控制，因此，从该理论模型分析可知，溶腔过程采用大排量进行溶腔不利于腔体最大半径的控制。由于该例是在 WinUbro 造腔数值模拟软件中建立的理想模型，未考虑实际地下盐岩存在不同深度、不同方向上的盐岩溶解速率不同、不溶物含量不同及隔夹层等因素的影响，因此，在不同排量下，实际腔体形态的变化规律尚需通过对溶腔井的形态进行监测对比分析而获知。

2 现场应用

一口溶腔井通常根据声呐测量次数而分成不同的阶段，一般一个溶腔阶段持续 3~4 个月，然后利用声呐测井对当前的腔体形状进行测量，得到实际的腔体形态。在实际溶腔过程中，每个阶段的各项溶腔参数都是根据上个阶段的声呐测井解释结果来设计的，因此，不能像理论分析一样使不同阶段的某些溶腔参数一致，而只研究注淡水排量的大小与排卤质量浓度及腔体形态之间的关系。为了在实际溶腔井中找到注淡水排量对排卤质量浓度及腔体形态的影响规律，对金坛储气库多口溶腔井进行数据整理，求取各个阶段的平均排量、平均排卤质量浓度、两管柱间距及平均不溶物含量，为降低其他变量对分析结果的影响，以平均排量及平均排卤质量浓度变化较大而两管柱间距及平均不溶物含量变化较小的阶段作为典型阶段，以金坛 A 井为例，对排量大小与排卤质量浓度及腔体形态之间的关系进行对比分析（图3、图4）。

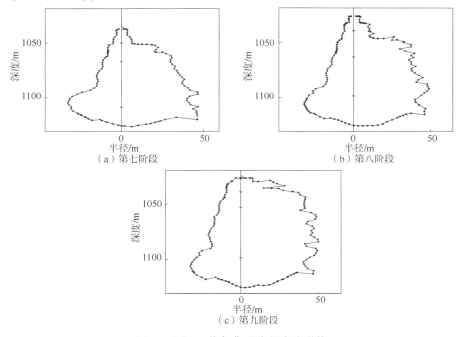

图 3　金坛 A 井各典型阶段溶腔形状

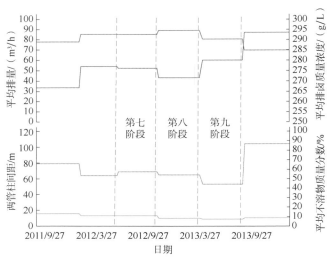

图 4 金坛 A 井典型阶段数据统计图

金坛 A 井已完成 10 次声呐测量，第七阶段与第八阶段两管柱间距与平均不溶物含量变化较小，对分析结基本不构成影响，而当排量变大时，排卤质量浓度明显下降，说明排量与排卤质量浓度间具有良好的相关性，而在腔体形态的发育上，第八阶段显然比第七阶段多了很多突起或凹陷，说明排量变大后，腔体形态不易控制，腔体形状发展不规则。对比第八阶段与第九阶段可知，平均不溶物含量变化依然很小，可忽略其对分析结果的影响，但两管柱间距比第八阶段缩短约 10m，可能对分析结果产生影响。两管柱间距越小，淡水与盐岩的接触面积越小，盐岩溶解体积越小，则排卤质量浓度应该越小，然而模拟显示，当排量下降后，排卤质量浓度非但没有变小，反而明显上升，说明排量下降与排卤质量浓度的上升具有良好的相关性（图 4）。对比两个阶段的腔体形态可知，第九阶段的突起或凹陷明显少于第八阶段，腔体形状的规则程度更高，说明排量越小，对腔体形态的控制越有利。由此得出排量对排卤质量浓度及腔体形态的影响规律：排量越小，则排卤质量浓度越高，腔体形态越规则，反之亦然。该例同时说明，腔体形态可以根据上一阶段的情况通过调整排量大小或其他溶腔参数来修正，即上一阶段的腔体形态在一定程度上影响下一阶段的腔体形态发育，因此，整个溶腔工作的初期腔体形态控制尤为重要。

为了更加深入地研究排量对腔体形态发育的影响，选择金坛 B 井和金坛 C 井的前两阶段进行数据整理及对比分析。这两口井都是刚开始溶腔不久的井，各阶段两管柱间距及不溶物含量均变化不大，其前两个阶段均符合排量越小、排卤质量浓度越高的规律（图 5），不同的是，金坛 B 井各阶段排量均高于金坛 C 井，因而使腔体的形态从一开始就受到了影响（图 6）。

显见，无论是金坛 B 井还是金坛 C 井，因第二阶段的排量均小于第一阶段，故第二阶段的腔体形态显得比第一阶段规则稳定；因金坛 C 井的排量小于金坛 B 井的排量，故无论是第一阶段还是第二阶段，金坛 C 井的腔体形态比金坛 B 井规则很多。由此，该例进一步说明排量越小，则排卤质量浓度越高，腔体形状越规则，并且初期溶腔形态越规则，对后面的溶腔形态控制越有利，反之亦然。对金坛多口井的数据进行整理及对比分析，发现该规律普遍存在，且相关性明显，可利用该规律作为调整溶腔参数的依据。

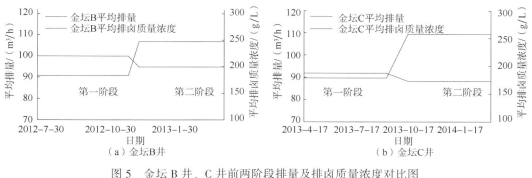

图 5　金坛 B 井、C 井前两阶段排量及排卤质量浓度对比图

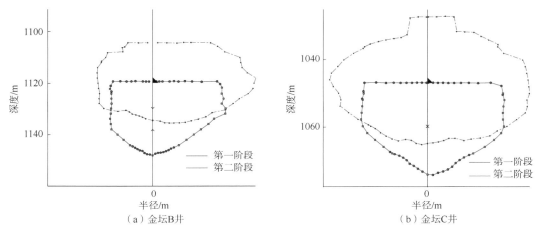

图 6　金坛 B 井、C 井前两阶段声呐测腔体积对比图

3　结论与认识

（1）盐穴储气库在溶腔过程中，注淡水排量与排卤浓度及腔体形态之间具有良好的相关性，排量越小，则排卤质量浓度越大，越容易符合盐企对排卤质量浓度的要求，腔体形状也越规则，形态越容易控制，并且初期溶腔形态越规则，对后面的溶腔形态控制越有利，反之亦然。因初期腔体形态的控制对后期腔体形态的发育影响较大，故对腔体初期形态的控制尤为重要，由此提出利用排量调节控制腔体的初期形态。

（2）实际溶腔时，在相同时间内，排量越小，腔体体积越小。为了得到相同的腔体体积，就需要更长的溶腔时间，因而影响储气库的经济效益。为此，并非排量越小越好，应该根据实际情况，制定合理的工作制度，有效匹配溶腔参数。

（3）对于盐穴储气库的溶腔工作，由于初期淡水与盐岩接触面积小，排卤质量浓度低，且后期腔体形态发育受初期腔体形状影响较大，因此，应该首先采取多开井、小排量的工作制度，这样既可以提高初期排卤质量浓度，又可以有效控制腔体的初期形状，到后期再采用大排量的造腔模式，以保证造腔时间及经济效益。

参 考 文 献

[1] 李建君,王立东,刘春,等.金坛盐穴储气库腔体畸变影响因素[J].油气储运,2014,33(3):269-273.
[2] 吴建发,钟兵,罗涛.国内外储气库技术研究现状与发展方向[J].油气储运,2007,26(4):1-3.
[3] 何爱国.盐穴储气库建库技术[J].天然气工业,2004,24(9):122-125.
[4] 赵志成,朱维耀,单文文,等.盐穴储气库水溶建腔机理研究[J].石油勘探与开发,2003,30(5):107-109.
[5] 田中兰,夏柏如.盐穴储气库造腔工艺技术研究[J].现代地质,2008,22(1):97-102.
[6] 班凡生,高树生,单文文,等.岩盐储气库水溶建腔排量和管柱提升优化研究[J].石油天然气学报,2005,27(2):411-413.
[7] Kunstman A S, Urbanczyk K M. A computer model for designing salt cavern leaching process developed at Chemkop[C]. Solution Mining Research Institute Meeting, Kracow, Poland, October 14T9, 1990.
[8] Kunstman A, Urbanczyk K. Application of WinUbro software to modelling of carven development in trona deposit[C]. Solution Mining Research Institute Technical Conference, Krakow, Poland, April 27-28, 2009.
[9] Maria K, Bartlomiej R. Analysing cavern geometry development based on sonar measurements from the point of view of geological deposit structure[C]. Solution Mining Research Institute Meeting, Kracow, Poland, May 11-14, 1997.

本论文原发表于《油气储运》2015年第34卷第2期。

盐穴储气库造腔过程动态监控数据分析方法

齐得山　巴金红　刘　春　汪会盟　王元刚　王晓刚　李建君

（中国石油西气东输管道公司储气库项目部）

【摘　要】 为保证储气库造腔正常进行，需持续对造腔过程中的各项生产动态数据进行监控分析，及时发现问题并采取相应措施。对造腔过程中监测到的压力排量、排卤质量浓度、形态体积等生产动态数据实际值与计算的理论值进行对比，通过分析两者的差异性与油垫深度、两管距、不溶物蓬松系数等造腔参数之间的关系，形成了一套根据生产动态数据判断造腔参数是否异常的方法。如果造腔过程中造腔内管脱落，会出现油垫压力实际值大于理论值，排卤质量浓度实际值小于理论值的现象；若油垫上升，则油垫压力实际值小于理论值，排卤质量浓度实际值大于理论值。结合声呐测腔数据，用生产实例对该方法进行了验证，对分析造腔过程是否偏离设计方案具有重要的指导意义。

【关键词】 盐穴储气库；水溶造腔；动态监控；声呐测腔

A dynamic monitoring and analysis method for the solution mining process of salt-cavern gas storage

Qi Deshan　Ba Jinhong　Liu Chun　Wang Huimeng
Wang Yuangang　Wang Xiaogang　Li Jianjun

(Gas Storage Project Department, PetroChina West-East Gas Pipeline Company)

Abstract To ensure the normal proceeding of gas storage solution mining, it is necessary to monitor and analyze continuously all production performance data in the process of solution mining so as to identify the problems and take the corresponding measures in time. In this paper, the actual production performance data measured in the process of solution mining (e. g., displacement pressure, brine concentration and morphologic volume) and their theoretical values were compared. After the relationship of the difference between actual values and theoretical values vs. oil blanket depth, tube spacing and insoluble fluff coefficient was investigated, a set of methods which can discriminate whether solution mining

---

基金项目：中国石油储气库重大专项"地下储气库关键技术研究与应用"子课题"盐穴储气库加快建产工程试验研究"，2015E-4008。

作者简介：齐得山，男，1987年生，工程师，2013年硕士毕业于中国石油大学（北京）地质工程专业，现主要从事盐穴储气库溶腔工艺优化设计及动态跟踪分析等研究工作。地址：江苏省镇江市润州区南徐大道商务A区D座中国石油西气东输管道公司储气库管理处，212000。电话：15862931327，E-mail：xqdsqideshan@petrochina.com.cn。

parameters are abnormal according to production performance data was established. If the inner pipe drops in the process of solution mining, the actual oil blanket pressure is higher than the theoretical value and the actual brine concentration is lower than the theoretical value. If the oil blanket moves upward, its actual pressure is lower than the theoretical value and the actual brine concentration is higher than the theoretical value. This method was verified based on sonar measurement data combined with production cases. It plays an instructive role in analyzing whether the solution mining process is in line with the design program or not.

Keywords salt-cavern gas storage; cavern-leaching; dynamic monitoring; cavern measuring with sonar

盐穴储气库造腔的实质是向盐层中注入淡水溶解盐岩，并采出近饱和卤水，从而在盐层中形成可储存天然气的腔体[1-6]。在造腔过程中，通过人为控制的方法可使整个腔体的形态体积最优化[7]。中国盐穴储气库的建设已有十多年的历史，研究者对盐穴储气库造腔方面的研究主要是关于溶腔工艺技术及腔体形态体积等造腔方案参数的优化分析[8-17]，而如何应用生产动态监测数据分析造腔过程是否偏离造腔方案设计方面的研究尚鲜见报道。

结合生产实例，论证了油垫深度、两管距等造腔参数的变化，会对造腔过程中监测到的压力排量、排卤质量浓度等生产动态数据产生影响。形成了一套根据生产动态数据判断造腔过程中设计参数是否变化的方法，可以实时监控造腔过程是否按照设计方案进行，对保障腔体形态体积的正常发展具有重要作用。

1 盐穴储气库单井造腔技术

目前国内外盐穴储气库建设普遍采用的是单井双管加油垫造腔[18-20]。钻完井后向盐层中下入造腔内管和造腔外管两层管柱，通过造腔管柱连续不断地向盐层中注入淡水对盐层进行溶蚀，直至形成最终的腔体。依据水的注入方式及水在腔体流动方向的不同，可将造腔方式分为正循环造腔和反循环造腔。正循环造腔是指从中心管注入淡水，水由腔体底部向上流动，卤水经中间管排出腔体，可以较好地控制腔体形态，但造腔速度较慢。反循环造腔是指从中间管注入淡水，水由腔体顶部向下流动，卤水经中心管排出腔体，造腔速度较快，但是腔体的形态不易控制。为了有效控制腔体形态，需要在生产套管中注入柴油（阻溶剂），以便在腔顶形成一个保护层，避免腔顶上溶过快。在造腔井井口和输水管道上，可以实时监控到注水压力、注水流量、排卤压力、排卤流量、油垫压力5种数据，通过化验分析可以测得排卤质量浓度。

造腔井中柴油与卤水的界面是油水界面，界面深度就是油垫深度；造腔内管与造腔外管的深度差是两管距，油垫界面与造腔内管底部出口深度之间的盐层段即造腔段（图1）。当正循环造腔时，油垫深度与外管深度一致；当反循环造腔时，外管深度大于油垫深度。正循环造腔时，如

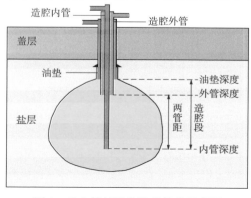

图1 盐穴储气库造腔井结构示意图

果油水界面保持到设计深度(外管深度),当向生产套管与造腔外管的环空注入柴油时,可在造腔外管的井口处观察到溢出的柴油,即注油见油。如果注油不见油,表明油水界面上升。

盐层中不溶于水的物质在卤水的浸泡下会发生膨胀,膨胀后的体积与原始体积的比值即为不溶物蓬松系数,一般通过室内实验得到。受盐层地质特征非均质性的影响,不同腔体及同一腔体不造腔段的不溶物蓬松系数是不同的,因此盐腔每次测完声呐后均要结合实际生产数据对不溶物蓬松系数进行校正,以使数值模拟所采用的不溶物蓬松系数最优化。

2 动态监控数据

盐穴储气库造腔过程中能够监测到的生产动态数据包括:井口压力流量、排卤质量浓度及腔体形态体积。这些数据受到油垫深度、两管距、不溶物蓬松系数等造腔方案参数的影响,通过分析生产动态数据可以推测出阶段造腔过程中关键的设计参数是否发生变化,从而判断实际的造腔过程是否偏离设计。

2.1 压力流量

在造腔井井口和输水管道上,可以实时监控到注水压力、注水流量、排卤压力、排卤流量及油垫压力。其中最关键的是油垫压力,注入压力排量、两管距及油垫深度的变化,均会对油垫压力产生影响。可以将造腔井井身结构看作一个 U 形管,造腔内管相当于 U 形管的一侧,造腔外管与生产套管的环空相当于 U 形管的另一侧,造腔内管下部的出口处就是 U 形管的底部连通点。

造腔内管井口压力与连通点处液柱压力关系为

$$p_{h_1} = \begin{cases} p_1 + \rho_1 g h_1 - p_f & \text{(正循环)} \\ p_1 + \rho_3 g h_1 + p_f & \text{(反循环)} \end{cases} \tag{1}$$

油垫压力与连通点处液柱压力关系为

$$p_{h_1} = p_2 + \rho_2 g h_2 + \rho_3 g h_3 \tag{2}$$

$$h_1 = h_2 + h_3 \tag{3}$$

根据 U 形管连通点处液柱压力相同的原理,可以推算出油垫压力与造腔内管井口压力之间的关系表达式为

$$p_2 = \begin{cases} p_1 - p_f + (\rho_1 - \rho_2) g h_2 - (\rho_3 - \rho_1) g h_3 & \text{(正循环)} \\ p_1 + p_f + (\rho_3 - \rho_2) g h_2 & \text{(反循环)} \end{cases} \tag{4}$$

式中:p_f 为管柱摩阻,造腔过程中管柱深度、注(排)水排量越大,该值就越大[21],Pa;p_{h_1} 为造腔内管下部出口处的压力,Pa;p_1 为造腔内管井口压力,Pa;p_2 为井口油垫压力,Pa;h_1 为造腔内管深度,m;h_2 为油水界面深度,m;h_3 为造腔段厚度,m;ρ_1 为注入淡水密度,取 1000 kg/m³;ρ_2 为柴油密度,金坛库区取 840 kg/m³;ρ_3 为排出卤水密度,kg/m³;g 为重力加速度,取 9.8 m/s²。

基于造腔方案设计的管柱深度和油垫深度，用监测到的井口流量、注入(排出)压力及排卤质量浓度的实际值可以计算理论上的井口油垫压力值 p_{2L}。造腔过程中若管柱深度无变化，则计算出的 p_f 与实际值相同或相近；若管柱脱落，管柱深度变小，则计算出的 p_f 小于实际值。将 p_{2L} 和同一时间监测的井口油垫压力实际值 p_{2S} 进行对比分析，如果两者的值差异明显或变化趋势不一致，可以排除压力流量和排卤质量浓度的影响，而只考虑管柱深度和油垫深度的变化。造腔过程中，造腔内管深度 h_1 不变，当油垫上升时，h_2 实际值就会小于理论值，h_3 实际值就会大于理论值；油垫深度 h_2 不变，腔内的造腔管柱发生脱落时，h_1 实际值就会小于理论值，h_3 实际值也会小于理论值。

由式(4)可知，在正循环造腔过程中：油垫上升而管柱深度不变，则 $p_{2L}>p_{2S}$；管柱脱落而油垫深度不变，则 $p_{2L}<p_{2S}$。在反循环造腔过程中：油垫上升而管柱深度不变，则 $p_{2L}>p_{2S}$；管柱脱落而油垫深度不变，则 p_{2L} 和 p_{2S} 的大小关系不定。

正常造腔过程中，由于造腔内管和造腔外管的直径、深度是已知且固定不变的，因此井口的注入排量和注入压力呈正相关系，即注入压力越大，注入排量也越大。但是，当排卤管柱被从卤水中结晶出而附着在管柱内壁上的盐严重堵塞时，注入排量和注入压力的变化趋势会不一致，即出现注入压力变化不大而注入排量却显著下降，或者注入压力显著增大而注入排量却变化不大的现象。

2.2 排卤质量浓度

采用已建立的溶腔数学模型，通过数值模拟可以计算出不同条件下的排卤质量浓度理论值。排卤质量浓度受到造腔段厚度、不溶物含量、循环方式及注入排量的影响：造腔段厚度越大，排卤质量浓度越大；不溶物含量越低，排卤质量浓度越大；反循环造腔，排卤质量浓度大；注入排量越小，排卤质量浓度越大[7,9,14]。

在阶段造腔过程中，不溶物含量和循环方式是已知且固定不变的，注入排量则可以通过井口流量计读取。而油垫上升或管柱脱落会造成造腔段的实际厚度和理论值(设计值)产生一定差异。如果实际造腔段厚度没有变化，那么基于实测数据计算得出的排卤质量浓度理论值和实际值就会相同或者相近。因此，当理论值和实际值的差异明显或变化趋势不一致时，可以首先排除不溶物含量、循环方式及注入排量 3 个已知确定值的影响，而只考虑造腔段厚度的变化。

当油垫上升而管柱深度不变时，实际的造腔段厚度就会大于设计值，那么排卤质量浓度的实际值就会大于计算的理论值。当管柱脱落而油垫深度不变时，实际的造腔段厚度就会小于设计值，排卤质量浓度的实际值就会小于计算的理论值。

2.3 形态体积

一口井从开始造腔到形成最终的腔体，要经历数个造腔阶段。每个阶段结束后要对腔体进行声呐检测，通过将测得的形态体积和模拟计算出的形态体积进行匹配，对造腔模拟所需的不溶物蓬松系数进行修正，以使每个腔体、每个造腔阶段模拟时所采用的不溶物蓬松系数最优化。

根据井口监测到的排卤流量和排卤质量浓度，可以计算出实际采出的盐岩体积。基于采盐体积、造腔段不溶物含量及声呐测得的腔体体积就可以计算出某造腔段的不溶物蓬松

系数，计算公式为

$$\beta = \frac{(V_c + V_1 - V_2)(1-\alpha)}{\alpha V_c} + 1 \qquad (5)$$

式中：β 为不溶物蓬松系数，一般取 1.2~1.9；α 为造腔段不溶物体积分数；V_c 为本阶段采盐体积，m^3；V_1、V_2 分别为上阶段、本阶段末声呐测腔体积，m^3。

由于每阶段的造腔段盐层厚度远大于阶段结束后油垫上升的高度，相邻两个造腔段盐层大部分重叠，其盐层内的不溶物含量和不溶物蓬松系数也是相近的。因此，可以将上阶段计算出的不溶物蓬松系数用于下阶段的造腔模拟，以使模拟出的结果更接近实际，为确定下阶段的造腔方案提供更精确的理论依据。

3 生产实例

3.1 管柱脱落

中国某盐穴储气库 A 腔体第 6 造腔阶段，正循环造腔，设计内管深度为 1146m，设计油垫和外管深度均为 1068m。第 6 造腔阶段内，每次从环空注油都见油，说明油水界面一直控制在设计深度。阶段造腔初期，模拟计算出的油垫压力和排卤浓度与监测到的实际值匹配良好，说明造腔过程正常（图 2）。随后造腔过程中，出现油垫压力实际值大于理论值、排卤质量浓度实际值小于理论值的现象，说明造腔段厚度变小，在确定油水界面深度不变的情况下，推断是管柱脱落造成的。

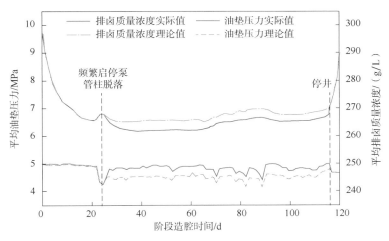

图 2 A 腔体第 6 造腔阶段生产数据分布图

A 腔体第 6 阶段结束，井场测声呐时发现造腔内管少了 3 节（30m）。而声呐测腔结果显示，第 6 造腔阶段内 1116m 以下的腔体直径基本无变化，说明 1116m 以下的盐层基本没有被淋滤，进而确定该阶段造腔内管深度在阶段初期就变成了 1116m，恰好比设计深度上移了 30m（图 3）。因此可推断出在第 6 造腔阶段的初期，A 腔体造腔内管底端的 3 节管柱发生脱落，从而导致模拟计算出的排卤质量浓度和油垫压力值产生异常。在生产监测数据发生异常的前几天，A腔体曾3次调整了注入压力排量并进行了启停泵，分析认为是造腔内管内频繁的压力变化导致底端3节管柱脱落。

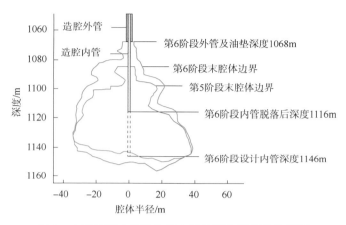

图3　A腔体第6造腔阶段末腔体形态及管柱位置图

3.2　油垫上升

中国某盐穴储气库B腔体第1造腔阶段，正循环造腔，设计内管深度1107m，设计油垫和外管深度均为1074m。阶段结束测声呐时，从B腔体中起出的造腔内外管长度都和阶段开始时下入的长度一致，说明第1阶段造腔过程中管柱并未发生脱落，深度保持在设计深度。阶段造腔的前期从环空注油见油，后期注油不再见油，并出现油垫压力实际值小于理论值、排卤质量浓度实际值大于理论值的现象（图4），说明油水界面上升致使造腔段厚度变大，最终致使模拟计算出的排卤浓度和油垫压力值产生异常。

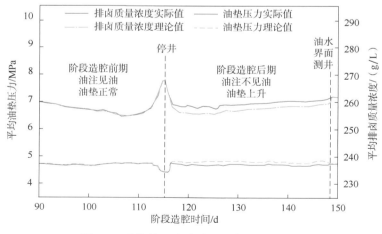

图4　B腔体第1造腔阶段生产数据分布图

油垫压力产生异常后，对该井进行了测井作业，测得油水界面值为1062m，比设计值上升了12m。阶段结束后声呐测得该井腔顶深度也为1062m（图5），再一次验证了上述推断的正确性。

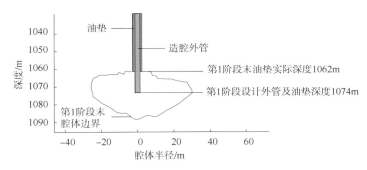

图 5　B 腔体第 1 造腔阶段末腔体形态及油垫位置图

4　结论

（1）在正循环造腔过程中，若油垫上升而管柱深度不变，则油垫压力理论值大于实际值；若管柱脱落而油垫深度不变，则油垫压力理论值小于实际值。

（2）在反循环造腔过程中，油垫上升而管柱深度不变，则油垫压力理论值大于实际值；管柱脱落而油垫深度不变，则油垫压力理论值和实际值的大小关系不定。

（3）油垫上升而管柱深度不变，排卤质量浓度的实际值大于计算出的理论值；管柱脱落而油垫深度不变，排卤质量浓度的实际值小于计算出的理论值。

参 考 文 献

[1] Seto M, Nag D K, Vutukur V S. In-situ rock stress measurement from rock cores using the emission method and deformation rate analysis[J]. Geotechnical and Geological Engineering, 1999, 17: 241-266.
[2] Bérest P, Brouard B, Durup J G. Tightness tests in saltcavern wells[J]. Oil & Gas Science and Technology, 2001, 56(5): 451-469.
[3] Liang W G, Yang C H, Zhao Y S, et al. Experimental investigation of mechanical properties of bedded salt rock[j]. International Journal of Rock Mechanics and Mining Sciences, 2007, 44(3): 400-411.
[4] 杨春和，梁卫国，魏东吼，等．中国盐岩能源地下储气储存可行性研究[J]．岩石力学与工程学报，2005，24(24)：4409-4417.
[5] 周学深．有效的天然气调峰储气技术——地下储气库[J]．天然气工业，2013，33(10)：95-99.
[6] 杨海军．中国盐穴储气库建设关键技术及挑战[J]．油气储运，2017，36(7)：747-753.
[7] 班凡生，高树生．岩盐储气库水溶建腔优化设计研究[J]．天然气工业，2007，27(2)：114-116.
[8] 丁国生，张昱文．盐穴地下储气库[M]．北京：石油工业出版社，2010：179-185.
[9] 肖学兰．地下储气库建设技术研究现状及建议[J]．天然气工业，2012，32(2)：79-82.
[10] 田中兰，夏柏如．盐穴储气库造腔工艺技术研究[J]．现代地质，2008，22(1)：97-102.
[11] 赵志成，朱维耀，单文文，等．盐穴储气库水溶建腔机理研究[J]．石油勘探与开发，2003，30(5)：107-109.
[12] 班凡生，耿晶，高树生，等．岩盐储气库水溶建腔的基本原理及影响因素研究[J]．天然气地球科学，2006，17(2)：231-266.
[13] 杨海军，于胜男．金坛地下储气库盐腔偏溶与井斜的关系[J]．油气储运，2015，34(2)：145-149.
[14] 李建君，王立东，刘春，等．金坛盐穴储气库腔体畸变影响因素[J]．油气储运，2014，33(3)：269-273.

[15] 王文权,杨海军,刘继芹,等.盐穴储气库溶腔排量对排卤浓度及腔体形态的影响[J].油气储运,2015,34(2):175-179.

[16] 刘继芹,焦雨佳,李建君,等.盐穴储气库回溶造腔技术研究[J].西南石油大学学报(自然科学版),2016,38(5):122-128.

[17] 郑雅丽,赵艳杰,丁国生,等.厚夹层盐穴储气库扩大储气空间造腔技术[J].石油勘探与开发,2017,44(1):1-7.

[18] 耿凌俊,李淑平,吴斌,等.盐穴储气库注水站整体造腔参数优化[J].油气储运,2016,35(7):779-783.

[19] 郭凯,李建君,郑贤斌.盐穴储气库造腔过程夹层处理工艺——以西气东输金坛储气库为例[J].油气储运,2015,34(2):162-166.

[20] 李建君,陈加松,吴斌,等.盐穴地下储气库盐岩力学参数的校准方法[J].天然气工业,2015,35(7):96-101.

[21] 李海伟,杨清玉.盐穴储气库溶腔过程中腔体净容积及油水界面计算实例[J].石油化工应用,2015,34(6):56-58.

本论文原发表于《油气储运》2017年第36卷第9期。

盐穴储气库造腔过程夹层处理工艺
——以西气东输金坛储气库为例

郭 凯[1] 李建君[2] 郑贤斌[1]

(1. 中国石油天然气股份有限公司天然气与管道分公司; 2. 中国石油西气东输管道公司)

【摘 要】 国内盐穴储气库造腔层段岩性复杂,不同深度存在不同厚度的夹层,夹层厚度从1m至10m以上,造腔结果相差较大。以金坛储气库8口造腔井为例,通过造腔参数分析,研究不同夹层处理工艺下的造腔效果,总结出多夹层盐岩段造腔工艺技术方案: 使油水界面在夹层上部且距离夹层至少3~4m; 当油水界面在夹层下部时,控制油垫厚度小于0.1m且腔顶直径不要太大; 当前阶段油水界面位于夹层下部且腔体需要扩容时,将油水界面调整至夹层上部; 当处理腔体下部夹层时,使内管位于夹层下部,至少在夹层位置。上述技术方案可为其他盐穴储气库建设提供参考和技术支持。

【关键词】 盐穴储气库; 夹层溶蚀; 油垫厚度; 油水界面

Interlayer treatment process in cavity building for salt cavern gas storage-A case study of Jintan Gas Storage of West-to-East Gas Pipeline

Guo Kai[1]   Li Jianjun[2]   Zheng Xianbin[1]

(1. PetroChina Company Limited Natural Gas & Pipeline Company;
2. PetroChina West-East Gas Pipeline Company)

Abstract In China, the cavity building interval for salt cavern gas storage often features complicated lithology, and interlayers with varying thickness (from 1m to 10m or more) exist at different depths, leading to distinct results of cavity building result. Taking the 8 cavity building wells of Jintan Gas Storage as an example, through analyzing the cavity building parameters, and studying the cavity building effect under different interlayer treatment process, the cavity building solution for multi-interbedded salt rock is proposed: (1) the OWC is kept at least 3-4m above the interlayer; (2) when the

---

基金项目: 中国石油天然气股份有限公司科技攻关项目"金坛盐穴地下储气库采卤造腔工程水处理及现场应用研究", K2011-31。

作者简介: 郭凯, 工程师, 1981年生, 2009年博士毕业于中国石油勘探开发研究院矿产普查与勘探专业, 现主要从事储气库运行管理与天然气运销工作。

OWC is under the interlayer, the oil cushion thickness is controlled less than 0.1m and cavern roof diameter should not be too big; (3) when the OWC is under the interlayer and the cavern needs to be expanded, the OWC is adjusted above the interlayer; and (4) when the interlayer beneath the cavern is treated, the inner pipe should be located under the interlayer or at least at the interlayer position. The solution can provide a reference and technical support for other salt cavern gas storage building.

Keywords salt cavern gas storage; interlayer dissolution; oil cushion thickness; OWC

国内盐穴储气库建设刚刚起步，建库资源缺乏，同时盐岩层段岩性变化较大，条件较差[1]，存在不同种类的夹层[2]，因此，造腔过程中夹层处理至关重要[3]，如果夹层未被溶蚀，会在夹层上部形成二次造腔，导致盐层段体积损失。金坛储气库作为国内首座盐穴储气库，已经形成一套完整的夹层处理工艺技术[4-6]，可为国内其他储气库建设中的夹层处理提供理论依据和技术支持。

金坛储气库造腔井段存在若干夹层，为使腔体体积与形态达到理想状态，采用合理的夹层处理工艺技术至关重要。不同夹层处理工艺下的造腔效果相差较大，最终形成的腔体形状也不同。金坛储气库夹层处理主要考虑油垫厚度，夹层与油水界面、造腔管柱的相对位置等因素。控制油水界面主要采用地面观察法，造腔过程中向环空注油，直至地面返油为止，该方法会注入过量柴油。如果油水界面在夹层下部较近位置，采用地面观察法注油，夹层下部会储存大量柴油，阻碍淡水上浮；若注油量较小，油水界面上移，夹层过早与泥岩接触，则影响造腔效果。影响夹层处理效果的另一个关键因素是造腔内管下入位置，注入水可影响内管下部一定距离的盐层，夹层距离内管下部较远时，因夹层中不溶物含量较高，阻碍淡水对管柱下部盐岩的溶蚀作用，使夹层不易溶蚀。

1 夹层处理不理想实例

金坛储气库造腔井较多，以 A 井、B 井为例说明油水界面与夹层相对位置对造腔效果的影响。

A 井造腔段 1048~1050m 为夹层，第四阶段将油水界面设定在 1052m，距离不溶物 2m。声呐测腔结果显示，腔体主体部分直径较小，仍有较大扩容空间，因此，造腔第五阶段保持油水界面位置不变，可对腔体进行扩容。第五次声呐测腔结果显示，腔体主体部分得到扩容。当造腔第五阶段结束时，腔内累计注油量为 109m³，退出油量为 40m³，因油水界面距离夹层较近，故约 70m³ 柴油存在于夹层下部，阻碍淡水上浮，经计算，油垫厚度为 0.11m；泥岩长时间(14 个月)大面积(660m²)与柴油接触，在泥岩夹层下部形成一层保护膜，进一步阻止夹层段的溶蚀。由于 A 井最终腔顶设定在 1028m，当前油水界面距离腔顶 24m，因此第六阶段体积较小，夹层附近腔体直径基本不变，残余油垫厚度未变薄导致夹层未垮塌[图 1(c)]。后续造腔阶段外管虽然上移，但大量柴油残留在腔内，泥岩夹层始终未溶蚀。

B 井造腔段 1137~1140m 为 3m 厚泥岩，第二次声呐测腔结果显示，腔体直径较小，因此，第三阶段对腔体进行扩容。第三次声呐测腔结果显示，腔顶面积为 306m²，腔内残余柴油量为 90m³，残余油垫厚度为 0.29m；后续造腔过程中该部分柴油未全部退出，第四造腔阶段结束时，夹层下部腔体面积为 1000m²，残余油厚度最大约 0.09m。因夹层下部存在大量残余油，阻碍淡水上浮，后续造腔过程中夹层未完全溶蚀，腔体形状不理想[图2(b)]。

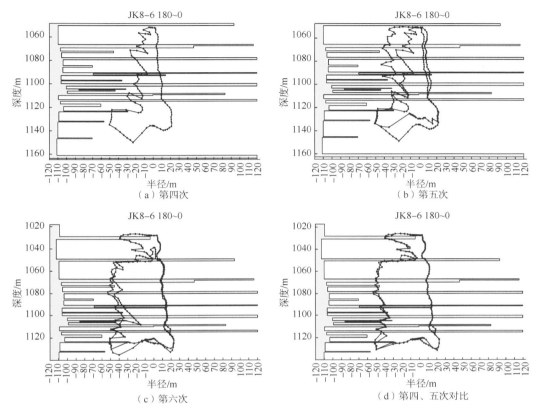

图1 A井声呐测腔结果

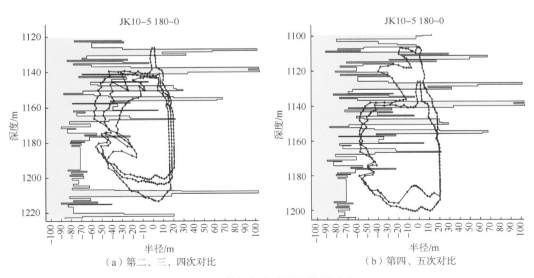

图2 B井各次声呐测腔结果对比

腔体下部夹层需要溶蚀时,内管与夹层相对位置对造腔效果有较大影响,以 C 井为例进行说明。

显见,在 1173~1175m 之间不溶物含量较高,第三阶段造腔管柱调整时,将造腔内管调整至 1171m,在夹层上部约 3m。由于造腔内管与夹层距离较远,注入淡水没有影响到夹层附近盐层,因此夹层未溶蚀。夹层附近腔体直径较小,腔体脖颈很快被上部不溶物堵塞,造成下部腔体体积损失,并在夹层上部形成二次溶腔(图 3)。

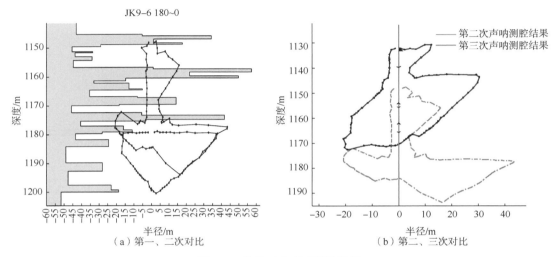

图 3　C 井前三次声呐测腔结果

通过对比 A 井、B 井、C 井夹层造腔过程可知,夹层不溶蚀情况有:(1)油水界面在夹层以下,且夹层下部残余油量过大;(2)泥岩夹层大面积与柴油接触,接触面位置腔体直径无法再扩大,油垫厚度无法变薄;(3)腔体中下部存在夹层需要溶蚀时,内管调整至夹层上部进行溶腔。

2　夹层处理成功实例

金坛储气库大部分井腔体形状较规则,以 D 井、E 井、F 井、G 井、H 井五口井夹层处理成功的井为例,分析夹层存在时的造腔工艺技术。

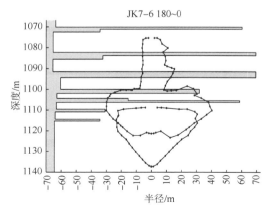

图 4　D 井第一、二次声呐测腔结果

D 井第一阶段油水界面 1108m,1105~1106m 为厚 1m 的夹层,因夹层厚度较小,且油水界面距离夹层 3m,腔内残余柴油量为 $50m^3$,油垫厚度为 0.02m。第二次声呐测腔结果显示,夹层已经充分溶蚀(图 4)。

E 井 1077~1079m 为厚 2m、平均不溶物含量为 50% 的含膏盐层,第八次声呐测腔结果显示,夹层下部腔体直径较大,因此,后续造腔时,内管置于夹层位置。第九次声呐测腔结果显示,夹层已经开始溶蚀(图5)。

F井第三阶段油水界面设定值为1080m,夹层位于1075m,夹层厚度为1.5m,平均不溶物含量为99.8%,且1080~1100m有三个夹层,平均不溶物含量较高。当第三阶段结束时,油水界面升至1078m,相对于设定值升高2m,但是距离夹层仍有3m。第三阶段腔内注油量为22m³,退油量为15m³,只有少量柴油留在腔内。第四次声呐测腔结果显示,1078m腔体直径由10m增至28m,1075m夹层溶蚀较好;在对腔体下部夹层处理的过程中,内管放置在夹层以下,故各夹层位置腔体直径变大,已经充分溶蚀(图6)。

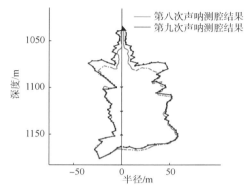

图5 E井第八、九次声呐测腔结果

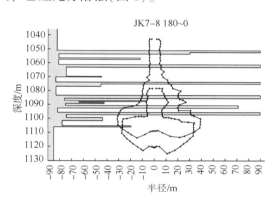

图6 F井第三、四次声呐测腔结果

G井1122~1125m存在厚度约4m的夹层,第一次声呐测腔结束之后,第二阶段将油水界面设定在1129m,第二次声呐测腔结果腔顶位置为1127m,油水界面上移2m,但仍与夹层有一定距离,第三阶段将油水界面上提至夹层以上,声呐测腔结果显示,夹层已经充分溶蚀(图7)。

H井1110m处夹层厚度约3m,第四造腔阶段结束后,因腔体下部直径较小,对腔体下部进行扩容。第四次声呐测腔结果显示,油水界面靠近腔顶位置,如果油水界面不调整,则夹层很可能浸泡在柴油中,导致后期阻碍夹层溶蚀。因此,第五阶段扩容过程中,将油水界面上移至夹层以上,由声呐测腔结果可知,夹层大部分被溶蚀,腔体形状比较规则(图8)。

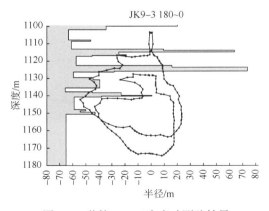

图7 G井第二、三次声呐测腔结果

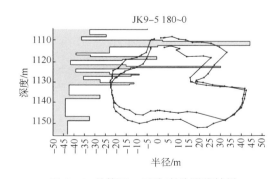

图8 H井第四、五次声呐测腔结果

3 结论

根据不同夹层的处理效果，在造腔过程中成功溶蚀夹层的方案有：

（1）造腔过程中油水界面可能会上移，油水界面距离夹层至少 3~4m，尽量在夹层上部。

（2）油水界面在夹层下部时控制注油量，保证油垫厚度小于 0.1m。

（3）油水界面在夹层下部时，控制腔顶直径不要太大，保证夹层附近在后续造腔阶段有扩容空间，使残余油垫厚度变薄。

（4）当前阶段油水界面位于夹层下部，且腔体需要扩容时，将油水界面调整至夹层上部。

（5）处理腔体下部夹层时，内管置于夹层下部一定距离，保证其至少在夹层位置。

金坛储气库作为中国第一座盐穴储气库，在储气库建设方面积累了宝贵经验，夹层处理工艺技术日臻成熟。国内其他储气库建库层段大部分存在夹层需要处理，金坛储气库夹层的处理案例在优化建库层段盐岩利用率方面具有借鉴意义，在加快储气库建设进程方面具有推进作用。

参 考 文 献

[1] Evan D J, Wiliams J D O, Hough E. The stratigraphy and lateral correlations of the Northwich and Preesall halites from the Cheshire Basin-East Irish Sea areas: implications for sedimentary environments, rates of deposition and the solution mining of gas storage caverns[C]. York: Solution Mining Research Institute Fall 2011 Technical Conference, 2011.

[2] Kenneth E, Mill S. Specialized techniques in solution cavern shape control[C]. Chicago: Solution Mining Research Institute Spring Meeting, 1981.

[3] Maria K, Bartlomiej R. Analysing cavern geometry development based on sonar measurements from the point of view of geological deposit structure[C]. Cracow: Solution Mining Research Institute Meeting, 1997.

[4] Kingdon A, Evans D J, Hough E, et al. Use of resistivity borehole imaging to assess the structure, sedimentology and insoluble content of the massively bedded Preesall Halite NW England[C]. York: Solution Mining Research Institute Fall 2011 Technical Conference, 2011.

[5] 李建君，王立东，刘春，等. 金坛盐穴储气库腔体畸变影响因素[J]. 油气储运，2014，33(3): 269-273.

[6] 肖学兰. 地下储气库建设技术研究现状及建议[J]. 天然气工业，2012，32(2): 79-82.

本论文原发表于《油气储运》2015 年第 34 卷第 2 期。

盐穴储气库天然气阻溶恒压运行技术

何 俊　井 岗　陈加松　付亚平　李建君

（国家管网集团西气东输分公司）

【摘　要】 盐穴储气库作为天然气管网的配套设施，其高效运行对于保证管网平稳供气作用巨大。恒压运行是盐穴储气库采用流体置换方式进行天然气注采的新模式，注采过程中腔内压力相对稳定，不需要垫底气，较为安全经济。但这种注采模式需要大量的饱和卤水置换天然气，注气阶段采出的卤水需要极大的存储空间或卤水处理设备，大大降低了该模式的经济性和可操作性。为此提出一种将天然气阻溶造腔与恒压运行有效结合的新工艺，其既能实现恒压运行，又无需对置换介质进行处置，具有良好的应用前景。2018 年，该技术在中国金坛盐穴储气库进行了先导试验，并取得成功。

【关键词】 盐穴储气库；恒压运行；天然气阻溶；方案研究；数据模拟

Constant-pressure operation technology for salt cavern gas storage through solution mining under gas

He Jun　Jing Gang　Chen Jiasong　Fu Yaping　Li Jianjun

（West-East Gas Pipeline Company, National Petroleum and Natural Gas Pipe Network Group Co. Ltd.）

Abstract The efficient operation of salt cavern gas storage, as an ancillary facility of the gas pipeline network, is of great significance to ensure the stable gas supply of the pipeline network. Constant-pressure operation is a new gas injection and production mode of salt cavern gas storage with fluid replacement, which is economic for the pressure in the cavern is stable, and no base gas is required during the process of gas injection and production. However, in this mode, a large amount of brine is required for the replacement of natural gas, and a very large storage space or treatment plant is required for the brine produced during gas injection. Thus, the economy and operability of the mode will be reduced greatly. Herein, a new technology effectively combining the solution mining under gas (SMUG) and constant-pressure operation is introduced, with which not only the constant-pressure operation can be realized, but also no treatment is required for the replacement medium. Therefore, the application of the new

---

基金项目：中国石油储气库重大专项资助项目"盐穴储气库加快建产工程试验研究"，2015E-4008。

作者简介：何俊，男，1987 年生，工程师，2012 年硕士毕业于中国石油大学（北京）石油与天然气工程专业，现主要从事地下储气库的造腔与综合评价工作。地址：上海市浦东新区世纪大道 1200 号，200122。电话：18651287636。Email：xqdshejun@petrochina.com.cn。

technology is very promising. The pilot test for the technology was carried out in Jintan Salt Cavern Gas Storage in 2018, having a great success.

Keywords salt cavern gas storage; constant-pressure operation; solution mining under gas (SMUG); plan research; data simulation

天然气阻溶造腔是盐穴储气库阻溶造腔的一种特殊造腔方式,即在地下盐腔达到一定体积或完腔后,使用天然气作为阻溶剂继续对腔体进行改造[1-8]。19世纪90年代,美国、德国已经开展了天然气阻溶工艺技术研究[9-10]。国家管网集团西气东输分公司在金坛储气库建设中,采用这种造腔方式对部分腔体特定层段的几何形态进行修复,取得了良好的效果[11]。若对该造腔工艺进行进一步优化,可利用已取得腔体的有效体积进行注采运行,并将储存的天然气作为阻溶剂继续造腔至完腔,这样不仅能够实现天然气恒压注采、减少垫底气投资,同时可以使盐穴提前投产注气,显著增加腔体的有效工作气量,从而提升建库效益。这种将天然气阻溶造腔和恒压运行相结合的新技术,称为盐穴储气库天然气阻溶恒压运行技术。

1 天然气阻溶恒压运行

1.1 造腔工艺

天然气阻溶恒压运行主要分为建槽、初始腔体构建、阻溶恒压运行3个阶段。

第1阶段:建槽阶段,即在主腔体的下部建造足够大的底坑。这一阶段的目的是确保底坑具有足够的体积和合适的形态,用于存放后期造腔产生的不溶物。合理的底坑形态对于后续的造腔影响极大,能否准确模拟腔底形态直接决定后续腔体的发展,然而目前大多数造腔设计软件未考虑不溶物沉降机理,底坑模拟形态往往与实际相差很大,极易造成底坑有效体积损失。由某井造腔过程中连续两个阶段的声呐形状(图1)可见,由于设计软件对不溶物沉降模拟不准确,依据数值模拟确定造腔管柱深度,导致该阶段沉降的不溶物隔断了右侧深坑,深坑内的空腔体积无法用于存储不溶物,进而造成右侧腔底迅速抬升,导致底坑体积及盐岩资源大量损失。

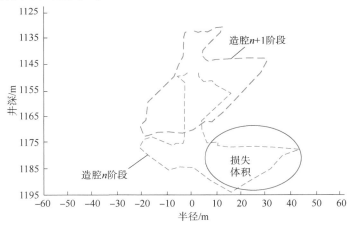

图1 某井造腔过程中底坑体积损失声呐形态图

第 2 阶段：构建初始腔体，作为天然气注采恒压运行的主体存储空间，这部分存储空间将在第 3 阶段不断增大直至完腔。在底坑建造完毕后，需进行声呐测腔，对底坑形状、体积等参数进行分析，并根据分析结果建造初始腔体。该阶段主要通过常规的正反循环造腔，打开腔体顶部及中部。理想的初始腔体应该满足两个条件：一是腔体顶部体积足够大，能够形成稳定的工作气量，满足经济性需求；二是腔体形态合理，能够保证盐层在后期造腔中得到充分利用（图2）。该阶段应注意：由于恒压运行对腔体稳定性有积极作用，腔体顶部可以适当放宽直径要求，以最大限度满足天然气存储需求；腔体中部应预留充足盐层在下一阶段造腔，因此初始腔体主体部分直径应小于腔体设计直径。

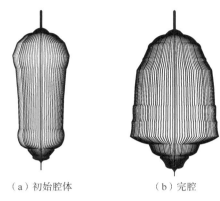

（a）初始腔体　　（b）完腔

图 2　天然气阻溶初始腔体及完腔形态图

第 3 阶段：阻溶恒压运行，初始腔体建造完毕后，依靠中心管进行淡水注入和卤水排出，环空注采天然气[10]，实现上部腔体注采运行，下部继续造腔。通过控制注水量和注入压力，实现恒压运行。该阶段的造腔不同于常规的自下而上造腔，而是自上而下造腔，随着下部腔体的扩大，腔体储气量不断增大，气水界面不断下压，直至达到最终的目标体积和腔体形态。天然气阻溶主要有双层管柱造腔和单层管柱造腔两种模式[12]。根据中国盐穴储气库井眼尺寸[13]，建议采用单层造腔管柱，通过注淡水排气和注气排卤流程切换，使气水界面在一定深度范围内浮动变化，达到对目标层段进行溶蚀的目的。该阶段主要包括注气排卤、注淡水排气、停井静溶 3 个过程(图3)：(1)注气排卤。通过生产管柱和排卤管柱环空注入天然气，将卤水从排卤管柱中排出，将气卤界面控制至设计深度的下部。(2)注淡水排气。通过排卤管柱注淡水，从生产管柱和排卤管柱环空采出天然气，将气水界面控制至设计深度的上部，在采气的过程中实施造腔。(3)停井静溶。该过程处于前两个过程中间，目的是使淡水与腔体充分接触，溶蚀目标层段。该过程并非必须，可根据腔内卤水浓度是否达到卤水消化标准来确定静溶时间。

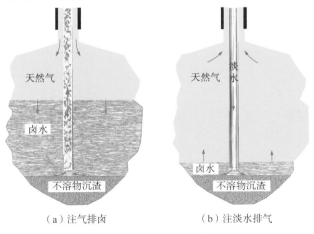

（a）注气排卤　　（b）注淡水排气

图 3　天然气阻溶恒压运行流程图

天然气阻溶恒压运行首先应根据第2阶段的腔体形态进行数值模拟，确定目标溶蚀层段及设计腔体形态，随后设定气水界面，进行注气排卤及注淡水排气，并循环此过程，依据生产数据的数值模拟或带压声呐测腔，判定是否达到腔体形状要求。直至腔体形状达到要求后，设定新的目标溶蚀层段开展下一轮次的天然气阻溶恒压运行。

1.2 技术优势

天然气阻溶恒压运行技术主要具有以下优势：(1)天然气可以全部采出，不需要垫底气投资；(2)初始腔体体积较小，投产时间相对较快，完成后即可投入注采运行，大大提前注气时间；(3)在常规注采模式下，每轮注采过程中，腔体都会受到交变应力作用[14]，对腔体稳定性造成损伤，而在恒压运行模式下，腔体受交变应力作用较小，有利于腔体延长寿命；(4)当恒压运行压力较高时，腔内压力可以抵抗腔体蠕变收缩，减少因腔体收缩造成的体积损失；(5)恒压运行下，腔体稳定性较强，可以适当扩大腔顶直径，增加上部盐层利用率。

1.3 技术不足及解决措施

天然气阻溶恒压运行技术存在以下缺点，建议采取相关解决措施（表1）。

表1 天然气阻溶恒压运行技术缺点及相应解决措施

缺点	预防措施
造腔井管理升级为气井管理，涉及可燃气体，造腔风险及成本增加	①排卤管安装紧急切断阀、井下安全阀(含环空安全阀)；②确保造腔过程中的工具、造腔井口和地面配套设施都达到气密性要求
造腔时，由于气水界面控制难度大，界面失控会影响腔体形态，严重时会溶穿腔顶，破坏套管鞋处密封性	①开发精准的SMUG模拟软件，精确计算界面位置变化；②采用光纤界面仪连续监测气水界面，中子测井辅助校核的方式控制油水界面
注采时，造腔管柱大段悬空，不溶物下落坍塌会增加造腔管柱损坏的风险[15]	①选用腔顶形状规则的腔体进行SMUG下恒压运行；②采用SMUG造腔的模拟软件，根据模拟情况合理设置管柱下深；③根据测井和监测结果，在采气期的时间窗口全面检测管柱泄漏情况，更换有问题的管柱；④利用造腔过程中的管柱检测和全库区的微地震监测，监测管柱损坏和夹层坍塌情况
运行时，可能由于气液界面控制不准发生超排，造成天然气进入排卤管柱，诱发井喷	①精确计量注气量、排卤量及管柱下深；②采用光纤界面仪连续监测气水界面，中子测井辅助校核的方式控制气水界面；③确保地面地下安全阀处于良好工作状态

2 数值建模

将天然气阻溶恒压运行简化（图4），建立模型。

根据压力平衡，可得

$$p_{wh}+\rho_w gh-\Delta p_{loss}=p_{sh}+\rho_g gh_g+\rho_b gh_b \tag{1}$$

式中：p_{wh}为井口注水压力，MPa；ρ_w为注入水的质量浓度，kg/m³；g为重力加速度，m/s²；h为排卤管柱深度，m；Δp_{loss}为压力损失，MPa；p_{sh}为套管鞋压力，MPa；ρ_g为天然气密度，kg/m³；h_g为套管鞋位置与气水界面的距离，m；ρ_b为排卤质量浓度，kg/m³；h_b为排卤管底部与气水界面的距离，m。

其中，

$$\Delta p_{\text{loss}} = \rho_w \frac{\lambda H v^2}{2D} \tag{2}$$

式中：λ 为沿程阻力系数；H 为排卤管的长度，m；v 为注水速度，m/s；D 为套管直径，m。

由于 $\rho_g g h_g$ 数值较小，可以忽略。对于式（1）两端求导，可得

$$\frac{\partial p_{\text{wh}}}{\partial t} + \frac{\partial p_w g h}{\partial t} - \frac{\partial \left(\rho_w \frac{\lambda H v^2}{2D} \right)}{\partial t} = \frac{\partial p_{\text{sh}}}{\partial t} + \frac{\partial \rho_b g h_b}{\partial t} \tag{3}$$

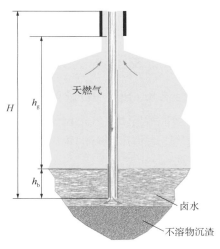

图4 天然气阻溶恒压运行简化示意图

流速 $v = \frac{4Q_w}{\pi D^2}$，同时，由于恒压运行，则有 $\frac{\partial p_{\text{sh}}}{\partial t} = 0$，综上可得

$$\frac{\partial p_{\text{wh}}}{\partial t} - \frac{4\lambda \rho_w H}{\pi D^3} \frac{\partial Q_w}{\partial t} = \rho_b g \frac{\partial h_b}{\partial t} + g h_b \frac{\partial \rho_b}{\partial t} \tag{4}$$

式中：Q_w 为注水流量，m³/h。

将腔体形状简化为圆柱，则有

$$\frac{\partial h_b}{\partial t} = \frac{Q_w - \frac{\partial V_s}{\partial t} + \frac{\partial (V_s a \eta)}{\partial t}}{A_b} = \frac{Q_w}{A_b} - \frac{1 - a\eta}{A_b} \frac{\partial V_s}{\partial t} \tag{5}$$

式中：V_s 为溶盐体积，m³；a 为不溶物含量（体积分数）；A_b 为腔体平均截面积，m²；η 为蓬松系数。

在单位时间内，卤水质量增量为注入淡水质量与溶盐质量之和，则有

$$V_b = \frac{\partial \rho_b}{\partial t} + \rho_b \frac{\partial V_b}{\partial t} = \rho_w Q_w + \frac{\partial V_s}{\partial t} \rho_s \tag{6}$$

式中：V_b 为排卤体积，m³；ρ_s 为盐岩密度，kg/m³。

腔内天然气质量变化率等于单位时间采气质量，则有

$$V_g \frac{\partial \rho_g}{\partial t} + \rho_g \frac{\partial V_g}{\partial t} = Q_g \rho_g \tag{7}$$

腔内天然气体积变化为注水量与不溶物膨胀体积之和再减去溶盐体积，则有

$$V_g \frac{\partial \rho_g}{\partial t} + \rho_g \left[\frac{\partial (V_s - V_s a \eta)}{\partial t} - Q_w \right] = Q_g \rho_g \tag{8}$$

式中：V_g 为腔内天然气体积，m³；Q_g 为采气流量，m³。

根据式(4)至式(6)、式(8)即可求得满足恒压运行条件下的注水量与注水压力。

3 软件研发及案例分析

3.1 CavSimu造腔设计软件

国家管网集团西气东输分公司在金坛盐穴储气库建设实践中，研发了一款适用于储气库天然气阻溶恒压运行的软件CavSimu，相比常规的盐穴储气库造腔软件，该软件主要有以下特色功能。

（1）实现气体阻溶造腔（SMUG）、单井单腔（SPSC）、双井单腔（DPSC）等多种造腔方式[16]的精准模拟。不同的造腔模式在同一腔体模拟过程中可以自由切换，能够很好地满足天然气阻溶恒压运行过程中常规造腔构建初始腔体、SMUG造腔完成后续阻溶恒压运行的需求。

（2）增加不溶物沉降模拟，将不溶物沉降作为造腔模拟的重要因素，更加准确地模拟腔底形态，指导排卤管柱下深。腔底形态的准确模拟直接决定腔体下部不溶物是否会堵塞和挤压排卤管柱，同时腔底形态的合理与否将直接影响恒压运行阶段的工作气量，对于天然气阻溶恒压运行尤为重要。

（3）增加能量消耗计算，计算每个造腔阶段的能量消耗，能够更加准确地辅助项目的经济性分析。

（4）集成注采运行模块，根据运行计划分析注采运行过程中腔内温度、压力及库存等参数。

（5）增加方案对比功能，在同一个模型下新建多个造腔方案，便于对比优化造腔方案。

3.2 恒压运行案例分析

天然气阻溶恒压运行主要运用软件的SMUG造腔功能。根据金坛储气库某井实际天然气阻溶试验结果，验证CavSimu相关造腔模拟模块的准确性。该井SMUG造腔试验采用单管SMUG造腔模式，试验目标溶蚀段为1034～1050m的畸形部位，通过8轮次的注气排卤和注淡水排气对目标段进行改造，利用光纤界面仪对气水界面进行跟踪测量。将模拟参数按照实际天然气阻溶试验设定后，进行SMUG模拟（图5、图6）。

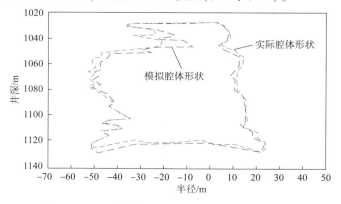

图5 SMUG造腔模拟结果与实际溶腔形态对比图

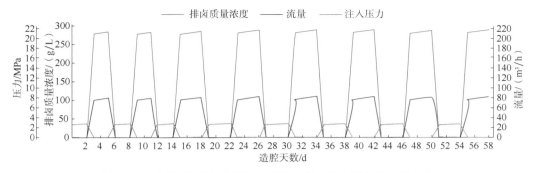

图6 恒压运行过程中模拟排卤质量浓度、注入压力及流量曲线

由模拟结果可见,每一轮次注气排卤时,排卤质量浓度均在300g/mL左右,造腔效率较高。对比模拟排卤浓度与实际排卤质量浓度发现(图7),各轮次天然气阻溶造腔排卤质量浓度与实际情况较吻合。对比腔体体积、平均能耗(表2)可知:新增体积模拟结果误差为6.3%,平均能耗模拟值误差为2.3%,CavSimu天然气阻溶造腔模拟模块能够精确模拟腔体发展及造腔能耗水平,可有效指导天然气阻溶恒压运行。

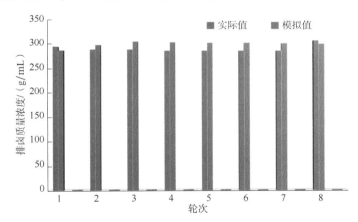

图7 各轮次天然气阻溶造腔模拟排卤质量浓度与实际排卤质量浓度对比图

表2 SMUG造腔模拟结果与实际情况对比表

对比项	新增体积/m³	平均能耗/(kW·h/m³)
模拟结果	10725.7	20.73
实际数据	11450.0	20.26

3.3 恒压运行模式分析

天然气阻溶恒压运行主要包括以下两种模式。

(1)按照常规腔体直径设计,直至达到设计体积转注采运行。使用CavSimu软件以每年注采1轮次进行30年恒压运行模拟,腔体平均直径增长29m,因此初始腔体直径控制在50m左右,可以保证30年恒压运行腔体不超溶。这种方式下,恒压运行不需要垫底气,腔体实现效益快,相对于常规造腔后注采运行,垫底气投资利用率(累计注采气量/垫底气)能

— 519 —

够得到大幅度提升。

（2）扩大单腔设计体积，在造腔中进行注采运行，完腔后直接废弃或储存其他流固体物质。充分利用恒压运行对运行中腔体稳定性的积极作用，经力学评价后扩大单腔设计直径，增加单腔体积，注采运行阶段与造腔阶段完全重合，完腔后对腔体进行报废或其他方式处置。这种模式不需要垫底气，还能扩大单腔体积，提高盐矿资源利用，同时，减少库区腔体数量，也能大大减少投资[17]。

第二种模式更能发挥天然气阻溶恒压运行优势。建议采用第一种模式进行单腔全生命周期现场试验，成功后采用第二种模式进行推广应用。

4 结论

结合天然气阻溶造腔和恒压运行优势，提出了盐穴储气库天然气阻溶恒压运行技术，分析了该技术的优点、缺点及解决方案，通过软件开发、数值模拟及现场试验，验证了该技术的可行性。

（1）恒压运行不需要垫底气维持腔内压力，可以减少30%以上的储气库建设投资，并且腔内维持较高的恒定压力对延长腔体寿命、减少腔体蠕变有积极作用。

（2）天然气阻溶恒压运行能够将造腔和注采运行良好结合，在大大减少造腔时间的同时，实现恒压运行。

（3）开发的CavSimu造腔模拟软件能够较为精确地模拟天然气阻溶造腔过程，可用于指导天然气阻溶恒压运行。

参 考 文 献

[1] 李铁，张永强. 地下储气库的建设与发展[J]. 油气储运，2000，19(3)：1-8.
[2] 杨春和，梁卫国，魏东吼，等. 中国盐岩能源地下储存可行性研究[J]. 岩石力学与工程学报，2005，24(24)：4409-4417.
[3] 丁国生，梁婧，任永胜，等. 建设中国天然气调峰储备与应急系统的建议[J]. 天然气工业，2009，29(5)：98-100.
[4] 王峰，王东军. 输气管道配套地下储气库调峰技术[J]. 石油规划设计，2011，22(3)：28-30.
[5] 罗天宝，李强，许相戈. 我国地下储气库垫底气经济评价方法探讨[J]. 国际石油经济，2016，24(7)：103-106.
[6] 成金华，刘伦，王小林，等. 天然气区域市场需求弹性差异性分析及价格规制影响研究[J]. 中国人口资源与环境，2014，24(8)：131-140.
[7] 吴洪波，何洋，周勇，等. 天然气调峰方式的对比与选择[J]. 天然气与石油，2009，27(5)：5-9.
[8] 周学深. 有效的天然气调峰储气技术——地下储气库[J]. 天然气工业，2013，33(10)：95-99.
[9] Stefan W. Releaching and solution mining under gas(SMUG) of existing caverns[C]. Basel：SMRI 2007 Spring Meeting，2007：62-67.
[10] Kunstman A，Urbńaczyk K. Designing of the storage caverns for liquid products，anticipating its size and shape changes during withdrawal operations with use of unsaturated brine[C]. Porto：SMRI 2008 Spring Meeting，2008：89-94.
[11] 刘继芹，刘玉刚，陈加松，等. 盐穴储气库天然气阻溶造腔数值模拟[J]. 油气储运，2017，36(7)：825-831.

[12] 李建君,陈加松,刘继芹,等.盐穴储气库天然气阻溶回溶造腔工艺[J].油气储运,2017,36(7):816-824.
[13] 付洪涛.盐穴储气库井钻井关键技术研究与应用[D].西安:西安石油大学,2013:21-29.
[14] 魏国齐,郑雅丽,邱小松,等.中国地下储气库地质理论与应用[J].石油学报,2019,40(12):1519-1530.
[15] 郑东波,黄孟云,夏焱,等.盐穴储气库造腔管柱损坏机理研究[J].重庆科技学院学报(自然科学版),2016,18(2):90-93.
[16] 班凡生.层状盐层造腔提速技术研究及应用[J].中国井矿盐,2015(6):16-18.
[17] 王元刚,李淑平,齐得山,等.考虑垫底气回收价值及资金时间价值的盐穴型地下储气库储气费计算方法[J].天然气工业,2018,38(11):122-127.

本论文原发表于《油气储运》2020年第39卷第11期。

盐岩储库腔底堆积物空隙体积试验与计算

任众鑫[1] 杨海军[1] 李建君[1] 刘建仪[2] 范 舟[3]

(1. 中国石油西气东输管道公司;2."油气藏地质及开发工程"国家重点实验室;
3. 西南石油大学材料科学与工程学院)

【摘 要】 受建库盐层本身地质特点影响,造腔结束后,底部一般会堆积大量不溶物。堆积物的存在既浪费了有效库容体积,又为整体建库施工带来了诸多不利因素。为科学评价该堆积物并进行有效处理或转变利用以增大库容体积,对其分布特征及堆积空隙体积进行了研究。通过室内溶腔实验获取了堆积不溶物样品,采用筛分法实验测定了其颗粒分布,并根据分形分布理论对其粒度分布特征进行了分析,进而基于可压缩堆积模型对堆积空隙体积进行了数值及模型研究。应用实例进行计算并与现场数据进行了对此评价验证,结果表明,在双对数坐标系下,腔底堆积物颗粒尺寸与累积数量呈线性相关,即其粒度分布可用分形分布函数进行表征,且分形维数作为分布特征参量能较好地描述颗粒均匀程度与集中性。

【关键词】 盐岩;储气库;堆积物;空隙体积;可压缩堆积模型

Testing and calculation of the pore volume of bottom deposits in the salt rock reservoir

Ren Zhongxin[1] Yang Haijun[1] Li Jianjun[1] Liu Jianyi[2] Fan Zhou[3]

(1. PetroChina West-East Pipeline Company; 2. State Key Laboratory of Oil and Gas Reservoir Geology and Exploration, Southwest Petroleum University;
3. School of Materials Scienee and Engineering, Southwest Petroleum University)

Abstract Affected by the geological characteristics of the reservoir-building salt layer, generally, after cavern creation is completed, a large amount of insoluble matter will accumulate at the bottom. The existence of deposits not only reduces the elective storage volume but also results in many unfavorable factors for the overall construction of the reservoir. In order to scientifically evaluate the deposits and effectively handle them or transform them to increase the storage volume, their distribution characteristics and deposition pore volume were studied. Samples of deposited insoluble matter were obtained by indoor cavern experiments. The particle size distribution was determined by sieving. The distribution characteris-

---

基金项目:中国石油天然气集团专业公司项目(2013B-3401-0503);中国石油天然气股份公司科技重大专项(2015E-40)。
作者简介:任众鑫,1987年生,男,汉族,山东菏泽人,工程师,硕士,主要从事储气库工艺技术研究。E-mail: zhongxren@petrochina.com.cn。

tics of particle size were analyzed according to the fractal distribution theory, and numerical modeling studies of the deposition pore volume were conducted based on the compressible packing model. Real cases were used in the calculations, and a comparison with field data was conducted. The results show that in the double-logarithmic coordinate system, the particle size of the bottom deposits is linearly correlated with the cumulative amount; that is, the distribution of particle size can be characterized by the fractal distribution function. The fractal dimension can be used as a distribution characteristic parameter to describe the uniformity and centrality of particles.

Keywords salt rock; gas storage; deposits; pore volume; compressible packing model

随着中国天然气事业及储运配套设施建设的步伐推进,盐岩储气库逐渐引起业界关注,其建设进度不断加快[1-3]。但由于盐岩本身所具有的层状分布、多泥质夹层、不溶杂质含量高等物性特征,腔体建成后在底部会堆积有一定厚度的不溶物[4-7]。相较国内,国外盐层纯度高,产生堆积物数量小,其研究也多集中在水溶建腔方面,对于堆积物的提及较少[8-9]。而国内盐层该问题较为突出,部分地区不溶物含量甚至可达 50%,由此产生的影响也较为明显[10-12]。堆积物的存在,不仅浪费了相当一部分库容体积,同时也为后续注气排卤及井下作业带来了安全隐患。因此有必要对堆积物分布特征及空隙体积进行研究,以为科学评价腔体库容体积、合理布置施工处理措施及提升储库整体建设质量提供理论基础与技术借鉴。

1 堆积物样品制备与测定

1.1 样品制备

由于腔底堆积物蓬松且浸满卤水,加之各方面条件限制,实现对其取样较为困难。根据相似性原理及现场工艺流程,自主设计模拟实验装置,利用腔体不同层位岩心分别进行相同条件下的室内水溶试验以获取堆积物样品。

实验装置如图 1(c)所示,对岩心进行中部钻孔(模拟井眼),分别下入同心圆管(模拟溶腔内、外管),在外管与孔壁间加注防护液(模拟溶腔油垫),控制溶蚀高度。在实验过程中,通过控制流量、调整油垫与管柱相对位置、改变循环方式,控制溶腔形态发展。

(a)试验前岩心　　　　　　(b)试验后岩心　　　　　　(c)试验模拟图

图 1　建腔模拟试验图

结合现场盐层与不溶物含量数据以及溶腔阶段发展动态，选用典型层位岩心进行试验。以金坛矿区某井为例，试验选用岩心截面约 10cm，对其进行处理打磨，使质量相等，为 1.17kg（以最小者为基准），层位及描述见表1。

表1 岩心层位及描述

编号	深度/m	岩性描述	编号	深度/m	岩性描述
1#	641.72~641.94	钙芒硝质盐岩	4#	695.15~695.35	泥质盐岩
2#	689.25~689.55	泥质钙芒硝岩	5#	712.95~913.30	泥质石膏盐岩
3#	708.70~708.89	灰黑色泥质盐岩			

水溶实验后，对形成的不溶堆积物烘干、分析，并进行分布测定。

1.2 粒度分布测定

粒度分布测定方法较多，其中利用筛分法既可测定粒度，还可以绘制累积粒度分布曲线[13]。对各组岩心所获取堆积物岩样分别采用筛分法进行测定，进而累加，得到堆积物样品总体粒径数据及各分组数据（以 3# 岩心样品为例）如图2、图3所示。

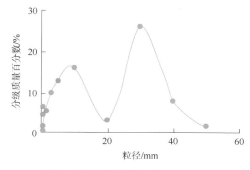

图2 样品总体粒径分级质量图

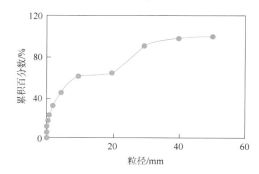

图3 样品总体粒径累积质量图

各分组所获取的堆积物样品粒度测定数据，其粒径范围及分级、累积质量百分数有所区别但差距有限，限于篇幅及数据重复性考虑，不将各组数据全部列出，仅以 3# 岩心获取的不溶物岩样为例（下文处理同），测定结果如图4、图5所示。由图4、图5可知，样品粒径多集中在 30~40mm 及 10~20mm，而各分组粒径集中范围有所差异，如 3# 样品粒径相对较大，有 30.8% 部分大于 40mm。

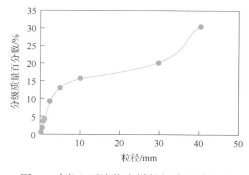

图4 3#岩心不溶物岩样粒径分级质量图

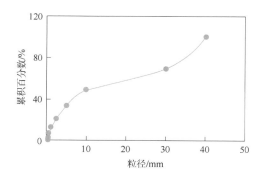

图5 3#岩心不溶物岩样粒径累积质量图

2 分布特征研究

2.1 分布模型选定

为进行后续特征描述、评价，研究与之相符的分布函数。

由粒径分级质量图(图4、图5)可看出，样品总体符合双正态分布，但分组粒径却各有差别。鉴于分形理论近年来的发展创新及独特优势[14-16]，采用分形函数对堆积物粒度特征进行描述分析。

根据 Mandelbrot 分形理论，多分散颗粒系统若符合分形分布，应满足[17]

$$n_{(>x)} = C \cdot x^{-D} \tag{1}$$

式中：$n_{(>x)}$ 为系统中粒度大于 x 的颗粒数，无量纲；x 为粒度，mm；C 为常数，无量纲；D 为分形维数，用以表征颗粒组成集中性和均匀性的特征参数。

定义粒度 x 到 $x+\mathrm{d}x$ 间颗粒数 $m(x)$

$$m(x) = \lim \frac{-\Delta n_{(>x)}}{\Delta x} = -\frac{\mathrm{d}n_{(>x)}}{\mathrm{d}x} \tag{2}$$

整理，得

$$m(x) = CDx^{-(D+1)} \tag{3}$$

根据分布矩概念，将累积分布与分形维数联系起来，累积质量为：

$$w_{(>x)} = \int_x^{x_{\max}} \rho C_V x^3 m(x) \mathrm{d}x = \rho C_V C \frac{D}{3-D}(x_{\max}^{3-D} - x^{3-D}) \tag{4}$$

式中：$w_{(>x)}$ 为累积质量，g；C_V 为体积形状系数，无量纲；ρ 为颗粒的密度，g/cm³。

注意到总累积质量

$$w_T = \int_0^{x_{\max}} \rho C_V x^3 m(x) \mathrm{d}x = \rho C_V C \frac{D}{3-D} x_{\max}^{3-D} \tag{5}$$

式中：w_T 为总累积质量，g。

则大于特定粒径值的累积质量，即粒度上累积分布，可用函数表示为

$$N(x) = \frac{w_{(>x)}}{w_T} \times 100 = 100\left[1 - \left(\frac{x}{x_{\max}}\right)^{3-D}\right] \tag{6}$$

式中：$N(x)$ 为粒度上累积分布函数。

令 $T = 3-D$，转换，可得下累积分布函数 $M(x)$

$$M(x) = 100\left(\frac{x}{x_{\max}}\right)^T \tag{7}$$

式中：$M(x)$ 为粒度上累积分布函数。

2.2 分布拟合及特征参量计算

将粒度值和累积颗粒数分别取对数，进行分布拟合。3#岩心不溶物岩样及样品总体拟

合结果图如图 6、图 7 所示。

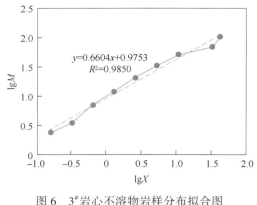

图 6 3#岩心不溶物岩样分布拟合图

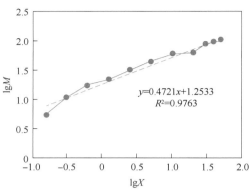

图 7 样品总体分布拟合图

对比实际数据线(实线)与拟合趋势线(虚线)可知,拟合偏差 R^2 在 0.97~0.99,两者在双对数坐标系下呈线性相关,即其分布具有分形特点。特征参量即分形维数 D 计算结果见表 2。

表 2 各组拟合分形维数计算数据表

不溶物编号	拟合斜率	分形维数
1#	0.8827	2.1173
2#	0.6627	2.3373
3#	0.6604	2.3396
4#	0.4173	2.5827
5#	0.3051	2.6949
整体	0.4721	2.5279

根据计算结果,对比样品粒径数据可知,分形维数越大,颗粒集中性和均匀程度越差;相反,分形维数越小时,颗粒均匀性越好,颗粒尺寸也越集中。

2.3 分形维数与溶解度关系

由于腔底堆积物是各岩层段不溶物沉积累加的结果,为避免采用某单一特征参数表征其分布特性,并考虑颗粒级配产生的协同效应[18],分析分形维数与溶解度数据之间关系。岩心各层分形维数和溶解度数据见表 3。

表 3 各组拟合分形维数计算数据表

不溶物编号	分形维数	溶解百分数/%
1#	2.1173	43.52
2#	2.3373	55.31
3#	2.3396	62.74
4#	2.5827	85.21
5#	2.6949	91.34
整体	2.5279	81.44

对分形维数与溶解度进行拟合,结果如图 8 所示。由图 8 可知,不溶物分形维数与溶解度呈线性相关。

为对两者关系的独立性做进一步分析研究,并验证与粒径组别范围的无关性,将 5 组样品不溶物中任意两组进行质量组合累加,得到 10 个新的组合,计算各组分形维数及溶解度并进行拟合(图 9)。

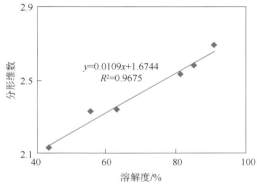 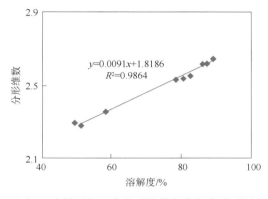

图 8 不溶物岩样分形维数与溶解度关系图　　　图 9 岩样累加组合分形维数与溶解度关系图

由图 9 可看出,不溶物分形维数与溶解度呈线性关系,且关系曲线可描述为

$$y = 0.0091x + 1.8186$$

式中:x 为溶解度,无量纲;y 为堆积物分形维数。

相关度在 0.98 以上。由此,堆积物分布分形维数可由其溶解度数据计算得到,且不溶物溶解度越大,形成的堆积物分形维数越大,亦即颗粒尺寸越不均匀,即堆积物分布特征可用分形函数进行表征。

3 空隙体积数值研究

3.1 空隙率数值模拟

受实验条件及因素限制,采用离散元仿真软件模拟不溶物堆积过程,并计算堆积空隙率。根据前述实验测定的堆积物分布结果,利用 Origin 拟合得出其分布特征参数,采用三维离散元方法模拟其堆积过程,并对模拟结果进行数值分析,从而得到各组不溶物岩样及总体堆积空隙率,结果见表 4。

表 4 不溶物岩样及总体堆积空隙率模拟数据表

不溶物编号	模拟空隙率/%	不溶物编号	模拟空隙率/%
1#	55.62	4#	42.28
2#	49.51	5#	39.21
3#	47.93	整体	43.11

3.2 堆积空隙率与分形维数关系

根据相关研究[19],随着颗粒粒度分形维数增加,其堆积空隙率线性减小,即其堆积结

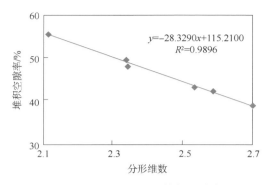

图 10 不溶物岩样分形维数与空隙率关系图

构越好,填充能力增强。据此对不溶物岩样数据进行空隙率与分形维数关系验证,验证结果如图 10 所示。

由图 10 可知,堆积物空隙率与其粒度分形维数呈线性关系,且随分形维数增加,堆积空隙率下降。分形维数越大即颗粒越分散,粒级越多,这时颗粒间会互相填隙,空隙率从而变小。

为进一步探索堆积空隙率与分形维数间线性关系的独立性,研究相同分形维数不同粒度范围的颗粒堆积空隙率变化。所用各组样品颗粒分布见表 5。

表 5 同分形维数、不同粒度范围的样品组数据($D=2.5584$)

12#样品组合		13#样品组合		14#样品组合		15#样品组合	
粒径/mm	分级质量/%	粒径/mm	分级质量/%	粒径/mm	分级质量/%	粒径/mm	分级质量/%
25.000	27.23	50.000	9.96	75.000	17.23	100.000	27.55
12.500	25.30	40.000	11.48	50.000	23.17	50.000	20.29
5.000	13.38	30.000	13.89	25.000	17.06	25.000	14.94
2.500	9.85	20.000	18.68	12.500	9.62	12.500	8.42
1.250	7.18	10.000	13.76	7.500	6.23	7.500	5.46
0.630	5.36	5.000	10.13	5.000	8.38	5.000	7.34
0.315	3.87	2.500	7.46	2.500	6.17	2.500	5.40
0.160	2.93	1.250	5.44	1.250	4.50	1.250	3.94
0.080	2.15	0.630	4.06	0.630	3.36	0.630	2.94
0.040	1.59	0.315	2.93	0.320	2.42	0.320	2.12
0.020	1.17	0.160	2.22	0.160	1.83	0.160	1.60

采用同样处理方法,模拟得到各组堆积空隙率值,见表 6。

由表 6 可知,分形维数相同粒度范围不同时,空隙率值也基本一致,即堆积空隙率与粒度范围无关,取决于其分形维数即颗粒分布特征。

表 6 空隙率模拟数据表

样品组合	最大粒径/mm	分形维数	模拟空隙率/%
12#	25	2.5584	43.12
13#	50	2.5584	43.11
14#	75	2.5584	43.12
15#	100	2.5584	43.12

4 空隙体积模型

4.1 可压缩堆积模型

腔底堆积物在形成过程中，颗粒间是相互影响的，小颗粒的存在会使大颗粒无法达到最紧密堆积，即松动效应；而大颗粒的存在会使小颗粒在靠近大颗粒壁面附近堆积，不能最密实，即附壁效应。同时，颗粒与颗粒间还存在互相压实的作用。

de Larrard F 等在考虑上述影响的基础上，为克服一般模型单一粒径假设的局限性，提出了可压缩堆积模型[20-22]。根据该研究，在多粒级颗粒堆积系统中，假设第 i 级颗粒占主导地位，则系统各堆积参数为

$$\begin{cases} \gamma_i = \beta_i / \left[1 - \sum_{j=1}^{i-1} \left[1-\beta_i+b_{ij}\beta_i(1-1/\beta_j) \right] y_j - \sum_{j=i+1}^{n} (1-a_{ij}\beta_i/\beta_j)y_j \right] \\ \gamma = \min(\gamma_i) \\ K = \dfrac{y_i/\beta_i}{1/\Phi - 1/\gamma} \end{cases} \quad (8)$$

式中：γ_i 为第 i 级颗粒为主时的虚拟堆积密实度，无量纲；β_i 为第 i 级颗粒的剩余堆积密实度，无量纲；y_j 为第 j 级颗粒所占体积分数，无量纲；γ 为系统实际堆积密实度，无量纲；K 为颗粒压实指数，无量纲；Φ 为系统实际堆积密实度，无量纲；a_{ij}，b_{ij} 为第 j 级颗粒对第 i 级颗粒产生的松动效应和附壁效应系数，无量纲，且 $a_{ij}=\sqrt{1-(1-d_j)^{1.02}}$，$j=i+1,\cdots,n$，$b_{ij}=1-(1-d_i)^{1.50}$，$j=1,\cdots,i-1$；$d_i$ 为第 i 级颗粒的粒径，mm。

4.2 模型计算分析

应用时，剩余堆积密实度可采用相关文献[23-24]所得数据。根据 de Larrard F 的试验校准研究，压实指数在简单倾倒形式堆积时可取 4.1，盐岩不溶物在堆积过程中主要受重力沉降作用并在此基础上压实形成[25-27]，故按该情况处理。

由此，在确定堆积物各粒级尺寸及质量分数后，通过为各粒级剩余堆积密实度取值，求出系统虚拟堆积密实度，进而根据压实指数公式求得系统实际堆积密实度，最终确定堆积空隙率。

4.3 计算模型

上述堆积空隙率计算方法所针对的是盐岩溶腔各层位，要计算整个腔体的堆积物空隙率，需对各分层数据进行累加求解。假设共 n 个溶腔层位，第 i 层位盐岩含量即溶解度为 x_i，层高 h_i，溶腔半径 l_i，则腔体不溶物总量 D_m 为

$$D_m = \sum_{i=1}^{n} (1-x_i) \cdot h_i \cdot \pi l_i^2 \quad (9)$$

对于溶腔各层位，溶解度可通过岩性测井数据转换求解，进而求得各层位堆积物空隙率，再进行累加求解，即可得到腔体总体空隙率，表达式如下

$$\phi = \frac{\sum_{i=1}^{n} \phi_i \cdot \lambda \cdot (1 - x_i) h_i \cdot \pi l_i^2}{D_m \cdot \lambda} = \frac{\sum_{i=1}^{n} \phi_i \cdot (1 - x_i) h_i \cdot \pi l_i^2}{D_m} \tag{10}$$

式中：λ 为不溶物堆积膨胀系数，无量纲。

由此，利用相关数据可对溶腔现场腔底堆积物空隙体积进行计算。

4.4 应用实例

实例井某溶腔段腔体测井解释数据见表7。

表7 实例井某溶腔阶段腔体层位测井解释数据表

序号	深度/mm	不溶物含量/%	序号	深度/mm	不溶物含量/%
1	680.0	47.77	12	685.5	27.42
2	680.5	47.77	13	686.0	27.42
3	681.0	47.77	14	686.5	27.42
4	681.5	47.77	15	687.0	27.42
5	682.0	16.95	16	687.5	32.51
6	682.5	24.28	17	688.0	32.51
7	683.0	24.28	18	688.5	32.51
8	683.5	19.54	19	689.0	24.50
9	684.0	28.79	20	689.5	24.50
10	684.5	28.79	21	690.0	49.89
11	685.0	28.79	22	690.5	17.99

采用可压缩堆积模型计算堆积空隙率，需首先确定堆积物各粒级组成，受条件限制，无法获取盐岩储库腔底不溶物具体颗粒组成。但根据前述分形研究结论，在分形维数相同时，空隙率值与粒度范围关系很小。故可根据上述基础数据，确定各层堆积后的分形维数，并选定50mm(此时筛下累积质量为100%)作为最大粒径，以此确定各盐岩层位溶解后粒径分布，具体数据见表8。

表8 各盐岩层位堆积粒度分布推导数据表

层位序号	分形维数	粒径分布/%									
		50.000mm	40.000mm	30.000mm	20.000mm	10.000mm	5.000mm	2.500mm	1.250mm	0.630mm	0.315mm	0.160mm
1	2.285	14.89	16.01	17.64	20.50	12.49	7.61	4.64	2.80	1.73	1.04	0.65
2	2.285	14.89	16.01	17.64	20.50	12.49	7.61	4.64	2.80	1.73	1.04	0.65
3	2.285	14.89	16.01	17.64	20.50	12.49	7.61	4.64	2.80	1.73	1.04	0.65
4	2.285	14.89	16.01	17.64	20.50	12.49	7.61	4.64	2.80	1.73	1.04	0.65
5	2.572	9.11	10.53	12.81	17.35	12.89	9.58	7.12	5.24	3.95	9.23	2.19
6	2.504	10.93	12.42	14.75	19.26	13.65	9.68	6.86	4.82	3.46	2.41	1.75
7	2.504	10.93	12.42	14.75	19.26	13.65	9.68	6.86	4.82	3.46	2.41	1.75
8	2.548	10.15	11.66	14.06	18.80	13.74	10.05	7.34	5.31	3.94	2.82	2.12

续表

层位序号	分形维数	粒径分布/%										
		50.000mm	40.000mm	30.000mm	20.000mm	10.000mm	5.000mm	2.500mm	1.250mm	0.630mm	0.315mm	0.160mm
9	2.462	11.68	13.13	15.37	19.63	13.52	9.31	6.41	4.37	3.05	2.06	1.46
10	2.462	11.68	13.13	15.37	19.63	13.52	9.31	6.41	4.37	3.05	2.06	1.46
11	2.462	11.68	13.13	15.37	19.63	13.52	9.31	6.41	4.37	3.05	2.06	1.46
12	2.475	11.45	12.92	15.19	19.52	13.56	9.42	6.55	4.50	3.17	2.16	1.54
13	2.475	11.45	12.92	15.19	19.52	13.56	9.42	6.55	4.50	3.17	2.16	1.54
14	2.475	11.45	12.92	15.19	19.52	13.56	9.42	6.55	4.50	3.17	2.16	1.54
15	2.475	11.45	12.92	15.19	19.52	13.56	9.42	6.55	4.50	3.17	2.16	1.54
16	2.427	12.31	13.71	15.86	19.89	13.37	8.99	6.04	4.02	2.74	1.81	1.25
17	2.427	12.31	13.71	15.86	19.89	13.37	8.99	6.04	4.02	2.74	1.81	1.25
18	2.427	12.31	13.71	15.86	19.89	13.37	8.99	6.04	4.02	2.74	1.81	1.25
19	2.502	10.97	12.45	14.78	19.28	13.65	9.66	6.84	4.80	3.44	2.39	1.74
20	2.502	10.97	12.45	14.78	19.28	13.65	9.66	6.84	4.80	3.44	2.39	1.74
21	2.265	9.90	11.41	13.83	18.64	13.76	10.16	7.50	5.48	4.10	2.97	2.25
22	2.562	18.90	19.21	19.63	20.29	10.60	5.54	2.90	1.50	0.80	0.41	0.22

采用前述计算方法，对以上数据进行实例井计算，该溶腔阶段腔体堆积物空隙率计算结果见表9。

表9 实例井某溶腔阶段堆积物空隙率计算数据表

序号	深度/mm	计算空隙率/%	序号	深度/mm	计算空隙率/%
1	680.0	51.70	13	686.0	48.37
2	680.5	51.70	14	686.5	48.37
3	681.0	51.70	15	687.0	48.37
4	681.5	51.70	16	687.5	49.37
5	682.0	43.27	17	688.0	49.37
6	682.5	47.70	18	688.5	49.37
7	683.0	47.70	19	689.0	47.74
8	683.5	46.57	20	689.5	47.74
9	684.0	48.65	21	690.0	51.96
10	684.5	48.65	22	690.5	46.18
11	685.0	48.65			

根据计算结果，实例井某溶腔阶段堆积物总体空隙率为48.23%，现场通过后续声呐测腔分析得出的空隙率44.26%，本计算方法结果值稍大，但偏差不超过8.97%，故计算方法可行。

5 结论

（1）通过室内溶腔实验获取了盐岩储气库腔底堆积物样品，采用筛分法对其粒度分布进行了实验测定，从而得到了样品粒径分级质量数据及总体累积质量数据。

（2）腔底堆积物粒度分布可用分形函数进行表征，且分形维数作为特征参量能很好地描述颗粒分布的集中性与均匀程度。

参 考 文 献

[1] 杨春和，梁卫国，魏东吼. 中国盐岩能源地下储存可行性研究[J]. 岩石力学与工程学报，2005，24(24)：4409-4417.

[2] 丁国生，李春，王皆明，等. 中国地下储气库现状及技术发展方向[J]. 天然气工业，2015，35(11)：107-112.

[3] 徐博，张刚雄，张愉，等. 我国地下储气库市场化运作模式的基本构想[J]. 天然气工业，2015，35(11)：102-106.

[4] 完颜祺琪，冉莉娜，韩冰洁，等. 盐穴地下储气库库址地质评价与建库区优选[J]. 西南石油大学学报（自然科学版），2015，37(1)：57-64.

[5] 李银平，葛鑫博，王兵武，等. 盐穴地下储气库水溶造腔管柱的动力稳定性试验研究[J]. 天然气工业，2016，36(7)：81-87.

[6] 王东旭，马小明，伍勇，等. 气藏型地下储气库的库存量曲线特征与达容规律[J]. 天然气工业，2015，35(1)：115-119.

[7] 胥洪成，王皆明，屈平，等. 复杂地质条件气藏储气库库容参数的预测方法[J]. 天然气工业，2015，35(1)：103-108.

[8] Yu L, Liu J. Stability of interbed for salt cavern gas storage in solution mining considering cusp displacement catastrophe theory [J]. Petroleum, 2015, 1(1)：82-90.

[9] Xiong J, Huang X, Ma H. Gas leakage mechanism in bedded salt rock storage cavern considering damaged interface [J]. Petroleum, 2015, 1(4)：366-372.

[10] 魏东吼. 金坛盐穴地下储气库造腔工程技术研究[D]. 青岛：中国石油大学（华东），2008.

[11] 李银平，杨春和，罗超文，等. 湖北省云应地区盐岩溶腔型地下能源储库密闭性研究[J]. 岩石力学与工程学报，2007，26(12)：2430-2436.

[12] 施锡林，李银平，杨春和，等. 盐穴储气库水溶造腔夹层垮塌力学机制研究[J]. 岩土力学，2009，30(12)：3615-3620.

[13] 王元磊. 粒度趋势分析方法的研究进展[J]. 山东师范大学学报（自然科学版），2008，23(2)：81-84.

[14] 蔡建超，胡祥云. 多孔介质分形理论与应用[M]. 北京：科学出版社，2016.

[15] 朱华，姬翠翠. 分形理论及其应用[M]. 北京：科学出版社，2011.

[16] 舒志乐，刘新荣，刘保县，等. 基于分形理论的土石混合体强度特征研究[J]. 岩石力学与工程学报，2009，28(S1)：2651-2656.

[17] Mandelbrot B. The Fractal Geometry of Nature[M]. San Francisco：W. H. Freeman, 1982.

[18] 韩涛. 矿渣粉粒度分布特征及其对水泥强度的影响[D]. 西安：西安建筑科技大学，2004.

[19] 吴成宝，段百涛. 中空玻璃微珠粒度分布分形特征及其与空隙率关系的研究[J]. 中国粉体技术，2008，14(1)：16-19.

[20] de Larrard F,Sedran T. Optimization of ultra-high-performance concrete by the use of a packing model [J]. Cement and Concrete Research,1994,24(6):997-1009.
[21] 陈瑾祥. 基于可压缩堆积模型的绿色混凝土材料性能研究[D]. 深圳:深圳大学,2015.
[22] 聂晶,基于可压缩堆积模型的水泥基复合材料性能研究[D]. 长沙:湖南大学,2008.
[23] 弗朗索瓦·德拉拉尔,混凝土混合料的配合[M]. 廖欣,叶枝荣译. 北京:化学工业出版社,2004.
[24] 周卫峰,赵可,王德群,等. 水泥稳定碎石混合料配合比的优化[J]. 长安大学学报(自然科学版),2006,26(1):24-28.
[25] 班凡生. 盐穴储气库水溶建腔优化设计研究[D]. 廊坊:中国科学院渗流流体力学研究所,2008.
[26] 汪蓬勃. 基于巨厚盐膏层以及碳酸盐储层的钻井技术研究[D]. 成都:西南石油大学,2015.
[27] 赵志成. 盐岩储气库水溶建腔流体输运理论及溶腔形态变化规律研究[D]. 廊坊:中国科学院渗流流体力学研究所,2008.

本论文原发表于《西南石油大学学报(自然科学版)》2018年第40卷第2期。

金坛盐穴储气库腔体畸变影响因素

李建君　王立东　刘　春　周冬林　王元刚　何　俊　薛　雨

（中国石油西气东输管道（销售）公司储气库管理处工艺技术研究所）

【摘　要】 金坛储气库在建设过程中，腔体不同程度地存在形状不规则的特点，分析腔体形态发育情况，找出腔体形态畸变的关键因素，并在后续造腔过程中加以控制，对于保障储气库经济效益和安全生产具有重要意义。基于金坛储气库的地震、测井和造腔数据分析，从地质和工艺两个方面出发，研究了多种因素对于盐穴形态变化的影响，结果表明：金坛储气库盐穴所处的构造位置、夹层的展布、不溶物的分布对盐穴的畸变无明显影响；而夹层数量和夹层厚度是影响盐穴形态的主要因素；溶腔速率的增大，会导致盐穴畸变；油垫层的位置和厚度会影响盐穴的形状；停井静溶期间，因卤水很快达到饱和，对腔体形态不会产生影响。目前，国内对于该问题的研究尚属空白，处于探索阶段。

【关键词】 盐穴储气库；腔体变形；地质；溶腔工艺

Factors affecting cavities distortion of Jintan Salt Cavern Gas Storage

Li Jianjun　Wang Lidong　Liu Chun　Zhou Donglin
Wang Yuangang　He Jun　Xue Yu

(Gas Storage Technology Research Institute of Gas Storage Dept,
PetroChina West-East Gas Pipeline Company (Marketing))

Abstract　During the construction of Jintan Gas Storage, the cavities are irregular in shape to a various degree. The development conditions of cavity morphology, the key factors affecting cavities distortion are figured out, and these factors are controlled in later cavity making process, which is of great significance for the economic benefit of gas storage and safe production. Based on the statistic analysis of Jintan Gas Storage's seismic, log and cavity making data, multiple factors affecting salt caverns' morphological changes are investigated from geology and technique aspects. Results show that the structural locations, interlayer layouts and distribution of insoluble substances of Jintan Salt Cavern Gas Storage do not have much effect on salt caverns' distortion, while the number and thickness of interlayers are the main factors affecting salt caverns' morphology. The increase in cavity dissolution rates

---

作者简介：李建君，工程师，1982年生，2008年硕士毕业于清华大学动力机械及工程专业，现主要从事盐穴储气库溶腔工艺及其优化、稳定性评价等研究工作。电话：13814795191；Email：cqklijianjun@petrochina.com.cn。

would lead to salt cavern's distortion. The location and thickness of oil cushion layers would affect salt cavern's shape, and during the period of well-resting and static solution, the condition in a rapid saturation of brine does not affect salt cavern's distortion. At present, research on this topic is still a blank in China, and still in exploring stage.

Keywords salt cavern gas storage; cavity distortion; geology; cavity solution technique

金坛储气库溶腔建库的目的层位于古近系阜宁组四段，属于陆相碎屑岩和蒸发岩沉积，富含盐岩并且发育泥岩石膏夹层。在储气库的建设过程中，多个腔体出现形态不规则等特点，称为盐穴畸变。盐岩体的强度和形态特点对于储气库的稳定和安全十分关键[1]，因此，研究腔体形态发育的影响因素对于控制腔体的形态，确保储气库的稳定和安全运行具有十分重要的意义。

1 盐穴畸变影响因素

造成盐穴畸变的因素很多，国外研究表明卤水温度、卤水浓度、不溶物含量等因素均对盐岩的溶蚀具有影响。腔体之间矿柱的宽度对于两个腔体的稳定性至关重要，模拟结果表明宽度与腔体直径之比应大于2.0[2]。卤水温度由40℃提高至60℃时，半径为3m厚度的盐岩溶蚀所需时间由42d减少至23d，溶蚀速度明显提高[3]。不溶物含量的增加以及分布的不均匀同样导致溶蚀速度的不均匀性，往往是造成腔体形态不规则的重要原因[4]。盐岩地层中不溶物夹层的分布也是造成腔体形态变形的危险因素[5]。构造活动和成岩过程中形成的裂缝不均匀分布也会导致溶蚀速度的明显变化[6]。

综合前人的研究成果，结合金坛地质条件和溶腔建设过程中遇到的问题，从地质因素和工艺因素两个方面分析金坛地区腔体形态变形的控制因素。地质因素主要包括：溶腔层段夹层的展布情况和夹层的岩性、夹层中岩盐的含量、腔体所处构造位置、溶腔层段的盐层中不溶物含量等。工艺因素主要是溶腔工艺中的各种参数。

2 地质因素

2.1 溶腔层段所处构造位置

根据金坛盐底构造图(图1)和盐层厚度分布图(图2)，盐穴E井、F井处于凹陷底部，均处于沉积中心位置，盐层厚度大。盐穴H井、G井位于斜坡上，盐层厚度也较大。盐穴E井、F井所处构造位置基本相同，但盐腔形状差别较大，盐穴E井为扁壶状，腔壁突出岩脊较多，盐穴F井的腔壁较光滑。盐穴A井、B井、C井、D井等完腔井腔壁光滑，突出岩脊较少，所处构造位置和盐层厚度相比盐穴E井、F井、H井、G井均较差。因此，金坛库区腔体所处的构造位置对盐穴壁面的形状影响不明确。

2.2 夹层的影响

2.2.1 夹层展布形状

腔体垂向上的形态特点受到垂向上夹层分布的影响，通过对比盐穴形态和地层剖面(图3)，可知盐脊出现的层段通常是厚夹层发育的层段，但夹层发育的层段不一定会出现盐脊影响腔体形态，表明垂向上的夹层分布对于腔体形态有影响，但不是绝对的。另外，

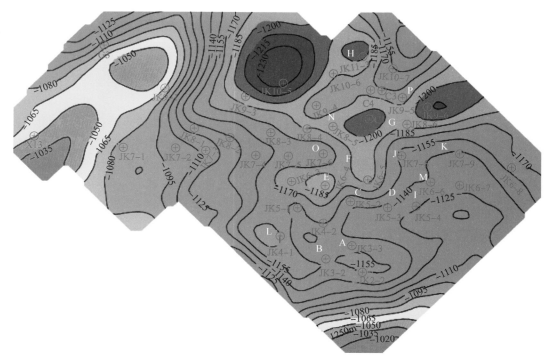

图 1 金坛储气库盐底构造图

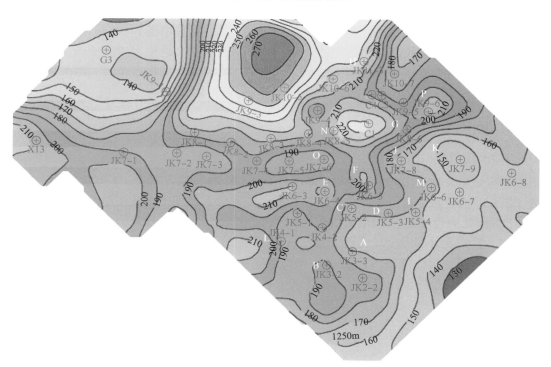

图 2 金坛储气库盐层厚度分布图

夹层的倾角对盐穴形状的影响也不确定，没有向倾角较大或较小方向发展的趋势。因此，针对金坛储气库而言，夹层展布形状对盐穴形状的影响也不确定。

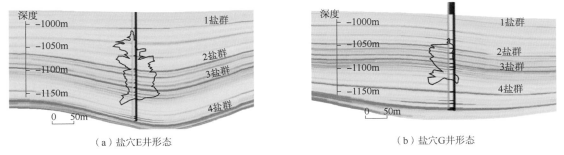

（a）盐穴E井形态　　　　　　　　　　（b）盐穴G井形态

图3　盐穴E井和G井夹层分布对腔体垂向形态发育的影响

2.2.2　夹层数量及厚度

统计各井夹层和盐穴发育形态，结果表明：溶腔层段内，夹层数量越多，对盐穴形状的负面影响越严重。盐穴在跨夹层段溶腔过程中，岩脊发育，盐穴壁面形状不规则，而在盐穴建槽期，由于溶腔层段岩盐较纯，夹层数量少，盐穴形状均为相对规则的陀螺状，壁面岩脊少（图4）。

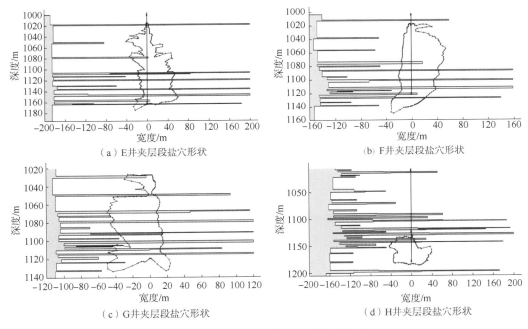

（a）E井夹层段盐穴形状　　　　　　　　（b）F井夹层段盐穴形状

（c）G井夹层段盐穴形状　　　　　　　　（d）H井夹层段盐穴形状

图4　单井夹层分布和盐穴形状对比图

3　工艺因素

影响盐穴形状和体积的工艺因素有井眼形状、管柱组合、注水排量、注水浓度、垫层位置、垫层厚度、溶腔阶段划分、溶腔速率、溶腔时间等。根据金坛储气库的溶腔数据和盐穴形状，选取溶腔速率、垫层厚度及位置、溶腔时间3个因素进行具体分析。

3.1 溶腔速率

统计相关井的平均溶腔速率(表1)可知，腔体形态较好井的溶腔速率也较小。为进一步分析溶腔速率对盐穴形状畸变的影响，对典型井进行分阶段详细统计分析(表2)。

表1 相关井平均溶腔速率统计表

井号	累计溶腔时间/d	累计净溶腔体积/m³	夹层数	平均溶腔速率/(m³/d)	备注
E	1831	214358	7	117	形状较差、有偏溶
F	2052	294591	6	144	形状较好、有偏溶
I	2017	169514	5	84	形状较好、无偏溶
J	2204	194446	8	88	形状较好、无偏溶
K	2016	169563	10	84	形状较好、无偏溶

表2 典型井阶段平均溶腔速率统计表

井号	阶段性溶腔体积/m³	净溶腔时间/d	平均溶腔速率/(m³/d)	备注
E	34969	215	162	第三次到第四次声呐之间，开始出现岩脊
F	45990	220	209	第三次到第四次声呐之间，开始出现岩脊
G	23258	133	175	第二次到第三次声呐之间，开始出现岩脊
H	43598	196	222	第三次到第四次声呐之间，出现严重的岩脊和岩谷
K	15957	322	50	第一次声呐到第二次声呐之间，形状较为规则
L	16410	377	43	第二次声呐到第三次声呐之间，形状较为规则

当溶腔速率较大时，容易导致岩脊和岩谷等不规则形状出现，采用畸变因数 δ 表征盐穴的不规则性，以进一步深入研究溶腔速率与不规则性的关系：

$$\delta = \frac{1}{n}\sum_{i=1}^{n}\left(\frac{D_i - \overline{D}}{\overline{D}}\right)^2$$

式中：D_i 为盐穴某一深度某一方向上的直径，m；$\overline{D}$ 为盐穴某一深度算术平均直径，m；n 为盐穴某一深度参与统计的直径数。

畸变因数越大，表面盐穴偏心越大，畸变因数能较好地表达盐穴的变形程度。

为确定水平溶腔速率与盐穴畸变因数的关系，必须剔除地质因素的影响，故应选择夹层数和平均不溶物含量基本相近的溶腔阶段进行分析(图5、图6)。根据所选井的水平溶腔速率与畸变因数的对比分析可知，金坛储气库在地质条件基本相同的情况下，溶腔速率与畸变因数呈现良好的线性关系，溶腔速率越大，盐穴壁面形状越容易出现不规则性。

3.2 垫层厚度及位置

分析盐穴历次声呐测量结果可知，垫层的厚度和位置也可能影响盐穴的形状。根据盐穴G井、K井、M井、P井邻近声呐检测结果(图7),盐穴G井在1050m深度位置盐层未得到充分溶蚀,盐穴P井在1179m深度位置盐层未得到充分溶蚀。根据现场数据,两个盐穴在形成该平顶形状的溶腔阶段,垫层柴油使用量较多,而溶腔阶段结束时退出柴油量较少,导致腔内有大量柴油剩余。盐穴M井、K井情况与盐穴G井、P井相似。根据溶腔情况分析,G井、P井腔内俘获柴油对后续溶腔造成影响,阻止了盐穴上溶,而M井、K井腔内俘获的大量柴油未阻止盐穴上溶,这是由于盐穴G井、P井第一次声呐腔顶距夹层较近,而盐穴M井、K井第一次声呐腔顶距夹层较远。对于盐穴M井、K井由于第一次声呐腔顶与夹层底有较厚的纯盐层,在后续溶腔过程中,这部分盐层被溶蚀,俘获的大量柴油被逐渐释放,因此未阻止盐穴上溶。

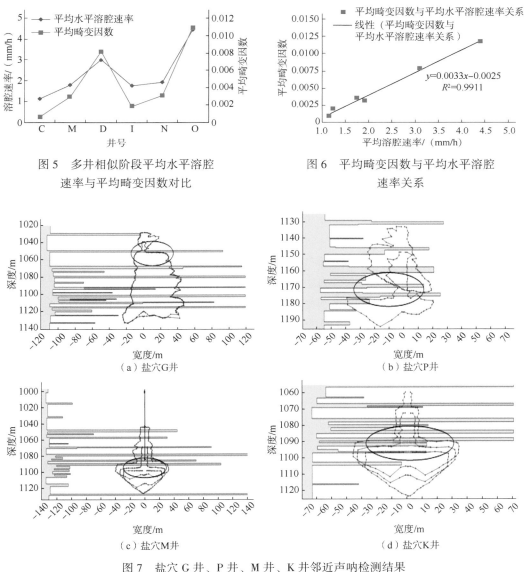

图5 多井相似阶段平均水平溶腔速率与平均畸变因数对比

图6 平均畸变因数与平均水平溶腔速率关系

图7 盐穴G井、P井、M井、K井邻近声呐检测结果

3.3 溶腔时间

溶腔时间包括净溶腔时间和停井时间，净溶腔时间对盐穴的影响最终可以归结到溶腔速率的影响。停井静溶时间，由于停井后卤水浓度上升较快，15d 左右则基本饱和，因此，长时间停井静溶对盐穴形状的影响不仅是溶腔的影响，还应包括化学动态平衡和其他一些未知作用的影响（图 8）。盐穴 E 井以上两次声呐测量间隔约 1a，而两次声呐形状几乎完全一样，第一次与第二次声呐测量时腔内卤水均已饱和。在一年的停井静溶中，岩脊未脱落，腔底深度未发生明显变化，因此腔内卤水饱和后，停井时间基本不会对腔体形状造成影响。

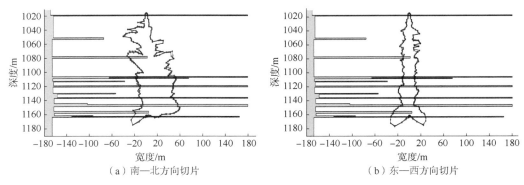

图 8　盐穴 E 井停井静溶前后声呐对比

4　结论

通过针对上述影响盐穴形状的地质因素和工艺因素的阶段分析，认为腔体形态发育的影响因素具有多元化的特点，就金坛储气库而言可以得出以下结论：

（1）盐穴的构造位置、夹层的展布、平均不溶物分布对盐穴畸变影响不明显。

（2）夹层数量和夹层厚度是影响盐穴形状的主要因素。

（3）溶腔速率越大，盐穴形状越易发生畸变。可以通过降低排量、调节注入工质或者调节两口距，降低水平溶腔速率，从而改善盐穴形状。

（4）垫层的深度位置和垫层厚度对盐穴形状也会产生不利影响，垫层的深度位置必须距离厚夹层一定距离，垫层和夹层之间有一定厚度的纯盐层，同时垫层厚度不宜过大。

（5）在溶腔周期内，停井静溶时间长短对盐穴形状无影响，长时间停井，盐穴形状基本保持不变。

参 考 文 献

[1] Kazarian V A, Smirnov V I, Remizov V V. The mass transfer processes intensification and improvement of cavern configuration formation at the construction of underground gas storages in rock salt [C]. Solution Mining Research Institute Spring Meeting, New Orleans, USA, April 19-22, 1998.

[2] Krainev B A, Kublanov A V, Baryakh A A. Geomechanical predictions and operational control of cavern convergence and subsidence due to the solution m inning procedure[C]. Solution Mining Research Institute Fall Meeting, Cleveland, USA, September 12-14, 1996.

[3] Andrej K, Kazimierz U. Application of winubro software to modelling of carven development in trona deposit

[C]. Solution Mining Research Institute Technical Conference, Krakow, Poland, April 27-28, 2009.

[4] Kenneth E, Mills. Specialized techniques in solution cavern shape control[C]. Solution Mining Research Institute Spring Meeting, Chicago, USA, April 27, 1981.

[5] Maria K, Bartlomiej R. Analysing cavern geometry development based on sonar measurements from the point of view of geological deposit structure[C]. Solution Mining Research Institute Meeting, Kracow, Poland, May 11-14, 1997.

[6] Evans D J, Heitmann N. Use of resistivity borehole imaging to assess the structure, sedimentology and insoluble content of the massively bedded Preesall Halite NW England[C]. Solution Mining Research Institute Technical Conference, New York, UK, October 3-4, 2011.

本论文原发表于《油气储运》2014年第33卷第3期。

盐岩储气库腔底堆积物扩容及工艺应用

任众鑫[1]　李建君[1]　巴金红[1]　蒋海涛[2]

（1. 中国石油西气东输管道公司；2. 中国石油渤海钻探工程院）

【摘　要】　针对国内在建的盐穴储气库其盐岩大都具有品位低、泥质夹层较多、不溶杂质含量较高的特点，对堆积物的形成及存在的影响进行了阐述，基于微积分原理采用线性插值方法对其堆积各项体积参数进行了分析和计算，得出了堆积物扩容体积的评价方法，并对相应扩容工艺进行了优选分析，最后结合现场生产和堆积物条件，在国内金坛某库区 A 井完成了扩容现场试验及效果分析，据分析结果充分利用堆积空隙体积多排出残余卤水近 5000m^3，腔体实现了有效扩容，扩容方案应用可行。

【关键词】　盐岩；储气库；腔底堆积物；扩容工艺；冲洗技术

地下储气库是天然气调峰的重要渠道，建设步伐正逐渐加快。盐岩储气库相较其他储气库，因具有注采率高、短期吞吐量大、垫气量低并可完全回收等优势而引起格外的重视[1-2]。但国内可供建库的盐岩地层大都存在品位低、夹层多、不溶杂质含量较高等劣势[3-4]，完腔后底部一般会堆积有大量不溶物，既浪费库容体积，又为后续施工带来了安全隐患。若通过对腔底堆积物进行工程处理，以避免该部分体积浪费，对单腔库容的增加无疑是非常明显的，尤其对盐层不溶杂质含量较高的地区，该问题的解决意义更为重大。

1　堆积物成因及影响

盐岩夹层中一般分布有泥岩、盐质泥岩、含钙芒硝泥岩，其中不溶水的杂质占比较高，在有些层位甚至可达 85% 或更多；泥质层位中盐岩含量很少，蒙脱石、伊利石、蒙脱石绿泥石混层等黏土矿物[5-6]含量较多。建库溶腔过程中，随着盐岩被注入淡水溶解并以卤水形式返至地面，其中不溶杂质部分会被释放并沉降，进而覆盖在未溶解盐岩表面或上阶段溶腔，加之黏土矿物遇水易软化膨胀，最后在腔底形成了大量堆积物，堆积示意图如图 1 所示。

堆积物存在首先会浪费库容体积，水不溶杂质部分堆积在腔底，本身会占据一定体积，再加上遇水膨胀，体积进一步扩大；后续进行注气排卤时，基于安全因素考虑会适当增大排卤管柱与堆积物顶面的距离，如此腔体库容因堆积物存在而损失了相当一部分的有效体

基金项目：本文为中国石油天然气集团公司项目"盐穴储气库腔底堆积物注气排卤扩容研究"（编号：2013B-3401-0503）的部分研究内容。

作者简介：任众鑫（1988—），硕士，工程师，从事储气库工艺技术研究工作。E-mail：zhongxren@petrochina.com.cn。

积；其次腔底存在堆积物还会使注气排卤管柱出现堵塞故障的风险大大增加；一方面上返卤水会携带部分不溶物颗粒进入排卤管，加上盐结晶及工艺因素影响，流通易产生不畅甚至堵塞；另一方面不溶物颗粒本身还会促进卤水结晶晶核的生成，进而加速卤水结晶，加快了管柱堵塞进程。另外，腔底堆积物还会对溶腔过程产生影响，其占用掉已成腔体积的同时也使腔体形状控制变得复杂。

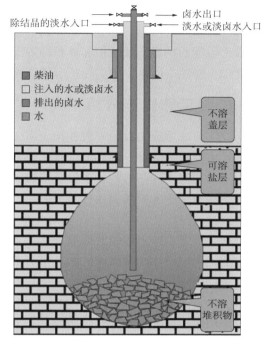

图 1 不溶物堆积示意图

2 堆积物扩容潜在体积评价

为对腔底堆积物进行有效处理或转变利用以实现腔体扩容，须对其扩容潜在体积完成科学评价。由于堆积物是盐层中不溶水部分沉降及膨胀所致，其分布势必会有一定的空隙体积，针对其堆积的体积参数进行分析和评价。造腔过程中，每阶段结束后会进行相应的声呐测腔，以获取腔体几何参数，为后续工程设计提供依据。将第 i 阶段声呐数据和测井数据中的井深进行合并、排序，去除重复数据，建立统一的井深序列表，得到井深系列下各层的层厚。根据微积分原理，采用线性插值方法得到任意井深 d_j 处空腔水平截面面积，对其在井深系列上进行积分并累加可得到目前成腔体积 V_{Ni}。公式如下：

$$V_{Nij} = \left[\frac{A_{(i,j)}+A_{(i,j+1)}}{2}\right] \times d_{zj} \tag{1}$$

$$V_{Ni} = \sum_{j=1}^{m} V_{Nij} \tag{2}$$

式中：$A_{(i,j)}$ 为任意井深 d_j 处空腔水平截面面积，m^2；$A_{(i,j+1)}$ 为任意井深 d_{j+1} 处空腔水平截面面积，m^2；d_{zj} 为 d_j 和 d_{j+1} 的深度差，m；m 为第 i 阶段声呐最大深度层数，1；V_{Nij} 为第 i 阶段声呐井深 d_j 处小层体积，m^3；V_{Ni} 为第 i 阶段声呐体积，m^3。

基于该方法，将历阶段声呐测腔数据进行合并、计算，得到造腔所溶掉的全部盐层原位体积 V，再得出最新成腔体积 V_N，两者之差即为堆积物分布体积 V_{st}，该体积值即为不溶物沉落并遇水膨胀后的分布体积。结合地层物性测井数据，由各层位不溶杂质含量，可计算得出腔体不溶物地层原位体积 V_{so}，则堆积物的空隙体积 V_s 可由 V_{st} 减去 V_{so} 得到。据相关研究，空隙中由于束缚水存在，有部分体积很难排出，因此求解可利用空隙体积 V_{sc} 时还要除去其束缚水的体积部分。

另外，由于排卤管柱下端与堆积物顶面间有一定距离，该深度范围内残留卤水体积 V_w 也可由上述方法求得。若通过循环携岩或取心等工艺将堆积物返至地面，则腔体体积可实现最大限度的扩容，此时扩容体积即等于堆积物分布体积 V_{st}。假定不携出堆积物，仅最大限度排出空隙中卤水，则实际扩容体积为其可利用空隙体积 V_{sc}。

3 扩容工艺优选分析

腔体形成后最大直径一般在中下部，70～100m。若通过钻井工程处理，采用取心或循环携岩将堆积物取出，预计上部携岩困难，施工效果不佳。鉴于堆积物空隙具有较好的连通性，可通过工具在堆积物中作业形成一定尺寸的坑孔，并增加排卤管柱下放的极限深度，尽可能多地将空隙中的卤水排至地面。

通过钻具钻进在腔底形成规则井眼，工艺环节上完全可行；但堆积物质地极为松软、胶结稳定性差，所形成的井眼易坍塌，并造成重复破碎或钻具埋卡，且作业管柱起出难重入。故可考虑采用冲洗技术方式，下入射流工具或连续油管，通过地面加压在腔底形成高速流场，将堆积物携离排卤管柱下端；在腔底形成坑孔后，下放排卤管柱以更多排出残余卤水，实现腔体扩容的目的。同时由于作业设备简单，施工费用预计较少。堆积物形成过程受多因素影响而成腔形状，故完腔后宜根据实测腔底形态优选扩容工艺方式，其中腔体底部中间高周围较低时更适宜采用冲洗作业方式进行扩容。

4 工艺应用及分析

4.1 试验准备

综合考虑现场生产及堆积物情况，在国内金坛某库区优选 A 井进行了工艺现场试验。对工艺流程及难点进行分析，明确试验相关安全隐患与风险，制定相应的应急措施预案。根据试验井具体情况及试验设计要求，检查与保养相关设备、工具与配件，按照相关作业 QHSE 标准开展井场准备工作，并完成试验前的地面配接和安装、调试，以及试验技术交底。根据历阶段声呐测腔数据，采用前述体积评价方法进行计算，作业前 A 井堆积物各体积参数见表 1。

表 1 作业前 A 井堆积物体积参数表

腔体层位/m	造腔阶段数	堆积物最大厚度/m	堆积物分布体积/m³	可利用空隙体积/m³
1025～1165	12	44.2	88000	21000

4.2 试验设备

结合试验井井身结构，考虑现场条件及施工情况，决定采用 4½in 油管柱。根据腔体底界深度和试验要求，组配连接试验管柱。基于作业提升载荷要求，优选作业设备。根据优选结果，试验采用 XJ-550 修井机，大钩载荷安全系数为 3.86。试验用冲洗工具如图 2 所示，工具底端有 3 个水眼，与轴线呈 10°夹角并均匀分布，工作液从水眼流出，破碎与冲蚀井底。

图 2　试验用冲洗工具图

4.3　试验过程

工具下放至距堆积物顶面以上一定距离，开泵循环，边冲洗边下放，注意观察管柱悬重变化，避免挂卡；单根冲洗过程中，缓慢上提、下放，活动作业管柱，并根据实时情况适当停泵分组间隔作业；下放至最大深度，加压反复遇阻无进尺后，起管柱进行声呐测腔，以评价试验效果。试验主要作业参数见表 2。

表 2　试验主要作业参数表

工具水眼尺寸	排量/(L/min)	泵压/MPa	工具水头压/MPa	比水功率/(W/mm^2)
8mm×3	1000	10	8.2	3.7

4.4　试验结果

根据声呐测腔数据，通过下入工具进行射流冲洗作业形成了有效的底部坑孔，使排卤管柱可多下放约 2m，据此进行试验效果分析。据分析结果，后续能充分利用堆积空隙体积并多排出残余卤水近 5000m^3，腔体实现了有效扩容。对于底部不溶物堆积呈严格倒 V 形的腔体，该工艺试验预计扩容效果更为明显。同时通过开展工艺试验，丰富工程实践经验，也为后续堆积物处理、腔体扩容建设提供了技术参考和借鉴。

4.5　成果分析

盐岩建库受制于多项因素，如地质、矿权、经济规划、建设成本及当地卤水消化能力等，在当前国内天然气调峰设施滞后、用气保供形势严峻的大环境下，建库思路的拓展提升和配套工艺技术的革新进步变得尤为重要。在不影响整体建设规划和既定工程质量的前提下，通过扩容工艺技术对已有腔体进行作业处理，实现单腔有效扩容，对于增加储库工作气量和满足国内用气需求具有一定的实际意义。

5　结论

（1）腔底堆积物的存在，对库容体积、建库施工和溶腔管理会产生影响，在一定程度上制约了工程建设。

（2）基于微积分原理和线性插值方法，对腔底堆积物的分布体积、空隙体积等进行了分析计算，得出了堆积物扩容体积的评价方法。

（3）在扩容工艺优选分析的基础上，基于冲洗技术思路，结合现场条件和实际情况，优选金坛某库区A井作为试验井进行了工艺现场试验，完成了试验效果和应用的分析。

参 考 文 献

[1] 杨毅，周志斌，李长俊．天然气地下储气库的经济性分析[J]．天然气技术，2007，1(1)：55-57．
[2] 李伟，杨宇，徐正斌，等．美国地下储气库建设及其思考[J]．天然气技术，2010，4(6)：3-5．
[3] 班凡生，耿晶，高树生，等．岩盐储气库水溶建腔的基本原理及影响因素研究[J]．天然气地球科学，2006，17(2)：261-266．
[4] 杜新伟．盐岩不溶物对储气库成腔影响[J]．石油化工应用，2016，35(1)：30-33．
[5] 姜德义，张军伟，陈结，等．岩盐储库建腔期难溶夹层的软化规律研究[J]．岩石力学与工程学报，2014，33(5)：865-873．
[6] 李国韬，霍永胜，郝国永．国外盐穴储气库盐腔残留水治理措施分析[J]．天然气技术与经济，2014，8(4)：48-50．

本论文原发表于《天然气技术与经济》2017年第11卷第5期。

金坛盐穴储气库现场问题及应对措施

李建君　巴金红　刘　春　何邦玉　李淑平　陈加松

（中国石油西气东输管道公司储气库项目部）

【摘　要】 金坛盐穴储气库采用单井双管油垫法分步式进行造腔。由于地下情况复杂，盐穴腔体在建造期间可能会遇到注水排卤井口泄漏、井筒泄漏、阻溶剂界面异常、造腔管柱故障及腔体泄漏等问题，若不能及时发现并采取应对措施，将会导致阻溶剂界面失控、造腔中断以及腔体体积损失等后果。金坛盐穴储气库通过在造腔实践中解决现场问题，形成了完善的造腔现场应对措施和监控体系，积累了处理造腔突发异常情况的经验，有效地保证了造腔的连续性，缩短了建腔周期，对后续盐穴储气库建设提供了理论依据和技术支持。

【关键词】 盐穴储气库；造腔；现场问题；应对措施；监控体系

Problems in the field of Jintan Salt-Cavern Gas Storage and their countermeasures

Li Jianjun　Ba Jinhong　Liu Chun　He Bangyu　Li Shuping　Chen Jiasong

(Gas Storage Project Department, PetroChina West-East Gas Pipeline Company)

Abstract In Jintan Salt-Cavern Gas Storage, caverns are built step by step by means of single-well dual-tubing oil cushion method. Due to the complicated underground conditions, salt cavities may encounter many problems during the construction of salt-cavern gas storage, such as wellhead leakage, wellbore leakage, abnormal blanket interface, leaching string failure and cavity leakage. If the problems cannot be found in time and measures are not taken correspondingly, some adverse consequences will occur, e.g. uncontrolled blanket interface, interrupted leaching and cavity volume loss. In Jintan Salt-Cavern Gas Storage, the on-site problems are solved in the leaching process. Thus, complete countermeasures and monitoring system for leaching sites are developed, the experience to deal with abrupt leaching anomaly is accumulated, the leaching continuity is ensured effectively and the leaching period is shortened. It provides the theoretical basis and technological support for the subsequent construction of salt cavern gas storages.

基金项目：中国石油储气库重大专项"地下储气库关键技术研究与应用"，2015E-40。

作者简介：李建君，男，1982年生，高级工程师，2008年硕士毕业于清华大学动力机械及工程专业，现主要从事盐穴储气库溶腔工艺及其优化、稳定性评价等研究工作。地址：江苏省镇江市润州区南徐大道商务A区D座中国石油西气东输管道公司储气库管理处，212000。电话：13814795191，E-mail：cqklijianjun@petrochina.com.cn。

Keywords salt-cavern gas storage; solution mining; on-site problems; countermeasures; monitoring system

盐穴储气库从开始建造到完腔需要经历一个相当长的过程[1]。金坛盐穴储气库主要采用单井双管油垫法分步式造腔，国内外学者对隔夹层垮塌机理[2-7]、腔体偏溶[8-10]、管柱变形损坏[11-12]等问题进行了相关研究，但对造腔过程中的现场问题研究很少。常见的现场问题有注水排卤井口泄漏、管柱堵塞和脱落、阻溶剂界面失控等。若造腔期间出现的问题不能及时被发现，并尽快找出原因采取相应的解决措施，则可能会导致造腔中断、腔体体积损失等后果。金坛盐穴储气库通过在造腔实践中解决现场问题，形成了完善的造腔现场应对措施和监控体系，可避免在后续造腔期间发生同类情况，最终达到加快造腔进度和尽快实现注气排卤的目的[13-19]，为后续储气库建设提供理论依据和技术支持。

1 造腔现场问题及应对措施

金坛盐穴储气库单井双管油垫法分步式造腔主要包括注水排卤井口、井筒、造腔管柱、阻溶剂界面及腔体5个主要部分(图1)。造腔期间任一环节出现问题，都可能会导致造腔中断。

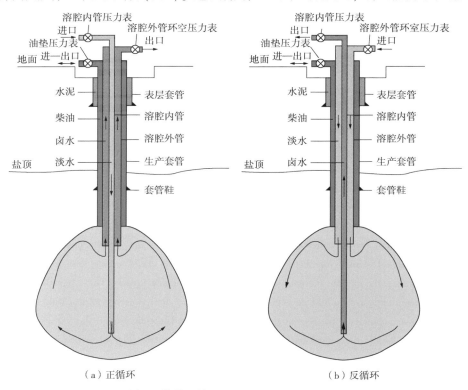

图1 单井双管油垫法分步式造腔示意图

1.1 注水排卤井口泄漏

造腔期间注水排卤井口除了注入淡水(或淡卤水)和排出卤水外，还有一个重要功能是向井下注入阻溶剂(如柴油)。为了确保造腔期间阻溶剂界面维持在设计界面附近，造腔时需要定期从井口向井下补注阻溶剂。通过对金坛储气库造腔井的井口密封情况进行跟踪，发现部分井在造腔期间注水排卤井口有阻溶剂泄漏情况，泄漏容易发生在177.8mm套管挂处。当井口发生泄漏时，现场应及时确认泄漏部位，对泄漏部位重新密封或更换井口相关元件。

1.2 井筒泄漏

井筒泄漏主要存在于两个时期：一是在造腔之前，主要是套管固井质量不合格或裸眼段漏失所致；二是在造腔期间，主要是生产套管鞋附近盐岩被溶蚀、固井套管局部损坏或造腔盐层段存在漏失层所致。造腔之前新钻井必须安排气密封试压测试，试压不合格造腔井将不允许开井造腔，金坛储气库新钻井A-2井就曾出现气密封试压不合格的情况，后经过多次气密封试压确认井筒不存在泄漏后才开井造腔；造腔期间一旦出现井筒泄漏，若发现不及时，阻溶剂将大量漏失导致阻溶剂界面失控引起腔体形状失控，现场可停井安排腔体密封性测试，确认井筒是否存在泄漏，测试不合格将不予进行造腔，试压介质采用卤水或氮气。

1.3 阻溶剂界面失控

阻溶剂界面控制是盐穴储气库建设非常关键的因素之一[20]。金坛储气库使用柴油作为阻溶剂，界面失控也是常见的问题。造腔期间虽定期补注阻溶剂并检测阻溶剂界面位置，但由于地下存在很多不确定因素[腔顶直径大、腔顶形状不规则、盐层(隔夹层)垮塌、阻溶剂漏失等]都会导致阻溶剂界面失控。金坛储气库虽已实施了阻溶剂界面监控措施，但目前不能实时监测，因此造腔期间阻溶剂界面失控时有发生。当阻溶剂界面上移达到10m以上时，应该立即停井，确认阻溶剂无泄漏并将界面调整至设计界面附近后才能开井，开井后需要增加阻溶剂界面检测频次。

1.4 造腔管柱故障

金坛盐穴造腔组合采用单井双管管柱结构，$\phi 114.3$ mm造腔内管一般下入腔体中下部，下部管柱悬空段长度为$30\sim 100$m，$\phi 177.8$mm造腔外管与造腔内管同心，位于造腔内管底端上部15m以上。造腔管柱故障主要出现在造腔内管[21-23]，如管柱弯曲变形、堵塞及脱落等。造腔外管泄漏将导致阻溶剂界面大量漏失而引起阻溶剂界面失控，目前金坛储气库尚未出现造腔外管泄漏，一旦出现造腔外管泄漏，应该立即停井并安排井下作业，更换泄漏管柱。

造腔内管弯曲变形较常见，尤其是在盐穴腔体建槽期，一般对腔体形状影响不大。建槽期间造腔管柱被腔底沉积不溶物掩埋$10\sim 20$m，管柱底端在注水期间长期受力不均或厚隔夹层垮塌会导致造腔管柱弯曲变形。在各个造腔阶段结束后的作业期间对变形管柱进行更换，根据管材使用年限将上提遇阻控制在$10\sim 20$t，当达到最大上提界限仍不能解决遇阻问题时，现场可考虑进行管柱割管作业，防止内管上提过程中脱落并卡在外管内。造腔内管堵塞多发生在反循环造腔过程中，当排卤浓度较高且长时间停井时，管柱内局部会出现盐结晶严重或管柱被不溶物掩埋较深的情况。卤水中携带的不溶物在管柱内积聚导致堵塞，内管堵塞严重时会导致造腔中断，当造腔内管发生严重堵塞且采用淡水反冲洗无效时，现场需要安排井下作业解堵，并更换遇阻管柱。造腔内管脱落在工程现场并不常见，但是一

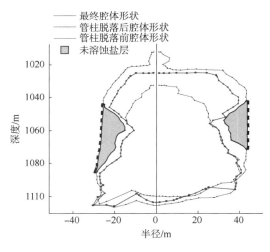

图 2 造腔内管脱落腔体形状发展示意图

且脱落将造成非常严重的后果，其主要原因是管柱丝扣上扣不牢或造腔时被垮塌不溶物砸中，大段管柱脱落会导致腔体形状发展不规则且腔体局部盐层溶蚀不充分，从而使得完腔体积达不到设计要求(图2)。可通过磁性定位(CCL)测井判断管柱落井情况，若确定发生管柱落井，则需要安排井下作业重新下入管柱。

1.5 腔体泄漏

建库盐层段存在断层、微裂缝等会导致腔体泄漏。盐岩密封性好，盐穴腔体建造期间很少发生腔体泄漏，泄漏会导致盐穴不能投入生产而被废弃，一旦出现腔体泄漏，应尽早处理，以免造成较大经济损失，目前金坛储气库还未出现腔体泄漏情况。

2 造腔现场问题预防和监控体系

2.1 筛查断层和漏失层

为预防造腔时腔体泄漏，应仔细筛查建库区断层和建库层段内漏失情况，并根据已有钻井、造腔井等资料不断完善整个库区地质资料，确保造腔时腔体避开断层，且建库层段无漏失。一般要求造腔井距离断层100m以上，并严格控制腔体与邻腔、断层之间的矿柱宽度不小于设计值。

2.2 检查密封性和完整性

为防止造腔期间注水排卤井口出现泄漏，现场维护人员应定期对井口装置进行维护保养(紧固螺栓、注脂等)，并安排井口取样。一旦发现井口存在泄漏，应及时对井口泄漏部位重新注脂密封或更换密封元件，以确保其密封性良好。实施造腔管柱下入作业时，检查每根管柱本体和丝扣是否完好，不合格管柱禁止下入井下，保证每根下入造腔管柱的密封性和完整性，并将每个造腔阶段的不溶物隔夹层数控制在3~5层，避免隔夹层大块垮塌损坏管柱。

2.3 定期提取动态参数

造腔时应尽量保持排量稳定，减少停井，及时掌握造腔井造腔动态，保证各井造腔正常有序进行。造腔期间每天安排现场井口卤水取样，并提取动态参数，如注入压力、油垫压力、排卤浓度、排量等，每周对现场提取数据进行造腔跟踪分析，以便发生造腔异常能尽早发现。通过跟踪现场造腔动态参数，可以及时发现阻溶剂界面失控、管柱堵塞、阻溶剂漏失等异常情况。

2.4 完善监控方法

目前金坛储气库常用阻溶剂界面监控方法主要包括地面观察和中子测井，其中地面观察仅局限于正循环造腔,且要求阻溶剂界面深度处于造腔外管管鞋处;中子测井费用较高,不能实时监测阻溶剂界面。光纤界面仪是一种新型的界面监测方法,将复合光缆沿造腔外管壁一并下入井中,能实时监测阻溶剂界面位置,且监测范围广、精度高,同时可减少阻溶剂使用量。通过光纤界面仪监控阻溶剂界面,能够及时发现造腔时阻溶剂界面是否异常,避免上部盐层过早打开,故应全面推广光纤界面仪在阻溶剂界面监控中的应用。

3 结束语

金坛盐穴储气库造腔现场遇到注水排卤井口泄漏、阻溶剂界面失控、造腔管柱故障等问题较多,在造腔生产实践中建立了完善的造腔现场预防和监控体系,及时发现并处理造腔期间的突发问题,有效地保证了造腔的连续性,减少了停井时间,缩短了造腔周期,积累了宝贵的现场经验。金坛盐穴储气库造腔工艺技术具有重要的借鉴意义,使单腔建库周期达到最优值,可大大加快盐穴储气库建设进程,可为后续盐穴储气库建设提供有意义的指导。

参 考 文 献

[1] 李铁,张永强,刘广文. 地下储气库的建设与发展[J]. 油气储运,2000,19(3):1-8.

[2] Charnavel Y,O'donnell J,Ryckelynck T. Solution mining at Stublach[C]. New York:SMRI Spring 2015 Technical Conference,2015:1-13.

[3] Devries K L,Mellegard K D,Callahan G D. Laboratory testing in support of a bedded salt failure criterion [C]. Wichita:SMRI Spring 2004 Technical Conference,2004:1-23.

[4] 袁光杰,班凡生,赵志成. 盐穴储气库夹层破坏机理研究[J]. 地下空间与工程学报,2016,12(3):675-679.

[5] 施锡林,李银平,杨春和,等. 卤水浸泡对泥质夹层抗拉强度影响的试验研究[J]. 岩石力学与工程学报,2009,28(11):2301-2308.

[6] 施锡林,李银平,杨春和,等. 盐穴储气库水溶造腔夹层垮塌力学机制研究[J]. 岩土力学,2009,30(12):3615-3620.

[7] 袁炽. 层状盐岩储库夹层垮塌的理论分析[J]. 科学技术与工程,2015,15(2):14-20.

[8] 杨海军,于胜男. 金坛地下储气库盐腔偏溶与井斜的关系[J]. 油气储运,2015,34(2):145-149.

[9] 郭凯,李建君,郑贤斌. 盐穴储气库造腔过程夹层处理工艺——以西气东输金坛储气库为例[J]. 油气储运,2015,34(2):162-166.

[10] 李建君,王立东,刘春,等. 金坛盐穴储气库腔体畸变影响因素[J]. 油气储运,2014,33(3):269-273.

[11] Bruno M,Dorfmann A. Well casing damage above compacting formation[C]. Texas:SMRI Spring 1996 Meeting,1996:1-9.

[12] 戴鑫,马建杰,丁双龙,等. 金坛盐穴储气库JT1井造腔异常情况分析[J]. 中国井矿盐,2015,46(1):26-29.

[13] 杨海军,李建君,王晓刚,等. 盐穴储气库注采运行过程中的腔体形状检测[J]. 石油化工应用,2014,33(2):22-25.

[14] 杨海军,霍永胜,李龙,等. TJ盐穴地下储气库注采气运行动态分析[J]. 石油化工应用,2010,29(9):61-65.

[15] 杨海军，郭凯，李建君. 盐穴储气库单腔长期注采运行分析及注采压力区间优化——以金坛盐穴储气库西2井腔体为例[J]. 油气储运，2015，34(9)：945-950.

[16] 李建君，陈加松，吴斌，等. 盐穴地下储气库盐岩力学参数的校准方法[J]. 天然气工业，2015，35(7)：96-102.

[17] 杨海军. 中国盐穴储气库建设关键技术及挑战[J]. 油气储运，2017，36(7)：747-753.

[18] 李建君，陈加松，刘继芹，等. 盐穴储气库天然气阻溶回溶造腔工艺[J]. 油气储运，2017，36(7)：816-824.

[19] 刘继芹，刘玉刚，陈加松，等. 盐穴储气库天然气阻溶造腔数值模拟[J]. 油气储运，2017，36(7)：825-831.

[20] 胡开君，巴金红，王成林. 盐穴储气库造腔井油水界面位置控制方法[J]. 石油化工应用，2014，33(5)：59-61.

[21] 李银平，杨春和，屈丹安，等. 盐穴储油(气)库水溶造腔管柱动力特性初探[J]. 岩土力学，2012，33(3)：681-686.

[22] 李银平，葛鑫博，王兵武，等. 盐穴地下储气库水溶造腔管柱的动力稳定性试验研究[J]. 天然气工业，2016，36(7)：81-87.

[23] 郑东波，黄孟云，夏焱，等. 盐穴储气库造腔管柱损坏机理研究[J]. 重庆科技学院学报(自然科学版)，2016，18(2)：90-93.

本论文原发表于《油气储运》2017年第36卷第8期。

Applications of numerical simulation in fault recognition during dissolving mining for gas storage caverns

Huiyong Song[1]   Jianjun Li [2]

(1. Department of Information Management, School of Economics and Management, Nanjing University of Chinese Medicine; 2. Department of Gas Storage Project, PetroChina West East Gas Pipeline Company)

Abstract   In the solution mining process of gas storage salt caverns, the insoluble layers collapse may lead to cavern leaching string damage. If the damage cannot be found in time, the string damage can make the effective volume loss and the shape unstable. Combining with field data analysis, a mathematical model to fast predict withdrawal brine density, shape and volume of cavern construction was built, and compiled simulation programs. Through numerical simulation, the withdrawal brine density, cavern shape and volume can be simulated fast and accurately, which can prevent great financial losses and potential safety hazard induced by the string damage.

Keywords   numerical simulation; salt cavern; underground gas storage; solution mining for gas storage cavern

1 Introduction

Salt cavern is internationally accepted as the optimal medium for oil and gas storage because of reliable leak-tightness and stable chemical property[1-3]. The reserve size of underground gas storage may reach several million or even hundreds of millions of cubic metres at a lower cost than other reserve ways. The mature technology of single well oil pad method is usually put into use during the solution mining for cavern construction, of which the cost soars and the shape and volume of the cavern is demanding[4-7]. All the ongoing and planned gas storages at home contain insoluble inter-layers, of which the thickness ranges from 2 to 12cm. These inter-layers may collapse, fracturing strings for cavern building, especially the inner ones. If not found in time, they may cause salt deposit losses below break points. The last phase of cavern building can even see a flat top, which may cause financial losses and potential safety hazard of the gas operation. Fig. 1 below is the sonar profile of a salt cavern. Since the inner strings fall off during the cavern building, the main part of the cavern(1,040-1,080m) doesn't meet the maximum volume, which affects the cavern shape to a degree as well.

Corresponding: Huiyong Song, njsonghuiyong@126.com; Jianjun Li, qklijianjun@petrochina.com.cn.

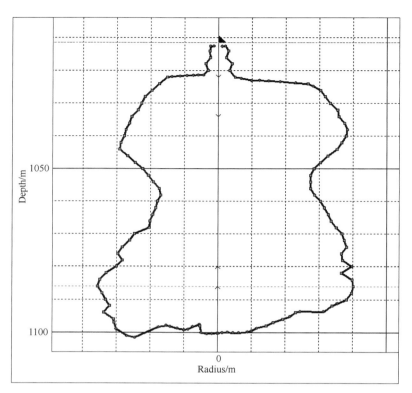

Figure 1　The Sonar Profile of A Salt Cavern

According to the long-term field data analysis, the density of withdrawal brine is a good indication of whether the down-hole strings fall off. However, thedensity anomaly is difficult to find and will lead to miscalculation once missed[8-13]. Therefore, if the density of withdrawal brine under normal circumstances can be forecasted fast and correctly and compared with those in fault, the accuracy of finding the down-hole fault will rise significantly, preventing great financial losses.

In response to the circumstances, what the article presents is a fast two-dimension model of solution mining for gas storage caverns, which calculates the effects of salt solution, convection and diffusion accurately, and takes the impact of the uneven distribution of insoluble substances into consideration as well. On this basis, the program, compiled by the use of computer numerical simulation, can stimulate the withdrawal brine density and cavern shape development fast and accurately.

2　The mathematical model deduction of cavern building

2.1　Basic underlying assumptions

To stimulate all the data with computer, we divide the whole cavern into several small circular tables with equal altitude, as shown in Figure 2 below.

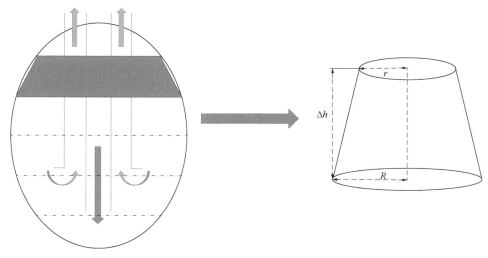

Figure 2 The discretization model of the cavern

Fresh water or weak brine enter the cavern through the injecting pipe end and dissolve the salt rock on the cavern walls, changing their shape. At the same time, as the brine density differs (low upside and high downside), the diffusion of the brine shall be seen. In addition, the water that entered the cavern will keep rising to the cavern top, forming the plume and forcing the brine to flow down to the outlet. Figure 3 shows the convection in the cavern, Q_i^p means the plume volume upward from Cell i, Q_i^d means the quantity of the flow shunted from plume Q_i^p to dissolve the salt rock on the cavern walls in Cell $i-1$, and Q_i^t means the quantity of the flow that goes down out of Cell i.

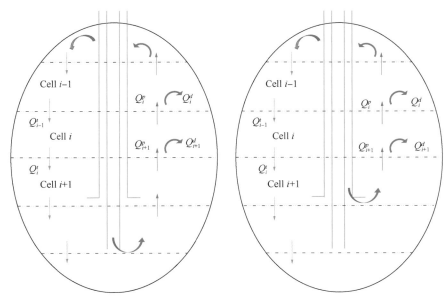

Figure 3 Convection term of the simulation model (Direct mode on the left and reverse on the right)

In conclusion, the salt drifting in and out in every cell could get divided into three kinds: that dissolved on the cavern walls, that drifting through convection and that through diffusion. Under the consumption that the parameters such as the cavern volume and temperature stand still during the period of dt, the calculation on the changes of the brine density in cells is as follows:

$$\underset{\text{Salt increment}}{\dfrac{\mathrm{d}V\mathrm{d}C}{}}=\underset{\text{Solution salt}}{\dfrac{M_s\mathrm{d}t}{}}+\underset{\text{Flow into salt}}{\dfrac{(C_i-C_o)Q\mathrm{d}t}{}}+\underset{\text{Diffusion salt}}{\dfrac{D\left(A\dfrac{\mathrm{d}C}{\mathrm{d}h}\right)_i\mathrm{d}t}{}}-\underset{\text{Spread out salt}}{\dfrac{D\left(A\dfrac{\mathrm{d}C}{\mathrm{d}h}\right)_o\mathrm{d}t}{}}$$

This equation is deduced as follows:

$$\dfrac{\mathrm{d}C}{\mathrm{d}t}=\dfrac{M_s}{\mathrm{d}V}+\dfrac{(C_i-C_o)Q}{\mathrm{d}V}+\dfrac{D\left(A\dfrac{\mathrm{d}C}{\mathrm{d}h}\right)_i-D\left(A\dfrac{\mathrm{d}C}{\mathrm{d}h}\right)_o}{\mathrm{d}V}$$

If dh is infinitesimal, the circular tables will be taken as approximate research units, so the equation turns out to be:

$$\dfrac{\mathrm{d}C}{\mathrm{d}t}=m_s+u\dfrac{\mathrm{d}C}{\mathrm{d}h}+D\dfrac{\mathrm{d}^2 C}{\mathrm{d}h^2} \tag{1}$$

Formula (1) means the equation of density changes in the salt caverns, which belongs to convection diffusion equations. In the equation, ms means the rate of salt solution (kg/s), u the flow velocity (m/s) and D the diffusion coefficient (m^2/s). Ulteriorly the discrete version of the salt balance in Cell i during the adjacent time comes as follows:

$$c_i^{t+1}-c_i^t=sol+con+dif=\dfrac{M_{si}^t}{\Delta V_i^{t+1}}+\dfrac{M_{ci}^t}{\Delta V_i^{t+1}}+\dfrac{M_{di}^t}{\Delta V_i^{t+1}} \tag{2}$$

The right side of the equal sign witnesses salt solution term, convection term and diffusion term successively. And here comes the calculation of the three terms.

2.2 Salt solution term

Point P_n on the cavern wall is chosen as the research unit and it has dissolution velocity and angle at both vertical and horizontal directions.

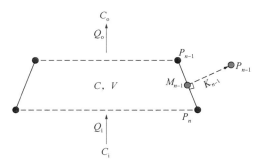

Figure 4 Simplified model of solution-mined cavities

In Fig. 4, P_{n-1} and P_n are the points on the cavern wall while M_{n-1} and M_n are corresponding midpoints. During the cavern construction, $P_{n-1}P_n$ is taken as a small unit with the same solution rate K_{n-1}. During the period of dt, the cavern wall moves to $P_{n-1}P_n$, as shown in Fig. 4.

K is taken as the solution rate on the cavern wall, and the fitting formula is as follows according to a large number of solution experiments:

$$K = k[1+0.0262 \cdot (T-20)] \cdot (C_s-C)^{\frac{3}{2}} \cdot \sqrt{C_s} \quad (3)$$

In which there is

$$k = \begin{cases} 0, & 0 \leq \alpha \leq \alpha_{min} \\ [k_h - (k_v-k_h) \cdot \sqrt{\cos\alpha}] \cdot \left[1-\left(\frac{\tan\alpha_{min}}{\tan\alpha}\right)^2\right], & \alpha_{min} < \alpha \leq 90° \\ k_h + (k_v-k_h) \cdot \sqrt{-\cos\alpha}, & 90° < \alpha \leq 180° \end{cases} \quad (4)$$

Where, K means the dissolving rate on the cavern wall m(m/h), T the temperature(℃), C the brine density(g/L), C_s the brine salinity(g/L), α the included angle between the cavern wall and the water level, α_{min} the lateral solution angle(the cavern wall lower than the angle will not be dissolved). The vertical dissolving rate k_v and horizontal solution rate k_h should get adjusted during the practical operation, and here is a set of reference value: $k_h = 2.22 \times 10^{-11}$ m/s/(g/L)^2; $k_v = 3.89 \times 10^{-11}$ m/s/(g/L)^2.

According to the analysis above, the calculation for salt dissolving term of every cell is as follows:

where, the salt dissolving rate is $\Delta m_s = (\Delta V_i^{t+1} - \Delta V_i^t) \cdot \rho_s$, ρ_s is the salt density(kg/m^3); α_i is the average insoluble content(%) on the stratum at the depth of Unit i, ΔV_i^{t+1} is the volume of Unit i at time $t+1$(m^3):

$$\Delta V_i^{t+1} = \frac{\pi \cdot \Delta h}{3} \cdot [(r_i^t + K \cdot \Delta t)^2 + (r_i^t + K \cdot \Delta t) \cdot (R_i^t + K \cdot \Delta t) + (R_i^t + K \cdot \Delta t)^2]$$

where, r_i^t and R_i^t are the radiuses of the upper surface and the lower surface of the circular tables respectively at time t(m^3).

2.3 Diffusion term

Brine drifts between these cells because of their salinity differences. The related calculation is as follows:

$$dif = D \cdot A_i^t \cdot \frac{(c_{i+1}^{t-1} - c_i^{t-1}) - (c_i^{t-1} - c_{i-1}^{t-1})}{\Delta h^2} \Big/ \left(\frac{\Delta V^{t+1}}{\Delta t}\right)$$

$$= D \cdot A_i^t \cdot \frac{(c_{i+1}^{t-1} - 2c_i^{t-1} + c_{i-1}^{t-1}) \cdot \Delta t}{\Delta h^2 \cdot \Delta V^{t+1}}$$

where D means the diffusion coefficient, c_i^t the brine salinity of unit i at time t under average environment(g/L), Δh the height of the unit(m), A_i^t the area of the lower section at time t(m^2), $A_i^t = \pi \cdot (R_i^t)^2$.

Specially, in the unit of cavern top the brine can't continue to spread upward, therefore the diffusion term comes as

$$dif = D \cdot A_i^t \cdot \frac{(c_{i+1}^{t-1} - c_i^{t-1}) \cdot \Delta t}{\Delta h^2 \cdot \Delta V^{t+1}}$$

Similarly, in the unit of cavern bottom the brine can't continue to spread downward, then the diffusion term comes as

$$dif = D \cdot A_i^t \cdot \frac{(c_{i-1}^{t-1} - c_i^{t-1}) \cdot \Delta t}{\Delta h^2 \cdot \Delta V^{t+1}}$$

2.4 Convection term

According to the convection flow model in Fig. 4 the convection term of Cell i is

$$con = \frac{(Q_{i+1}^p c_{i+1}^p - Q_i^p c_i^p) + (Q_{i-1}^t c_{i-1}^t - Q_i^t c_i^t)}{\Delta V_{i+1} / \Delta t}$$

First, based on the flow conservation in the cells here comes the equation set:

$$\begin{cases} Q_i^d + Q_{i-1}^t - Q_i^t = \Delta V_i / \Delta t \\ Q_i^d + Q_i^p = Q_{i+1}^p \end{cases}$$

And the equation set is solved as

$$Q_i^t = Q_{i+1}^p - Q_i^p + Q_{i-1}^t - \Delta V_i / \Delta t$$

The factor of Local Mixing Ratio is used in the calculation of plume Q^p, and the upward recursion starting from the cell at the injecting pipe end shows the relationship between LMR and Q^p as below:

$$LMR_i = \frac{Q_i^p - Q_{i+1}^p}{Q_{i+1}^p} = \Gamma(\bar{\rho}_{i+1} - \rho_{i+1}^p) \cdot (Q_{i+1}^p)^m$$

Thereinto, $\bar{\rho}_{i+1}$ means the brine density of unit i at time t under average environment and ρ_{i+1}^p the plume density, $m = -0.27$, so Γ can get calculated through the formula below:

$$\Gamma = C_i(\bar{\rho}_{inlet} - \rho_{inlet}^p) \cdot (Q_{in})^n$$

Thereinto, $C_1 = 220$, $\bar{\rho}_{inlet}$ means the density of injection fluids, ρ_{inlet}^p the density of the fluids floating up through the cells at the injecting pipe end, Q_{in} the volume of injection fluids and $n = -0.10$. The brine density in the two formulas above can be got through table lookup, so the formula of the plume is

$$Q_i^p = (1 + LMR_i) Q_{i+1}^p$$

Under the initial situation there is

$$Q_i^p \Big|_{i=0} = (1 + LMR) Q_{in} = [1 + \Gamma(\bar{\rho}_{i+1} - \rho_{i+1}^p) \cdot Q_{in}^m] \cdot Q_{in}$$

Then the distribution of propellant flow Q^t can get calculated, and so the convection term can be got as well.

This section content combined, the accurate cavern shape and the accurate distribution of brine in the cavern can get calculated at any time during the cavern construction (the radiuses of the upper surface and the lower surface of the circular tables of every cell), and so do the changes of the cavern volume and the density of the withdrawal brine.

3 Numerical simulation and model verification

Based on the above model, using Visual Studio 2010 and C# computer language, computer simulation program was compiled.

The model validation compares the truthful data from three wells (Well 1, Well 2 and Well 3) with numerical simulation results in order to verify the validity and veracity of the model of dissolving mining for gas storage caverns mentioned in this article.

3.1 Leached cavern shape

Figure 5 shows the shapes of the three wells after the dissolving mining process, in which black lines mean the actual cavern walls and red means the simulated ones. According to Fig. 6, the two results are closed to each other.

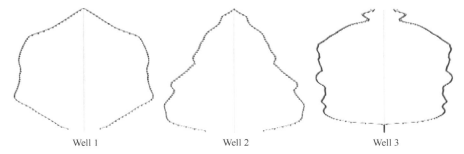

Figure 5 Real and simulated shapes of the 3 cavities after dissolving-mined

Figure 6 is the photo of falling-off inner leaching pipe during some salt cavern construction process. Under the knocking and extruding of the collapsing insoluble interlayers, the pipe screw threads get crumpled seriously, leading pipe-columnsto fall in cavern. The brine density is got stimulated under regression simulation with author-developed numerical program and the difference is easy to find when compared with the real brine density (See Fig. 7). The real brine density sees a sharp fall compared with the stimulated brine density after sampling point 13, while other factors witness no changes, which means that the damaged or falling-off down-hole pipe-columns diminish the distance between the two ends of the pipe, lowering the withdrawal brine density.

Figure 6 Photo of damaged inner leaching pipe

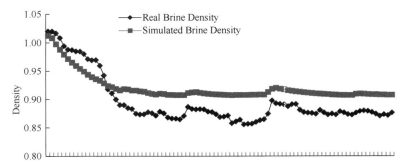

Figure 7　Comparison of the simulated brine density and the real brine density

4　Conclusions

The research above shows that the stimulated shape of the model during the dissolving mining for gas storage caverns and the results tested by sonar match, which means the mathematical model of cavern construction is scientific and reasonable. This article sees a real down－hole fault under regression simulation based on the numerical program that comes from the related mathematical model. The result comes out that down－hole faults can get estimated accurately by the use of this program, proving that the research here holds significant guidance for down－hole fault recognition.

Acknowledgments

The research work was supported by National Natural ScienceFoundation of China under Grant No. 51174170 and 51074137.

References

[1] Chen Weizhong, Wu Guojun, Dai Yonghao, et al. Stability Analysis of Abandoned Salt Caverns Used for Underground Gas Storage. Chinese Journal of Rock Mechanics and Engineering, 2006, 25(4): 848-854.
[2] Li Jianzhong. Salt Cavern Underground Gas Storage for Petroleum Stategic Reserve[J]. Chinese Journal of Rock Mechanics and Engineering, 2004, 23(S2): 4787-4789.
[3] He Aiguo. Building Technology for Salt Cavern Gas Storage[J]. Natural Gas Industry 2004, 24(9): 122-125.
[4] Jianjun Liu, Rui Song, and Mengmeng Cui, "Numerical Simulation on Hydromechanical Coupling in Porous Media Adopting Three-Dimensional Pore-Scale Model," The Scientific World Journal, vol. 2014, Article ID 140206, 8 pages, 2014. doi: 10.1155/2014/140206
[5] Ban Fansheng, Gao Shusheng(2007). Salt Water Soluble Gas Sorage Built Cavern Optimization Design Research. Natural Gas Industry, (2), pp. 114-116.
[6] Jianjun Liu, Zheming Zhu, and Bo Wang, "The Fracture Characteristic of Three Collinear Cracks under True Triaxial Compression," The Scientific World Journal, vol. 2014, Article ID 459025, 5 pages, 2014. doi: 10.1155/2014/459025
[7] Li Qingyang, Wang Nengchao, Yi Dayi. Numerical Analysis[M]. Tsinghua University Press, 2008.
[8] Jianjun Liu, Linzhi Zhang, and Jinzhou Zhao, "Numerical Simulation on Open Wellbore Shrinkage and Casing Equivalent Stress in Bedded Salt Rock Stratum," The Scientific World Journal, vol. 2013, Article ID 718196,

5 pages, 2013. doi: 10.1155/2013/718196
[9] Jianjun Liu and Qiang Xiao, "The Influence of Operation Pressure on the Long-Term Stability of Salt-Cavern Gas Storage," Advances in Mechanical Engineering, vol. 2014, Article ID 537679, 7 pages, 2014. doi: 10.1155/2014/537679
[10] Liu Jianjun, Li Quanshu. Numerical Simulation of Injection Water Flow through Mudstone Interlayer in Low Permeability Oil Reservoir Development. Disaster Advances, 2012, 5(4): 1639-1645
[11] Liang Weiguo, Yang Chunhe, Zhao Yangsheng(2008). Physical-Mechanical Properties and Limit Operation Pressure of Gas Deposit in Bedded Salt Rock. Chinese Journal of Rock Mechanics and Engineering, (1), pp. 22-27

本论文原发表于《EJGE》2014年第19期。

注采运行与安全监测

盐穴储气库运行损伤评价体系

敖海兵　陈加松　胡志鹏　齐得山　付亚平　任众鑫　刘继芹　张新悦

（中国石油西气东输管道公司储气库项目部）

【摘　要】 腔体损伤评价是盐穴储气库安全运行涉及的重要问题之一。通过对盐岩力学特性开展室内系统试验，利用损伤力学、岩石流变学理论及有限元数值模拟方法，研究了盐穴储气库运行过程中腔体损伤评价体系。盐岩损伤评价方法包括最大拉应力准则、最大剪应力准则、最大畸变能准则、膨胀强度准则、蠕变强度准则。利用 Abaqus 软件对所关注盐穴的评价指标进行数值模拟计算，并将计算结果与声呐实测结果对比，表明：所采用的腔体损伤评价体系能够很好地对腔体注采运行过程中的损伤破坏进行评价，可为今后盐岩损伤评价预测及控制提供参考依据。

【关键词】 储气库；盐岩；损伤；声呐；评价

Study on the damage assessment system of salt-cavern gas storage

Ao Haibing　Chen Jiasong　Hu Zhipeng　Qi Deshan　Fu Yaping
Ren Zhongxin　Liu Jiqin　Zhang Xinyue

（Gas Storage Project Department, PetroChina West-East Gas Pipeline Company）

Abstract　The cavity damage evaluation is one of the important problems involved in safe operation of salt-cavern gas storage. By virtue of damage mechanics, rock rheology theory and finite element numerical simulation, the evaluation system of cavity damage during the operation of salt-cavern gas storage was investigated by performing laboratory system tests on the mechanical properties of salt rocks. The methods for evaluating salt rock damage include maximum tensile stress criterion, maximum shear stress criterion, maximum distortional energy criterion, expansion strength criterion and creep strength criterion. The evaluation index of the concerned salt caverns was numerically simulated and calculated using Abaqus software, and the calculation results were compared with the measured sonar values. It is shown that the cavity damage evaluation system adopted in this paper can be well used to evaluate the damage in the process of gas injection and withdraw, and it provides the basis for the assessment,

---

基金项目：中国石油储气库重大专项子课题"地下储气库关键技术研究与应用"，2015E-4008。

作者简介：敖海兵，男，1988年生，工程师，2013年硕士毕业于西南石油大学矿产普查与勘探专业，现主要从事盐穴地下储气库稳定性评价的相关研究。地址：江苏省镇江市中国石油西气东输管道公司储气库项目部，212000。电话：13775532364，E-mail：xqdsahb@petrochina.com.cn。

prediction and control of salt rock damage in the future.

Keywords　gas storage；salt rock；damage；sonar；assessment

　　盐岩具有稳定的力学性能、较低的渗透性、良好的蠕变性、较强的损伤自愈性，能够保证储气库的密闭性，适应储气库运行压力的不断变化，因此被国内外公认为天然气储存的理想介质[1]。但是，在储气库建造和运行过程中涉及的安全性问题却不容忽视，否则可能引发重大的生产事故。例如，法国 Tersanne 盐穴储气库由于运行压力不合理，腔体运行10 年后因体积减小 30% 而失效；阿尔及利亚 Berkaoui 盐丘由于地表沉降等原因，在地表发生塌陷，形成一个直径为 350m 的天坑；美国 Brenham 盐丘由于储存在盐穴内的 LPG 通过井泄漏到地表，引发严重的爆炸事故[2]。

　　盐穴储气库的安全性主要包括腔体围岩无损伤破坏和腔内天然气无泄漏两个方面。盐岩损伤破坏是由其内部各种缺陷(微裂缝、微孔洞)相互作用、扩展，最终贯通成宏观断裂的过程，研究盐岩损伤破坏对于保证腔体安全性很有必要。盐岩损伤需要通过一些本构关系来描述，随着研究的深入，盐岩的本构模型从早期的弹性模型到塑性模型再到流变模型，更加真实地反映出岩盐的实际力学性质[3-13]。SMRI(Solution Mining Research Institute)等国际知名学术机构和国内外专家针对模拟岩盐力学性能的本构模型做过大量研究[14-24]，目前比较流行的本构模型有一般指数模型、Lemaitre 模型及 MDCF 模型。Chan 等[25]提出盐岩蠕变、损伤断裂多机制耦合 MDCF 本构模型；陈卫忠等[26]提出基于 Burges 模型的盐岩非线性蠕变损伤本构方程和损伤演化方程；韦立德等[27]利用对应性原理建立了基于细观力学的盐岩蠕变损伤本构模型；胡其志等[28]在 Bingham 蠕变模型中引入一个非线性函数，通过引入损伤到岩石的加速蠕变阶段，建立了考虑温度损伤的蠕变本构方程；丁靖洋等[29]通过构建 Abel 黏壶本构关系替代西原模型中牛顿黏壶的方法，求解获得了盐岩分数阶流变本构模型。

　　目前，对于盐岩损伤的研究主要集中在本构模型和数值模拟方法等方面，盐岩损伤评价体系尚未见报道。通过对盐岩力学特性开展室内系统试验和数值模拟评价，利用损伤力学、岩石流变学理论及有限元数值模拟的方法，开展对盐穴储气库运行过程中腔体损伤评价预测方法的研究，同时结合盐穴储气库多次声呐带压测腔实测的盐穴腔体形状和体积数据，对评价结果进行回归分析，总结出了一套科学合理的盐穴储气库注采运行过程中腔体损伤评价预测及控制体系。

1　声呐带压测腔

　　盐穴储气库声呐带压测腔为近年来国内应用的一项新兴技术，声呐带压测腔结果能够真实有效地反映腔体当前体积和形状状态。腔体投产注采运行后，为了检测地下储气库腔体形状，监测腔体的稳定性，进行声呐带压测腔。由于施工难度大、危险系数高、成本费用高等原因，到目前为止，金坛盐穴储气库仅完成 4 个盐穴的声呐带压测腔，下面选取 2 个具有代表性的腔体进行分析。

　　盐穴 A 是中国金坛盐穴储气库的第一个盐腔，整个建库过程经历了钻井、造腔、注气排卤、注采运行 4 个阶段。2005 年 5 月 27 日开始造腔，2008 年下半年完成地下盐层的腔体建造，该腔体完腔后声呐实测体积为 116508m^3。2010 年 11 月 28 日注气投产，经历了多个注采循环，由于采气最主要是满足季节性调峰的需求，因此采气大多在冬夏两季进行，根据用户对气量的需求不同，腔体的采气率及最低压力也不同，故每个注采循环不同。腔体采气操作完成后通常立刻进行注气操作，采气率一般比注气率要大，因此腔体压力停留在循环最低压的时间很短，而停留在最高压的时间较长，且闲置期一般保持在最高压。腔体载荷加载过程中，最高压为16.81MPa，最低压为8.94MPa(图1)。注采运行5年后，于2015年11月25日进行了声呐带压测腔，将测得结果与完腔声呐测腔结果对比发现腔顶位置发生大面积垮塌，有4m的位移差，并且在盐腔中部也发生了局部垮塌(图2)。

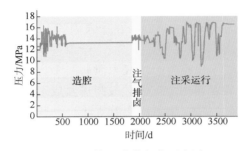

图1 腔体A载荷加载历史图

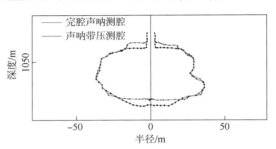

图2 腔体A声呐带压测腔与完腔声呐测腔结果对比图

盐穴B由盐化老腔改造而来，整个建库过程与盐穴A相比少了钻井、造腔2个阶段。腔体改造完成后，进行了声呐测腔，腔体体积为105664m³。2007年1月30日注气排卤，2007年7月2日注气投产。盐穴B为最早投产的一批腔体，其注采频繁，一年内注采周期可达4次，腔内最高压力为14.68MPa，最低压力为8.95MPa(图3)。在投产过程中，于2013年11月2日进行了带压声呐测腔，将测得结果与完腔声呐测腔结果对比发现腔体体积、形状基本未变(图4)。

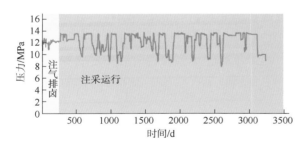

图3 腔体B载荷加载历史图

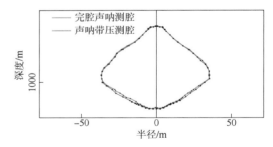

图4 腔体B声呐带压测腔与完腔声呐测腔结果对比图

2 运行损伤评价方法

盐岩属于弹—黏性体岩石，应力—应变曲线先是有一小段弹性变形，然后为非弹性变形部分，并继续不断地蠕变，故盐穴储气库盐岩损伤评价与其他岩体有所差别。目前，世界范围内对盐岩损伤评价方法并没有统一的规范，在材料力学强度准则的基础上，结合盐岩蠕变等性质，对盐岩损伤进行评价。对于岩盐在各种应力状态下发生损伤破坏的强度理论有第一强度理论(最大拉应力理论)、第三强度理论(最大剪应力理论)、第四强度理论(最大畸变能理论)、膨胀强度理论、蠕变强度理论。

2.1 最大拉应力准则

根据材料力学最大拉应力理论可知,盐岩不发生拉应力破坏的条件为

$$\sigma_1 \leqslant \sigma_t \tag{1}$$

式中:σ_1 为最大主应力,MPa;σ_t 为岩盐抗拉强度,可由巴西劈裂试验获得,MPa。

对金坛地区盐岩进行抗拉强度测定,包括顶、底板共9个层位(表1)盐岩最大抗拉强度为1.62MPa,最小抗拉强度为0.84MPa,平均抗拉强度为1.1MPa。当盐岩受到拉张应力作用时,变得很脆,非常低的拉张强度即可使岩盐发生破碎。因此,为了保证腔体注采运行过程的安全,盐穴绝对不允许出现拉张应力。

表1 金坛地区盐岩抗拉强度测定结果

深度/m	岩性	抗拉强度/MPa
981.98~982.09	盐岩	1.55
1023.30~1023.41	盐岩	1.10
1035.97~1036.09	盐岩	1.00
1054.90~1054.96	含膏盐岩	1.00
1066.04~1066.10	含泥盐岩	0.85
1088.23~1088.31	含泥盐岩	1.20
1103.61~1103.67	含泥盐岩	0.84
1124.90~1125.00	盐岩	0.98
1138.33~1138.38	盐岩	1.62

2.2 最大剪应力准则

目前,在岩石工程实践中被广泛应用的最大剪应力准则为莫尔—库伦破坏准则,该准则能够比较全面地反映岩石的强度特征,对塑性岩石和脆性岩石的剪切破坏均适用,并能解释岩石的抗拉强度远小于岩石的抗压强度特征。该准则的表达式为

$$\tau = c + \sigma_n \tan\varphi \tag{2}$$

式中:τ 为剪切面上的剪应力(剪切强度),MPa;σ_n 为剪切面上的正应力,MPa;c 为黏结力,MPa;φ 为内摩擦角,(°)。

经过转换,该表达式可由主应力表示为

$$\sigma_1 = \frac{1+\sin\varphi}{1-\sin\varphi}\sigma_3 + \frac{2c\cos\varphi}{1-\sin\varphi} \tag{3}$$

式中:σ_3 为最小主应力,MPa。

式(2)、式(3)可以用莫尔强度包络线表示,为了得到盐岩的破裂包络线,需要在不同围压下进行一组三轴实验,每个实验在不同条件下进行,直到盐岩破坏。记录盐岩破坏时

的轴压和围压,绘制莫尔应力圆,取所得的一系列莫尔应力圆的公切线,最后连接成莫尔强度包络线。通常将盐岩的莫尔强度包络线简化为直线形式。若盐岩任意一点处的应力落在包络线以下,则盐岩不会发生破裂,若刚好在包络线上则处于破裂极限状态,若在包络线以上则盐岩将发生破裂。

对金坛地区盐岩进行剪切强度测定,包括顶、底板共9个层位(表2)最终得到盐岩的黏结力和内摩擦角数据,黏结力最大值为5MPa,最小值为3MPa,平均值为4.07MPa;内摩擦角最大值为43.6°,最小值为31°,平均值为39.07°。因此,金坛地区盐岩剪切强度准则可由包络线表示为

$$\tau = 4.07 + 0.81\sigma_n \tag{4}$$

表2 金坛地区盐岩剪切强度测定结果

深度/m	岩性	黏结力/MPa	内摩擦角/(°)
981.00~981.60	盐岩	4.0	40.6
1019.91~1023.30	盐岩	4.6	38.0
1036.27~1038.44	盐岩	4.6	40.0
1054.54~1054.74	含膏盐岩	3.3	42.0
1065.84~1066.45	含泥盐岩	5.0	34.4
1088.30~1090.72	含泥盐岩	3.0	43.6
1103.65~1104.30	含泥盐岩	4.8	31.0
1122.94~1123.70	含泥盐岩	3.7	40.0
1138.80~1139.15	盐岩	3.6	42.0

2.3 最大畸变能准则

盐岩在外力作用下所发生的变形既包括盐岩的体积改变,也包括盐岩的形状改变,最大畸变能理论认为当盐岩任意一点的应力偏量第二不变量J_2达到某一定值时,该点就开始进入塑性状态,可表示为

$$(\sigma_1-\sigma_2)^2+(\sigma_2-\sigma_3)^2+(\sigma_3-\sigma_1)^2=2\sigma_s^2 \tag{5}$$

式中:σ_2为中间主应力,MPa;σ_s为盐岩初始屈服应力,MPa。

利用有限元软件模拟计算时,常采用Von-Mises屈服准则。Von-Mises等效应力表达式为

$$\sigma_{eq}=\sqrt{\frac{3}{2}\sigma'_{ij}\sigma'_{ij}} \tag{6}$$

$$\sigma'_{ij}=\sigma_{ij}+p\delta_{ij} \tag{7}$$

$$p=-\frac{1}{3}(\sigma_1+\sigma_2+\sigma_3) \tag{8}$$

式中:σ_{eq}为Von-Mises等效应力,MPa;σ'_{ij}为偏应力张量,MPa;σ_{ij}为应力张量,MPa;p为等效压应力,MPa;δ_{ij}为单位矩阵。

由于最大畸变能准则考虑了σ_1、σ_2、σ_3这3个主应力的共同影响,因此该准则比最大剪应力准则更符合测试结果,在工程中也得到了广泛应用。

采用盐岩三轴压缩试验进行最大畸变能强度测定，对圆柱形盐岩试样施加与地应力相仿的围压，轴压逐渐加大，直到盐岩试样破裂，得到抗压强度(表3)。

表3 金坛地区盐岩最大畸变能强度测定结果

深度/m	岩性	抗压强度/MPa
981.00~981.60	盐岩	46.0
1019.75~1019.91	盐岩	40.0
1037.27~1038.44	盐岩	43.0
1055.17~1055.82	含膏盐岩	40.0
1066.30~1066.45	含泥盐岩	37.0
1090.50~1090.72	含泥盐岩	41.0
1104.10~1104.30	含泥盐岩	32.8
1123.50~1123.70	含泥盐岩	39.0
1138.28~1138.82	盐岩	41.0

对盐岩的初始屈服应力进行统计计算，得到盐岩在20MPa围压下的变形强度为19.98MPa，故盐岩不发生变形破坏的条件为：$\sigma_{eq} \leq 19.98\text{MPa}$。

2.4 膨胀强度准则

膨胀损伤是一些岩石材料特有的性质，在单轴和三轴压缩试验条件下，体应变随轴向应力增加到一定程度后突然减小，这时试件内部产生微裂缝，导致试件体积增加，通常情况下肉眼看不到试件发生损坏，但对于盐穴储气库长期运行来说是不允许发生膨胀损伤的。膨胀准则有线性和非线性两种[30]，简单的线性膨胀准则可以表示为

$$\eta = \frac{\sqrt{J_2}}{aI_1 + b} \tag{9}$$

$$I_1 = \sigma_1 + \sigma_2 + \sigma_3 \tag{10}$$

$$J_2 = \frac{1}{6}\left[(\sigma_1-\sigma_2)^2 + (\sigma_2-\sigma_3)^2 + (\sigma_3-\sigma_1)^2\right] \tag{11}$$

式中：η为膨胀指数；a、b为盐岩系数；I_1为应力张量第一不变量，MPa。

由于膨胀损伤与加载路径有关，因此考虑加载路径的线性膨胀准则为

$$\eta = \sqrt{J_2}\frac{\cos\theta - \frac{1}{\sqrt{3}}\sin\theta\cos\varphi}{c\cos\theta - \frac{1}{3}I_1\sin\varphi} \tag{12}$$

$$\theta = \arctan\frac{2\sigma_2 - \sigma_1 - \sigma_3}{\sqrt{3}(\sigma_1 - \sigma_3)} \tag{13}$$

式中：θ为洛德角，(°)。

非线性膨胀准则：

$$\eta = \sqrt{J_2}\frac{\sqrt{3}\cos\theta - D_2\sin\theta}{D_1\left(\dfrac{I_1}{\sigma_0}\right)^n + T_0} \quad (14)$$

式中：D_1、D_2、T_0、n 为盐岩材料参数；σ_0 为标准应力，MPa。

非线性膨胀准则对于某些岩盐可以更好地反映其强度性质，但对于实际工程问题，线性准则更容易使用。

通过盐岩三轴应力试验，可以获得平均应力和膨胀应力（表4）。对 I_1 和 $\sqrt{J_2}$ 进行拟合（图5），通过测定岩盐试件得到 $a=0.274$、$b=-10.8$，代入式(9)可知：当 $\eta>1$ 时，即在直线上方区域，盐岩发生膨胀损伤；当 $\eta<1$ 时，即在直线下方区域，盐岩不会发生膨胀损伤。

表4　金坛地区盐岩膨胀强度测定结果

平均应力/MPa	膨胀应力/MPa	I_1/MPa	$\sqrt{J_2}$/MPa
33.0	23.9	98.80	13.78
28.2	24.7	84.70	14.30
25.8	17.5	77.50	10.00
39.3	42.9	117.90	24.80
39.1	27.3	117.30	15.76
30.8	32.4	92.40	18.70
27.4	22.2	82.20	12.80
37.9	38.7	113.74	22.40
28.9	26.6	86.60	15.40
30.1	15.5	90.50	8.90

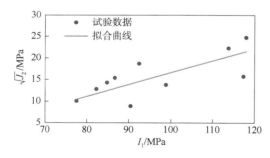

图5　金坛地区盐岩线性膨胀回归图

2.5　蠕变强度准则

当盐岩受到三轴压缩作用时，不管偏应力值有多小，都将发生蠕变变形。盐岩在受到压应力作用时，为高韧性材料，具有良好的蠕变特性，可以在蠕变时承受很大的变形而不断裂，能适应储存压力的反复变化。但是，为了设计目的和经济效益的需要，蠕变应变每年不得超过1%，整个生命周期不得超过10%，其公式为

$$\varepsilon_{eq}(t) = 10^{-6}\left\{\int_0^t \left(\dfrac{\sigma_{eq}}{K}\right)^{\frac{\beta}{\alpha}}\exp\left[-\dfrac{Q}{R}\left(\dfrac{1}{T}-\dfrac{1}{T_{ref}}\right)\right]^{\frac{1}{\alpha}}dt\right\}^{\alpha} \quad (15)$$

式中：$\varepsilon_{eq}(t)$ 为等效蠕变应变；t 为时间，a；Q/R 为 Arrhénius 系数；T 为温度，℃；T_{ref} 为基准温度，℃；α、β、K 为材料蠕变参数。

通过盐岩蠕变试验，可获得盐岩蠕变强度（表5）。盐岩蠕变阶段可以分为3个阶段：

初始蠕变、稳态蠕变及加速蠕变。经历很短的初始蠕变阶段，盐岩就进入稳态蠕变阶段，为了得到稳定的盐岩稳态蠕变过程，整个试验一般持续一周，试验过程中计算机实时采集数据。在偏应力、温度、岩性相同的情况下，蠕变速率有时不同，可能是盐岩各矿物含量和矿物内部结构不同造成的。通过对不同应力状态下稳态蠕变期蠕变速率与应力进行拟合得到蠕变速率公式：

$$\dot{\varepsilon}_{vp} = 5 \times 10^{-8} (\sigma_1 - \sigma_3)^{3.75} \tag{16}$$

式中：$\dot{\varepsilon}_{vp}$ 为蠕变速率，h^{-1}。

表5 金坛地区盐岩蠕变强度测定结果

深度/m	岩性	σ_3/MPa	σ_1/MPa	稳态蠕变期的蠕变速率/h^{-1}
979.42~979.92	盐岩	5	30.0	1.002×10^{-4}
1021.11~1021.41	盐岩	5	35.0	2.298×10^{-4}
1037.96~1038.27	盐岩	5	35.0	4.756×10^{-4}
1053.93~1054.47	含膏盐岩	5	35.0	1.635×10^{-3}
1065.58~1066.05	含泥盐岩	5	35.0	3.201×10^{-4}
1088.91~1089.27	含泥盐岩	5	35.0	1.990×10^{-4}
1103.61~1104.10	含泥盐岩	5	30.0	1.671×10^{-4}
1124.65~1125.08	盐岩	5	40.0	1.594×10^{-3}
1137.01~1137.29	盐岩	5	31.5	2.000×10^{-4}

3 数值模拟评价

基于实验室单轴和三轴试验获得的力学参数，结合现场测试结果，利用有限元分析软件对所关注盐穴的以上评价指标进行计算，并将计算结果与实际声呐测试结果进行对比，分析其有效性。尽管使用三维模型计算可以更好地反映盐穴在运行过程中的变形和损伤区域，但是计算数据庞大、计算速度缓慢，根据大量的盐穴储气库地质力学评价分析实例，选择盐穴的关注剖面建立轴对称模型进行分析计算。因此采用轴对称模型对盐穴进行地质力学计算，静力计算使用 Mohr-Coulomb 模型，蠕变计算使用 Lemaiitre 模型。由于温度和热应力对腔体稳定性评价具有较大的影响，故引入温度和热应力影响因素，将温度场和已建立模型进行顺序耦合，更真实地还原地层实际情况。

3.1 最大拉应力指标

考虑温度影响因素，计算蠕变模型中腔内最低压力工况下的最大主应力（图6）。腔体A在顶部、中部、底部都存在大面积拉张应力区域，极有可能发育裂缝，从而形成相对独立的小岩块。当顶部和中部的小岩块自身重力大于与岩层的黏结力时，小岩块将会垮塌，掉落到腔体底部。腔体B未出现拉应力，完全处于压应力状态，因此不会有垮塌现象发生。

3.2 最大剪应力指标

最大剪应力的计算采用摩尔—库伦准则，判断静力模型是否存在塑性区（图7）。腔体A、B均无塑性区域产生，不会发生剪切破坏。

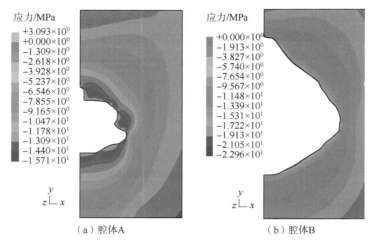

图 6 蠕变模型腔体最大主应力云图

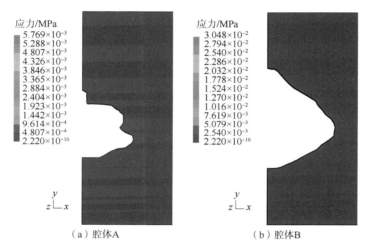

图 7 静力模型腔体最大剪应力云图

3.3 最大畸变能指标

运用 ABAQUS 软件对 Von-Mises 等效应力 σ_{eq} 进行计算(图 8),Von-Mises 等效应力 σ_{eq} 小于材料初始屈服应力 σ_s,盐岩不会发生变形破。

3.4 膨胀强度指标

采用线性膨胀准则,计算最低压力工况下的膨胀指数(图 9),可以判断两个腔体围岩的膨胀指数 $\eta \leqslant 1$,由此可见,采用线性膨胀准则,两个腔体均不会有膨胀损伤区域的产生。

3.5 蠕变强度指标

计算 2 个腔体在最危险工况下蠕变应变(图 10),基于设计目的,以蠕变应变每年不得超过 1%,整个生命周期不得超过 10% 进行判断,两个腔体均不会发生损伤破坏。

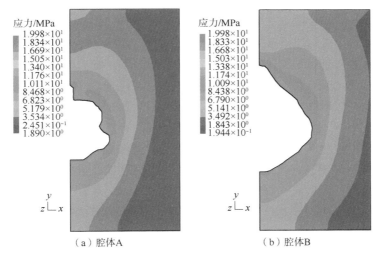

图 8 蠕变模型腔体最大畸变能云图

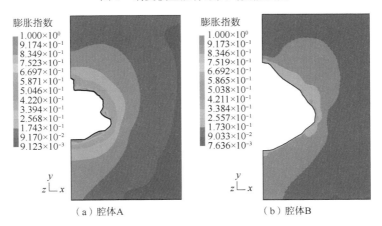

图 9 蠕变模型腔体膨胀强度云图

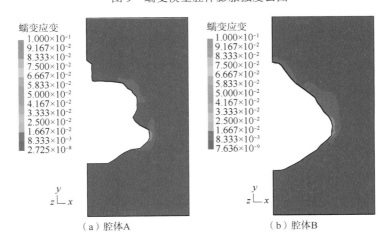

图 10 蠕变模型腔体蠕变强度云图

3.6 评价结果与实测结果对比

对腔体 A、B 评价指标的模拟结果进行统计，同时与声呐实测结果进行对比（表6）。数值模拟评价考虑每个可能造成盐穴储气库运行损伤的因素，综合以上5个评价指标，只要其中任何一个指标超出规定标准，则盐穴储气库处于运行损伤中。模拟结果表明：腔体 A 最大剪应力指标、最大畸变能指标、膨胀强度指标、蠕变强度指标都在无损伤规定范围内，而在腔体的顶部、中部、底部存在大面积区域拉应力，超出了最大拉应力指标，极有可能发生损伤破裂，模拟评价结果与腔体 A 声呐带压实测结果能够很好吻合。同时可知造成腔体 A 腔顶垮塌的原因是有拉应力的存在，拉应力的出现可能是采气速率过快使腔体温度大幅下降，从而在腔壁围岩形成了热拉张应力。腔体 B 模拟结果和声呐结果均无损伤，故腔体 B 可作为腔体 A 有无损伤的一个参照。综上所述，采用的腔体损伤评价体系能够很好地评价腔体注采运行过程中的损伤破坏，评价结果与腔体实际状态吻合。

表6 声呐实测结果与模拟评价结果对比

腔体	声呐实测结果	数值模拟评价结果				
		最大拉应力指标	最大剪应力指标	最大畸变能指标	膨胀强度指标	蠕变强度指标
A	顶部垮塌	$\sigma_1>0$	无塑形区域	无变形区域	$\eta \leq 1$	$\varepsilon_{eq} \leq 10\%$
B	无损伤	$\sigma_1<0$	无塑形区域	无变形区域	$\eta \leq 1$	$\varepsilon_{eq} \leq 10\%$

4 结论

（1）针对盐岩弹—黏性性质，在材料力学强度准则的基础上，提出适用于盐岩的损伤评价准则：第一强度理论（最大拉应力理论）、第三强度理论（最大剪应力理论）、第四强度理论（最大畸变能理论）、膨胀强度理论、蠕变强度理论。

（2）针对5个盐岩损伤评价准则进行盐岩室内力学试验，获得盐岩力学参数，得到金坛地区盐穴储气库运行损伤评价标准，同时利用软件对以上评价指标进行数值模拟计算，并将计算结果与声呐实测结果对比，分析其有效性。

（3）声呐实测结果与数值模拟评价结果对比表明：所采用的腔体损伤评价体系能够很好地评价腔体注采运行过程中的损伤破坏，造成腔体 A 腔顶垮塌的原因是有拉应力存在。

参 考 文 献

[1] 杨海军. 中国盐穴储气库建设关键技术及挑战[J]. 油气储运，2017，36(7)：747-753.

[2] 宋桂华，李国韬，温庆河，等. 世界盐穴应用历史回顾与展望[J]. 天然气工业，2004，24(9)：116-118.

[3] 杨春和，陈锋，曾义金. 盐岩蠕变损伤关系研究[J]. 岩石力学与工程学报，2002，21(11)：27-29.

[4] 李建君，陈加松，吴斌，等. 盐穴地下储气库盐岩力学参数的校准方法[J]. 天然气工业，2015，35(7)：96-102.

[5] 高小平，杨春和，吴文，等. 温度效应对盐岩力学特性影响的试验研究[J]. 岩土力学，2005，26(11)：1775-1778.

[6] 邵保平,赵阳升,赵金昌,等.层状盐岩温度应力耦合作用蠕变特性研究[J].岩石力学与工程学报,2008,27(1):90-96.

[7] Staudtmeister K, Rokahr R B. Rock mechanical design of storage caverns for natural gas in rock salt mass [J]. International Journal of Rock Mechanics and Mining Sciences, 1977, 34(3-4):300.

[8] 唐明明,王芝银,丁国生,等.含夹层盐岩蠕变特性试验及其本构关系[J].煤炭学报,2010,35(1):42-45.

[9] 陈剑文,杨春和,高小平,等.盐岩温度与应力耦合损伤研究[J].岩石力学与工程学报,2005,24(11):1986-1991.

[10] 刘建平,姜德义,陈结,等.盐岩水平储气库的相似模拟建腔和长期稳定性分析[J].重庆大学学报,2017,40(2):45-51.

[11] 任众鑫,李建君,汪会盟,等.基于分形理论的盐岩储气库腔底堆积物粒度分布特征[J].油气储运,2017,36(3):279-283.

[12] 耿凌俊,李淑平,吴斌,等.盐穴储气库注水站整体造腔参数优化[J].油气储运,2016,35(7):779-783.

[13] 任松,白月明,姜德义,等.温度对盐岩疲劳特性影响的试验研究[J].岩石力学与工程学报,2012,31(9):1839-1845.

[14] Berest P, Djizanne H, Brouard B, et al. Rapid depressurizations:can they lead to irreversible damage? [C]. Regina:Solution Mining Research Institute Spring 2012 Technical Conference,2012:1-20.

[15] Avdeev Y, Vlorobyev V, Krainev B, et al. Criteria for geomechanical stability of salt caverns[C]. El Paso: The Fall 1997 Meeting,1997:1-11.

[16] 郭建强,刘新荣,王景环,等.基于能量原理盐岩的损伤本构模型[J].中南大学学报(自然科学版),2013,44(12):5045-5050.

[17] 金犇,夏焱,袁光杰,等.盐穴地下储气库排卤管柱盐结晶影响因素实验研究[J].天然气工业,2017,37(4):130-134.

[18] 郑雅丽,赵艳杰,丁国生,等.厚夹层盐穴储气库扩大储气空间造腔技术[J].石油勘探与开发,2017,44(1):137-143.

[19] 李银平,葛鑫博,王兵武,等.盐穴地下储气库水溶造腔管柱的动力稳定性试验研究[J].天然气工业,2016,36(7):81-87.

[20] 李杭州,廖红建,盛谦.基于统一强度理论的软岩损伤统计本构模型研究[J].岩石力学与工程学报,2006,25(7):1331-1336.

[21] 徐卫亚,韦立德.岩石损伤统计本构模型的研究[J].岩石力学与工程学报,2002,21(6):787-791.

[22] 王军保,刘新荣,郭建强,等.盐岩蠕变特性及其非线性本构模型[J].煤炭学报,2014,39(3):445-451.

[23] 杨友卿.岩石强度的损伤力学分析[J].岩石力学与工程学报,1999,18(1):23-27.

[24] 谢和平,彭瑞东,周宏伟,等.基于断裂力学与损伤力学的岩石强度理论研究进展[J].自然科学进展,2004,14(10):1086-1092.

[25] Chan K S, Brodsky N S, Fossum A F, et al. Damage-induced non-associated inelastic flow in rock salt [J]. International Journal of Plasticity,1993,10(6):623-642.

[26] 陈卫忠,王者超,伍国军,等.盐岩非线性蠕变损伤本构模型及其工程应用[J].岩石力学与工程学报,2007,26(3):467-472.

[27] 韦立德, 杨春和, 徐卫亚. 基于细观力学的盐岩蠕变损伤本构模型研究[J]. 岩石力学与工程学报, 2005, 24(23): 4253-4258.
[28] 胡其志, 冯夏庭, 周辉. 考虑温度损伤的盐岩蠕变本构关系研究[J]. 岩土力学, 2009, 30(8): 2245-2248.
[29] 丁靖洋, 周宏伟, 陈琼, 等. 盐岩流变损伤特性及本构模型研究[J]. 岩土力学, 2015, 36(3): 769-776.
[30] Karimi-jafari M, Gatelier N, Brouard B, et al. Multi-cycle gas storage in salt caverns[C]. York: SMRI Fall 2011 Technical Conference, 2011: 1-18.

本论文原发表于《油气储运》2017年第36卷第8期。

金坛盐穴储气库上限压力提高试验

井 岗 何 俊 陈加松 杨 林 李建君

(中国石油西气东输管道公司储气库项目部)

【摘 要】 为了精确获取造腔层段最小主应力以确定盐穴储气库上限压力，确保气腔安全稳定运行且最大限度地发挥盐穴储气库的功能。在金坛储气库进行小型水力压裂地应力测试，测量5个层段的最小主应力，选取注采B井套管鞋处最小主应力22.5MPa的80%，即18MPa作为理论上限压力。上限压力的确定需进行数值模拟，研究表明，注采B井在上限压力18MPa下满足气腔稳定性和密闭性的要求。根据注采站压缩机技术参数和安全考虑，金坛盐穴储气库确定了上限压力提高0.5MPa的方案，且提压过程中利用实时微地震技术监测腔体及围岩的稳定性。经过现场提压试验，注采B井上限压力达到17.5MPa，库容从3481.06×10⁴m³增加到3590.39×10⁴m³，工作气量从2137.02×10⁴m³增加到2246.15×10⁴m³，增加了5.11%。研究认为：虽然盐穴储气库行业普遍采用最小主应力的80%~85%的经验值作为盐穴储气库上限运行压力，但采取的上限压力需要进行数值模拟。通过理论模拟研究和现场试验，金坛盐穴储气库上限压力可以从当前的17.0MPa提高到17.5MPa。

【关键词】 盐穴储气库；上限压力；最小主应力；稳定性；微地震

Maximum pressure threshold increase test for Jintan Salt Cavern Gas Storage

Jing Gang　He Jun　Chen Jiasong　Yang Lin　Li Jianjun

(Gas Storage Project Department, PetroChina West-East Gas Pipeline Company)

Abstract The objective of the present study was to accurately ascertain the minimum principal stress of the gas chamber layer of salt cavern gas storages in order to determine the maximum pressure threshold, thereby ensuring the safe and continuous operation of the gas chamber and the maximization of the function of salt caverns. Small-scale hydraulic fracturing tests were carried out in the Jintan salt cavern gas storage to measure the minimum principal stress in five layers. Eighty percent of the minimum principal stress of 22.5MPa (i.e., 18MPa) at the casing shoe of the injection-production well B was selected as the theoretical maximum pressure threshold. The maximum pressure threshold was also determined using numerical simulation. The results showed that the injection-production well B met the gas

---

基金项目：中国石油储气库重大专项子课题(2015E-4008)。

作者简介：井岗，1986年生，男，汉族，陕西渭南人，工程师，硕士，主要从事盐穴储气库地质研究及稳定性监测方面的工作。E-mail：xqdsjing-gang@petrochina.com.cn。

chamber stability and airtightness requirements at the maximum pressure threshold of 18MPa. A 0.5MPa increase in the maximum pressure threshold was determined for the Jintan salt cavern gas storage based on the technical parameters and safety considerations of the injection-production station compressor. Real-time microseismic technology was used to monitor the stability of the chamber and the surrounding rock during the pressure increase process. After the on-site pressure test, the maximum pressure threshold of the injection-production well B reached 17.5MPa, the storage capacity increased from $3,481.06\times10^4m^3$ to $3,590.39\times10^4m^3$, and the functional gas volume increased by 5.11% from $2,137.02\times10^4m^3$ to $2,246.15\times10^4m^3$. Although the industry generally uses the empirical value of 80%~85% of the minimum principal stress as the maximum pressure threshold for salt cavern gas storages, our study showed that the maximum pressure threshold needs to be numerically simulated. Based on the theoretical simulation studies and the field test results, we have determined that the maximum pressure threshold of the Jintan salt cavern gas storage can be increased from the current 17.0MPa to 17.5MPa.

Keywords salt cavern gas storage; maximum pressure threshold; minimum principal stress; stability; microseism

为了存储更多的天然气，盐穴储气库运行上限压力越高越好，但上限压力不能太高，否则易造成天然气泄漏[1]，地下腔体的稳定性是盐穴储气库安全平稳运行的重中之重。对于盐穴储气库运行压力限值的确定国内外学者进行了大量研究。过去20年，上限压力确定的经验方法是采用上覆岩石测井密度曲线积分获取垂直应力，采取垂直应力的80%~85%作为上限压力。Pierre认为，盐穴储气库上限压力确定必须进行机械完整性测试，并基于地质学、岩石力学、几何学以及气腔的实际运行条件等方面进行数值模拟研究[1]。Reinhard认为，盐穴储气库上限压力需要从三轴主应力状态、腔体几何参数、矿柱比、运行历史数据及盐岩蠕变特征等方面考虑并进行岩石原应力测试[2]。Fritz在德国Ruhrgas盐穴储气库通过选取新钻井K6井11个层段进行水力压裂测试以获取原位应力数据，进而为溶腔设计及腔体运行上限压力提供基础数据[3]。刘飞等运用解析方法和数值模拟方法确定中国含夹层的薄盐层中盐穴储气库运行压力，认为上限压力与盐层和夹层的破裂压力有关[4]。赵艳杰等基于盐岩的物理力学参数实验数据和现场地层资料，利用FLAC 3D软件进行恒压流变分析，模拟了淮安库区典型井的注采循环安全运行压力限制[5]。杨海军等利用带压声呐测腔修正盐岩蠕变参数进行上限压力的优化，增加了单腔工作气量，提高了经济效益[6]。

1 上限压力确定因素

1.1 避免破裂

水力压裂主要用于油气井增产或者原位应力测量，当井口压力足够大时，在井筒裸眼段就会产生裂缝。盐穴储气库设计的上限压力必须小于岩层初始开裂压力，甚至必须小于关井压力。在盐岩地层破裂测试中，井筒测试裸眼段用两个膨胀紧密的封隔器上下隔离，并被注入卤水或气体，当压力上升到初始开裂压力时岩石产生裂缝。此时停泵并经一定时间后压力会稳定在一个定值，这个数值称为裂缝传播压力或者关井压力[7]（图1）。初始开裂压力和关井压力之间的差值称为岩石抗拉强度，在盐岩地层中，岩石抗拉强度为1~3MPa。

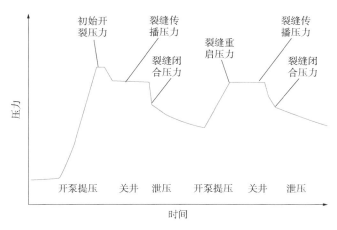

图 1 经典的水力压裂测试过程中压力随时间演化图

1.2 机械完整性测试

通常认为，套管鞋处的固井水泥环是一个密封薄弱点，固井水泥环处存在两个界面，水泥环与套管的界面和水泥环与岩层的界面，在盐穴储气库周期性较大的压力变动过程中，水泥环会发生膨胀或收缩，基于这个原因，认为固井水泥环比盐岩地层本身相对薄弱。国内外 400 多口盐穴储气注采井运行多年，只有很少量的盐穴注采井发生泄漏。这些发生泄漏井的泄漏点绝大多数为注采管柱，而非固井水泥环。为了保证盐穴储气库安全运行，国内外都会在盐穴气腔投产前进行机械完整性测试（MIT，Mechanical Integrity Testing）或被称为密封性测试，以检验注采管柱密封性。过程中要求腔内压力持续升高到上限压力或者更高并且稳定一定时间。MIT 通常的做法是在生产套管和测试管柱的环空中注入氮气，把气水界面压到生产套管固井水泥环以下 15m 处停止注气，并对气水界面和井口压力进行观测，过快的气水界面变动可能预示着较差的密封性。国外一些国家在盐穴气腔的整个生命周期里都把密封性测试作为一个常规手段用来监测气腔安全性并且发布了相关标准，当气腔达不到标准要求时，会利用技术手段对套管水泥环进行检测及更换注采管柱甚至进行修井作业，完成这些工序后再进行第二次 MIT 测试阶[8-9]。对于测试压力需要慎重选取，测试压力必须大于或者等于上限压力，但必须小于最小主应力。一些盐穴储气库公司认为，测试压力必须等于上限压力，较大的测试压力会对井筒完整性产生危害[10]。而另一些盐穴储气库公司会在技术论证后采用较大的测试压力，优点在于盐穴气腔安全运行若干年后可进行上限压力的提高以增加工作气量。在金坛储气库采用 1.1 倍上限运行压力进行腔体完整性测试，以此保证上限运行压力下气腔的密封性和安全性。

1.3 最大运行压力梯度

盐穴储气库运行上限压力的选取是一个复杂的问题，在过去几十年进行了充分的研究。盐穴储气库的拥有者希望尽可能地扩大上限压力和下限压力之间的范围以提高工作气量在库容中的比例。上限压力在一些国家会受到政府部门的强制规定，所以对于这些储气库拥有者为了提高经济效益就只能对下限压力进行优化研究。

加拿大国家盐穴储气库标准 Z341 Series-14"Storage of hydrocarbons in underground formations"中对上限压力规范两条：（1）上限压力不能超过上覆岩层压力(垂向应力)或最小主应力的80%；（2）当缺少破裂压力资料时，上限压力通过套管鞋深度乘以上限压力梯度18.1kPa/m获取。对于盐穴储气库运行上限压力梯度，一些发达国家的规定见表1。

表1 主要盐穴储气库国家运行上限压力梯度规定统计

国家	省份	盐层类型	最大运行压力梯度/(kPa/m)
美国	Kansas	层状盐层	17.0
美国	Louisiana	盐丘	20.4
美国	Mississippi	盐丘	20.4
美国	Texas	层状盐层和盐丘	19.2
加拿大		层状盐层和盐丘	18.1
法国		层状盐层和盐丘	单独规定（一般为16.0~19.0）
德国		层状盐层和盐丘	单独规定
英国		层状盐层	单独规定

加拿大艾伯塔省政府从盖层稳定的角度对上限压力进行规范，认为张裂缝破坏和剪裂缝破坏是引起盖层失效的两种机制，基于此规定两项：（1）为了避免张裂缝破坏上限压力通过式（1）确定，其中盖层裂缝闭合压力梯度通过经典的小型水力压裂获取；（2）为了避免盖层剪切破坏导致库容损失，必须进行地质力学的数值模拟。

$$p_{\text{mop}} = 0.8 G_{\text{CFC}} D \tag{1}$$

式中：p_{mop}为上限压力，kPa；0.8为安全系数，无量纲；G_{CFC}为盖层裂缝闭合压力梯度，kPa/m；D为腔顶垂深，m。

2 水力压裂地应力测试

2.1 水力压裂地应力测试原理

加拿大国家盐穴储气库标准Z341 Series-14规定盐穴储气库上限压力不超过上覆岩层压力或最小主应力的80%，获取造腔段精准的最小主应力数据对于确定盐穴储气库上限压力至关重要。水力压裂地应力测试是目前工业界广泛应用的岩层地应力测试手段，被认为是目前最可靠和有效的深层地应力测试方法。通过小体积、高压流体注入，小型水力压裂测试，在测试层位产生一张性破裂，并将破裂扩展到远离井筒影响范围之外的原始地层中，然后停止流体注入，破裂将随压力下降而闭合。通过地质力学与瞬态渗流的理论方法分析压降曲线，获取裂缝闭合压力。裂缝闭合压力等效于地层的最小主应力[11]。

2.2 地应力测试结果分析

金坛储气库新钻A井选取5个层段进行小型水力压裂，每次测试中裂缝张开与闭合进行4~5次，使得裂缝经过多次重张与扩展，以获得多个裂缝闭合压降曲线，采用均方根法、回流刚度法和G函数法计算裂缝闭合压力，多种方法表明，水压致裂法测试的数据具有较好的一致性，确定了盐层最小主应力值（表2）。利用密度测井资料确定上覆岩层垂向应力值[图2(a)为钻井井径测井曲线，井径测井曲线上的红色虚线标注小型压裂测试的层位，

图2(b)为建库段伽马测井曲线,图2(c)方框标注的最小主应力是小型压裂的测试结果,蓝色斜线为由密度测井资料确定的垂向应力值],对比表明,垂向应力与最小主应力值相等。利用阵列声波岩石力学参数计算得到了全井段的水平主应力值,即最大主应力、垂向应力、最小主应力,3个主应力的值十分接近,反映了盐岩蠕变地层随时间推移,地应力趋于各向同性的典型特征。

表2 金坛储气库 A 井地应力参数表

井号	岩性	测试垂深/m	最小主应力/MPa	最小主应力梯度/(kPa/m)	垂向主应力/MPa	垂向主应力梯度/(kPa/m)
A 井	盐岩	1042.0	24.496	23.509	23.770	22.812
A 井	泥岩	1074.5	24.529	22.828	24.530	22.829
A 井	泥岩	1085.5	25.127	23.148	24.780	22.830
A 井	盐岩	1104.0	25.528	23.123	24.970	22.618
A 井	盐岩	1121.0	25.309	22.577	25.590	22.825
平均值				23.037		22.783

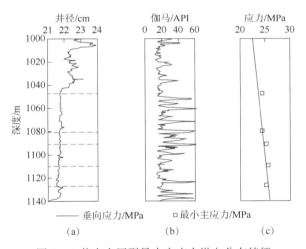

图2 A 井水力压裂最小主应力纵向分布特征

根据小型水力压裂实验结果可知,最小主应力即为垂向应力,同时,由于盐岩典型蠕变特征三向主应力基本相等,同一深度位置处三向主应力测试误差值小于5%。如图3所示,利用表2所列举的测试深度和地应力数值进行交会制图,并且进行线性拟合,可以得到 A 井最小主应力与深度关系为

$$y = 0.02154x + 1.03424 \qquad (2)$$

式中:y 为最小主应力,MPa;x 为垂直深度,m。

根据拟合结果可知,该线性方程的方差为0.97,说明拟合程度较好,可以满足实际工程中预测最小主应力的要求。

2.3 盐穴注采 B 井上限压力理论值

盐穴注采 B 井实钻盐层为 976.4~1145.0m，平均不溶物含量为 12.6%。技术套管下深 996.18m，造腔盐层段为 1011.4~1132.0m，厚度为 121m。造腔工艺全部使用正循环造腔模式，累计采卤 225.98×10⁴m³，完腔声呐体积为 198720.10m³。从图 4 所示声呐测量剖面图可知，腔顶深度为 1011.62m，腔体净高度为 76m。2011 年 12 月 5 日投产，注气排卤完后声呐体积为 193828.63m³。按照式（2）计算得到的最小主应力值为 22.78MPa。根据加拿大国家盐穴储气库标准 Z341 Series-14，盐穴储气库上限压力不超过上覆岩层压力或最小主应力的 80%，金坛储气库盐穴注采 B 井上限压力理论值为 18.0MPa。因此，依据地应力测试结果将现行设计上限压力 17.0MPa 进行提压从理论上讲是可行的，但还需要从腔体稳定性和密封性的角度考虑进行数值模拟论证。

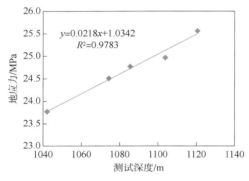

图 3 A 井地应力与深度关系曲线

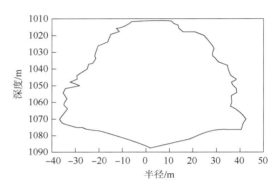

图 4 盐穴注采 B 井腔体剖面图

3 地质力学数值模拟

3.1 评价指标

通过对盐岩力学特性开展室内研究和数值模拟评价，利用地质力学、岩石流变学理论及有限元数值模拟，开展盐穴储气库运行过程中腔体稳定性和密封性预测研究[12-16]。通过分析，选取变形量、体积收缩率、剪胀安全系数为评价指标，并对指标安全临界值进行确定。各个指标含义及临界值如下：(1)变形量作为可以直接量化的指标，一直作为地下空间结构安全性评价指标之一。考虑到盐岩具有典型的蠕变特征，且其变形能力较强。根据前期理论和室内实验研究成果，提出盐穴储气库运行 30a 最大蠕变变形量不大于其最大直径的 5%作为盐穴储气库变形安全的临界判别值。(2)体积收缩率定义为溶腔体积减小量与溶腔原体积之比。根据金坛盐岩地质特征及国外行业标准，提出运行 30a 体积收缩率不大于 10%作为其安全评价的临界值。(3)盐岩在失效破坏时，表现出较为显著的剪胀破坏特征。主要表现为，当盐岩开始发生破坏时，其渗透率将会显著增加。剪胀破坏在盐岩试样实验过程中其体积达到最小值时表现最为突出。在这个点之后，试样中的微裂缝增加，将会导致试样体积增加。盐岩发生剪胀破坏主要与两个主应力不变量相关：平均应力不变量和偏应力张量不变量的平方根。该准则已经被很多实验结果验证，并在很多盐穴油气储库工程设计和安全评价中得到广泛应用并取得很好的应用效果。采用该判别准则对盐穴储气库的

安全性进行评价，计算公式为

$$F_{vs} = \frac{0.27I_1}{\sqrt{J_2}} \tag{3}$$

式中：F_{vs}为剪胀安全系数，无量纲；I_1为第一主应力不变量，MPa；J_2为第二主应力不变量偏差，MPa^2。

该判别准则的安全临界值定义为：当$F_{vs}<1.5$，盐穴围岩将会发生局部破坏；当$F_{vs}<1.0$，盐穴围岩将会发生破坏；当$F_{vs}<0.6$，盐穴围岩将会发生垮塌。

3.2 盐穴注采B井数值模拟

盐穴储气库数值分析一般采用有限元分析软件对所关注的评价指标进行计算。本次研究通过建立注采B井三维模型分析盐穴在运行过程中的变形和损伤，具体为开展盐穴注采B井变形量、体积收缩率及剪切安全系数3个评价指标在运行上限压力18MPa下运行30a数值模拟。从盐穴注采B井溶腔围岩中变形量分布（图5）可知，变形量整体偏小，在距离腔体附近最大的变形量为0.07%，距离腔体越远，变形量越小，远低于安全临界值5%。同时计算结果表明，底部隆起变形量要比顶部下沉的变形量大，符合地下空间变形经验认识。从盐穴注采B井在上限压力为18.0MPa时运行30a的体积收缩率随时间变化关系曲线（图6）可知，运行30a后体积收缩率为0.6%，远低于10%的安全临界值，从体积收缩率角度考虑，上限运行压力18.0MPa可满足稳定性要求。从盐穴注采B井围岩中剪胀安全系数分布（图7）可知，围岩的剪胀安全系数整体偏高，一般都在3左右，只有在夹层位置处的局部区域为2左右，可以保证盐穴的运行安全。通过对注采B井上述指标的分析可知，当上限运行压力提到18MPa时，盐穴储气库注采B气腔可以满足稳定性要求。

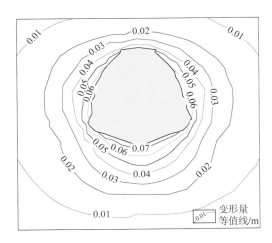

图5 盐穴注采B井内压18.0MPa
运行30a变形量分布图

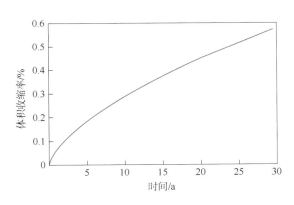

图6 盐穴注采B井内压18.0MPa
运行30a体积收缩率

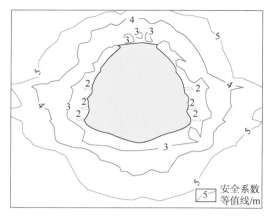

图 7 盐穴注采 B 井内压 18.0MPa
运行 30a 剪胀安全系数分布图

4 微地震监测

对于地下腔体的稳定性研究，目前国内外的主流技术是地表沉降监测和力学数值模拟，这两种技术都属于间接方式并且具有滞后性。国外曾经利用实时微地震提前 3 周预测了采卤盐腔的垮塌，这使得微地震技术在地下腔体监测方面声名鹊起。微地震技术起源于地热开发行业，后引入石油天然气行业进行地下原生裂缝、诱发裂缝几何特性分析及储层有效改造体积估算。微地震监测的目的是实时预警盐穴储气库生产过程中天然裂缝活动、腔壁垮塌、顶板破裂信号，进而采取有效措施控制和防范盐腔垮塌，以及对盐穴储气库的生产运行技术参数进行安全稳定性评价[17-20]。监测井 A 井为盐穴储气库新完钻井，完钻方式为二开裸眼完钻，盐顶埋深 982.4m，钻井生产套管下深 998.4m，为了防止盐层溶蚀完钻后井筒灌满饱和卤水。根据现场条件采取井中观测方式，选定 12 级 Maxwave 三分量检波器接收，级距为 10m，检波器的位置尽量靠近监测目的层段深度，并且从仪器安全考虑检波器不能出生产套管进入裸眼段，所以，12 级三分量检波器设计位置为 870~980m，仪器采样间隔为 0.5ms，记录长度为 10s，增益为 40dB。

4.1 提压方案

针对金坛盐穴注采 B 井理论上限压力为 18.22MPa 的计算结果，考虑注采站压缩机实际情况和安全生产的要求，本次提压方案制定了提高 0.5MPa 的提压目标。盐穴注采 B 井初步设计的上限压力为 17MPa，因技术套管 996.16m 的埋深产生的气柱压力为 1.2MPa，故盐穴注采 B 井当前的井口注气压力上限为 15.8MPa。因此，提压方案要求的提压目标对应的井口上限压力为 16.3MPa。根据提压目标将整个微地震监测分为注采井静止期、注采井正常注气升压期、注采井提压期、提压稳定期 4 个阶段。具体微地震提压监测方案见表 3。

表 3 金坛盐穴注采 B 井微地震提压监测表

阶 段	压力变动区间/MPa	压力变化值/(MPa/d)	微地震监测期/d
注采井静止期	13.98~13.98	0	5
正常注气升压期	13.98~15.80	0.2~0.3	8
注采井提压期	15.80~16.30	0.1	5
提压稳定期	16.30~16.30	0	7

4.2 微地震监测成果

盐穴注采 B 井从 2017 年 10 月 13 日进入冬季注气期，2017 年 10 月 25 日完成注气，井口压力从 13.98MPa 上升到 16.30MPa。为期 25d 的监测过程中，共监测到 172 个微地震事件，震级范围在 -2.68~-0.39。其中微地震事件主要集中在 C1 井、C2 井两井之间（图 8），从两井联井地震剖面上看存在断层，推测为天然断层位置处地应力调整导致发生小的岩石破裂所致。从时间上，看微地震事件出现在提压试验各个阶段，微地震事件数量没有随着压力的升高集中出现，证明了提压过程中及提压后腔顶和围岩力学性质稳定，没有出现破裂或者腔壁掉块现象(图9)。经过现场提压试验，注采B井上限压力达到17.5MPa，库容从 $3481.26×10^4m^3$ 增加到 $3590.39×10^4m^3$，工作气量从 $2137.02×10^4m^3$ 增加到 $2246.15×10^4m^3$，工作气量增加5.11%。

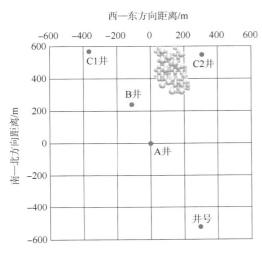

图8 金坛盐穴注采B井提压试验微地震事件平面分布图

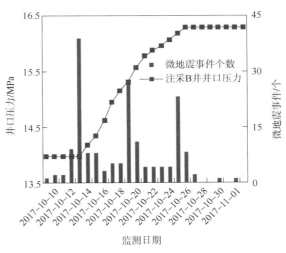

图9 金坛盐穴注采B井提压试验微地震事件个数与井口压力统计图

5 结论

（1）金坛储气库建库层状盐岩段地应力具有各向同性的应力状态，结合地应力特征和行业规范，认为金坛盐穴储气库上限压力具有提压可行性。

（2）开展了盐穴注采B井理论上限压力18MPa下运行30a的数值模拟，分析选取的评价指标，研究认为在上限压力18MPa下腔体满足稳定性要求。

（3）开展了盐穴注采B井上限压力提高0.5MPa的试验，并利用实时微地震技术监测提压腔体稳定性。经过现场提压试验，注采B井上限压力达到17.5MPa，库容从 $3481.26×10^4m^3$ 增加到 $3590.39×10^4m^3$，工作气量从 $2137.02×10^4m^3$ 增加到 $2246.15×10^4m^3$，工作气量增加5.11%。

参 考 文 献

[1] Pierre B. Benoît B, Fabien F, et al. Maximum pressure in gas storage caverns[S]. New York: Solution Mining Research Institute, 2015: 1-17.

[2] Reinhard R, Staudtmeister K, Zander-schi-ebenhofer D. Rock mechanical determination of the maximum internal pressure for gas storage caverns in rock salt[S]. Hannover: Solution Mining Research Institute, 1998: 1-20.

[3] Fritz R, Klaus B, Helmut D. Hydraulic fracturing stress measurements in the krumhorn gas storage field, Northwestern Germany[S]. Houston: Solution Mining Research Institute, 1996: 1-20.

[4] 刘飞, 宋桂华, 李国韬, 等. 含有夹层的薄盐层中盐穴储气库运行压力的确定[J]. 天然气工业, 2004, 24(9): 133-135.

[5] 赵艳杰, 马纪伟, 郑雅丽, 等. 淮安盐穴储气库注采循环运行压力限值确定[J]. 油气储运, 2013, 32(5): 526-531.

[6] 杨海军, 郭凯, 李建君. 盐穴储气库单腔长期注采运行分析及注采压力区间优化——以金坛盐穴储气库西2井腔体为例[J]. 油气储运, 2015, 34(9): 945-950.

[7] Horvath P L, Sven E W. Determination of formation pressures in rock salt with regard to cavern storage[S]. Krakow: Solution Mining Research Institute, 2009: 83-90.

[8] Crotogin F R. SMRI reference for external well mechanical integrity testing/performance, data evaluation and assessment[S]. Krakow: Solution Mining Research Institute, 1994: 95-100.

[9] Thiel W R. Precision methods for testing the integrity of solution mined underground storage cavern[S]. Krakow: Solution Mining Research Institute, 1993: 377-383.

[10] Gillhaus A. Natural gas storage in salt caverns-present status, developments and future trends in Europe[S]. Germany: Solution Mining Research Institute, 2007: 1-19.

[11] 周冬林, 杨海军, 李建君, 等. 盐岩地层地应力测试方法[J]. 油气储运, 2017, 36(12): 1385-1390.

[12] 敖海兵, 陈加松, 胡志鹏, 等. 盐穴储气库运行损伤评价体系[J]. 油气储运, 2017, 36(8): 910-917, 936.

[13] 杨春和, 陈峰, 曾义金. 盐岩蠕变损伤关系研究[J]. 岩石力学与工程学报, 2002, 21(11): 1602-1604.

[14] 高小平, 杨春和, 吴文, 等. 温度效应对盐岩力学特性影响的试验研究[J]. 岩土力学, 2005, 26(11): 1775-1778.

[15] 任松, 白月明, 姜德义, 等. 温度对盐岩疲劳特性影响的试验研究[J]. 岩石力学与工程学报, 2012, 31(9): 1839-1845.

[16] 徐卫亚, 韦立德. 岩石损伤统计本构模型的研究[J]. 岩石力学与工程学报. 2002, 21(6): 787-791.

[17] Emmanuelle N, Eric B, Guillaume B. Real time microseismic monitoring dedicated to forecast collapse hazards by brine production[C]. Solution Mining Research Institute Fall Technical Meeting, 2007.

[18] 赵博雄, 王忠仁, 刘瑞, 等. 国内外微地震监测技术综述[J]. 地球物理学进展, 2014, 29(4): 1882-1888.

[19] Fortier E, Maisons C, Valette M. Contribution to a better understanding of brine production using a long period of microseismic monitoring[C]. Solution Mining Research Institute Fall Technical Meeting, 2005.

[20] Lilian D, Emmanuelle N, Giovanni G. Contribution to salt leaching and post-leaching monitoring issues feedback form a long term mnicroseismic survey[C]. Solution Mining Research Institute Spring Technical Meeting, 2006.

本论文原发表于西南石油大学学报(自然科学版)2018年第40卷第6期。

金坛盐穴储气库 JZ 井注采运行优化

陈加松　程　林　刘继芹　李建君　李淑平　汪会盟　井　岗

（中国石油西气东输管道公司储气库项目部）

【摘　要】 JZ 井是金坛盐穴储气库第一口先导性试验井，目前已投产 5 年。基于最新声呐测腔及力学特征分析，JZ 井目前注采运行方式无法满足腔体稳定性要求，腔顶曾发生部分垮塌。利用有限元数值模拟软件对 JZ 井最新腔体形状进行注采运行方式优化和力学稳定性评价分析，结果表明：注采运行压力由原来的 7~17MPa 优化至 10~17MPa；最低运行压力的停留时间实现量化表征，不超过 10 天；每年进行 6 次调峰可以保证腔体运行安全；腔体运行单日压降控制在 0.5MPa 以内，采气速率控制在 $50\times10^4 m^3/d$。针对不同腔体形状优化相应运行压力范围及条件，对于保证金坛盐穴储气库安全运行及精细化管理具有重要意义。

【关键词】 盐穴储气库；注采优化；参数量化；稳定性；盐岩蠕变；数值模拟

Injection and production optimization of Well JZ in Jintan Salt Cavern Gas Storage

Chen Jiasong　Cheng lin　Liu Jiqin　Li Jianjun
Li Shuping　Wang Huimeng　Jing Gang

(Gas Storage Project Department, PetroChina West-East Gas Pipeline Company)

Abstract Well JZ is the first pilot test well in Jintan Salt Cavern Gas Storage, and so far, it has been put into operation for 5 years. It is shown from the latest cavity measuring with sonar and the mechanical characteristics analysis that the current injection and production pattern of Well JZ can't meet the demand of cavity stability and the roof of the cavity partially collapsed. In this paper, finite element numerical simulation software was applied to conduct injection and production mode optimization and mechanical stability evaluation and analysis on the latest cavity shape of Well JZ. It is indicated that the operation pressure range of injection and production is optimized from the initial 7-17MPa to the current 10-17MPa. The dwell time of the minimum operation pressure is quantified, and it shall be not longer than 10 days. The operation of the cavity can be kept safe by conducting 6 peaking shavings each year. It is necessary to control the daily cavity operation pressure drop below 0.5MPa and keep the gas production

---

基金项目：中国石油重大科技专项资助项目"地下储气库关键技术研究与应用"，2015E-40。

作者简介：陈加松，男，1986 年生，工程师，2013 年硕士毕业于中国石油大学（北京）矿产普查与勘探专业；现主要从事盐穴储气库建设及稳定性评价工作。地址：江苏省镇江市中国石油西气东输管道公司储气库项目部，212000。电话：18605253656，E-mail：chenjiasong@petrochina.com.cn。

rate at $50×10^4$ m^3/d. In order to ensure the safe operation and fine management of Jintan Salt Cavern Gas Storages, it is of great significance to optimize the operation pressure range and condition based on different cavity shapes.

Keywords salt cavern gas storage; injection and production optimization; parameter quantification; stability; salt rock creep; numerical simulation

地下储气库是国家能源战略安全、城市用气稳定的重要保障,盐穴储气库以其利用率高、注采气时间短、垫层气用量低的优点受到各国政府及石油企业的重视[1-7]。虽然盐岩具有致密渗透率极低、裂缝自愈合的特点,但其注采运行的稳定性仍不容忽视。徐素国等[8-9]对盐穴储气库的注采运行方式及腔体稳定性评价进行了相关研究。目前,常利用有限元软件模拟盐穴腔体注采运行过程中腔体周围的应力场分布,并采用岩石力学方法分析其力学性质,从而确定储气库的注采运行方式[10-15]。中国盐穴储气库建设较晚,投入运行腔体数量较少,关于盐穴储气库注采方式的研究尚处于方案设计阶段,注采运行的稳定性跟踪分析较少。而随着盐穴储气库应用规模的不断扩大,针对单腔注采方式的优化将更加重要。

根据金坛盐穴储气库运行经验,腔体的盐岩地质条件、腔体形状、临腔矿柱宽度等对腔体的力学性质有较大影响,需要根据腔体自身的力学条件制定注采运行方案。JZ 井作为金坛盐穴储气库先导性试验井,由于受到当时造腔工艺技术的限制,腔体形状未按照预先设计的"倒梨形"发展,完腔后在注采运行过程中腔顶发生过部分垮塌,腔体稳定性存在隐患。基于金坛地区盐岩力学参数试验分析与模拟回归,利用 Abaqus 有限元软件对 JZ 井最新声呐形状进行力学稳定性评价和注采运行方案优化,确定 JZ 井安全运行压力范围和注采频率,并将最低运行压力的停留时间量化。

1 地质及生产概况

1.1 地质概况

JZ 井位于金坛盆地直溪桥凹陷陈家庄—上白塘次洼,地质构造简单。盐岩埋深为 978.6~1169.8m,厚度为 191.2m,NaCl 纯度达 90%以上,可作为天然气良好储层,上部为泥岩覆盖,可作为天然气的良好盖层。JZ 井由于其良好的地质条件,而成为西气东输金坛储气库第一口先导性试验井。

1.2 生产概况

JZ 井于 2009 年 9 月完成造腔,2010 年 11 月完成注气排卤开始储气。从 2011 年 1 月开始注采运行,井口运行压力范围为 7.94~15.81MPa(对应腔内运行压力约为 8.5~17MPa)。调峰规模为早期注 1 个月高压停留 3 个月,采 2 个月低压停留 1 个月的快注慢采运行方式;之后改为注 1 个月高压停留 3 个月,采 2 个月低压停留 1 个月的慢注快采运行方式,单日的压降基本控制在 0.3MPa 以下,同时伴有多次一般规模的调峰(图1)。

2015 年 12 月,JZ 井运行满 5 年,对其进行带压声呐测腔(图2),腔体顶部发生约 5m 的垮塌。由于 JZ 井脖颈预留较厚(约 25m),虽然目前的腔体形状对腔体今后的注采运行不构成实质性威胁,但仍需要根据目前监测获得的数据信息,针对 JZ 井的腔体现状重新制定注采运行方案,以保证腔体顶部在今后的运行过程中不再发生坍塌,确保注采安全。

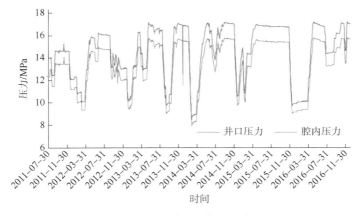

图 1　JZ 井注采运行井口压力图

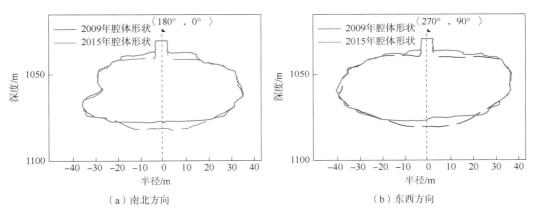

（a）南北方向　　　　　　　　　　　　　（b）东西方向

图 2　JZ 井两期声呐形状对比图

中国盐穴造腔技术尚处于起步探索阶段,造腔新工艺、新方法以及腔体形状控制技术尚待开发研究,JZ 井所获得的实际腔体形状、体积与初设均有较大差别,腔体的力学性能存在一定缺陷,尤其是腔顶的大平顶在后期注采运行过程中存在垮塌风险。

2　确定注采运行方式

2.1　采气速率

根据初步设计腔体稳定性要求,腔体压降不可以超过 0.5MPa/d。溶腔的日采气能力取决于注采管柱尺寸及结构、溶腔压力、井口压力及井口冲蚀流量。目前,金坛储气库采用 7in(1in=2.54cm)注采气管柱,经计算当腔内压力为 7~17MPa 时,单腔采气能力可达 $210×10^4$~$342×10^4 m^3/d$。考虑到采气压降为 0.5MPa/d,对于腔体体积约 $10×10^4 m^3$ 的腔体,每天采气量不超过 $50×10^4 m^3$。因此 JZ 井的单日采气量应尽量控制在 $50×10^4 m^3$ 以内。

2.2　注采运行压力

关于腔体运行压力区间的确定,国内外根据模拟研究和实际运行得出一些经验方法,但是具体方法需要结合现场实际情况和地质力学模拟计算确定。

2.2.1 高压初步确定方法

关于盐穴储气库最高运行压力的确定,CSA Z341—2003《地下碳氢储库》建议取生产套管鞋深度处岩层最小主应力的80%,最小主应力由地应力测试获取;若库区内尚未进行地应力测试,最小主应力数据未知的情况下,最大运行压力可取生产套管鞋深度处上覆岩层载荷的75%~85%;根据金坛储气库初步设计,目前金坛储气库最高运行压力梯度为0.017MPa/m,腔内最高运行压力可取该最大运行压力梯度乘以生产套管鞋处深度,JZ井生产套管鞋深度约为1015m,因此,该井最高运行压力可取17MPa,但是该最高运行压力是否可行需要经过地质力学数值模拟验证。

2.2.2 低压初步确定方法

关于盐穴储气库最低运行压力的确定,CSA Z341—2003《地下碳氢储库》建议套管鞋处最小运行压力梯度为0.0034MPa/m;部分学者认为可取生产套管鞋深度处上覆载荷压力的20%~30%;根据金坛储气库初步设计,金坛储气库最小运行压力的梯度为0.006MPa/m,因此,该井最低运行压力可取6MPa,但是该最低运行压力是否可行需要经过地质力学数值模拟验证。

2.3 注采频次

金坛储气库是按照每年2次调峰进行设计的,但是随着天然气市场需求急剧增加,管道运输距离延长,自然环境的不确定性,每年2次调峰已无法满足市场需求,根据历年金坛储气库的调峰统计,金坛储气库每年的调峰次数达2~6次。国外同类型的盐穴储气库的调峰次数根据文献记载可达6~10次[4]。

2.4 最低运行压力停留时间

盐穴储气库最低压力的停留时间对于盐穴稳定性具有较大影响,若低压运行时间过长,会加剧盐穴腔体的蠕变损伤,使腔体体积收缩加快。目前,关于最低压力的停留时间尚无准确定论,只有"尽量缩短最低压力停留时间"的模糊概念。目前金坛储气库按照每年2次调峰、最低压力停留不超过1个月进行初步设计。

3 数值模型

3.1 力学参数选取

根据泥岩与盐岩单轴、三轴蠕变试验结果和现场实际数据进行分析校核[16],采用以下基本力学参数(表1)。盐岩的蠕变本构模型采用简化后的 Lemaître 模型:

$$\dot{\varepsilon}_{vp}^{eq} = A\sigma^n t^m \tag{1}$$

式中:$\dot{\varepsilon}_{vp}^{eq}$为盐岩稳态蠕变应变率;$\sigma$为盐岩试样受到的应力差,MPa;$t$为时间,d;$A$、$n$、$m$分别为盐岩的蠕变参数,其主要与盐岩样品的结构组分有关,A还与盐岩试样试验温度相关。

表1 金坛储气库基本力学参数

岩性	弹性模量/GPa	泊松比	黏聚力/MPa	内摩擦角/(°)	抗拉强度/MPa
盐岩	6	0.35	1	30	1
泥岩	10	0.27	1	35	1

根据江苏金坛的盐岩蠕变试验及现场试验数据，通过反演的方法最终确定盐岩的主要蠕变参数：$A=3.0\times10^{-7}$，$n=3.75$，$m=-0.525$。

3.2 模型建立

根据JZ井腔体2015年12月的声呐数据及相关地质数据，利用Abaqus软件建立腔体的二维轴对称几何模型，选取腔体某纵剖面的1/2作为研究对象，计算模型为1000m×1600m的长方形，平面为xy坐标系平面，竖直方向为y轴，坐标原点为模型顶部左侧顶点，整个模型考虑了0～1600m的地层信息，由于泥岩夹层较少且厚度较薄，因此暂不考虑夹层影响（图3）。根据腔体受力情况，分别施加地应力和内部拉张力，地应力场采用3向等压自重应力场，模型底边加以纵向约束，左边和右边加以横向约束。

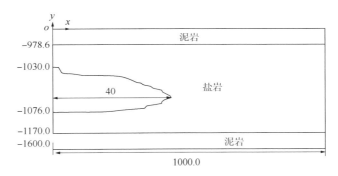

图3 JZ井计算模型示意图（单位：m）

4 计算结果

4.1 静力计算

腔体静力分析利用莫尔库仑准则，从弹塑性的角度计算了运行压力分别为6～14MPa时9种方案腔体周围塑性区的情况（图4）。当腔内压力为6MPa时，腔壁的支撑压力远小于原岩应力，腔体周围存在较大塑性区，腔脖部位在重力作用下易出现掉块；随着内压增大，塑性区逐渐减小，当内压达到11MPa时，塑性区很小，且腔体顶部的塑性区消失；当内压为14MPa时，基本无塑性区。因此，尽量减少溶腔在内压小于11MPa的情况下运行，以免腔体顶部岩块进一步垮塌，损坏溶腔。建议腔体最小内压为10MPa，且停留时间不宜过长。

4.2 恒压蠕变

腔体恒压蠕变分析是从黏弹性的角度分别计算了溶腔内压为6～17MPa时流变30年腔体周围的稳定性评价指标情况。根据计算结果，利用稳定性评价指标进行分析（表2），在

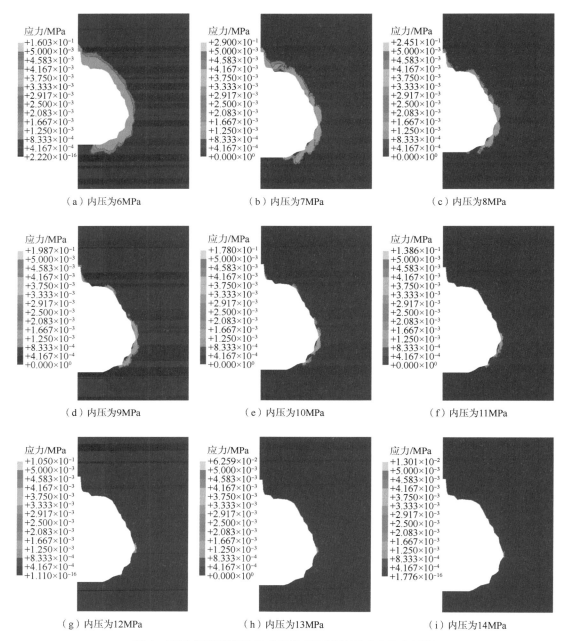

图 4 不同内压工况下 JZ 井腔体周围塑性区应力分布云图

内压为 6~17MPa 的恒压流变情况下,腔体周围均未出现拉张应力区和穿刺区,当内压为 9MPa 及以下时,腔体前 5 年的体积收缩率超过了 5%;当内压为 6MPa 时,腔体会形成膨胀损伤区,大于 6MPa 时,不会形成膨胀损伤区;当内压为 6~8MPa 时,会形成剪切塑性破坏区,大于 8MPa 时,不会形成塑性破坏区。综上,建议腔体最小内压确定为 10MPa。

表2 JZ井恒压流变30年稳定性评价指标

压力/MPa	运行时间/a	腔体收缩率/%	套管鞋处应变/%	压力/MPa	运行时间/a	腔体收缩率/%	套管鞋处应变/%
6	5	10.49	2.56	12	5	3.17	0.68
6	10	14.11	3.29	12	10	4.53	0.94
6	20	18.49	4.19	12	20	6.25	1.28
6	30	21.27	4.69	12	30	7.39	1.50
7	5	8.99	2.14	13	5	2.34	0.49
7	10	12.16	2.77	13	10	3.43	0.70
7	20	16.02	3.56	13	20	4.76	0.97
7	30	18.50	4.08	13	30	5.75	1.14
8	5	7.56	1.76	14	5	1.61	0.34
8	10	10.32	2.31	14	10	2.45	0.51
8	20	13.70	2.99	14	20	3.54	0.70
8	30	15.88	3.44	14	30	4.36	0.84
9	5	6.32	1.46	15	5	1.02	0.21
9	10	8.68	1.92	15	10	1.63	0.32
9	20	11.58	2.50	15	20	2.46	0.48
9	30	13.46	2.88	15	30	3.10	0.58
10	5	5.14	1.15	16	5	0.56	0.11
10	10	7.14	1.55	16	10	0.97	0.19
10	20	9.62	2.03	16	20	1.55	0.29
10	30	11.23	2.36	16	30	2.04	0.37
11	5	4.10	0.91	17	5	0.26	0.05
11	10	5.77	1.23	17	10	0.49	0.09
11	20	7.85	1.64	17	20	0.85	0.16
11	30	9.21	1.91	17	30	1.20	0.20

4.3 注采运行蠕变

4.3.1 单循环设计

按照最灵活的随机注采方式设计单循环。假设连续注气、注满停腔、连续采气、采完停腔。注采压力梯度设为0.5MPa极端条件，设定不同的最高压力、最低压力、停腔时间，按每年6次循环设计(表3)。

表3 一年时间内腔体注采气内压变化方案(运行压力为10~17MPa)

方案1（最低压力停留0天）		方案2（最低压力停留10天）		方案3（最低压力停留20天）		方案4（最低压力停留30天）	
运行天数/d	腔体内压/MPa	运行天数/d	腔体内压/MPa	运行天数/d	腔体内压/MPa	运行天数/d	腔体内压/MPa
0	17	0	17	0	17	0	17
14	10	14	10	14	10	14	10

续表

方案1 (最低压力停留0天)		方案2 (最低压力停留10天)		方案3 (最低压力停留20天)		方案4 (最低压力停留30天)	
运行天数/d	腔体内压/MPa	运行天数/d	腔体内压/MPa	运行天数/d	腔体内压/MPa	运行天数/d	腔体内压/MPa
28	17	24	10	34	10	44	10
60	17	38	17	48	17	58	17
74	10	60	17	60	17	60	17
88	17	74	10	74	10	74	10
120	17	84	10	94	10	104	10
134	10	98	17	108	17	118	17
148	17	120	17	120	17	120	17
180	17	134	10	134	10	134	10
194	10	144	10	154	10	164	10
208	17	158	17	168	17	178	17
240	17	180	17	180	17	180	17
254	10	194	10	194	10	194	10
268	17	204	10	214	10	224	10
300	17	218	17	228	17	238	17
314	10	240	17	240	17	240	17
328	17	254	10	254	10	254	10
360	17	264	10	274	10	284	10
		278	17	288	17	298	17
		300	17	300	17	300	17
		314	10	314	10	314	10
		324	10	334	10	344	10
		338	17	348	17	358	17
		360	17	360	17	360	17

4.3.2 最优单循环筛选

使用盐岩蠕变模型,在50年生命周期对不同单循环进行计算。分析稳定性评价指标,筛选最优单循环。注采运行稳定性评价指标[17-21]主要包括:(1)注采运行过程中腔壁无拉张应力形成;(2)注采运行过程中腔壁没有或仅局部区域出现膨胀损伤区;(3)注采运行过程中不出现或仅局部区域出现塑性区;(4)腔体最大蠕变应变区域5年运行周期内低于5%,整个生命周期内低于10%;(5)腔体体积收缩率平均每年低于1%,整个生命周期内低于20%;(6)由于注采运行导致的地面沉降速率低于0.044mm/d;根据JZ井稳定性评价结果(表4),运行压力区间为10~17MPa时,运行压力区间基本能够满足JZ井的安全运行要求，设计的 4 种运行方案，在运行 50 年中均未出现拉张应力区、膨胀损伤区和塑性破坏区，仅在局部区域出现应变较大，腔体整体保持安全稳定。但是随着最低运行压力持续时间延长，腔体收缩显著加快，蠕变应变增加，穿刺区域范围加大，地面沉降速率加大。根据 JZ 井 4 种注采运行模拟方案注采流变 50 年腔体体积收缩率（图 5）对比可以得出，方案 2、方案 3、方案 4 相比方案 1 具有明显的体积收缩加快现象，特别是从方案 1 到方案 2，腔体稳定性评价的力学指标发生显著变化，最低压力停留 10 天具有明显的稳定性评价指标的临界点特征，因此建议尽量缩短腔体最低运行压力停留时间，最长停留时间不要超过 10 天。

表 4　JZ 井注采运行 50 年稳定性评价指标（运行压力为 10~17MPa）

注采方案	时间/a	刺穿区域	腔体收缩率/%	蠕变应变（套管鞋）/%	腔体最大蠕变应变	地面沉降/m	地面沉降速率/(mm/d)
方案 1	5	无	4.44	0.92	尖角局部大于 5%	0.0023	0.1686
	10	无	5.57	1.14	—	0.0028	0.0329
	20	有（少许）	7.01	1.42	—	0.0033	0.0212
	30	有（少许）	8.01	1.62	—	0.0038	0.0150
	40	有（局部）	8.79	1.77	—	0.0041	0.0120
	50	有（局部）	9.45	1.90	尖角局部大于 10%	0.0044	0.0102
方案 2	5	有（少许）	5.84	1.21	尖角局部大于 5%	0.0028	0.2073
	10	有（局部）	7.32	1.50	—	0.0034	0.0438
	20	有（局部）	9.14	1.86	—	0.0042	0.0270
	30	有（局部）	10.39	2.11	—	0.0047	0.0197
	40	有（局部）	11.37	2.30	—	0.0052	0.0161
	50	有（较大范围）	12.19	2.47	尖角局部大于 10%	0.0055	0.0124
方案 3	5	有（局部）	6.68	1.38	局部大于 5% 扩大趋势	0.0032	0.2307
	10	有（局部）	8.33	1.70	—	0.0038	0.0489
	20	有（局部）	10.35	2.10	—	0.0047	0.0307
	30	有（较大范围）	11.73	2.39	—	0.0053	0.0219
	40	有（较大范围）	12.81	2.61	—	0.0058	0.0175
	50	有（较大范围）	13.71	2.79	局部大于 10% 扩大趋势	0.0062	0.0146
方案 4	5	有（局部）	7.27	1.50	局部大于 5% 扩大趋势	0.0034	0.2475
	10	有（局部）	9.04	1.85	—	0.0041	0.0533
	20	有（较大范围）	11.22	2.29	—	0.0050	0.0332
	30	有（较大范围）	12.71	2.59	—	0.0057	0.0237
	40	有（较大范围）	13.88	2.83	—	0.0062	0.0190
	50	大范围	14.84	3.03	局部大于 10% 扩大趋势	0.0066	0.0161

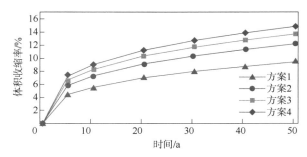

图 5 JZ 井 4 种注采运行模拟方案注采流变 50 年腔体体积收缩率(运行压力为 10~17MPa)

5 结论

通过跟踪 JZ 井的注采运行数据,结合其声呐测腔形状变化,利用有限元软件进行仿真模拟,并且进行力学分析,优化了 JZ 井的注采运行方式。将盐穴储气库最低运行压力停留时间量化,针对单个腔体形状优化相应运行压力范围及条件,对保障盐穴储气库安全运营与精细化管理提供了借鉴与参考。

参 考 文 献

[1] 周学深. 有效的天然气调峰储气技术——地下储气库[J]. 天然气工业, 2013, 33(10): 95-99.

[2] 吴忠鹤, 贺宇. 地下储气库的功能和作用[J]. 天然气与石油, 2004, 22(2): 1-4.

[3] 苏欣, 张琳, 李岳. 国内外地下储气库现状及发展趋势[J]. 天然气与石油, 2007, 25(4): 1-4.

[4] Barron T F. Regulatory, technical pressures prompt more US salt-cavern gas storage[J]. Oil & Gas Journal, 1994, 92(37): 55-67.

[5] 丁国生, 梁婿, 任永胜, 等. 建设中国天然气调峰储备与应急系统的建议[J]. 天然气工业, 2009, 29(5): 98-100.

[6] 郑雅丽, 赵艳杰. 盐穴储气库国内外发展概况[J]. 油气储运, 2010, 29(9): 652-655.

[7] 丁国生, 谢萍. 利用地下盐穴实施战略石油储备[J]. 油气储运, 2006, 25(12): 16-19.

[8] 徐素国. 层状盐岩矿床油气储库建造及稳定性基础研究[D]. 太原: 太原理工大学, 2010: 51-105.

[9] 赵克烈. 注采气过程中地下盐岩储气库可用性研究[D]. 武汉: 中国科学院武汉岩土力学研究所, 2009: 14-35.

[10] Schmidt U, Staudtmeister K. Determining minimum permissible operating pressure for a gas cavern using the finite element method[J]. Petroleum Abstracts, 1989, 30(22): 103-113.

[11] Adams J B. Natural gas salt cavern storage operating pressure determination[J]. Petroleum Abstracts, 1997, 38(25): 97-107.

[12] 丁国生, 谢萍. 西气东输盐穴储气库库容及运行模拟预测研究[J]. 天然气工业, 2006, 26(10): 120-123.

[13] 丁国生. 金坛盐穴储气库单腔库容计算及运行动态模拟[J]. 油气储运, 2007, 26(1): 23-27.

[14] 王同涛, 闫相祯, 杨恒林, 等. 多夹层盐穴储气库最小允许运行压力的数值模拟[J]. 油气储运, 2010, 29(11): 877-879.

[15] 时文, 申瑞臣, 徐义, 等. 盐穴储气库运行压力对腔体稳定性的影响[J]. 石油钻采工艺, 2012, 34(4): 89-92.

[16] 李建君,陈加松,吴斌,等.盐穴地下储气库盐岩力学参数的校准方法[J].天然气工业,2015,35(7):96-102.

[17] 杨海军,郭凯,李建君.盐穴储气库单腔长期注采运行分析及注采压力区间优化——以金坛盐穴储气库西2井腔体为例[J].油气储运,2015,34(9):945-950.

[18] Rokahr R B, Staudtmeister K, Schiebenhofer D K. Development of a new criterion for the determination of the maximum permissible internal pressure for gas storage in caverns in rock salt[C]. Hannover: SMRI Research and Development Project Report 1997:145-155.

[19] 敖海兵,陈加松,胡志鹏,等.盐穴储气库运行损伤评价体系[J].油气储运,2017,36(8):910-915.

[20] 丁国生,张保平,杨春和,等.盐穴储气库溶腔收缩规律分析[J].天然气工业,2007,27(11):94-96.

[21] 屈丹安,杨春和,任松.金坛盐穴地下储气库地表沉降预测研究[J].岩石力学与工程学报,2010,29(增刊1):2705-2711.

本论文原发表于《油气储运》2018年第37卷第8期。

盐穴储气库单腔长期注采运行分析及注采压力区间优化
——以金坛盐穴储气库西 2 井腔体为例

杨海军[1] 郭 凯[2] 李建君[1]

(1. 中国石油西气东输管道公司储气库管理处；2. 中国石油天然气与管道分公司)

【摘 要】 西气东输金坛盐穴储气库建设于地质条件复杂的层状盐层中，耗资巨大且有一定的使用期限，因此，确保腔体在安全稳定运行的前提下，尽量增加其工作气量，意义重大。以西气东输金坛盐穴储气库的盐矿老井西 2 井腔体为例，通过声呐数据对比分析，发现在目前的注采运行方式下存在一定程度的效益浪费，因而利用数值模拟方法校核了盐岩的蠕变参数，并利用修正后的参数对注采最小气压和最大气压进行了优化，增加了单腔工作气量，提高了经济效益。

【关键词】 盐穴储气库；数值模拟；蠕变；注采运行；注采压力区间

Analysis on long-term operation and interval optimization of pressure for single cavity injection/production in underground salt cavern gas storage-Taking the cavity of Well Xi-2 in salt cavern gas storage in Jintan as an example

Yang Haijun[1] Guo Kai[2] Li Jianjun[1]

(1. Gas Storage Project Department, PetroChina West-East Gas Pipeline Company;
2. PetroChina Company Limited Natural Gas&Pipeline Company)

Abstract Gas storage in Jintan salt cavern for West-to-East gas transmission project is constructed in laminar salt layer with complicated geological conditions, which costs a lot and has a certain service life. Therefore, under the premise of ensuring safe and stable operation of the cavity, it is of great importance to increase the gas production. Taking the old Well Xi-2 in salt mine of gas storage in Jintan cavern for West-to-East gas transmission project as an example, it is found through the comparative analysis of sonar data that there are some benefit waste under the current mode of injection/production, so the approach of numerical simulation is used to verify and check the creep parameters of rock

---

作者简介：杨海军，高级工程师，1961 年生，1983 年毕业于大庆石油学院开发系钻井工程专业，现主要从事地下储气库的运营管理工作。电话：13951201923，E-mail：weisz@petrochina.com.cn。

salt. And the corrected parameters are used to optimize the injection/production minimum pressure and maximum pressure, so as to increase the working gas production of single cavity and improve the economic benefit.

Keywords salt cavern gas storage; numerical simulation; creep deformation; injection/production operation; interval of injection/production pressure

地下油气储库作为油气管道的配套设施,在安全平稳供气和战略储备方面具有重要作用,特别是西气东输1~3线陆续建成投产,加快储气库建设势在必行。西方发达国家建设了大量油气储库,技术和经验相对成熟,我国尚处于初期建设和前期摸索阶段[1]。盐岩因其极低的渗透率、良好的蠕变性能和损伤自我恢复能力成为能源储备的良好场所,2003年中国石油西气东输管道公司在江苏金坛开始投资建设亚洲第一个盐穴储气库。与西方国家在巨厚的盐丘中建设储气库相比,我国在层状盐岩中建设储气库,具有埋深浅、夹层多、盐岩纯度低等不利因素[2-3]。

盐穴储气库虽然相比其他储存设备和储存方式具有较好的安全性,但近几十年来在国外也曾发生各类灾难性事故,如油气渗透、库区塌陷等,危害生命、财产、环境等的安全。我国在层状盐穴中建储气库,既缺乏经验又缺少成功案例作参考,而复杂的地质条件更是增加了施工风险。因此,在储气库建设前和运行过程中对盐穴腔体进行力学稳定性评价和研究尤为重要。关于盐岩的力学性能,国内外专家开展了大量研究工作,形成了一些盐穴腔体的稳定性判据,数值模拟主要使用FLAC3D、Abaqus、ANSYS等有限元软件[4-5]。

盐穴储气库建设投资巨大,具有一定的时效性,通常规划30~50年,使其在规划时间内、安全运行条件下工作气量最大化,方才符合经济效益最大化的原则[6]。工作气量的大小通常与腔体的几何参数(包括腔体深度、形状、高度、半径和矿柱宽度等)及运行参数(包括腔内最大气压p_{max}、最小气压p_{min}、循环周期等)有关。腔体设计阶段使用的力学参数、造腔结束后最终的几何参数、实际运行参数等因素导致腔体设计与现实情况存在一定程度的误差。因此,需要在腔体实际运行过程中,将腔体实际工作情况与模拟结果对比,利用有限元数值模拟方法反演并修正相关力学参数,重新调整运行参数。为此,以金坛储气库西2井为例,在不考虑其他腔体影响的情况下,利用Abaqus软件进行有限元数值模拟,以稳定性评价标准[7]为依据对腔体的生产运行方式进行评价与优化。

1 金坛储气库西2井腔体工程概况

西2井为盐矿老井,采盐结束后处于废弃状态,为了建设地下储气库,中国石油西气东输管道公司储气库项目部对一批盐矿老井(约40口)进行评估并从中选取6口改建地下盐穴储气库[8]。其中,西2井腔体自2005年开始施工检修,2007年1月开始注气排卤,2007年10月开始注采运行至今。

金坛储气库在构造位置上处于金坛盆地直溪桥凹陷,西2井位于盆地西部,属于陈家庄次凹。该井盐岩层深度为920.3~1065.2m,盐厚144.9m,盐岩顶部和底部主要为泥岩(图1)。根据相邻井的岩性解释结论,盐岩层段的主要岩性有盐岩19层/105.68m,其盐岩纯度较高，一般在90%左右；含泥盐岩11层/16.5m；泥质盐岩10层/6.4m；含钙芒硝盐岩9层/11.52m；盐质泥岩2层/0.8m；泥岩5层/8.1m，含3个厚度超过1.5m的泥岩夹层。

2005年1月11日对西2井腔体进行第一次声呐测腔，结果显示：腔体形状较规则，近似轴对称(图1)，腔体顶底深度为937.0~1010.8m，高为73.8m，最大半径为41m，高径比为0.925，腔体体积为12.98×10⁴m³，套管鞋深度为931.2m，腔体顶部距离盐层顶为16.8m，套管鞋距离盐穴顶仅5.8m(脖颈)，该腔体的几何参数与现今设计的理想腔体几何参数具有较大差距。

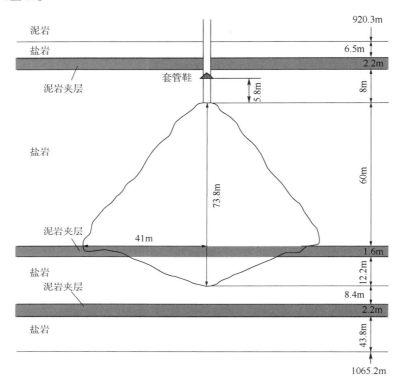

图1　金坛储气库西2井地质剖面及腔体示意图

盐矿老井主要采用自下而上分层开采、择优先用的方法采矿，且采矿时间较长，一般为5~10年，甚至更长，西2井缺少造腔数据，遂根据腔体的声呐体积为12.98×10⁴m³及盐矿的溶腔速度，估测其溶腔时间为6~8年。

2005年开始对西2井原始的井眼、固井、套管等进行修缮及更换，2007年1月31日开始注气排卤，6月30日结束，为期5个月，约150天。排卤完毕后，平压近4个月，2007年10月22日开始注采气运行，工作气压8.0~13.5MPa，运行方式为注气2个月，采气2个月，最高压停留3个月，期间伴有应急采气等突发状况。工作5年后，于2013年6月再次进行声呐测腔，发现腔体顶板下沉0.3m，底板上鼓0.6m，腔体体积约12.83×10⁴m³，损失约1.2%。说明在目前运行方式下，腔体体积收缩率较小，在规定使用期限内尚有一定的扩大调峰能力，即改善运行参数，增加工作气量。

2 计算模型与稳定性判断依据

2.1 计算模型

中国科学院武汉岩土力学研究所针对金坛储气库的岩性开展相关力学试验,根据泥岩、盐岩和泥岩夹层的单轴试验、三轴试验、蠕变试验结果[9],得到基本力学参数(表1)。盐岩流变本构模型为 $\dot{\varepsilon} = A_0 \sigma^n$(其中,$\dot{\varepsilon}$ 为蠕变率,σ 为偏应力,A_0、n 均为材料参数),根据金坛储气库盐岩蠕变试验及西2井的检测数据,通过反演方法最终确定的盐岩蠕变参数为:$A_0 = 3.6 \times 10^{-6} \mathrm{MPa}^{-3.5}/\mathrm{a}$,$n = 3.5$。

表1 金坛储气库基本力学参数

岩性	弹模/MPa	泊松比	黏聚力/MPa	内摩擦角/(°)	密度/($10^3 \mathrm{kg/m^3}$)
盐岩	13000	0.30	2.5	30	2.16
泥岩	10000	0.27	2.5	35	2.46
泥岩夹层	4000	0.30	2.0	30	2.35

根据西2井腔体2005年1月的声呐数据及相关地质数据,利用Abaqus软件建立腔体的二维轴对称几何模型,选取腔体的一个纵剖面的一半作为研究对象,计算模型为一个2000m×2000m正方形,平面为XY坐标系平面,竖直方向为Y轴,坐标原点为模型顶部左侧顶点。整个模型考虑了0~2000m的地层信息,因泥岩夹层较少且厚度较薄,故暂不考虑夹层的影响(图2)。根据腔体的受力情况,分别施加地应力和内部拉张力,地应力场采用三向等压自重应力场,模型底边加以纵向约束,左边和右边加以横向约束。采用有限元数学方法对建立的物理模型进行模拟计算,需要对几何模型划分网格,网格划分质量的好坏直接决定结果的准确度和精度,通常腔体周围的网格需要加密以分析腔体周围的物理过程。

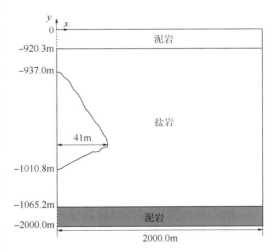

图2 金坛储气库西2井腔体物理模型示意图

2.2 稳定性判断依据

(1) 无张应力分布准则。盐岩在拉张应力作用下较脆,抗拉张强度极低,因此,盐穴周围腔壁中任何时候都不允许有拉张应力的存在,否则将导致盐岩层的破坏。通常腔内在较高气压下及长期盐岩蠕变过程中会形成拉张应力。

(2) 膨胀或损伤判断依据。储气库溶腔和运行导致盐岩层中初始的各相同性的应力发生改变,从而形成盐层中的偏应力,当盐岩受到偏应力的作用且超过某个最大值时,因微观裂缝的形成和扩展,岩石体积增大而发生扩容。这种现象通常被认为是膨胀,会造成腔体不稳定(剥落)及相对较高渗透性区域的气体刺穿盐岩层。膨胀指数 η 是表征岩石损伤的一个极好的指标。

在检查膨胀或损伤判据之前,要先创建新的输出变量指标 Lode Angle(θ):

$$\theta = \arctan\left(\frac{1}{\sqrt{3}} \frac{2\sigma_{mid} - \sigma_{max} - \sigma_{min}}{\sigma_{max} - \sigma_{min}}\right) \quad (1)$$

式中:θ 角的取值范围应该介于 $-\pi/6 \sim \pi/6$ 之间,若模型中大多数区域在此范围内则通过检查,否则检查不合格;σ_{min}、σ_{max}、σ_{mid} 分别为最小主应力、最大主应力、中间主应力。

检查膨胀或损伤,膨胀指数 η 的计算公式如下:

$$\eta = \sqrt{J_2} \frac{\cos\theta - \frac{1}{\sqrt{3}}\sin\theta \cdot \cos\phi}{C \cdot \cos\phi + p \cdot \sin\phi} \quad (2)$$

$$J_2 = \frac{\delta}{\sqrt{3}} \quad (3)$$

$$p = \frac{1}{3}(\sigma_1 + \sigma_2 + \sigma_3) \quad (4)$$

$$\delta = \sqrt{\frac{3}{2}(\sigma_{min}^2 + \sigma_{mid}^2 + \sigma_{max}^2)} \quad (5)$$

式中:C 和 ϕ 为岩石内部参数,分别为黏聚力和内摩擦角;σ_1、σ_2、σ_3 分别为第一主应力、第二主应力和第三主应力。

当膨胀指数 $\eta < 1$ 时,不会发生膨胀或损伤;当腔内气压在最小气压时,最易发生膨胀或损伤。

(3) 蠕变应变准则。盐穴在工作过程中会发生收缩变形[10],盐岩蠕变应变不能超过给定的限值,腔体的体积收缩率一般要求每年蠕变率小于1%,整个运行周期内小于10%,套管鞋处的盐岩蠕变率一般要求小于0.3%,否则套管鞋会发生塑性变形而损伤。

(4) 最大气压或气体渗漏准则。当切向应力(即平行于穴壁)小于作用于盐穴壁的拉张力(即气压)时,需要考虑气体渗透的风险。即墙壁周围水平应力的大小大于垂直应力的大小时,腔体无渗透风险,反之则有渗透风险,腔体内在最大气压时更易发生渗透。

3 腔体数值反演及生产运行方式优化

利用现有腔体形状和注采方式以及实验力学参数模拟得到的结果与实际情况可能差别较大,在其他条件确定不可改变的情况下,需要通过数值模拟方法反演相关实验力学数据,使拟合模拟结果与实际结果相匹配,从而确保今后模拟结果的准确性。

3.1 腔体数值反演拟合蠕变参数

盐岩具有蠕变性,盐穴腔体在工作过程中可能出现顶板下沉、底板上鼓及体积收缩等现象,根据声呐测腔数据及原始模拟结果的对比分析,发现模拟结果与实际情况存在一定差距,主要原因可能是地下温度对盐岩的蠕变参数有一定影响,因此,对盐岩蠕变参数进行修正,使模拟结果尽量与实际情况接近(表2)。

表2 盐岩蠕变参数修正表(腔体工作5年数据)

试验对比	蠕变参数 A_0	蠕变参数 n	顶板位移/m	底板位移/m	腔体体积收缩率/%
声呐数据			0.3	0.6	1.20
原始模拟结果	3.6×10^{-6}	3.5	0.6	1.5	10.21
反演修正结果	3.6×10^{-7}	3.5	0.2	0.5	1.90

3.2 生产运行方式的数值模拟优化

根据修正的模拟结果,西2井腔体以目前的注采方式在规划的使用期限内(30年)蠕变应变较慢,体积收缩率不足5%,且腔体未产生任何膨胀损伤,说明该腔体在安全运行条件下存在一定程度的经济效益浪费,因此需要调整其最大工作气压 p_{max} 和最小工作气压 p_{min},以增加工作气量,优化生产方式。

3.2.1 最小工作气压 p_{min} 的确定

通过经验公式可以确定腔内最小内压,如要求套管鞋深度的 p_{min} 梯度介于上覆岩层重力估测的垂直应力的0.2~0.3倍或者更高[11],恒压流变5年周期内腔体体积收缩小于5%。根据国外经验计算腔内最小内压为4.58MPa,为验证其准确性进行了6次模拟,最小内压范围为4~9MPa,模拟腔体工作30年的情况(表3)可知,随着工作时间变长,腔体顶板下沉、底板上鼓位移逐渐增加,腔体体积损失越来越大,且随着腔体最小内压的增加,腔体体积损失越来越小,腔壁位移渐渐变小。腔体最小内压为4MPa时会形成损伤区,其他最小内压均未产生损伤、拉应力区和渗漏。腔体工作内压为6MPa时第1个工作5年周期内体积收缩超过5%,工作内压为6MPa和7MPa时在工作周期内体积损失率变化较大(图3),因此选7MPa作为腔体最小内压。

表3 金坛储气库西2井腔体恒压流变30年后稳定性判断数据

工作压力/MPa	时间/a	顶板下沉位移/m	底板上鼓位移/m	最大半径/m	套管鞋处应变/%	体积损失率/%	损伤区	拉应力区	渗漏风险
4	5	0.167	0.455	0.377	0.41	1.92	无	无	无
4	10	0.412	1.192	1.031	0.94	9.85	有	无	无
4	15	0.489	1.458	1.268	1.10	12.67	有	无	无
4	20	0.544	1.646	1.436	1.21	14.62	有	无	无
4	25	0.587	1.796	1.568	1.28	16.15	有	无	无
4	30	0.624	1.921	1.678	1.35	17.44	有	无	无
5	5	0.167	0.455	0.377	0.41	1.92	无	无	无
5	10	0.364	1.051	0.906	0.84	8.41	无	无	无
5	15	0.432	1.286	1.115	0.98	10.85	无	无	无
5	20	0.479	1.450	1.261	1.07	12.56	无	无	无
5	25	0.516	1.581	1.377	1.15	13.92	无	无	无
5	30	0.548	1.691	1.475	1.21	15.05	无	无	无

续表

工作压力/MPa	时间/a	顶板下沉位移/m	底板上鼓位移/m	最大半径/m	套管鞋处应变/%	体积损失率/%	损伤区	拉应力区	渗漏风险
6	5	0.167	0.455	0.377	0.41	1.92	无	无	无
	10	0.323	0.933	0.800	0.75	7.22	无	无	无
	15	0.381	1.133	0.979	0.88	9.88	无	无	无
	20	0.421	1.274	1.105	0.96	11.74	无	无	无
	25	0.453	1.387	1.205	1.02	13.25	无	无	无
	30	0.480	1.482	1.289	1.08	14.55	无	无	无
7	5	0.167	0.455	0.377	0.41	1.92	无	无	无
	10	0.286	0.824	0.704	0.67	5.88	无	无	无
	15	0.334	0.990	0.852	0.77	7.66	无	无	无
	20	0.367	1.110	0.959	0.85	8.92	无	无	无
	25	0.394	1.206	1.044	0.90	9.93	无	无	无
	30	0.417	1.288	1.117	0.94	10.78	无	无	无
8	5	0.167	0.455	0.377	0.41	1.92	无	无	无
	10	0.252	0.723	0.614	0.60	4.84	无	无	无
	15	0.292	0.864	0.740	0.68	6.29	无	无	无
	20	0.320	0.964	0.829	0.74	7.35	无	无	无
	25	0.342	1.045	0.901	0.79	8.20	无	无	无
	30	0.361	1.114	0.962	0.83	8.92	无	无	无
9	5	0.167	0.455	0.377	0.41	1.92	无	无	无
	10	0.224	0.644	0.545	0.54	3.94	无	无	无
	15	0.254	0.752	0.641	0.60	5.09	无	无	无
	20	0.277	0.833	0.713	0.65	5.95	无	无	无
	25	0.295	0.899	0.771	0.69	6.65	无	无	无
	30	0.310	0.956	0.822	0.72	7.41	无	无	无

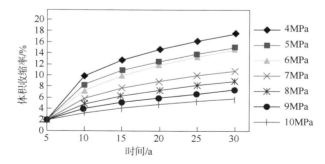

图3　金坛储气库西2井腔体4~9MPa内压下流变30年体积收缩率变化曲线

3.2.2 最大工作气压 p_{max} 的确定

关于腔体最大工作气压，国外认为套管鞋深度的 p_{max} 梯度应该介于上覆岩层重力估测的垂直应力的 0.75~0.85 倍或者更小[12]，最大压力梯度为 0.017MPa/m，计算可得西 2 井腔体的最大压力为 15.83MPa。因此选取 5 种注采运行方式进行模拟，其内压变化值为 7~14MPa，7~15MPa，7~16MPa，8~15MPa，8~16MPa（表 4），在该 5 种工况下腔体体积收缩率均较小，流变 30 年后体积收缩率均小于 10%，提高最小内压和最大内压均可以减小腔体体积收缩率，但提高最小内压更有利于减小腔体体积收缩率（图 4）。5 种工况模拟均未形成损伤区，但只有 7~14MPa 和 8~15MPa 在流变过程中未形成拉应力区，其他 3 种工况流变过程中局部区域形成了拉应力区，存在气体渗漏的风险。因此，腔体的最高内压确定为 15MPa，注采循环内压可取为 8~15MPa，应急采气情况下，最小内压可以降为 7MPa，但停留时间不宜过长。

表 4　金坛储气库西 2 井腔体注采流变 30 年后稳定性判断数据

工作压力/ MPa	时间/ a	顶板下沉位移/ m	底板上鼓位移/ m	最大半径/ m	套管鞋处应变/ %	体积损失率/ %	损伤区	拉应力区	渗漏风险
7~14	5	0.167	0.455	0.377	0.41	1.92	无	无	无
	10	0.204	0.569	0.481	0.49	3.17	无	无	无
	15	0.227	0.646	0.552	0.54	4.01	无	无	无
	20	0.245	0.707	0.608	0.59	4.68	无	无	无
	25	0.260	0.758	0.655	0.62	5.23	无	无	无
	30	0.273	0.804	0.696	0.65	5.71	无	无	无
7~15	5	0.167	0.455	0.377	0.41	1.92	无	无	无
	10	0.200	0.558	0.471	0.48	3.06	无	有	有
	15	0.222	0.630	0.539	0.53	3.84	无	有	有
	20	0.239	0.687	0.589	0.57	4.46	无	有	有
	25	0.253	0.736	0.633	0.61	4.98	无	有	有
	30	0.266	0.778	0.672	0.63	5.43	无	有	有
7~16	5	0.167	0.455	0.377	0.41	1.92	无	无	无
	10	0.198	0.549	0.462	0.48	2.96	无	有	有
	15	0.218	0.616	0.524	0.52	3.69	无	有	有
	20	0.234	0.670	0.572	0.56	4.27	无	有	有
	25	0.247	0.716	0.614	0.59	4.77	无	有	有
	30	0.259	0.756	0.650	0.62	5.21	无	有	有
8~15	5	0.167	0.455	0.377	0.41	1.92	无	无	无
	10	0.188	0.526	0.442	0.45	2.70	无	无	无
	15	0.204	0.581	0.493	0.49	3.29	无	无	无
	20	0.217	0.626	0.534	5.15	3.78	无	无	无
	25	0.228	0.665	0.570	0.55	4.21	无	无	无
	30	0.238	0.700	0.602	0.57	4.58	无	无	无

续表

工作压力/MPa	时间/a	顶板下沉位移/m	底板上鼓位移/m	最大半径/m	套管鞋处应变/%	体积损失率/%	损伤区	拉应力区	渗漏风险
8~16	5	0.167	0.455	0.377	0.41	1.92	无	无	无
	10	0.185	0.518	0.435	0.45	2.62	无	有	有
	15	0.200	0.568	0.481	0.48	3.16	无	有	有
	20	0.212	0.610	0.519	0.51	3.62	无	有	有
	25	0.223	0.647	0.553	0.54	4.01	无	有	有
	30	0.232	0.680	0.582	0.56	4.36	无	有	有

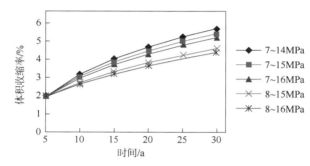

图4 金坛储气库西2井腔体5种注采方式流变30年体积收缩变化曲线

4 结论

通过对金坛盐穴储气库西2井腔体两次声呐数据进行对比分析，发现其在5年的注采运行中腔体体积收缩率较小且在规划期限内及安全运行条件下，现行的运行方式存在一定程度的经济浪费，遂利用Abaqus软件对该腔体进行数值模拟，反演并修正了盐岩的蠕变参数，进而对今后25年的注采方式进行模拟，确定了腔体的最小气压为7MPa，最大气压为15MPa，腔体在7~15MPa压力区间范围内运行是安全的，建议采用8~15MPa方案，应急情况下最小气压可以降至7MPa，但停留时间不宜过长，从而将该腔体之前的8~13.5MPa运行方式进行了优化，增大了工作气量。

参 考 文 献

[1] 谭羽非,陈家新,余其铮,等. 国外盐穴地下储气库的建设与研究进展[J]. 油气储运,2001,20(1):6-8.

[2] 杨春和,梁卫国,魏东吼,等. 中国盐岩能源地下储存可行性研究[J]. 岩石力学与工程学报,2005,24(24):4410-4417.

[3] 杨春和,李银平,陈锋. 层状盐岩力学理论与工程[M]. 北京:科学出版社,2009.

[4] Asgari A, Ramezanzadeh A, Mohammad S, et al. Stability analysis of natural-gas storage caverns in salt formations[C]. Solution Mining Research Institute Fall 2012 Technical Conference, Bremen, Germany, October 1-2, 2012.

[5] Adams J B. Determination of salt cavern operating pressures using rock mechanics and finite element analysis[C]. Solution Mining Research Institute Fall 1996 Meeting, Cleveland, Ohio, USA, October 20-23, 1996.

[6] 赵克烈. 注采气过程中地下盐岩储气库可用性研究[D]. 武汉：中国科学院武汉岩土力学研究所, 2009.

[7] 吴文, 侯正猛, 杨春和. 盐岩中能源(石油和天然气)地下储存库稳定性评价标准研究[J]. 岩石力学与工程学报, 2005, 24(14)：2497-2505.

[8] Qu D, Wei D H. Use of sonar surveys for planning conversion of brine wells to gas storage caverns[C]. Solution Mining Research Institute Spring 2007 Technical Conference, Basel, Switzerland, April 29 – May 2, 2007.

[9] 梁卫国, 徐素国, 赵阳升, 等. 盐岩蠕变特性的试验研究[J]. 岩石力学与工程学报, 2006, 25(7)：1386-1390.

[10] 丁国生, 杨春和, 张宝平, 等. 盐岩地下储库洞室收缩形变分析[J]. 地下空间与工程学报, 2008, 4(1)：80-84.

[11] Rokahr R, Staudtmeister K, Schiebenhofer D Z. Rock mechanical determination of the maximum internal pressure for gas storage caverns in rock salt[C]. Solution Mining Research Institute Fall 1998 Meeting, Rome, ltaly, October 4-7, 1998.

本论文原发表于《油气储运》2015 年第 34 卷第 9 期。

盐穴储气库注采运行对邻近铁路安全影响分析

张 幸[1] 李文斌[1] 谢 楠[1] 贲建华[1] 王成林[1] 黄盛隆[1] 叶良良[2]

（1. 国家管网集团西气东输公司；2. 中国科学院武汉岩土力学研究所）

【摘 要】 盐穴储气库作为地下天然气储存的一种主要方式，在我国油气储运行业中的重要作用日益突出。但是由于盐岩的流变特性，在储气库的运营过程中盐穴腔体体积会随着运营年限逐渐减少，从而可能引发地面沉降等一系列环境问题。针对某盐穴储气库建设工程，通过适用盐岩蠕变变形的幂指数模型对储气库运营中盐岩腔引发的地面沉降进行模拟，并对邻近铁路运营可能产生的影响进行分析。分析显示，储气库运行30年后，储气库区域沉降表现为整体的均匀沉降，无明显的地表倾斜，区域最大沉降196mm，最大纵横向倾斜率未超过0.1mm/m，储气库群运行所产生的区域地面沉降对铁路的运行无明显影响。

【关键词】 盐穴储气库；地面沉降；铁路安全；数值分析；注采运行

Influence Analysis of Injection-production Operation of Salt Cavern Gas Storage on the Safety of Adjacent Railways

Zhang Xing[1] Li Wenbin[1] Xie Nan[1] Ben Jianhua[1] Wang Chenglin[1]
Huang Shenglong[1] Ye Liangliang[2]

(1. PipeChina West East Gas Pipeline Company;
2. Institute of Rock and Soil Mechanics, Chinese Academy of Sciences)

Abstract Salt cavern gas storage is a main method of underground natural gas storage. With the construction and development of salt cavern gas storage in China, its position in the oil and gas storage and transportation industry has become very prominent. At the same time, due to the rheological characteristics of salt rock, the volume of the salt cavern will gradually decrease with the operating time during the operation of the gas storage, which may cause a series of environmental problems such as surface subsidence. For the construction project of salt cavern gas storage in Pingdingshan, the surface subsidence caused by salt cavern in the operation of gas storage was simulated by using the power exponential model suitable for salt rock creep deformation, and the possible influence on the operation of adjacent railways was analyzed. It shows that after 30 years of operation, the subsidence of the gas storage

---

作者简介：张幸，男，1986年生，汉族，安徽宿州人，硕士，高级工程师，主要从事储气库工程技术研究。E-mail：zhangxing@pipechina.com.cn。

area is uniform as a whole, with no obvious surface tilt. The maximum subsidence in the area is 196mm, and the maximum vertical and horizontal tilt rates do not exceed 0.1mm/m. The regional ground subsidence caused by the operation of gas storage group has no obvious influence on the operation of the adjacent railway.

Keywords salt cavern gas storage; surface subsidence; railway safety; numerical analysis; injection-production operation

由于盐岩的流变性与重结晶特性，深部地质结构体中盐岩地层具有完善的密封性，其在我国及世界能源和天然气地下储存中获得了广泛应用[1-3]，我国第一座盐穴储气库——金坛储气库一期已建成投入使用。国内外已建成盐岩储气库的相关分析与监测结果显示，盐岩储气库在运行过程中，其腔体会随着时间逐渐缩小，与此同时，会导致一定程度的地面沉降，严重时可能危及地面建筑和人员财产安全[4-6]。因此，在盐岩储气库建设过程中有必要开展地面沉降对地面构筑物的安全影响评价[7]。

针对盐岩储气库建设引起的地面沉降，国内外已进行了大量的研究，研究方法有概率积分法、拟合函数法、数值模拟方法等，并结合实际的地面沉降监测数据进行综合分析[8-13]。德国 Bernburg 储库群埋深为 500~650m，存储体积约为 $10^5 m^3$，根据5年多的观测记录，其中心区域沉降量逐年递增，累积达到42mm，影响范围约为450m[14]。美国 Mont Belvieu 储库包括124个液化烃存储盐腔，埋深为750~1700m，运行期间地表沉降速率为20~40mm/a，影响范围为1500m[15]。与国外的盐岩层相比，我国盐岩层多为陆相沉积，泥质夹层及泥岩层盐岩层互层较多，盐岩层的力学性质较为复杂[16]。

我国某盐岩储气库井位分布区存在既有货运铁路，针对这一工程实际，通过数值模拟方法分析了该地区由于盐穴储气库注采运行而产生地面沉降和地表倾斜值，对铁路运行的安全影响进行了分析评价。

1 项目概况

该盐穴储气库是西气东输二线工程地下储气库建设库址，已开始进行前期建设。根据建设规划，该储气库的井位分布区是当地工业企业货运铁路穿越区，井位布设与铁路相对位置如图1所示。由于盐穴储气库具有很强的蠕变特性，在长期运营过程中，必然会产生一定量的地面沉降，为了保证货运铁路的运营安全，有必要对储气库运营产生的地面沉降和地表倾斜值进行分析。

根据《铁路运输安全保护条例》第十八条规定，在铁路线路两侧路堤坡脚、路堑坡顶、铁路桥梁外侧起1000m范围内，及在铁路隧道上方中心线两侧各1000m范围内，禁止从事采矿、采石及爆破作业。按照上述规定，铁路两侧各1000m总计2000m宽度范围内是不允许进行采矿作业的。考虑到储气库建设需要，在不影响铁路安全运营的前提下，有必要开展盐穴储气库建设对铁路安全的影响分析，确定保护带范围内的储气库建设不影响铁路的安全运行。

根据《Ⅲ、Ⅳ级铁路设计规范》5.6.4条规定，在松软地基上填筑路基应进行工后沉降分析。沉降量应满足以下要求：Ⅲ级铁路基的工后沉降不应大于30cm，Ⅳ级铁路不应大于

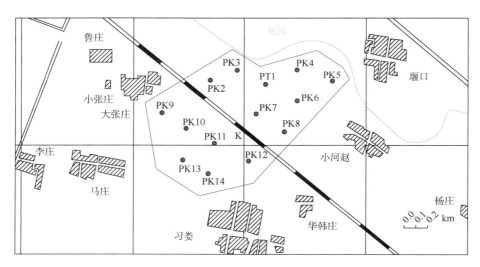

图 1 储气库井位分布

40cm。依据《铁路技术管理规程》，速率小于 120km/h 的线路要求钢轨横向倾斜差值不得超过 4mm，依据我国铁轨标准轨距 1.435m，垂直于线路方向的地表倾斜值应小于 4mm/1.435m，即 0.278%。

综合以上规范的各种规定，穿越该储气库区域铁路线路为低于 80km/h 低速工业企业货运铁路。依据国家铁路分级标准，可定义为该铁路为Ⅲ级铁路，其应有的沉降变形许可值可参考该等级铁路标准或高一级等级标准。据此，该铁路最大沉降允许值可确定为 300mm，年最大沉降速度为 40mm/a，最大横向倾斜值为 0.278%。

2 储气库注采运行地面沉降分析

根据该盐矿建库区域地层条件和腔体形态设计，盐穴储气库顶深为 1710m，底深为 1930m。储气库盐穴腔体形状分为三部分，下部为圆锥，高度为 40m，中部为圆柱，高度为 110m，顶部为半椭圆，高度为 70m，腔体最大直径为 80m，腔体模拟形状图如图 2 所示。针对以上腔体形状和实际的地层，建立如图 3 所示的盐腔数值分析模型，数值模拟采用 FLAC 2D 软件，能较好地处理岩石的黏弹塑性问题，建立的模型是二维轴对称数值模拟地质模型，模型建立中考虑了大于 2m 的泥岩夹层。模型大小为 3000m×2200m，上边界为地面，为自由面，其他三个边界面的法向位移是固定的。下边界深度为 2200m，水平方向为 3000m，地下深度 1355m 以上为泥岩层，其下为盐岩和泥岩夹层。图 4 所示为网格剖分图，在腔体周围的网格密度大，外部边界附近网格密度小，网格单元数量约为 22100 个，具有较好的计算精度。

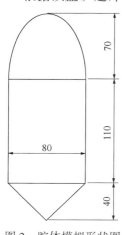

图 2 腔体模拟形状图（单位：m）

 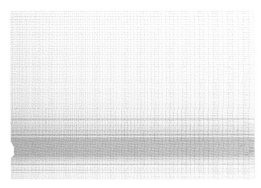

图 3 沉降分析地质模型图　　　　　图 4 沉降分析网格模型图

地质体的本构模型采用适用于盐岩的稳态蠕变的幂指数蠕变模型，表达式为

$$\dot{\varepsilon}=A\bar{\sigma}^n \tag{1}$$

式中：$\dot{\varepsilon}$ 为稳态蠕变率，a^{-1}；A 为材料参数，$MPa^{-n}\cdot a^{-1}$；n 为材料参数；$\bar{\sigma}$ 为偏应力，MPa。

根据该盐穴储气库力学试验结果，并结合国内外盐岩力学特性的研究成果，确定计算参数，模拟参数见表 1。

表 1　蠕变参数值表

试样类别	$A/(MPa^{-n}\cdot a^{-1})$	n
盐岩	2×10^{-7}	3.3
泥岩夹层	2.8×10^{-8}	2

分析模拟考虑储气库注采运行工况，运行压力为 12~28MPa，每年运行两个周期。
第一注采周期：
采气期：12 月 1 日—2 月 28 日，共 90 天，压力由上限至下限均匀递减；
平衡期：3 月 1 日—3 月 15 日，保持下限压力；
注气期：3 月 16 日—6 月 30 日，共 107 天，压力由下限至上限均匀递增；
平衡期：7 月 1 日—7 月 15 日，保持上限压力。
第二注采周期：
采气期：7 月 16 日—9 月 15 日，共 62 天，压力由上限至下限均匀递减；
注气期：9 月 16 日—11 月 30 日，共 76 天，压力由下限至上限均匀递增。
运行方案如图 5 所示。
通过对平顶山储气库运行工况下 30 年的蠕变变形分析，30 年腔体蠕变体积收缩 18.6%。该条件下，储气库区地表沉降形成以井口为中心的凹陷沉降区，储气库最大沉降发生的井口，约为 16.8mm，地表以井口为圆心，半径 0~500m 的范围内为主沉降区，500~2000m 范围为沉降急速减少区，2000m 以外为转微沉降区，图 6 为储气库腔周沉降分布图。

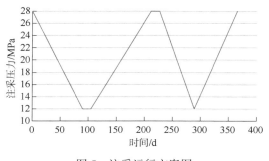

图 5　注采运行方案图

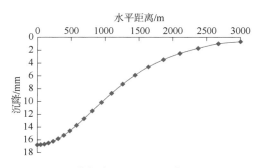

图 6　储气库腔周地表沉降分布图

3　储气库腔群地面沉降对铁路影响分析

盐岩储气库建设运营引发的地面沉降是一个动态与长期的过程，其沉降量随着时间逐步增加，地面上某一点的沉降也受到区域内多个储气库腔体共同作用的影响[17]。为考虑储气库腔群共同作用的影响，运用叠加原理，设腔群各腔体第 i 井口与地面某点的距离分别为 r_i，该腔体对该点产生的沉降用 $S(r_i)$ 表示，则该点受腔群影响产生的总沉降可以用各个单腔产生的沉降和计算：

$$S = \sum_{i=1}^{n} S(r_i) \tag{2}$$

储气库井区与铁路相对位置如图 7 所示，分别沿铁路横轴线和纵轴线建立坐标系，K 为坐标原点。为分析储气库群运行引发地面沉降对铁路安全运行的影响，分别对垂直于铁路线的 A-A1、B-B1、C-C1、D-D1、E-E1 等五条横剖面及沿铁路线的 F-F1 纵剖面进行分析。所有储气库运行产生的沉降通过单腔运行沉降线性叠加的方法获得储气库群的整体沉降。

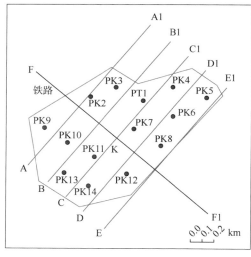

图 7　储气库井区与铁路相对位置图

图 8 为 A-A1 剖面、B-B1 剖面、C-C1 剖面、D-D1 剖面、E-E1 剖面的地面沉降图，图 9 为 F-F1 铁路纵轴线剖面地面沉降曲线。A-A1 剖面、B-B1 剖面、C-C1 剖面、D-D1 剖面、E-E1 剖面的最大沉降分别为 177.6mm、192.6mm、196.0mm、190.2mm、171.1mm。区域沉降以铁路纵轴线和 C-C1 为中心，最大沉降位于坐标原点 K 附近，为 196.0mm，在储气库运行的 30 年期间产生的地面沉降，未超过Ⅲ级铁路最大许可工后沉降 300mm，年平均沉降速率为 6.6mm/a，远低于许可沉降速率 40mm/a。储气库区域的地面沉降表现为整体沉降，无明显的地表倾斜，铁路横纵线的纵横向最大倾斜率均小于 0.01%，未超过铁路许可横向倾斜率 0.278%，储气库群运行

所产生的区域地面沉降对铁路的运行影响很小。

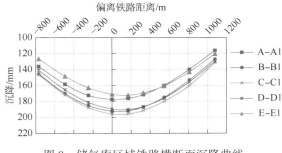

图8 储气库区域铁路横断面沉降曲线　　图9 铁路纵轴线 F-F1 剖面沉降曲线

4 结论

通过对储气库建设的地面沉降进行模拟及对铁路运营影响分析,获得如下结论:

(1)储气库单腔注采运行 30 年,其形成了明显的地面沉降,地表凹陷以井口为圆心,半径 0~500m 的范围内为主沉降区,500~2000m 范围为沉降急速减少区,2000m 以外为转微沉降区,井口中心最大沉降为 16.8mm。

(2)以单腔运行产生的沉降进行叠加,储气库区域的地面沉降表现为整体沉降,无明显的地表倾斜,储气库腔群产生的地面沉降最大为 196.0mm,铁路横纵线的纵横向最大倾斜率均未小于 0.01%,未超过铁路许可横向倾斜率 0.278%,储气库群运行所产生的区域地面沉降对铁路的运行影响很小。

由于盐穴地表沉降是一种长期且缓慢的过程,为防范风险,在储气库的建设过程中应建立地面沉降监测网,且在临近铁路区域布设加密测点,保证储气库建设过程中的铁路运行安全。

参 考 文 献

[1] 杨春和,梁卫国,魏东吼,等. 中国盐岩能源地下储存可行性研究[J]. 岩石力学与工程学报,2005,(24):4409-4417.

[2] Shi X L, Chen Q L, Ma H L, et al. Geomechanical investigation for abandoned salt caverns used for solid waste disposal [J]. Bulletin of Engineering Geology and the Environment, 2021, 80: 1205-1218.

[3] Yin H W, Yang C H, Ma H L, et al. Study on damage and repair mechanical characteristics of rock salt under uniaxial compression [J]. Rock Mechanics and Rock Engineering, 2019, 52: 659-671.

[4] Bérest P, Brouard B. Safety of salt cavern used for underground storage [J]. Oil & Gas Science and Technology, 2003, 3(58): 361-384.

[5] 李银平,孔君凤,徐玉龙,等. 利用 Mogi 模型预测盐岩储气库地表沉降[J]. 岩石力学与工程学报,2012,31(9):1739-1745.

[6] 陈雨,李晓,侯正猛,等. 不同腔形下盐岩储库区地表最大变形预计新方法[J]. 岩土工程学报,2012,34(5):826-833.

[7] 井文君,杨春和,孔君凤,等. 盐岩地下储备库引发地表沉陷事故的风险分析[J]. 岩土力学,2011,32(S2):544-550.

[8] 陈雨,李晓. 盐岩储库区地面沉降预测与控制研究现状与展望[J]. 工程地质学报, 2010, 18(2): 252-260.

[9] 任松,姜德义,杨春和,等. 岩盐水溶开采沉陷新概率积分三维预测模型研究[J]. 岩土力学, 2007, 28(1): 134-138.

[10] 孔君凤,李银平,杨春和,等. 盐穴储气库群地表铁路路基变形及运输安全评价[J]. 岩石力学与工程学报, 2012, 31(9): 1776-1784.

[11] 李二兵,方秦,谭跃虎,等. 层状盐岩中天然气地下储存库地表沉降分析[J]. 解放军理工大学学报(自然科学版), 2011, 12(6): 629-634.

[12] Eickemeier R. A new model to predict subsidence above brine fields[C]. In: Proceedings of SMRI Fall Meeting, Nancy, France, 2005.

[13] Liu W, Li Y P, Yang C H, et al. A new method of surface subsidence prediction for natural gas storage cavern cavern in bedded salt rock[J]. Environmental Earth Sciences, 2016, 75(9): 1-17.

[14] Menzel W, Schreiner W. Results of rock mechanical investigations for establishing storage cavern in salt formations[C]. Proc. 6th Symp. on Salt. Alexandia: The Salt Institute, 1985.

[15] Ratigan J. Ground subsidence at Mount Belvieu, Texas (panel discussion-surface subsidence)[C]. In: Proceedings of SMRI Spring Meeting, Atlanta, 1991.

[16] 杨春和,李银平,陈锋. 层状盐岩力学理论与工程[M]. 北京:科学出版社, 2009.

[17] 屈丹安,杨春和,任松. 金坛盐穴地下储气库地表沉降预测研究[J]. 岩石力学与工程学报, 2010, 29(S1): 2705-2711.

本论文原发表于《岩土工程技术》2023年第37卷第1期。

盐穴储气库稳定性的实时微地震监测

井岗 何俊 陈加松 王元刚 肖恩山 姚威
薛雨 闫凤林

(中国石油西气东输管道公司储气库项目部)

【摘 要】 盐穴地下储气库常面临地表沉降、盐岩破坏、气体渗漏及腔体收缩过快等安全稳定性问题。为了监测盐穴储气库注水排卤和注气排卤过程中的腔体稳定性,确保储气库安全平稳运行,利用实时微地震技术对金坛盐穴储气库连续监测12天,通过在井下布置高精度检波器,接收生产过程中的微地震事件,研究储气库造腔及注采气过程中微地震事件诱发因素,综合分析不同微地震信号特征。研究表明:检波器位于3层套管中导致储气库微地震信号衰减严重,特别是套管间环空液体对横波的阻碍;观察到的一些微地震事件可能为腔壁垮塌或破裂诱发所致,注水排卤工艺参数变化会引起微地震信号突然增多,统计微地震信号数量认为当前的生产工艺参数可以满足储气库安全稳定运行。

【关键词】 盐穴储气库;微地震;注水排卤;注气排卤;稳定性

Application of real time micro-seismic technique to the stability monitoring of salt cavern gas storage

Jing Gang He Jun Chen Jiasong Wang Yuangang Xiao Enshan
Yao Wei Xue Yu Yan Fenglin

(Gas Storage Project Department, PetroChina West-East Gas Pipeline Company)

Abstract Salt cavern gas storages are often faced with the problems of safety and stability, such as surface subsidence, salt rock failure, gas leakage and fast cavity shrinkage. In order to monitor the stability of the cavity in the process of water or gas injection and brine discharge, so as to ensure the stable and safe operation of gas storage, a real-time micro-seismic technique was used to monitor the Jintan salt cavern gas storage continuously for 12 days. A high-precision geophone was placed downhole to receive the micro-seismic events in the process of production. The factors inducing the micro-seismic events in the process of solution mining and gas injection/production were investigated and the character-

---

基金项目:中国石油储气库重大专项子课题"地下储气库关键技术研究与应用",2015E-4008。
作者简介:井岗,男,1986年生,工程师,2012年硕士毕业于中国石油大学(北京)矿产普查与勘探专业,现主要从事地下储气库相关技术的研究工作。地址:江苏省镇江市润州区南徐大道60号商务A区D座中国石油西气东输储气库管理处,212050。电话:15665115327,Email:xqdsjinggang@petrochina.com.cn。

istics of different micro-seismic signals were analyzed comprehensively. It is shown that the micro-seismic signals of the gas storage are attenuated seriously if the geophone is set in the three-layer casing and the liquid in the casing annulus is especially fatal to transverse wave propagation. It is indicted that some observed micro-seismic events may be caused by the collapse or rupture of the cavity wall and the quantity of micro-seismic signals is increased sharply due to the variation of water injection and brine discharge parameters. It is demonstrated from the current micro-seismic signal statistics that the current production parameters can guarantee the stable and safe operation of gas storage.

Keywords　salt cavern gas storage; micro-seismic; water injection and brine discharge; gas injection and brine discharge; stability

目前，中国地下储气库的腔体稳定性监测方法主要有地表沉降监测[1-5]和力学稳定性评价，但是这两种方法均不能实现盐穴储气库的实时监测。

在国外，微地震监测[6]采卤盐穴发展较早，意大利曾经提前3周利用实时微地震预测了采卤腔体的垮塌，促进了实时微地震监测地下腔体稳定性的发展。目前，埃克森美孚在加拿大艾伯塔省、壳牌加拿大公司在阿曼Fahud、道达尔公司在委内瑞拉均进行过微地震监测研究。

中国的呼图壁储气库进行了长达两年的实时微地震监测试验，监测废弃油气藏型储气库在注采运行过程中断层活化情况。

实时微地震监测的目标是预警储气库生产过程中的岩层破裂、腔体垮塌信号，从而提前采取有效措施，控制和防范地下腔体垮塌，并对生产运行工艺参数进行安全评价。

1　数据采集

在大地构造位置上，金坛盆地位于扬子准地台的东北部，是苏南隆起区常州坳陷带中的次一级构造。金坛盆地位于茅山推覆带和上黄—大华隆起带之间，北临丹阳盆地、陵口盆地，南临南渡盆地，是一个北东走向的新生代盆地，盆地北东向长33km，北西向宽22km。钻遇地层从浅到深为第四系东台组（Qd）、古近系三垛组（Es）、戴南组（Ed）、阜宁组（Ef），造腔层段为阜宁组四段（Ef_4），是厚度约200m稳定分布的层状盐岩。

1.1　监测井、目标井

监测井A为金坛盐穴储气库造腔井，采用二开裸眼完钻方式，盐顶深度为988.5m。钻井技术套管（直径244.5mm）下深1008.77m，造腔管柱中间管（直径177.8mm）下深1085.71m，中心管（直径114.3mm）下深1115.93m。监测井A腔顶深度为1055m，腔体体积为$25×10^4m^3$。注水排卤井B为储气库造腔井，采用二开裸眼完钻方式，盐顶深度为984m。钻井技术套管下深999.46m，造腔管柱中间管下深1075.45m，中心管下深1095m。注水排卤井B腔顶深度为1040m，腔体体积为$26.7×10^4m^3$，与监测井A距离为332m。注气排卤井C为储气库完腔注气排卤井，完钻方式为二开裸眼完钻，盐顶深度为999.8m。钻井技术套管下深1020m，注采管下深1016m，排卤管下深1106m，腔顶深度为1034m，腔体体积为$16.4×10^4m^3$，与监测井A距离为311m（图1）。

1.2 观测系统

以监测井 A 的井口为坐标原点,建立注水排卤井 B 轨迹和注气排卤井 C 轨迹的统一储气库监测坐标系,确立检波器位置与储气库的相对坐标。微地震监测有井中监测、地面监测及浅井长期埋置 3 种方式。地面监测接收排列布设快,水平方向定位精度高,震源机制分析相对准确,其缺点是高度控制差,研究区存在玄武岩屏蔽层,不利于精细分析。浅井长期埋置具有丛式井成本低和分辨率较高的特点,其缺点是单井成本高,高度控制差,更适用于微地震长期监测。井中监测的信号信噪比高、微地震事件多、高度控制好,但是选井困难、较远事件无法监测。根据现有条件采取井下监测方式,采用 8 个 slimwave 三分量检波器接收,检波器间距定为 10m,根据储气库目的层垂深,检波器位置应该尽可能靠近监测井套管鞋(图 2)。监测井 A 的 8 个三分量检波器设计放置深度为 970~1040m(表 1),单流阀深度建议在 1055m 以下(仪器下井之前需提出中心管,在中心管底部安装单流阀后再次下放中心管,用泵车打清水置换卤水,确保中心管内全部为清水,防止卤水对仪器及电缆造成腐蚀)。

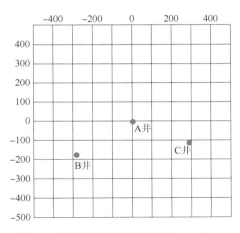

图 1 微地震监测井位图(单位: m)

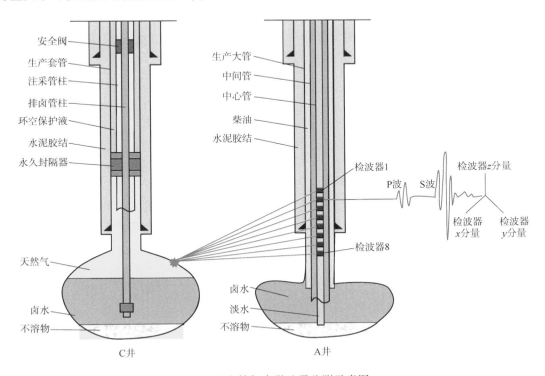

图 2 盐穴储气库微地震监测示意图

表1 监测井 A 的 8 个检波器位置分布

检波器号	测深/m	北偏移量/m	东偏移量/m	垂深/m
1	970	0	0	970
2	980	0	0	980
3	990	0	0	990
4	1000	0	0	1000
5	1010	0	0	1010
6	1020	0	0	1020
7	1030	0	0	1030
8	1040	0	0	1040

数据采集利用地面炮对井下检波器进行三分量定向，记录地面炮信号，已知地面炮位置，依据地震波传播原理确定检波器三分量方位。三分量检波器记录有效信号的初至波，通过水平分量旋转和垂直分量旋转实现 P 波、SH 波、SV 波分离。在整个反演、正演过程中，对比计算得出的地面炮位置和实际位置，越接近实际位置，说明检波器三分量定位越准确，速度模型越精准（依据监测井、目标井的声波测井曲线，对储气库地质分层，建立 P 波、S 波二维层状速度模型），重复该过程，反复调整速度模型，使其达到最理想的状态，并以此定位储气库中岩石破裂产生的微地震事件。

通过现场分析可控源井下检波器方向的校正信号，初至信号比较弱，分析原因为：(1) 库区 500~600m 埋深范围内稳定分布一套玄武岩，其对地震波信号有很强的屏蔽作用；(2) 3 层套管对震源信号造成了一定程度的衰减。

2 数据处理

2.1 微地震事件识别

微地震信号能量较弱，从背景噪声中识别有效微地震信号是实时微地震监测的关键。通过分析微地震信号的幅度谱、频谱、能量包络、频带范围，从而进行有效事件的确定[7-9]。通过实际数据的处理分析，在背景噪声的基础上识别 3 类微地震信号：(1) 微地震事件有效波，视速度为 4500~5000m/s 的上行波，只能观察到纵波观察不到横波，视为垮塌或破裂形成的微地震事件，推测不能接收到横波的原因为 3 层套管之间存在 2 个卤水环空，横波在液体中是不传播的；(2) 视速度为 5300~5800m/s 的上行波，只能观察到纵波观察不到横波，可能为地震波沿套管向上传播造成的；(3) 地面干扰波，视速度为 5000~5500m/s 的下行波，只能观察到纵波观察不到横波，波形持续时间短，可能是不明机械造成的，背景噪声以低频干扰为主，干扰频率约为 20Hz。

2.2 微地震事件自动拾取

微地震信号初至拾取是基础性处理工作[10-12]，在长达 12 天的实时微地震监测过程中可能存在大量的有效事件。因此需要在人工干预、识别的基础上进行自动拾取。初至重拾取是指对初至波形进行 2 次拾取，初步确定微地震震源位置，在直线传播路径的假设下将每个三分量转换成 P 波、SH 波、SV 波，将微地震群在纵向进行分组，按位置重排每组里的 P 波、SH 波、SV 波，最后根据重排后同向轴的相似性重新拾取初至事件。

2.3 微地震事件定位

利用射线追踪法和 Geiger 算法等定位方法联合确定微地震事件位置。检波器接收破裂能量信号并传输至地面仪器，再送入处理软件接口进行数据自动处理，经筛选后得到有效破裂事件，之后通过反演定位找到有效事件发生的位置，对微地震事件进行能量分析确定震级，分析每个微地震事件的 P 波、S 波数量，将定位误差控制在 10m 以内，从而实时判断裂缝的空间展布与走向以及裂缝方位等信息。微地震事件的定位流程为：(1) 人工交互优化参数，得到连续的波形数据；(2) 自动拾取 P 波、S 波初至事件；(3) 根据 P 波确定震源方向、事件方位及相关参数，用扫描的方式得到震源的最佳位置，即在三维空间中利用射线追踪技术，以误差最小为原则确定震源位置；(4) 事件检查并对其重新拾取初至时间、重新处理，更换不同的速度模型，重复以上过程，确定事件的最准确定位(图 3)。

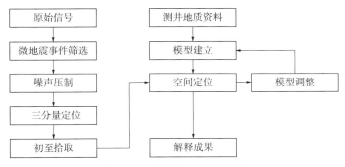

图 3 微地震数据处理流程图

2.4 微地震事件解释

根据生产过程中注气、造腔情况，实时监测可分为 3 个阶段。注气排卤和注水排卤同时进行阶段(9月22日8:00—9月27日18:00；9月27日18:00—9月29日8:00)；单独注气排卤阶段(9月27日6:00—9月27日18:00；9月29日8:00—9月30日19:00)；注气排卤和注水排卤同时停止阶段(9月30日19:00—10月3日5:00)。

由 3 个阶段微地震数据记录(图 4)可见：注气、造腔同时进行阶段监测到的震动信号最多，当注水排卤停井时微地震信号随之减少，随着造腔和注气排卤先后停井时监测到的震动信号明显减少，即 9月29日8:00 注水排卤井 B 停井后，自动识别个数明显减少 9月30日19:00 注气排卤井 C 停井后，自动识别个数又下降(图 5)。9月27日、9月28日两天中自动识别出的微地震信号较多，虽然不全是微地震有效信号，包括套管波、地面干扰波等噪声，但可以反映微地震事件相对活动剧烈程度。

分析注气排卤井 C 的工艺参数(图 6)，包括注气排量和注气压力，9月27日、9月28日这两天压力和注气量变化很小，因此自动识别个数增多可能与井 C 无关。

通过对注水排卤井 B 的工艺参数(图 7)分析可知：9月27日6:00—18:00 停井，重新开井之后一直到 9月29日8:00 停井，井口压力和注水量相对关井前较大。自动识别微地震事件数量突然增多可能与井口压力和注水量增大有关。

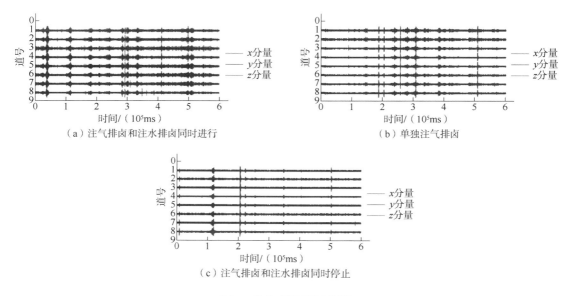

图 4 微地震数据记录图

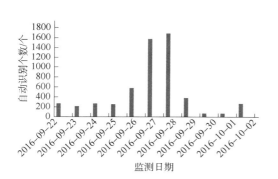

图 5 实时微地震事件自动识别柱状图

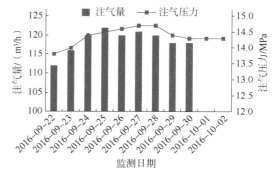

图 6 注气排卤井 C 井的工艺参数柱状图

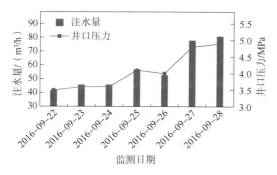

图 7 注水排卤井 B 井的工艺参数柱状图

Emmanuele Naymn 在 Solvay SA 地区利用实时微地震预测因采卤产生的垮塌[13-14]，监测数据显示：在垮塌前 3 周接收到的微地震事件数量突然增多，从平时每天 20~30 个微地震事件突变到每天 80~100 个，持续了 3 周从量变到质变的微地震事件爆发增长后盐腔垮塌。排除因造腔井开井引起微地震事件增多因素外，监测期内微地震事件数量平稳，因此认为目前的生产工艺参数满足储气库安全稳定要求[15-20]。

3 结论

针对国内盐穴储气库缺少实时监测方法的现状,采用实时微地震监测技术对储气库注水排卤和注气排卤连续监测12天,通过井下观测的方式进行数据采集和微地震事件识别、自动拾取,分析生产工艺参数与微地震事件数量之间的关系。

(1)针对研究区存在一套稳定分布的玄武岩屏蔽层,库区微地震监测采取井下监测方式最佳。

(2)3层套管井身结构对微地震信号衰减严重,建议在后续研究中将检波器下入技术套管内,避免多层套管对信号能量的衰减,特别是套管之间环空液体对横波的阻碍。

(3)自动识别微地震事件,可以判断地下活动的相对剧烈程度,通过分析生产工艺参数认为注水排卤造腔比注气排卤引起微地震事件剧烈。

参 考 文 献

[1] 李建君,陈加松,吴斌,等. 盐穴地下储气库盐岩力学参数的校准方法[J]. 天然气工业,2015,35(7):96-102.

[2] 屈丹安,杨春和,任松. 金坛盐穴地下储气库地表沉降预测研究[J]. 岩石力学与工程学报,2010,29(增刊1):2705-2711.

[3] 王同涛,闫相祯,杨秀娟,等. 考虑盐岩蠕变的盐穴储气库地表动态沉降量预测[J]. 中国科学:技术科学,2011,41(5):687-692.

[4] 任松,姜德义,杨春和. 盐穴储气库破坏后地表沉陷规律数值模拟研究[J]. 岩土力学,2009,30(12):3595-3606.

[5] 张新生,常晓云,蒋丽云,等. 盐穴地下储气库稳定性评价系统及其应用[J]. 天然气工业,2015,35(11):83-90.

[6] 赵博雄,王忠仁,刘瑞,等. 国内外微地震监测技术综述[J]. 地球物理学进展,2014,29(4):1882-1888.

[7] 张晓林,张峰,李向阳,等. 水力压裂对速度场及微地震定位的影响[J]. 地球物理学报,2013,56(10):3552-3560.

[8] 段银鹿,李倩,姚韦萍. 水力压裂微地震裂缝监测技术及其应用[J]. 断块油气田,2013,20(5):644-648.

[9] 严永新,张永华,陈祥,等. 微地震技术在裂缝监测中的应用研究[J]. 地学前缘,2013,20(3):270-274.

[10] 刘劲松,王赟,姚振兴. 微地震信号到时自动拾取方法[J]. 地球物理学报,2013,56(5):1660-1666.

[11] 宋维琪,冯超. 微地震有效事件自动识别与定位方法[J]. 石油地球物理勘探,2013,48(2):303-307.

[12] 刘振武,撒利明,巫芙蓉,等. 中国石油集团非常规油气微地震监测技术现状及发展方向[J]. 石油地球物理勘探,2013,48(5):843-853.

[13] Emmanuelle N. Contribution to salt leaching and postleaching monitoring issues-feedback from a long term microseismic survey[C]. Brussels:SMRI Fall 2006 Technical Conference,2006:2-19.

[14] Emmanuelle N,Eric B,Guillaume B. Real time micro-seismic monitoring dedicated to forecast collapse hazards induced by brine production[C]. HAlifax:SMRI Spring 2007 Technical Conference,2007:1-12.

［15］Fortier E, Maisons C, Sainte T, et al. Contribution to better understanding of brine production using a long period of microseismic monitoring［C］. Nancy：SMRI Fall 2014 Technical Conference, 2014：10-23.

［16］Hosseini Z, Collins D, Shumila V, et al. Induced microseismic monitoring in salt caverns［C］. Francisco：US Rock Mechanics/Geomechanics, 2015：15-24.

［17］Mercerat D, Berenard P, Ineris N, et al. Induced seismicity monitoring of an underground salt cavity under a transient pressure experiment［C］. Basel：SMRI Spring 2017 Technical Conference, 2017：102-113.

［18］Patrick R, Eric F, Christophe M, et al. Microseismicity induced within hydrocarbon storage in salt cavern, Manosque, France［C］. Avignon：SMRI Fall 2015 Technical Conference, 2015：87-98.

［19］Julie S, Mark L, Dario B, et al. Passive seismic observations at Grand Bayou［C］. Avignon：SMRI Fall 2015 Technical Conference, 2015：15-27.

［20］Rodrigo P, Sacha D, Vincent R, et al. Statistical analyses of microseismicity in salt cavern storages［C］. Salzburg：SMRI Fall 2016 Technical Conference, 2016：1-11.

本论文原发表于《油气储运》2018年第37卷第7期。

热应力对盐穴储气库稳定性的影响

李建君　敖海兵　巴金红　陈加松　汪会盟　李淑平　胡志鹏

（中国石油西气东输管道公司储气库项目部）

【摘　要】 盐腔注采周期的变化导致腔内温度周期性变化，进一步导致围岩热应力的周期性变化。由于盐岩抗拉强度低，当这种周期性热应力为拉应力且数值较大时，盐岩将破碎从而导致腔体垮塌。结合声呐带压测腔实测的盐穴腔体形状和体积数据，采用有限元数值模拟方法，将温度场与已建力学模型进行耦合建立热应力模型，并对常规模型和热应力模型进行对比，以分析热应力对腔体稳定性的影响。结果表明：在常规模型中表现稳定的部分区域，在热应力模型中却发生垮塌现象，热应力模型计算结果与声呐实测带压测腔结果吻合，由此推断温度变化所产生的热应力是影响腔体稳定性的重要因素，严重时可能导致腔体垮塌。因此，腔体热应力影响因素的提出，进一步完善了腔体稳定性评价体系，使评价结果更加接近于地层的真实情况。

【关键词】 盐穴储气库；热应力；温度；稳定性；数值模拟

Influence of thermal stress on the stability of salt-cavern gas storage

Li Jianjun　Ao Haibing　Ba Jinhong　Chen Jiasong
Wang Huimeng　Li Shuping　Hu Zhipeng

(Gas Storage Project Department, PetroChina West-East Gas Pipeline Company)

Abstract The change of salt-cavern injection-production cycle leads to the periodic change of cavern temperature, and then the periodic change of thermal stress of surrounding rock. The tensile strength of salt rock is low, so when the periodic thermal stress is tensile stress and its value is higher, salt rock will be broken, resulting in cavity collapse. In this paper, cavern shape and volume data measured by sonar under pressure were used. By virtue of finite-element numerical simulation, a thermal stress model was established by coupling the temperature field with the existing mechanical model. Then, the conventional model and the thermal stress model were compared to analyze the influence of thermal stress on cavern stability. It is indicated that some stable zones in the conventional model suffer collapse in the thermal stress model. The calculation results of thermal stress model are in accordance with the values measured by sonar under pressure, so it is inferred that the thermal stress generated by temperature change is the main factor influencing cavity stability and may even lead to cavity collapse. To sum up, the proposal of thermal stress as the influential factor makes the cavity stability evaluation system more complete and the evaluation results closer to the real strata situations.

Keywords salt-cavern gas storage; thermal stress; temperature; stability; numerical simulation

盐穴储气库凭借其良好的密封性、经济性越来越受到油气储运行业的青睐[1]，其运行的安全性也逐渐受到重视，盐穴储气库运行的安全性主要涉及盐腔垮塌和泄漏两方面。盐腔壁面大面积垮塌主要由岩体损伤引起，严重时可能导致地质灾难，因此，能否准确预测盐穴腔体损伤，找出相应的控制方法，对于盐穴储气库安全高效运行至关重要，国内外学者对此进行了大量研究。Avdeev 等[2-4]预设多个稳定性评价标准，采用数值模拟的方法对储气库的稳定性进行评价。吴文等[5]系统地提出盐穴储气库的稳定性研究的内容和评判标准。任松等[6]建立了盐穴储气库稳定性多层次、多指标综合评价体系，并定义了稳定性等级标准。李建君等[7]应用造腔和注采运行回归法建立了盐穴储气库岩石蠕变参数校准试验方法。

虽然国内外专家学者已从不同的角度、层次、指标对盐穴储气库稳定性进行了评价，但是考虑温度因素的研究很少，均侧重于对力学特性影响的研究[8-14]。陈剑文等[15]通过对不同围压和温度下的盐岩进行力学特性试验，分析围压和温度对其损失的影响。高小平等[16]对不同温度下盐岩蠕变特性进行了实验研究，分析了应力水平和温度对盐岩蠕变特性的影响。任松等[17]通过对不同温度下盐岩试件进行循环载荷实验，分析其疲劳损失。关于热应力对盐穴储气库稳定性影响的研究更是鲜见报道。但是相关研究资料[18]表明，每1℃的温差最高能产生 1MPa 的热应力。对于抗拉强度(1~2MPa)很低的盐岩来说，足以使其发生垮塌。

在此，基于单轴试验和三轴试验获得的力学参数，结合现场声呐带压测腔的结果，利用有限元分析软件对腔体进行数值模拟，将热应力耦合到已有的模型中，与原有模型进行对比分析，以检验在腔体稳定性评价过程中考虑热应力影响因素的必要性。

1 腔体热学问题

我国盐穴储气库埋藏深度大，具有高温、高压的特点，目的层段盐岩的温度与地表温度相差甚大。在天然气注采运行过程中，受腔内压力、腔内气体体积、注气温度等因素的影响，腔内的温度不断变化。温度变化对盐岩的影响包括温度对盐岩力学特征的影响和温度变化产生的热应力作用两个方面。

1.1 温度影响

在不同温度下盐岩各基本物理力学特性差异很大，总的表现为随着温度升高，盐岩的力学性能明显劣化。温度增加，弹性模量和峰值应力明显下降，但峰值应变却明显增长，塑性变形特征变得更加明显，蠕变是盐岩的一个重要力学特征。腔体注采气运行过程中，在温度和压力的作用下由于蠕变原因可能产生井眼闭合，造成卡钻、套管被挤毁等事故，

甚至可能造成腔体体积收缩率过大而导致提前报废。蠕变应变既是应力的函数，也是温度的函数，稳定蠕变应变本构方程为

$$\varepsilon = \frac{\exp\left[-\frac{Q}{R}\left(\frac{1}{T}-\frac{1}{T_r}\right)\right]}{K^n}\sigma^n t^m \tag{1}$$

式中：ε 为蠕变；σ 为应力，MPa；t 为时间，a；K、m、n 为盐岩蠕变常数；Q 为激活自由能，J/mol；R 为普适气体常数，J/(mol·K)；T 为实际温度，K；T_r 为参考温度，K。

通过式(1)可以计算出盐岩在不同温度下的蠕变应变数值。在腔体深度1000 m左右，温度50 ℃下的蠕变应变是常温20 ℃下蠕变应变的2.6倍。因此，温度对盐岩弹性、塑形、蠕变性质影响都很大。在进行盐岩样品试验时，如果仅仅在常温下进行试验，未考虑岩样实际环境的温度，最终得到试验力学参数与现场实际盐岩力学参数往往相差很大。

1.2 热应力

随着温度的变化，如果盐岩是均匀材料，且各向同性、没有约束，那么在岩体内部将不会有热应力产生。如果盐岩不满足这种理想假设，则必然会产生热应力[19]。产生热应力的基本条件是在约束下有温度的变化，约束条件可以分为外部变形约束、互相变形约束、内部变形约束[20-23]。热应力将导致盐岩内部介质破裂，微裂纹萌生、扩展、贯通，直至宏观裂纹出现，从而改变盐岩的力学性质和力学行为。

物体的热胀冷缩属性是其产生热应力的根本原因[24]。腔体产生热应力有3种形式：(1)由于存在地温梯度，整个地层由浅到深热胀冷缩程度均不一样，于是受到互相变形的约束，在地层中也将产生热应力。(2)对于盐岩来说，由于含有泥质、石膏等杂质，当温度发生变化时，内部不同矿物颗粒热膨胀系数不同，产生的热膨胀不同，但是各矿物之间互相固结，造成内部颗粒不能随温度变化而自由变形，因此盐岩由于内部变形约束而产生热应力。(3)在注采气过程[25-26]中，腔内温度不断变化，当腔内温度比围岩温度高时，围岩接触气体被直接加热，表层围岩温度上升较快，而盐岩热传导性差，温度向深部围岩传递很慢，因此深部围岩温度较低，在盐岩的内外之间出现温差，分别以其各自的体胀系数膨胀，但由于矿物颗粒彼此牵制，表层围岩的膨胀被温度较低的深部围岩约束，结果使表层围岩产生压应力。相反，当腔内温度比围岩低时，表层围岩首先受到冷却开始快速降温，体积收缩，结果在表层围岩产生拉应力。随着注采周期的变化，腔内温度周期性变化，热应力也呈周期性变化。这种周期性热应力作用对盐岩内部介质连接造成破坏，使其力学性质劣化，并产生微裂缝。随着时间的延长，微裂缝不断扩展、贯通，最终形成宏观裂缝，被裂缝裂开的盐岩表皮开始层层剥落。这也是在腔体突出岩脊处更容易发生岩体剥落、垮塌的原因。

采用有限元方法，对一点处应力、应变等物理量的分布状态进行求解。尽管使用三维模型计算可以更好地反映盐穴在运行过程中应力和应变情况，但是三维计算庞大、速度慢，根据大量的盐穴储气库地质力学评价分析实例，选择盐穴的关注剖面建立轴对称模型进行分析计算也能说明问题，因此采用轴对称模型对盐穴进行地质力学计算。对于腔壁处盐岩，在考虑热应力的情况下，应变分量包括由于自由热膨胀引起的应变分量和物体内各部分之

间的互相约束引起的应变分量两部分，即

$$\boldsymbol{\xi} = \boldsymbol{D}^{-1}\boldsymbol{\sigma} + \boldsymbol{\xi}_0 \quad (2)$$

式中：$\boldsymbol{\xi}$ 为总应变矩阵；$\boldsymbol{\xi}_0$ 为由温度引起的初应变矩阵；$\boldsymbol{D}$ 为弹性矩阵；$\boldsymbol{\sigma}$ 为考虑温度作用的应力矩阵。

对于轴对称物体，若各点都能无约束地自由膨胀，其初应变有轴向应变 ξ_{z0}、径向应变 ξ_{r0}、环向应变 $\xi_{\theta0}$ 及剪切应变 γ_{zr0} 共 4 个分量，可记为

$$\boldsymbol{\xi}_0 = \begin{Bmatrix} \xi_{z0} \\ \xi_{r0} \\ \xi_{\theta0} \\ \xi_{zr0} \end{Bmatrix} \quad (3)$$

另外，对于各向同性的材料，$\xi_{z0} = \xi_{r0} = \xi_{\theta0} = \alpha t$，$\gamma_{rz0} = 0$。

因此，热应力 $\boldsymbol{\sigma}$ 的表达式为

$$\boldsymbol{\sigma} = \begin{Bmatrix} \sigma_z \\ \sigma_r \\ \sigma_\theta \\ \sigma_{zr} \end{Bmatrix} = \boldsymbol{D}(\boldsymbol{\xi} - \boldsymbol{\xi}_0)$$

$$= \frac{E}{1+\nu} \begin{Bmatrix} \xi_z + \dfrac{\nu}{1-2\nu}(\xi_z + \xi_r + \xi_\theta) \\ \xi_r + \dfrac{\nu}{1-2\nu}(\xi_z + \xi_r + \xi_\theta) \\ \xi_\theta + \dfrac{\nu}{1-2\nu}(\xi_z + \xi_r + \xi_\theta) \\ \dfrac{1}{2}\gamma_{zr} \end{Bmatrix} - \begin{Bmatrix} \dfrac{E\alpha}{1-2\nu}\Delta T \\ \dfrac{E\alpha}{1-2\nu}\Delta T \\ \dfrac{E\alpha}{1-2\nu}\Delta T \\ 0 \end{Bmatrix} \quad (4)$$

式中：E 为弹性模量，MPa；ν 为泊松比；α 为膨胀系数，℃$^{-1}$；ΔT 为温度变化量，℃。

由于腔体内压及地静压力的存在，因此不能单纯地将热应力作为判断腔壁围岩是否处于拉应力状态的标准，需要综合考虑有效应力。有效应力为腔体内压、地静压力、热应力的合力，若有效应力为拉应力，则腔体可能产生裂缝，发生垮塌。

2 模型建立

盐穴 A 位于江苏省金坛地区，是中国第一个盐穴储气库的第一个盐腔，整个建库过程经历了钻井、造腔、注气排卤、注采运行 4 个阶段。2008 年下半年正式完成了地下盐层腔

体建造,并进行声呐测腔。2010年11月正式注气投产,经过5年的注采运行,2015年底进行了带压声呐测腔。将测得结果与完腔声呐进行对比,发现腔顶位置发生大面积垮塌,有4m的位移差,且在腔体中部突出岩脊处也有局部垮塌(图1)。为了分析造成垮塌的原因,基于声呐带压测腔实测盐穴腔体形状和体积数据,建立腔体常规数值模拟模型和热应力数值模拟模型,并进行对比分析。

金坛地区腔体普遍埋深约1000m,其温度变化形式有以下两种:(1)地温梯度。随着深度增加,地层温度逐渐升高,正常情况下地温梯度为(2~3)℃/(100m)。金坛地区地温梯度为2.55℃/(100m),埋深1000m处的盐层温度在51.5℃左右。(2)注采运行温度变化。在注采运行过程中,井口注气温度通常不是一个定值,由于调峰的需要,腔内压力、气体体积季节性变化,因此腔内温度在实际注采运行过程也不断地变化。

根据腔体A在某一深度点实测的温度和金坛地区地温梯度情况,同时从岩石热力学实验中获取盐岩的比热容、密度、传导率、弹性模量数据,运用有限元数值模拟软件Abaqus模拟地层初始温度场(图2)。

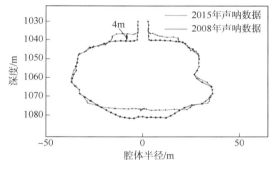 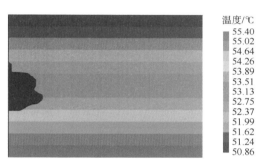

图1 2008年与2015年盐穴A声呐数据对比图

图2 金坛储气库腔体A盐岩层初始温度场模拟结果

目前,由于技术条件、成本费用等原因,无可以直接测量腔内实时温度的仪器。为了获取腔内的实时温度,自主研究开发了一套温度—压力模拟软件。根据理想气体状态方程、气体与地层对流换热等原理,将压力作为一个参考标准,通过对模拟压力与实测压力进行反复匹配,二者吻合良好,保证模拟结果准确性,最终获得注采运行过程中腔内的温度数据(图3)。将腔体内温度模拟数据导入Abaqus软件进行加载,进而模拟出腔体在注采运行过程中腔体—地层的完整温度场。整体上,围岩温度服从地温梯度分布规律;腔体附近,围岩温度受腔内温度传热影响,影响范围可达70m,但完全受其控制的区域仅限腔壁5~7m围岩内(图4)。

3 数值模拟

当盐岩受到压应力作用时是一种可塑性很好的材料,能够承受很大的变形。但当受到拉张应力作用时,盐岩却变得很脆,非常低的拉张强度即可使盐岩发生破碎。因此,盐穴在注采运行过程中不允许出现拉张应力,特别是腔顶区域。对比常规数值模

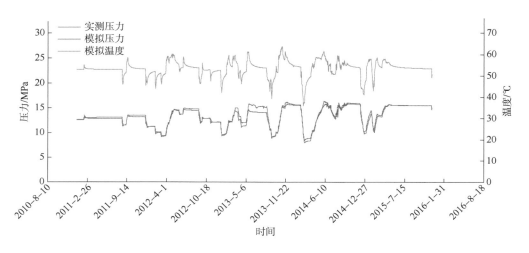

图 3 腔体 A 温度—压力模拟曲线

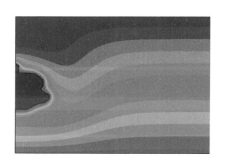

图 4 金坛储气库腔体 A 腔体—地层温度场模拟结果

型与热力学数值模型,观察是否有拉张应力区域。

在 Abaqus 模型计算结果中,根据腔体内最低压力时围岩的最大主应力进行拉张应力的判定。由于 Abaqus 规定压应力为负值,拉张应力为正值,因此观察最大主应力云图是否有正值区,若有正值区则存在拉应力。在此,基于常规模型模拟出腔体 A 最大主应力云图[图 5(a)],腔体 A 最大主应力分布范围为 $-24.6 \sim -8.3$ MPa,皆为压应力区域,没有拉应力存在。然而,实测声呐结果显示腔体 A 发生了垮塌现象,存在拉应力,模拟结果与实际情况不符,因此需要进行热应力模型分析。

热应力模型是在已有腔体数学—力学模型的基础上,考虑温度影响因素,将腔体温度场进行耦合而建立的。根据腔内最低压力下其最大主应力云图[图 5(b)],腔体 A 在腔体顶部、中部、底部岩块处均存在大面积拉张应力区域,这些区域为盐岩破碎区域。对于腔体顶部岩块、中部突出岩脊来说,由于重力作用,将发生垮塌,掉落到腔体底部。将腔体 A 两次实测声呐数据进行对比发现,声呐测量结果与数值模拟结果能够很好吻合:在实测声呐数据中腔体顶部发生大范围垮塌,模拟结果同样预测顶部将发生垮塌,垮塌区域基本一致;实测声呐数据中腔体中部岩脊突出区域发生小范围垮塌,模拟结果也同样预测到了。

如果没有热应力的作用,腔体处于稳定状态,热应力是导致腔体围岩破碎,促使其垮塌的重要因素。当腔体温度较围岩低时,温度变化所产生的热应力为拉应力,其值很大,可以使有效应力最大主应力从压应力 8MPa 变为拉应力,不可忽略。在今后的腔体稳定性评价中,应当充分考虑热应力作用的影响。

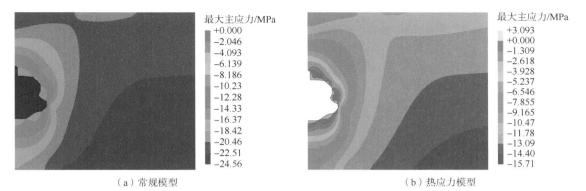

(a) 常规模型　　　　　　　　　　　　　　(b) 热应力模型

图 5　基于两种模型的腔体 A 最大主应力云图

4　结论

（1）温度的变化导致热应力产生，盐腔中产生的热应力有 3 种形式：地温梯度的存在，地层受到互相变形约束产生热应力；盐岩含有杂质，不同矿物颗粒热膨胀系数不同，存在内部变形约束而产生热应力；随盐腔注采周期变化温度不断变化，表层围岩受到互相变形约束产生热应力。

（2）在实际的注采运行过程中，调峰、应急气量所占总库容比例通常很高，因此盐腔体积变化很大，导致盐腔压力、温度变化很大。通过热应力公式计算可知，对于盐岩 1℃ 的温差最大可产生 1MPa 的热应力。当热应力为拉应力时，若温差足够大，则盐腔稳定性将受到影响。

（3）将温度场耦合到已有模型中，建立热应力模型。热应力模型计算结果能够与声呐实测带压测腔结果很好地吻合，可以解释在注采运行过程中腔体发生垮塌现象的原因，在今后的腔体稳定性评价中，需要充分考虑热应力作用的影响。

参 考 文 献

[1] 杨海军. 中国盐穴储气库建设关键技术及挑战[J]. 油气储运，2017，36(7)：747-753.

[2] Avdeev Y, Vlorobyev V, Krainev B, et al. Criteria for geomechanical stability of salt caverns[C]. Texas: the Fall 1997 Meeting EIPaso, 1997：1-11.

[3] Staudtmeister K, Rokahr R B. Rock mechanical design of storage caverns for natural gas in rock salt mass[J]. International Journal of Rock Mechanics and Mining Sciences, 1977, 34(3-4)：300-313.

[4] Mohanty S, Vandergrift T. Long term stability evaluation of an old underground gas storage cavern using unique numerical methods[J]. Tunnelling and Underground Space Technology, 2012, 30(4)：145-154.

[5] 吴文，侯正猛，杨春和. 盐岩中能源（石油和天然气）地下储存库稳定性评价标准研究[J]. 岩石力学与工程学报，2005，24(14)：2497-2505.

[6] 任松，李小勇，姜德义，等. 盐岩储气库运营期稳定性评价研究[J]. 岩土力学，2011，32(5)：1465-1472.

[7] 李建君，陈加松，吴斌，等. 盐穴地下储气库盐岩力学参数的校准方法[J]. 天然气工业，2015，35(7)：96-102.

[8] 邰保平，赵阳升，赵金昌，等. 层状盐岩温度应力耦合作用蠕变特性研究[J]. 岩石力学与工程学报，

[9] 刘建平,姜德义,陈结,等.盐岩水平储气库的相似模拟建腔和长期稳定性分析[J].重庆大学学报,2017,40(2):45-51.

[10] 任众鑫,李建君,汪会盟,等.基于分形理论的盐岩储气库腔底堆积物粒度分布特征[J].油气储运,2017,36(3):179-283.

[11] 耿凌俊,李淑平,吴斌,等.盐穴储气库注水站整体造腔参数优化[J].油气储运,2016,35(7):779-783.

[12] 金虓,夏焱,袁光杰,等.盐穴地下储气库排卤管柱盐结晶影响因素实验研究[J].天然气工业,2017,37(4):130-134.

[13] 郑雅丽,赵艳杰,丁国生,等.厚夹层盐穴储气库扩大储气空间造腔技术[J].石油勘探与开发,2017,44(1):137-143.

[14] 李银平,葛鑫博,王兵武,等.盐穴地下储气库水溶造腔管柱的动力稳定性试验研究[J].天然气工业,2016,36(7):81-87.

[15] 陈剑文,杨春和,高小平,等.盐岩温度与应力耦合损伤研究[J].岩石力学与工程学报,2005,24(11):1986-1991.

[16] 高小平,杨春和,吴文,等.盐岩蠕变特性温度效应的实验研究[J].岩石力学与工程学报,2005,24(11):2054-2059.

[17] 任松,白月明,姜德义,等.温度对盐岩疲劳特性影响的试验研究[J].岩石力学与工程学报,2012,31(9):1839-1845.

[18] Berest P, Djizanne H, Brouard B. Rapid depressurizations: Can they lead to irreversible damage? [C]. Saskatchewan: Solution Mining Research Institute Spring 2012 Technical Conference,2012:1-20.

[19] Kingery W D. Factors affecting thermal stress resistance of ceramic materials[J]. Journal of the American Ceramic Society,1955,38(1):3-17.

[20] 唐世斌,唐春安,朱万成,等.热应力作用下的岩石破裂过程分析[J].岩石力学与工程学报,2006,25(10):2071-2078.

[21] 朱兴吉,程旭东,彭文山.热应力作用下液化天然气储罐球形罐顶应力分布及裂缝形态[J].石油学报,2014,35(5):993-1000.

[22] 王贤能,黄润秋.引水隧洞工程中热应力对围岩表层稳定性的影响分析[J].地质灾害与环境保护,1998,9(1):43-48.

[23] 李静,林橙焰,杨少春,等.套管—水泥环—地层耦合系统热应力理论解[J].中国石油大学学报(自然科学版),2009,33(2):63-69.

[24] 李维特,黄保海,毕仲波.热应力理论分析及应用[M].北京:中国电力出版社,2004:55-62.

[25] 王彬,陈超,李道清,等.新疆H型储气库注采气能力评价方法[J].特种油气藏,2015,22(5):78-81.

[26] 殷代印,何超,董秀荣.储气库调峰能力数值模拟研究[J].特种油气藏,2015,22(1):95-98.

本论文原发表于《油气储运》2017年第36卷第9期。

金坛盐穴地下储气库地表沉降预测研究

屈丹安[1,2]　杨春和[1,3]　任　松[1]

(1. 重庆大学　西南资源开发及环境灾害控制工程教育部重点实验室；
2. 中国石油西气东输管道公司；3. 中国科学院　武汉岩土力学研究所)

【摘　要】 金坛盐穴地下储气库作为西气东输的重要配套工程，其主要功能是为西气东输管道的季节调峰和安全稳定运行提供用气，确保长江三角洲地区的天然气用气安全。地表沉降的准确预测是其保证储气库建设和生产运行安全的一种重要手段。先应用盐岩水溶开采沉降新概率积分三维预测模型，基于开采沉降的分层传递原理，对盐穴20年变形收缩引起的地表沉降及盐穴报废垮塌可能引起的地表沉降进行预测。然后应用地表沉降动态预测模型计算出金坛盐穴地下储气库地表沉降的时间影响因素，获得地表任意点任意时刻的沉降预测值。对比分析表明预测值与实测值十分接近，证实传递预测模型较传统预测方法具有更高的预测精度。

【关键词】 岩石力学；地表沉降；盐穴；储气库；传递模型；动态预测

Study and prediction of surface subsidence of salt rock caves used for gas storages in Jintan salt mine

Qu Dan'an[1,2]　Yang Chunhe[1,3]　Ren Song[1]

(1. Key Laboratory for the Exploitation of Southwestern Resources and the Environmental Disaster Control Engineering, Ministry of Education, Chongqing University;
2. PetroChina West-East Gas Pipeline Company; 3. Institute of Rock and Soil Mechanics, Chinese Academy of Science)

Abstract　The salt rock caves used for gas storages in Jintan salt mine are the important auxiliary engineering of West-East Gas Transmission. It's used for gas peak shaving and ensures the safety of supplying gas in the Yangtze River Delta. The accurately predicting the surface subsidence is the important method for ensuring the safety of the gas storages. Based on the rule of the transferring if one terrane by

---

基金项目：国家杰出青年基金项目(E50725414)，国家自然科学基金创新群体项目(50621403)；国家重点基础研究发展计划项目(2009CB7246003, 2009CB724602)。

作者简介：屈丹安(1965—)，男，1985年毕业于西北大学化工机械专业，现为博士研究生，主要从事盐穴地下储气库工程建设等方面的研究工作。E-mail: qudanan@petrochina.com.cn。

one terrane, the surface subsidence caused by the shrinking of salt rock caves in 20 years and the surface subsidence caused by the collapse of salt rock caves are predicted by the new probability integral 3D model. Then the time factor of the surface subsidence of the Jintan salt mine is obtained by the dynamic forecasting model and the surface subsidence can be achieved. The predicted value is very close to the measurement value, which proves that the transfer model has more forecasting precision.

Keywords rock mechanics; surface subsidence; salt rock caves; gas storage; transfer model; dynamic forecasting

金坛盐穴地下储气库作为西气东输的重要配套工程，其主要功能是为西气东输管道的季节调峰和安全稳定运行提供用气，确保长江三角洲地区的天然气用气安全。地表沉降的准确预测是其保证储气库建设和生产运行安全的一种重要手段。为保障金坛储气库的安全生产运行，在首批储气的老腔区域进行了地表沉降监测网的布设，结合预测理论掌握整个片区的地表沉降情况。

本文先应用盐岩水溶开采沉降新概率积分三维预测模型[1]，基于开采沉降的分层传递原理，对盐穴20年变形收缩引起的地表沉降及盐穴报废垮塌可能引起的地表沉降进行预测。然后应用地表沉降动态预测模型计算出金坛盐穴地下储气库老腔区域地表沉降的时间影响因素，获得地表任意点任意时刻的沉降预测值。对比分析表明实际监测值与预测值十分接近。

在随机介质概率积分模型及推导过程中有这样的假设：随机介质颗粒尺寸一致；整个上覆岩层的下沉系数由一个地表下沉系数来表示。然而，如果将上覆岩层看成一个整体，实际情况与其有较大的差别。很显然，上覆岩层破坏后的颗粒大小并不一致，而与岩层的性质有很大的关系，具有较强的分层特性[2-6]。没有考虑岩土体的层状结构、岩土体的破碎和断裂尺寸是导致概率积分法下沉预计缺陷的根源[7-12]。本文通过对上覆岩层变形传递规律的研究，提出开采沉降分层传递预测模型来解决这一问题。

1 预测理论研究

1.1 单盐穴引起的地表沉降预测模型

将上覆岩层按岩性分成若干层，从采空区顶板直至地表从小到大依次编号。采动对第一层岩层的移动变形，可以这样考虑：将第二层至地表的岩层移走，用相应的荷载加在第一层岩层上，对于第一层岩层来说其变形和原模型是一致的。这样，在第一层岩层的上表面将产生下沉空间，其形状由第一层岩层的预测参数决定；然后再将第二层岩层还原，在第二层岩层上加上等效的荷载，那么由于第一层岩层的下沉空间，将导致第二层岩层产生相应的下沉空间2。依此类推，直至地表（图1）。采用开采沉降新概率积分三维预测模型[1]对单层岩层下沉变形进行预测，得到开采沉降分层传递

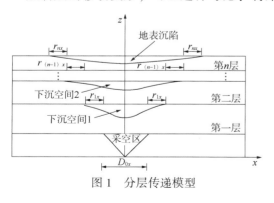

图1 分层传递模型

地表下沉预测公式为

$$W_n(x, y) = q_n \iiint_{V_n} \frac{h_n}{2(h_n - s_z)} \left(\frac{1}{r_{xn}^2} + \frac{1}{r_{yn}^2} \right)$$

$$\exp\left\{ -\pi \frac{h_n}{h_n - s_z} \left[\frac{(x - s_x)^2}{r_{xn}^2} + \frac{(y - s_y)^2}{r_{yn}^2} \right] \right\} \mathrm{d}s_x \mathrm{d}s_y \mathrm{d}s_z \tag{1}$$

式中：r_{xn}，r_{yn} 分别为第 n 层岩层 x 和 y 方向的地表影响半径；q_n 为第 n 层岩层的下沉系数；s_x，s_y，s_z 分别为对 V_n 积分时 x，y，z 三个方向的变量。其中，当 $n=1$ 时，V_n 取盐穴的参数，r_{xn}，r_{yn}，q_n 取第一层岩层的参数；当 $n>1$ 时，$V_n = W_{n-1}(x, y)$；r_{xn}，r_{yn}，q_n 均取第 n 层岩层的参数。

r_{xn}，r_{yn}，q_n 的计算公式[13-14]如下：

$$r_{xn} = \sqrt{\frac{h_n^2 \pi \sigma_{x\max}}{6 \sum_{i=n}^{N} \gamma_i h_i}} \tag{2}$$

$$r_{yn} = \sqrt{\frac{h_n^2 \pi \sigma_{y\max}}{6 \sum_{i=n}^{N} \gamma_i h_i}} \tag{3}$$

$$q_n = 0.991 - 0.238 \frac{E_n}{3600} - 0.224 \frac{\rho_n h_n}{100 \times 3600} \tag{4}$$

式中：N 为上覆岩层总数；γ_i 为第 i 层岩层的容重；h_i 为第 i 层岩层的有效厚度；$\sigma_{x\max}$，$\sigma_{y\max}$ 分别为第 n 层岩层 x，y 方向的岩石强度；E_n 为第 n 层岩体弹性模量；ρ_n 为第 n 层岩体的容重。

式(1)即为水溶开采沉降基于新概率积分的三维分层传递地表下沉预测公式。然后基于倒数第二层岩层产生的下沉空间，应用以下公式计算地表水平、倾斜等变形值。

地表点 $A(x, y)$ 沿地表任意方向倾角 ϕ 的倾斜曲率计算公式分别为

$$i(x, y, \phi) = \frac{\partial W(x, y)}{\partial x} \cos\phi + \frac{\partial W(x, y)}{\partial y} \sin\phi \tag{5}$$

$$K(x, y, \phi) = \frac{\partial^2 W(x, y)}{\partial x^2} \cos^2\phi + \frac{\partial^2 W(x, y)}{\partial x \partial y} \sin(2\phi) + \frac{\partial^2 W(x, y)}{\partial y^2} \sin^2\phi \tag{6}$$

地表点 $A(x, y)$ 沿地表任意方向倾角 ϕ 的水平移动计算公式为

$$U(x, y, \phi) = \frac{(b_x + b_y)(r_x + r_y)}{4} \cdot \left[\frac{\partial W(x, y)}{\partial x} \cos\phi + \frac{\partial W(x, y)}{\partial y} \sin\phi \right] \tag{7}$$

式中：b_x，b_y 分别为 x 和 y 方向的水平移动系数。

地表点 $A(x, y)$ 沿地表任意方向倾角 ϕ 的水平变形计算公式为

$$\xi(x,y,\phi)=\frac{(b_x+b_y)(r_x+r_y)}{4}\left[\frac{\partial^2 W(x,y)}{\partial x^2}\cos^2\phi+\frac{\partial^2 W(x,y)}{\partial x\partial y}\sin(2\phi)+\frac{\partial^2 W(x,y)}{\partial y^2}\sin^2\phi\right]$$
(8)

式(1)至式(8)即为开采沉降分层传递预测模型。

1.2 多盐穴引起的地表沉降预测模型

当多个盐穴分布较近时，对地表造成的变形将相互影响，导致严重的沉降问题。因此，必须对多盐穴共同作用下的地表沉降进行预测。

数值模拟和模型试验证明了在这种情况下可以采用叠加的方法来进行处理。因此，多盐穴引起的地表沉降分层传递预测模型如下：

$$W_n(x,y)=\sum_{i=1}^{n}W'(x-l_{ix},y-l_{iy}) \tag{9}$$

$$W_n(x,y)=\sum_{i=1}^{n}W'(x-l_{ix},y-l_{iy}) \tag{10}$$

$$K_n(x,y,\phi)=\sum_{i=1}^{n}K'(x-l_{ix},y-l_{iy},\phi) \tag{11}$$

$$U_n(x,y,\phi)=\sum_{i=1}^{n}U'(x-l_{ix},y-l_{iy},\phi) \tag{12}$$

$$\xi_n(x,y,\phi)=\sum_{i=1}^{n}\xi'(x-l_{ix},y-l_{iy},\phi) \tag{13}$$

式中：l_{ix}，l_{iy} 分别表示第 i 个盐穴中点坐标距原点的距离，$i=1\sim N$。

1.3 地表沉降动态预测模型

吴侃等[7]的研究表明，地下开采引起的地表移动变形是一个复杂的四维空间问题。在煤田开采条件下，地表移动过程可从 6 个月延续到数年，而在盐矿床开采条件下，移动持续时间甚至达到 100 年以上[8]。地表各点处的移动变形值在开采期间变化明显，移动终止时发生压缩变形的区域，在移动期间可能遭受拉伸，反之亦然。因此在进行开采设计和选择地表建筑物保护措施时，不仅要考虑移动过程稳定后的终止状态，还必须考虑地表移动变形随时间的发展过程。

假设地表下沉速率 $\dfrac{\mathrm{d}W(t)}{\mathrm{d}t}$ 与地表某点最终下沉值 W_0 和某一时刻 t 的动态下沉值 $W(t)$ 之差成正比[15]：

$$\frac{\mathrm{d}W(t)}{\mathrm{d}t}=c[W_0-W(t)] \tag{14}$$

式中：c 为时间影响因素，与上覆岩层的力学性质有关，量纲为 $1/t$。

初始时刻：$t=0$，$W(t)=0$，对式(14)积分，可得

$$W(t) = W_0(1-e^{-ct}) \tag{15}$$

式(15)即为地表动态移动过程中的时间函数，令

$$\phi(t) = 1-e^{-ct} \tag{16}$$

式(16)称为时间影响函数，且有

$$W(t) = W_0\phi(t) \tag{17}$$

式(17)即为开采沉降分层传递动态预测模型。

2 金坛盐穴地下储气库地表沉降预测

2.1 盐穴及覆岩参数

东1井和东2井、岗1井和岗2井距离较近，互相影响较大，其平面位置如图2所示。以点(4 044+5 239，35+23 506)为中心坐标点，则4井的坐标及其他基本参数见表1。

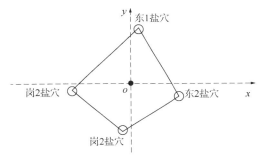

图 2　4井平面位置图

表 1　4井盐穴基本参数表

盐穴名称	坐标/m		海拔/m	容积/($10^4 m^3$)	盐层厚度/m
	x	y			
东1	24.03	73.60	3.51	12.66	148.54
东2	93.49	-24.19	3.44	12.66	148.45
岗1	-92.73	-15.91	3.90	9.75	132.04
岗2	-4.59	-72.95	4.16	9.75	134.65

图3是东1井和东2井、岗1井和岗2井的声呐测井及简化模型图。传递模型计算需要上覆岩石物理力学参数，见表2。

表 2　计算模型所用的岩石物理力学参数

土层编号	E_i/MPa	γ_i/(MN/m^3)	f_t/MPa	h_i/m
1	2520	0.0224	0.50	400
2	3510	0.0267	4.03	170

续表

土层编号	E_i/MPa	γ_i/(MN/m^3)	f_t/MPa	h_i/m
3	3815	0.0273	5.80	140
4	3125	0.0270	4.80	190
5	20	0.0180	0.02	40

注：f_t 为岩石抗拉强度。

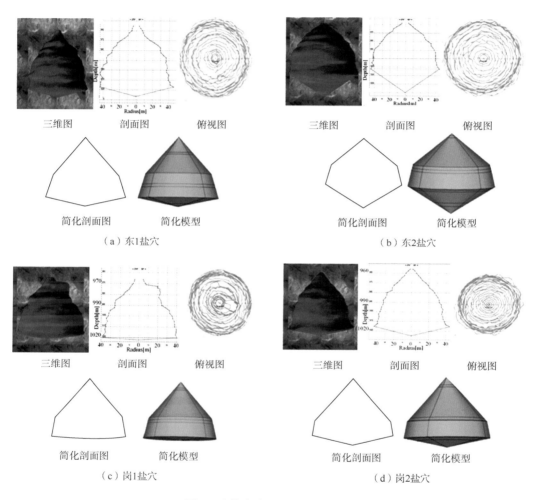

图 3　腔体声呐测试及简化模型

2.2　盐穴 20 年收缩变形导致的地表沉降预测

应用多盐穴引起的地表沉降预测模型[式(9)至式(13)]进行计算，需要盐穴 20 年变形收缩的容积损失。这部分地层损失主要是由于盐岩蠕变特性造成，本文采用"金坛盐矿已有盐穴可用性评估研究报告"的计算结果，该研究报告应用 FLAC 3D 软件对盐穴营运期的变形进行了数值计算，结果见表 3。

表 3　各井 20 年腔体收缩变形量

盐穴名称	变形量/%	盐穴名称	变形量/%
东 1	18.09	岗 1	19.30
东 2	18.78	岗 2	21.03

知道容积损失还必须知道损失空间的形状才能进行积分，本文采用以下方法得到积分空间，以东 2 井为例，其余三井计算方法一样。

设盐穴容积在一定时间内由 V_0 收敛为 V_1，容积损失率为 ρ，则

$$\rho = V_1/V_0 \tag{18}$$

设盐穴边界总收敛值为 Δr，如图 4 所示。

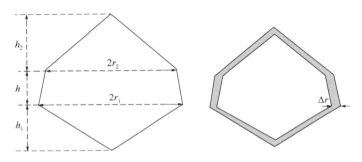

图 4　东 2 盐穴剖面图及收缩变形图

盐穴容积可以看成是 2 个圆锥和一个圆台的体积，则

$$V_0 = \frac{\pi}{3} r_2^2 h_2 + \frac{\pi}{3} r_1^2 h_1 + \frac{\pi h}{3}(r_1^2 + r_2^2 + r_1 r_2) \tag{19}$$

$$V_1 = \frac{\pi}{3}(r_2 - \Delta r)^2 (h_2 - \Delta r) + \frac{\pi}{3}(r_1 - \Delta r)^2 (h_1 - \Delta r) + \\ \frac{\pi h}{3}[(r_1 - \Delta r)^2 + (r_2 - \Delta r)^2 + (r_1 - \Delta r)(r_2 - \Delta r)] \tag{20}$$

将 $\rho = 18.78\%$，$r_1 = 40.0\text{m}$，$r_2 = 33.5\text{m}$，$h = 20\text{m}$，$h_1 = 2\text{m}$，$h_2 = 36\text{m}$。代入式(19)，式(20)可解得 $\Delta r = 3.171\text{m}$，$V_0 = 235625.6\text{m}^3$。

预测模型积分空间为 $V_1 - V_0$，如图 4 中充填部分。用同样的方法可以得到其他 3 个腔体变形积分空间。

应用多盐穴引起的地表沉降预测模型[式(9)至式(13)]进行数值求解得到的预测结果如下：图 5 为 4 盐穴 20 年收缩变形导致的地表下沉和水平变形等值线图，其中最大地表下沉量值为 1.153m，最大水平变形量为 0.417m，整个地表变形波及范围为东西长 900m，南北宽 870m。

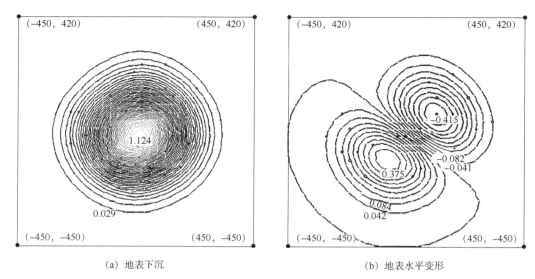

(a) 地表下沉　　　　　　　　　　(b) 地表水平变形

图 5　20 年收缩变形导致的地表等值线图(单位：m)

2.3 盐穴破坏失稳导致的地表沉降预测

盐穴破坏，积分空间为腔体本身，直接应用多盐穴引起的地表沉降预测模型进行数值求解得到的预测结果如下：

图 6 为 4 盐穴破坏失稳导致的地表下沉和水平变形等值线图，其中最大地表下沉量值为 6.791m，最大水平变形量为 2.473m，整个地表变形波及范围为东西长 990m，南北宽 930m。

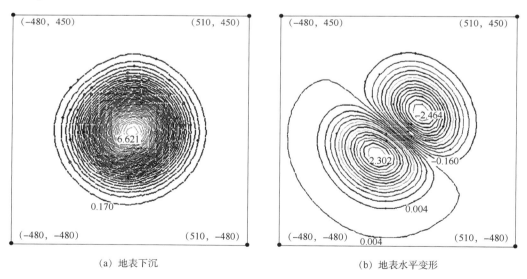

(a) 地表下沉　　　　　　　　　　(b) 地表水平变形

图 6　盐穴破坏失稳导致的地表等值线图(单位：m)

2.4 地表沉降动态预测

应用动态预测模型 $W(t) = W_0(1-e^{-ct})$ 对盐穴油气储库地表沉降进行动态预测。从预测模型可知，动态预测的关键是时间影响因素 c 的确定。前面计算得到了地表 20 年的下沉量 $W(20)$，和地表最终下沉量 W_0，因此可解得相应的 c 值。

地表最终下沉最大值为 6.791m，20 年地表下沉最大值为 1.153m，代入动态预测公式可解得：$c = 0.0094$。

东 1 和东 2、岗 1 和岗 2 盐穴共同作用下地表动态变形计算公式为

$$W(t) = W_0(x, y)(1 - e^{0.0094t}) \tag{21}$$

更换地表不同点的下沉值 $W_0(x, y)$，可以得到该点 (x, y) 的动态预测曲线。

图 7 为地表最大下沉最大值点的动态变化曲线图。至此，得到了整个片区的地表动态变形情况。

近两年来，共在东 1 井口位置进行了 3 次地表沉降测量，测量数据如图 8 所示，可以看到实测值与该点的动态曲线 $W(t) = 6.387(1 - e^{-0.0094t})$ ($t \approx 1 \sim 2a$) 十分接近。

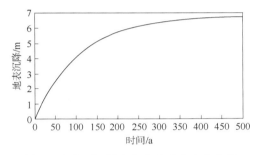

图 7 地表下沉最大值点的动态变化曲线

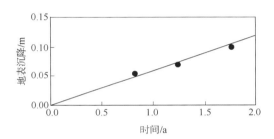

图 8 地表下沉动态变化曲线与实测值

3 结论

本文先应用盐岩水溶开采沉降新概率积分三维预测模型，基于开采沉降的分层传递原理，对金坛盐穴地下储气库 20 年变形收缩引起的地表沉降及盐穴报废垮塌可能引起的地表沉降进行预测。

（1）分层传递预测模型能充分考虑不同岩层的力学特性。

（2）分层传递预测模型良好反映了上覆岩层各向异性对地表变形的影响。

（3）分层传递预测模型较传统预测方法具有更高的预测精度。

（4）金坛盐穴地下储气库地表动态沉降时间影响因素 $c = 0.0094$。

（5）金坛盐穴地下储气库破坏失稳将会导致十分严重的地表沉陷问题，地表最大下沉值将达 6.791m，地表 45°方向最大水平位移值将达 2.473m，整个地表变形波及范围将达东西长 990m，南北宽 930m。

参 考 文 献

[1] 任松, 姜德义, 杨春和. 岩盐水溶开采沉陷新概率积分三维预测模型研究[J]. 岩土力学, 2007, 28(1): 133-138.

[2] 高大钊. 岩土工程的回顾与前瞻[M]. 北京: 人民交通出版社, 2001.

[3] Torano J, Rodriguez R. Probabilistic analysis of subsidence-induced strains at the surface above steep seam

mining[J]. International Journal of Rock Mechanics and Mining Sciences,2000,37(7):1 161-1 167.

[4] Singh R P,Yadav R N. Prediction of subsidence due to coal mining in Taniganj coalfield[J]. Engineering Geology,1995,39(1):103-111.

[5] 崔希民,陈至达. 非线性几何场论在开采沉陷预测中的应用[J]. 岩土力学,1997,18(4):24-29.

[6] Reddish D J,Yao X L,Benbia A,et al. Modeling of caving over the Lingan and Phallen mines in the sidney coalfield cape breton[C]// Proceedings of 5th International Mine Water Congress. Nottingham:[s. n.],1994:105-124.

[7] 吴侃,靳建明,戴仔强. 概率积分法预计下沉量的改进[J]. 辽宁工程技术大学学报,2003,22(1):19-22.

[8] Brown C B,King I P. Automatic embankment analysis:Equilibrium and instability conditions[J]. Geotechnique,1966,16(3):209-219.

[9] Kwinta A,Hejmanowski R,Sroka A. A time function analysis used for the prediction subsidence[J]. Mining Science and Technology,1996:419-424.

[10] 高明中,余忠林. 急倾斜煤层开采对地表沉陷影响的数值模拟[J]. 煤炭学报,2003,28(6):578-582.

[11] Baek J,Kim S W,Park H J,et al. Analysis of ground subsidence in coal mining area using SAR interferometry[J]. Geosciences Journal,2008,12(3):277-284.

[12] Singh R,Mandal P K,Singh A K,et al. Optimal underground extraction of coal at shallow cover beneath surface subsurface objects:Indian practices[J]. Rock Mechanics and Rock Engineering,2008,41(3):421-444.

[13] Booth C J. Confined-unconfined changes above long-wall coal mining due to increases in fracture porosity[J]. Environmental and Engineering Geoscience,2007,13(4):355-367.

[14] 邹友峰. 地表下沉系数计算方法研究[J]. 岩土工程学报,1997,19(3):109-112.

[15] 刘立民,刘汉龙,连传杰,等. 半解析方法用于房柱式开采沉陷的三维计算[J]. 煤炭学报,2005,27(5):462-467.

本论文原发表于《岩石力学与工程学报》2010年第29卷增1。

气密封检测技术在储气库井应用研究

李海伟[1]　李梦雪[2]　孟凡琦[1]　任众鑫[1]　王元刚[1]

(1. 中国石油管道有限责任公司西气东输分公司储气库管理处；
2. 中国石油管道有限责任公司西气东输分公司杭州管理处)

【摘　要】　国内油气田天然气井随着天然气开采出现大量环空带压现象，造成严重安全隐患，据资料统计环空带压现象90%左右由螺纹泄漏引起[1]。特殊扣油套管受到螺纹类型、加工误差、材质选择、运输磕碰、不规范操作、密封脂的选择等方面的影响，都可能导致螺纹出现气体泄漏[2]。在金坛盐穴储气库已投产注采井运行过程中，已经有4口井出现环空带压，给储气库的安全运行带来一定的安全隐患。利用气密封检测技术对金坛储气库注采完井管柱螺纹进行密封检测，有效剔除泄漏管柱入井，避免因螺纹泄漏引发环空带压。目前该技术在金坛储气库4口井注采完井作业中成功应用，为储气库的长期注采运行提供了安全保障。

【关键词】　储气库；完井；气密封检测；螺纹；泄漏

The application of gas seal detection technology in gas storage well

Li Haiwei[1]　Li Mengxue[2]　Meng Fanqi[1]　Ren Zhongxin[1]　Wang Yuangang[1]

(1. Gas Storage Department, PetroChina Pipeline Company Ltd-west-east Gas Pipeline Branch Company; 2. Hangzhou Department, PetroChina Pipeline Company Ltd-west-east Gas Pipeline Branch Company)

Abstract　With the occurrence of large number of annulus pressure phenomena in natural gas wells in domestic oil and gas fields, serious security risks are caused. According to the data, the leakage of the annulus is about 90%, which is caused by thread leakage. Special oil casing thread type, processing error, material selection, transportation knock against, non-standard operation, the influence of the choice of sealing grease, could lead to a thread a gas leak. During the operation of injection wells in Jintan salt underground gas storage, there are 4 wells showing annulus pressure, which brings certain security risks to the safe operation of gas storage. Sealing detection of the pipe column thread of injection and

---

基金项目：中国石油天然气股份有限公司重大科技专项"地下储气库关键技术研究与应用"子课题"盐穴储气库加快建产工程试验研究"，项目编号：2015E-400801。

作者简介：李海伟，男(1986—)，工程师，在读硕士生，2010年毕业于中国石油大学(华东)资源勘查工程专业，现主要从事地下盐穴储气库钻井、溶腔工程方向的研究工作，邮箱：cqklihaiwei@ petrochina. com. cn。

production completion of Jintan gas storage by gas seal detection technology, effectively eliminate leakage of pipe column into well and avoid ring pressure caused by leakage of thread. At present, the technology has been successfully applied to the injection and recovery operation of 4 wells in Jintan gas storage, which provides security for the longterm injection and production operation of the gas storage.

Keywords　　gas storage; completion; gas seal detection; thread; leakage

美国在20世纪80年代首先提出特殊扣螺纹气密封检测技术,并将该技术成功应用到油气井中,目前对油套管螺纹的气密封检测已成为国外石油行业强制性要求[3]。四川油气田自20世纪90年代最早引进使用特殊扣套管,但在生产过程中一直存在套管螺纹泄漏现象,某区块30口井有超过半数环空带压[4]。为确保金坛盐穴储气库注采井长期安全运行,注采完井管柱近期均引进特殊扣型气密套管;同时,并先后组织专业套管队进行下套管作业、上扣扭矩检测、涂抹专用密封脂,目的通过现场措施尽量保证螺纹密封性完好[5]。尽管在金坛储气库注采井完井作业中采取一系列措施确保螺纹密封,但目前还是有4口井出现环空带压,给储气库安全运行带来一定隐患[6]。为确保金坛储气库后期注采气安全、可靠运行,特引进螺纹气密封检测技术对注采管柱螺纹连接进行气密封检测。现场应用表明,该技术能有效对管柱螺纹密封性进行检测,有效剔除泄漏油套管,保证了注采气井井筒的密封性[7]。

1　检测原理

油套管螺纹气密封检测技术主要利用氦气分子直径比天然气小、更易渗漏的特点,在油套管下井之前,从油管或套管管柱内下入有双封隔器的测试工具,向测试工具内注入氦氮混合气,工具坐封,加压至规定值,稳压一定时间后泄压,利用氦气(表1)分子直径很小、在气密封扣中易渗透的特点[8],通过高灵敏度的氦气探测器在螺纹连接外探测氦气有无泄漏,从而来判断螺纹连接的密封性。

表1　常温下不同介质的渗透率表

气体	氢气	氦气	水蒸气	氖气	氮气	空气	氩气
相对分子质量	2	4	18	10	28	29	40
渗透率/mD	1.41	1	0.47	0.45	0.37	0.37	0.32

2　设备简介

气密封检测设备(图1)主要包括液气动力系统、增压系统及检测气源、检测执行系统、控制系统及其他辅助系统。

液气动力系统为增压储能器提供高压水源,为绞车及操作台提供液压源、气源,并携带相关驱动装置的高压设备。增压储能器则利用来自液气动力系统的高压水源,对进入储能器的氦氮混合气体进行增压的压力容器。检测系统主要包括液气动力系统,由原动机及其辅助装置,高压柱塞泵、液压泵、气泵、各种阀件和气路、液路高低压管线等组成。增压系统及检测气源由增压储能器及进出口阀件、氦氮混气装置及氦气瓶组成。检测执行系

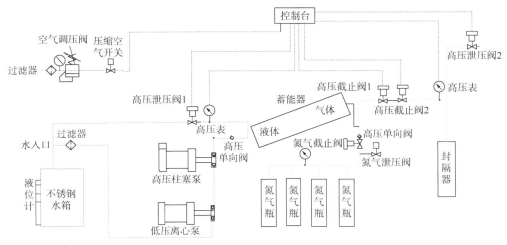

图 1 气密封检测设备示意图

统由检测工具及其高压管线、集气套和检漏仪组成[9]。控制系统及辅助系统由绞车及操作台、井架操作台、井架滑轮组成。

3 现场应用

3.1 井况简介

金坛储气库注采井统一采用外径 177.8mm、壁厚 9.19mm、气密螺纹、钢级 N80 注采管柱，管柱配套工具有流动短节、井下安全阀、封隔器、坐落接头及引鞋[10,11]。排卤管柱使用 API NU 螺纹的普通油管，管柱配套有坐落接头，管柱结构为：引鞋+短油管+坐落接头+油管[12]（图2、图3）。

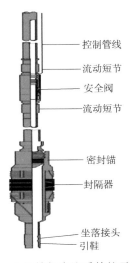

图 2 金坛储气库注采管柱示意图

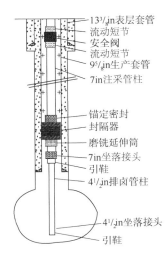

图 3 金坛储气库注采完井示意图

— 643 —

3.2 施工操作步骤

3.2.1 准备阶段

启动设备,将高压水泵与供水管线连接好,打开氮气、氦气气瓶,运转发动机,按标准检查设备各部分,确认设备正常。把检测工具在待测套管外下放,让检测工具的双封隔器中间点对应于待测套管接箍中部,然后在绞车钢丝绳上做好记号,对钢丝绳位置进行标定[13]。待测套管上扣完成后,将检测工具放入到待测套管内,上提套管直到待测接箍高出钻台面一定高度,调整检测工具位置定位在套管内接箍的上下位置,将检测集气套扣在套管接箍上,将检漏仪探测头插入检测护套的通孔。

3.2.2 氦气增压

储能器中加载检测气体,待储能器内压力达到 4~6MPa 后,关闭气控气源供气阀,将工具放空阀打开,坐封测试阀、储能排液阀关闭,启动高压水泵将储能器内的混合气体打压,让检测工具坐封。

3.2.3 加压、检测

高压水泵继续对工具和储能器打压,直至达到检测压力,停泵,关闭坐封阀,稳压20s后泄压,根据检漏仪探测到的氦气泄漏率来判断螺纹密封性能是否合格。若泄漏率超过 2×10^{-7} bar·mL/s,则表明螺纹密封性不合格,需采取相应的整改措施;若泄漏率低于 2×10^{-7} bar·mL/s,则表明螺纹密封性能合格[14]。

3.2.4 泄压、解封

检测完毕后将坐封阀、储能排液阀打开,泄掉储能器内压力,然后关闭坐封测试阀,打开工具放空阀,将卡封工具压力泄掉,解封,取出测试工具,继续检测下一螺纹密封性[15]。重复以上操作步骤,直到所有待测管柱螺纹全部检测完毕(图4)。

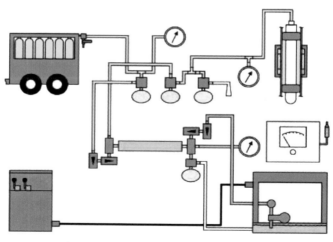

图 4 设备加压、坐封检测阶段示意图

3.3 应用实例

2017年在金坛储气库4口井先后进行5次螺纹气密封检测施工作业,套管扣型主要包括 API NU 和 BGT2 两种扣型。经过现场施工作业数据统计分析,两种扣型油套管螺纹均存在泄漏,单井最大泄漏率为3.51%,平均泄漏率为2.23%(表2)。现场采取的整改措施如下:(1)对油套管螺纹进行二次上扣;(2)卸开对扣并对密封台阶面和螺纹进行再次清理,重新上扣;(3)更换油套管母螺纹接头;(4)清理不密封油套管,更换新油套管。

表2 金坛盐穴储气库气密封检测数据统计表

井号	作业管柱	扣型	钢级	套管数量/根	泄漏套管数量/根	泄漏百分比/%
A井	114.3mm 排卤管柱	API NU	J55	114	4	3.51
A井	177.8mm 注采管柱	BGT2	N80	94	1	1.06
B井	177.8mm 注采管柱	BGT2	N80	91	3	3.30
C井	177.8mm 注采管柱	BGT2	N80	89	1	1.12
D井	177.8mm 注采管柱	BGT2	N80	93	2	2.15
平均						2.23

4 主要结论

(1)油套管螺纹气密封检测技术,可以有效地剔除螺纹密封不严的套管入井,保证了井筒密封性。

(2)造成油套管螺纹密封不严泄漏的原因是多方面的,不能单靠入井前通径、螺纹清理、涂抹密封脂、扭矩检测等措施,来避免和判别油套管螺纹连接处是否存在泄漏。

(3)油套管螺纹密封检测设备目前已经国产化,检测费用也大幅降低,建议在盐穴储气库注采完井作业中全面推广使用,为盐穴储气库长期注采气运行提供可靠的安全保障。

参 考 文 献

[1] 吴建发,钟兵,罗涛. 国内外储气库技术研究现状与发展方向[J]. 油气储运, 2007, 26(4):1-3.
[2] 杨海军. 中国盐穴储气库建设关键技术及挑战[J]. 油气储运, 2017, 36(7):747-753.
[3] 马晓伟. 套管气密封检测技术在深层采气井中应用[J]. 石油矿场机械, 2016, 45(7):72-74.
[4] 高宝奎,高德利. 高温高压井测试对套管安全的特殊影响[J]. 天然气工业, 2002, 22(4):40-42.
[5] 王保辉,闫相祯,杨秀娟,等. 衰竭气藏型地下储气库库存量动态预测研究[J]. 科学技术与工程, 2012, 12(10):2286-2289.
[6] 耿凌俊,李淑平,吴斌,等. 盐穴储气库注水站整体造腔参数优化[J]. 油气储运, 2016, 35(7):779-783.
[7] 李建中,李奇,胥洪成,等. 盐穴地下储气库气密封检测技术[J]. 天然气工业, 2011, 31(5):90-92.
[8] 胡茂中,白国云. 低充氦浓度氦质谱检漏技术应用研究[J]. 真空科学与技术学报, 2011, 31(2):208-211.
[9] 张富臣. 高压气密封检测系统回收卸压装置及控制技术研究[D]. 大庆:东北石油大学, 2011.
[10] 李朝霞,何爱国. 砂岩储气库注采井完井工艺技术[J]. 石油钻探技术, 2008, 36(1):16-19.
[11] 张宝岭,王西录,徐兴平. 高压封隔器密封胶筒的改进[J]. 石油矿场机械, 2009, 38(1):85-87.
[12] 李建中,徐定宇,李春. 利用枯竭油气藏建设地下储气库工程的配套技术[J]. 天然气工业, 2009,

29(9):97-99.

[13] 李国韬,刘飞,宋桂华,等.大张坨地下储气库注采工艺管柱配套技术[J].天然气工业,2004,24(9):156-158.

[14] 阳小平,王凤田,陈俊,等.地下储气库注采井多层管柱电磁探伤技术[J].天然气技术,2008,2(6):43-46.

[15] 刘啸峰,陈实.气密封检测技术在高压深井中的应用[J].内蒙古石油化工,2010,36(23):81-82.

本论文原发表于《石油化工应用》2018年第37卷第3期。

气体示踪技术在盐穴地下储气库微泄漏监测中的应用

王建夫[1]　张志胜[2]　安国印[3]　王文权[1]　巴金红[1]　尹　浩[4]　康延鹏[1]

(1. 中国石油储气库分公司；2. 国家石油天然气管网集团有限公司西气东输分公司；
3. 中国石油华北油田公司；4. 中国科学技术大学地球和空间科学学院)

【摘　要】　常规的监测手段无法满足盐穴地下储气库盐穴、井筒等注采系统的微泄漏监测要求。为此，基于油气藏监测中常用的气体示踪检测技术，依托高分辨质谱仪建立了SF_6痕量气体示踪剂检测方法(检测设备主要由在线冷冻大气采样器、在线多气氛反应热解/热脱附炉等组成)，在中国石油金坛盐穴储气库开展了盐腔、腔体间连通、环空带压井筒和井口及地面管线微泄漏监测现场试验。研究结果表明：(1)目标盐腔不存在天然气泄漏情况，3口相邻老腔井间也无连通现象；(2)4口目标井均存在微泄漏现象，检测到的SF_6含量介于$1\times10^{-13} \sim 5\times10^{-13}$，由于估算的天然气泄漏速率为1mL/min，认为该储气库井口及地面管线不存在完整性破坏；(3)环空带压井筒微泄漏监测试验结果表明，A井、F井均不存在井筒泄漏，其环空带压应该是因油套间温度/压力升高引起。结论认为：建立的SF_6痕量气体示踪剂检测方法能够实现对盐穴储气库微泄漏监测和评价的要求，下一步应扩大监测区域，制定覆盖全库区的实时监测方案，建立一套有效的盐穴储气库微泄漏长期监测系统和安全预警系统。

【关键词】　盐穴储气库；气体示踪；SF_6；痕量检测；微泄漏监测；腔体连通；环空带压；井口泄漏；现场试验

Application of gas tracer technology to micro-leakage monitoring of salt-cavern underground gas storage

Wang Jianfu[1]　Zhang Zhisheng[2]　An Guoyin[3]　Wang Wenquan[1]
Ba Jinhong[1]　Yin Hao[4]　Kang Yanpeng[1]

(1. PetroChina Gas Storage Company; 2. West East Pipeline Company, Pipeline China;
3. PetroChina Huabei Oilfield Company; 4. School of Earth and Space Sciences,
University of Science of Technology of China)

Abstract　Conventional monitoring methods cannot realize the required micro-leakage monitoring of

---

基金项目：中国石油天然气集团有限公司科技攻关项目"盐穴储气库双井造腔关键技术研究"(编号：2019B-3205)。
作者简介：王建夫，1989年生，工程师，硕士；主要从事盐穴储气库造腔工程设计及管理工作。地址：(212000)江苏省镇江市润州区南徐大道60号商务中心A区D座0513。ORCID：0000-0001-9696-4485。E-mail：wjf_jscqk@petrochina.com.cn。

salt caverns, wellbores and other injection and production systems of salt-cavern underground gas storages. In this paper, the SF_6 trace gas tracer detection method is developed by using the gas tracer detection technology commonly used for oil and gas reservoir monitoring, combined with a high-resolution mass spectrometer. Its detection equipment mainly includes an online frozen atmosphere sampler and online multi-atmosphere reaction pyrolysis/thermal desorption ovens. The method is tested on site in the microleakage monitoring of salt caverns, cavern connection, wellbores and wellheads with sustained casing pressure and surface pipeline in the Jintan Salt-Cavern Gas Storage. And the following research results are obtained. First, there is no phenomenon of gas leakage in the target salt caverns and there is no connection between three adjacent old cavern wells. Second, micro-leakage occurs in the four target wells, and the detected SF_6 content is in the range of $1×10^{-13}–5×10^{-13}$, but the leakage rate of natural gas is estimated to be 1 mL/min, so it is inferred that there is no integrity damage at the wellheads and surface pipelines of the gas storage. Third, micro-leakage monitoring results of the wellbores with sustained casing pressure show that Wells A and F have no wellbore leakage, and their sustained casing pressure is resulted from the increase of tubing-casing annulus temperature/pressure. In conclusion, the SF_6 trace gas tracer detection method developed in this paper can realize micro-leakage monitoring and evaluation of salt-cavern gas storages. What's more, it is recommended to enlarge the monitoring area, formulate a real-time monitoring plan covering the whole area of the gas storage, and develop a set of effective long-term monitoring system and safety early warning system for the micro-leakage of salt-cavern gas storages.

Keywords Salt-cavern gas storage; Gas tracer; SF_6; Trace detection; Micro-leakage monitoring; Cavern connection; Sustained casing pressure; Wellhead leakage; Field test

盐穴地下储气库(以下简称盐穴储气库)作为一类重要的储气设施,在季节调峰和长输管道应急方面发挥了重要作用[1-2]。由于盐穴储气库存储气量巨大且一般位于人口较为稠密的地区,一旦发生天然气泄漏爆炸事故,将造成巨大的人员伤害和经济损失。例如,2001年1月美国堪萨斯州Yaggy储气库泄漏爆炸,损失约$600×10^4 m^3$的天然气,导致2死1伤,数百居民被疏散[3]。2021年6月,湖北省十堰市燃气泄漏爆炸事故造成重大人员伤亡,更是给储气库安全生产敲响了警钟。

目前,盐穴储气库开展的安全监测手段主要有地表沉降监测、微地震监测和带压声呐测井等[4-7]。2006年至今,中国石油金坛盐穴储气库已连续进行14次地面沉降监测,库区沉降基本稳定,未发生超量沉降及突发性灾害沉陷[5]。2016年利用微地震监测技术对造腔和注气过程进行了监测,结果表明造腔活动未产生腔体较大的破裂或垮塌,注气压力上升可导致断层活动[6]。对投产盐腔每5年进行一次带压声呐测腔监测,2015年JZ井声呐测腔发现腔顶发生约5m垮塌[7]。但以上监测手段主要是针对盐腔形态、裂缝、腔体垮塌等,无法监测腔体或井筒等注采系统的微泄漏情况。

气体示踪检测技术是将气体示踪剂加入密闭容器内,通过物理或化学检测手段获取容器外部示踪剂的体积分数及分布,来评价容器的泄漏状态,可有效监测盐穴、井筒或地面管线的密封性。该技术已经广泛应用于油气藏监测中,主要用于二氧化碳、空气泡沫等气驱井组间的连通情况,判断受效方向等[8-10],国内尚未见该技术在盐穴储气库微泄漏监测中的研究和应用情况。

为此,笔者依托高分辨质谱仪建立了 SF_6 痕量气体示踪剂检测方法(检测设备主要由在线冷冻大气采样器、在线多气氛反应热解/热脱附炉等组成),并经现场试验验证了技术的可行性。进而利用该监测方法在金坛盐穴储气库开展了盐腔微泄漏监测、腔体间连通性监测、环空带压井筒微泄漏监测和井口及地面管线微泄漏监测现场试验,试验结果及现场经验可为其他储气库提供借鉴。

1 气体示踪微泄漏监测原理

基于气体示踪的微泄漏监测技术是通过在密闭容器外检测示踪剂含量来判断容器是否发生微泄漏。示踪气体通常选择六氟化硫(SF_6),其为一种人工合成的惰性气体,无色、无味、无毒、无腐蚀性、不燃、不爆炸,具有良好的化学稳定性和热稳定性,作为安全性保护气被广泛应用于电力和电器工业中[11-12],可作为盐穴储气库微泄漏监测技术的气体示踪剂。

现有的气体示踪剂检测方法检测对象是气体样本,检测能力较低,无法满足大气中痕量示踪剂的检出限要求。本文提出了一种新的检测方法:(1)建立示踪剂气体吸附再解脱附的检测流程,提高检测能力;(2)对 SF_6 进行定性定量分析,获得满足现场微泄漏监测要求的极低 SF_6 含量的检出限和标准气体定量曲线;(3)进行现场试验,验证该方法的可行性。

1.1 痕量气体示踪剂检测流程

痕量气体示踪剂检测主要依托于气相色谱—傅里叶变换静电场轨道阱高分辨质谱仪(型号:赛默飞 Q Exactive GC Orbitrap)。该仪器采用先进的静电场轨道阱(Orbitrap)检测技术,分辨率超过 160000FWHM($m/z=127$)且具有良好的灵敏度,满足示踪剂痕量气体检测要求。根据现场检测要求研制了与之配套的在线冷冻大气采样器和在线多气氛反应热解/热脱附炉。检测流程如图1所示:(1)采用自主研发的在线冷冻大气采样器低温采集监测点的空气样本,将采集后的吸附柱两端密封,放置于-20℃便携式冷阱中保存;(2)采用多气氛反应热解/热脱附炉对低温保存的吸附柱进行大体积、快速热脱附,释放吸附的气体示踪剂;(3)脱附后的气体载入气相色谱—傅里叶变换静电场轨道阱超高分辨质谱仪进行检测、记录。

1.2 建立示踪剂 SF_6 检测方法

1.2.1 实验步骤

(1)配置标准气体:用 10μL 气体进样器准确抽取高纯 SF_6 气体 0.4μL,注入 2.0mL 高纯氮气空瓶中,制成体积分数为 $2.0×10^{-5}$ 的一级 SF_6 标准气体。再分别抽取一级标准气 1.0μL、2.0μL、4.0μL、5.0μL、8.0μL 注入同样 2.0mL 高纯氮气空瓶中,制成体积分数分别为 $1.0×10^{-8}$、$2.0×10^{-8}$、$4.0×10^{-8}$、$5.0×10^{-8}$ 和 $8.0×10^{-8}$ 的二级 SF_6 标准气体。

(2)使用在线冷冻大气采样器采样:在-20℃采样温度下,以 300mL/min 抽取室内空气,将二级 SF_6 标准气体各取 1μL 注入抽气管口,混合气通过填充固相吸附剂 Matrix Car-

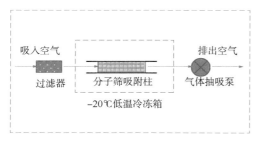
(a)在线冷冻大气采样器

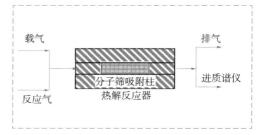
(b)在线多气氛反应热解/热脱附炉

(c)气相色谱—傅里叶变换静电场轨道阱超高分辨质谱仪

图1 痕量气体示踪剂检测流程图

boxen 1000 的分子筛吸附柱吸附 SF_6，在抽取 40L 室内空气后停止，将吸附柱两端密封即可获得标准气体吸附浓缩样品。

（3）吸附样品热脱附：将吸附柱去除封口后，放入在线多气氛反应热解/热脱附炉中，高纯氦气作为载气，以流速 1.2mL/min 通过吸附柱，热解/热脱附炉初始加热温度 30℃，保持 0.5min 后以 100℃/min 快速升温至 300℃，保持 3min 进行在线气体热脱附。

（4）示踪剂气体检测：脱附后气体载入气相色谱—傅里叶变换静电场轨道阱高分辨质谱仪进行检测。

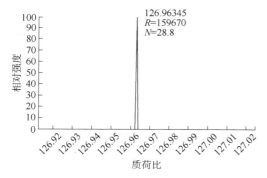

图2 实测气体示踪剂定性、定量质谱峰图

1.2.2 示踪剂 SF_6 定性定量分析

经检测，SF_6 用于定性定量的特征碎片离子（SF_5）质谱峰的理论分子量为 126.96354u，气体示踪剂定性准确度实验实际测得的分子量为 126.96345u（图2），实际误差值为 0.09mmu，定性分辨率为 159670 FWHM，定性精度为 0.71×10^{-6}，完全满足储气库微泄漏监测对气体示踪剂精确定性分析的要求。

由于使用的高分辨质谱仪具有优秀的降噪抑噪能力，其待测物的响应背景噪声往往为 0，因此必须采用统计学意义上的检出限（DL，Detection Limit），见式（1）。由统计学形式指定的检出限和定量限采用相对标准偏差（RSD，Relative Standard Deviation）见式（2），可间接测量质谱仪所检测到的定量物质的离子计数值,避免了当选择无基线噪声区域的测量峰面积去推断灵敏度时造成的不确定性,确保无论在高背景噪声下,还是在低背景噪声下定量检出限均严格有效。

$$DL = RSD \times t_a \times N_i \quad (1)$$

$$RSD = \frac{\sqrt{\frac{\sum(X_i - \overline{X})^2}{n-1}}}{\overline{X}} \times 100\% \quad (2)$$

式中:DL 为示踪剂检出限;RSD 为重复 n 次进样所测得的响应值的相对标准偏差;t_a 为 T 检验下单侧置信度 99% 自由度 $n-1$ 时的置信因子;N_i 为标准物质的体积分数;X_i 为第 i 次检测标准物质峰面积;$\overline{X}$ 为 n 次检测的标准物质峰面积的平均值;n 为检测次数。

经过 7 次重复检测,获得不同浓度下的 SF_6 标准气体检测峰面积(X_i),并计算获得平均峰面积($\overline{X}$)、标准偏差(S)、相对标准偏差(RSD)(表 1)。检出限是按照多次测定痕量体积分数为 2.5×10^{-16} 的 SF_6 标准气体之后,在被测定的 SF_6 标准气体痕量体积分数下所产生的信号能以 95% 置信度区别于空白样品而被测定出来的最低分析的量,通过式(1)和式(2)计算获得 SF_6 检出限为 $DL = 4.26 \times 10^{-17}$,该检测体积分数完全满足现场痕量示踪剂的检测要求。同时,建立了 SF_6 标准气体的定量曲线(图 3), $n_i = 3.014 \times 10^{-19} X_i + 8.331 \times 10^{-18}$, $R_2 = 0.9947$, 曲线拟合程度较高,可准确定量分析 SF_6 的体积分数。

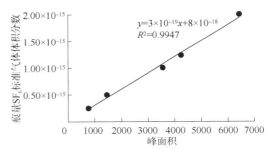

图 3 痕量 SF_6 标准气体定量拟合曲线图

表 1 不同痕量体积分数 SF_6 标准气体的峰面积表

痕量体积分数	不同检测序次峰面积							平均峰面积	标准偏差	相对标准偏差
	第1次	第2次	第3次	第4次	第5次	第6次	第7次			
2.50×10^{-16}	750	724	825	739	774	795	775	768.86	34.54	0.045
5.00×10^{-16}	1463	1466	1586	1475	1613	1237	1570	1487.14	126.64	0.085
1.00×10^{-15}	3576	3434	3718	3792	3221	3683	3140	3509.14	252.74	0.072
1.25×10^{-15}	4816	4005	4838	3961	4060	4059	3919	4236.86	406.32	0.096
2.00×10^{-15}	6040	6256	7263	6799	6490	6380	5909	6449.29	462.72	0.072

1.3 现场验证

基于气体示踪技术的储气库微泄漏需要超高灵敏度的检测技术,因此,需要验证该技术对天然气微泄漏监测的可行性。现场验证试验是在给定的井场放置示踪气体泄漏源,设定示踪气体 SF_6 的泄漏速度为 $1.0 mL/min$,持续泄漏 5d 后在无风或微风天气分别在距泄漏

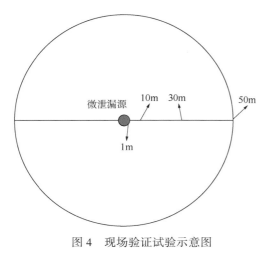

图4 现场验证试验示意图

源1m、10m、30m、50m处抽取地表空气40L进行检测(图4)，检测结果见表2。从图4、表2可看出，示踪气体体积分数和泄漏点距离有一定的负相关性，距离越远，检测到的示踪剂体积分数越低，表明该方法可以指示泄漏点位置区域。由于距泄漏点50m的检测点是位于农田田埂下方的凹陷处，该处示踪剂体积分数是距泄漏点30m处的3.8倍，表明示踪气体SF_6在低洼地方可以形成聚集，有更强的示踪性。试验结果表明：可以通过监测示踪气体SF_6在地表大气中的含量来监测盐穴储气库天然气的泄漏情况，并获取泄漏点位置，评价腔体或库区天然气的泄漏状态。

表2 试验场地的地表大气中SF_6的检测体积分数表

检测点距离/m	实际检测积分面积	检测SF_6体积分数
1	5700264265	1.11×10^{-7}
10	82689837	1.62×10^{-9}
30	638510	1.25×10^{-11}
50	2464408	4.82×10^{-11}

2 现场应用

中国石油金坛盐穴储气库位于江苏省金坛市直溪镇，建库盐层区域面积11.2km^2，建库深度约1000m，盐层最厚区域达180~230m[13-14]。经过10余年建设，截至2020年，已累计建成盐腔超过40个，形成库容约$12\times10^8m^3$，是亚洲规模最大的盐穴储气库。针对金坛盐穴地下储气库的特点，笔者利用基于气体示踪的微泄漏监测技术开展了盐腔微泄漏监测、腔体间连通性监测、环空带压井筒微泄漏监测和井口及地面管线微泄漏监测试验。

现场试验分为以下3个阶段：(1)试验前检测目标井示踪剂背景值；(2)选择注入井注入示踪剂；(3)检测目标井示踪剂含量。第一阶段测量结果显示目标井示踪剂背景值均为0，可认为储气库矿区内自然条件下无示踪剂SF_6。第二阶段注入示踪剂，根据注气特点设计井场示踪剂注入流程如图5所示，试验前所有闸阀均处于关闭状态。步骤为：(1)打开闸阀2，采用井场天然气放空装置放空闸阀1与闸阀2之间的管线；(2)待放空后，关闭闸阀2和放空装置，从闸阀1和闸阀2间的压力表旋塞阀处连接示踪剂钢瓶，注入SF_6示踪剂；(3)示踪剂注入完成后，关闭压力表阀门，打开闸阀2及与井口之间的所有注入闸阀，闸阀1保持关闭状态；(4)启动空气压缩机，待天然气进口压力高于井口压力后，缓慢打开闸阀1，推动示踪剂进入盐腔。结合监测计划，本次选择了6口井注入示踪剂，其中A井~E井为新腔井，F井为老腔井，注入结果见表3。第三阶段根据监测内容，开展了目标井示踪剂含量检测和分析。

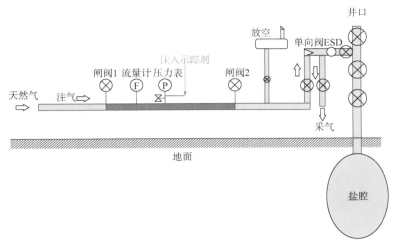

图 5 井场注示踪剂示意图

表 3 示踪剂注入情况表

井号	注入日期	示踪剂注入量/kg	腔体体积/m³	示踪剂体积分数
A	2020-10-20	29	206713	$1.709×10^{-6}$
B	2020-10-20	27	206953	$1.589×10^{-6}$
C	2020-10-20	32	198720	$1.962×10^{-6}$
D	2020-10-27	34	226430	$1.829×10^{-6}$
E	2020-10-29	41	215810	$2.314×10^{-6}$
F	2020-10-30	40	136903	$3.559×10^{-6}$

2.1 盐腔微泄漏地面监测

盐腔具有极低的渗透率、良好的蠕变行为，是理想的石油和天然气等碳氢化合物的地下储备场所[15]。但也存在天然气泄漏风险，主要有以下 4 种类型（图 6）：（1）泥岩夹层密封性不足引起气体近水平渗漏；（2）盐腔与断层连通引起断层泄漏；（3）盖层被突破失效导致气体上窜；（4）井筒完整性失效导致气体逃逸。而天然气泄漏可能导致灾难性事故发生，为此，开展了盐腔微泄漏地面监测试验。

经现场勘查后，根据现场实际情况，选取 A 井和 E 井这 2 口井进行地面区域监测，腔体形态如图 6 所示。监测网格为中国结形状[图 7（a）]，以井口作为坐标中心点，每隔 50m 设置 1 个采集点，共计 25 个采集点，整个现场采集覆盖面积为 $1.2×10^{5} m^{2}$。采集气体时从 GPS 测量仪上测得各个采集点坐标，记录工作时的风力、风向和地表温度。采集时根据现场实际情况进行了调整，实际采集点分布如图 7（b）、（c）所示。

2020 年 9 月 19—20 日、2020 年 9 月 28—29 日在风力较小的天气分别进行了 A 井、E 井背景值测量，均未检测出 SF_6。2020 年 10 月 20 日在 A 井注入示踪剂 29kg，腔内示踪剂体积分数为 $1.709×10^{-6}$。2020 年 10 月 29 日在 E 井注入示踪剂 41kg，腔内示踪剂体积分数为 $2.314×10^{-6}$。时隔约 1 个月后，2020 年 11 月 21—22 日，同一检测网格下分别进行了这两口井的示踪剂体积分数区域性检测，结果均未检测出 SF_6。示踪剂注入前后均未检测到

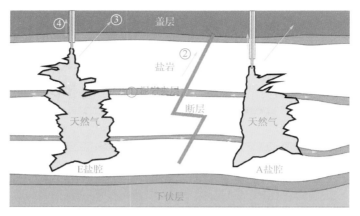

图 6　盐穴储气库天然气泄漏示意图

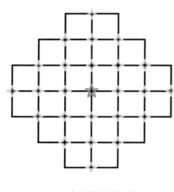

(a) 中国结监测网格

(b) A井实际监测网格

(c) E井实际监测网格

图 7　地面采样点分布图

SF_6，表明两个盐腔气密封性均较好，不存在天然气泄漏情况，这也与现场多年安全运行情况相符。

2.2　腔体间连通性监测

当相邻腔体间矿柱宽度过低时，由于蠕变破坏或泥岩夹层穿刺漏失等原因，可能引起储气库间串库现象[16]。此时各个盐腔之间将会失去独立性，严重影响储气库稳定性和注采气功能。为此，开展了腔体间连通性监测试验。

F井、G井、H井为独立的采卤老腔改建的储气库，F井处于G井、H井的中间位置，距离G井、H井较近且均为22.1m(图8)。老腔形态均比较规则，但由于腔体间距离较近，存在串库风险，所以选择该井组作为试验井组。监测方案为F井作为示踪剂注入井，G井、H井作为示踪剂检测井。2020年9月28日在G井、H井进行了示踪剂背景值测量，为了安全，并考虑到该井组刚经历过采气，决定采取收集地面管线内放空的天然气来代替腔内天然气样本的方法，结果未检测到SF_6。2020年10月30日将40kg示踪剂注入F井中。2020年12月4日进行了G井、H井示踪剂检测，采集天然气样本时，在井口泄压口边释放天然气边抽取，最终抽取体积达到40L，两井均未检测到SF_6。示踪剂注入F井前后，G井、H井内均未检测到示踪剂，试验结果表明F井与G井、H井之间未发生腔体间连通。虽然腔体间距较小，但3口老腔改造井一直采用同注同采的方式运行，运行压力也比较合理[14]，所以并未发生腔体间连通。

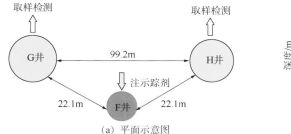
（a）平面示意图

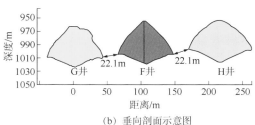
（b）垂向剖面示意图

图8 腔体流通性监测示意图

2.3 环空带压井筒微泄漏监测

环空带压是盐穴储气库注采气井的一个突出问题，较大的环空压力可能存在生产安全隐患，但环空带压的原因一直难以判断[17]。2018年金坛5口老腔改造井出现不同程度的环空带压现象，A环空（图9）压力分别达到了3.8MPa、4.5MPa、3.8MPa、6.6MPa、10.0MPa[18]。2019年金坛储气库现场曾开展过分布式光纤测环空带压井油套管泄漏试验，但并未检测出泄漏位置[19-20]。为了判断环空带压是否由井筒泄漏产生，开展了新腔D井和老腔F井环空带压井筒微泄漏监测试验，此时该两口井A环空压力较低，分别为1.0MPa和0.2MPa。

2020年9月29日对D井和F井开展了A空示踪剂背景值测量，均未检测出SF_6。检测时

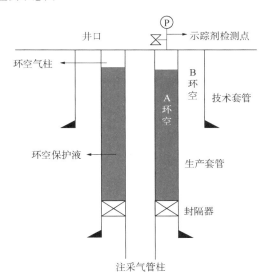

图9 储气库注采气井环空示意图

打开井口A环空压力表针型阀，边释放环空气体边采集样本。10月27日对D井注入示踪剂34kg；10月30日对F井注入示踪剂40kg。11月23日对D井、F井进行环空气体示踪剂检测，此时，环空压力较低，仅释放出少量气体，测试结果为SF_6含量为0，说明两井井筒和封隔器处不存在泄漏。根据金坛储气库实践经验，注采气时A环空压力会升高，曾发生注气时某井的A环空压力由2MPa升至8MPa的现象，而非注采气期间A环空泄压后则一直无环空压力。综上，认为上述2口井环空带压不太可能是注采气管柱泄漏引起的，很有可能因生产过程中注采气管柱与生产套管间温度/压力升高导致[17]。

2.4 井口及地面管线微泄漏监测

井口及地面管线是盐穴储气库注采气系统重要的组成部分，一旦发生破裂或失效，将导致天然气泄漏甚至大火或爆炸[21]。例如，2004年8月美国得克萨斯州Moss Bluff储气库泄漏爆炸起火，大火燃烧6天，波及半径120m，方圆3mile(1mile=1609.34m)居民撤离，损失天然气$1.7×10^8 m^3$[22]。金坛盐穴储气库已经运行超过10年，为了检验井口及地面管线

的完整性，开展了井口及地面管线的微泄漏监测试验。

2020年9月29日对A井、B井、C井、E井进行了井口附近示踪剂背景值检测，未检测到SF_6。10月20日进行了A井、B井、C井示踪剂注入，10月29日进行了E井示踪剂注入。11月23日在B井井口进行了示踪剂检测，检测到SF_6体积分数为1.797×10^{-13}，表明B井井口或地面管线存在微泄漏。为了验证结果的准确性，12月4日对A井、B井、C井、E井井口进行了示踪剂检测，检测到SF_6体积分数分别为1.441×10^{-13}、9.052×10^{-14}、1.299×10^{-13}、4.724×10^{-13}，表明上述4口井井口或地面管线确实存在微泄漏现象。原因为该4口井在注入示踪剂之后均进行过采气作业。例如，B井在注入示踪剂后第一次检测时正在进行采气，腔内示踪剂随着天然气流经至井口及地面管线，在某些闸门或仪表与管线连接处发生微量泄漏。可以看出泄漏的SF_6体积分数较低，介于$1\times10^{-13}\sim5\times10^{-13}$，计算结果表明，示踪剂泄漏量约$1.0\times10^{-6}$ mL/min，天然气泄漏量初步估计为1 mL/min，泄漏速率极低。根据天然气爆炸极限浓度计算，当天然气泄漏量小于1 L/min，不存在爆炸风险，处于安全可控范围内，所以认为井口及地面管线不存在完整性破坏。

试验结果表明：气体示踪技术可在盐腔微泄漏监测、腔体间连通性监测、环空带压井筒微泄漏监测和井口及地面管线微泄漏监测方面发挥重要作用，可实现盐穴储气库微泄漏监测和评价。但是现场施工过程中也存在几点问题，有待完善解决：（1）现场采集时吸附柱需要在-20℃下低温保存，需要质量大的携带式冷阱、移动电源等配套设备，采集不便，后期需要研发可移动便携式采集装置；（2）示踪剂注入前要选择一段地面封闭管道进行放空，浪费了大量的天然气资源，后期应研发示踪剂高压注入装置，在高压下注入示踪剂。

3 结论

（1）依托于高分辨质谱仪建立了由在线冷冻大气采样器、在线多气氛反应热解/热脱附炉组成的痕量气体示踪剂检测方法。实验建立了示踪剂SF_6定性定量检测方法，得到了SF_6检出限$DL=4.26\times10^{-17}$和SF_6标准气体的定量曲线，并通过现场试验验证了该方法的可行性。

（2）金坛储气库腔体微泄漏监测试验结果表明目标盐腔不存在天然气泄漏现象；腔体间连通性监测试验表明3口相邻老腔间不存在连通现象；环空带压井筒微泄漏监测试验结果表明目标井不存在井筒泄漏，环空带压的原因不太可能是注采气管柱泄漏引起，很有可能因油套间温度/压力升高引起；井口及地面管线微泄漏监测试验结果表明4口目标井存在微泄漏现象，检测到SF_6体积分数基本在$1\times10^{-13}\sim5\times10^{-13}$，估算天然气泄漏速率为1 mL/min，认为井口及地面管线不存在完整性破坏。

（3）本次试验对气体示踪技术在盐穴储气库微泄漏监测中的应用进行了初步尝试，下一步的研究重点是进一步扩大监测区域，制定覆盖全库区大面积监测方案。同时，设计自动采集装置，对库区实现实时监测。最终，开发微泄漏监测评价软件，建立一套有效的盐穴储气库微泄漏长期监测系统和安全预警系统。

参 考 文 献

[1] 周淑慧，王军，梁严. 碳中和背景下中国"十四五"天然气行业发展[J]. 天然气工业，2021，41（2）：171-182.

［2］郑雅丽，完颜祺琪，邱小松，等. 盐穴地下储气库选址与评价新技术［J］. 天然气工业，2019，39（6）：123-130.

［3］Bérest P, Brouard B. Safety of salt caverns used for underground storage［J］. Oil & Gas Science and Technology—Revue d'IFP, 2003, 58(3)：361-384.

［4］杨海军. 中国盐穴储气库建设关键技术及挑战［J］. 油气储运，2017，36（7）：747-753.

［5］何春海，姚威，刘玉刚，等. 盐穴储气库地面沉降监测与分析［J］. 土工基础，2017，31（4）：476-478.

［6］魏路路，井岗，徐刚，等. 微地震监测技术在地下储气库中的应用［J］. 天然气工业，2018，38（8）：41-46.

［7］陈加松，程林，刘继芹，等. 金坛盐穴储气库JZ井注采运行优化［J］. 油气储运，2018，37（8）：922-929.

［8］陈龙龙，白远，云彦舒，等. 气体示踪在非均质特低渗透油藏二氧化碳驱气窜监测中的应用［J］. 钻采工艺，2020，43（3）：56-59.

［9］李补鱼，褚万泉，龚山华，等. 气体示踪剂在中原油田油气藏监测中的应用［C］//2012油气藏监测与管理国际会议暨展会论文集. 西安：西安石油大学，2012.

［10］郭文敏，李治平，吕爱华，等. 气体示踪表征CO_2驱渗流特征实验研究［J］. 西南石油大学学报(自然科学版)，2015，37（1）：111-115.

［11］田杰，羊衍秋，杨亮，等. SF_6示踪剂远程泄漏监测系统设计及性能［J］. 原子能科学技术，2014，48（9）：1686-1692.

［12］龚山华，李补鱼，杜辉，等. 天然气中SF_6等多种氟化物气体示踪剂的气相色谱检测［J］. 西安石油大学学报(自然科学版)，2015，30（6）：97-100.

［13］杨海军，于胜男. 金坛地下储气库盐腔偏溶与井斜的关系［J］. 油气储运，2015，34（2）：145-149.

［14］尹雪英，杨春和，陈剑文. 金坛盐矿老腔储气库长期稳定性分析数值模拟［J］. 岩土力学，2006，27（6）：869-874.

［15］陈祥胜，李银平，施锡林，等. 地下盐穴储气库泄漏原因及防治措施研究［J］. 岩土力学，2019，40（增刊1）：367-373.

［16］Wang T T, Li J J, Jing G, et al. Determination of the maximum allowable gas pressure for an underground gas storage salt cavern: A case study of Jintan, China［J］. Journal of Rock Mechanics and Geotechnical Engineering, 2019, 11(2)：251-262.

［17］王兆会，曲从锋，周琛洋，等. 基于环空压力的储气库井风险分级及管控程序［J］. 钻采工艺，2020，43（6）：17-20.

［18］岳春林. 盐穴储气库老腔改造井环空带压影响因素分析［J］. 环球市场，2019（12）：387.

［19］王文权，贾建超，赵明千，等. 盐穴储气库井筒泄漏监测室内模拟实验［J］. 石油钻采工艺，2020，42（4）：513-517.

［20］安国印，王文权，陈春花，等. 盐穴储气库井筒泄漏监测技术现场试验研究［C］//第31届全国天然气学术年会(2019)论文集. 合肥：中国石油学会天然气专业委员会，2019.

［21］李丽锋，罗金恒，赵新伟. 盐穴地下储气库井口破裂火灾事故危险分析［J］. 油气储运，2013，32（10）：1054-1057.

［22］Rittenhour T P, Heath S A. Moss Bluff Cavern 1 blowout［C］//Proc. SMRI Fall Technical Conference, Bremen, Germany, 2012：119-130.

本论文原发表于《天然气工业》2021年第41卷第12期。

Geomechanical investigation of roof failure of China's first gas storage salt cavern

Tongtao Wang[1]　Chunhe Yang[1]　Jiasong Chen[2]　J. J. K. Daemen[3]

(1. State Key Laboratory of Geomechanics and Geotechnical Engineering, Institute of Rock and Soil Mechanics, Chinese Academy of Sciences; 2. Department of Gas Storage Project, West-East Gas Pipeline Company, PetroChina; 3. Mackay School of Earth Sciences and Engineering, University of Nevada)

Abstract　JK-A cavern is the first operating cavern of Jintan underground gas storage, which is also the first operating salt cavern gas storage in China. In 2015, the sonar survey carried out under working conditions indicated that a massive roof collapse of the JK-A cavern had taken place. To prevent similar accidents from happening again, finding the causes and the collapse time of JK-A cavern would be valuable. By using the target salt formation information, rock properties, and monitored gas pressure, a 3D geomechanical model of JK-A cavern has been built. The responses of the rock mass around JK-A cavern under leaching, sealing tests, debrining, and gas injection/delivery have been investigated. A safety evaluation system consisting of cavern volume shrinkage, dilatancy safety factor, displacement, vertical stress, and equivalent strain is proposed to find the likely collapse time and the reason for the JK-A cavern roof collapse. Results show that there are two main causes for the cavern roof failure. (1) A large-span flat roof, detrimental for bearing loads. Once local damage takes place, massive collapse may be triggered by the self-weight of the loosened rock. (2) The decrease speed of the internal gas pressure is too fast. The loads applied to the cavern roof cannot be transferred in a timely fashion, which causes a stress concentration zone to form, and local damage develops. The roof collapse of JK-A cavern took place 1.3 years after its gas injection/delivery started. Numerical analysis results show that dilatancy safety factor and vertical stress have high accuracy and sensitivity in the prediction of the cavern roof collapse. There is a high potential for a roof collapse of JK-A cavern to happen again. We propose that the gas pressure and pressure decrease rate are to be controlled strictly in the later operating period to prevent any collapses in the future.

Keywords　3D geomechanical model; Bedded rock salt formation; Salt cavern; Underground gas storage; Failure analysis; Numerical simulation

1　Introduction

Jintan salt cavern underground gas storage (UGS) is the first salt cavern gas storage of China. Its construction began in 2002. The first cavern, JK-A cavern, was completed in November 2010. This marked China entering a rapid development stage for the construction of salt cavern UGS. Jintan rock salt is a typical bedded rock salt; it contains many interlayers with high insoluble contents. This causes many challenges for cavern construction and operation (Wang et al., 2018a, b, c; Li et al., 2018; Wang et al., 2017; Wang et al., 2015), such as cavern stability, tightness, debrining, and operating parameters design. To cope with these challenges and to ensure the cavern safety, many methods are used during the UGS construction and operation, such as surface subsidence monitoring, micro-seismic monitoring, sonar survey, cavern temperature monitoring, gas flow and gas pressure monitoring. According to the sonar survey of 2015, a massive roof collapse of JK-A cavern had taken place. The collapse of the cavern roof may threaten the tightness of the casing shoe and even cause more roof collapse. A similar accident had been reported by Crossley (1998). At Regina South Gas Storage Cavern No. 5, with a large-span flat roof, a massive roof collapse occurred after operating for five years. A second roof collapse happened two years later. This accident caused gas leakage and the cavern was operated with a smaller internal gas pressure after some treatments. Finding the cause of the JK-A cavern roof collapse would be valuable. It can assist in avoiding similar accidents from happening again in Jintan UGS and provide a reference for dealing with similar problems for other places.

Salt caverns used for the storage of natural gas, petroleum, and compressed air, have been widely developed. Related technological and academic problems have been extensively studied by many scholars. Van Sambeek et al. (1993) found the dilatancy failure of rock salt associated with the mean stress invariant and the square root of the stress deviator invariant, and proposed a criterion to predict the failure of rock salt. Bérest and Brouard (2003) analyzed the failure of Kiel 101 gas storage salt cavern, Germany, and confirmed that low pressure and high decrease rate of gas pressure were the main reasons for the failure. Bruno (2005) studied the effects of different parameters on the failure zone and displacement of the rock mass around caverns in the Permian, Michigan, and Appalachian Basins. He pointed out that a formation above the cavern roof with a small stiffness was bad for the cavern stability. Evans (2008) classified and analyzed 228 accidents in salt caverns used for energy storage, but confirmed salt cavern was still one of the safest ways to store gas and petroleum. Lux (2009) summarized his studies about the energy storage in rock salt over the last 30 years, and indicated that salt caverns for energy storage should satisfy the requirements of static stability, tightness, acceptable surface subsidence, and environmentally safe abandonment. He particularly pointed out that roof structure has a notable effect on cavern safety. Bérest (2013) found the failure of salt cavern used for gas and oil storage showing typical plastic (cavern volume shrinkage and cavern bottom heave) and brittle (roof fall, spalling, and sluffing) damage characteristics. Liu et al. (2015) tested the permeability of the mudstone cap rock and interlayers in bedded salt formations of Jintan, China.

Zhang et al. (2015) used a time dependent mathematical model to predict the surface subsidence induced by an UGS salt cavern in a bedded rock salt formation and investigated the influence factors on the subsidence. Sobolik and Lord (2015) investigated the issues related to the operation and abandonment of large-diameter caverns and their long-term implications for oil storage facilities in domal salt of West Hackberry, Louisiana and Bryan Mound sites, Texas, USA. Wang et al.

(2016) proposed a safety evaluation system for UGS salt caverns and investigated the safety of a salt cavern close to a fault. Khaledi et al. (2016) simulated the responses of the rock mass around a salt cavern by simulating the cyclical loads. Wang et al. (2017) analyzed the failure of an overhanging block on the wall of an UGS salt cavern, and indicated that the decrease of buoyant weight causing by debrining was the main reason for the failure. Zhang et al. (2017) built a physical model to study the deformation and stress of a rock mass around a salt cavern, and confirmed that a rapid pressure change had notable effects on cavern displacement and local damage. Belzer and DeVries (2017) used numerical simulation to predict the failure of the casing of a salt cavern UGS, and summarized several typical failures of UGS casing. Wang et al. (2018b) studied the cavern shape, dimensions, operating parameters, and safety of salt cavern UGS in Jianghan, Hubei, China, ultra-deep formation. Labaune et al. (2018) compared the dilatancy criteria based on the strain and stress, and indicated the strain-based criterion is more conservative than the stress-based criterion. From the above literature review we can conclude: How to evaluate and predict the safety of salt caverns is challenging, and is still of interest. This is also the reason why so many scholars carry out studies in the fields related to salt caverns use for energy storage.

The motivation for this paper is to find the reason for and the time when the roof failure happened in China's first salt cavern UGS. A 3D geomechanical model of JK-A cavern is built based on the sonar survey data and the information about the target formation where the cavern is located. The responses of the rock mass around the cavern are simulated during leaching, sealing test, debrining and operating. A new index system is proposed to predict the cavern roof failure time and to identify the causes of the failure. Research results can provide a reference for Jintan salt cavern UGS to avoid similar accidents, and for other places with similar conditions.

2　3D geomechanical model and boundary conditions

2.1　Background of JK-A cavern

JK-A cavern is located in the Jintan salt mine district, Changzhou city, Jiangsu province, China. It is the first completed cavern of Jintan UGS, and is also the first UGS salt cavern in China. It is also an information cavern and a test cavern. The drilling, leaching and debrining of JK-A were completed in November 2003, September 2009 and November 2010 respectively (Yuan et al., 2006; Li et al., 2011). Five sonar surveys were carried out during the cavern leaching, during October 2005, July 2006, June 2007, September 2007, and September 2008. To ensure the debrining was carried out smoothly, a sonar survey was carried out in July 2009, after the cavern leaching was completed. Fig. 1 presents the shape of JK-A obtained by the sonar survey of 2009. The cavern has a tabular shape. Its height, excepting the chimney, is about 40 m and its maximum diameter is about 80 m. JK-A has a largespan flat roof with a span of about 50 m. Many overhanging blocks have been mapped on the cavern wall. The rock salt maintained above the cavern roof (the chimney) has a thickness of about 52 m(-978 to -1030 m). An interlayer immediately above the cavern roof has a thickness of about 2 m. The rock salt between the interlayer and the cavern bottom has a salt content of about 90%. Constrained by the cavern leaching technology at that time, the cavern shape, volume and dimensions greatly differ from the design values. This increases the risk of accidents during the later operation. The depth of the casing shoe is about −1016 m; the depths of cavern roof and bottom are about −1040 and −1081 m; and the effective volume is about 11.65×10^4 m^3.

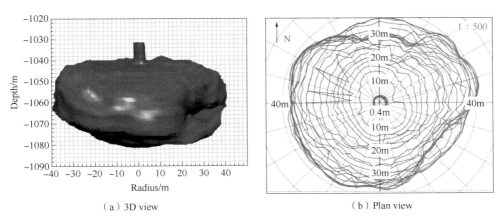

Fig. 1 Shape and dimensions of JK−A obtained by the sonar survey of July 2009

After the completion of debrining in November 2010, JK−A cavern was formally put into operation. Its operating parameters were: gas pressure ranged from 7 to 15.8 MPa; maximum storage capacity was about 2000×10^4 m^3; working gas capacity was 850×10^4 m^3, and was 1100×10^4 m^3 for extreme condition; maximum injection/delivery gas capacity was 125×10^4 m^3/day. The production casing used for gas injection/delivery has a dimension of ϕ 177.8 mm × 9.19 mm (outer diameter and wall thickness). To evaluate the cavern safety, a sonar survey was carried out on JK−A cavern under working condition in November 2015. The sonar survey showed that the depth of the casing shoe was still −1016 m, the roof and bottom depths were −1036 and −1076 m respectively, and cavern volume was about 12.0×10^4 m^3.

Fig. 2 presents the average profiles of JK−A cavern obtained by sonar surveys of 2009 and 2015. The cavern roof obtained in 2015 shows that notable collapse had taken place compared with that obtained in 2009. The failed zone has a length of about 50 m, and a thickness ranging from 2 to 4 m. By using the sonar surveys, we calculate the volume of the collapsed zone to be about 3300 m^3. The shape of the cavern middle parts has changed slightly. The cavern bottom has heaved about 5 m from 2005 to 2009. The sonar survey results show that the caverns of Jintan

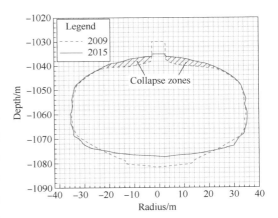

Fig. 2 Average profiles of JK−A cavern obtained by the sonar surveys of 2009 and 2015

have only small bottom upheaving displacements (Yang et al., 2015b). We believe that the bottom upheaving is caused by the accumulation of blocks collapsed from the cavern roof. The volume of the bottom heave is about 3900 m^3, calculated using the sonar data, which is basically equal to that of the collapsed roof zone. This indicated that no significant volume expansion of these collapsed blocks took place, and that they were not dissolved. Considering the loads acting on the cavern wall during the sealing tests and the debrining are basically constant (Wang et al., 2017), we assume that the roof collapse may have taken place during the gas injection/delivery, not during the sealing tests or the debrining. The chimney top depth has increased by about 5 m (from -1030 to -1035 m), which is much larger than that of the other locations except for the collapsed zone and the cavern bottom. This maybe because the large-span flat roof has a poor capacity to resist the vertical creep deformation, and a zone with a large displacement is formed in the cavern roof. However, these large deformation areas, except for the chimney bottom, have collapsed. Ultimately, the sonar survey data show the chimney top depth increasing significantly.

The sonar survey results show an interesting phenomenon: the volume of JK-A obtained in 2015 increased by about 3200 m^3, an increase of about 2.7%, compared with that obtained in 2009. This seems contradictory with the expected situation. Due to the creep of rock salt, the volume of a salt cavern should decrease with time (Langer, 1999). The cavern volume increase may have been caused by two reasons. (1) The unsaturated brine continues to dissolve rock salt during the sealing test and the debrining, which increases the cavern volume. To increase the cavern leaching speed, the brine, with a salt content of about 300 g/L, required by the brine process factory is expelled from the cavern rather than the fully saturated brine with a salt content of about 330 g/L. JK-A cavern was the first cavern of Jintan. The sealing test and the debrining took more than two years, and workovers were carried out several times. The unsaturated brine may have changed into fully saturated brine during these two years. We roughly calculate the volume of JK-A cavern can increase by about 1600 m^3 by the brine becoming saturated. (2) The measuring error of the sonar survey. According to engineering experience, the sonar survey error is about 2% of the cavern volume, viz., about 2000 m^3 for JK-A cavern. These two reasons may lead to this deviation observation. It also indicates that Jintan rock salt has a high creep deformation resistance, and the cavern has only a small volume shrinkage.

Gas was injected in JK-A cavern in November 2010, when it started operating. By November 2015, when the sonar survey under working condition was carried out, it had operated for five years. Fig. 3 presents the monitored gas pressure of JK-A cavern during 2010-2015. The gas pressure significantly differs from the previous design value (Wang et al.,

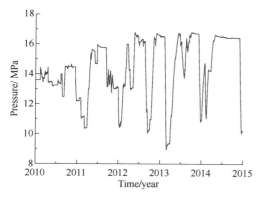

Fig. 3 Field monitored gas pressure in JK-A cavern during 2010-2015

2016). This is mainly because the salt cavern UGS has advantages of good flexibility and high peak-shaving capacity, so that it is not only used for seasonal but also for time interval peak-shaving and even for pressure control of pipeline grids. This leads to the internal gas pressure changing sharply, which is bad for the cavern stability.

2.2 Rock property parameters

JK-A cavern is the first core information cavern of Jintan salt cavern UGS. Cores with a total length of about 280 m of caprock, rock salt, and substratum formations were obtained. First, all the cores were photographed and logged carefully. Then, they were wrapped in plastic film to prevent air-slaking. Systematic lab tests were carried out on these cores based on the requirements of different projects. The tests included regular mechanical tests, creep tests, permeability tests, breakthrough pressure tests, and x-ray diffraction analysis, etc. Numerous experimental data have been obtained (Yang et al., 2015a, b; Yang et al., 2009). The experimental methods and observations are not repeated in this paper. Based on previous research (Wang et al., 2017), three types of rocks will be used during the construction of the 3D geomechanical model of JK-A cavern, mudstone, rock salt and interlayer. Their property parameters are listed in Table 1 (Yang et al., 2009, 2015a, b). The main components of the interlayer are anhydrite, argillaceous dolomite, and glauberite, and their water solubility is poor.

Table 1 Mechanical parameters used in the numerical simulations

Property	Materials		
	Rock salt	Interlayer	Mudstone
Young's modulus/GPa	3.99	3.80	4.72
Poisson's ratio	0.24	0.28	0.19
Cohesion/MPa	5.45	5.70	8.19
Friction angle/(°)	30.51	30.43	39.57
Tensile strength/MPa	1.04	1.08	1.67

The Norton-Hoff model provides a good fit for the creep deformation of Jintan rock salt, and has been proved by the field monitoring data to have a high reliability (Yang et al., 2015b). Therefore, it is used in the later numerical simulation. It is expressed mathematically as

$$\dot{\varepsilon} = A \cdot (\bar{\sigma})n \tag{1}$$

where $\dot{\varepsilon}$ is the steady creep rate; A is the material constant; $\bar{\sigma}$ is the normal stress, $\bar{\sigma} = \left(\frac{3}{2}\right)^{1/2} (\sigma_{ij}^d \sigma_{ij}^d)^{1/2}$; σ_{ij}^d is the deviator part of σ_{ij}; n is the stress exponent, and is usually valued as 3-6 (Bérest et al., 2001).

Based on available experimental results (Yang et al., 2015b), A and n of rock salt and interlayer of the target formation are: $A = 2.996 \times 10^{-9}$ MPa$^{-4.480}$/a, $n = 4.480$ for rock salt; and $A = 12.0 \times 10^{-6}$ MPa$^{-3.5}$/a, $n = 3.5$ for interlayer. The mudstone does not have the creep deformation characteristics.

2.3 3D geomechanical model and boundary conditions

To investigate the reasons and factors causing the roof collapse of JK−A cavern, a 3D geomechanical model of JK−A cavern is established based on the sonar survey data of 2009 and the geological features of the target formation. Fig. 4 presents the 3D geomechanical model and partial enlarged view of JK−A cavern constructed using ANSYS software (ANSYS Inc, 2012). The model is a cuboid with a length of 600 m, a width of 300 m, and a height of 700 m. According to the geological exploration, the depths of the roof and bottom of the bedded rock salt formation are about −978 and −1170 m respectively. The bedded rock salt formation has a thickness of about 192 m, and is located at the middle of the model. Two compacted mudstone layers with thicknesses of more than 300 m are located at the upper and lower reaches of the rock salt formation. JK−A cavern is at the center of the model. Its shape and dimensions are determined on the basis of the 2009 sonar survey data. Considering its symmetry, 1/2 cavern is modeled. The height and diameter of JK−A cavern are about 40 and 80 m respectively. The distances between the cavern and the model boundaries are all about 10 times the cavern dimensions along theirrespective directions. This ef-fectively minimizes the influence of boundary effects on the numerical results. There are four interlayers in the model. Their top depths are −1036, −1050, −1059, and −1076 m, and their thicknesses are 2.1, 3.2, 2.5, and 2.8 m respectively from top to bottom. Three interlayers intersect the cavern. The other interlayers have large distances to the cavern and are not modeled. The model bottom is fixed, viz., no displacements of the bottom are permitted. A horizontal fixed boundary is applied on the four vertical surfaces of the model. Overlying pressure of −16.1 MPa (compressive stress is defined as negative) is applied to the model top boundary, calculated by the depth and average density of the overlying formation. Due to the typical creep deformation capacity of rock salt (Langer, 1999), the tectonic stresses of the rock salt formation are released completely. Therefore, the distribution of the current initial in−situ stresses of the rock salt formation are considered satisfying the lithostatic pressure distribution, viz., the in−situ stress and its gradient are equal to each other along three directions at any given depths (Bruno and Dusseault, 2002). The cavern leaching is a continuous and gradual change process, which cannot be simulated by step excavation. We assume the cavern is constructed instantaneously. To eliminate the adverse influence of this assumption, an internal pressure acting on the cavern wall that gradually decreases from the initial in−situ stress to brine hydraulic pressure is used in the numerical calculation to simulate the cavern leaching. The construction of JK−A cavern includes drilling, leaching, sealing testing, debrining, and operating, which makes that the loads to which the cavern wall are subjected become very complicated. How these loads are simulated in the calculations will be given in Section 2.4.

ANSYS software (ANSYS Inc, 2012) is used to build the 3D geomechanical model. FLAC^{3D} is used for the calculations for its notable advantage in dealing with large deformation problems related

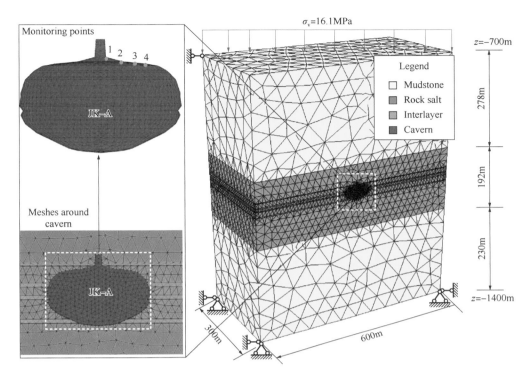

Fig. 4 3D geomechanical model and partial enlarged views of JK-A cavern constructed using ANSYS software

to rock and soil mechanical engineering (Itasca Consulting Group, 2005). Due to the model having large dimensions and only the zones close to the cavern being of concern, radial grids are used in the model. The sizes of the elements increase with increased distance to the cavern. The cavern has the smallest element size. This ensures that the study zone has a high accuracy with a good efficiency. Three kinds of elements are used: tetrahedral, hexahedral, and pyramidal. To avoid a poor mesh quality impact on numerical results, Meshtool, embedded in ANSYS software, is used to check the mesh quality (ANSYS Inc, 2012). Moreover, elements with several different sizes have been simulated to find the best element sizes. The element type and size used in this paper ensures the numerical simulation with high reliability and calculating efficiency. The model includes 31,597 nodes and 161,666 elements. To investigate the stress and deformation of the roof of JK-A cavern, four monitoring points are set, as shown in Fig. 4. They are numbered 1, 2, 3, and 4, and their Cartesian coordinates are (2.5, 0, −1038), (12.5, 0, −1040.3), (15, 0, −1040.5), (20, 0, −1041) (unit: m). Tecplot Inc (2013) is used for the result post-processing. To ensure the convergence of the numerical calculation, several other points are also monitored, and we found that the element type and size ensure the convergence requirement. Numerical results have a good independence relation with element size.

2.4 Operating condition simulation

The drilling for JK-A cavern began in October and was completed in November 2003, about 58 days. Due to the diameter of a well being much smaller than that of a salt cavern, and the drilling time being short, the drilling is not simulated in the numerical simulation. The cavern leaching lasted about 3.6 years (from January 2005 to September 2008). During the cavern leaching, the in-situ stress of the location where the cavern is planned decreased from the initial in-situ stress to the brine hydraulic pressure. Based on the depth of the casing shoe of JK-A cavern, the initial in-situ stress is calculated as about -23 MPa while the brine pressure is about -12 MPa. Therefore, the cavern leaching process can be simulated as the cavern operating with an internal pressure that gradually decreases. From September 2008 to October 2009, JK-A cavern was full of brine, and a sealing test was carried out. During that period, the pressure at the casing shoe was kept constant, at about -12 MPa. The debrining began in October 2009. Due to it being the first debrining in China, many accidents happened, and the debrining was completed on November 2010. The entire debrining took more than a year, which is much longer than the time the later caverns took (Wang et al.,2018a). The debrining uses compressed gas to expel the brine, and the pressure in the cavern basically does not change. Therefore, the cavern is subjected to a constant pressure (-12 MPa) during the debrining. The loads applied to JK-A cavern from November 2010 to November 2015, are given in Fig. 3. To accurately simulate the internal pressure in the calculations, segmented functions are used. About forty segmented functions are used to depict the gas pressure. Based on above analysis of the internal loads to which JK-A cavern is subjected during the leaching, sealing test, debrining and operating, Fig. 5 gives the internal loads applied to JK-A cavern for the numerical simulations. It is shown that the simulating internal loads applied to JK-A cavern have good agreement with the actual monitoring data, which can depict the actual situations.

3 Results and discussion

To find the likely failure time and the reasons that caused the cavern roof failure, the responses of JK-A are simulated using the model in Fig. 4 and the internal loads in Fig. 5. Rock salt shows typical creep and damage self-healing characteristics while the damage of salt cavern shows both plastic features (such as cavern volume shrinkage and cavern bottom heave) and brittle features (such as roof fall, spalling, and sluffing) (Bérest, 2013). Those characteristics lead to the failure prediction of salt caverns becoming complicated. Therefore, new failure prediction indexes are proposed, and they are used to find the potential collapse time and to analyze the causes of the JK-A cavern roof failure. The applicability of the proposed index is verified by the numerical results.

3.1 Evaluation criteria

To evaluate the stability of JK-A cavern roof, a new failure prediction index is introduced, consisting of volume shrinkage, dilatancy safety factors, displacement, vertical stress, and equiv-

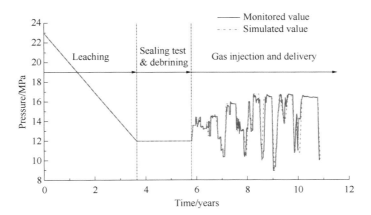

Fig. 5 Internal loads applied on JK-A cavern for the numerical simulations

alent strain. The definitions and critical values of the five indicators are explained and discussed as following.

(1) Volume shrinkage. The ratio between the volume reduction and cavern original volume, it is used mainly to evaluate the plastic damage of salt caverns. Published research shows that the volume shrinkage ranges from several percent to about 60% (Bérest and Brouard, 2003). According to the sonar survey results (Yang et al., 2015a, b), Jintan salt caverns have a small volume shrinkage. 30% for 30 years is proposed as the critical value of volume shrinkage for Jintan (Wang et al., 2016).

(2) Dilatancy safety factor. It was proposed by Van Sambeek et al. (1993), and has been proved by many actual engineering applications (Sobolik and Ehgartner, 2006). Its mathematical expression is written as (Sobolik and Ehgartner, 2006)

$$SF = \frac{0.27 I_1}{\sqrt{J_2}} \quad (2)$$

where SF is the safety factor for dilatancy; I_1 is the first invariant of the stress tensor, $I_1 = \sigma_1 + \sigma_2 + \sigma_3$; J_2 is the second invariant of the deviatoric stress tensor, $J_2 = \frac{1}{6}[(\sigma_1 - \sigma_2)^2 + (\sigma_2 - \sigma_3)^2 + (\sigma_3 - \sigma_1)^2]$. σ_1, σ_2, σ_3 are the first, second, and third principal stresses.

When the numerical simulations are completed, the stresses required by Eq. (2) are extracted and input into Tecplot Inc (2013) for the post-processing, and the SF contours are obtained. According to mechanical properties of Jintan rock salt, these thresholds are used: dilatancy safety factor smaller than 0.6 indicates collapse, ranging from 0.6 to 1.0 indicates failure, and ranging from 1.0 to 1.5 indicates local damage (Wang et al., 2016, 2017, 2018b).

(3) Displacement. Serves as one of the most widely used indicators in rock and soil engineering and has the advantages of having a simple physical meaning and of being monitored eas-

ily. Based on the previous study about Jintan salt cavern UGS (Wang et al., 2016), the maximum displacement should not exceed 5% of the cavern maximum diameter to avoid damage, such as roof fall, spalling, and sluffing.

(4) Vertical stress. JK-A cavern had a large-span flat roof. Moreover, the roof is not surrounded by a rock mass, but contacts gas directly. We speculate that the roof instability may be sensitive to the vertical stress. Moreover, low vertical stress zones indicate where the rock salt is suffering larger shear stress, and where failure takes place more easily. Therefore, the vertical stress is selected as one of the stability indicators. The depth of JK-A cavern roof is about −1000 m, and the initial in-situ stress is about −23 MPa. There are no related criteria about how to determine the critical safe value of vertical stress.

(5) Equivalent strain. Considering that the damage of rock salt shows the damage characteristic of ductile metal (such as copper), equivalent strain is adopted to evaluate the salt cavern safety. It is an indicator to assess the plastic creep safety, and its mathematical equation is (Wang et al., 2016)

$$ES = \sqrt{\frac{2}{3}\varepsilon^{dev} : \varepsilon^{dev}} \tag{3}$$

where ES is the effectivestrain; ε^{dev} is the deviatoric strain tensor.

Based on the mechanical properties of Jintan rock salt and engineering experience (Wang et al., 2016), 3% for 30 years is proposed as the critical value.

These indicators and their thresholds are used in Section 3.2 to find the likely collapse time and the reasons that caused the collapse of JK-A cavern roof.

3.2 Results analysis and discussion

Fig. 6 presents the volume shrinkage of JK-A vs. time. The volume shrinkage increases with time. When the time comes into the operating period, volume shrinkage changes slightly with the change of internal gas pressure. This is mainly because the steady creep rate of Jintan rock salt is small, and the internal pressure at current level prevents accelerated creep of the rock salt. Therefore, the volume shrinkage is mainly determined by time, not by the change of gas pressure. Based on the threshold of volume shrinkage given in Section 3.1, JK-A cavern is safe.

Moreover, the results show that thecollapse time of the JK-A cavern roof cannot be determined by the relation between the volume shrinkage vs. time. The volume shrinkage of JK-A cavern after operating five years is about 1.5%. By comparing the sonar survey data of 2009 and 2015, the volume shrinkage of JK-A cavern is about −2.7%. The reason causing this unusual phenomenon has been discussed in Section 2.1. The volume shrinkage of JK-A cavern is small, and may be smaller than the error of the sonar survey. We calculate that the maximum error between the volume shrinkages obtained by numerical simulation and by field monitoring data is about 4.2% under extreme conditions, which is a high accuracy for an actual engineering application. This indicates that the 3D geomechanical model, mechanical properties, boundary conditions, and loads used in the paper are reliable and accurate. Therefore, we have reasons to believe that the other results (Figs. 7–11) obtained by the numerical simulation have high reliability and accuracy.

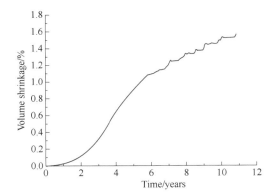

Fig. 6 Volume shrinkage of JK–A vs. time

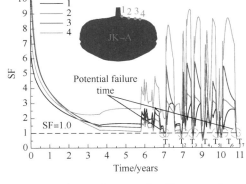

Fig. 7 Relations between the dilatancy safety factors (SFs) of monitoring points and time using Eq. (2)

Fig. 7 presents the relations between the dilatancy safety factors of monitoring points and time using Eq. (2), where the numbers 1–4 indicate the monitoring points. Based on the threshold of the dilatancy safety factor given in Section 3.1, the safety factor with value of 1.0 is used here to predict the potential collapse time of JK–A cavern roof. Because three different stages, viz., leaching, sealing test and debrining, and gas injection–delivery are simulated, the relations between the dilatancy safety factors of monitoring points and time are discussed separately. (1) Cavern leaching. The dilatancy safety factors of the monitoring points decrease with time. This is mainly because the loads applied to the cavern wall gradually decrease from the initial in-situ stress (about −23 MPa) to the brine hydraulic pressure (about −12 MPa) while the far-field stress remains constant. This makes that the deviatoric stress acting in the rock mass around the cavern increases, which causes the decrease of the cavern safety. The minimum values of the dilatancy safety factors of the monitoring points are still in the safe range during cavern leaching. (2) Sealing test and debrining. The dilatancy safety factors of the monitoring points 1–3 decrease slowly while that of point 4 increases slightly. This is because during this period the internal loads applied to the cavern remain constant with a relative large value of about −12 MPa (accounting for about 50% of the initial in-situ stress). The creep of rock salt is slow and slightly changes the stress distribution in the rock mass. Due to the difference in location, the stress change may be different. As a result the dilatancy safety factors of the monitoring points have different variations with time. (3) Operating period. The dilatancy safety factors of all monitoring points change sharply. Generally, they all decrease with time. This indicates that the dilatancy safety factor is sensitive with time and the change of internal gas pressure when the internal pressure is low. Results show that failure takes place at the monitoring points 1–3 prior to that at point 4. This may be caused by the fact that JK–A cavern has a large-span flat roof, and the roof center is the weakest part, followed by the roof edge.

Numerical simulation shows that there are seven potential times when the roof collapse may have taken place. These potential collapse times are marked as T_1-T_7, and their value are $T_1 = 7.15$, $T_2 = 7.95$, $T_3 = 8.45$, $T_4 = 9.25$, $T_5 = 9.75$, $T_6 = 9.95$ and $T_7 = 10.75$ years. They are all in the operating period. From Fig. 5, we find that the internal pressures are all low and their change rates are steep at these potential collapse times. This indicates that low gas pressure and rapid gas pressure change may be the likely main factors causing the roof collapse. Actually, the roof collapse could only take place one time. However, the current numerical software cannot accurately simulate discontinuous damage, such as the cavern roof collapse. To find the potential collapse time, the responses of the rock mass around JK-A cavern roof at $T_1 = 7.15$, $T_2 = 7.95$, $T_3 = 8.45$, and $T_4 = 9.25$ years are investigated first.

Fig. 8 presents the dilatancy safety factors in the rock mass around the JK-A cavern on the vertical section at different times by using Eq. (2). The thresholds for the dilatancy safety factor are given in Section 3.1. The zone with a dilatancy safety factor smaller than 0.6 (red zone) mainly is located at the cavern roof and interlayers. The red zone in the cavern roof forms a large connected area. This indicates that the cavern roof is the weakest zone, where failure takes place most easily. By comparing the results in Fig. 8(a)-(d), the areas of the red zone show an increase at first and then a decrease with time. This may be caused by the changes of internal gas pressure, the creep load and the structure of the cavern roof. It indicates that the dilatancy safety factor is sensitive to gas pressure and creep time. By comparing the areas of low dilatancy safety factor zone, we conclude that there is a high possibility of the JK-A cavern roof collapse having taken place after 7.15 years. From the dilatancy safety factor viewpoint, the roof collapse of the JK-A cavern will take place again during the later operating period until a stable dome structure is formed. The depth of the casing shoe of JK-A cavern is -1016 m, and the depth of the cavern roof after collapse is about -1035 m. The thickness of the roof salt maintained is about 19 m. If a new collapse takes place, the thickness of the roof reserved salt may become smaller than the critical value of 15 m for Jintan. This may cause a tightness failure of JK-A cavern. Crossley (1998) reported a similar salt cavern UGS accident, as well as a second roof collapse that caused leakage of the wellbore. We propose that the gas pressure and especially its decrease speed should be strictly controlled to avoid the roof collapsing again.

Fig. 9 presents the displacement of the rock mass around JK-A cavern on the vertical section at different times. The displacements increase with time. The bottom heave displacements are much larger than the displacements at the other locations. However, the displacements all are small (only about 5 cm), much smaller than the thresholds of displacement given in Section 3.1. This is well in accordance with the volume shrinkage in Fig. 6. The sonar survey data obtained in 2009 and 2015 also indicate that JK-A cavern only had a small deformation over five years (2010-2015) of operating. The results also confirmed the reliability of the 3D geomechanical model presented in the paper from the viewpoint of displacement.

Numerical simulations indicate the collapse of JK-A cavern is not induced by excessive deformation but by excessive stress, which indicates a brittle failure. By comparing the results in Figs. 9(a) - (d), the displacements of the rock mass around the cavern do not show notable changes. Therefore, we cannot judge the potential time of the roof collapse using the displacement criterion.

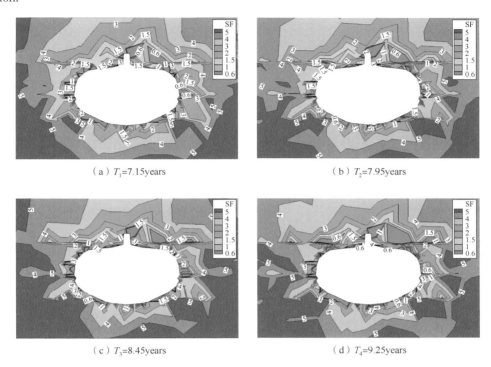

(a) T_1=7.15years (b) T_2=7.95years

(c) T_3=8.45years (d) T_4=9.25years

Fig. 8 Dilatancy safety factor (SF) in the rock mass around JK-A cavern on the vertical section at four different times using Eq. (2)

Fig. 10 presents the vertical stress in the rock around JK-A cavern on the vertical section at different times. The original in-situ stress at the location of the casing shoe is about -23 MPa. Due to the presence of the cavern, the in-situ stress redistributes. The red zones (low vertical stress) indicate where the rock salt suffers larger shear stress, and where failure takes place more easily. Therefore, the red areas are considered the potential failure zones. The red areas mainly are located at the cavern roof, followed by the zones around interlayers. The roof structure and the property difference at interlayers have notable effects on the stress redistribution. Due to the cavern roof not being surrounded by rock, once local damage take places the failed zone will fall from the roof and may trigger more collapse. The failed zones around interlayers, constrained by adjacent rock, will remain in their original locations. That is the reason why the sonar survey only found the collapse of the roof and not also the collapse of the cavern walls. By comparing the results of Fig. 10 (a)-(d), the roof collapse of JK-A cavern most probably took place at the time of 7.15 years, viz., 1.3 years after the operation started. This is because at that time the red zone around the cavern roof is large and connects together indicating a high possibility of collapse. By analyzing the mo-

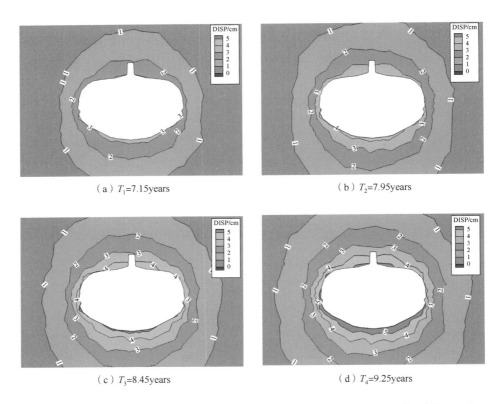

(a) T_1=7.15years (b) T_2=7.95years

(c) T_3=8.45years (d) T_4=9.25years

Fig. 9 Displacement of rock mass around JK-A cavern on the vertical section after different times

nitored gas pressure in Fig. 5, the gas pressure was 10.4 MPa at the time of 7.15 years, which is the lowest gas pressure over that period. Moreover, the gas pressure decreased from 15.8 to 10.4 MPa within 18 days at that period. The rapid gas pressure decrease leads to a significant and fast stress increase in the rock mass around the cavern roof. Due to the cavern having a large-span flat roof, stresses in the cavern roof cannot be released timely and stress concentration is formed. Local damage takes place in the cavern roof and may trigger additional collapses by the self-weight of the loosened rock. Moreover, there is an interlayer immediately above the cavern roof. Its strength and deformation are smaller than those of the rock salt, and stress concentration and deformation incompatibility are produced in the interface between the interlayer and rock salt. By reacting to dramatically changing gas pressure, stress concentration effect and deformation incompatibility, the rock salt between the interlayer and the cavern roof may weaken, fracture and fall. Similar failure mechanisms had been reported in the roof collapse analysis of cavern No. 5 in Regina South UGS (Crossley, 1998). Santolo et al. (2018) also confirmed that the roof structure was critical to the stability of a shallow man-made cavern. Ultimately, these factors together caused the roof collapse. In conclusion, the main reasons that caused the failure from the viewpoint of vertical stress are (ⅰ) the cavern roof has large-span flat structure, and (ⅱ) the internal gas pressure decrease speed is too fast.

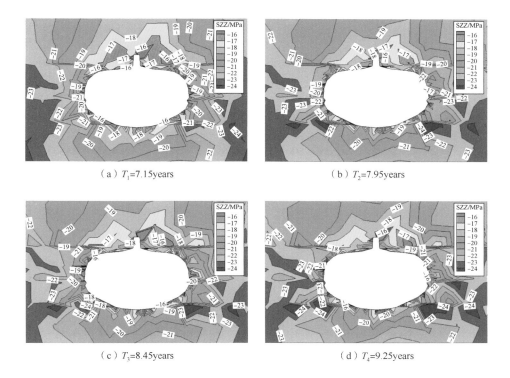

Fig. 10 Vertical stress in the rock mass around JK-A cavern on the vertical section at four different times

Fig. 11 presents the equivalent strain in the rock mass around JK-A cavern on the vertical section at different times using Eq. (3). The equivalent strain increases with time. The areas with high equivalent strain are located mainly around interlayers. Due to the short time simulated, the overall equivalent strain is still small, much smaller than the critical value of 3% for 30 years. Therefore, creep deformation damage will not happen around JK-A cavern. By comparing the results in Fig. 11 (a)-(d), the equivalent strain increases smoothly without any sudden changes or without exceeding the critical value. Therefore, the potential roof collapse and its reason of JK-A cavern roof cannot be identified by using the equivalent strain.

From the above analysis, the JK-A cavern roof collapse took place after 7.15 years (viz., 1.3 years after the operation started). The reasons that caused the collapse are (i) the cavern has a large-span flat roof, which is bad for bearing the loads; (ii) the internal gas pressure decrease speed is too fast, which causes a stress concentration in the cavern roof; and (iii) the cavern roof contacts gas directly without being surrounded by rock. When the local damage takes place the failed zone falls under its self-weight. This aggravates the roof collapse. Due to the collapse of the cavern roof confirmed as having taken place after 7.15 years, the safety of the rock mass around JK-A cavern at times of $T_5 = 9.75$ years, $T_6 = 9.95$ years, and $T_7 = 10.75$ years are not investigated.

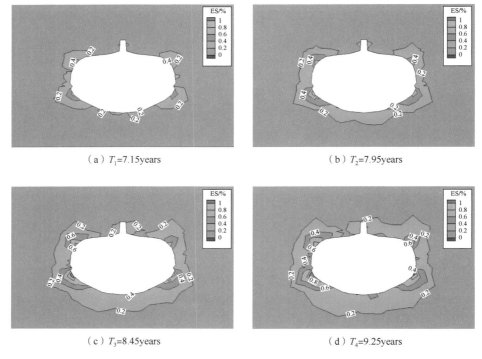

Fig. 11　Equivalent strain (ES) in the rock mass around JK-A cavern on the vertical section at four different times using Eq. (3)

4　Summary and conclusions

(1) By comparing the sonar survey results obtained in 2009 and 2015, we observed that the roof of China's first salt cavern gas storage (JK-A cavern) collapsed. A 3D geomechanical model has been built using sonar survey data, geological features, and monitoring gas pressure to analyze the potential time of the collapse and the causes of the collapse.

(2) A new index system is proposed to predict the roof failure of JK-A cavern. It consists of volume shrinkage, dilatancy safety factor, displacement, vertical stress, and equivalent strain. Results show dilatancy safety factor and vertical stress have a good sensitivity and reliability to predict the cavern roof collapse.

(3) JK-A cavern roof collapse took place at 7.15 years (viz., 1.3 years after the operating started). The main reasons for the collapse are that the cavern has a large-span flat roof, the internal gas pressure decrease speed is too fast, and the self-weight aggravates local damage leading to massive collapse.

(4) JK-A cavern has a high risk of roof collapse taking place again during the future operating period. The gas pressure and its decrease speed should be strictly controlled to avoid the roof collapsing again, in particular because it may threaten the tightness of the cavern.

Acknowledgements

The authors wish to acknowledge the financial supports of National Natural Science Foundation of China (Grant Nos. 41502296), Youth Innovation Promotion Association CAS (Grant No. 2016296), National Natural Science Foundation of China Innovative Research Team (Grant No. 51621006), and Natural Science Foundation for Innovation Group of Hubei Province, China (Grant No. 2016CFA014). We thank the editor Prof. Janusz Wasowski and reviewers for providing constructive suggestions and valuable references.

References

ANSYS Inc, 2012. ANSYS 14.5 Mechanical APDL Verification Manual. Canonsburg, Pennsylvania, USA. http://www.ansys.com.

Belzer, B., DeVries, K., 2017. Numerical prediction of tensile casing failure by salt creep for evaluating the integrity of cemented casings of salt caverns. In: SMRI Spring 2017 Technical Conference, 23-26 April 2017, Albuquerque, New Mexico.

Bérest, P., 2013. The mechanical behavior of salt and salt caverns. In: Proceedings of the ISRM International Symposium EUROCK 2013, Wroclaw, Poland, 23-26 October 2013, pp. 17-30.

Bérest, P., Brouard, B., 2003. Safety of salt caverns used for underground storage blow out; mechanical instability; seepage; cavern abandonment. Oil Gas Sci. Technol. 58 (3), 361-384.

Bérest, P., Bergues, J., Brouard, B., Durup, J., Guerber, B., 2001. A salt cavern abandonment test. Int. J. Rock Mech. Min. 38 (3), 357-368.

Bruno, M., 2005. Geomechanical Analysis and Design Considerations for Thin-bedded Salt Caverns. Office of Scientific & Technical Information Technical Reports.

Bruno, M., Dusseault, M., 2002. Geomechanical analysis of pressure limits for thin bedded salt caverns. In: SMRI Spring 2002 Technical Meeting April 29-30, 2002, Banff, Alberta.

Crossley, N., 1998. Sonar surveys used in gas-storage cavern analysis. Oil Gas J. 96 (18), 98-108.

Evans, D., 2008. Accidents at UFS sites and risk relative to other areas of the energy supply chain, with particular reference to salt cavern storage. In: Solution Mining Research Institute SMRI Fall 2008 Technical Conference, 13-14 October 2008.

Itasca Consulting Group, 2005. Inc FLAC^{3D} Version 3.0 Users' Manual (2005). Minneapolis, Minnesota, USA.

Khaledi, K., Mahmoudi, E., Datcheva, M., Schanz, T., 2016. Stability and serviceability of underground energy storage caverns in rock salt subjected to mechanical cyclic loading. Int. J. Rock Mech. Min. 86, 115-131.

Labaune, P., Rouabhi, A., Tijani, M., Blancomartín, L., You, T., 2018. Dilatancy criteria for salt cavern design: a comparison between stress-and strain-based approaches. Rock Mech. Rock. Eng. 51 (1), 1-13.

Langer, M., 1999. Principles of geomechanical safety assessment for radioactive waste disposal in salt structures. Eng. Geol. 52 (3-4), 257-269.

Li, L., Yang, H., Liu, Y., Fang, L., 2011. Debrining case study on the newly built cavern in Jintan. In: Ningxia Young Scientist Forum, 14-16 July 2011, Yinchuan, Ningxia, China, pp. 452-455.

Li, J., Shi, X., Yang, C., Li, Y., Wang, T., Ma, H., 2018. Mathematical model of salt cavern leaching for gas storage in high-insoluble salt formations. Sci. Rep. 8, 372.

Liu, W., Li, Y., Yang, C., Daemen, J., Yang, Y., Zhang, G., 2015. Permeability characteristics of mudstone cap rock and interlayers in bedded salt formations and tightness assessment for underground gas storage caverns. Eng. Geol. 193, 212-223.

Lux, K., 2009. Design of salt caverns for the storage of natural gas, crude oil and compressed air: geomechanical aspects of construction, operation and abandonment. Geol. Soc. Lond. Spec. Publ. 313, 93–128.

Santolo, A., Forte, G., Santo, A., 2018. Analysis of sinkhole triggering mechanisms in the hinterland of Naples (southern Italy). Eng. Geol. 237, 42–52.

Sobolik, S., Ehgartner, B., 2006. Analysis of Shapes for the Strategic Petroleum Reserve. Sandia National Laboratories, Albuquerue, NM, USA, pp. 2006 (No. SAND 2006–3002).

Sobolik, S., Lord, A., 2015. Operation Maintenance and Monitoring of Large-Diameter Caverns in Oil Storage Facilities in Domal Salts. Proc Mech Beh Salt VII. Taylor & Francis Group, London.

Tecplot Inc, 2013. Tecplot 360 User's Manual (2013). Bellevue, Washington, USA. https://www.scc.kit.edu/downloads/sca/tpum.pdf.

Van Sambeek, L., Ratigan, J., Hansen, F., 1993. Dilatancy of rock salt in laboratory test. Proceedings, 34th U.S. Symposium on Rock Mechanics, University of Wisconsin–Madison, Madison, WI, June 27–30, B. C. Haimson (ed.). Int. J. Rock Mech. Min. Sci. Geomech. Abstr. 30, 735–738.

Wang, T., Ma, H., Yang, C., Shi, X., Daemen, J., 2015. Gas seepage around bedded salt cavern gas storage. J. Nat. Gas Sci. Eng. 26, 61–71.

Wang, T., Yang, C., Ma, H., Li, Y., Shi, X., Li, J., Daemen, J., 2016. Safety evaluation of salt cavern gas storage close to an old cavern. Int. J. Rock Mech. Min. 83, 95–106.

Wang, T., Yang, C., Li, J., Li, J., Shi, X., Ma, H., 2017. Failure analysis of overhanging blocks in the wall of a gas storage salt cavern: a case study. Rock Mech. Rock. Eng. 50 (1), 125–137.

Wang, T., Ding, S., Wang, H., Yang, C., Shi, X., Ma, H., Daemen, J., 2018a. Mathematic modelling of the debrining for a salt cavern gas storage. J. Nat. Gas Sci. Eng. 50, 205–214.

Wang, T., Ma, H., Shi, X., Yang, C., Zhang, N., Li, J., Ding, S., Daemen, J., 2018b. Salt cavern gas storage in an ultra-deep formation in Hubei, China. Int. J. Rock Mech. Min. 102, 57–70.

Wang, T., Yang, C., Wang, H., Ding, S., Daemen, J., 2018c. Debrining prediction of a salt cavern used for compressed air energy storage. Energy 147, 464–476.

Yang, C., Li, Y., Chen, F., 2009. Bedded Salt Rock Mechanics and Engineering. Science Press, Beijing (in Chinese).

Yang, C., Li, Y., Zhou, H., 2015a. Failure Mechanism and Protection of Salt Caverns for Large-scale Underground Energy Storage. Science Press, Beijing (in Chinese).

Yang, C., Wang, T., Li, Y., Yang, H., Li, J., Xu, B., Daemen, J., 2015b. Feasibility analysis of using abandoned salt caverns for large-scale underground energy storage in China. Appl. Energy 137, 467–481.

Yuan, G., Shen, R., Tian, Z., Yuan, J., Gao, Y., Lu, L., Wei, D., Yang, H., 2006. Research and field application of quick-speed solution mining technology. Acta Pet. Sin. 27 (4), 139–142.

Zhang, G., Wu, Y., Wang, L., Zhang, K., Daemen, J., Liu, W., 2015. Time-dependent subsidence prediction model and influence factor analysis for underground gas storages in bedded salt formations. Eng. Geol. 187, 156–169.

Zhang, Q., Duan, K., Jiao, Y., Xiang, W., 2017. Physical model test and numerical simulation for the stability analysis of deep gas storage cavern group located in bedded rock salt formation. Int. J. Rock Mech. Min. 94, 43–54.

本论文原发表于《Engineering Geology》2018年第243卷。

Failure analysis of overhanging blocks in the walls of a gas storage salt cavern: A case study

Tongtao Wang[1] Chunhe Yang[1] Jianjun Li[2] Jinlong Li[1]
Xilin Shi[1] Hongling Ma[1]

(1. State Key Laboratory of Geomechanics and Geotechnical Engineering,
Institute of Rock and Soil Mechanics, Chinese Academy of Sciences;
2. West-to-East Gas Pipeline Company Gas Storage Project Department,
PetroChina Company Limited)

Abstract Most of the rock salt of China is bedded, in which non-salt layers and rock salt layers alternate. Due to the poor solubility of the non-salt layers, many blocks overhang on the walls of the caverns used for gas storage, constructed by water leaching. These overhanging blocks may collapse at any time, which may damage the tubing and casing string, and even cause instability of the cavern. They are one of the main factors threatening the safety of caverns excavated in bedded rock salt formations. In this paper, a geomechanical model of the JJKK-D salt cavern, located in Jintan salt district, Jintan city, Jiangsu province, China, is established to evaluate the stability of the overhanging blocks on its walls. The characters of the target formation, property parameters of the rock mass, and actual working conditions are considered in the geomechanical model. An index system composed of stress, displacement, plastic zone, safety factor, and equivalent strain is used to predict the collapse length of the overhanging blocks, the moment the collapse will take place, and the main factors causing the collapse. The sonar survey data of the JJKK-D salt cavern are used to verify the reliability and accuracy of the proposed geomechanical model. The results show that the proposed geomechanical model has a good reliability and accuracy, and can be used for the collapse prediction of the overhanging blocks on the wall of the JJKK-D salt cavern. The collapse length of the overhanging block is about 8 m. We conclude that the collapse takes place during the debrining. The reason behind the collapse is the sudden decrease of the fluid density, leading to the increase of the self-weight of the overhanging blocks. This study provides a basis for the collapse prediction method of the overhanging blocks of Jintan salt cavern gas storage, and can also serve as a reference for salt cavern gas storage with similar conditions to deal with overhanging blocks.

Keywords Rock mechanics; Rock salt; 3D geomechanical model; Salt cavern gas storage; Numerical simulation

1 Introduction

Jintan bedded rock salt hosts the fourth salt section of the Funing group of the tertiary formation, which belongs to continental clastic rock salt deposits. The overall core body has an east-north strike, with a length of 11.75 km and a width of 5.6 km, showing bedded structure or bed-like structure, composed of many layers with different properties. The layers with the same property are continuous, and have stable plane distribution with a dip of less than 10°. Generally, the core body is composed oftens of rock salt and nonsaltlayers. The thickness of single layers usually ranges from about 5 to 10 m. The total thickness of the core body is about 143-173 m, and its average is about 150 m. The main components of the non-salt layers are anhydrite, mudstone, and glauberite, which have poor solubility in water. This causes many irregular overhanging blocks to form on the cavern wall. These non-salt layers have high strengths even after having been immersed for a long time in brine, which leads these overhanging blocks to have a large length. These overhanging blocks are not always stable and their stability is affected by many factors. When they collapse, these blocks can cause damage and even failure of the tubing and casing. Under some extreme conditions, the sudden collapse of the overhanging block can cause cavern wall massive chain fall offs and even cavern failure.

Figure 1 presents the 3D shape and average radius of the JJKK-D salt cavern obtained by sonar survey. There are many irregular overhanging blocks on the cavern wall. Notable partial solution is taking place as a result of the presence of the non-salt layers during the leaching. Therefore, study on the safety evaluation of the overhanging blocks, identifying the conditions under which these blocks fall and the factors affecting their collapse, are important and valuable for ensuring the safety of gas storage caverns constructed in bedded rock salt formations.

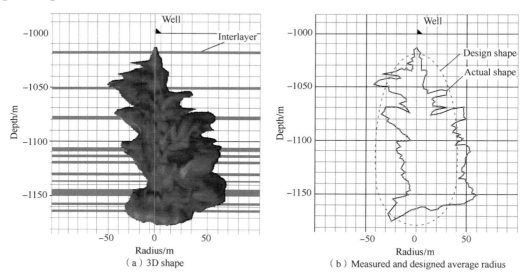

Fig. 1 3D shape and average radius of the JJKK-D salt cavern obtained by sonar survey in Jintan typical bedded rock salt formation

Many studies focus on the stability assessment of salt caverns used for natural gas storage, waste disposal, and compressed air storage. Staudtmeister and Rokahr(1997) built a 2D numerical model to study the long-term stability of gas storage salt caverns, and proposed the stress intensity factor to determine whether wall spalling takes place. Langer and Heusermann (2001) investigated the safety of a salt mine to serve for the disposal of radioactive waste and pointed out that the mechanical stability and integrity were the two most important assessment factors. They indicated that the geology investigations, laboratory and field experiments, and geomechanical modeling were the necessary means to accurately predict the cavern stability. Heusermann et al. (2003) indicated that the effects of cavern depth, cavern shape and dimension, pillar width, distance between cavern and salt boundary, and internal gas pressure should be highlighted during the design of salt cavern gas storage. They used the LUBBY2 constitutive model to depict the non-linear creep of rock salt and discussed the effects of the above factors on the cavern stability. Swift and Reddish (2005) established a numerical model of Bostock No. 5 salt cavern, Winsford, England, to predict the stability of the abandoned cavern used for the disposal of waste materials and confirmed that the old cavern would be stable for along time. Han et al. (2007) developed a 3D geomechanical model of salt cavern gas storage located in a bedded rock salt formation and investigated the effects of the roof salt thickness, Young's modulus of overlying formation, interlayer properties, and cavern dimension on the cavern longterm stability. Wang et al. (2011) used the strength reduction method and cusp catastrophe model to quantitatively study the effects of depth, internal gas pressure, pillar width, and creep time on the stability of the pillar between adjacent salt cavern gas storages. Bérest (2013) analyzed the reason behind the fall of overhanging blocks on the walls of a salt cavern used for compressed air storage and indicated that the blocks failed during the first gas injection. He pointed out that this could be caused by the density decrease of the fluid in the cavern during the debrining, which led to the self-weight increasing significantly, and then the failure happened. Djizanne et al. (2014) built a 2D model of a gas storage salt cavern with irregular shape to study the stability of the overhanging blocks under the debrining and speedy gas production, and indicated that buoyancy was the main reason causing the fall of the overhanging blocks. Deng et al. (2015) used the deformation reinforcement theory to predict the stability evaluation and failure of rock salt gas storage caverns, and concluded that the unbalanced force can be used as an effective index to specify the local failure position and failure mode of gas storage caverns. Wang et al. (2016) built a 3D geomechanical model of a new cavern close to an old one and introduced the displacement, plastic zone areas and volumes, equivalent strain, safety factor, and volume shrinkage to assess the safety of the cavern close to an old cavern and to optimize the cavern dimensions and running parameters. Based on the above literature reviews, at least two conclusions can be reached: (1) stability assessment is one of the most critical factors during salt cavern design and there is no effective index (e.g., stress, displacement, plastic zone, safety factor) to accurately evaluate the stability of a gas storage salt cavern; (2) using the geomechanical model built based on the target formation characteristics and property parameters and actual running parameters to predict the responses (e.g., stress, strain, and displacement) of the rock masses around the salt cavern is an effective and reliable research method.

The JJKK-D salt cavern gas storage is located in the Jintan salt district, Jintan city, Jiangsu province, China. The drilling of it was completed on 30 August, 2005. The leaching began on 2 July, 2006 and was finished on 12 September, 2011. The leaching consisted of 11 stages. A sonar survey was carried out 12 times. The sonar survey results of 13 September, 2011 showed that the total volume of the cavern was 214, 358 m^3 after the completion of leaching. The debrining was started on 21 June, 2013 and finished on 20 September, 2014. To monitor the safety of the JJKK-D salt cavern, a sonar survey was carried out on 29 November, 2015, under working conditions. The cavern total volume was 217, 801 m^3. The shapes of the cavern obtained in 2011 and 2015 show little change, except for the parts at the depth of about -1070 m, where spalling took place. There is no clear conclusion about the time the spalling happened and the reasons for the failure.

The main motivation for this paper is to find the time when the overhanging blocks fall off from the cavern wall and the reasons for the falling off. A 3D geomechanical model of the JJKK-D salt cavern is built based on the sonar survey data of 13 September, 2011 and the information about the target formation where the cavern is located. The responses of the rock masses of the cavern are simulated during the leaching, debrining, and running. A comprehensive index composed of stress, displacement, plastic zone, safety factor, and equivalent strain is used to reveal when and why the overhanging blocks fail. The results can provide the basis for the collapse prediction and the control methods of the overhanging blocks of Jintan salt cavern gas storage, and can also serve as a reference for gas storage salt caverns with similar conditions to deal with overhanging blocks.

2　3D Geomechanical model and its boundary conditions

2.1　Mechanical properties and creep constitutive model

The Jintan salt cavern gas storage is the first gas storage cavern of China. The authors' group members have carried out many tests on the rock salt samples from this site, and rich and reliable experimental data of Jintan rock salt mechanical properties have been published (Yang et al. 2009, 2015a, b; Wang et al. 2016). Therefore, the tests on the rock salt are not included in this paper. Based on the characteristics of the formation where the JJKK-D cavern is located, three kinds of rocks are used in the geomechanical model, viz., mudstone, rock salt, and interlayer. The mudstones are distributed in the overlying and underlying strata. Rock salt and interlayer alternate in the bedded rock salt formation where the cavern is located. Based on the previous tests' results (Yang et al. 2009, 2015a, b; Wang et al. 2016), the mechanical properties of the three kinds of rocks are listed in Table 1.

Due to the typical creep characteristic of the rock salt (Yang et al. 1999), which has great influence on the stresses and deformations around the cavern, the effects of the rock salt creep are considered in the numerical simulation. Based on the previous tests' results (Guo et al. 2012; Li et al. 2014), the initial creep stage of Jintan rock salt lasts for a short time only, usually a few to over ten hours. The steady-state creep stage lasts a long time. The accelerated stage is seldom observed in the tests under simu-lating the actual in situ stress condition of the target formation. Therefore, only the steady-state creep stage parameter is used in the design, which has been accepted, widely used, and proved by the salt cavern gas storage designers of China (Yang et al. 2015a, b). The Norton-Hoff law is used to depict the steady-state creep characteristic of Jintan rock salt and mathematically expressed as (Bérest et al. 2001; Wang et al. 2010):

$$\dot{\varepsilon} = A \cdot (\overline{\sigma})^n \tag{1}$$

where $\dot{\varepsilon}$ is the steady-state creep rate; A is a material constant; $\overline{\sigma}$ is the deviatoric stress, $\overline{\sigma} = \sigma_1 - \sigma_3$, σ_1 and σ_3 are the maximum and minimum principal stresses respectively; n is the stress index, and is usually valued as 3-6 for rock salt (Bérest et al. 2001).

Table 1   Mechanical parameters used in the numerical simulations

Property	Materials		
	Rock salt	Interlayer	Mudstone
Young's modulus/GPa	3.99	3.8	4.72
Poisson's ratio	0.24	0.277	0.185
Cohesion/MPa	5.45	5.70	8.19
Friction angle/(°)	30.51	30.43	39.57
Tensile strength/MPa	1.04	1.08	1.67
Density/(kg/m³)	2200	2400	2700

Based on the previous experimental results (Yang et al. 2015a, b; Wang et al. 2016), the creep parameters of the rock salt and interlayer of the target formation used in Eq. (1) are:

Rock salt: $A = 2.996 \times 10^{-9} \mathrm{MPa}^{-4.480}/a$, $n = 4.480$;

Interlayer: $A = 12.0 \times 10^{-6} \mathrm{MPa}^{-3.5}/a$; $n = 3.5$.

Detailed information about the experimental conditions and procedures are included in the references Yang et al. (2015a, b) and Wang et al. (2016).

2.2 Geomechanical model and boundaries

According to geological exploration and the sonar survey results, the top and bottom depths of the bedded rock salt formation where the JJKK-D cavern is located are -983 and -1217.8 m, respectively, and those of the cavern are -1014.6 and -1176.4 m, respectively. Figure 2 presents the geomechanical model and boundary of the JJKK-D cavern. The model is a cuboid with a length of 800 m, a width of 600 m, and a height of 800 m. The JJKK-D cavern is located at the center of the model. Its dimensions are based on the sonar survey results of 2011, with reasonable simplifications. The height of the cavern is about 162 m, its maximum diameter is about 60 m, and its total volume is about 214,358 m³. Considering that the tiny interlayers have no notable effects on the stability of the cavern but greatly decrease the calculating efficiency and may lead to non-convergence of the calculation, a proper simplification is adopted. The interlayers with a thickness less than 1.5 m are not included in the model. Three interlayers are included in the model. Their thicknesses are 2.2, 1.8, and 2.0 m from upper to lower, respectively. An overlying pressure is applied to the model top, 16.1 MPa, based on the depth and density of the overlying strata. Considering the creep of rock salt, we assume that the in situ stresses are equal along three directions at a given depth. The gradient of the in situ stress along the depth is 2.3 MPa/100 m. Hinge supports are applied on the four vertical surfaces along the horizontal direction, to constrain the horizontal displacement of these boundaries. The horizontal and vertical displacements of the bottom boundary are all constrained. Brine/gas pressure is applied on the cavern internal surface, which changes with time as shown in Fig. 3.

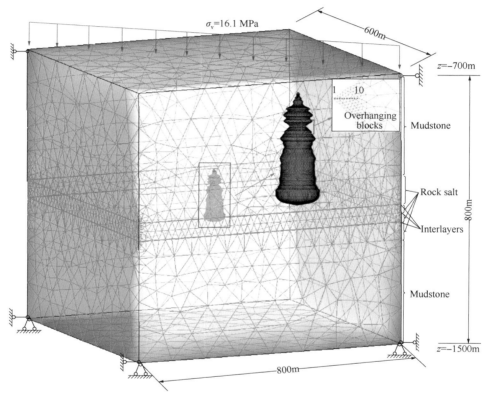

Fig. 2 Geomechanical model and boundaries of the JJKK-D cavern. Points 1, 2, 3, ⋯, 10 are the monitoring points used to evaluate the stability of overhanging blocks

Due to the large size of the geomechanical model and only the strain and deformation of the rock masses around the cavern being of concern in this paper, a radial mesh is used, viz., the mesh size increases with the distance to the cavern. The Meshtool embedded in the ANSYS software (ANSYS Inc. 2012) is used to check the mesh quality. To eliminate the effects of the mesh size on the results, different mesh sizes have been simulated. The mesh size used in the paper ensures calculation accuracy and high efficiency. Quadrilateral, hexagonal, and pyramidal elements are used. There are 751, 700 elements and 126, 144 nodes. The geomechanical model is imported into

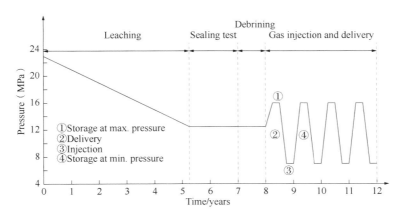

Fig. 3 Pressure applied to the internal surface of the JJKK-D cavern used in the numerical simulation

the FLAC^{3D} software (Itasca Consulting Group Inc. 2005) for the calculation. Tecplot software (Tecplot Inc. 2013) is used for the post-processing. The entire calculation for each condition takes about 180 h, carried out on a standard Lenovo D30 workstation.

To investigate the response of the overhanging blocks during the debrining and operation, ten monitoring points are set. The first point is located at the front edge, numbered as 1, and then one point is set per meter, as shown in Fig. 2. The points are numbered 1, 2, 3, ⋯, 10.

2.3 Internal pressure

The construction of gas storage salt caverns includes drilling, leaching, debrining, and gas injection/production stages. Due to the small size of the initial well and the drilling lasting for a short time only, we assume that the drilling has no notable effects on the cavern stability. Therefore, the drilling phase is not included in the simulation. During the leaching, the stress on the cavern wall is decreased with the increase of time and cavern size, and trends to the hydrostatic pressure of the brine. The internal pressure applied to the cavern in the leaching stage is equal to the hydrostatic pressure of the brine, and the effect of the brine gradient is included. Based on the casing shoe depth and overlying formation density, the in situ stress at the casing shoe is about 23 MPa. Similarly, the brine hydrostatic pressure at the casing shoe is about 12 MPa. After the completion of the leaching, the well completion and cavern tightness tests are carried out. During these operations, the cavern is full of brine. Subsequently, the debrining is carried out to form the gas storage. The gas is used to expel the brine during the debrining, which means that the gas pressure basically equals the brine pressure. According to the design parameters of the available gas storage salt cavern of Jintan (Yang et al. 2009, 2015a, b; Wang et al. 2016), the maximum and minimum gas pressures are 17 and 6 MPa, respectively. An operational period can be divided into four stages, viz., (1) storage at max pressure; (2) delivery; (3) injection; (4) storage at min pressure (Fig. 3). The duration of each stage is approximately 3 months (Wang et al. 2015). As stated in the introduction, the whole leaching time is about 5.2 years. The duration of the cavern tightness test and debrining is about 3 years. After running for about 4 years, a sonar survey is carried out under working conditions. Figure 3 presents the pressure applied to the internal surface of the JJKK-D cavern, which is used in the numerical simulation. A brine gradient is accounted for in simulating the leaching, tightness testing, and debrining process. The gas gradient is not included in simulating the gas storage running.

3 Numerical results and analysis

Using the formation parameters, 3D geomechanical model, and internal pressure shown in Fig. 3, the response of the rock mass around the JJKK-D cavern is simulated. The motivation for this study is to analyze the failure of the overhanging blocks at a depth of about-1070 m. Mainly, the stability of the block shown in Fig. 2 is discussed. That of the rock masses at the other locations is not included in this section. The durations of the leaching completion (5.2 years), before and after the debrining (8 and 8.5 years), and after running for 4 years (12.5 years) are selected as the study time frame to identify the possible time when the overhanging blocks fail and the reason that explains the failure. Because $FLAC^{3D}$ cannot directly simulate the spalling of the overhanging blocks, we propose an index system to evaluate the block's stability during the post-processing of the results. The index consists of the vertical stress, displacement, plastic zone, safety factor, and equivalent strain, which can decrease the risk caused by using only one single indicator. A coordinate grid is developed to locate the failed zone conveniently.

3.1 Vertical stress

Due to the fact that overhanging blocks are not surrounded by rock but by a fluid, the stresses in the overhanging blocks are mainly affected by the buoyant weight. The buoyant weight has a vertical direction. We postulate that the vertical stresses are more sensitive to the change of the buoyant weight of the overhanging blocks under different conditions (e.g., the leaching, well completion and cavern tightness tests, debrining, and running). Therefore, the vertical stress is chosen as the indicator to evaluate the stability of the overhanging blocks. Figure 4 presents the vertical stress contours in the overhanging blocks at different times. The tensile stress is defined as negative and compressive stress as positive. The high stress zones (red) appear at the right upper zone of the blocks and increase with time in the beginning and then decrease with time. This is mainly because the overhanging blocks can be approximately assumed to act as a cantilever beam, and the high tensile stresses first appear at the top surface of the fixed end of the beam under its buoyant weight. The fluid density decreases during the debrining. This greatly reduces the buoyancy of the overhanging blocks, viz., the selfweight increases notably, which leads to the increase of the vertical stress in the blocks. As time increases, the creep of rock salt transfers the stresses from the high stress zone to the low stress zone, and causes the stresses in the blocks to become uniform. Ultimately, the areas of the high stress zone decrease. By comparing the results in Fig. 4b, c, the stresses are increased greatly after the debrining, and the high stresses zone (red) penetrates nearly the entire overhanging blocks. This indicates that the overhanging blocks may collapse during the debrining. The reason for this is that the brine is replaced by natural gas, which causes the buoyancy acting on the blocks to decrease and causes the blocks to fall. Based on the results in Fig. 4c, we can deduce the length of the failed blocks to be about 8-10 m.

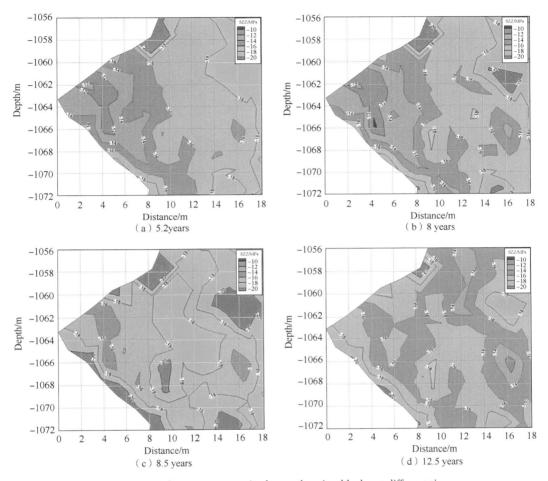

Fig. 4　Vertical stress contours in the overhanging blocks at different times

3.2　Displacement

Figure 5 presents the displacement contours in the overhanging blocks over time. The displacement increases with time. Large displacement mostly occurs in the tip areas of the overhanging blocks, which conforms to the displacement rule of a cantilever beam. By comparing the results in Fig. 5b, c, the debrining is not the key factor but the time affecting the displacement of the overhanging blocks. This maybe caused by the shortcoming of FLAC^{3D} that treats the rock salt as a continuous material. Moreover, the results indicate that using only one indicator to evaluate whether the overhanging blocks fail or not and to find the time at which the failure takes place is not sufficient. Several assessing indicators should be used. This is also the reason that the combination of the stress, displacement, plastic zone, safety factor, and equivalent strain are selected in this study.

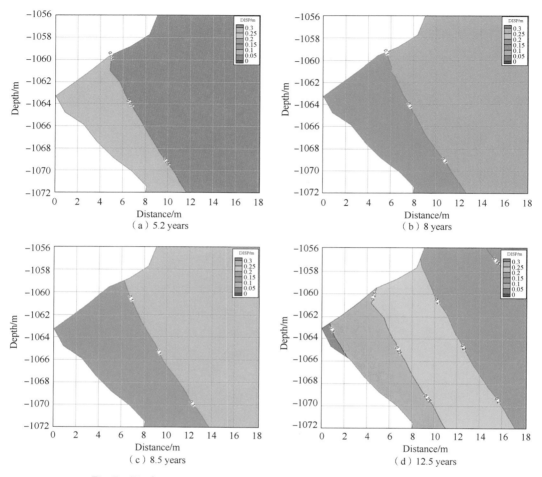

Fig. 5 Displacement contours in the overhanging blocks at different times

3.3 Plastic zone

Due to the special characteristics of rock salt, such as creep, self-healing, and large deformation capacity, its damage prediction in actual engineering remains challenging. There is still no widely accepted and approved criterion for rock salt damage prediction. Considering its simple expression, easily carrying out the calculation with a computer, and wide use in rock and soil engineering, the Mohr-Coulomb criterion is used as an indicator to evaluate whether failure of the overhanging blocks takes place. The plastic zone caused by the tensile failure is also included, viz., the plastic zone used in the paper is equal to the sum of the parts caused by shear and tensile stresses. Figure 6 presents the plastic zone contours of the overhanging blocks at different times. The red area indicates the plastic zone. The plastic zones increase with time and show a sudden increase after the debrining. As shown in Fig. 6c, the plastic zone penetrates nearly the entire blocks after debrining, which indicates that the failure of the overhanging blocks has taken place. From the viewpoint of the plastic zone, we can deduce that the failure of the overhanging block takes place during the debrining or at the time just after the debrining has been completed. The results in Fig. 6c show that the length of the failed zone is about 8 - 10 m. Considering the shortcoming of FLAC^{3D} mentioned above, the results in Fig. 6d may not appear in the actual cavern. Therefore, the results in Fig. 6care more suitable to serve as the basic data to analyze the failure of the blocks.

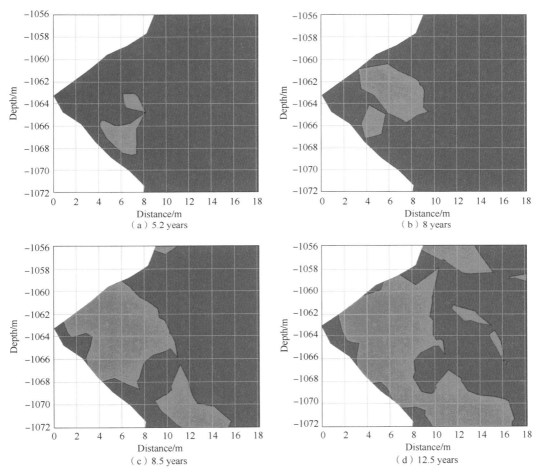

Fig. 6 Plastic zone contours of the overhanging blocks at different times.
The red areas indicate the plastic zones(color figure online)

3.4 Safety factor

To qualitatively evaluate the stability of the overhanging blocks, a safety factor is used, first proposed by Van Sambeek et al. (1993) to predict the dilatancy of rock salt. They indicated that the dilatancy of rock salt is mainly affected by the first principal stress invariant and the second principal stress deviator, and proposed the corresponding criterion. The criterion has beenverified by many experimental results, and has been used in the design and safety assessment of salt caverns serving for the storage of gas and petroleum. Good results are obtained(Sobolik and Ehgartner 2006). The criterion is mathematically expressed as:

$$SF_{vs} = \frac{0.27I_1}{\sqrt{J_2}} \quad (2)$$

where SF_{vs} is the safety factor for dilatancy; I_1 is the first invariant of the stress tensor, $I_1 = \sigma_1 + \sigma_2 + \sigma_3$; J_2 is the second invariant of the deviatoric stress tensor, $J_2 = \frac{1}{6}[(\sigma_1-\sigma_2)^2 + (\sigma_2-\sigma_3)^2 + (\sigma_3-\sigma_1)^2]$; σ_1, σ_2, and σ_3 are the maximum, intermediate, and minimum principal stresses, respectively.

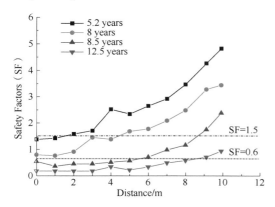

Fig. 7 Relations between the safety factors and distance at different times

By using Eq. (2), we developed a program using the FISH language embedded in FLAC^{3D} to obtain the safety factor of the rock masses. The safety factors of the points shown in Fig. 3 are monitored. According to Sobolik and Ehgartner (2006), a safety factor less than 0.6 indicates general collapse and one less than 1.5 indicates local damage. These thresholds are used in this paper to assess the block safety. Figure 7 presents the relations between the safety factors and distance at different times. The safety factors increase with increase of the distance and decrease with time. By comparing the results of the 8th and 8.5th year, the safety factors are changed greatly before and after the debrining. If the threshold of the safety factor is selected as 1.5, the length of failed blocks in the 8th year is about 5 m, while that in the 8.5th year is about 9 m, increasing by about 80 %. This confirms that the buoyancy has notable effects on the safety of the overhanging blocks. If the threshold is selected as 0.6, the lengths of the failed blocks at the 8.5th and 12.5th years are about 6 and 9 m, respectively. Based on the above results, we consider the overhanging blocks to have failed during the debrining or just after the debrining completion from the safety factor viewpoint, and the length of the failed zone is about 8 m.

By comparing the results in Figs. 6 and 7, the possible length of the failed zone obtained by the Mohr-Coulomb criterion is about 8-10 m, while that obtained by the dilatancy criterion is about 6-9 m (at the 8.5th and 12.5th years). This indicates that the dilatancy takes place before the plastic failure, viz., the dilatancy criterion is more severe than the Mohr-Coulomb criterion. The length of the failed zone obtained by the dilatancy criterion is conservative compared with that obtained by the Mohr-Coulomb criterion.

3.5 Equivalent strain

Because of the typical creep characteristic of rock salt, creep failure becomes a critical issue for the cavern safety. To evaluate the creep safety of the overhanging blocks, the equivalent strain is used, which should be less than 3 % over the design life time (Wu et al. 2005) to avoid creep failure. The equation used to calculate the equivalent strain is written as (Im and Atluri 1987):

$$\varepsilon_{eq} = \sqrt{\frac{2}{3}\varepsilon^{dev} : \varepsilon^{dev}} \qquad (3)$$

where ε_{eq} is the equivalent strain; ε^{dev} is the deviatoric strain tensor.

Figure 8 presents the equivalent strain contours of the overhanging blocks at different times obtained by Eq. (3). The equivalent strain increases with time and decreases with the distance to the block tip. By comparing the results in Fig. 8b, c, the equivalent strain does not change much before and after the debrining. This shows that the equivalent strain is not sensitive to the density change of the fluid stored in the cavern. This may be because this indicator is mainly affected by the creep deformation, while the creep deformation is mainly determined by time. Moreover, the results show that the equivalent strain of the overhanging blocks has a large safety margin. It cannot be used as an effective indicator to evaluate the dimensions of the failed zone and the time at which the failure takes place.

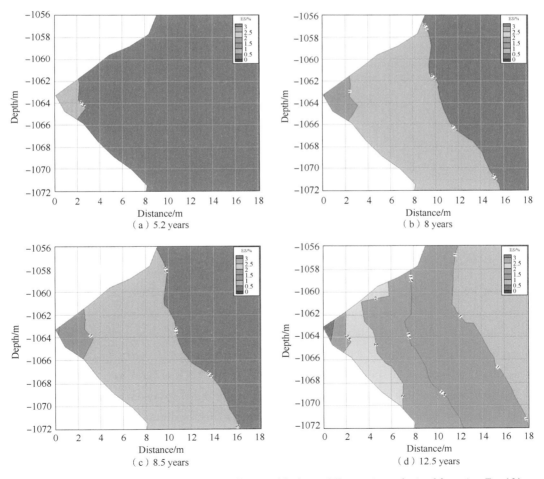

Fig. 8 Equivalent strain contours of the overhanging blocks at different times obtained by using Eq. (3)

4 Monitoring results analysis and verification

Figure 9 presents the average radius of the JJKK-D salt cavern obtained in 2011 and 2015. It shows that failure of the overhanging blocks takes place at a depth of about −1060 to −1080 m. The length of the failed blocks is about 8 m, which is the same as the results obtained by the numerical simulation. This indicates that the proposed geomechanical model is reliable.

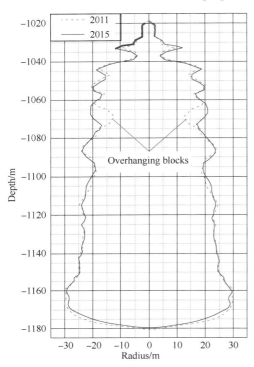

Fig. 9 Average radius of the JJKK-D salt cavern obtained in 2011 and 2015

By using the sonar survey results obtained in 2011 and 2015, the volumes of the parts with depth ranging from −1060 to −1080 m are 19, 762 and 24, 959 m^3. As shown in Fig. 9, the average radius of the cavern in the part with depth from −1060 to −1080 m almost does not change, except at the location of the overhanging blocks. Therefore, we can estimate the initial volume of the overhanging blocks to be about 5200 m^3. Considering the expansion of the broken rock, the volumes of the sediment deposits formed by the broken blocks are larger than the initial volume of the overhanging blocks. The expansion coefficient of broken Jintan rock salt is usually valued as 2.0, which is verified by the field monitoring data. We find that the volume of the sediment deposits formed by the broken blocks is 10, 400 m^3. If the overhanging blocks fail during the gas injection/delivery stage, the total volume of the JJKK-D cavern obtained in 2015 should be less than that obtained in 2011 by about 5200 m^3 at least. However, the volume of the JJKK-D cavern obtained in 2015(217, 801 m^3) is increased by about 3443 m^3 compared with that obtained in 2011 (214, 358 m^3). It is a very interesting phenomenon. Usually, the volume of a salt cavern shrinks gradually because of the creep of rock salt, especially as the collapse causes about another 5000 m^3 volume to be lost. The volume increase of the JJKK-D cavern looks unreasonable. By analyzing the sonar data, we deduce two reasons that may explain the apparent volume increase of the JJKK-D cavern.

The first reason is that the collapse of the overhanging blocks does not take place during the running stage, but happens after the completion of the leaching or during the debrining. To improve the cavern construction rate, the brine is not saturated. Its salt content is about 285 g/L, and that of the Jintan saturated brine is about 320 g/L, which means that the rock salt can continuously dissolve in the brine after the completion of leaching. The failed blocks will form a not compacted sediment deposit, which increases the surface areas of the salt contacting the unsaturated brine and improves the dissolving of the broken salt in the brine. This means that these salts contained in the failed blocks are preferentially dissolved into the brine. Ultimately, these dissolved salts of the failed blocks are expelled out of the cavern by the brine during the debrining. This is one possible reason behind the volume increase of the JJKK-D cavern.

The second reason is the error of the sonar survey. By comparing the salt production with the sonar data, the error of the sonar survey is usually about 1% of its measuring results. Taking the JJKK-D cavern as an example, the measuring error is about ±2000 m^3, which accounts for only about 30-40% of the volumes of the sediment deposits formed by the broken blocks. It also confirms that the collapse of the overhanging blocks does not take place during the running stage.

The cavern is full of brine, without any disturbance for the duration after the completion of leaching and before the debrining. Moreover, a pressure is applied to the well head to test the tightness of the cavern, which increases the pressure applied to the cavern wall. Therefore, the loading condition at this stage is better than that during the leaching stage, which is more favorable for the stability of the overhanging blocks. We are inclined to believe that the collapse taking place at this stage has a low probability. The debrining lasts for about a year, while the density of the fluid stored in the cavern is decreased from about 1200 kg/m^3 (brine) to 100 kg/m^3 (compressed natural gas). This increases the self-weight of the overhanging blocks by about 6000 t. Therefore, we deduce that the change of buoyant weight caused by the density change is the main reason leading to the collapse. This is also confirmed by the numerical simulation.

By comparing with the monitoring data, the proposed geomechanical model and stability evaluation indicators have a good accuracy and reliability, and can be used to analyze the reason for the failure of the overhanging blocks in the wall of the JJKK-D cavern, predict the dimensions of the failed zone, and find the main factor causing the failure. Based on the above results, we suggest that more attention should be given to these overhanging blocks during the leaching and redissolutioning should be used to eliminate these blocks based on the sonar survey results.

5 Summary and conclusions

(1) Based on the formation characteristics, sonar survey data, and actual loading condition of the JJKK-D salt cavern gas storage, a geomechanical model is built. A stability assessment index system composed of the stress, displacement, plastic zone, safety factor, and equivalent strain is proposed. The failure of the overhanging blocks in the JJKK-D cavern is analyzed.

(2) The results show that the length of the failed zone in the overhanging blocks is about 8 m. The failure takes place during the debrining, and the reason for the failure is the sudden density change of the fluid stored in the cavern.

(3) By comparing with and analysis of the sonar survey data, the results obtained by the proposed geomechanical model are accurate and reliable. The monitoring of these overhanging blocks should be highlighted during the debrining to eliminate their adverse effects.

Acknowledgments

The authors wish to acknowledge the financial supports of the National Natural Science Foundation of China (grant nos. 41502296, 41472285, 51404241, 51304187) and Youth Innovation Promotion Association CAS (grant no. 2016296). Thanks go to Professor J. J. K. Daemen, Mackay School of Earth Sciences and Engineering, University of Nevada (Reno), for the careful proofreading and constructive suggestions. We would like to thank an anonymous reviewer for their helpful and constructive comments and suggestions.

List of symbols

A Material constant

I_1 First invariant of the stress tensor, $I_1 = \sigma_1 + \sigma_2 + \sigma_3$

J_2 Second invariant of the deviatoric stress tensor, $J_2 = \dfrac{1}{6}[(\sigma_1 - \sigma_2)^2 + (\sigma_2 - \sigma_3)^2 + (\sigma_3 - \sigma_1)^2]$

SF_{vs} Safety factor for dilatancy

n Stress index, and is usually valued as 3–6 for rock salt

$\bar{\sigma}$ Deviatoric stress, $\bar{\sigma} = \sigma_1 - \sigma_3$

σ_1 Maximum principal stress

σ_2 Intermediate principal stress

σ_3 Minimum principal stress

$\dot{\varepsilon}$ Steady-state creep rate

ε^{dev} Deviatoric strain tensor

ε_{eq} Equivalent strain

References

[1] ANSYS Inc. (2012) ANSYS 14.5 mechanical APDL verification manual. Canonsburg, Pennsylvania, USA. Home page at: http://www.ansys.com.

[2] Bérest P (2013) The mechanical behavior of salt and salt caverns. In: Proceedings of the ISRM International Symposium EUROCK 2013, Wroclaw, Poland, 23–26 October 2013. ISRM-EUROCK-2013-002. International Society for Rock Mechanics, pp 17–30.

[3] Bérest P, Bergues J, Brouard B, Durup JG, Guerber B (2001) A salt cavern abandonment test. Int J Rock Mech Min Sci 38(3): 357–368.

[4] Deng JQ, Yang Q, Liu YR, Pan YW (2015) Stability evaluation and failure analysis of rock salt gas storage caverns based on deformation reinforcement theory. Comput Geotech 68: 147–160.

[5] Djizanne H, Bérest P, Brouard B (2014) The mechanical stability of a salt cavern used for compressed air energy storage (CAES). In: Proceedings of the 2014 Spring SMRI Technical Conference, San Antonio, Texas, USA, 4–7 May 2014.

[6] Guo YT, Yang CH, Mao HJ (2012) Mechanical properties of Jintan mine rock salt under complex stress paths. Int J Rock Mech Min Sci 56: 54–61.

[7] Han G, Bruno MS, Lao K, Young J, Dorfmann L (2007) Gas storage and operations in single-bedded salt

caverns: stability analyses. SPE Prod Oper 22(03): 368-376.

[8] Heusermann S, Rolfs O, Schmidt U(2003) Nonlinear finite-element analysis of solution mined storage caverns in rock salt using the LUBBY2 constitutive model. Comput Struct 81(8): 629-638.

[9] Im S, Atluri SN(1987) A study of two finite strain plasticity models: an internal time theory using Mandel's director concept, and a general isotropic/kinematic-hardening theory. Int J Plast 3: 163-191.

[10] Itasca Consulting Group Inc. (2005) FLAC^{3D} version 3.0 users' manual. Itasca Consulting Group, Inc., New York, USA.

[11] Langer M, Heusermann S (2001) Geomechanical stability and integrity of waste disposal mines in salt structures. Eng Geol 61(2-3): 155-161.

[12] Li YP, Liu W, Yang CH, Daemen JJK (2014) Experimental investigation of mechanical behavior of bedded rock salt containing inclined interlayer. Int J Rock Mech Min Sci 69: 39-49.

[13] Sobolik SR, Ehgartner BL (2006) Analysis of cavern shapes for the strategic petroleum reserve. No. SAND2006-3002. Sandia National Laboratories, Albuquerue, NM, USA.

[14] Staudtmeister K, Rokahr RB (1997) Rock mechanical design of storage caverns for natural gas in rock salt mass. Int J Rock Mech Min Sci 34: 300.e1-300.e13.

[15] Swift GM, Reddish DJ(2005) Underground excavations in rock salt. Geotechn Geol Eng 23: 17-42.

[16] Tecplot Inc. (2013) Tecplot 360 user's manual. Tecplot Inc., Bellevue, Washington, USA. Available online at: https://www.scc.kit.edu/downloads/sca/tpum.pdf. Accessed 21 July 2016.

[17] Van Sambeek LL, Ratigan JL, Hansen FD(1993) Dilatancy of rock salt in laboratory tests. Int J Rock Mech Min Sci Geomech Abstr 30: 735-738.

[18] Wang TT, Yan XZ, Yang XJ, Yang HL(2010) Dynamic subsidence prediction of ground surface above salt cavern gas storage considering the creep of rock salt. Sci China Tech Sci 53(12): 3197-3202.

[19] Wang TT, Yan XZ, Yang HL, Yang XJ(2011) Stability analysis of the pillars between bedded salt cavern gas storages by cusp catastrophe model. Sci China Tech Sci 54: 1615-1623.

[20] Wang TT, Yang CH, Yan XZ, Daemen JJK(2015) Allowable pillar width for bedded rock salt caverns gas storage. J Petrol Sci Eng 127: 433-444.

[21] Wang TT, Yang CH, Ma HL, Li YP, Shi XL, Li JJ, Daemen JJK(2016) Safety evaluation of salt cavern gas storage close to an old cavern. Int J Mech Min Sci 83: 95-106.

[22] Wu W, Hou ZM, Yang CH (2005) Investigations on evaluating criteria of stabilities for energy (petroleum and natural gas) storage caverns in rock salt. Chin J Rock Mech Eng 24: 2497-2505(in Chinese).

[23] Yang CH, Daemen JJK, Yin JH(1999) Experimental investigation of creep behavior of salt rock. Int J Rock Mech Min Sci 36: 233-242.

[24] Yang CH, Li YP, Chen F(2009) Mechanics theory and engineering of bedded salt rock. Science Press, Beijing(in Chinese)

[25] Yang CH, Li YP, Zhou HW (2015a) Failure mechanism and protection of salt caverns for large-scale underground energy storage. Science Press, Beijing(in Chinese).

[26] Yang CH, Wang TT, Li YP, Yang HJ, Li JJ, Qu DA, Xu BC, Yang Y, Daemen JJK(2015b) Feasibility analysis of using abandoned salt caverns for large-scale underground energy storage in China. Appl Energy 137: 467-481.

本论文原发表于《Rock Mechanics and Rock Engineering》2016年10月5日。

地面工程

盐穴储气库造腔地面工艺技术

杨清玉[1]　庄清泉[2]

(1. 中国石油西气东输管道公司储气库项目部；2. 大庆油田工程有限公司)

【摘　要】 金坛盐穴储气库造腔地面注水、采卤的介质环境对地面工艺和设备提出了更高的要求。玻璃钢管道的应用降低了大排量、长距离输送过程中的摩阻损失，解决了常规钢质管道输送河水及卤水介质时的结垢及腐蚀带来的运行成本增加。给出了计算溶腔过程中有效体积的理论计算方法。

【关键词】 储气库；盐穴；造腔；工艺

1　地面工艺流程简介

金坛盐穴地下储气库作为西气东输管道工程的配套设施，在管道调峰、处理应急事故等方面将具有重要的商业价值及社会意义。

盐穴造腔用水量大，对水质要求不高，水源为就近的河水和湖水。地表水经沉降池沉降由泵增压输送至注水站的 2 座 2000m³ 淡水罐进行缓冲、沉降，然后靠罐本身静水压输至高压注水泵进口，泵进口采用双吸流程(可根据需要取自淡水罐和卤水罐)，注水泵升压至单井高压阀组并经分配控制、调节和计量送至站外，经站外注水管网到单井井口，根据造腔工艺要求采用正循环或反循环将水注进盐腔内溶腔。

井口返出卤水经回水管网，回到注水站内的低压阀组间，按卤水所含介质、组分及含量等不同分别流向不同去处。

含盐量不小于 285g/L 的近饱和卤水根据卤水中硫酸根离子质量浓度的高低分别经过一条卤水汇管外输至盐厂缓冲罐和外销卤水成品罐；未饱和卤水进入 2 座 2000m³ 未饱和卤水储罐，经卤水储罐缓冲，再次由注水泵升压至高压阀组分配，经注水管网到另一组井口；井口返出液经回水管网，回到注水站内，重复上述流程。

2　注采工艺设备的选择

在盐穴造腔初期，由于腔内体积小，注入水在腔体内的滞留时间短且与腔壁接触面积小，采出的卤水难以达到盐厂或直接外销的含盐质量浓度要求。金坛储气库造腔注水采卤站通过选用与恒速电机相配套的液力偶合器来驱动离心泵，液力偶合器的调速范围为 1~1/5，可根据不同时期的工艺要求大范围调整单个注水泵的注入量，单井注入量可通过高压阀组间的调节阀调节。另外，通过控制高压阀组间汇管上的阀门可实现一组腔体注淡水而另一组腔体注卤水。以上设计原则确保了系统和单井的注采灵活调配。

3　站外单井玻璃钢注水、采卤管线

根据当地盐业公司的管道运行经验，钢质管道的主要缺点是寿命短，经简单沉降处理的地面淡水和从盐腔返回的卤水介质导致管壁腐蚀、结垢严重、摩阻大，生产运行能耗大，对焊接和防腐质量要求高，因腐蚀穿孔等导致卤水泄漏带来的环境污染不容忽视。针对钢质管道的缺点，金坛储气库第一批14口新井注水、排卤管道采用了玻璃钢管道。其中高压注水管道规格为 HVF-12DN150，压力等级为 12MPa；低压注水管道规格为 HVF-3.5DN150，压力等级为3.5MPa，设计单井注水量小于120m³/h。目前已有8口井的地面管线投入溶腔生产，总长度达13860m，没有发生过任何的管线渗漏或破损。注水过程中管路摩阻较低，以某井为例，注水管线长达1400m，在100m³/h的流量下，管线水力摩阻为0.31MPa，仅为同等直径钢质管道的1/2。

4 造腔体积的计算方法

根据每天腔体采出卤水的量和实测得到的含盐质量浓度，可依据式（1）计算地下腔体的有效容积，用来指导生产和造腔工程的其他作业。

$$V_f = V_s \frac{1-\alpha\beta}{1-\alpha} \tag{1}$$

式中：V_f 为腔体有效容积，m³；V_s 为溶解的固体盐体积，m³；α 为盐岩中不溶物比例，小数；β 为盐岩中不溶物崩落以后的膨胀系数，常数。

式（2）为注入1m³淡水所溶解固体盐的体积公式。

$$V_s = \frac{1}{\rho_s\left(\dfrac{100}{m}-1\right) - \rho_{bo}\left(\dfrac{m_o}{m}-1\right)} \tag{2}$$

其中

$$m = \frac{C_s}{10 \times \rho_b} \tag{3}$$

式中：ρ_s 为纯盐的密度，kg/L；ρ_{bo} 为腔体内条件下（腔体温度）饱和卤水的密度，kg/L；ρ_b 为地面条件下（采出卤水温度）采出卤水的密度，kg/L；m_o 为腔体内条件下（腔体温度）饱和卤水的质量百分数，%；m 为地面条件下（采出卤水温度）采出卤水的质量百分数，%；C_s 为卤水中盐的质量浓度（单位体积卤水中溶解盐的质量），g/L。

5 结语

（1）金坛盐穴储气库造腔地面工程建成投运后一直运行良好，地面玻璃钢管线摩阻没有明显增加，地面管网没有发生渗漏现象。

（2）通过应用液力偶合器调整泵的排量以及优化高、低压阀组间工艺设计，可以满足多种工况条件下的快速造腔和盐厂的卤水要求。

（3）为确保溶腔过程中因腔体超压或欠压造成腔体结构性损坏，需增设油垫压力监测报警和超压保护装置。

（4）为确保因地面管线破损后腔体静溶时不欠压，应备用一个可移动式小型泵车，以便及时向欠压的腔体补充淡水。

（5）玻璃钢管线受到机械外力后易破损，一旦出现意外事故，将会造成卤水泄漏污染环境，今后在高效、安全输送卤水管道的选择方面还需进一步研究。

（6）西气东输金坛造腔注水采卤站是国内第一个盐穴储气库造腔地面工程，目前仅运行 8 个月，其适应性还有待于进一步评价。

本论文原发表于《油气田地面工程》2007 年第 26 卷第 10 期。

盐穴储气库注采集输系统优化

吕亦瑭[1]　黄　坤[1]　方　亮[2]　杨海军[2]

(1. 西南石油大学石油工程学院；
2. 中国石油西气东输管道分公司)

【摘　要】 储气库作为能源储备的一种重要方式，越来越多地应用到中国各地。M储气库作为中国第一个盐穴储气库，是中国建设其他储气库经验的重要来源。研究分析了M储气库地面注采集输系统工艺，在分析目前运行情况基础上，结合理论及现场实际对注采集输系统工艺，主要包括单井中的分离器、调压装置、流量计以及注采站中的分离器及阀门进行了优化。通过优化，M储气库的单井和注采站工艺流程更简单，更便于维护，解决了遗留问题，为M储气库后期建设提供了建议。

【关键词】 储气库；集输系统；优化；单井；注采站

Optimization of injection and production system in salt cavity gas storage

Lv Yitang[1]　Huang Kun[1]　Fang Liang[2]　Yang Haijun[2]

(1. College of Petroleum Engineering, Southwest Petroleum University;
2. PetroChina West-East Gas Pipeline Company)

Abstract As an important way of energy reserves, underground gas storages are increasingly being used in China. Construction, put-into-production and operation of M salt cavern gas storage, as the first one in China, can provide valuable reference for future gas storage construction. Described is the surface injection and production process system in M gas storage, analyzed are its operation parameters in recent period and optimized are configuration, installation and parameters of the surface injection and production process system, mainly including separators, pressure regulators and flow meter in single well and separators and valves in injection and production station. The technological processes of single well and injection and production station after optimization become more simple and easier to maintain, accordingly, solved are many problems left over from the past and suggestions are proposed for later construction of M gas storage.

Keywords Gas storage; Gathering and transportation system; Optimization; Single well; Injection and production station

---

作者简介：吕亦瑭(1991—)，女，四川遂宁人，硕士研究生，主要从事油气储运研究工作。

M储气库于2007年正式投入运营,作为亚洲的第一座地下盐穴储气库[1],M储气库借鉴了很多国外建设盐穴储气库的经验[2-3]。由于每个盐穴的情况都不尽相同,因此在运营的过程中,M储气库逐渐暴露出了很多问题。基于M储气库后期建设需要,对现有的集输系统的优化显得特别重要。针对运营中出现过滤器超压截断关闭、节流区工艺烦琐、分离器精度不够、设备冗杂等问题分别做出了优化。

1 M储气库注采集输系统现状

M储气库工程集输管网井场均匀分布在M盐矿拟建库区内,围绕东西注采气站建设。共包括注采气站站外63座井场设施。其中老腔6口井、新腔57口井。东站进站井数共计40口,集配气阀组8个,西站进站井数共计23口(包括老腔6口井),集配气阀组5个。

M储气库主要工艺包括注气工艺和采气工艺。

注气的一般流程:A分输站来的干气经超声波流量计计量,通过输气管道输至M储气库的西注采气站的进出站阀组,分成两路,一路进西站注气装置,另一路经干气联络线进东站注气装置。在注气装置中,干气经压缩机增压后,经进出站阀组的分配器分别输送至各集配气阀组,由集配气阀组经集输管网分别送至各注采气井井场,计量后注入盐穴中储存。注气流程框图如图1所示。

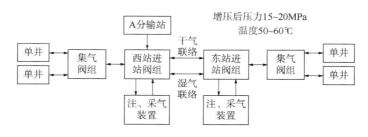

图1 M储气库注气流程框图

采气的一般流程:各井口来气在井场经节流、分离、超声波流量计计量后出井场,通过集输管网分别送至集配气阀组,由集配气阀组再输至东、西两站的进出站阀组,然后进入采气装置。在采气装置中,天然气先经TEG脱水或应急加药后外输,东站干气经干气联络线送入西站阀组与西站干气统一通过输气干线输送至A分输站进入西气东输管道。采气流程框图如图2所示。

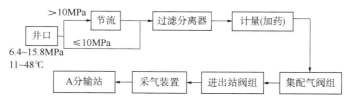

图2 M储气库采气流程框图

2 M 储气库注采集输系统优化

M 储气库现有的集输系统存在的主要问题有:

(1) 储气库在建设过程中存在井口注采气设施的施工周期过长,现场工作量大,配备施工人员过多等问题。

(2) 由于单井设有过滤分离器,在注气排卤后的溶腔中残留有卤水,在采气作业的初期,会带出较多的水分,当气流通过过滤组件时,液体微粒就被聚结成较大的液滴聚集在滤网上,堵塞滤网,引起分离器进出口压差超限,从而触发井口紧急切断阀动作,关闭单井,影响正常生产。

(3) 单井和注采站工艺流程较复杂。

因此,建议根据 M 储气库运行情况,可以通过井口装置橇装化设计和施工,缩短施工周期,保证装置的安装质量;进一步优化流程,优选设备。

2.1 M 储气库单井工艺优化

地下储气库的单井主要工艺可以划分为井口区、节流区、调压区、分离区、污水区、计量区、注醇区,井口注采集输示意图如图 3 所示。

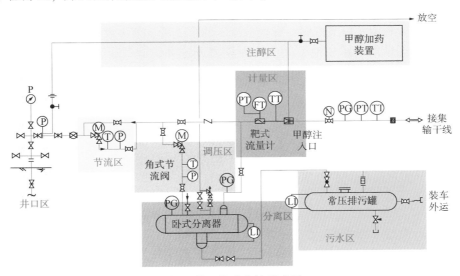

图 3 井口注采集输示意图

2.1.1 节流调压区

优化前单井节流区的工艺流程为:当天然气压力>10MPa 时,节流至 10MPa,进入过滤分离器;当天然气压力≤10MPa 时,直接进入过滤分离器。为调节采出气压力,设置了两路电动调压阀,一用一备。

可对该调节工艺进行简化,充分利用井口节流阀调节和控制流量,为便于调节,推荐将目前井口手动节流阀更换为电动节流阀。

节流阀的外形结构与截止阀并无区别,只是它们启闭件的形状有所不同。节流阀的启闭件大多为圆锥流线型,通过它改变通道截面积实现调节流量和压力的功能[4]。

井场装置第一级节流阀的功能是调控气井产量,将井口采出气体根据压力大小(通常以 10MPa 为界)进行节流,故应在临界状态下操作,计算公式为

$$d=\left(\frac{q_v}{156p_1}\right)^{0.476}(\Delta ZT)^{0.238} \tag{1}$$

第二级和第三级节流阀的功能是调控采气管线起点压力,计算公式为

$$d=\left(\frac{q_v}{324}\right)^{0.476}\left[\frac{\Delta ZT}{p_2(p_1-p_2)}\right]^{0.238} \tag{2}$$

式中:p_1 为阀前压力,100kPa(绝);p_2 为阀后压力,100kPa(绝);d 为节流阀流通直径,mm;q_v 为天然气流量($p=101.325$kPa,$t=20$℃),m³/d;Δ 为天然气相对密度;Z 为阀前气体压缩因子,$Z=0.7660$;T 为阀前气体绝对温度,K。

节流阀通常应在半开状态操作以满足调节要求,故计算得到的 d 应乘以 2。

气体在节流区节流与不节流高压输送相比有两点意义:节流后尽量以保证不形成水合物为原则,既满足了下游压力要求又提高了输送的安全性和可靠性;节流后单井井场设备设计压力较不节流低,降低了设备投资,提高了经济性[5]。

优化节流区工艺后增加了设备利用率和设备自动化程度。将井口的手动节流阀更换为电动节流阀,充分利用井口节流阀,配合节流区的第二级、第三级节流阀达到降压的目的。

为满足调峰气量的需求,地下储气库需频繁开停井,储气压力高,在井口需要节流,井流物节流后易形成水合物,但单井管线的冻堵具有间歇性。目前,该储气库采取在井口节流前注入水合物抑制剂来防止节流后水合物的形成,在单井出站管线上注入水合物抑制剂防止冬季地温低时集输管道中水合物的形成。从目前运行情况看,该方案较好地满足了储气库的生产需求,有效防止了水合物的形成。运行生产中只需对抑制剂的比例及用量结合 HYSYS 软件模拟分析的结果进行适当的调整即可。

2.1.2 分离区

为降低集输管道和设备的腐蚀速率、确保管道的输送能力、防止各单井采出天然气集输过程中生成水合物,避免游离水对下游工艺的影响,在单井设置了过滤分离器。

分离区的分离器类型较多,但在此处设置分离器的主要目的是进行气液分离,要求精度并不高,因此建议选用重力分离器即可。重力分离器分为卧式和立式两种。一般来讲,M 储气库在采气时,分离器需要处理大产量的气体。当卧式和立式分离器的直径相同时,在相同的操作条件下,卧式分离器的处理能力为立式分离器的 4 倍。卧式分离器对气体所携液滴的运动方向与液滴所受重力的方向垂直,有利于沉降分离,其液面波动小、稳定性好,处理单位气量的成本低于立式分离器[6]。因此推荐采用卧式重力式分离器对储气库井场采出气进行气液分离。

根据《分离器规范》(SY/T 0515—2014)的要求:仅靠重力沉降作用能除去气相中直径大于 100μm 的液滴;若安装有捕集器,则要求能除去气相中直径大于 10μm 的液滴[7]。优化后将井口的过滤分离器替换成卧式分离器,再根据气量大小进行设备选型。

调整分离器布局和优选分离器类型后,分离器大小较原方案明显减小,可大幅节省投资。

2.1.3 计量区

计量区工艺与常规天然气计量工艺基本一致，主要区别是根据注采工艺要求，考虑计量是否需要双向计量。一般而言，现目前的储气库都要求一套系统同时满足双向注采，因此流量计必须满足双向计量的要求。满足该要求的流量计有外夹式超声流量计、靶式流量计和质量流量计。各流量计的计量方案对比，见表1[8]。

表1 单井流量计对比

项目	外夹式超声流量计	靶式流量计	质量流量计
量程比	1:80	1:15	1:10
压损	无	很小	较小
测双向流	可以	可以	可以
准确度/%	±1	±1	±1
通道数	1	—	—
管径影响	管径越小，准确度越低	对准确度无影响	缩径较大
承压情况	常压	高压	高压
雷诺数要求	$1000<Re<5000$	$Re>1000$	$Re>1000$
介质影响	被测气体中杂质和管壁结垢对计量准确度影响很大	被测气体中的杂质对计量准确度影响很小	被测气体中的杂质一起计量(质量)
流量范围/(m^3/h)	0~62500	0~62500	0~31000(需2台并联)
工艺管路	直管段20D/5D	直管段10D/5D	直管段5D/2D
现场安装	复杂，准确度受安装误差影响较大	简单	简单
维修	可在线拆卸	不能在线拆卸	不能在线拆卸
业绩	气田开采应用较少	国内已建储气库大量使用	国外储气库有应用
投资	较高	较低	较高

综合考虑流量计的可靠性、本工程工况的适应性和相关业绩等因素，推荐采用靶式流量计。

2.1.4 注醇区

为防止水合物的生成，在井口区和计量区后注入抑制剂，各种抑制剂选用标准见表2[9-10]。

表2 抑制剂类型及适用条件

抑制剂类型	适用条件
甲醇	低产、气流温度低于-40℃、出水量大且压力较高的场合，季节性或临时性局部解冻(甲醇剧毒，使用时应注意安全)
乙二醇	气流温度不低于-25℃、高产、不出地层水的场合
THI(热力学抑制剂)	气流温度-40~25℃，可根据气体组分配置

可采用与注采气管线同沟铺设的方式将抑制剂由专用泵注入气井的套管，同时在站场内设注醇泵房。如果抑制剂里有甲醇，则鉴于甲醇的中等毒性与空气混合易爆炸的特性，在泵房内设置排风系统、配套净化设施，确保操作区的安全。

2.2 M 储气库注采工艺优化

M 储气库设置 2 座注采站（注气站和采气站合并），总体工艺如下：

A 分输站来的干气进入西站的进站阀组，通过清管接收装置之后分成两路，一路进西站注气装置，另一路经干气联络线进入东站的进出站阀组区，再进入东站注气装置。注气时，干气经注气装置增压后，经进出站阀组区的分配器分别输送至各集配气阀组，由集配气阀组经集输管网分别送至各注采气井井场，各井场经单井计量后注入储气井中储存。采气时，各井场来气接至进出站阀组区，再进入采气装置，脱水后送入输气干线。当出采气装置的气体压力低于输气干线压力时，将其送入注气装置增压后外输。东西注采气站间设一条湿气联络线和一条干气联络线，满足东西注采气站与各自所辖井场互相匹配操作。

目前，东西注采站在压缩机进口前均设置了旋风分离器和过滤分离器，单井来气首先通过汇管分配后进入旋风分离器，旋风分离器出口天然气再汇合进入汇管，再经汇管分配给过滤分离器。即采用并联的旋风分离器与并联的过滤分离器串联的方式。该分离工艺存在如下问题：

（1）选用旋风分离器作为过滤分离器的第一级分离，旋风分离器分离精度较多管干式除尘器低，只适用于处理气量不大、粉尘粒径大于 5μm、压力和流量较稳定、对分离精度要求不很高的站场。而 M 储气库东西注采站均设置有压缩机，且站场处理量大，压力和流量并不稳定，因此此处选取旋风分离器欠妥。

（2）旋风分离器和过滤分离器进出口均设置截断阀，因此需在旋风分离器和过滤分离器上分别设置一套安全泄压阀；推荐采用多管干式除尘器与过滤分离器一一对应，即将多管干式除尘器与过滤分离器串联。

（3）在旋风分离器和过滤分离器之间设有一套汇管，增加了压力设备。

（4）该流程较为复杂，不便于生产运行和管理。

从目前国内外输气站场除尘设备的设置，同时考虑东西注采站压缩机对来气气质的要求，结合东西注采站注气量波动范围较大、来气压力不稳定的特点，推荐采用多管干式除尘器。多管干式除尘器分离原理与旋风分离器相同，但由于旋风子旋转半径小，是一种高效的除尘设备，其分离效率较旋风分离器高，噪音小。

考虑到过滤分离器滤芯需定期更换的需求，原东西注采站分别设置了 2 套过滤分离器，可实现一用一备，因此推荐东西注采站也各设置 2 套多管干式除尘器与过滤分离器一一对应，实现一用一备。

目前采用并联的旋风分离器与并联的过滤分离器串联的方式进行过滤分离，增加了截断阀和安全阀，推荐采用多管干式除尘器与过滤分离器一一对应，即串联的方式进行过滤分离，且多管干式除尘器与过滤分离器之间不再设置截断阀，将多管干式除尘器和过滤分离器视为同一个压力容器，多管干式除尘器上不再设置安全阀和手动放空阀，只需保留过滤分离器上的安全阀和手动放空阀即可。优化后的工艺流程见图 4。

从图 4 可看出，优化后的工艺更简单，减少了原旋风分离器出口和过滤分离器进口管

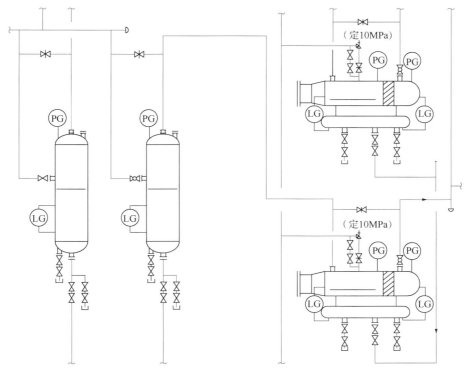

图 4　优化后的注采站过滤分离区工艺流程

道截断阀，同时由于多管干式除尘器和过滤分离器之间的连接管道中无任何截断阀，因此可将其视为一个压力容器，只设置一套安全泄压阀，即将原旋风分离器上的安全阀和手动放空阀取消，只保留过滤分离器上的安全阀和手动放空阀即可满足工艺要求。同时将原旋风分离器更换为多管干式除尘器，提高分离粒径，减小了原分离器的外径，更好地适应来气流量和压力的波动。

3　结论

本文研究分析了 M 储气库地面注采集输系统工艺，在分析目前运行情况基础上，结合现场实际对注采集输系统工艺进行了优化，得到：单井注采集输工艺，可采用重力分离器取代过滤分离器实现注采系统气液分离，从而减小现场定期更换过滤分离器滤芯的工作量；取消单井一用一备的调压装置，利用单井节流阀实现流量和压力控制，同时给原节流阀配置电动执行机构或将原手动节流阀更换为电动节流阀，确保节流可靠性；根据单井流量计选型比选结果，建议单井计量采用靶式流量计，既满足生产需要同时又节省投资。注采站工艺，可对东西注采站分离工艺进行优化，建议选择外径更小、分离精度更高、适应性更强的多管干式除尘器取代旋风分离器。同时推荐采用多管干式除尘器和过滤分离器一一对应的组合分离方式，即一套多管干式除尘器串联一套过滤分离器，同时多管干式除尘器和过滤分离器之间不再设置截断阀，为此可将多管干式除尘器和过滤分离器视为一个整体，只设置一套安全泄压阀。

参 考 文 献

[1] 李国韬,郝国永,朱广海,等.盐穴储气库完井设计考虑因素及技术发展[J].天然气与石油,2012,30(1):52-54.
[2] 苏欣,张琳,李岳.国内外地下储气库现状及发展趋势[J].天然气与石油,2007,25(4):1-4.
[3] 吴忠鹤,贺宇.地下储气库的功能和作用[J].天然气与石油,2004,22(2):1-4.
[4] 陈晋市,刘昕晖,元万荣,等.典型液压节流阀口的动态特性[J].西南交通大学学报,2012,47(2):325-332.
[5] 张建业,伍藏原,黄兰,等.异常高压气井井口节流阀开度控制方法研究[J].石油钻采工艺,2010,32(增刊):115-117.
[6] 曾自强,张育芳.天然气集输工程[M].北京:石油工业出版社,2001.78-84.
[7] SY/T 0515—2014,分离器规范[S].
[8] 应后民,肖国祥,周成义,等.超声波流量计选用的注意事项[J].计量与测试技术,2013,40(6):43-44.
[9] 张文.油田污水处理技术现状及发展趋势[J].油气地质与采收率,2010,17(2):108-110.
[10] 毕曼,贾增强,吴红钦,等.天然气水合物抑制剂研究与应用进展[J].天然气工业,2009,29(12):75-78.

本论文原发表于《天然气与石油》2015年第33卷第1期。

电磁流量计在盐穴储气库造腔过程中的应用

成 凡 焦雨佳 张青庆 刘 春

(中国石油管道有限公司西气东输分公司储气库管理处)

【摘 要】 盐穴储气库在造腔过程中,电磁流量计用于计量注入和排出的液体体积,其可靠性直接影响到造腔工程师对造腔进度的监控,及相应施工计划的安排。除定期检测外,文章根据岩盐中泥质不溶物膨松系数在一定范围这一特征,统计分析生产数据,评价其可靠性。结果表明,电磁流量计可靠性能满足盐穴储气库造腔生产需要。

【关键词】 电磁流量计;盐穴储气库;造腔

电磁流量计由流量传感器和转换器两部分组成,与其他类型的流量计相比,电磁流量计具有无压损、口径大、量程比宽、耐腐蚀、适用于高黏度介质、性价比高等特点[1-3],可充分减小管道内的阻力,符合节能降耗的要求。目前,电磁流量计已广泛应用于造腔生产中,虽然经过定期检测,但由于被测介质性质及流速等的影响,电磁流量计在造腔中的应用效果并未评价。调研结果表明[4-6],岩盐中不溶物膨松系数在特定地区是常数,且往往在一定范围内,可用生产中溶盐体积、腔体实际体积、不溶物含量等参数表达。本文通过介绍电磁流量计的测量原理,造腔地面工艺,确定电磁流量计在造腔站的适用性,造腔站内电磁流量计的选型与安装,明确该类流量计的使用符合减小测量误差及稳定性的要求,再根据日常生产中的注入液体、排出液体体积以及相应的质量浓度,对溶盐体积进行计算,结合阶段造腔实际体积及不溶物含量,计算不溶物膨松系数,评价流量计的可靠性。

1 电磁流量计测量原理

电磁流量计是利用法拉第电磁感应定律测量,如图1所示,用于测量电极与运动的流体构成的回路中产生的电参数,并根据公式换算成流体速度,再测量流通截面液位高度得到流通面积,两者相乘得到所测流量。

$$U_e = \kappa B L v \quad (1)$$

$$Q = \pi \frac{L^2}{4} v \quad (2)$$

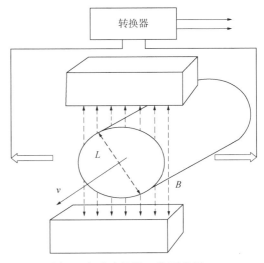

图1 电磁流量计工作原理图

作者简介:成凡(1989—),男,工程师。

$$Q = \pi \frac{U_e L}{4\kappa B} \tag{3}$$

式中：Q 为被测介质体积流量，m³/s；U_e 为直流式电磁流量计用来测量流体横越磁场时所感生的电动势，V；B 为磁感应强度，T；L 为电极间距，即测量管内径，m；v 为液体平均流速，m/s；k 为修正系数。

2 造腔站电磁流量计选型

电磁流量计选型与安装应根据工艺条件来确定，主要包括被测介质电导率、温度、压力、介质流量范围、工艺管道的管径、介质状态，介质的腐蚀性，工艺管道材质等因素，保证测量的准确度，并根据环境条件选择防爆及防护等级。

2.1 造腔地面工艺

造腔地面工艺分为注水工艺和排卤工艺。注水过程中，淡水或淡卤水通过离心泵增压至汇管，然后通过各井管线流量计计量，进入盐层。排卤过程为，溶盐产生的卤水依靠自身压力通过管道计量后，最终进入汇管外输。

2.2 选型及安装

造腔站电磁流量计测量介质主要为卤水或者淡卤水，电导性较强，但具有较强的腐蚀性。采用的衬里材料为聚四氟乙烯（PTFE），电极材料选用哈氏合金（HC），耐腐蚀性能好，能有效预防注入和排出液体含有颗粒杂质、饱和卤水结晶等问题。防护等级选用 IP65 用以完全防止外物侵入并防喷射水进入。

在安装位置方面，各单井注水排卤流量计在高低压阀室，与注水泵机组隔离，接地良好。单井注水排卤管道为 DN150，流量计的前直管段长度为 87cm，后直管段 55cm，同时，在回卤的低压阀室，电磁流量计安装在介质的上流处，符合电磁流量计安装要求[7-8]。

3 造腔站电磁流量计可靠性分析

根据岩盐中泥质不溶物膨松系数在一定范围这一特征，通过对金坛储气库的造腔井生产数据进行分析，计算腔体溶盐量，结合声呐测腔体积，计算不溶物膨松系数，检验电磁流量计计量的可靠性。

3.1 盐层物理化学性质

金坛盐岩层段岩性主要为盐岩、含泥盐岩夹含盐泥岩、钙芒硝泥岩、云质泥岩、泥岩、粉砂岩等。对于占盐岩段 80% 以上的盐岩层，主要矿石以石盐为主，其次为钙芒硝、石膏，局部出现无水芒硝，杂质主要为黏土矿物，其次为白云岩、碳酸盐岩等。石盐的化学组成主要是 NaCl，占 74.9%~90.8%，其次是 Na_2SO_4、$CaSO_4$，其他盐类甚微。

3.2 造腔运行中溶盐量计算

由于日常造腔中，主要注入淡水或者淡卤水，离子质量浓度非常低，且日常化验中测量了注入液体的氯离子质量浓度，其他离子未详细测量，所以在进行溶解盐体积计算时，注入液体只考虑氯离子。

$$V_{溶} = V_{采} + V_{剩余} \tag{4}$$

$$V_{采} = (V_{排} \cdot C_{排} - V_{注} \cdot C_{注}) / \rho_{氯化钠} \tag{5}$$

$$V_{剩余} = V_{剩余液} \cdot C_{排} / \rho_{氯化钠} \tag{6}$$

$$V_{剩余液} = m_{剩余水} / (\rho_{排} - C_{排}) \tag{7}$$

$$m_{剩余水} = (V_{注} \cdot \rho_{注} - V_{注} \cdot C_{注}) - (V_{排} \cdot \rho_{排} - V_{排} \cdot C_{排}) \tag{8}$$

由于每次测腔都有等待和准备时间，所以腔内卤水质量浓度会继续升高，腔体体积扩大。故该阶段溶盐体积应按照式(9)计算：

$$阶段溶盐体积 V_s = \sum V_{采} + \sum m_{剩余水} \div (\rho_{声} - C_{声}) \times C_{声} \div \rho_{氯化钠} \tag{9}$$

式中：$V_{溶}$ 为日溶盐体积，m^3；$V_{采}$ 为日采盐体积，m^3；$V_{剩余}$ 为腔体扩大后未排出液体的溶盐体积，m^3；$V_{剩余液}$ 为腔体扩大后未排出液体体积，m^3；$m_{剩余水}$ 为腔体扩大后未排出的液体中等效淡水溶剂质量，kg；$V_{排}$ 为日排出液体体积，m^3；$V_{注}$ 为日注入液体体积，m^3；$C_{排}$ 为日排出液体氯化钠质量浓度，g/L；$C_{注}$ 为日注入液体氯化钠质量浓度，g/L；$\rho_{排}$ 为日排出液体密度，kg/m^3；$\rho_{注}$ 为日注入液体密度，kg/m^3；$\rho_{氯化钠}$ 为氯化钠密度，kg/m^3；$\rho_{水}$ 为水的密度，kg/m^3；V_s 为阶段溶盐体积，m^3；$C_{声}$ 为声呐测量时腔体内氯化钠质量浓度，g/L；$\rho_{声}$ 为声呐测量时腔体内液体密度，kg/m^3。

由于2016年前并未进行注水质量浓度化验，所以本文选择了2016—2017年进行造腔并在该时间段内完成设定造腔阶段的井进行统计分析，结果见表1。

表1 金坛储气库2016—2017年完成设定造腔阶段的井溶盐量及不溶物膨松系数计算结果

序号	井名	阶段声呐体积/m³	阶段不溶物含量	阶段溶盐体积/m³	不溶物膨松系数
1	T1	51491.0	0.176	63037.00	1.9
2	T2	42439.7	0.208	84122.00	2.9
3	T3	38477.3	0.227	49772.34	1.8
4	T4	28774.4	0.208	43190.34	2.3
5	T5	14954.4	0.147	19537.28	2.4
6	T6	39465.0	0.181	44591.82	1.5
7	T7	24276.3	0.269	28561.64	1.4
8	T8	33216.5	0.203	41432.30	1.8
9	T9	28783.7	0.281	36815.59	1.6
10	T10	24649.4	0.282	44977.06	2.2
11	T11	23079.3	0.25	44361.37	2.4

3.3 膨松系数

盐岩矿床不论品位多高，都普遍存在着水不溶物，主要分布在盐岩层内及夹层中。由于金坛盐矿不溶物中的一大部分属于泥岩，主要以黏土矿物中的伊利石为主，其次为伊/蒙混层，无高岭石，这些黏土矿物的强吸水性及吸水后极易蓬松性，会减少腔体有效体积，所以实际溶解盐的体积并不是腔体体积，它们之间存在以下关系[9-10]：

$$\beta = \frac{1}{\alpha}\left(1 - \frac{1-\alpha}{V_s}V_f\right) \qquad (10)$$

式中：V_f 为阶段腔体有效体积，m^3；V_s 为阶段溶解的固体盐体积，m^3；α 为阶段盐岩中不溶物比例，小数；β 为阶段盐岩中不溶物崩落以后的膨松系数，常数。

根据实验，金坛储气库不溶物膨松系数在 1.5～3.0 范围内。由于盐穴造腔高度近 150m，含多个夹层，夹层含黏土矿物含量不同，所以膨松系数不同，同时夹层的厚度不同，所以利用该方法模拟的腔体体积的评价应根据不同造腔阶段来进行。表1列举了2017年测腔井的阶段造腔体积及该阶段不溶物含量，并计算该阶段的不溶物膨松系数。

从表1可以看出，除T7井外，其他井不溶物膨松系数均在 1.5～3.0 范围内，说明造腔站内的电磁流量计的准确性和稳定性能够满足生产的需要。

3.4 误差分析

根据该方法对盐穴储气库电磁流量计效果进行评价时，不溶物膨松系数计算误差主要受下列因素影响。

（1）注入液体溶液质量浓度化验误差。受盐厂清罐等因素影响，当日供应的淡卤水质量浓度可能会出现上午高，下午低的情况，而每天淡水供应化验的质量浓度只有上午一个参数，所以对计算结果有一定影响。该误差可通过增加质量浓度化验次数得到有效解决。

（2）由注入液体中硫酸钠、硫酸钙等含量的不确定引起的误差，但这些矿物在溶液中含量非常少，在一个造腔阶段内其影响可忽略。

（3）流量计计量误差。现场选用的是准确度等级为 0.5 级的流量计。检测报告显示，流量计实际相对示值误差较大，这可能导致计算的不溶物膨松系数偏离正常范围。以 T7 井为例，见表2，注入端流量计 FT1 在 $30m^3/h$ 时，相对示值误差达 6%，在 $90m^3/h$ 时，相对示值误差为 3%，明显大于排出端流量计 FT2 在对应检测点的相对示值误差。同时，生产数据显示，T7 井在该造腔阶段，FT1 示值在 30～$80m^3/h$ 范围，这直接导致了计算的注入盐量偏大，而溶盐量偏小，最终使不溶物膨松系数偏低。考虑到注入液体溶液质量浓度化验误差对于所有同期造腔井的不溶物膨松系数影响结果相近，而除 T7 井外，其他井不溶物膨松系数都在正常范围内，所以流量计 FT1 实际测量误差大是导致 T7 井不溶物膨松系数偏小的主要原因。

表2 T7井电磁流量计检测结果

流量计	压力/MPa	温度/℃	介质	投产年份	标准流量/(m^3/h)	实测流量/(m^3/h)	相对示值误差/%
FT1	6.2	25	淡卤水	2006	30.34	32.30	6
					90.04	92.92	3
					180.84	182.41	0.8
FT2	0.3	29	卤水	2014	30.12	31.08	3
					90.32	90.85	0.6
					180.23	180.44	0.1

从表2中可以看出，随流速的增大，相对示值误差逐渐减小。这可能是由于测量的流体内部杂质在管道内易形成沉淀，虽然采用了聚四氟乙烯作为衬里材料保护电极，但是在低流速状态下，这些沉淀附着于内衬的表面使测量结果发生偏差，所以电磁流量计实际测量误差必将随着时间的推移而逐渐增大。为提高电磁流量计在盐穴储气库造腔过程中的测量精度，可采用以下方法。

（1）调整工艺，在可控范围内提高流速，减少杂质沉淀，提高测量精度。

（2）定期清理流量计。

（3）对于检测过程中发现的误差值超过最大允许误差的流量计，仪表系数偏离原出厂标定的仪表系数时，可根据检测机构的检测结果进行仪表系数校准。

4　结束语

综上所述，电磁流量计的准确性和稳定性能够满足盐穴储气库造腔的需要。但是，电磁流量计在盐穴储气库使用中，受待测液体不纯的影响，实际的测量误差随时间的推移而逐渐增大，特别是在低流速的情况下，相对示值误差明显较大。造腔设计者可根据特定地区岩盐中不溶物膨松系数在一定范围内这一特征，进行生产分析，评价流量计计量在使用中的可靠性。对于不溶物膨松系数不在特定范围内的井，排除质量浓度检验影响后，可通过提高流速，定期清理流量计，仪表系数校准等方式提高测量的数据的准确性，以便于合理安排造腔计划，指导造腔施工。

参　考　文　献

[1] 张皓，杨金城.电磁流量计的测量原理及其在工程设计中的应用[J].石油化工自动化，2014，50(6)：55-58.

[2] 赵保生，吴蓉，刘志森，等.电磁流量计发展及趋势[J].自动化仪表，2017，38(5)：67-71.

[3] 韩军伟，徐明，张军阳，等.电磁流量计在西北油田钻井现场应用试验[J].科学管理，2017(8)：183-184.

[4] 杜新伟.盐岩不溶物对储气库成腔的影响[J].石油化工应用，2016，35(1)：30-32.

[5] 田中兰，夏柏如.盐穴储气库造腔工艺技术[J].现代地质，2008，22(1)：97-102.

[6] 杨春和，李银平，陈锋.层状盐岩力学理论与工程[M].北京：科学出版社，2009.

[7] 赵淞江，刘小红.浅谈电磁流量计对城市供水管道流量测量的应用及测量不确定度评定[J].计量与测试技术，2017，44(10)：28-29.

[8] 陈洁.影响电磁流量计误差因素的分析[J].内蒙古石油化工，2014(24)：82-83.

[9] 杨清玉，李祥.盐穴造腔腔体净容积的计算方法[J].中国井矿盐，2010，41(1)：22-23.

[10] 陈晓源，张蕾，李应芳.多夹层盐穴储库沉渣碎胀——膨松系数试验研究[J].矿业研究与开发，2013，33(2)：34-38.

本论文原发表于《工业计量》2018年第28卷第5期。

节能技术在储气库地面工程中的应用

柳 雄 云少闯 黄 玮

(中国石油西气东输管道公司储气库项目部)

【摘 要】 天然气地下储气库作为长距离输气管道工程中的储气设施，一般是用于对下游用户不均衡用气进行调峰。坚持石油开采的经济性，坚持节能降耗，已经成为我国石油企业可持续发展的重要举措。因此，在石油开采中，应当积极进行优化工艺设计和配套设备，实现绿色生产。文章以油气藏型地下储气库的地面设施为例，对节能减耗的实施进行了探讨。

【关键词】 储气库；地面工程；节能

一般而言，天然气的地下储气是经过对废弃的油气藏进行改造后而完成的，少数天然气地下储气库是经过对水层、岩穴和矿井进行改造后建设而成的。在地下储气库中，一般由地下储气库、注气井、地面集注管网、露点控制装置和注气装置等组成。

1 储气库的地面工艺

储气库的注气工艺一般比较简单，气源为干气，由长输管道输送而来，只需要对气源进行过滤和分离，即可对其进行增压，将其注入底层进行储存。在油气藏的地下储气库建设中，还存在一些没有采出的原油，并且在油气藏储气库的建设中，一般都有边水存在，底层温度高，所以，需要对其地下储气库的采出气进行脱水、脱烃后，才能输入输气管网，在其地面工程的建设中对采出气的脱水和脱烃是其工艺的核心。

2 地面工程的流程优化设计

地面工程的流程完善主要是用于保障工程的安全正常运作，而通过对流程进行物料分析平衡、和热平衡的优化改造，可以到达工程的节能减耗。

2.1 物料的分析与平衡

通过对储气库的采气系统进行分析，发现采出水、杂质、甲醇、乙二醇、缓蚀剂、污水排放、放空天然气、废弃脱硫剂等属于无效的流量，可以进行减少或者取消。通过分析发现，由于采出水是由地质条件决定的，因此无法取消或减少；甲醇、乙二醇及缓蚀剂也不能取消，但可以对其注入的位置进行优化；对于污水，可以采用采用密闭排放工艺从而实现回收；对于天然气的放空，可以将脱硫装置泄压的一部分进行回收利用，将乙二醇闪蒸气大部分回收利用，并减少事故放空气体。

作者简介：柳雄(1986—)男，湖北武汉，中国石油西气东输管道公司储气库项目部，助理工程师，本科学历，现从事储气库建设项目管理工作。

2.2 热平衡分析

在储气库的采气过程中，一般包括放出热量和吸收热量的过程，因此对系统的热量进行分析和平衡，可以对热量进行合理回收、优化换热流程。技术人员通过对温度和热量进行对比分析，从而来确定哪些热量和冷量可予以回收，哪些过程可实现热量交换。比如，进站天然气的冷却过程属于放热的过程，虽然可以进行热交换，但是由于温位相差比较大，无法满足梯级利用原则，而且其释放的热量非常大，因此可以考虑采用空气冷却方法。

3 地面工程的参数优化

实行参数优化往往具有不错的节能效果。在储气库的注采系统中，参数优化可以从以下几个方面进行改进。

3.1 注采气规模和烃水露点控制温度

合理的注采气规模可以降低工程的投资，也可以提高设备运行的效率，降低能耗。大部分储气库一般应用于季节调峰，有的则应用于应急调峰。储气库应注意以"均采均注"为原则，来确定设计的规模，并且按着设计规模的120%作为其规模的上限。

在储气库的采出气烃水露点方面，烃露点一般要不低于最低输送环境下的温度，水露点一般要至少低于最低制冷温度5℃，实际中，制冷温度可以取2~3℃的裕量。

3.2 注气压缩机入口压力、排量、排气温度

压缩机入口压力与增压所消耗的功率成反比，在对压缩机操作中应当使其入口压力保持在允许的较高压力下，使其拥有较小的压比和较小的压缩功率需求。在储气库注气工艺中，出口压力会随着注气的进行而逐步升高，所以，压缩机的排压就需要达到最高注气压力的目标才行。在实际中，如果按照最高注气压力和平均注气量来对压缩机组进行配置，则会出现注气初期机组低效运行和综合注气能力可能不足的问题。要解决这些问题，既要对压缩机组的排量参数进行优化，还需要采用小排量的多机组并联的方法运行。此外，可以通过减少机械损失来提高压缩机组的效率。

3.3 空冷器设计参数优化

在空冷器设计参数中，热负荷、工艺气进口温度、介质物性是由地面工艺和采出气条件决定的，一般不会进行调整。但是可以对工艺气出口温度和设计空气温度进行调整。在进站空冷器中，采出气进站后温度一般比较高，需要对其进行冷却，这样一方面能降低后续制冷的包袱，一方面降低了乙二醇循环量、制冷所需的冷量或压差。所以来说，采出气经过冷却后温度越低越好。一般而言，常见的进站空冷器位于生产分离器气相出口，这样能降低空冷气的负荷，但是会对烃水露点控制产生不利的影响，如果进站温度过高，空冷器冷却后的凝液就会增多，携液对下游设施会产生影响，那么就可以采用在空冷器后设置分离器。如果底层采出的天然气携液量很少，不会对空冷器造成的负荷不大，则可以将进站空冷器的位置设于生产分离器气相入口。

4 结束语

在储气库地面工程的施工设计中，除了应用以上措施来实现节能减耗外，还可以通过

对换热器和低温分离器进行优化来实现。

参 考 文 献

[1] 先智伟,谢箴. 地下储气库的地质条件要求和选型[J],天然气与石油,2004,22(2):5-7.
[2] 吴忠鹤,贺宇. 地下储气库的功能和作用[J],天然气与石油,2004,22(2):1-4.

本论文原发表于《资源节约与环保》2013年第7期。

盐穴储气库地面工程技术要点研究

刘 岩[1]　程 林[2]

（1. 大庆油田工程有限公司；2. 中国石油西气东输管道公司）

【摘　要】　金坛储气库是我国仅有的 1 座已建成投产的盐穴储气库，对其设计、建设和运行经验进行总结，可为平顶山和淮安等盐穴储气库的建设设计提供指导。盐穴储气库是利用地下已有的盐层，采用人工的方式在盐层中水溶形成腔体，用于天然气回注、储存和采出的系统体系。盐穴储气库地面配套工程须额外增加造腔地面配套系统和注气排卤系统；注采气工艺中压缩机的选型也须兼顾注气排卤工况；盐穴采出气含有卤水，且各腔体独立，压力会有所不同，采气脱水工艺须能够适应腔体温度、压力变化。处理装置设计规模须根据盐穴储气库的功能定位、注采周期和注采能力等因素确定。多级离心泵作为造腔注水泵是最适宜的注气排卤设备，可满足各种工况条件下的快速造腔和盐厂的卤水回收要求。采出气处理工艺的选择须避免水合物生成并合理地控制水露点，三甘醇脱水和 J-T 阀+乙二醇技术是常用的脱水工艺。压缩机厂房的降噪是储气库建设的重要设计之一，采用吸声、隔声、隔振和通风消声等噪声控制措施可以实现整体降噪的目的。

【关键词】　盐穴储气库；注采工艺；注气排卤；降噪

Key point study on the surface engineering technology of salt cavern gas storages

Liu Yan[1]　Cheng Lin[2]

(1. Daqing Oilfield Engineering Co., Ltd;
2. PetroChina West East Gas Pipeline Company)

Abstract　Jintan Gas Storage is the only salt cavern gas storage that has been built and put into production in China. The design, construction and operation experiences are summarized to provide guidance for the construction design of other salt cavern gas storages such as Pingdingshan Salt Cavern Gas Storage and Huaian Salt Cavern Gas Storage. The salt cavern gas storage is a system that utilizes existing underground salt beds to form caverns in salt beds by the water-soluble artificial method and is applied for natural gas injection, storage and recovery. The gas injection and brine discharge as well as cavity making system should be added to the surface supporting system of the salt cavern gas storage. The

---

作者简介：刘岩：高级工程师，硕士，2004 年毕业于吉林大学高分子化学与物理专业，从事油气储配和净化处理工作，0459-5902019，1324064773@qq.com，黑龙江省大庆市让胡路区大庆油田工程有限公司，163712。

selection of compressors in gas injection and production process should also take into account the working conditions of gas injection and brine discharge. Salt cavern produced gas contains brine, and the every chamber is independent, so their pressure will be different. The gas production and dehydration process should be able to adapt to the temperature and pressure changes of the chambers. The design scale of the processing device should be determined according to the function orientation, injection-production cycle and injection-production capacity of the salt cavern gas storage. The multi-stage centrifugal pump, as the water injection pump for cavity making, is the optimal device for gas injection and brine discharge, which can meet the requirements of rapid cavity making and brine recovery in salt plants under various working conditions. The treatment process for produced gas needs to avoid hydrate formation and control dew point reasonably. Triglycol dehydration and J-T valve+ ethylene glycol technology are commonly used dehydration processes. The noise reduction of compressor workshops is one of the important measures in the construction of gas storages. Noise control measures such as sound absorption, sound insulation, vibration isolation and ventilation noise elimination can be used to achieve the overall noise reduction design.

Keywords　salt cavern gas storage；injection and production process；gas injection and brine discharge；noise reduction

建设储气库是最经济有效的天然气储存和调峰手段,在平衡天然气生产和市场需求、提高管网输送效率、保障安全平衡供气等方面发挥了重要作用。盐穴储气库宜用于季节调峰、应急调峰、战略储备,具有灵活存储和短期吞吐量大的特点。盐穴储气库可根据市场调峰需求的逐步增加来分期设计、建设,操作的机动性强、利用率高,一年中可多次注采气循环;在应急情况下,盐穴储气库采气速率快,应急能力强[1]。

1　盐穴储气库建库基本原理

盐穴储气库是指利用地下已有的盐层或盐丘,采用人工的方式在盐层或盐丘中水溶形成腔体,用于天然气回注、储存和采出的系统体系。盐穴储气库一般采取水溶建库方式:先将采卤管柱下放至盐层,将淡水或不饱和卤水用泵通过采卤管柱注入盐层中,水洗溶盐后返回地面,再连续循环注入淡水或不饱和卤水,返出卤水。水洗溶盐造成盐层中的空间逐渐扩大,最终形成设计要求的地下盐穴。腔体形成后,再下入注采气油管柱,进行注气排卤作业。注气排卤结束后,从井内起出排卤油管柱,再安装注采天然气井口,进而完成盐穴储气库地下工程建设,转入注采气运行阶段。

金坛盐矿盐岩层位于地下约1100m深,盐腔温度为53℃,腔体运行压力为6~17MPa。图1为金坛盐矿地质剖面和拟建盐腔示意图。图2为造腔后盐穴腔体结构图。

盐穴储气库建库特性和运行特性区别于其他类型的储气库:地面配套工程须额外增加造腔地面配套系统和注气排卤系统;注采气工艺中压缩机的选型也须兼顾注气排卤工况[2];盐穴采出气含有卤水,且各腔体独立,压力会有所不同,采气脱水工艺须能够适应腔体温度、压力变化。

2　注采工艺

采用工况分析的方式对盐穴储气库的运行进行全面模拟,根据工况分析的结果来确定合理的盐穴储气库建设分期实施方案，进一步优化注采气的规模、配置以及集输系统的能力。

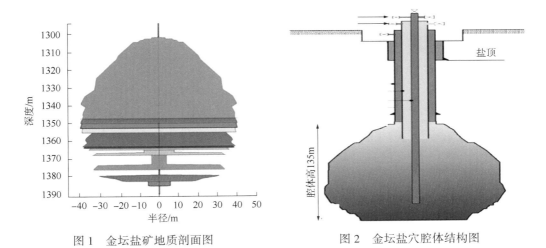

图 1　金坛盐矿地质剖面图　　　　图 2　金坛盐穴腔体结构图

储气库注采周期应根据市场调峰需求情况确定。注采气装置设计规模须根据盐穴储气库的功能定位、注采周期和注采能力，并结合长输管道供气能力和调峰需求等合理确定。注采气装置应根据造腔实际进度，分期进行建设。

注气装置的设计须满足盐穴储气库运行周期内的各种工况条件要求，并应同时适应注气排卤周期内的工况条件要求。以季节调峰为主时，采气装置的规模宜按高月高日峰值计算，采气装置不宜设置备用；兼顾季节调峰以及应急供气时，采气装置总规模应按应急供气量确定，多套采气装置宜并联设置。

天然气水化物是水与烃类作用生成的结晶体，外表冰和致密的雪类似，是一种笼形晶状包络物。一般而言，$C_1 \sim C_4$ 的烃类物质可形成水化物，C_5 以上的烃类不能形成水化物。水化物的性质不稳定，一旦稳定存在的条件破坏，就会迅速分解为烃和水。天然气水化物是采输气中经常遇到的难题之一。水化物生成于井筒中时会造成堵塞，并减少气体流动断面、降低采气量、损坏井筒内部件，甚至造成气井停产等危害。因此水化物生成于井口或地面管线中时会导致下游压力降低，妨碍正常输气，甚至完全堵塞管道并造成停气。

盐穴储气库在造腔时，一方面由于采用了温度低的水进行洗盐滤提，使盐穴储库具有冷型储槽的特性。随着天然气的开采，虽然储库周围盐层能供给部分热量，但远不能补偿降压带来的"焦耳-汤姆逊效应"，使天然气迅速冷却，再加上在经过生产作业线上的压力和摩擦等损耗，最终导致井口天然气温度低于储库温度。如果天然气中存在游离水，且温度降低到某特定值时，大部分采气将发生在水化物的生成区。另一方面，如果井口在一定温度下进行高压注气，也容易发生水化物沿井筒的冻堵问题。因此，在储库的设计和运行过程中，一方面须连续预测采出过程水化物的生成范围，另一方面还须适时判定井口天然气是否生成水化物，从而为指导现场实际生产提供可靠依据。

3　脱水工艺分析

盐穴储气库的采出气处理以控制水露点为目的，处理后的天然气应满足输气管线的输送气质要求。采气装置处理工艺的设计应满足启停方便、调节灵活和操作弹性大等要求[3-4]。采气原理流程如图3所示。

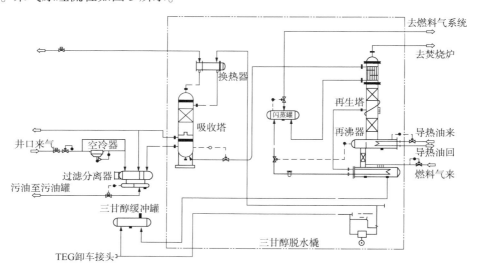

图3 采气原理流程

调峰采气装置推荐采用三甘醇脱水技术，仅在须满足应急供气的采气装置中，推荐采用J-T阀制冷+注甲醇工艺来满足外输要求。

采出气的组成与输气管道来的天然气基本相同，但含有一定量的卤水。在采气初期，盐腔内为高温高压气体，采出气在集配气阀组进行节流，温度、压力降低，析出的液体经分离器分离后，气相进入集注站，经过滤分离后进入脱水装置。连续采气时，随着采出气压力降低，采出气温度也逐步降低，此时，主要控制在集输过程中无水合物形成。当采出气降低到一定压力，则须停止采气一段时间，此时盐腔内处于低压高温状态。因为盐腔内温度较高，需考虑采出气温度是否能满足三甘醇脱水装置进塔温度要求，若温度高于40℃，须在脱水装置前增加空冷或水冷设备[5]。

金坛储气库高压采气和低压高温采气工况见表1。

表1 金坛储气库高压采气和低压高温采气工况

条件	节流前		节流后		水合物形成温度/℃
	压力/MPa	温度/℃	压力/MPa	温度/℃	
工况1	17	45	9.8	24.54	16.83
		50	9.8	29.24	
工况2	11	45	9.8	40.60	
		50	9.8	45.50	

4 造腔及注气排卤工艺

造腔及注气排卤地面工艺(图4)设计主要包括：(1)盐穴造腔地面工艺技术选择；

(2)盐穴造腔注水泵机组调速技术方案的比选;(3)注气排卤工艺技术方案选择。

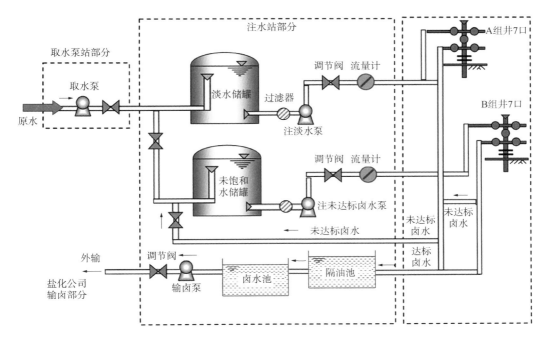

图 4 盐穴储气库造腔排卤工艺流程

4.1 注水泵类型选择及组合

造腔淡水注入及采卤应用的注水泵主要分为离心式和柱塞式两类。离心式泵的主要特点是排量大、流量平稳、压力波动小,利于注采卤管柱的长期运行,缺点是产生的压力相对较低;柱塞泵虽然能产生较高的压力,但是其流量不均衡,压力波动范围较大,工作压力不平稳,易导致注采卤管柱的振动和疲劳破坏,不利于管柱的长期运行[8-9]。因此,选择了多级离心泵作为造腔注水泵,电动机驱动。盐穴造腔注水泵机组以无级变速形式满足造腔井周期性不断变化的注水压力以及注水量的要求,节能降耗达到 10%~20%。

国内第一次将液力调速技术应用在造腔注水泵机组上,通过调整泵的排量和扬程来优化的高、低压阀组间工艺技术,满足各种工况条件下的快速造腔和盐厂的卤水回收要求。

4.2 造腔工艺流程

国内首次优化组合出盐穴造腔注水和采卤系统的"A+B"等工艺技术。为加快造腔进度,对造腔井进行分组,每阶段同时两井组进行造腔,一组注入淡水,另一组注入未饱和卤水。如第一批井为 A、B 两组,A 组井注入淡水,B 组井注入 A 组井采出的未饱和卤水。主要工艺流程包括以下步骤:

(1)从水源地来的淡水依次经储罐缓冲、沉降后,经注水泵升压并输送至高压阀组进行分配、控制、调节和计量,再由高压注水管路输送至 A 组注水井口。

(2)从 A 组井口返出的卤水经低压回水管网输送至造腔注水站内,经过取样分析后,未饱和卤水进入储罐,饱和卤水输往下游接收企业。

(3) 未饱和卤水经注水泵升压至高压阀组进行分配、控制、调节和计量后由高压注水管路输送到 B 组注水井口，从 B 组井口返出的卤水再经低压回水管网输至造腔站，计量后输往下游卤水接收企业。

4.3 布站方案

盐穴为单腔单井，并且井与井间距离较远，每口井宜单独设置 1 座井场。简化井口设计，造腔期间井场除造腔管线外不设置辅助设施，造腔结束后需在井口安装温度检测、压力检测、紧急切断阀及远程操作的电动球阀等设施。注采系统和造腔系统均采用两级布站，注采集配气阀组与造腔集配气阀组应合并建设[6]。

天然气集输管网呈枝网状结构，注采管线合一设置，单井进集配气阀组。集配气阀组的管线串接进集注站；造腔注水系统管线敷设至集配气阀组后采用枝网状结构至各井场；单井卤水管线进集配气阀组，再串接进集注站；卤水外输管线均单独铺设至各下游用户[7]。

5 压缩机厂房和空冷器区降噪

金坛储气库工程批复的环评报告书中要求的噪声控制目标为《工业企业厂界环境噪声排放标准》(GB 12348—2008)的 1 类厂界标准，即厂界处噪声等效声级不超过 45dB(A)。噪声治理难度极大，通过与清华大学声学研究所等公司合作研究，噪声治理达到了环保要求。

采取吸声、隔声、隔振和通风消声等噪声控制措施进行整体降噪的设计方案。压缩机厂房的屋面和墙体均安装高隔声量轻质泄爆降噪体；在纵墙上安装采光隔声窗；压缩机厂房内配置隔声门(声闸结构)；厂房顶部和四周安装进、排风风机消声器；压缩机四周基础安装使用隔振材料。空冷器区的屋面和墙体均安装轻质吸隔声降噪体；空冷器设备上部安装排风导流消声筒；空冷器进风口安装消声片，进风消声片外围再设置吸隔声屏障[10]。

(1) 厂房墙体和屋面安装超高隔声量轻质泄爆吸隔声降噪体。压缩机厂房为甲类生产厂房，耐火等级为二级，其设计结构需要满足相应的防火和防爆要求。降噪体由吸、隔声不燃板材现场复合拼装组成，以确保优良的降噪性能。

(2) 安装隔声门和隔声采光窗。压缩机厂房的大门和逃生门全部采用隔声门(声闸结构)。根据等透射量原则，单层隔声门设计隔声量≥35dB，钢制、防火防爆型，隔声门上安装防爆、逃生推杠锁。压缩机厂房采用在侧墙安装隔声采光窗的方式采光，每个窗户共两道，每道采用双层玻璃。

(3) 安装进、排风风机及配套消声器。压缩机厂房四周墙体安装进风消声器，进风消声器的设计消声量≥40dB(A)。

(4) 压缩机基础进行隔振处理。压缩机组在运行时会产生较强的振动，其通过基础使厂房结构产生振动传声。为减小设备基础刚性接触导致的结构传声，在基础外围地面以下安装高强隔振板，此措施可以有效地减少振动的传播，从而降低厂房结构的振动传声。

(5) 排烟消声围护。压缩机组末端配套安装有燃驱尾气排放装置，若压缩机排烟管道安装的消声器消声量达不到设计要求，需在烟囱周围安装吸声围护，使其满足厂区噪声排放要求。

6 结束语

目前我国仅有1座盐穴储气库——金坛储气库建成投产。随着天然气管网建设日益完善，气源更加多样化，盐穴储气库更能适应LNG气源的快速注采特性，在沿海地区建设盐穴储气库，可解决卤水排放处理问题，更具有投资优势。目前中国石油平顶山和淮安等盐穴储气库的建设即将启动，盐穴储气库建设的高峰即将到来。将我国首座盐穴储气库的设计、建设和运行经验进行总结，可对后续工程设计起到指导作用，进一步提高盐穴储气库的设计水平。

参 考 文 献

[1] 丁国生，张昱文.盐穴地下储气库[M].北京：石油工业出版社，2010：1-10.
[2] 郑雅丽，赵艳杰.盐穴储气库国内外发展情况[J].油气储运，2010，29(9)：652-655.
[3] 朱荣强，于连兴，王进军，等.盐穴地下储气库地面工程工艺技术[J].煤气与热力，2015，35(3)：47-52.
[4] 张瑞军.天然气三甘醇脱水系统工艺技术研究[J].中国石油和化工标准与质量，2011，31(7)：44.
[5] 冯叔初，郭揆常.油气集输与矿场加工[M].2版.东营：中国石油大学出版社，2006：420-429.
[6] 徐洋.天然气地下盐穴储气库技术研究[D].成都：西南石油大学，2005：10-18.
[7] 班凡生.深层盐穴储气库造腔及注采技术分析[J].重庆科技学院学报(自然科学版)，2016(2)：62-64.
[8] 解恺.盐穴储气库注气排卤优化设计[D].成都：西南石油大学，2016：23-28.
[9] 庄清泉.注气排卤技术在盐穴造腔中的应用[J].油气田地面工程，2010，29(12)：65-66.
[10] 贾宇，张巍.天然气压缩机降噪工艺研究[J].油气田环境保护，2011，21(1)：30-33.

本论文原发表于《油气田地面工程》2019年第38卷第2期。

盐穴储气库注水站整体造腔参数优化

耿凌俊[1]　李淑平[2]　吴　斌[2]　刘　春[2]　刘继芹[2]
王元刚[2]　何　俊[2]　王英杰[3]

(1. 江苏省常州市国土资源局金坛分局；2. 中国石油西气东输管道公司储气库管理处；3. 西南石油大学)

【摘　要】 针对盐穴储气库水溶建腔的优化问题，综合考虑水溶建腔过程涉及的诸多因素，结合非线性规划理论和水溶建腔的生产实际，建立了多井造腔工艺参数优化数学模型，计算得到最大造腔量和最小回罐量两种多井造腔工艺参数优化方案。研究表明：节约能耗不仅需要降低回罐量，而且需要保持卤水罐较低质量浓度；在淡水注入量一定的情况下，采用反循环的井越多，造腔量越大，配合高注入量可以显著增加造腔量，但卤水缓冲罐需维持较高质量浓度，能耗较大；较多地采用正循环，配合较低注入量，可以在满足淡水消耗量和卤水外输量要求下，减少回罐量，使卤水缓冲罐维持较低质量浓度，降低能耗。采用该模型，合理配置造腔井数并合理分配各井的循环模式、注入工质和注入量等参数，可以得到最优多井工艺参数配置方案，具有较好的指导意义。

【关键词】 盐穴储气库；溶腔进度优化；非线性规划；水溶建腔；溶腔模拟

Optimization of the solution mining parameters on salt-cavern gas storages water-injection station

Geng Lingjun[1]　Li Shuping[2]　Wu Bin[2]　Liu Chun[2]　Liu Jiqin[2]
Wang Yuangang[2]　He Jun[2]　Wang Yingjie[3]

(1. Land and Resources Bureau of Jintan City; 2. Gas Storage Management Department of West-East Gas Pipeline Company, CNPC; 3. Southwest Petroleum University)

Abstract In order to optimize the solution mining of salt-cavern gas storages, a mathematical model was established for optimizing the parameters of multi-well solution mining technology on the basis of non-linear programming theories and solution mining practices after comprehensive analysis was made on factors related with solution mining. Based on the calculation, two multi-well solution mining parameter optimization programs were prepared by considering the maximum cavern volume program and the minimum flowing back rate to the tank program. It is shown that, for purpose of energy conservation, the

---

作者简介：耿凌俊，男，助理工程师，1985年生，2008年毕业于南京工业大学数学与应用数学专业，现主要从事国土资源管理工作。地址：江苏省常州市金坛区东门大街158号，213200。电话：13776386641，E-mail: 760851205@qq.com。

volume of brine flowing back to the tank shall be reduced, but also the concentration in the brine tank shall be kept at low level. When fresh water injection rate is given, the more the wells for reverse circulation are adopted, the larger the caverns are. And the volume of the caverns can be increased remarkably if the injection rate is high, but the energy consumption is higher because the concentration in the brine buffer tank is kept at high level. Generally, the program of normal circulation with low injection rate is used. Based on this program, the backflow is reduced on condition that fresh water consumption and brine delivery are satisied. In this way, the concentration in the brine buffer tank is kept at lower level, so energy consumption drops. With this model, a practically optimal multi-well mining parameter arrangement program is developed by deploying the number of wells for solution mining rationally and arranging the circulation mode, injection working medium and injection rate for each well scientiically. The program plays an instructive role for ield operation.

Keywords salt cavern gas storages; leaching schedule optimization; non-linear programming; solution mining; solution mining simulation

随着天然气的普及,天然气用量迅速增加,随之而来的天然气用量季节性不均衡问题日趋突显。盐穴储气库因其使用寿命长、适合强注强采等特点,已经成为保障天然气稳定供应的重要设施。金坛储气库作为西气东输的配套工程,建设持续时间长,水溶建腔过程涉及诸多因素,包括淡水供应量、卤水外输质量浓度、卤水外输量、注入量及注入质量浓度等。为了经济高效地完成储气库建设任务,需要综合考虑溶腔过程涉及的诸多因素,对水溶建腔这一长期过程进行合理规划和工艺参数优化。

1 问题阐述

水溶建腔即向盐层注入未饱和工质(淡水或淡卤水),排出较高浓度的卤水,依靠盐层在未饱和工质中的溶解,实现扩大腔体体积的目的。金坛盐穴储气库东注水站在水溶建腔过程中(图1),还需要考虑为每口井配置注入量、循环模式、注入工质类型等。由于淡水供应来自盐化,而盐化的淡水供应量有限,且要求东注水站外输与淡水供应量等量的饱和卤水。因此,在安排造腔进度时,淡水注入总量、外输卤水总量及盐化淡水供应量三者要趋近等量(以小时为单位计算),且外输卤水的混合质量浓度也应满足盐化要求。同时,为了保持卤水缓冲罐的存量,所有井的回罐总量与注淡卤水总量也要保持平衡。

可见,需要协调以下3个条件约束实现溶腔进度最优化:(1)淡水消耗;(2)与供应量平衡,回罐量与注淡卤水量平衡;(3)外输混合质量浓度达标。满足第3个约束时,淡水消耗量与外输量也会达到平衡。

2 数学模型及求解

对于溶腔进度优化,可采用非线性规划[1-2]模型求解。分别以最大造腔量和最佳回罐量作为目标函数(f),附加约束条件为非线性约束(G, H_i)。

金坛储气库东注水站总计有30个注水口,而溶腔过程中,最优情况下实际使用的接口数可能少于30,将实际使用的接口数记为N。对于第i个注水口,其循环模式为x_{i1},注入量为x_{i2},注入工质质量浓度为x_{i3},内外管深度分别为x_{i4}、x_{i5},排出卤水去向为x_{i6};另外,

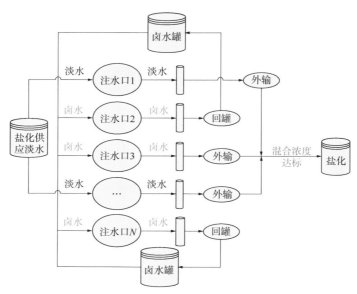

图 1　金坛储气库东注水站水溶建腔工艺流程示意图

对于第 i 个注水接口，对应的排卤质量浓度为 $c_i = C(X_i)$，对应的排量为 $p_i = P(X_i)$，X_i 为一维向量。盐矿为东注水站提供的淡水量限值为 W，外输到盐化的最低质量浓度限值为 S，注入的淡卤水质量浓度为 U，W 和 S 均为常量。同时，当 $x_{i1}=1$ 时，表示正循环，当 $x_{i1}=0$ 时，表示反循环；当 $x_{i3}=1$ 时，表示注淡卤水，当 $x_{i3}=0$ 时，表示注淡水；当 $x_{i6}=1$ 时，表示去向为外输；当 $x_{i6}=0$ 时，表示去向为回罐。

2.1　约束条件

首先，所有外输井的卤水混合质量浓度 G 均需达到盐化对卤水的要求：

$$G = \frac{\sum_{i=1}^{N} c_i p_i x_{i6}}{\sum_{i=1}^{N} p_i x_{i6}} \geqslant S \tag{1}$$

由于单位时间内盐化供应的淡水量有限，因此，所有井的淡水注入总量 H_1 均需与淡水供应量平衡：

$$H_1 = \sum_{i=1}^{N}\left(1 - \frac{x_{i3}}{U}\right) x_{i2} \tag{2}$$

为了保持卤水缓冲罐中的持液量，所有井的回罐量总量 H_2 均需与淡卤水注入总量平衡：

$$H_2 = \sum_{i=1}^{N}(1 - x_{i6}) p_i - \sum_{i=1}^{N} \frac{x_{i3}}{U} x_{i2} = 0 \tag{3}$$

其他附加约束包括：

$$H_{3i} = (1-x_{i1})x_{i1} \quad (i=1,\cdots,N) \tag{4}$$

$$H_{4i} = (U-x_{i1})x_{i3} \quad (i=1,\cdots,N) \tag{5}$$

排卤质量浓度的计算涉及水溶建腔机理[3]和溶腔模拟[4]，可以采用简化的二维溶腔模拟[5]方法计算排卤质量浓度 $C(X_i)$ 和排量 $P(X_i)$。排卤质量浓度根据循环模式、注入工质及注入量等参数，利用经验方法计算，排量则与注入量相同。经验方法可归纳为4种情况：(1) 正循环和注淡水，排卤质量浓度为 200~280g/L，随注入量增加而降低；(2) 正循环和注卤水，排卤质量浓度为 280~310g/L，随注入量增加而降低，随卤水质量浓度增加而增加；(3) 反循环和注淡水，排卤质量浓度为 280~310g/L，随注入量增加而降低；(4) 反循环和注卤水，根据生产经验，排出卤水一般已经达到饱和，质量浓度为 310g/L，但对于注入卤水质量浓度较低的情况，排卤质量浓度为 280~310g/L。

单位时间造腔体积的计算[6]未考虑不溶物堆积损失的体积以及地面和地下温度差异造成的误差。卤水罐中的卤水质量浓度，采用回罐卤水的混合质量浓度近似替代，并在计算过程中不断变化。

2.2 目标函数

从整体上看，溶腔过程中减少的能耗与最大化造腔量是对立的。要取得最大化的造腔量，必将导致更多的能耗。从系统能量平衡的角度来看，减少回罐量可以降低系统整体能耗，但也会导致造腔量的减少。因此，分别以最佳回罐量和最大造腔量为目标函数，建立多元非线性规划数学模型[5]。

最大造腔量模型如下：

$$\begin{cases} \max f_1 = \dfrac{\sum\limits_{i=1}^{N} c_i p_i - x_{i3} x_{i2}}{\rho} \\ G, H_1, H_2, H_{3i}, H_{4i} \quad (i=1,\cdots,N) \end{cases} \tag{6}$$

最佳回罐量模型如下：

$$\begin{cases} \max f_1 = \sum\limits_{i=1}^{N}(1-x_{i6})p_i \\ G, H_1, H_2, H_{3i}, H_{4i} \quad (i=1,\cdots,N) \end{cases} \tag{7}$$

2.3 模型求解

对于溶腔进度优化数学模型的求解，常采用罚函数法[7]将问题转化为无约束[10]的最优化问题，进而采用直接法求解。由于溶腔进度优化模型中包含等式约束[11]，因此，选择 SUMT 外点法[12]构造的罚函数：

$$F(X,M) = M\left[(G-S)^2 + (H_1-W)^2 + H_2^2 + \sum_{i=1}^{N}H_{3i}^2 + \sum_{i=1}^{N}H_{4i}^2\right] - f \tag{8}$$

式中：M 为罚因子，是一个较大的整数；与 M 相乘的部分称为罚项。

对于罚函数 $F(X,M)$，只有变量在可行域内，罚项趋近于 0 时，罚函数趋近于目标函数。由于罚因子可以有多个值，不同的 M 值产生的误差不同，因此还需要迭代，直至误差符合要求，即可确定最佳 M 值和最优解。

构造罚函数后，罚函数的最优解即溶腔进度优化的最优解，问题也变为对无约束的罚函数求极值。可以按照以下步骤迭代求解：(1) 任意给定初始点 X_0，取 $M_1>1$，给定允许误差 $\varepsilon>0$，令 $k=1$；(2) 求 $F(X,M)$ 的最优解，设为 $X^k=X(M_k)$，即 $\min F(X,M)=F(X^k,M_k)$；(3) 若 $S-G(X^k)>\varepsilon$，则取 $M_{k+1}=\alpha M_k$（其中 $\alpha=10$），令 $k=k+1$，返回(2)，否则，终止迭代，得最优解 $X^*\approx X^k$；(4) 对于 $F(X,M)$ 的求解，采用直接法[8-9]，给定步长，依次沿各坐标系轴方向进行步长式探索，寻找下降方向直至逼近函数极小点。

3 实例分析

根据金坛储气库的生产情况，假设淡水供应量为 1100m³/h，外输质量浓度限值为 290g/L，从生产数据中筛选出 29 口井作为模型的初始值(表 1)。实际上，这些井并不是同一时期设计和建腔的，其中有些井早已完腔，并用于储气调峰；另外一些井，可能尚处于某个溶腔阶段。为了便于分析计算，认为这些井可以在同一时间建腔，也可以处在各自不同的设计阶段。

表 1 金坛储气库各井初始状态

井名	模式	注入速度/(m³/h)	工质质量浓度/(g/L)	排卤质量浓度/(g/L)	去向
CAV105	正循环	79.6	0	295.7	外输
CAV106	正循环	99.1	0	268.2	回罐
CAV107	正循环	102.7	0	296.9	外输
CAV113	正循环	109.1	0	292.7	外输
CAV22	正循环	89.8	0	243.4	回罐
CAV41	正循环	97.5	0	274.8	外输
CAV42	正循环	97.6	0	239.2	回罐
CAV54	正循环	80.1	0	264.8	回罐
CAV63	正循环	97.1	0	173.2	回罐
CAV67	正循环	97.2	0	288.5	外输
CAV68	正循环	104.4	0	158.1	回罐
CAV71	正循环	104.4	0	207.5	回罐
CAV72	正循环	111.7	0	304.7	外输
CAV73	正循环	106.5	0	185.1	回罐
CAV74	正循环	107.1	0	255.3	回罐
CAV75	正循环	107.7	0	300.6	外输
CAV76	正循环	83.1	0	293.2	外输

续表

井名	模式	注入速度/(m³/h)	工质质量浓度/(g/L)	排卤质量浓度/(g/L)	去向
CAV78	正循环	86.4	0	290.7	外输
CAV79	正循环	87.5	0	260.5	回罐
CAV81	正循环	103.8	0	164.8	回罐
CAV83	正循环	105.3	0	243.1	回罐
CAV85	正循环	69.6	0	300.2	外输
CAV86	正循环	69.1	0	296.3	外输
CAV91	正循环	113.5	0	149.9	回罐
CAV93	正循环	80.2	0	211.7	回罐
CAV94	正循环	80.1	0	208.7	回罐
CAV95	正循环	74.4	0	222.3	回罐
CAV96	正循环	76.4	0	196.5	回罐
CAVCH5	正循环	98.9	0	315.3	外输

将初始值输入程序后，计算得到最大造腔量和最小回罐量溶腔进度规划方案（表2、表3）。在追求最大造腔量的情况下，大部分井的循环模式是反循环，并且注淡水的井均采用反循环，这是因为在反循环时，排卤质量浓度高，造腔效率高，符合实际生产规律。部分采用正循环的井均注入淡卤水，是为了平衡卤水罐液量。另外，卤水罐质量浓度维持在高位，导致较多能量损耗。在追求经济节能的情况下，大部分井采用正循环的溶腔模式，且注入量均较低，尽量保证较高的排卤质量浓度。而卤水罐中的卤水质量浓度则维持在较低水平。

表2 金坛储气库各井最大造腔量规划方案

井名	模式	注入速度/(m³/h)	工质质量浓度/(g/L)	排卤质量浓度/(g/L)	去向
CAV105	反循环	40.09	0.00	298.32	外输
CAV106	正循环	67.54	222.67	288.74	外输
CAV107	反循环	19.83	0.00	306.70	外输
CAV113	反循环	40.08	0.00	298.32	外输
CAV22	反循环	51.03	222.67	310.00	外输
CAV41	反循环	20.01	222.67	310.00	外输
CAV42	反循环	20.00	222.67	310.00	外输
CAV54	反循环	20.00	222.67	310.00	外输
CAV63	反循环	20.00	222.67	310.00	外输
CAV67	反循环	20.00	222.67	310.00	外输
CAV68	反循环	20.00	222.67	310.00	外输

续表

井名	模式	注入速度/(m³/h)	工质质量浓度/(g/L)	排卤质量浓度/(g/L)	去向
CAV71	反循环	20.00	222.67	310.00	外输
CAV72	反循环	20.00	222.67	310.00	外输
CAV73	反循环	20.00	222.67	310.00	外输
CAV74	反循环	20.00	222.67	310.00	外输
CAV75	反循环	20.00	222.67	310.00	外输
CAV76	正循环	41.38	222.67	298.10	外输
CAV78	正循环	100.00	222.67	278.33	外输
CAV79	正循环	20.02	222.67	306.66	外输
CAV81	反循环	100.00	0.00	278.33	外输
CAV83	反循环	100.00	0.00	278.33	回罐
CAV85	反循环	100.00	0.00	278.33	外输
CAV86	反循环	100.00	0.00	278.33	外输
CAV91	反循环	100.00	0.00	278.33	回罐
CAV93	反循环	100.00	0.00	278.33	回罐
CAV94	反循环	100.00	0.00	278.33	外输
CAV95	反循环	100.00	0.00	278.33	回罐
CAV96	反循环	100.00	0.00	278.33	回罐
CAVCH5	反循环	100.00	0.00	278.33	外输

表3 金坛储气库各井最小回罐量规划方案

井名	模式	注入速度/(m³/h)	工质质量浓度/(g/L)	排卤质量浓度/(g/L)	去向
CAV105	反循环	90.00	0.00	285.00	外输
CAV106	正循环	30.00	87.10	300.00	外输
CAV107	反循环	37.85	0.00	298.69	外输
CAV113	反循环	32.19	0.00	299.64	外输
CAV22	正循环	60.00	87.10	290.00	外输
CAV41	正循环	20.07	87.10	306.65	外输
CAV42	正循环	24.60	87.10	305.90	外输
CAV54	正循环	24.38	87.10	305.94	外输
CAV63	正循环	24.68	87.10	305.89	外输
CAV67	正循环	23.38	87.10	306.10	外输
CAV68	正循环	23.06	87.10	306.16	外输
CAV71	正循环	23.88	87.10	306.02	外输

续表

井名	模式	注入速度/(m³/h)	工质质量浓度/(g/L)	排卤质量浓度/(g/L)	去向
CAV72	正循环	23.81	87.10	306.03	外输
CAV73	正循环	23.27	87.10	306.12	外输
CAV74	正循环	22.54	87.10	306.24	外输
CAV75	正循环	22.51	87.10	306.25	外输
CAV76	正循环	21.69	87.10	306.39	外输
CAV78	正循环	90.00	87.10	280.00	外输
CAV79	正循环	21.22	87.10	306.46	外输
CAV81	反循环	74.07	0.00	287.66	外输
CAV83	正循环	82.54	0.00	147.46	回罐
CAV85	反循环	91.99	0.00	279.67	外输
CAV86	反循环	97.19	0.00	278.80	外输
CAV91	正循环	97.05	0.00	102.95	回罐
CAV93	正循环	100.00	0.00	100.00	回罐
CAV94	反循环	98.43	0.00	278.59	外输
CAV95	正循环	99.51	0.00	100.49	回罐
CAV96	正循环	100.00	0.00	100.00	回罐
CAVCH5	反循环	99.18	0.00	278.47	外输

将两种规划方案进行对比(表4)可以看出,两种方案在造腔量和回罐量上,差别并不明显。但由于造腔量为小时造腔量,而每个腔体造腔过程需要历时数年,因此,小时造腔量差值的长时间累计值将比较可观。

表4 金坛储气库各井最大造腔量与最小回罐量规划方案对比

方案	造腔量/(m³/h)	注淡水量/(m³/h)	注卤水量/(m³/h)	回罐量/(m³/h)	总注入量/(m³/h)	工质质量浓度/(g/L)
最大造腔量	160.20	1100.00	500.01	500.01	1600.01	222.87
最经济节能	152.14	1099.99	479.10	479.09	1579.09	87.10

由于卤水罐中的盐含量是通过消耗能量在地下盐层溶解并排入卤水罐,属于无效的耗能,因此,将卤水罐中的盐含量作为评价能量损耗的标准是合适的。回罐量与卤水罐质量浓度的乘积可以用来评价两种方案的能量损耗情况,乘积越大,能量损耗越大,相反,乘积越小,能量损耗也越小。

4 结论

结合金坛储气库水溶建腔的生产实际和非线性规划理论,建立了溶腔进度优化数学模型,并给出了数值求解方法。利用该模型可以对溶腔进度方案进行量化计算,通过对比分析得到最佳方案。讨论了最大造腔量和最小回罐量两种方案,得出以下结论:

(1)通过减少回罐量和降低卤水罐质量浓度,可以减少溶腔过程中能量的无效损耗。当回罐量和卤水罐质量浓度均达到较低水平时,溶腔方案最经济节能。

(2)采用反循环模式造腔,排卤质量浓度高,造腔效率高。在追求最大造腔量的情况下,势必要求更多的井采用反循环模式。采用反循环造腔,并配注高流量的淡水,能够显著增加造腔量。该情况下,回罐的卤水质量浓度和排量普遍偏高,因而使卤水罐中卤水质量浓度维持在高位,导致较多的无效能量损耗。

(3)通过优化数学模型,可以确定复杂情况下的最经济节能方案。方案中大部分井采用正循环的溶腔模式,并且注入量均较低,应该尽量保证较高的排卤质量浓度,使卤水罐中卤水质量浓度维持在较低水平。

参 考 文 献

[1] 张义森. 实用非线性规划[M]. 北京:科学出版社,1981:110-115.
[2] 席少霖,赵凤治. 最优化计算方法[M]. 上海:上海科学技术出版社,1983:99-101.
[3] 赵志成,朱维耀,万玉金,等. 盐穴储气库水溶建腔机理研究[J]. 石油勘探与开发,2003,30(5):107-109.
[4] Saberian A. Numerical simulation of development of solutionmined storage cavities[D]. Austin:The University of Texas at Austin,1974:4-5.
[5] Saberian A,Podio A L. A numerical model for development of solution mined cavities[J]. SMRI,1976,11(4):303.
[6] 李晓颖. 盐穴储气库造腔注水站设计要点[J]. 油气田地面工程,2008,27(3):3-4,15.
[7] 白春阳,石东伟. 非线性规划在数学建模中的应用[J]. 科技信息,2011(29):171,213.
[8] 宋士吉,张玉利,贾庆山. 非线性规划[M]. 2 版. 北京:清华大学出版社,2013:305-311.
[9] 王一铁. 无约束求极值的一种直接法[J]. 临沂师专学报,1999,21(6):10-11.
[10] 邓乃扬. 无约束最优化计算方法[M]. 北京:科学出版社,1982:105.
[11] 史秀波,李泽民. 用非线性方程组求解等式约束非线性规划问题的降维算法[J]. 经济数学,2007,24(2):208-212.
[12] 龙腾,刘莉,李怀建,等. 可行方向 SUMT 外点法的研究及应用[J]. 系统工程与电子技术,2011(3):685-689.

本论文原发表于《油气储运》2016 年第 35 卷第 7 期。

盐穴储气库造腔管理分析系统设计与应用

李淑平　刘继芹　齐得山　陈加松　敖海兵　王成林

(中国石油西气东输管道公司储气库管理处)

【摘　要】 在盐穴储气库建设过程中，由于经验不足，在管理和效率方面存在较多问题。为了提高盐穴储气库建库的工作效率和管理水平，自主设计和开发了盐穴储气库造腔管理分析系统。该系统利用关系数据库组织和维护造腔生产数据，设计和实现造腔分析、声呐分析、注采运行以及系统管理等功能模块，实现造腔进度快速计算和监控、油垫压力监测以及造腔参数优化等许多科学实用的功能，大大降低了人工计算量，确保造腔设计和进度监测的准确性，有效地提高了管理水平，具有非常好的推广价值。

【关键词】 盐穴储气库；造腔分析；造腔优化；油垫压力监测

金坛盐穴储气库，是中国盐穴储气第一库。在建库之初，面对中国特有地质特点和国情，由于经验不足，在造腔管理和效率方面遇到了较多问题。概括来讲，主要包括三个方面的问题。首先，随着建库任务的推进，生产数据日趋庞杂，对生产数据的整理分析变成一项烦琐的任务。不仅数据的筛选困难，为了获取某项造腔参数，还需要大量的手工计算，计算效率和准确率都难以保证。其次，造腔异常和井下故障难以发现，没有科学有效的检测手段，仅能依靠现场检查和井下作业，成本高。另外，库区多井同时造腔方案，没有科学的优化方法，导致淡水供应和卤水外输不均衡的矛盾突出；同时协调困难，导致库区造腔高能耗、低效率的问题严重。因此，急需一套用于盐穴储气库造腔管理和分析的软件。

1　系统架构

盐穴储气库造腔管理分析系统(简称GSDMAS)采用客户端/服务器模式(C/S模式)。服务器端采用SQL SERVER数据库。根据造腔过程中涉及的诸多因素和实际需求，对数据库关系模式[1]进行设计，规范数据存储规则，降低数据冗余。客户端采用模块化设计，根据生产需要和问题，设计多个功能模块，并且随着新的生产问题不断发现，新的功能模块也随之增加，使系统功能不断丰富和完善。

综合需求分析，该系统架构设计图如图1所示。

为了快速开发，客户端选择基于.net平台开发，采用C#语言编写，要求编码规范，注释详细。模块功能必须经过全面的测试。

2　模块功能

GSDMAS系统客户端设计开发6大功能模块，包括盐穴地图、造腔设计、造腔分析、声呐分析、注采运行和系统管理。

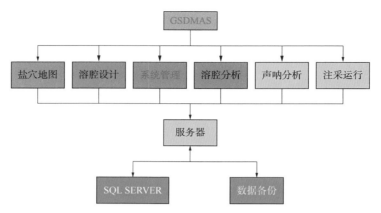

图 1 盐穴储气库造腔管理分析系统架构

2.1 盐穴地图

该模块主要功能是基于各个井位的大地坐标，展示库区各井相对位置以及各个腔体当前横截面形状。

2.2 造腔分析

该模块包含众多子模块，功能丰富，是 GSDMAS 的核心模块。该模块包含的子模块主要有：造腔监控模块，用于监控各个腔体的阶段进度和工程进度，也具有数据统计功能；数据选择模块，也即数据查询模块，用于查询和编辑各个腔体详细的生产数据；报告生成模块，用于从数据库中导出各个腔体的综合月报和日报数据；声呐测量计划模块，可优化安排各个腔体的声呐测量计划；图表分析模块，是造腔分析模块的核心模块，功能包括排卤浓度监测、地面地下声呐三体积、排量压力对比、油垫压力监测、注入量监测以及造腔效率监控等。

2.3 造腔设计

它是造腔设计的辅助模块，可以生成岩性表和批量生成造腔阶段数据等，可供造腔模拟软件使用。

2.4 声呐分析

用于声呐数据分析，可展示二维和三维腔体形状。即可展示腔体某个阶段的二维形状，也可以展示多个阶段叠加的二维形状，为腔体偏溶分析和造腔设计提供参考。

2.5 注采运行

负责储气库注采运行监控和注采数据查询。注采运行监测内容包括腔体压力状态、运行气量和库容以及最近 30d 压力波动等。

2.6 系统管理

主要用于生产数据维护、用户管理、系统公告发布和系统日志等。其中，生产数据维护包括单井信息维护、日报数据更新、声呐数据维护、溶腔阶段维护和注采气数据更新等子模块。用户管理模块可以添加删除系统用户、设置用户权限和编辑用户信息等功能。

3 技术创新

在 GSDMAS 系统的开发过程中,通过技术创新解决了 3 大技术难题,包括排卤浓度监测、油垫压力监测和造腔参数优化。

3.1 排卤浓度监测

在水溶造腔过程中,通过生产日报采集各个腔体的排卤质量浓度,利用每日排卤质量浓度数据绘制排卤质量浓度曲线。由于缺少参考依据,通过排卤质量浓度曲线,仅能看到排卤质量浓度波动情况,无法监测造腔过程是否存在异常。

为了解决该问题,自主设计开发了造腔模拟器[2]。利用造腔模拟器,就可以为每个腔体的每日排卤质量浓度提供一个理论值,利用排卤质量浓度的理论值绘制一条与实际排卤质量浓度同步的理论排卤质量浓度曲线,通过对比两条曲线的差异便可以及时发现潜在的造腔异常。GSDMAS 内置的造腔模拟器采用的网格模型如图 2 所示。

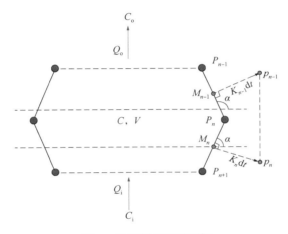

图 2 造腔模拟网格图解

网格质量浓度变化方程:

$$\frac{dC}{dt} = m_s + u\frac{dC}{dh} + D\frac{d^2C}{dh^2}$$

式中:C 为网格质量浓度,g/L;m_s 为溶盐速率;u 为流速,m/s;D 为扩散常数,m²/s;h 为网格高度,m。

3.2 溶腔参数优化

金坛盐穴储气库东注水站水溶建腔过程如图 3 所示。在实际水溶造腔的生产中,还需要考虑为每口井配置注入量、循环模式、注入工质类型等。由于淡水供应来自盐化,并且盐化的淡水供应量有限,同时,盐化要求东注水站外输与淡水供应量等量的饱和卤水。因此,在安排造腔进度时,淡水注入总量、外输卤水总量和盐化淡水供应量三者要趋近等量(本文以小时为单位计算)。与此同时,外输卤水的混合质量浓度也要满足盐化要求。另外,为了保持卤水缓冲罐的存量,所有井的回罐总量与注淡卤水总量也要保持平衡。

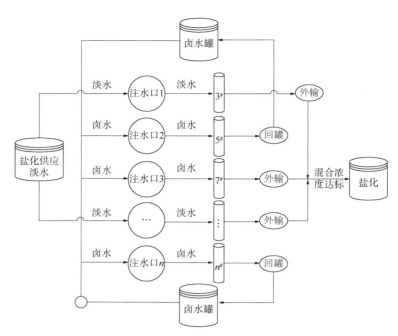

图 3　金坛储气库水溶造腔示意图

综上所述，需要协调以下 3 个条件约束来实现溶腔参数最优化。

（1）淡水消耗量与供应量平衡；（2）回罐量与注淡卤水量平衡；（3）外输混合浓度达标。

在满足第（3）个约束的情况下，淡水消耗量与外输量也会达到平衡。

考虑上述问题和约束，采用非线性规划理论[3]，建立该问题的数学模型，如下。

$$H_1 = \sum_{i=1}^{N} \left(1 - \frac{X_{i3}}{U}\right) X_{i2} = W$$

$$G = \frac{\sum_{i=1}^{N} C(X_i) P(X_i) X_{i6}}{\sum_{i=1}^{N} P(X_i) X_{i6}} \geqslant S$$

$$H_2 = \sum_{i=1}^{N} (1 - X_{i6}) P(X_i) - \sum_{i=1}^{N} \frac{X_{i3}}{U} X_{i2} = 0$$

$$H_{3i} = (1 - X_{i1}) X_{i1} \quad (i = 1, \cdots, N)$$

$$H_{4i} = (U - X_{i3}) X_{i3} \quad (i = 1, \cdots, N)$$

$$\max f_1 = \frac{\sum_{i=1}^{N} C(X_i) P(X_i) - X_{i3} X_{i2}}{\rho}$$

式中：N 为注水站注水口数量；X_{i1} 为第 i 个注水口的循环模式；X_{i2} 为第 i 个注水口的注入

量，m³/h；X_{i3} 为第 i 个注水口的注入工质质量浓度，g/L；X_{i4}，X_{i5} 为第 i 个注水口的内、外管深度，m；X_{i6} 为第 i 个注水口的排出卤水去向；c_i 为第 i 个注水口的排卤质量浓度，g/L，$c_i=C(X_i)$，X_i 是一维向量；p_i 为第 i 个注水口的排量，m³/h，$p_i=P(X_i)$，X_i 为一维向量；W 为盐化供应淡水量限值，常量，m³/h；S 为外输到盐化的最低质量浓度限值，常量，g/L；U 为注入的淡卤水质量浓度，常量，g/L；ρ 为纯盐密度。

对于以上溶腔参数优化数学模型的求解，常见的思路是采用罚函数法[4]将问题转化为无约束[5]的最优化问题，采用直接法对其求解，进而得到最优的溶腔参数。

3.3 油垫压力监测

油垫压力监测遇到的问题与排卤质量浓度监测类似，需要为实际油垫压力曲线匹配一条理论油垫压力曲线作为判断依据。创新性地将水动力学管流理论与盐岩蠕变理论相结合，准确计算油垫压力。相关理论比较成熟，本文不再赘述。

4 现场应用

从 2010 年开始至今，主要应用于金坛盐穴储气库。借助 GSDMAS，生产数据的管理和维护更加规范，数据整理更加简单高效，显著提高管理水平；利用系统的功能模块，使得造腔跟踪分析一键即可完成，准确快速，避免手工计算，降低劳动强度；通过数据分析诊断造腔故障，降低现场作业量，节约成本；利用造腔优化模块，实现多井造腔排量得以优化，可以更好地协调淡水注入量与卤水外输量的矛盾，更好地控制能耗和造腔效率。

GSDMAS 系统的辅助功能模块也非常实用。利用其进度跟踪及声呐计划功能，优化井下作业和声呐检测，降低施工成本；利用其用户权限管理，将库区各个造腔任务具体到人，责任到人，既有利于提高工作效率，也有利于规范管理，确保生产任务保质保量地完成。

GSDMAS 的应用取得了非常好的经济效益和社会效益，有效地保障了金坛库区造腔任务的顺利推进。截至目前，金坛库区已经顺利完成造腔 $683.5 \times 10^4 \mathrm{m}^3$，完腔 14 口井，有效库容达 $5 \times 10^8 \mathrm{m}^3$。

5 结论

（1）为了解决金坛盐穴储气库建设中造腔管理和效率的问题，结合造腔生产实际，设计和开发盐穴储气库造腔管理分析系统。系统架构设计良好，数据库数据冗余较少，采用模块化设计，系统功能丰富，且具有一定的可扩展性。

（2）借助 GSDMAS 系统，金坛库区的造腔管理形成一套规范的工作制度，使库区生产数据的管理和维护更加规范高效，造腔跟踪分析更加快速便捷。

（3）通过技术创新，解决了水溶建腔 3 大技术难题，利用排卤质量浓度监测和油垫压力监测功能，不需要现场作业，即可监测造腔异常和故障，有效地减少了施工作业量，降低造腔成本；利用造腔参数优化功能，优化配置注水站造腔参数，进而节约能耗，提高造腔效率。

（4）GSDMAS 系统在金坛库区的应用取得了非常好的经济和社会效益，在盐穴储气库建设方面具有非常好的推广价值。

参 考 文 献

[1] 周定康,许婕,李云洪,等.关系数据库理论及应用[M].武汉:华中科技大学出版社,2005.
[2] Saberian A,Podio A L. A numerical model for development of solution mined cavities[J]. SMRI,1976.
[3] 席少霖,赵凤治.最优化计算方法[M].上海:上海科学技术出版社,1983.
[4] 王一铁.无约束求极值的一种直接法[J].临沂大学学报,1999,21(6):10-11.
[5] 邓乃扬.无约束最优化计算方法[M].科学出版社,1982.

本论文原发表于《石油化工应用》2016年第35卷第8期。

经济评价

考虑垫底气回收价值及资金时间价值的盐穴型地下储气库储气费计算方法

王元刚　李淑平　齐得山　李建君

(中国石油管道有限责任公司西气东输分公司)

【摘　要】 我国的地下储气库(以下简称储气库)与油气管道捆绑运营,没有单独的定价机制,计算储气库储气费时也未充分考虑资金的时间价值及油气藏型、盐穴型储气库垫底气的回收价值,导致计算结果的准确性欠佳。未来储气库实行独立、市场化运营是必然趋势,因而需要建立一种符合我国储气库运营模式的储气费定价机制。为此,以国内某盐穴型储气库建设投资项目为例,采用二分法建立了一种考虑垫底气可回收的储气费计算模型,计算出该储气库在不同内部收益率下的储气费,并分析了影响储气费的主要因素。结果表明:(1)当储气费为 1.02 元/m^3 时,可满足内部收益率 8%的要求;(2)在盐穴储气库工作气量确定的情况下,年储转次数(储气库年实际注采气量与年设计工作气量的比值)是影响储气费的最重要因素,地下及地面工程等建设投资的影响次之,而经营成本的影响最小;(3)在储转次数大于 1.4 时,盐穴储气库注采运行的工作效率达到最大,建议将盐穴储气库的储转次数设定为 1.4。结论认为:该储气费计算方法在保证能获得一定利润的前提下,充分考虑了资金的时间价值以及垫底气的回收价值,计算得到的储气费较为合理,可推广到类似盐穴型储气库的应用计算。

【关键词】 盐穴地下储气库;储气费;垫底气回收;二分法;资金时间价值;敏感性分析;年储转次数;建设投资

A methodology for calculating the gas storing price of salt cavern UGSs considering the recycle value of cushion gas and the time value of capital

Wang Yuangang　Li Shuping　Qi Deshan　Li Jianjun

(West-East Gas Pipeline Company, PetroChina Pipeline Company)

Abstract At present, domestic underground gas storages (UGSs) is treated only as an auxiliary

---

基金项目:中国石油天然气集团有限公司储气库重大专项子课题"盐穴储气库加快建产工程试验研究"(编号 2015E-4008)。

作者简介:王元刚,1986 年生,工程师,硕士;主要从事管道建设及盐穴地下储气库工程建设工作。地址:(230031)安徽省合肥市政务区怀宁路 288 号置地广场 C 座 28 层。ORCID:0000-0001-6745-9121。
E-mail:515935924@qq.com。

facility of oil and gas pipeline companies without an independent pricing system. And when its gas storing price is calculated, the recycle value of cushion gas of oil/gas reservoir UGSs and salt cavern UGSs and the time value of capital are not taken into full consideration, so its calculation accuracy is not good enough. Independent and market-oriented operation is the inevitable development trend of UGSs in the future, so it is necessary to establish a gas storing pricing system suitable for domestic UGS operation mode. In this paper, the construction investment project of one salt cavern UGS in China was taken as the research object. A gas storing price calculating model considering the recycle of cushion gas was established by the dichotomy method. Then, the gas storing price of this case UGS under different internal rates of return (IRR) was calculated and the main factors influencing the gas storing price were analyzed. And the following research results were obtained. First, the IRR of 8% can be satisfied when the gas storing price is CNY 1.02 per m^3. Second, when the working gas volume of a salt cavern UGS is determined, annual storage withdrawal frequency (the ratio between the actual and designed annual working gas volume) is the most important factor influencing gas storing price, and the construction investment of underground and ground engineering takes the second place, while the effect of operation cost is the least. Third, the working efficiency of a salt cavern UGS is the maximum when the storage-withdrawal frequency is over 1.4, the value of which is thus recommended here. In conclusion, this gas storing price calculating method takes full consideration of the time value of capital and the recycle valve of cushion gas while ensuring a certain profit, so its calculation result is more reasonable. This method shall be popularized and applied to the calculation of similar salt cavern UGSs.

Keywords Salt cavern UGS; Gas storing price; Recycle of cushion gas; Dichotomy; Time value of capital; Sensitivity analysis; Annual storage-withdrawal frequency; Construction investment

我国的地下储气库经过约20年的发展，形成了一定的规模。但由于储气库作为管道的辅助设施，与管道捆绑运营[1-4]，相应的储气费计入管输费，与管输费一并收取，没有单独、成熟的定价机制，也未在天然气价格体系中单独设立"储气费"科目[3]。随着天然气应用范围越来越广以及储气库的大规模建设并投入使用，储气库独立运营已经成为一种趋势，储气费独立定价被逐渐提上日程。

美国及欧洲的储气库运营模式较为成熟，前者在联邦能源监管委员会管理下，其储气库普遍按市场需求定价[5]，而后者则采用协商定价[6]。欧美以及我国储气环节的定价方式表明，采用何种定价方式必须与本国天然气产业发展情况相适应[7-10]。

储气费定价是储气库建设的重要评价指标，国内外很多学者对其进行了深入的研究[11-12]，通过对储气库建设的投资进行分析，研究注采气量及电力成本等对储气费的影响规律[13-15]，求解出一种储气费计算的基本模型。由于储气库建设周期较长，投资回收期较长，在进行经济评价时需要充分考虑资金的时间价值；确定储气费时需要保证企业所耗费的全部成本得到补偿，并在正常情况下能获得一定的利润。目前我国在计算储气库储气费时，未考虑油气藏型储气库垫底气可部分回收及盐穴储气库垫底气可完全回收等特点，计算得出的地下储气库储气费评价指标偏高，项目盈利能力等指标偏低，影响和制约了我国地下储气库的发展[16]。为此，笔者尝试建立一种考虑资金时间价值以及储气库废弃后垫底气可全部回收的储气费计算方法，并对储气费的影响因素进行敏感性分析。

1 储气费计算模型

1.1 储气费计算方法

为保证储气库的正常运营与发展,储气库的建设与注采运行需要满足一定的盈利能力。运营期内,在满足一定内部收益率的条件下,使累计净现值为 0 时的储气费用为市场参考值。笔者采用二分法[17-19],在设定区间内计算出合理的储气费,计算流程如图 1 所示。

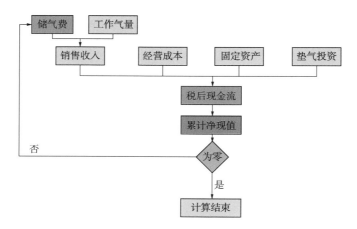

图 1 储气费计算流程图

(1) 计算出储气库的建设投资,包括地下建设、地面建设、垫气费等,其中地下与地面投资均为固定资产投资。

(2) 计算出盐穴储气库每年的工作气量,设定储气费初始值,计算出销售收入。

(3) 根据工作气量计算整个评价期内的运营成本。

(4) 根据固定资产投资及垫气费等费用计算折旧与摊销费用。

(5) 根据销售收入及经营成本等费用,计算出税金及附加费用。

(6) 根据计算出的折旧、摊销等费用,结合企业所得税、银行贷款利率等经济评价参数,计算出储气库的调整所得税以及税后现金流。

(7) 根据建设运营期内累计净现值(NPV)调整储气费的设定值,当 NPV = 0 时的价格为市场参考值。

1.2 储气费计算模型

基于上述分析,设定储气费上下限 $[p_{storemin}, p_{storemax}]$,令 $p_{store} = p_{storemin}$(p_{store} 表示储气费,元/m³),进行评价期内 NPV 的初始计算。

$$NPV = \sum_{i=1}^{T_{life}} NCF(i)/(1 + IRR)^i \tag{1}$$

式中:T_{life} 为评价期,a;IRR 为内部收益率;$NCF(i)$ 为第 i 年所得税后净现金流量,元。

评价期结束时流动资金可全部回收,与油气田开发以及含水层储气库不同,盐穴储气库废弃时,垫底气可大部分回收甚至全部回收,因此净现金流计算公式为

$$\text{NCF}(i) = \begin{cases} s(i) - c_{\text{fix}}(i) - c_{\text{cus}}(i) - c_{\text{flow}}(i) - \\ c_{\text{ope}}(i) - t_{\text{bus}}(i) - t_{\text{adj}}(i), \ i < T_{\text{life}} \\ s(i) + \sum\limits_{1}^{T_{\text{life}}-1}[c_{\text{flow}}(i) + c_{\text{cus}}(i)] - c_{\text{fix}}(i) - \\ c_{\text{ope}}(i) - t_{\text{bus}}(i) - t_{\text{adj}}(i), \ i = T_{\text{life}} \end{cases} \quad (2)$$

式中：$s(i)$ 为销售收入，元；$c_{\text{fix}}(i)$ 为固定资产投资，元；$c_{\text{cus}}(i)$ 为垫气费，元；$c_{\text{flow}}(i)$ 为流动资金，元；$c_{\text{ope}}(i)$ 为经营成本，元；$t_{\text{bus}}(t)$ 为税金及附加，元；$t_{\text{adj}}(j)$ 为调整所得税，元。

现金流计算涉及费用项较多，公式中各费用项计算公式如下。

（1）销售收入、固定资产、垫气费以及经营成本等基本投资计算。盐穴储气库由于注采灵活，可实现一年多轮次注采运行，则

销售收入为

$$s(i) = V_{\text{work}}(i) \times \text{cynum} \times p_{\text{store}} \quad (3)$$

垫气费为

$$c_{\text{cus}}(i) = V_{\text{cus}}(i) \times p_{\text{gas}} \quad (4)$$

式中：$V_{\text{word}}(i)$ 为工作气量，m³，cynum 为年储转次数，$V_{\text{cus}}(i)$ 为新增垫气量，m³，p_{gas} 为垫气价格，元/m³。

固定资产投资 $c_{\text{fix}}(i)$ 以及经营成本 $c_{\text{ope}}(i)$ 根据储气库建设规模、投产井数、运行条件等进行计算；流动资金 $c_{\text{flow}}(i)$ 表示维持生产运营所必需的资金，该部分可以按照经营成本乘以一定比例系数进行计算。

（2）税金及附加。城市建设维护税及教育费附加计算方法如下：

$$t_{\text{bus}}(i) = \text{VAT}(i) \times r_{\text{vat}} = [s(i) \times r_{\text{out}} - c_{\text{ope}}(i) \times r_{\text{in}}] \times r_{\text{vat}} \quad (5)$$

式中：VAT(i) 为增值税，元；r_{vat} 为附加税税率；r_{out} 为销项税额比例；r_{in} 为进项税比例。

（3）折旧计算。定义二维函数 $f_{\text{dep}}(i,j)$ 表示第 i 年的固定资产在第 j 年产生的折旧费用。储气库在第一口井完腔后开始运营，在运营过程中所有固定资产需要进行折旧计算，采用年限平均法折旧，建设期的资产以及产生的利息自投产后第一年开始折旧，该部分费用每年产生的折旧费用为

$$f_{\text{dep}}(i,j)_{j=T_{\text{con}}+1}^{T_{\text{con}}+T_{\text{dep}}} = \sum_{k=1}^{T_{\text{con}}}[c_{\text{int}}(k) + c_{\text{fix}}(k)]/T_{\text{dep}}, \ i = T_{\text{con}} \quad (6)$$

式中：T_{con} 为建设期，a；T_{dep} 为折旧年限，a；$c_{\text{int}}(k)$ 为第 k 年建设期利息，元。

运营期发生的固定资产投资从使用后第二年开始折旧，运营期第 i 年的投资在折旧期内每年的折旧费用如下：

$$f_{\text{dep}}(i,j)_{j=i+1}^{i+T_{\text{dep}}}=c_{\text{fix}}(i)/T_{\text{dep}}, \quad i>T_{\text{con}} \tag{7}$$

将式(6)与式(7)合并，可得第 i 年的投资在第 j 年产生的折旧费用为

$$f_{\text{dep}}(i,j)_{j=i+1}^{i+T_{\text{dep}}}=\begin{cases}\sum_{k=1}^{T_{\text{con}}}[c_{\text{int}}(k)+c_{\text{fix}}(k)]/T_{\text{dep}}, & i=T_{\text{con}} \\ c_{\text{fix}}(i)/T_{\text{dep}}, & i>T_{\text{con}}\end{cases} \tag{8}$$

则每年的总折旧费用为

$$c_{\text{dep}}(j)=\sum_{i=1}^{T_{\text{life}}}f_{\text{dep}}(i,j) \tag{9}$$

建设期利息 $c_{\text{int}}(i)$ 的计算方法如下：

设定二维函数 $f_{\text{int}}(i,j)$ 为第 i 年贷款在第 j 年产生的利息。每年新贷款时间按年中计算，则第 i 年贷款当年产生的利息为

$$f_{\text{int}}(i,j)=[c_{\text{fix}}(i)+c_{\text{cus}}(i)]\times p_{\text{loan}}\times r_{\text{bank}}/2, \quad j=i \tag{10}$$

式中：p_{loan} 为建设投资中贷款额所占的比例；r_{bank} 为银行贷款利率。

第 i 年贷款的后续新增利息为

$$f_{\text{int}}(i,j)=[c_{\text{fix}}(i)+c_{\text{cus}}(i)]\times p_{\text{loan}}\times(1+r_{\text{bank}}/2)\times(1+r_{\text{bank}})^{(j-i-1)}\times r_{\text{bank}}, \quad j>i \tag{11}$$

综合式(10)、式(11)，第 i 年贷款在第 j 年产生的利息为

$$f_{\text{int}}(i,j)=\begin{cases}[c_{\text{fix}}(i)+c_{\text{cus}}(i)]\times p_{\text{loan}}\times r_{\text{bank}}/2, & j=i \\ [c_{\text{fix}}(i)+c_{\text{cus}}(i)]\times p_{\text{loan}}\times(1+r_{\text{bank}}/2)\times(1+r_{\text{bank}})^{(j-i-1)}\times r_{\text{bank}}, & j>i\end{cases} \tag{12}$$

则建设期每年的利息总额计算公式如下：

$$c_{\text{int}}(j)=\sum_{i=1}^{i=j}f_{\text{int}}(i,j) \tag{13}$$

(4) 摊销。摊销费与经营期内的资产折旧计算类似，用年限平均法进行摊销，第 i 年垫气费在后续每年的摊销额为

$$f_{\text{amo}}(i,j)_{j=i+1}^{i+T_{\text{amort}}}=c_{\text{cus}}(i)/T_{\text{amort}} \tag{14}$$

式中：$f_{\text{amo}}(i,j)$ 为第 i 年新增垫气费在第 j 年单独产生的摊销费用，元；T_{amort} 表示摊销年限，a。

评价期内所有的摊销费为

$$c_{\text{amo}}(j)=\sum_{i=1}^{T_{\text{life}}}f_{\text{amo}}(i,j) \tag{15}$$

式中：$c_{\text{amo}}(j)$ 为第 j 年摊销总额，元。

（5）调整所得税计算。在弥补息税前利润亏空后，如果息税前利润仍有结余，则结余部分需要缴纳调整所得税。

$$\begin{aligned} t_{adj}(i) &= f_{adj}(i) \times r_{adj} \\ &= [\text{ebit}(i) - f_{def}(i)] \times r_{adj} \\ &= [f_{pro}(i) + f_{in-gr}(i) - f_{def}(i)] \times r_{adj} \\ &= [s(i) - c_{tot}(i) - t_{bus}(i) + f_{in-gr}(i) - f_{def}(i)] \times r_{adj} \end{aligned} \quad (16)$$

式中：$f_{adj}(j)$为应纳调整所得税总额，元；r_{adj}为所得税税率；$\text{ebit}(i)$为息税前利润，元；$f_{def}(i)$为弥补息税前利润亏空额，元；$f_{pro}(i)$为利润总额，元；$f_{in-gr}(i)$为总利息，元；总成本$c_{tot}(i)$计算公式如下：

$$c_{tot}(i) = c_{dep}(i) + f_{amo}(i) + c_{ope}(i) + f_{in-gr}(i) \quad (17)$$

式中：$c_{tot}(i)$为总成本，元。

因此，调整所得税简化为

$$t_{adj}(i) = [s(i) - c_{dep}(i) - f_{amo}(i) - c_{ope}(i) - t_{bus}(i) - f_{def}(i)] \times r_{adj} \quad (18)$$

根据储气库建设规模，确定出建设投资以及经营成本等费用后，NPV与p_{store}为一元一次函数，如果根据设定的p_{store}，计算出NPV表示负值，则在$[p_{store}, p_{storemax}]$区间内采用二分法继续计算NPV，如果NPV表示正直，则在$[p_{storemin}, p_{store}]$区间内重新计算，直至NPV=0。

2 计算实例

2.1 投资数据

以我国某储气库为例，该储气库建设期5年，寿命为30年，其进度安排以及全部建设完成后的投资数据见表1。

储气费计算参数见表2。

表1 某储气库各项投资费用表

项目	建设期					投入运行期					
	第1年	第2年	第3年	第4年	第5年	第1年	第2年	第3年	第4年	第5年	第6年
工作气量/$10^4 m^3$	0	0	0	0	0	4173	8346	34143	59940	84346	100564
垫气量/$10^4 m^3$	0	0	0	0	0	3192	6384	9576	29748	48856	65836
垫气费/万元	0	0	0	0	0	7310	14620	21930	68126	111885	150771
固定资产/万元	98852	21819	23921	24984	23444	18259	18259	20106	18004	12443	2961
经营成本/万元	1764	1764	1764	1764	1764	2162	2560	3778	5317	6786	7912

表 2　储气费计算费用表

项目	数值	项目	数值
贷款比例	60%	进项税额比例	5.5%
银行利率	4.9%	税及附加比例	10%
所得税比例	25%	法定公积金所占比例	10%
内部收益率	8%	折旧年限/a	20
销项税额比例	11%	摊销年限/a	30

2.2　储气费计算结果

将基础数据代入计算模型，得到储气费为 1.02 元/m³ 时，可满足内部收益率 8% 的要求。

2.3　敏感性分析

影响储气费的因素较多，为了找出影响储气费的主要因素，对年储转次数、经营成本、建设投资等不确定因素进行了敏感性分析，计算出各不确定因素变化范围由 -20% 变化到 20% 时储气费的变化情况（表 3）。

表 3　国内某储气库敏感性分析表　　　　　　　　　　　　　　单位：元/m³

变动因素	各因素由-20%变化到20%时的储气费				
	20%	10%	0	-10%	-20%
年储转次数	0.86	0.93	1.02	1.13	1.26
垫气费①	1.07	1.04	1.02	0.99	0.97
地面投资①	1.06	1.04	1.02	1.00	0.98
地下投资①	1.08	1.05	1.02	0.99	0.96
经营成本①	1.05	1.03	1.02	1.01	0.99

① 按我国某盐穴型储气库投资估算，垫气费投资为 374642 万元，地面投资为 113221 万元，地下投资为 169831 万元，经营成本为 37335 万元。

将计算结果绘制成敏感性分析图（图 2），从图 2 中可以看出，年储转次数对储气费影响较大，其余依次为地下投资、垫气费、地面投资，相对而言，经营成本的敏感性不强。

2.4　不同年储转次数分析

盐穴型储气库不同于气藏型储气库，每年可以进行多周期注采和应急采气。从实际运行情况来看，平均每年多次注采气量可能大于同年工作气量。所以，设计年储转次数为 0.6~2.0 倍不同情况，分析年储转次数对储气费的影响。

通过对比不同年储转次数时的储气费（表 4）可以看出，随着年储转次数的增加，储气费逐渐降低，当年储转次数为 2.0 时，储气费降低为 0.53 元/m³，为 1 倍年工作气量的 52%。

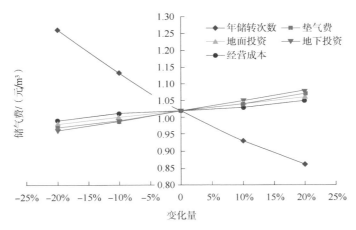

图 2 国内某储气库敏感性曲线图

表 4 不同年储转次数与储气费情境分析表

年储转次数	0.6	0.8	1.0	1.2	1.4	1.6	1.8	2.0
储气费/(元/m³)	1.67	1.26	1.02	0.86	0.74	0.66	0.59	0.53

通过分析储气费与注采周期变化曲线(图3)可以看出,年储转气量对储气费的影响较大,在年储转次数小于 1.0 时,即年采气量小于工作气量时,随年储转次数的增加储气费大幅度下降;而随年储转次数的增加,储气费下降幅度有逐渐变小的趋势;在年储转次数大于 1.4 时,注采运行的工作效率达到最大,储气费与年储转次数近似成直线对应关系,随着年储转次数的继续增加,储气库运行将大幅度超过设定工作载荷,相关运行维护费用也大幅度增加。因此,要保证盐穴储气库能达到最大经济效益,盐穴储气库最优年储转次数应为 1.4。

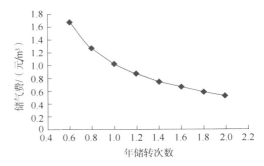

图 3 储气费与年储转次数对应关系图

3 结论

(1)储气库建设周期时间较长,本文储气费计算模型在保证能获得一定的利润的基础上,充分考虑了资金的时间价值以及垫底气的回收价值。

(2)在盐穴储气库工作气量确定的情况下,年储转次数对储气费影响较大,其次为地下及地面工程等建设投资,经营成本对储气费影响较小;

(3)年储转次数小于 1.0 时,随年储转次数的增加储气费大幅度下降;随年储转次数的继续增加,储气费下降幅度有逐渐变小的趋势;在年储转次数大于 1.4 时,注采运行的工作效率达到最大,盐穴储气库最优年储转次数建议为 1.4。

(4)储气费计算模型以盐穴储气库建设运营为依据,计算得到的储气费较为合理,可推广到类似盐穴储气库的应用计算。

参 考 文 献

[1] 陈思源,张奇,王歌,李彦.基于博弈分析的我国天然气储气库开发策略及运营模式研究[J].石油科学通报,2016,1(1):175-182.
[2] 刘剑文,孙洪磊,杨建红.我国地下储气库运营模式研究[J].国际石油经济,2018,26(6):59-67.
[3] 丛威.中国天然气储气库发展现状及问题[EB/OL].(2015-07-21)[2018-07-21].http://gas.in-en.com/html/gas-2302458.shtml.
[4] 王震,任晓航,杨耀辉,等.考虑价格随机波动和季节效应的地下储气库价值模型[J].天然气工业,2017,37(1):145-152.
[5] 孟浩.美国储气库管理现状及启示[J].中外能源.2015,20(1):18-24.
[6] 洪波,丛威,付定华,等.欧美储气库的运营管理及定价对我国的借鉴[J].国际石油经济,2014,22(4):23-29.
[7] 田静,魏欢,王影.中外地下储气库运营管理模式探讨[J].国际石油经济,2015,23(12):39-43.
[8] 姚莉,肖君,吴清,等.地下储气库运营管理及成本分析[J].天然气技术与经济,2016,10(6):50-54.
[9] 李圣彦,李博,张立恒,等.天然气管网冬季保障供应措施探讨[J].天然气技术与经济,2016,10(2):67-69.
[10] 张刚雄,李彬,郑得文,等.中国地下储气库业务面临的挑战及对策建议[J].天然气工业,2017,37(1):153-159.
[11] Boogert A,De Jong C.Gas storage valuation using a Monte Carlo method[J].Birkbeck Working Papers in Economics & Finance,2007,15(3):81-98.
[12] Denholm P,Sioshansi R.The value of compressed air energy storage with wind in transmission-constrained electric power systems[J].Energy Policy,2009,37(8):3149-3158.
[13] Drury E,Denholm P,Sioshansi R.The value of compressed air energy storage in energy and reserve markets[J].Energy,2011,36(8):4959-4973.
[14] Chen Z,Forsyth P A.A semi-Lagrangian approach for natural gas storage valuation and optimal operation[J].Society for Industrial and Applied Mathematics,2007,30(1):339-368.
[15] 胡奥林,何春蕾,史宇峰,等.我国地下储气库价格机制研究[J].天然气工业,2010,30(9):91-96.
[16] 罗天宝,李强,许相戈.我国地下储气库垫底气经济评价方法探讨[J].国际石油经济,2016,24(7):103-106.
[17] 周新,吴晓峰.二分法与迭代法处理化学工程中的气体计算[J].山东化工,2014,43(5):155-159.
[18] 黄胙翰,李占松.优化二分法计算天然河道水面曲线[J].人民黄河,2018,40(8):113-115.
[19] 游宇堃,王淳.基于解析优化二分法的配电网分布式电源准入容量计算[J].南昌大学学报(工科版),2013,35(2):184-186.

本论文原发表于《天然气工业》2018年第38卷第11期。

我国地下储气库垫底气经济评价方法探讨

罗天宝[1] 李 强[2] 许相戈[3]

(1. 中国石油西气东输管道公司；2. 中国石油天然气与管道分公司；3. 新疆石油工程设计有限公司)

【摘　要】 地下储气库垫底气根据回收释放性质的不同，分为基础垫底气和补充垫底气。目前我国不区分地下储气库类别和垫底气性质统一采取摊销处理的评价方法，未考虑油气藏型储气库垫底气可部分回收及盐穴储气库垫底气可完全回收等特点，计算得出的地下储气库储气费评价指标偏高，项目盈利能力等指标偏低，影响和制约了我国地下储气库的发展。结合不同地下储气库特点及垫底气回收性质，提出基础垫底气采取摊销，补充垫底气采取期末回收并计算资金占用费的评价方法，为正确评价地下储气库项目经济效益以及下一步储气服务价格机制改革提供理论参考，为我国储气库实现商业化运营，促进天然气市场繁荣创造条件。

【关键词】 天然气；地下储气库；垫底气；经济评价

The economic evaluation methodology of base gas in China's underground gas storage

Luo Tianbao[1]　Li Qiang[2]　Xu Xiangge[3]

(1. PetroChina West-East Pipeline Company; 2. PetroChina Natural Gas & Pipeline Company; 3. Xinjiang Petroleum Engineering Design Company Ltd. ,)

Abstract　Base gas of underground gas storage can be divided into foundation base gas and supplement base gas with the different properties of recovery and release. The current unified evaluation method by means of amortization in which the categories of domestic underground gas storage and base gas properties are not distinguished and the characteristics of base gas of reservoirs underground gas storage (can be partly of the recycling) and base gas of salt-cavern gas storage (can be completely recycled) are not considered. The index of gas storage fee is high and project profitability index is accordingly low, which influences and restricts the development of underground gas storage in China. The paper proposes a new evaluation method adopting amortization for foundation base gas and final recovery and calculating the financial cost for supplement base gas considering the different features of the underground gas storage and recovery properties of base gas to provide theoretic references for evaluating economic benefits of underground gas storage project and the next gas storage service price mechanism reform, create conditions for the commercial operation of domestic gas storage, and promote the natural gas market come into prosperity.

Keywords　natural gas; underground gas storage; base gas; economic evaluation

1 我国地下储气库发展概况

地下储气库是天然气供应链中调峰储备的重要组成部分，具有库容量大、调峰能力强、安全性好、储气费低等特点，已成为世界上应用最广、最直接、最经济有效的调峰方式。

美国和俄罗斯这两个天然气消费生产大国的地下储气库总工作气量分别占其天然气年消费量的17.4%和17%(不包括战略储备气量)。部分发达国家和地区的调峰应急储备达到年消费量的17%~27%。我国储气库建设规模与国外相比存在较大差距。截至2014年底，我国已建成地下储气库群10座，其中气藏型储气库群9座，盐穴型储气库1座，形成调峰气量$30×10^8m^3$，占天然气年消费量的1.7%，距世界10%的平均水平有较大差距。根据预测，2020年我国天然气需求量可达到$3200×10^8m^3$，储气库作为最主要的调峰方式，其规模至少应达到消费量的10%以上才能满足调峰和保供的需求。"十三五"期间，我国地下储气库建设将掀起高潮，濮阳油田、华北油田、辽河油田以及新疆油田将逐步形成有效工作气量，预计2020年可形成$300×10^8m^3$的有效工作气量。

按地质条件不同，地下储气库可分为油气藏储气库、水层储气库和盐穴储气库三类。其中油气藏储气库是利用枯竭的气层或油层而建设，将天然气储存在天然的岩石孔隙中。油气藏储气库是目前最常用、最经济的一种地下储气形式，具有造价低、运行可靠、垫底气可部分回收等特点。水层储气库是将高压气体注入含水层的空隙中将水排走，在非渗透性的含水层盖层下直接形成储气场所。与油气藏储气库相比，水层储气库需要更多的垫底气，注气和采气时需要更严密的监控，其储气和调峰能力相对较低。盐穴地下储气库是采取水溶开采方式，在地下较厚的盐层或盐丘中形成人造地下洞穴，形成一定的地下空间，用于储存天然气。盐穴储气库具有构造完整，密封性好，垫底气量低且可完全回收，注采灵活，短期吞吐量大等特点，虽然建库成本较高，但是每年多次注采循环可降低储气库单位注采成本。

2 地下储气库垫底气现行经济评价方法

地下储气库在投产初期，需要提前向储层中注入一定量的天然气或其他非可燃气体(目前国内主要是天然气)，使储层的地层压力上升到最低工作压力，以便为储气库正常储存或释放天然气提供能量保证，称之为垫底气。根据国内外数据分析，油气藏储气库垫底气量占最大库容量的35%~60%，垫底气投资占总投资的30%~45%；盐穴储气库垫底气量占最大库容量的35%~48%，垫底气投资占总投资的26%~43%。

垫底气根据回收释放性质的不同，可分为基础垫底气(不可回收释放)和补充垫底气(可回收释放，或称附加垫底气)两类。基础垫底气是指当气库压力下降到气藏废弃压力时，因地质结构的复杂性导致气库内无法采出的气量，也称之为"死气"。该垫底气在气藏作为储气库利用时已经事先存在气库内，它是衡量气库闲置资源量的重要指标。补充垫底气是指在基础垫底气的基础上后续为了升高气库压力，保证采气井在下限压力能够达到最低调峰能力时所需另外注入的气量，这部分气量在储气库报废时可以作为商品气回收释放。

根据《中国石油天然气集团公司建设项目经济评价参数〔2016〕》规定，进行储气库建设项目经济评价时，不区分基础垫底气和补充垫底气，均按20年摊销处理。这种评价方法认为，从油气田企业转让得到的废弃油气藏（含基础垫底气），类似采矿权，是企业拥有和控制的，具备无形资产的特征，应作为无形资产核算；补充垫底气在储气库投运初期注入地层中，形成后续投资列入无形资产投资支出，与基础垫底气共同作为无形资产核算。

这种评价方法，一方面不能真实地反映垫底气的资产属性，不能客观地反映基础垫底气和补充垫底气取得渠道不同、报废时可利用的价值不同以及投资补偿渠道不同的特点；另一方面，这种评价方法没有客观反映出油气藏储气库垫底气（包括基础垫底气和补充垫底气）可部分回收和盐穴储气库垫底气（全部为补充垫底气）可完全回收的实际运行特点，计算得出的地下储气库储气费评价指标偏高，项目盈利能力指标、清偿能力指标、财务生存能力指标偏低，从而导致在现行低油价和经济新常态下较其他调峰方式（尤其是LNG）竞争力偏低，项目论证难以通过，影响和制约了我国地下储气库的发展。

3 地下储气库垫底气经济评价新方法探讨

3.1 垫底气经济评价新思路

垫底气的经济评价既要考虑投资回收，满足价值补偿的要求，又要满足现行会计准则的要求。本文按照实质重于形式的原则，充分考虑基础垫底气和补充垫底气取得渠道的不同、将来可利用价值不同以及投资补偿渠道不同等因素，正确区分资本性支出和经营性支出的界限，采取差别化评价方法进行处理。

首先，根据现行评价方法，基础垫底气按摊销处理。补充垫底气因在项目报废时可采出利用，可在项目计算期末作为商品气销售，补偿其占用投资，从而增加项目的现金流入量，因此补充垫底气可采取期末回收并计算资金占用费的评价方法进行处理。其中，作为商品天然气的销售单价按成本加成法计算确定，成本利润率可按行业成本利润率或项目要求达到的报酬率计取。

其次，补充垫底气因在项目报废时才采出利用，购买补充垫底气所使用的资金在整个计算期内被占用，无法再用于其他用途获取资金的保值增值。因此，对于垫底气投资占用资金可通过流动资金占用费的形式予以补偿，即按补充垫底气投资、流动资金贷款利率和占用期限计算流动资金占用费，计入当期财务费用。

再次，补充垫底气是为了维持地下储气库在最低运行压力时能够正常发挥调峰功能，在项目投运初期被注入地层中的天然气，在正常运行情况下不会作为工作气被采出利用，仅在项目报废时才予以回收采出。根据这一特点，在进行项目评价时，可在计算期末年将补充垫底气量计入当年工作气量，以评价地下储气库项目经济效益及计算天然气储气费指标。

3.2 案例应用分析

某盐穴地下储气库项目共部署井位60口，分二期建设，一期工程建设27口新盐腔和6个老腔，设计库容量为$16.7 \times 10^8 m^3$，工作气量为$9.5 \times 10^8 m^3$；二期工程建设27口盐腔作为战略储备，设计库容量为$18.9 \times 10^8 m^3$，工作气量为$10.6 \times 10^8 m^3$。项目总投资1125512万元，其中补充垫底气投资357583万元，建设期利息29286万元，铺底流动资金2911万元。

项目建设投资的40%由企业自有资金筹措，其余60%由银行贷款解决。项目评价期30年，财务基准收益率8%，分别按储气费1.2元/立方米计算项目效益和按项目达到8%的内部收益率反算储气费两种方法进行评价。现将该项目分别按两种不同的垫底气评价方法进行分析计算。

（1）现行垫底气评价方法。补充垫底气投资357583万元，按20年摊销计算。根据有关规定，计算评价结果。

（2）垫底气评价新方法。假设其他数据不变、参数不变，计算期末年将补充垫底气 $15.6075 \times 10^8 m^3$ 作为商品气销售，计算销售收入，计入现金流入量；补充垫底气 $15.6075 \times 10^8 m^3$ 并入当年工作气量；计算期内垫底气投资357583万元，按3.92%的流动资金贷款利率和每年周转一次计算资金占用费。根据有关规定，计算评价结果见表1。

表1 主要经济评价数据及指标对比

序号	项目名称	单位	现行评价方法	新的评价方法	备注
1	项目总投资	万元	1125512	1125512	
	其中：补充垫气费	万元	357583	357583	
2	建设期利息	万元	29286	29286	
3	铺底流动资金	万元	2911	2911	
4	总成本费用	万元	54164	53168	年均
	其中：折旧	万元	14791	14791	年均
5	经营成本	万元	20329	20453	年均
6	营业收入	万元	195076	201319	年均
7	利润总额	万元	138288	145430	年均
8	总投资收益率		12.70%	14.28%	
9	资本金利润率		23.56%	24.78%	
10	平均利息备付率(ICR)		29.41%	33.11%	
11	平均偿债备付率(DSCR)		3.42%	3.42%	
12	财务内部收益率(税后)		11.30%	11.16%	
13	财务内部收益率(税前)		13.52%	13.71%	
14	财务净现值(税后)	万元	232975	242056	
15	财务净现值(税前)	万元	430701	473224	
16	储气费	元/立方米	0.8754	0.8675	

从以上评价指标对比可以看出，对基础垫底气采用摊销核算、补充垫底气采用期末回收核算的评价方法，充分反映了基础垫底气和补充垫底气的不同流转特点，计算得出的项目盈利能力指标、偿债能力指标和单位储气费等指标均优于按现行评价方法得出的指标。这种经济评价方法更真实客观地反映了地下储气库建设项目经济效益水平和作为调峰设施的竞争能力水平，可解决部分地下储气库建设项目因经济效益差而被否定的问题，有利于

加快我国地下储气库投资建设步伐，为引导地下储气库建设行业持续健康发展奠定良好的基础。

本文合理区分基础垫底气的资产属性和补充垫底气的存货属性。作为天然气供应链中的重要组成部分，未来地下储气库将成为专业化的市场主体，采用独立运营的商业模式。合理确定储气库服务价格有益于上游管道公司、下游销售公司或城市燃气公司、储气库运营者等商业主体利益的公平分配，为我国储气库实现商业化运营，促进天然气市场繁荣创造条件。

参 考 文 献

[1] 宋东昱，田静. 中国天然气储气调峰面临的挑战与对策[J]. 国际石油经济，2014(6)：39-45.

[2] 魏欢，田静，李建中，等. 中国天然气地下储气库现状及发展趋势[J]. 国际石油经济，2015(6)：57-62.

[3] 龚继忠，朱静，况中文，等. 关于地下储气库垫底气会计核算方法的探讨[J]. 财会研究，2012(2)：35-38.

[4] 罗天宝，程凤华，孟少辉. 盐穴地下储气库建设项目经济评价应注意的几个问题[J]. 国际石油经济，2015(12)：92-95.

[5] 郑德文，赵堂玉，张刚雄，等. 欧美地下储气库运营管理模式的启示[J]. 天然气工业，2015(11)：97-101.

[6] 丛威，洪波，裴国平，等. 欧美储气库独立运营商业模式相关经验借鉴[J]. 中国能源，2014(5)：29-33.

本论文原发表于《国际石油经济》2016年第24卷第7期。

盐穴地下储气库建设项目经济评价应注意的几个问题

罗天宝[1]　程风华[2]　孟少辉[1]

(1. 中国石油西气东输管道公司；2. 新疆油田分公司风城油田作业区)

【摘　要】 目前石油建设项目经济评价方法与参数中关于盐穴地下储气库建设项目评价方法与参数的选取还有很多不完善之处。盐穴储气库项目通常以分期分批的模式建设，投资的分摊应按工程项目投资与功能一致的原则进行。注气排卤标志着从建设阶段向生产运行阶段的过渡，是明确划分建设期和生产运行期的界限。盐穴储气库天然气损耗率按 0.15%~0.25% 考虑比较合适。根据外输管道所属工程范围的不同，垫底气价的确定也不相同。盐穴储气库安全生产费用计提按管道运输企业安全生产费用提取标准计算较为合适。盐穴储气库建设项目作为独立项目进行经济效益评价时，选取 6% 的增值税税率较为合适。应急储备天然气占用的资金可在储气库项目运行期内平均摊销。应多方面多角度充分论证建设盐穴储气库调峰的必要性和经济性，为地下盐穴储气库项目投资决策提供科学依据。

【关键词】 盐穴；储气库；建设项目；经济评价

Related problems of economic evaluation for the salt cavern gas storage reservoir construction project

Luo Tianbao[1]　Cheng Fenghua[2]　Meng Shaohui[1]

(1. PetroChina West East Pipeline Company; 2. Windy City Oilfield Operation District of Xinjiang Oilfield Company)

Abstract　For the economic evaluation method and parameter of the oil construction project, there are some imperfections in the method and parameter selection for salt cavern gas storage reservoirs. Usually the construction mode of salt caverns is done by stages and in batches, and the cost allocation should be consistent with the function in different stages and batches. The process of injecting gas to eject brine not only symbolizes the transition period from construction to production, but also can be the boundary. The rate of natural gas loss ranges from 0.15% to 0.25%. The lowest price of natural gas depends on the engineering type of the exporting pipelines. The safety production expense of salt caverns is estimated according to the standard of pipeline transportation enterprises. When evaluating the economic benefit evaluation of the salt cavern gas storage reservoir independently, the VAT rate is set as 6% properly. The funds for emergency system of natural gas can be shared in production. The necessity and economy of building salt cavern gas storage reservoir for peakshaving should be demonstrated with different aspects, to provide scientific basis for investment decisions on the salt caverns.

Keywords　salt caverns; gas storage reservoir; construction project; economic evaluation

盐穴地下储气库是利用水溶开采方式在地下较厚的盐层或盐丘中形成人造地下洞穴而建成的地下空间，用于储存天然气。与其他类型储气库相比，盐穴地下储气库具有建库周期长、建库成本高、构造完整、密封性好、可塑性大、实际运行寿命长、综合成本低等特点，尤其是注采能力强、注采速度快、垫底气利用率高(在需要时，垫底气可以完全回收)的优点。越来越多的国家重视盐穴储气库建设。目前石油建设项目经济评价方法与参数中关于盐穴地下储气库建设项目评价方法与参数的选取还有很多不完善之处，这些问题对盐穴地下储气库建设项目经济评价存在一定的影响，有时可能会影响评价的结果乃至项目决策。笔者结合盐穴地下储气库的特点和中国首座盐穴地下库(金坛储气库)的建设运行情况，探讨盐穴地下储气库建设项目经济评价时应该注意的几个问题。

1 投资分摊

储气库建设遵循的原则是在总体规划的指导下滚动建设，通常将储气库和天然气干线管道作为一个整体，总体规划，配套建设，分步实施。由于盐穴储气库建设受盐化工企业卤水消化能力的制约，项目建设周期一般比较长，导致项目达到设计经济规模、实现预期经济效益所经历的时间也很长。但是盐穴储气库工程具有单腔建成后能独立运行发挥功效的特点，因此，为了最大限度地实现项目建设的经济效益，盐穴储气库项目通常以分期分批的模式建设。例如，平顶山储气库工程分两期建设，配套的地面工程和外输管道工程需要结合项目总体规划统筹考虑。对于先期实施的一期工程，如果将该阶段全部资金投入计入一期工程的话，评价得出的一期工程投资效益指标会降低，影响项目的决策，也不能正确反映项目投入产出匹配的效益。

在进行项目经济评价时，投资的分摊应按工程项目投资与功能一致的原则进行。如果设计内容和投资中考虑了后期扩建的功能，前期和后期投资应该分摊计入经济评价模型。投资分摊的比例应根据投资项目的功能情况确定，通常采用规模指数法和规模比例法确定分摊的比例。

2 建设期与生产期划分界限

盐穴储气库的建造大致经历以下几个阶段或过程：建库目标库址的确定(包括区域普查、建库目标库址、建库区块和层段的确定)，建库方案设计(包括储气库建设地质方案及钻井工程、造腔工程、注气排卤工艺和地面工程方案等)，储气库施工建设。盐穴储气库一般由多个单腔构成，每个单腔需要按钻完井、造腔、注气排卤等步骤进行建设。注气排卤是盐穴储气库首次注入天然气的作业，标志着从建设阶段向生产运行阶段的过渡，是明确划分建设期和生产运行期的界限，是合理确定项目建设期的基础。对于整体建设的盐穴储气库来说，建设期和生产运行期的划分界限可按第一批井注气排卤结束时间确定；对于分期分批建设的盐穴储气库，项目建设期和生产运行期的划分界限可按一期工程第一批井注气排卤结束时间确定。

3 天然气损耗

天然气损耗是盐穴储气库建设项目经营成本的重要组成部分，不仅关系到盐穴储气库运行的经济特性，也是评价盐穴储气库经济效益的重要因素。盐穴地下储气库天然气损耗量主要由地质构造泄漏损耗、卤水携带气量损耗和地面系统天然气损耗构成。(1)地质构造泄漏损耗。盐穴储气库是利用水溶开采方式在地下较厚的盐层或盐丘中形成的人造地下洞穴，高温高压下的盐具有一定的塑性，在产生一定的裂缝时能够自动愈合，地下盐穴就形成了很好的密封储存库。因此，可以不考虑从构造溢出点逃逸的天然气损耗。(2)卤水携带气量损耗。注气排卤过程中，采出的卤水在分离外输时会携带一部分天然气而产生损耗。(3)地面系统天然气损耗。主要包含注采井泄漏、地面管道和设备泄漏、注采气系统放空等各种损耗。

根据天然气损耗的以上构成，以金坛盐穴储气库为跟踪对象，采用统计分析方法，分别计算了卤水携带气量损耗和地面系统天然气损耗。经计算分析，盐穴储气库储气天然气损耗率大约为 0.05%。

目前我国天然气工业是上中游一体化经营，储气库在天然气产业链上未独立成为一个功能性组成环节，而是作为干线管网的功能组成部分，与管道输气系统形成一个相互关联的整体。因此，作为天然气管网配套设施建设的盐穴储气库，天然气损耗应综合考虑输气损耗和储气损耗。根据近几年统计资料，盐穴储气库天然气损耗率按 0.15%~0.25%考虑比较合适。

4 垫底气价

储气库在投产初期，需要提前向储层中注入一定量的天然气或其他非可燃气体，使储层的底层压力上升到最低工作压力，以便为储气库正常存储和释放天然气提供能力保证，这种提前注入的气体(目前国内主要是天然气)称之为垫底气。根据对金坛地下盐穴储气库、平顶山地下储气库、淮安赵集地下盐穴储气库、淮安楚州地下盐穴储气库等项目数据的分析，垫底气量占库容量的 35%~48%，垫底气投资占总投资的 26%~43%。垫底气价格是合理确定垫底气投资的关键因素，也是影响盐穴储气库经济效益的重要因素，垫底气价格是否合理，将对投资项目的决策起着关键作用。

根据外输管道所属工程范围的不同，垫底气价的确定也不相同。若连接天然气干线管网和盐穴储气库的外输管道属于天然气管网建设范围时，根据《国家发改委关于调整天然气价格的通知》(发改价格〔2013〕1246号)要求，垫底气价为储气库建设项目所在省份天然气门站价加上天然气注入费；若外输管道属于储气库项目配套工程，依托储气库项目建设时，垫底价应为储气库建设项目所在省份天然气门站价加上外输管道管输成本和天然气注入费。其中，为合理确定外输管道管输成本，外输管道投资和输送天然气时所发生的成本费用与储气库项目主体投资和天然气储转发生的成本费用应分别计列和归集；天然气注入费应综合考虑垫底气注入时所消耗压缩机台班和注入气量计算。

5 安全生产费用计提

为建立企业安全生产投入长效机制，加强安全生产费用管理，保障企业安全生产资金投入，维护企业、职工以及社会公共利益，2012年2月和12月，财政部、安全监管总局及中国石油天然气集团公司分别下发了《关于印发〈企业安全生产费用提取和使用管理办法〉

的通知》(财企[2012]16号)和《关于集团公司安全生产费用提取标准和使用范围的通知》,明确中国石油天然气集团公司所属在中华人民共和国境内从事勘探生产、危险品生产和存储、工程技术服务、建设工程施工、机械制造、交通运输的企业以及其他经济组织均应提取安全生产费用。安全生产费用的计提根据企业所提供的生产服务内容的不同,采取不同的计取标准。

盐穴地下储气库作为天然气管道网络的功能性组成部分之一,依托输气干线管道建设,与管道输气系统形成一个相互关联的整体,由天然气管道公司拥有和运营,用以优化管网系统运行,提高供气的可靠性和安全性,并满足用气调峰需求。因此,盐穴储气库安全生产费用计提按照管道运输企业安全生产费用提取标准计算较为合适。

6 增值税税率选取

根据国务院进一步扩大交通运输业和部分现代服务业营业税改征增值税试点的要求,2013年5月,财政部、国家税务总局下发财税[2013]37号《关于在全国开展交通运输业和部分现代服务业营业税改征增值税试点税收政策的通知》,明确自2013年8月1日起,在全国范围内开展交通运输业和部分现代服务业营改增试点,要求在我国境内提供交通运输业和部分现代服务业服务等应税服务的单位和个人,改缴增值税,不再缴纳营业税。增值税纳税人根据所提供应税服务内容和性质的不同,分别适用不同的税率,其中交通运输服务(含管道运输服务)适用税率为11%,仓储服务(指利用仓库、货场或者其他场所代客存放、保管货物的活动)适用的税率为6%。储气库作为天然气管道网络的功能性组成部分,其主要功能是将从天然气田采出的天然气再注入可以保存气体的空间,以解决管道企业的季节调峰和应急保安用气,为管道输送企业提供仓储服务。因此,盐穴储气库建设项目作为独立项目进行经济效益评价时,模拟地下盐穴储气库单独运营,按提供仓储服务考虑,选取6%的增值税税率较为合适。

7 应急储备天然气评价

盐穴地下储气库储存的天然气主要有三方面用途:调峰用气(用于解决供气区和用户之间年供需不平衡)、应急保安供气(长输管道维修或发生意外情况时保证用户需求)、应急储备天然气(作为国防战略能源应对战争等突发事件,保障国民经济安全)。

战略储备天然气作为保障国民经济安全发展的支柱能源,在储气库项目建成投运初期加注到盐腔中,在储气库正常运行情况下不会被作为工作气采出利用,仅在影响国民经济安全的重大事件发生时或储气库项目报废时才被采出利用。因此,这部分天然气具有储存周期长、资金占用量大且占用时间长的特点。根据谨慎性原则要求,为尽早回收投资,取得收益,减少风险,应急储备天然气占用的资金可在储气库项目运行期内平均摊销。考虑到应急储备天然气储存周期长和重大事件发生偶然性的特点,这部分天然气可在项目评价期最后或项目终止年计算一次工作气量,以合理评价盐穴储气库项目经济效益及计算天然气储转费指标。

8 盐穴储气库项目经济评价方法探讨

管道天然气的调峰方式主要有两大类,一类是通过建设调峰设施满足调峰需求,调峰方式包括地下储气库调峰、LNG 调峰、液化石油气调峰、上游气源调峰、管道调峰、管束调峰和储气罐调峰等；另一类是通过对用户用气量进行调节来满足调峰需求，调峰方式包括选择可中断用户和实行峰谷气价等。

地下储气库根据性质和用途的不同，可分为调峰储气库和战略储备储气库。其中调峰储气库项目建设以实现项目经济效益最大化为目标，投资决策的重要参考依据包括项目盈利能力、偿债能力和财务生存能力等评价指标，以及相比其他调峰设施储气库调峰价格竞争能力分析指标。因此，在进行调峰储气库建设项目经济评价时，一方面按要求达到的目标基准收益率"反算"出单位储气费指标；另一方面按集团公司要求的最高储气费指标计算项目范围内的效益和费用，分析项目的盈利能力、清偿能力。《中国石油天然气集团公司建设项目经济评价参数》[2015 版]规定，不分地下储气库类型储气费指标均为 1.2 元/立方米。但根据国外储气库建设经验及数据统计分析，不同类型的地下储气库其储气费价格水平是不同的。建议根据已建不同类型的地下储气库实际情况，确定不同的"标杆"储气费控制指标或采用"净回值法"测算的储气费价格空间作为控制指标。

从技术角度，采用 SWOT 分析原理分析评价盐穴地下储气库调峰相对于其他类型地下储气库和 LNG 调峰的竞争态势，分析判断建设盐穴储气库调峰设施的必要性和可行性。从经济角度，结合国际石油天然气行业现状及未来发展趋势、国家天然气利用政策等因素，分析计算不同原油价格对应的天然气价格下盐穴地下储气库调峰相对其他类型地下储气库和 LNG 调峰的价格竞争力，分析判断建设盐穴储气库调峰设施的经济性。价格竞争力分析应综合考虑调峰设施周转费用和调峰设施至用户的管输费用，多方面多角度充分论证建设盐穴储气库调峰的必要性和经济性，为地下盐穴储气库项目投资决策提供科学依据。

财务评价是在现行财税制度和价格体系的条件下，计算项目范围内的效益和费用，分析项目的盈利能力、清偿能力，以考察项目在财务上的可行性。合理选取建设项目效益费用参数及采用适当的评价方法是正确计算评价盐穴储气库建设项目经济效益的基础，有助于提高盐穴地下储气库建设项目投资决策的科学性。

参 考 文 献

[1] 王秀芝，刘传喜，曹艳，等. 气田开发项目经济评价中有关问题的探讨[J]. 当代石油石化，2006，14(8)：30-33.
[2] 胡奥林，何春蕾，史宇峰，等. 我国地下储气库价格机制研究[J]. 天然气工业，2010(9)：91-96.
[3] 吴玉国，陈保东，郝敏，等. 城市供气调峰措施[J]. 油气储运，2008，27(12)：13-19.
[4] 高发连. 地下储气库建设的发展趋势[J]. 油气储运，2005，24(6)：15-18.
[5] 宋杰，刘双双，李巧云，等. 国外地下储气库技术[J]. 内蒙古石油化工，2007(8)：209-212.
[6] 陶卫方，刘旭，王起京. 地下储气库的损耗构成与变化规律[J]. 油气储运，2010，29(4)：258-259.
[7] 龚继忠，朱静，况中文，等. 关于地下储气库垫底气会计核算方法的探讨[J]. 财会研究，2012(2)：35-38.
[8] 丁国生，张昱文. 盐穴地下储气库[M]. 北京：石油工业出版社，2010.

本论文原发表于《国际石油经济》2015 年第 12 期。

成果篇

获得省部级及以上科技奖励

序号	成果名称	获奖类型	获奖等级	颁奖机构	获奖时间
1	深部盐矿采卤溶腔大型地下储气库建设关键技术及应用	国家科学技术进步奖	二等奖	中华人民共和国国务院	2011年
2	长三角地区刘庄地下储气库建库方案及工程实施	科学技术奖	三等奖	上海市人民政府	2015年
3	地下储气库运行安全保障技术研究	科学技术进步奖	二等奖	中国石油天然气集团公司	2016年
4	层状盐岩油气地下储存岩石力学理论、评估体系及造腔关键技术	科技进步奖	一等奖	湖北省人民政府	2010年
5	盐穴型天然气地下储气库风险评估方法研究	科学技术进步奖	二等奖	中国石油天然气集团公司	2013年
6	盐岩溶腔油气储库建造技术研究与应用	科技进步奖	三等奖	山西省科学技术奖励委员会	2011年
7	已有采卤老腔改建储气库工程技术	科学技术进步奖	三等奖	中国石油天然气集团公司	2008年
8	地下储气库设施完整性管理研究	科学技术进步奖	三等奖	中国石油天然气集团公司	2011年
9	西气东输储气库(金坛)造腔过程腔体形状控制和检测技术研究	科学技术进步奖	二等奖	中国石油天然气集团公司	2010年
10	西气东输储气库(金坛)含盐层系三维精细地质建模及地质综合研究	科学技术进步奖	二等奖	中国石油天然气集团公司	2012年
11	金坛储气库几何设计参数优化研究	科学技术进步奖	三等奖	中国石油天然气集团公司	2017年
12	复杂地质条件下层状盐岩储气库建设技术与应用	科技进步奖	一等奖	湖北省人民政府	2017年
13	盐穴储气库天然气阻溶造腔技术	能源创新奖	三等奖	中国能源协会	2017年
14	盐穴储气库精细建库技术及配套设计系统	科学技术进步奖	二等奖	中国石油天然气集团公司	2018年
15	复杂地质条件下层状盐岩水溶造腔关键技术与应用	科技奖	二等奖	中国石油和化工自动化应用协会	2021年

续表

序号	成果名称	获奖类型	获奖等级	颁奖机构	获奖时间
16	储气库技术研究团队	优秀创新团队	/	国家石油天然气管网集团有限公司	2022年
17	盐穴储气库高效建造与安全运行保障技术及产业化团队	科技促进发展奖	/	中国科学院	2023年
18	超深大型盐穴储气库建库关键技术及工程应用	科学技术进步奖	二等奖	湖北省人民政府	2023年

授权专利

序号	专利号	专利名	授权日期	专利类型
1	ZL 2015 1 1024387.5	一种盐穴储气库的注气排卤方法及装置	2019-05-07	发明专利
2	ZL 2016 1 1190014.X	一种应力接头	2019-06-11	发明专利
3	ZL 2016 1 0681172.9	盐穴储气库造腔过程气水界面控制模拟实验系统	2020-06-09	发明专利
4	ZL 2016 1 0320212.7	力学参数测定方法及装置	2021-01-01	发明专利
5	ZL 2017 1 0976200.4	一种起排卤管的方法	2021-04-30	发明专利
6	ZL 2018 1 0928198.8	盐穴储气库运行压力上限确定方法	2021-06-01	发明专利
7	ZL 2016 1 1141795.3	应用于盐穴储气库注气排卤阶段的气水界面监测方法	2021-08-03	发明专利
8	ZL 2018 1 0814294.X	盐穴储气库建造方法	2021-08-03	发明专利
9	ZL 2018 1 1409466.1	盐穴储气库的造腔方法及装置	2021-08-03	发明专利
10	ZL 2016 1 1157733.1	盐穴储气库造腔过程中不溶物堆积形态预测方法及装置	2022-03-01	发明专利
11	ZL 2018 1 1010321.4	地下储气库最高运行压力确定方法及装置	2022-08-30	发明专利
12	ZL 2018 1 1465153.8	对流连通老腔筛选方法及装置	2022-10-04	发明专利
13	ZL 2019 1 1111854.6	监测地下储气库泄漏的方法及装置	2022-11-01	发明专利
14	ZL 2010 2 0625438.6	便携式计算机控制采集井下油水界面检测仪	2011-09-07	实用新型专利
15	ZL 2012 2 0371077.6	井下介质界面监测装置	2013-02-27	实用新型专利
16	ZL 2017 2 0706873.3	一种连续油管管柱装置	2018-04-17	实用新型专利
17	ZL 2018 2 0399255.3	一种盐穴储气库井筒腐蚀的监测实验装置	2018-11-09	实用新型专利
18	ZL 2018 2 0717812.1	一种盐穴储气库造腔管柱防堵塞装置	2019-01-04	实用新型专利
19	ZL 2018 2 1445480.2	一种注氮设备	2019-06-11	实用新型专利

发布标准

序号	标准号	标准名	标准级别	发布时间
1	SY/T 6806—2019	盐穴地下储气库安全技术规程	行业标准	2019
2	Q/SY 1416—2011	盐穴储气库腔体设计规范	企业标准	2011
3	Q/SY 1417—2011	盐穴储气库造腔技术规范	企业标准	2011
4	Q/SY 1418—2011	盐穴型储气库声呐检测技术规范	企业标准	2011
5	Q/SY 1599—2013	在役盐穴地下储气库风险评价导则	企业标准	2013
6	Q/SY 06025—2017	盐穴储气库造腔系统地面工程设计规范	企业标准	2017
7	Q/SY XQ 192—2016	盐穴储气库带压测腔技术规范	公司标准	2016
8	Q/SY XQ 193—2016	盐穴储气库井下作业管理规定	公司标准	2016
9	Q/SY XQ 194—2016	盐穴型储气库运行管理规范	公司标准	2016
10	Q/SY XQ 195—2016	油气藏型地下储气库运行管理规范	公司标准	2016
11	Q/SY XQ 215—2018	盐穴储气库注油、退油作业管理规定	公司标准	2018
12	Q/SY XQ 216—2018	盐穴储气库注气排卤操作技术规程	公司标准	2018
13	Q/SY XQ 217—2018	盐穴储气库注采运行监测规范	公司标准	2018
14	Q/SY XQ 218—2018	热媒加热系统操作维护规程	公司标准	2018

出版专著

序号	书名	作者	出版时间	出版社
1	盐穴储气库造腔工程	杨海军　李龙　李建君	2018年	南京大学出版社
2	盐穴储气库安全监测技术	李龙　李建君　等	2023年	石油工业出版社
3	盐穴储气库项目建设风险管控指南	李龙　李建君　程林	2023年	石油工业出版社

获得软件著作权

序号	软件名	获得时间
1	气垫阻溶造腔界面控制咨询软件【简称：GBI】V1.0	2016 年
2	地下盐穴储气库工作平台【简称：CavBen】V1.0.0	2017 年
3	基于 Cavsim 盐穴储气库设计软件 V1.0	2018 年
4	地下储气库天然气微泄漏监测大数据管理系统【简称：天然气微泄漏监测软件】V1.0	2022 年
5	盐穴储气库小间距对井造腔模拟软件【简称：小间距对井造腔软件】V1.0	2023 年